P9-CDL-625

ENCYCLOPEDIA OF science technology AND ethics

EDITORS AND CONSULTANTS

ENCYCLOPEDIA OF science technology AND ethics

EDITED BY

CARL MITCHAM

volume

a–c

MACMILLAN REFERENCE USA

An imprint of Thomson Gale, a part of The Thomson Corporation

Detroit • New York • San Francisco • San Diego • New Haven, Conn. • Waterville, Maine • London • Munich

Encyclopedia of Science, Technology, and Ethics

Carl Mitcham, Editor in Chief

For more information, contact
Macmillan Reference USA
An imprint of Thomson Gale
27500 Drake Rd.
Farmington,
Hills, MI 48331-3535
Or you can visit our Internet site at
http://www.gale.com

LIBRARY OF CONGRESS CATALOGING-IN-PUBLICATION DATA

Encyclopedia of science, technology, and ethics / edited by Carl Mitcham.
p. cm.
Includes bibliographical references and index.
ISBN 0-02-865831-0 (set, hardcover : alk. paper)—ISBN 0-02-865832-9 (v. 1) —
ISBN 0-02-865833-7 (v. 2)—ISBN 0-02-865834-5 (v. 3)—ISBN 0-02-865901-5 (v. 4)
1. Science—Moral and ethical aspects—Encyclopedias.
2. Technology—Moral and ethical aspects–Encyclopedias.
I. Mitcham, Carl. Q175.35.E53 2005
503—dc22 005006968

This title is also available as an e-book.
ISBN 0-02-865991-0
Contact your Thomson Gale representative for ordering information.

Printed in the United States of America
10 9 8 7 6 5 4 3 2 1

CONTENTS

EDITORIAL AND PRODUCTION STAFF

PREFACE

The *Encyclopedia of Science, Technology, and Ethics* has had multiple origins. It was when contributing an article on the philosophy of technology to the pioneering first edition of the *Encyclopedia of Bioethics* (1978), that I began to dream of a more general encyclopedic introduction to issues of technology and ethics. Inspired by the perspective of scholars as diverse as Jacques Ellul and Hans Jonas, bioethics appeared only part of a comprehensive need to grapple intellectually with the increasingly technological world in which we live. This idea was pursued in a state-of-the-field chapter on "Philosophy of Technology" in *A Guide to the Culture of Science, Technology, and Medicine* (1980) edited by one of my mentors, Paul T. Durbin. Thus when Stephen G. Post, the editor of the third edition of the *Encyclopedia of Bioethics* (2004), suggested to Macmillan the idea of a more general "Encyclopedia of Technoethics," with me as potential editor, I was primed to be enthusiastic—although I also argued that the field should now be expanded to include ethics in relation to both science and technology.

A high-school attraction to philosophy as critical reflection on how best to live had early morphed into the critical assessment of scientific technology. In contemporary historical circumstances, what has a more pervasive influence on the way we live than modern technology? My initial scholarly publications thus sought to make philosophy and technology studies a respected dimension of the academic world. Over the course of my curriculum vitae this concern further broadened to include science, technology, and society (STS) studies. Given the narrow specializations of professional philosophy, STS seemed better able to function as a home base for philosophy of technology. In fact, in the mid-1980s, George Bugliarello, George Schillinger, and I (all colleagues at Brooklyn Polytechnic University) made a proposal to Macmillan Reference for an Encyclopedia of Science, Technology, and Society." That proposal was declined, but a version eventually found truncated expression in *The Reader's Adviser*, 14th edition, vol. 5, *The Best in Science, Technology, and Medicine* (1994), co-edited with William F. Williams, a colleague at Pennsylvania State University, where I served for a period during the 1990s as director of the Science, Technology, and Society Program. Thus when the opportunity arose to edit an encyclopedia on science, technology, and ethics, I also wanted not to limit such a reference work to ethics in any narrow sense.

Other associations that broadened my perceptions in both philosophy and STS in ways that have found modest reflections here should also be mentioned. One was the collegiality of two professional associations, the Society for the Philosophy of Technology (founded 1980) and the Association for Practical and Professional Ethics (founded 1991), with members from both becoming contributors. Service as a member of the Committee on Scientific Freedom and Responsibility of the American Association for the Advancement of Science, 1994–2000, was one of the most professionally rewarding experiences of my career, and contributed its own perspective. Finally, the critical fellowship of Ivan Illich introduced me to friends and ideas with whom I might not always agree though they seldom failed to inspire.

Developmental Process

When the possibility for the present encyclopedia finally emerged in the Fall of 2002, my initial desire was not only to work with previous colleagues but to seek the collaboration of others who had become leaders in institutionalizing discussions of science, technology, and ethics. Obvious candidates for associate editors were philosopher Deborah Johnson, whose work on computer and engineering ethics during the 1980s and 1990s had helped define both fields, and Stephanie Bird and Raymond Spier, the editors of *Science and Engineering Ethics* (founded 1995), the leading journal in this area of interdisciplinary discourse. It was also desirable to make sure that the project had representation not just from the scientific and technical community (which neuroscientist Bird and biochemical engineer Spier clearly brought to the team) but also from different points on the ethical and political spectrum. Fortunately, political scientist Larry Arnhart, with whom I had recently become acquainted, was willing to bring to the table a conservative philosophical perspective that might otherwise have been inadequately represented, and to go beyond the call of editorial duty in many respects.

The first editorial meeting place in New York City in January 2003, hosted by Hélène Potter of Macmillan Reference USA. This two-day workshop established the general framework for the *Encyclopedia of Science, Technology, and Ethics* and became the basis for collegial productivity over the next two years. During the Spring and Summer 2003 we set up an Editorial Advisory Board which included Durbin, Bugliarello, and Schillinger as well as more than twenty other representatives of important disciplinary and regional perspectives. Commissioned articles began to be submitted in August 2003 and continued over the next eighteen plus months.

For the first year—during a portion of which I served as a Fulbright Scholar at the University of the Basque Country in Spain (where Nicanor Ursua was a supportive host)—the editors worked with authors to refine article definitions, learn from their contributions about new topics that needed to be covered, and thereby deepened and broadened the content of the encyclopedia. Four scholars who played especially important roles in these regards were Robert Frodeman, Valerie Miké, Roger Pielke Jr., and Daniel Sarewitz.

Self Assessment

As the first edition of a reference work, some important topics remain missing from *ESTE*, because of problems with schedule, author availability, or simple oversight. Indeed, because the themes of science, technology, and ethics are so broad, the *Encyclopedia of Science, Technology, and Ethics*, despite its four-volume length, is necessarily selective. Yet in an effort to not let perfection become the enemy of the good, the project has been pursued in a belief that it might advance in its own modest way a contemporary social process in the ethical assessment of science and technology.

This encyclopedia is thus a work in progress. It aims to synthesize, but does not claim to be final or complete. Indeed, all reference works today have to contend with a knowledge production industry that makes it difficult to secure any stable orientation. Despite its efforts, the project cannot hope to please all scientists, engineers, and ethicists—or other scholars and general readers. But the hope is to have pleased sufficient numbers that those who see opportunities for improvement will consider offering to make a second edition better. Critical comments and recommendations are welcome.

Acknowledgments

Beyond those already mentioned, all Associate Editors and members of the Editorial Advisory Board deserve special thanks for their contributions. I should nevertheless single out Stephen H. Cutcliffe, Paul T. Durbin, Helen Nissenbaum, and Nancy Tuana for extra work in identifying authors and reviewing articles. Adam Briggle, as the most qualified and hard-working research assistant I could imagine, functioned during many months of the second year as an editorial assistant. A graduate seminar in STS at the Center for Science and Technology Policy Research during Fall 2004 contributed critical perspectives that measurably improved the project. But primary credit must go to the editors and contributors, many of whom worked well beyond what would have been appropriate for their modest honoraria.

Finally, in the background, my spouse, Marylee, and family cheered on the project whenever they found me available—and when they found me absent, simply suffered a work schedule that for more than two years seldom let me come up for air. Colleagues in the Division of Liberal Arts and International Studies at the Colorado School of Mines similarly tolerated with good nature a tendency to commandeer more work space than was rightly mine; and the Division Director, Laura J. Pang, was generous in directing toward the project modest but not insignificant resources from very limited funds. In the foreground, the daily work of managing the encyclopedia preparation process depended on a production team at Macmillan to efficiently commission articles, maintain contact with authors, coordinate reviews, copyedit manuscripts, secure illustrations, check revisions and bibliographies, and prepare all materials for publication. In particular, I am fortunate to have had Monica Hubbard, Senior Editor with Macmillan Reference USA, as a guide through the process. This project would not have come to fruition without her good work, good humor, and persistence over the previous two years.

CARL MITCHAM
EDITOR IN CHIEF
MAY 2005

INTRODUCTION

Human beings are in the midst of creating a new world through science and technology. But what kind of world we create will not be decided by science and technology alone. It will depend even more significantly on our views, implicit or explicit, about the nature of the good life—about good and bad, right and wrong, and our abilities to enact ideals in the face of limited knowledge and temptations to ease or arrogance.

Virtually all sciences and technologies today have implications for ethics and politics, and ethics and politics themselves increasingly influence science and technology—not just through law, regulation, and policy initiatives, but through public discussions stimulated by the media, public interest organizations, and religious concerns. According to Alan Leshner, CEO of the American Association for the Advancement of Science, the largest interdisciplinary scientific society in the world, a new science-society relationship has emerged in the public realm and within the scientific community. As he wrote in the lead editorial in *Science* (February 11, 2005):

> We've been used to having science and technology evaluated primarily on the basis of potential risks and benefits. However, our recent experience suggests that a third, values-related dimension will influence the conduct and support of science in the future.

In response, Leshner called on members of the technoscientific community to engage others in discussing the meaning and usefulness of science, engineering, and technology. But such engagement cannot be a one-way street; it must also stimulate scientists and engineers in self-examinations of the social character of their professions and the proper roles of science and technology in society. Additionally, the non-scientific public would do well to eschew any easy criticism or naive enthusiasm in the pursuit of informed consideration. Such multi-path assessment is precisely what science, technology, and ethics is all about, and the present encyclopedia aims to contribute in the broadest possible way to this on-going process of promotional and critical reflection.

To this end the *Encyclopedia of Science, Technology, and Ethics* has three objectives:

- To provide a snapshot of emerging bodies of work in the co-construction of an ethical, scientific, and technological world;
- To design and build bridges between these not always collaborative efforts;
- To promote further reflection, bringing ethics to bear on science and technology, and science and technology to bear on ethics.

Background: The Encyclopedic Idea

The term "encyclopedia" comes from the Greek, *enkyklios* (general) + *paideia* (education), and thus alludes to the classical conception of *paideia* as character formation that transmits a level of cultural achievement from one generation to the next among the educated few. In this classical form education came to include the liberal arts of logic, grammar, rhetoric, arithmetic, geometry, astronomy, and music. As achievements in these fields accumulated and became more extensive, explicit efforts were naturally undertaken to summarize them. Early examples of such summaries were the *Antiquitates rerum humanarum et divinarum* and *Disciplinae* of Marcus Terentius Varro (116–27 B.C.E.), neither of which survives. The oldest extant work in this tradition is the *Historia naturalis* of Pliny the Elder (23–

79 C.E.). The *Etymologiarum* of Isadore of Seville (560–636) became a work of standard reference that helped transmit classical learning into the Middle Ages. Medieval and Renaissance encyclopedias continued this tradition in, for example, the *Speculum majus* of the thirteenth century Dominican scholar Vincent of Beauvais and the *Encyclopaedia seu orbis disciplinarum tam sacrarum quam prophanarum epistemon* of the sixteenth century German scholar Paul Scalich, the latter being the first to use the term "encyclopedia" in its title.

The work with which the term is most commonly associated, the Enlightenment *Encyclopédie ou Dictionnaire raisonné des sciences, des arts et des métiers* (1751–1772) marked a three-fold change in the encyclopedia idea. First, the French encyclopedia was written to educate the many as well as the few; the aim was to popularize or democratize knowledge. Second, the knowledge summarized in the French encyclopedia included technical craft traditions as well as learned or intellectual knowledge, thus building a bridge between intellectual and workshop traditions of knowing and making. Third, the French encyclopedia proposed not simply to summarize existing cultural achievements but to produce new ones. In the project of the *philosophes* Denis Diderot, Jean d'Alembert, and others, the modern idea of education as going beyond the transmission of previous cultural achievements to produce new cultural formations found one of its paradigmatic cultural expressions.

As the modern project of knowledge production took hold it proceeded by means of disciplinary specialization. In this context the encyclopedic idea also became a kind of counter movement to the creation of more and more specialized knowledge in the physical sciences, the social sciences, the humanities, and the arts. Projects that exemplified efforts at synthesis range from G. W. F. Hegel's *Encyclopedia of the Philosophical Sciences* (1817) to Otto Neurath, Rudolf Carnap, Charles W. Morris's *International Encyclopedia of the Unified Sciences* (1938–1969).

It is on all three of these visions that the present *Encyclopedia of Science, Technology, and Ethics* (*ESTE*) seeks to draw. *ESTE* aims to summarize, in however provisional a manner, emerging bodies of knowledge bearing on the co-construction of an ethical, scientific, and technological world; to promote new collaborative efforts in this interdisciplinary field of thinking and acting; and to stimulate new cross-fertilizations and syntheses between science, technology, and ethics.

The *ESTE* Idea

Moral teachings and ethical inquiries regarding the creation and use of science and technology have been part of religious and philosophical traditions from the earliest periods. Repeated cautions about over reliance on science and technology occur in the primary texts of many religious traditions (see the Tower of Babel, the myth of Daedalus, and the tales of Chuang Tzu) and in classic Western philosophy (Plato's *Gorgias*). By contrast, modern European history displays a rejection of the tradition of caution in favor of a commitment to science and technology as the best means to improve the human condition—even as restatements of caution have appeared especially in the Faust story, *Frankenstein*, *Brave New World*, and some popular science fiction.

Since their rise in the sixteenth and seventeenth centuries, science and technology have nevertheless been increasingly involved with a series of ethical and political challenges. During the eighteenth century, the Enlightenment and Romanticism sparred over the ethical character of the scientific view of the world in both epistemological and metaphysical terms. The nineteenth century witnessed the rise of major political reactions against the evils of the Industrial Revolution, reactions that influenced the military and ideological conflicts of the twentieth century. During the last half of the twentieth century whole new fields of ethical reflection emerged to deal with the technoscientific world of nuclear weapons (nuclear ethics), chemical transformation of the environment (environmental ethics), biomedical advances (bioethics), and computers and information technology (computer ethics). Additionally, the ethics of scientific research and of the engineering practice became specialized areas of study.

As the twenty-first century begins, ethical and political challenges have become global in scope and intensified by the terrorist use of technology and science. Science, technology, and ethics interactions thus transcend disciplinary and cultural boundaries—and promise to play ever more prominent roles in human affairs for the foreseeable future. *ESTE* thus aims to integrate more specialized work in the applied ethics of particular technologies, in the professions of engineering and science, and in science and technology policy analyses, to point toward general themes and grapple with contemporary issues, while including articles that provide historico-philosophical background and promote cross-cultural comparative reflection. Had *ESTE* needed a subtitle, it might well have been "Toward Professional, Personal, and Political Responsibility in the Technoscientific World." The goal is to help us all practice a more informed seeking of the good in the high-tech, deeply scientific world in which we progressively live.

Building Bridges to Promote Reflection

The field of science, technology, and ethics is not mature. As a result this encyclopedia seeks to exercise as much a creative or formative role as it does a reporting or summary one. *ESTE* is an experiment in synthesis. Although it is clear that advances in science and technology are insufficient in and of themselves to constitute true human progress, previous encyclopedic efforts to survey the ethical challenges involved with both advancing and responding to such advances have focused only on specific areas such as biomedical ethics, computer ethics, or environmental ethics—or provided synthesis at the higher level of ethics in general. The present encyclopedia is the first to attempt a mid-level synthesis of the various specializations of applied ethics as they deal especially with science, technology, engineering, and medicine in order to promote interactive scholarly reflection, practical guidance, informed citizenship, and intelligent consumerism.

To meet these diverse but overlapping purposes *ESTE* coverage aims to include (although not exhaust) four themes: (1) types of science and technology; (2) approaches to ethics; (3) types of science, technology, and ethics interactions; and (4) historical and cultural contexts.

(1) The terms "science" and "technology" are somewhat flexible. In the present context "science" indicates the modern sciences of physics, chemistry, biology, and geology—and their numerous extensions: psychology, nuclear physics, biochemistry, cosmology, and more. "Technology" refers primarily to the modern activities of making and using artifacts, especially in applied science, engineering, medicine, decision-making, and management. The merging of science and technology in science that is highly dependent on advanced engineering instrumentation (cyclotrons, electron microscopes, advanced computers) and major capital investments, and in technology that is highly dependent on scientific knowledge or theory (designer materials, computers, biotechnology, genetic engineering, etc.) is sometimes referred to as "technoscience." None of these understandings of science and technology are excluded from *ESTE*, although the encyclopedia has not been able to include everything equally.

(2) Ethics is likewise understood broadly to be concerned with all questions of right and wrong, good and bad, in science, engineering, and technology. Although science provides descriptive knowledge of the world, on its own it is not able to interpret the human meaning of this knowledge, nor to provide full guidance for distinguishing between proper and improper processes in the acquisition of knowledge. Likewise, engineering and technology provide increasingly powerful means, but tell us little about the ends to which they should be dedicated. Ethics, generally speaking, is concerned with identifying proper means and distinguishing good and bad ends. Traditions or schools of ethical reflection and analysis include those of consequentialism, deontology, virtue ethics, natural law, and more.

Adapting a working definition from the *Encyclopedia of Bioethics*, 2nd edition (1995), *ESTE* is concerned with the *multiple moral dimensions—from vision and conduct through decision and policy making at the personal, professional, and governmental levels— of science and technology broadly construed, and employing a diversity of methods in interdisciplinary settings*. This description emphasizes the unity of ethics and politics both within technoscientific communities and in the technoscience-society relationship.

(3) Science, technology, and ethics interactions can take place within technoscientific communities and outside of such communities. Furthermore, interactions outside professional communities may take place at the personal or public levels, thus suggesting the following matrix:

	Professional	**Personal**	**Public**
Science	Professional ethics of doing science	Personal interpretations and uses of science by non-scientists	Political and policy issues raised by science in relation to society
Technology	Professional ethics of doing technology, especially engineering and clinical medicine	Personal interpretations and uses of technology by non-engineers and non-physicians	Political and policy issues raised by engineering and technology in relation to society

External (personal and public) issues may further be divided into those that stress the social-political adjustment to accommodate scientific and technological change or questions about how society should promote, support, or regulate science, engineering, and technology. Science policy (both science for policy and policy for science) and technology policy are specialized approaches.

Each of the six matrix boxes further interact: professional ethics of science and engineering can overlap and influence each other; the social impacts of science and technology are sometimes difficult to distinguish; internalist ethics often has implications for external issues and vice versa. *ESTE* aspires to be cognizant of the full spectrum of this complex diversity in possible relationships.

(4) Science, technology, and ethics interactions in these broad senses have, furthermore, been examined from multiple historical and cultural perspectives: The Continental European tradition, for instance, tends to focus more globally on science and technology as a whole, whereas in the Anglo-American tradition the ethics of particular technologies (as in medical ethics or computer ethics), areas of professional practice (engineering ethics, business ethics), or issues (equity, privacy, risk) dominate. In *ESTE* perspectives from different philosophical schools are to be further complemented by those from diverse religious, political, and cultural or linguistic traditions.

Types of Articles

The Editorial Board considered these four themes in writing scope notes for *ESTE*'s more than 670 articles, using the following four-part categorization scheme:

1. Introductions and overviews
    - 1.1 Specialized introductions
    - 1.2 Overviews
2. Concepts, case studies, issues, and persons
    - 2.1 Key concepts
        - 2.1.1 Concepts, Ethical and Political
        - 2.1.2 Concepts, Scientific or Technological
    - 2.2 Case studies
    - 2.3 Issues
        - 2.3.1 Issues, Historical and Social
        - 2.3.2 Issues, Scientific or Technological
        - 2.3.3 Issues, Phenomena
    - 2.4 Persons and narratives
        - 2.3.1 Persons and figures, premodern
        - 2.3.1 Persons and figures, modern to World War I
        - 2.3.2 Persons and figures, post-World War I
    - 3. Sciences, technologies, institutions, and agencies
        - 3.1 Particular sciences and technologies
        - 3.2 Social institutions
        - 3.3 Organizations and agencies
4. Philosophical, religious, and related perspectives
    - 4.1 Philosophical perspectives
    - 4.2 Religious perspectives
    - 4.3 Political and economic perspectives
    - 4.4 Cultural and linguistic perspectives

The Topical Outline presents the full list of articles organized by these categories.

INTRODUCTIONS AND OVERVIEWS. As this categorization framework indicates, there are two types of introductory articles in *ESTE*. One consists of the thirty-three specialized introductions to existing applied ethics fields such as "Agricultural Ethics," "Bioethics," "Computer Ethics," and "Engineering Ethics." The second is a set of more than a dozen Overview entries that serve two kinds of purpose. In the first instance they are stand-alone articles to review a few central concepts such as Science, Technology, and Ethics themselves. In the second they provide introductions to composite articles. In both instances, unlike all other *ESTE* entries, they give internal references to closely related articles.

CONCEPTS, CASE STUDIES, ISSUES, AND PERSONS. The bulk of *ESTE* articles, as is appropriate in an emerging dialogue, deal with concepts, case studies, issues, and persons. In relation to concepts, the distinction between those classified as Ethical and Political in character (such as "Plagiarism" and "Trust") and those classified as Scientific or Technological ("Efficiency" and "Networks") could in many instances be contested. Why is "Aggression" ethical but "Ethology" scientific? Is not "Human Nature" as much ethical as scientific? But the interest here is simply to make a rough distinction between those more closely associated with ethics or politics and those more easily associated with science or technology. Ethics concepts also tend to have a longer history than scientific or technological ones. In each instance, however, articles aim to bring out both ethical and scientific or technological dimensions.

The distinction between Case Studies and Issues is likewise somewhat arbitrary, since along with such clear instances as the "DC-10 Case" and "DDT" are included the "Apollo Program" and the "Asilomar Conference." But the intuition is that the case studies are modestly more closely tied to historical particulars than are issues. It is also important to note that *ESTE* has avoided attaching the names of persons to cases, at least in article titles, opting instead for more generic descriptors. Since there are an indefinite number of cases, there has also been an attempt to group some kinds of cases together, as in the three entries on "Misconduct in Science." Among the case studies some are more expansive than others, often reflecting a sense that other material relevant to the case is provided elsewhere, but sometimes just as a result of the accidents or oversights that inevitably find their way into such a large compilation.

The separation of Issues into three types is again not meant to be hard and fast but suggestive. But some issues are more Historical and Social than Scientific or Technological. Then there are some Phenomena that have an issue-like dimension related to science, technology, and ethics. For instance, although the notion of elements is covered in the entry on "Chemistry," to provide some historical and phenomenological perspective articles are included on what in the European tradition have served as the four traditional elementary phenomena: "Air," "Earth," "Fire," and "Water."

The classification of Persons and Figures is divided into Premodern, Modern to World War I, and Modern since World War I. The ancient/modern division is quite common. Using World War I as a divide in the modern period recommended itself because of the role the Great War played in stimulating recognition of the destructiveness of modern science and technology, and thus ethical discussion.

SCIENCES, TECHNOLOGIES, INSTITUTIONS, AND AGENCIES. Articles on sciences, technologies, institutions, and agencies are not comprehensive. For instance, although there is an article on "Chemistry" there is none on physics or biology. The reason is that chemistry tends to be an overlooked science when it comes to ethics, whereas physics and biology are dealt with in numerous other articles such as "Nuclear Ethics" and "Bioethics". At the same time, because of the profound significance of the mathematical discipline of probability and statistics, together with its under-appreciation in ethical and political discussions, this topic has been given a somewhat more extensive treatment. The length of this treatment, which includes introductory technical material, reflects a belief in the importance of this new form of thinking that demands both attention and comprehension especially in ethical assessment. In like manner, there might have been articles on a host of social institutions as well as organizations and agencies. The goal was simply representation and illustration of the importance that these realities must play in ethical reflection and practical action that engages the world transforming character of science and technology.

PHILOSOPHICAL, RELIGIOUS, AND RELATED PERSPECTIVES. Finally, the four sets of Perspectives articles—Philosophical, Religious, Political–Economic, and Cultural–Linguistic—aim to give *ESTE* a breadth that would otherwise be lacking. Here special efforts have also been made to secure contributors from throughout the world. *ESTE* represents authors from 28 countries, reflecting the growing interest of scholars worldwide in these important issues.

Organization of the Encyclopedia

Entries vary in length from 250 to 5000 words and are arranged alphabetically. In general structure they begin with a statement of how the topic relates to the theme of the encyclopedia, followed by some background of a historical or developmental character. The main body aims to provide an authoritative exposition of its particular theme, concept, case, issue, person, science or technology, or perspective, and to conclude with critical application or comments.

In selective instances entries are composed of more than one article. For example,

RESPONSIBILITY

Responsibility: Overview

Responsibility: Anglo-American Perspectives

Responsibility: German Perspectives

Since any article is going to exclude as well as include, and this kind of composite occurs only occasionally, references to Related Articles at the end of each entry provide another means for broadening a reader's knowledge. In a synthetic, interdisciplinary encyclopedia like *ESTE* topics invariably have tendrils that reach out into multiple entries.

Bibliographies for each article are another important feature, often complemented by a few Internet Resources. They were prepared by the contributors and verified by a bibliographic editor. Although brief, bibliographies nevertheless serve different purposes from article to article. Seldom are primary sources listed. Some bibliographic items refer readers to sources used or cited by the contributor with internal reference, for example: (Jones 2000, p. 100). In cases where a bibliographic entry is not explicitly used in the text it is often briefly annotated for significance.

The article bibliographies are supplemented by two appendices: a selective, annotated general bibliography; and selective, annotated list of Internet resources. Written entries are further enhanced with more than 300 graphics that range from tables to photographs.

SPECIAL FEATURES. The main body of alphabetical entries is complemented by eight introductory essays. Given the constructive character of the encyclopedia, these essays present selective but fundamental perspectives on the dialogue among science, technology, and ethics. These range from science and technology studies scholar Sheila Jasanoff's argument for new forms of citizen participation in technoscientific governance to engineer-inventor Ray Kurzweil's argument for the ethical responsibility to promote scientific research and technological development. Historian Ronald Kline compares and contrasts developments in research ethics and engineering ethics, while philosophers Deborah Johnson and Thomas Powers set out a new program for research in ethics and technology that would help bridge the divide Kline observes. Computer science philosopher Helen Nissenbaum argues for new practices in science and engineering that would complement the Johnson-Powers program in scholarship. Mathematician Valerie Miké proposes a new ethical use of scientific evidence in the promotion and utilization of both science and technology. Science, technology, and society scholar Carl Mitcham and philosopher and environmental scientist Robert Frodeman note some ethical challenges associated with the expansion of knowledge, both scientific and technological. Philosopher of science and technology Hans Lenk calls attention to a range of emerging, ethically relevant special features in contemporary technologies themselves.

These introductory essays, which are an unusual feature in an encyclopedia, are especially recommended to readers seeking synthetic perspectives. Although they are necessarily limited in their scope, they point the way toward the kinds of interdisciplinary reflection that is crucial to further enhancement of the science, technology, and ethics dialogue.

The Appendices are another special *ESTE* feature. Along with the "Selective, Annotated General Bibliography on Science, Technology, and Ethics," and the annotated list of "Internet Resources on Science, Technology, and Ethics", there is a "Glossary of Terms" often found in discussions of science, technology, and ethics, and a "Chronology of Historical Events Related to Science, Technology, and Ethics." Finally, a set of ethics codes from around the world enhances appreciation of the truly transnational character of the science, technology, and ethics interactions at the levels of both theory and practice.

Comments and Qualifications

As will be immediately obvious to any reader, some topics are treated at greater length than others; some articles are more argumentative or polemical than others; and some articles contain more overlaps than others. Across all such variations, however, the goal has been a balance that would provide an index to emerging bodies of work contributing to the co-construction of an ethical, scientific, and technological world, enhance links between not always collaborative efforts, and further theoretical and practical engagements between science, technology, and ethics. Of course, in making such decisions there is never any one perfect way; there is always room for improvement.

With regard to length: Often less well known topics are treated at greater length than more well known. *ESTE* has, for instance, made no effort to replace other more specialized synthetic works such as the *Encyclopedia of Bioethics* (1978, 1996, 2004), the *Encyclopedia of Applied Ethics* (1998) and its offshoots, or the *Encyclopedia of Ethical, Legal, and Policy Issues in Biotechnology* (2000)—although it has tried to pick up many of the themes and issues found in such works and place them in a distinct and broader perspective. Additionally, in some cases contributors simply submitted articles longer than specified, but that were just so good it would have been a mistake to cut them.

With regard to polemics: There has been a serious effort to allow contributors when appropriate to express their views in stimulating, thought-provoking arguments rather than insist on rigid adherence to uniformly balanced reports that could come across as dull or pedantic. At the same time, efforts have also been made to complement arguments in one article with arguments in others.

With regard to overlaps: It has been judged a positive feature when, for instance, similar themes occur in entries on "Acupuncture," "Confucian Perspectives," and "Chinese Perspectives." Similarly, the importance of the idea of social contract for science justifies related treatments in entries on "Social Contract for Science," "Social Contract Theory," "Governance of Science, and Rawls, John."

The fields of economics and statistics presented special challenges. Ethics today cannot be seriously pursued without appreciation for the achievements in these disciplines, which themselves overlap. Contemporary economics is heavily mathematical, involving extensive use of probability and statistics, and it is for the latter an important area of application. Relations between a number of entries related to economics are highlighted in "Economics: Overview," but a number of approaches were nevertheless slighted. There are two articles each for probability and statistics, with one containing a brief introduction to basic concepts in terms of elementary mathematics. The goal was to include sufficient technical detail and symbolism to serve as a point of entry to further study, but there are many illustrations and adequate narrative text to convey the main concepts to those who may prefer to skip over any unfamiliar mathematics. These technical articles provide useful background for more applied entries based on statistics, such as "Biostatistics," "Epidemiology," and "Meta-Analysis," as well as for the implicit use of statistics in many other articles. They are further complemented by biographical entries on, for example, "Nightingale, Florence" and "Pascal, Blaise."

Conclusion

In the world of high-intensity science and technology, how does one lead the good life? What is the form of the just state? Is it sufficient to practice the traditional virtues in traditional ways? To apply received moral principles to new technological opportunities? Or is it not necessary to rediscover ethical and political practice in forms equal to the radical re-founding of knowledge and power that itself has constituted modern science and technology? Without in any way suggesting the end of tradition or of scientific and technological progress, *ESTE* seeks to make common cause with all persons of good will who see a need for critical ethical reflection in the midst of the new world we are creating—remembering that questions can be asked in order to seek the good with greater diligence. In a pluralistic world it is, in addition, no mean feat to practice such questioning with a tolerance and pursuit of principled compromise that avoids the failures of relativism or self-righteousness. The aspiration here is to provide common ground for scholars in the various disciplines who would place their work in broader perspectives, students desiring to deepen their knowledge of complex issues, scientists and engineers sharing their expertise with a participating public, and citizens who aspire to make intelligent decisions in the increasingly scientific and technological world in which we all now live.

CARL MITCHAM
EDITOR IN CHIEF

INTRODUCTORY ESSAYS

TECHNOLOGIES OF HUMILITY: CITIZEN PARTICIPATION IN GOVERNING SCIENCE

SHEILA JASANOFF

• • •

In his prescient 1984 book, the sociologist Charles Perrow forecast a series of "normal accidents" in high-risk technologies. The term applied with precision to events that were strung like dark beads through the later years of the twentieth century—most notably, the 1984 chemical plant disaster in Bhopal, India; the 1986 loss of the *Challenger* shuttle and, in the same year, the nuclear plant accident in Chernobyl, USSR; the contamination of blood supplies with the AIDS virus in Europe and North America; the prolonged crisis over BSE ("mad cow disease") in the United Kingdom; and the U.S. space program's embarrassing, although not life-threatening, mishaps with the Hubble telescope's blurry lens, and several lost and extremely expensive Mars explorers. To these we may add the discovery of the ozone hole, climate change, and other environmental disasters as further signs of disrepair. Occurring at different times and in vastly different political environments, these events nonetheless served collective notice that human pretensions of control over technological systems need serious reexamination.

American theorists like Perrow chalked up these failings of technology to avoidable error, especially on the part of large organizations (Clarke 1989, Short and Clarke 1992, Vaughan 1996), but some European analysts suggested a more troubling scenario. Passionately set forth by the German sociologist Ulrich Beck (1992), the thesis of "reflexive modernization" argued that risks are endemic in the way that contemporary societies conduct their technologically intensive business. Scientific and technical advances bring unquestioned benefits, but they also generate new uncertainties and failures, so that doubt continually undermines knowledge and unforeseen consequences confound faith in progress. The risks of modernity, Beck suggested, cut across social lines and operate as a great equalizer of classes. Wealth may increase longevity and improve the quality of life, but it offers no certain protection against the ambient harms of technological societies. This observation was tragically borne out when the collapse of the World Trade Center on September 11, 2001 ended the lives of some 3,000 persons, not discriminating among corporate executives, stock market analysts, computer programmers, secretaries, firefighters, policemen, janitors, and restaurant workers. In many other contexts, however, vulnerability remains closely tied to socioeconomic circumstances, inequalities persist in the ability of groups and individuals to defend themselves against risk.

"Risk," on this account, is not a matter of simple probabilities, to be rationally calculated by experts and avoided in accordance with the cold arithmetic of cost-benefit analysis (Graham and Wiener 1995). Rather, it is part of the modern human condition, woven into the very fabric of progress. The problem we urgently face is how to live well with the knowledge that our societies are inevitably "at risk." Critically important normative questions of risk management cannot be addressed by technical experts with conventional tools of prediction. Such questions determine not only whether we will get sick or die, and under what conditions, but also who will be affected and how we should respond to uncertainty and ignorance. Is it sufficient, for instance, to assess technology's consequences, or must we also seek to evaluate its aims? How should we choose when the values of science appear to conflict with other fundamental values? Has our ability to innovate in some areas run unacceptably ahead of our powers of control? Will some of our most revolutionary technologies increase inequality, promote violence, threaten cultures or harm the environment? And are our institutions, national or supranational, up to the task of governing our dizzying technological capabilities? (Never far from the minds of philosophers and authors of fiction, some of these concerns were also famously articulated in recent times by Bill Joy, co-founder and chief scientist of Sun Microsystems.)

To answer these questions, the task of managing technologies has to go far beyond the model of "speaking truth to power" that once was thought to link knowledge to political action (Price 1965). According to this template, technical input to policy problems must be developed independently of political influences; the "truth" so generated adequately constrains subsequent exercises of political power. The accidents and troubles of the late twentieth century, however, have called into question the validity of this model: both as a descriptively accurate rendition of ways in which experts relate to policy-makers (Jasanoff 1990), and as a normatively acceptable formula for deploying specialized knowledge within democratic political systems. There is growing awareness that even technical policy-making needs to get more political—or, more accurately, to recognize its political foundations more explicitly. Across a widening range of policy choices, technological cultures must learn to supplement the expert's narrow preoccupation with measuring the risks and benefits of innovation with greater attentiveness to the politics of science and technology.

But how can this expansion in the expert's role be reconciled with well-entrenched understandings of the relations between knowledge and power or expertise and public policy? How should these understandings be modified in response to three decades of research on the social dimensions of science? Can we imagine new institutions, processes, and methods for restoring to the playing field of governance some of the normative and political questions that were too long side-lined in assessing the risks and benefits of technology? And are there structured means for cultivating the social capacity for deliberation and reflection on technological change, much as expert analysis of risks has been cultivated for many decades?

There is a growing need, to this end, for what we may call "technologies of humility." These are methods, or better yet institutionalized habits of thought, that try to come to grips with the ragged fringes of human understanding—the unknown, the uncertain, the ambiguous, and the uncontrollable. Acknowledging the limits of prediction and control, technologies of humility confront "head-on" the normative implications of our lack of perfect foresight. They call for different expert capabilities and different forms of engagement between experts, decision-makers, and the public than were considered needful in the governance structures of high modernity. They require not only the formal mechanisms of participation but also an intellectual environment in which citizens are encouraged to bring their knowledge and critical skills to bear on the resolution of common problems.

The Social Contract between Science and the State

In the United States the need for productive working relations between science and the state was famously articulated not by a social theorist or sociologist of knowledge but by the quintessential technical expert: Vannevar Bush, the distinguished Massachusetts Institute of Technology (MIT) engineer and presidential adviser. Bush foresaw the need for major institutional changes following the intense mobilization of science and technology during the Second World War. In 1945 he produced a report, *Science: The Endless Frontier*, that laid the basis for American policy towards science and technology. Science, in Bush's vision, was to enjoy government patronage in peacetime as in war. Control over the scientific enterprise, however, would be wrested from the military and lodged with the civilian scientific community. Basic research, uncontaminated by industrial application or state ambitions, would thrive in the free air of universities. Scientists would establish the substantive aims as well as the intellectual standards for their research. Bush firmly believed that the bountiful results flowing from scientists' endeavors would be translated into beneficial technologies, contributing to the nation's prosperity and progress. Although his design took years to materialize, and even then was only imperfectly attained, the U.S. National Science Foundation (NSF) eventually emerged as the primary state funder of basic research. (The creation of the National Institutes of Health [NIH] to sponsor biomedical research divided U.S. science policy in a way not contemplated in Bush's original design. In the recent politics of science, NIH budgets have proved consistently easier to justify than appropriations for other branches of science.) The exchange of government funds and autonomy in return for discoveries, technological innovations and trained personnel came to be known as America's "social contract for science."

Signs of wear and tear in the "social contract" appeared in the 1980s. A spate of highly publicized cases of alleged fraud in science challenged the reliability of peer review and, with it, the underlying assumptions concerning the autonomy of science. The idea of science as a unitary practice also broke down as it became clear that research varies from one context to another, not only across disciplines, but—even more important from a policy standpoint—across institutional settings. It was recognized, in particular, that regulatory science, produced to support governmental efforts to manage risk, was fundamentally different from research driven by scientists' curiosity. At the same time, observers of science in society began questioning whether the categories of basic and applied research held meaning in a world where the production and uses of science were densely connected to each other, as well as to larger social and political consequences (Jasanoff, Markle, Petersen, and Pinch 1995).

Rethinking the relations of science with other social institutions generated three major streams of analysis. The first stream takes the "social contract" essentially for granted but points to its failure to work as its proponents had imagined. Many have criticized science, especially university-based science, for deviating from idealized norms of purity and disinterestedness. Despite (or maybe because of) its simplicity, this critique has seriously threatened the credibility of researchers and their claims to autonomy. Others have tried to replace the dichotomous division of *basic* and *applied* science with more differentiated categories, calling attention to the particularities of science done in different settings to meet different objectives. Still others have sought to respecify from the ground up how scientific knowledge is actually produced. This last line of analysis seeks not so much to correct or refine Vannevar Bush's vision of science as to replace it with a more complex account of how knowledge-making fits into the wider functioning of society.

DEVIANT SCIENCE. Scientific fraud and misconduct appeared on the U.S. policy agenda in the 1980s. Political interest reached a climax with the notorious case of alleged misconduct in an MIT laboratory headed by Nobel laureate biologist David Baltimore. He and his colleagues were exonerated after years of inquiry, which included investigations by Congress and the FBI (Kevles 1998). This and other episodes heightened the tendency for policy-makers and the public to suspect that all was not in order in the citadels of basic science and greatly increased federal powers for the supervision of research. Some saw the Baltimore affair as a powerful sign that legislators were no longer content with the old social contract's simple *quid pro quo* of money and autonomy in exchange for technological benefits (Guston 2001). Others, like the science journalist Daniel Greenberg (2001), accused scientists of profiting immoderately from their alliance with the state, while failing to exercise moral authority or meaningful influence on policy. American science, at any rate, was asked to justify more explicitly the public money spent on it. A token of the new relationship between science and government came with the reform of NSF's peer review criteria in the 1990s. The Foundation now requires reviewers to assess proposals not only on grounds of technical merit, but also with respect to their wider implications for society—thus according greater prominence to science's social utility. In effect, the fraud investigations of the previous decade opened up other taken-for-granted aspects of scientific autonomy, and forced scientists to account for their objectives as well as their honesty.

To these perturbations may be added a steady stream of challenges to the supposed disinterestedness of academic science. In areas ranging from climate change to biotechnology, critics have charged researchers with having sacrificed their objectivity in exchange for grant money or, worse, equity interests in lucrative start-up companies (Boehmer-Christiansen 1994). These allegations have been especially damaging to biotechnology, because that industry benefits significantly from the rapid transfer of skills and knowledge from universities. Since most western governments are committed to promoting such transfers, biotechnology is caught on the horns of a particular dilemma: how to justify its promises of innovation and progress credibly, when the interests of most scientists are aligned with those of industry, government or, occasionally, public interest advocates.

While financially motivated, pro-industry bias has attracted the most criticism, academic investigators have also come under scrutiny for alleged pro-environment and anti-technology biases. In several cases involving biotechnology—in particular, that of the monarch butterfly study conducted by Cornell University scientist John Losey (1999) in the United States, and Stanley Ewen and Arpad Puzstai's (1999) controversial rat-feeding study in the United Kingdom—industry critics questioned the quality of university-based research and implied that political orientations had prompted premature release or over-interpretation of results. In April 2002 a controversy erupted over an article in *Nature* by a University of California scientist, Ignacio Chapela, who concluded that DNA from genetically modified corn had contaminated native species in Mexico. Philip Campbell, the journal's respected editor, did not retract the paper, but stated that "the evidence available is not sufficient to justify the publication of the original paper," and that readers should "judge the science for themselves" (*Washington Times* 2002). As in the Losey and Ewen and Puzstai cases, critics charged that Chapela's science had been marred by non-scientific considerations. Environmentalists, however, viewed all these episodes as pointing to wholesale deficits in knowledge about the long-term and systemic effects of genetic modification in crop plants.

CONTEXT-SPECIFIC SCIENCE. The second line of attack on the science–society relationship focuses on the basic-applied distinction. One attempt to break out of that dualism was proposed by Donald Stokes (1997), whose quadrant framework, using Louis Pasteur as the prototype, suggested that "basic" science can be done within highly "applied" contexts. Historians and sociologists of science and technology have long observed that foundational work can be done in connection with applied problems, just as applied problem-solving is often required for resolving theoretical issues (for example, in designing new scientific instruments). To date, formulations based on such findings have been slow to take root in policy cultures.

Another example of the contextualing approach can be found in the work of Silvio Funtowicz and Jerome Ravetz (1992). They proposed to divide the world of policy-relevant science into three nested circles, each with its own system of quality control: (1) "normal science" (borrowing the term from Thomas Kuhn), for ordinary scientific research; (2) "consultancy science," for the application of available knowledge to well-characterized problems; and (3) "post-normal science," for the highly uncertain, highly contested knowledge needed for many health, safety, and environmental decisions. These authors noted that, while traditional peer review may be effective within "normal" and even "consultancy" science, the quality of "post-normal" science cannot be assured by standard review processes. Instead, they proposed that work of this nature be subjected to *extended peer review*, involving not only scientists but also the stakeholders affected by the use of science. Put differently, they saw accountability, rather than mere quality control, as the desired objective when science becomes "post-normal." (A problem with this analysis lies in the very term "post-normal science." When scientific conclusions are so closely intertwined with social and normative considerations as in Funtowicz and Ravetz's outermost circle, one may just as well call the "product" by another name, such as "socially relevant knowledge" or "socio-technical knowledge.")

Sheila Jasanoff's 1990 study of expert advisory committees in the United States provided another perspective on this issue by noting that policy-relevant science (also referred to as "regulatory science")—such as science done for purposes of risk assessment—is often subjected to a special kind of "peer review." Regulatory science is reviewed by multidisciplinary committees rather than by individually selected specialists. The role of such bodies is not only to validate the methods by which risks are identified and investigated, but also to confirm the reliability of the agency's interpretation of the evidence. Frequently, regulatory science confronts the need to set standards for objects or concepts whose very existence was not previously an issue for either science or policy: "fine particulate matter" in air pollution control; the "maximum tolerated dose" (MTD) in bioassays; the "maximally exposed person" in relation to airborne toxics; or the "best available technology" in programs of environmental regulation. In specifying how such terms should be defined or characterized, advisory committees have to address issues that are technical as well as social, scientific as well as normative, regulatory as well as metaphysical. What *kind* of entity, after all, is a "fine" particulate or a "maximally exposed" person, and by what markers can we recognize them? Studies of regulatory science have shown that the power of advisory bodies to definitively address such issues depends on their probity, representativeness, transparency, and accountability to higher authorities—such as courts and the public. In other words, the credibility of regulatory science rests upon factors that have more to do with democratic accountability than with the quality of science as assessed by peer scientists.

NEW MODES OF KNOWLEDGE PRODUCTION. Going beyond the quality and context-dependency of science, some have argued the need to take a fresh look at the structural characteristics of contemporary science in order to make it more socially responsive. Michael Gibbons and his co-authors (1994) concluded that the traditional disciplinary science of Vannever Bush's "endless frontier" has been largely supplanted by a new mode of knowledge production. The salient properties of this new mode, in their view, include the following:

- Knowledge is increasingly produced in contexts of application (i.e., *all* science is to some extent "applied" science);
- Science is increasingly transdisciplinary—that is, it draws on and integrates empirical and theoretical elements from a variety of fields;
- Knowledge is generated in a wider variety of sites than ever before, not just universities and industry, but also in research centers, consultancies, and think-tanks;
- Participants in science have grown more aware of the social implications of their work (i.e., more "reflexive"), just as publics have become more conscious of the ways in which science and technology affect their interests and values.

The growth of this new mode, as Gibbons et al. note, has necessary implications for quality control. Besides old questions about the intellectual merits of their work, scientists are being asked new questions about its marketability, and its capacity to promote social harmony and welfare.

In other work, Helga Nowotny, Peter Scott, and Michael Gibbons (2001) have grappled with the implications of these changes for knowledge production in public domains. Nowotny et al. propose the concept of "socially robust knowledge" as the solution to problems of conflict and uncertainty. Contextualization, in their view, is the key to producing science for public ends. Science that draws strength from its socially detached position is too frail to meet the pressures placed upon it by contemporary societies. Instead, they imagine forms of knowledge that gain robustness from their very embeddedness in society. The problem, of course, is how to institutionalize polycentric, interactive, and multipartite processes of knowledge-making within institutions that have worked for decades at keeping expert knowledge away from populism and politics. The question confronting the governance of science is how to bring knowledgeable publics into the front-end of scientific and technological production—a place from which they have historically been excluded.

The Participatory Turn

Changing modes of scientific research and development provide at least a partial explanation for the current interest in improving public access to expert decision-making. In thinking about research today, policy-makers and the public frequently focus on the accountability of science rather than its quality. As the contexts for science have become more pervasive, dynamic and heterogeneous, concerns about the integrity of peer review have transmuted into demands for greater public involvement in assessing the costs and benefits, as well as the risks and uncertainties, of new technologies. Such demands have arisen with particular urgency in the case of biotechnology, but they are by no means limited to this field.

The pressure for accountability manifests itself in many ways, including demands for greater transparency and participation. One notable example came with U.S. federal legislation in 1998, requiring public access, pursuant to the Freedom of Information Act, to all scientific research generated with public funds (Omnibus Consolidated and Emergency Supplemental Appropriations Act of 1999, P.L. 105–277, 1998). The provision was hastily introduced and scarcely debated. Its sponsor, Senator Richard Shelby (R-Alabama), tacked it on as a last-minute amendment to an omnibus appropriations bill. His immediate objective was to force disclosure of data from a controversial study by the Harvard School of Public Health of the health effects of human exposure to fine particulates. This Six Cities Study provided key justification for the U.S. Environmental Protection Agency's stringent ambient standard for airborne particulate matter, issued in 1997. This sweeping enactment showed that Congress was no longer willing to concede unchecked autonomy to the scientific community in the collection and interpretation of data. Publicly funded science, Congress determined, should be available at all times for public review.

Participatory traditions are less thoroughly institutionalized in European policy-making, but in Europe, too, recent changes in the rules and processes governing expert advice display a growing commitment to involving the public in technically-grounded policy decisions. In announcing the creation of a new Directorate General for Consumer Protection, for example, the European Commission observed in 1997 that, "Consumer confidence in the legislative activities of the EU is conditioned by the *quality and transparency* of the scientific advice and its use on the legislative and control process" (emphasis added). The commitment to greater openness is also evident in the strategies of several new United Kingdom expert bodies, such as the Food Standards Agency, created to restore confidence in the wake of the BSE crisis. Similarly, two major public inquiries—the Phillips Inquiry on BSE and the Smith inquiry on the Harold Shipman murder investigation—set high standards for public access to information through the Internet. All across Europe, opposition to genetically modified foods and crops prompted experiments with diverse forms of public involvement, such as citizen juries, consensus conferences, and referenda (Joss and Durant 1995).

Although admirable, formal participatory opportunities cannot by themselves ensure the democratic and deliberative governance of science. There are, to start with, practical problems. People may not be engaged enough or possess enough specialized knowledge and material resources to take advantage of formal procedures. Participation may occur too late to identify alternatives to dominant or default options; some processes, such as consensus conferences, may be too *ad hoc* or issue-specific to exercise sustained influence on policy. Even timely participation does not necessarily improve decision-making. Empirical research has consistently shown that transparency may exacerbate rather than quell controversy, leading parties to deconstruct each other's positions instead of deliberating effectively. Indeed, the Shelby Amendment reflects one U.S. politician's conviction that compulsory disclosure of data will enable challenges to researchers' own interpretations of their work. It is in this sense an instrument that can be used for fomenting scientific dissent. By contrast, participation constrained by established formal discourses, such as risk assessment, may not admit novel viewpoints, radical critique, or considerations lying outside the taken-for-granted framing of a problem.

Technologies of Humility

Participation alone, then, does not answer the problem of how to democratize technological societies. Opening the doors to previously closed expert forums is a necessary step—indeed, it should be seen by now as a standard operating procedure of democratic politics. But the formal mechanisms adopted by national governments are not enough to engage the public effectively in the management of global science and technology. What has to change is the *culture* of governance, nationally as well as internationally, and for this we need to address not only the mechanics but also the substance of participatory politics. The issue, in other words, is no longer whether the public should have a say in technical decisions, but how to promote more meaningful interaction among policy-makers, scientific experts, corporate producers, and the informed public.

The analytic ingenuity of modern states has been directed for many decades toward refining what we may call the "technologies of hubris." To reassure their publics, as well as to keep the wheels of science and industry turning, national governments have developed a series of predictive methods (e.g., risk assessment, cost-benefit analysis, climate modeling) that are designed, on the whole, to facilitate management and control, even in areas of high uncertainty (e.g. Porter 1995). These methods achieve their power through claims of objectivity and a disciplined approach to analysis, but they suffer from three significant limitations. First, they show a kind of peripheral blindness toward uncertainty and ambiguity. Predictive methods focus on the known at the expense of the unknown, producing overconfidence in the accuracy and completeness of the pictures they produce. Well-defined, short-term risks command more attention than indeterminate, long-term ones. At the same time, technical proficiency conveys the impression that analysis is not only rigorous, but complete—in short, that it has adequately taken account of all possible risks. Predictive methods tend in this way to downplay what falls outside their field of vision, and to overstate whatever falls within (Irwin and Wynne 1996).

Second, the technologies of predictive analysis tend to preempt political discussion. Expert analytic frameworks create high entry barriers against legitimate outsider positions that cannot express themselves in terms of the dominant discourse (Winner 1986). Claims of objectivity hide the exercise of judgment, so that the normative presuppositions of studies and models are not subjected to general debate. The boundary work that demarcates the space of "objective" policy analysis is carried out by experts, so that the politics of making demarcations remains locked away from public review and criticism (Jasanoff 1990).

Third, predictive technologies are limited in their capacity to internalize challenges that come from outside their framing assumptions. Techniques develop and grow more sophisticated, to be sure, but not necessarily in ways that revisit the values on which they were founded. For example, techniques for assessing chemical toxicity have become ever more refined, but they continue to rest on the demonstrably faulty assumption that people are exposed to one chemical at a time. Synergistic effects, long-term exposures, and multiple exposures are common in normal life but have tended to be ignored as too messy for analysis. Even in the aftermath of catastrophic failures, modernity's predictive models are often adjusted only to take on board lessons that are compatible with their initial assumptions. When a U.S.-designed chemical factory in Bhopal released the deadly gas methyl isocyanate, killing thousands, the international chemical industry made many improvements in its internal accounting and risk communication practices. But no new methods were developed to assess the risks of technology transfer between radically different cultures of industrial production.

At the beginning of the twenty-first century, the unknown, unspecified and indeterminate aspects of scientific and technological development remain largely unaccounted for in policy-making; treated as beyond reckoning, they escape the discipline of analysis as well as politics. What is lacking is not just the knowledge to help fill the gaps, but the processes and methods for eliciting what the public wants and for using what is already known. To bring these dimensions out of the shadows and into the dynamics of democratic debate, they must first be made concrete and tangible. Scattered and private knowledge has to be amalgamated, perhaps even disciplined, into a dependable civic epistemology. The human and social sciences of previous centuries undertook just such a task of translation. They made visible the social problems of modernity—poverty, unemployment, crime, illness, disease, and, lately, technological risk—often as a prelude to rendering them more manageable, using what I have termed the "technologies of hubris." Today, there is a need for "technologies of humility" to complement the predictive approaches: to make apparent the possibility of unforeseen consequences; to make explicit the normative that lurks within the technical; and to acknowledge from the start the need for plural viewpoints and collective learning. How can these aims be achieved?

From the abundant literature on technological disasters and failures, as well as from studies of risk analysis and policy-relevant science, we can abstract four focal points around which to develop the new technologies of humility. They are *framing*, *vulnerability*, *distribution*, and *learning*. Together, they generate the questions we should ask of almost every human enterprise that intends to alter society: what is the purpose; who will be hurt; who benefits; and how can we know? On all these points, we have good reason to believe that wider public engagement would improve our capacity for analysis and reflection. Participation that pays attention to these four points promises to lead to richer deliberation on the substance of decision-making.

FRAMING. It is an article of faith in the policy literature that the quality of solutions to perceived social problems depends on the adequacy of their original framing (Schon and Rein 1994). If a problem is framed too narrowly, too broadly, or simply wrongly, then the solution will suffer from the same defects. To take a simple example, a chemical testing policy focused on single chemicals cannot produce knowledge about the environmental health consequences of multiple exposures: the framing of the regulatory issue is more restrictive than the actual distribution of chemical-induced risks, and hence is incapable of delivering the optimal management strategies. Similarly, a belief that violence is genetic may discourage the search for controllable social influences on behavior. A focus on the biology of reproduction may delay or impede effective policies for curbing population growth. When facts are uncertain, disagreements about the appropriate frame are virtually unavoidable and often remain intractable for long periods. Yet, few policy cultures have adopted systematic methods for revisiting the initial framing of issues, despite calls to do so (Stern and Fineberg 1996). Frame analysis thus remains a critically important, though neglected, tool of policy-making.

VULNERABILITY. Risk analysis treats the "at-risk" human being as a passive agent in the path of potentially disastrous events. In an effort to produce policy-relevant assessments, human populations are often classified into groups (e.g., most susceptible, maximally exposed, genetically predisposed, children or women) that are thought to be differently affected by the hazard in question. Based on physical and biological indicators, these classifications tend to overlook the social foundations of vulnerability and to subordinate individual experiences of risk to aggregate numerical calculations (e.g. Irwin and Wynne 1996). Recent efforts to analyze vulnerability have begun to recognize the importance of socio-economic factors, but assessment methods still take populations rather than individuals as the unit of analysis. These approaches not only disregard differences within groups but reduce individuals to statistical representations. Such characterizations leave out of the calculus of vulnerability such factors as history, place, and social connectedness, all of which may play crucial roles in determining human resilience. Through participation in the analysis of their vulnerability, ordinary citizens might regain their status as active subjects rather than remain undifferentiated objects in yet another expert discourse.

DISTRIBUTION. Controversies over such innovations as genetically modified foods and stem cell research have propelled ethics committees to the top of the policy-making ladder in several countries. Frequently, however, these bodies are used as "end-of-pipe" legitimation devices, reassuring the public that normative issues have not been omitted from deliberation. The term "ethics," moreover, does not cover the whole range of social and economic realignments that accompany major technological changes, nor their distributive consequences, as technology unfolds across global societies and markets. Attempts to engage systematically with distributive issues in policy processes have not been altogether successful. In Europe, consideration of the "fourth hurdle"—the socioeconomic impact of biotechnology—was abandoned after a brief debate. In the United States the congressional Office of Technology Assessment, which arguably had the duty to evaluate socio-economic impacts, was dissolved in 1995 (Bimber 1996). President Clinton's 1994 injunction to federal agencies to develop strategies for achieving environmental justice produced few dramatic results (Executive Order 12298, 1994). At the same time, episodes like the rebellion against Monsanto's "terminator gene" demonstrate a deficit in the capacity for ethical analysis in large corporations, whose technological products can fundamentally alter people's lives. Sustained interactions between decision-makers, experts and citizens, starting at the upstream end of research and development, could do much to expose the distributive consequences of innovation.

LEARNING. Theorists of social and institutional learning have tended to assume that what is "to be learned" is never a part of the problem. A correct, or at least a better, response exists, and the only issue is whether actors are prepared to internalize it. In the real world, however, learning is complicated by many factors. The capacity to learn is constrained by limiting features of the frame within which institutions act. Institutions see only what their discourses and practices permit them to see. Experience, moreover, is polysemic, or subject to many interpretations, no less in policy-making than in literary texts. Even when the fact of failure in a given case is unambiguous, its causes may be open to many different readings. Just as historians disagree over what caused the rise or fall of particular political regimes, so policy-makers may find it impossible to attribute their failures to specific causes. The origins of a problem may look one way to those in power, and quite another way to the marginal or the excluded. Rather than seeking monocausal explanations, then, it would be fruitful to design more avenues through which societies can collectively reflect on the ambiguity of their experiences and to assess the strengths and weaknesses of alternative explanations. Learning, in this modest sense, is a suitable objective of civic deliberation.

Conclusion

The enormous growth and success of science and technology during the last century has created difficult contradictions for institutions of governance. As technical activities have become more pervasive and complex, so too has the demand grown for more complete and multivalent evaluations of the costs and benefits of technological progress. It is widely recognized that increased participation and interactive knowledge-making would improve accountability and lead to more credible assessments of science and technology. Such approaches would also be consistent with changes in the modes of knowledge production, which have made science more socially embedded and more closely tied to contexts of application. Yet, modern institutions still operate with conceptual models that seek to separate science from values and emphasize prediction and control at the expense of reflection and social learning. Not surprisingly, the real world continually produces reminders of the incompleteness of our predictive capacities.

To move public discussion of science and technology in new directions, there is a need for "technologies of humility," complementing the predictive "technologies of hubris" on which we have lavished so much of our past attention. These *social technologies* would give combined attention to substance and process, and stress deliberation as well as analysis. Reversing nearly a century of contrary development, these approaches to decision-making would seek to integrate the "can do" orientation of science and engineering with the "should do" questions of ethical and political analysis. They would engage the human subject as an active, imaginative agent in making policy, as well as a source of knowledge, insight, and memory.

BIBLIOGRAPHY

Beck, Ulrich. (1992). *Risk Society: Towards a New Modernity*. London: Sage Publications.

Bimber, Bruce. (1996). *The Politics of Expertise in Congress: The Rise and Fall of the Office of Technology Assessment*. Albany: State University of New York Press.

Boehmer-Christiansen, Sonja. (1994). "Global Climate Protection Policy: The Limits of Scientific Advice, Parts 1 and 2." *Global Environmental Change* 4(2): 140–159; 4(3): 185–200.

Bush, Vannevar. (1945). *Science: The Endless Frontier*. Washington, DC: U.S. Government Printing Office.

Clarke, Lee. (1989). *Acceptable Risk? Making Decisions in a Toxic Environment*. Berkeley: University of California Press.

Ewen, Stanley W. B., and Arpad Pusztai. (1999). "Effect of Diets Containing Genetically Modified Potatoes Expressing Galanthus Nivalis Lectin on Rat Small Intestine." *Lancet* 354: 1353–1354.

Executive Order 12298. (1994). "Federal Actions to Address Environmental Justice in Minority Populations and Low-Income Populations." Washington, DC: February 11.

Funtowicz, Silvio O., and Jerome R. Ravetz. (1992). "Three Types of Risk Assessment and the Emergence of Post Normal Science." In *Social Theories of Risk*, eds. Sheldon Krimsky and D. Golding. New York: Praeger.

Gibbons, Michael; Camille Limoges; Helga Nowotny, et al. (1994). *The New Production of Knowledge*. London: Sage Publications.

Graham, John D., Jonathan B. Wiener, eds. (1995). *Risk versus Risk: Tradeoffs in Protecting Health and the Environment*. Cambridge, MA: Harvard University Press. A preeminent example of the calculative approach.

Greenberg, Daniel S. (2001). *Science, Money, and Politics: Political Triumph and Ethical Erosion*. Chicago: University of Chicago Press.

Guston, David H. (2001). *Between Politics and Science: Assuring the Integrity and Productivity of Research*. Cambridge: Cambridge University Press.

Irwin, Alan, and Brian Wynne, eds. (1996). *Misunderstanding Science? The Public Reconstruction of Science and Technology*. Cambridge: Cambridge University Press.

Jasanoff, Sheila. (1990). *The Fifth Branch: Science Advisers as Policy-makers*. Cambridge, MA: Harvard University Press.

Jasanoff, Sheila; Gerald E. Markle; James C. Petersen; and Trevor Pinch, eds. (1995). *Handbook of Science and Technology Studies*. Thousand Oaks, CA: Sage Publications.

Joss, Simon, and John Durant, eds. (1995). *Public Participation in Science: The Role of Consensus Conferences in Europe*. London: Science Museum.

Kevles, Daniel J. (1998). *The Baltimore Case: A Trial of Politics, Science, and Character*. New York: Norton.

Losey, John E.; L. S. Rayor; and M. E. Carter. (1999). "Transgenic pollen harms monarch larvae" *Nature* 399: 214.

Nowotny, Helga; Peter Scott; and Michael Gibbons. (2001). *Re-Thinking Science: Knowledge and the Public in an Age of Uncertainty*. Cambridge, UK: Polity.

Omnibus Consolidated and Emergency Supplemental Appropriations Act of 1999, P. L. 105-277 (codified at 42 U.S.C. § 1395 *et. seq.*). (1998). The Office of Management and Budget in the Clinton administration controversially narrowed the scope of the law to apply not to *all* publicly-funded research, but only to research actually relied on in policy-making. The issue was not completely resolved as of May 2005.

Perrow, Charles. (1984). *Normal Accidents: Living with High Risk Technologies*. New York: Basic Books.

Porter, Theodore M. (1995). *Trust in Numbers: The Pursuit of Objectivity in Science and Public Life*. Princeton: Princeton University Press.

Price, Don K. (1965). *The Scientific Estate*. Cambridge, MA: Harvard University Press. The locus classicus of the view of the right relations between knowledge and power.

Schon, Donald A., and Martin Rein. (1994). *Frame/Reflection: Toward the Resolution of Intractable Policy Controversies*. New York: Basic Books.

Short, James F., and Lee Clarke, eds. (1992). *Organizations, Uncertainties, and Risk*. Boulder: Westview Press.

Stern, Paul C., and Harvey V. Fineberg, eds. (1996). *Understanding Risk: Informing Decisions in a Democratic Society*. Washington, DC: National Academy Press.

Stokes, Donald E. (1997). *Pasteur's Quadrant: Basic Science and Technological Innovation*. Washington, DC: Brookings.

Vaughan, Diane. (1996). *The Challenger Launch Decision: Risky Technology, Culture, and Deviance at NASA*. Chicago: University of Chicago Press.

Winner, Langdon. (1986). "On Not Hitting the Tar Baby." In his *The Whale and the Reactor*. Chicago: University of Chicago Press.

INTERNET RESOURCES

European Commission. *1997 Communication of the European Commission on Consumer Health and Safety*. COM (97) 183 fin. Available from http://europa.eu.int/comm/food/fs/sc/index_en.html.

Joy, Bill. (2000). "Why the Future Doesn't Need Us." *Wired*. http://www.wired.com/wired/archive/8.04/joy.html.

Washington Times. "Nature Regrets Publication of Corn Study." April 5, 2002. Available from http://www.washingtontimes.com/national/20020405-9384015.htm.

ETHICS AND TECHNOLOGY: A PROGRAM FOR FUTURE RESEARCH

DEBORAH G. JOHNSON
THOMAS M. POWERS

• • •

In this paper we present a program for future study of ethics and technology. Most generally, the analysis involves understanding the role of technology in moral action. On the one hand, technology shapes and is shaped by moral thought and action; on the other, this shaping is rarely acknowledged, let alone understood, by moral philosophers. Thus the program sketched here is aimed at making technology visible as an element of moral philosophy. We lay out a line of reasoning that uncovers the intentionality of the design of technological artifacts, and then we compare human moral action to features of the design and use of technological artifacts. This line of reasoning provides the groundwork for extensive future research. The program description is both a plan of study for our own research as well as a call for other scholars to turn their attention to the issues outlined.

In thinking about the nature of a technology, we argue that traditional philosophical theories of human action and ethics can be usefully extended to technology. Contemporary action theory has suggested a causal model of intentional behavior in humans, and we believe that (with modification) this model is applicable to technology. Indeed, when technology is viewed in relation to a causal model of intentional behavior, the moral nature of technological agency becomes apparent. Similarly, traditional notions from ethics, such as goodness, responsibility, and accountability, can be extended in order to understand technology in a new light.

The Artifactual Platform

The world in which humans act and live is a world filled with human-made objects. In addition to the objects of the natural world, these human-made objects provide an enabling and inhibiting background for human thought and action, and for all of the arrangements of human life. This background influences and informs what we think, how we act, and how we arrange ourselves into units, organizations, and institutions.

By noting the presence of human-made objects, we introduce a distinction between the human-made and the natural world, though we readily admit the two are intertwined. Indeed, they are often so intertwined that it is difficult to separate them. The natural world has been dramatically affected by human activity, and technology is, at least in part, the manipulation of natural potential. Scientific research from the late twentieth and early twenty-first centuries suggests there is very little left of a natural world that is untouched by human agency; the balance, over human history, has clearly shifted toward a relatively larger class of human-made objects. In other words, we are living in the anthropocene, on an increasingly anthropogenic planet (Allenby 2004).

Even though, as a matter of ontology, it will be increasingly difficult to maintain a distinction between the classes of human-made and natural objects, the difference remains significant. The human-made world could be otherwise, and the future human-made world is, to some extent, a matter of human choice and human action. Indeed, work in "normative" design and engineering, seen in the universal design, green engineering, and appropriate technology movements, presupposes that there are morally better (and worse) ways to create the future human-made world. The analysis herein provides these normative enterprises with a philosophical footing.

Moral philosophy has always presumed the natural world as the background for human action and morality, but has failed to recognize the powerful role of the human-made world in moral thought and behavior. Rather than focusing on the background, moral philosophy has concentrated attention on human agency, and the presumption has been that moral action (through human beings) is part of the embodied world. The embodied world has been understood to consist both of natural things and human bodies, though, to be sure, some ethicists have acknowledged that morality might be different if humans had different sorts of bodies or acted in a natural world ordered in a different way. Moral philosophers have considered a typical action to consist of an agent (an embodied being) moving his or her body in some way, even if only in a very small way—a wink, a bit of pressure on a trigger, and so on. If the agent does not move his or her body in some way, then there is no action. Even speech acts require movement of the speech organs, and most philosophers have recognized that humans can commit moral wrongs with mere words.

So our starting place is the idea that human agency operates in an embodied world, noting how the embodied world includes both human-made and natural objects. But we want to call attention to the *normative* features of the human-made part and come to grips with the moral importance of technology in constituting the background for human action. We will call the human-made part of the embodied world, as far as it concerns human action, the *artifactual platform*. This platform is the class of constructed objects and systems of objects that are created by and come to influence human action.

Often, descriptions of action incorporate human-made objects into the action. For instance, when we say "John shot Bill," use of a gun is implicit; when we say "Mary flew to London," use of an airplane is presumed; and so on. This feature of descriptive language is what Joel Feinberg (1970) has called the "accordion effect." We can choose an expanded description that includes the artifact, or a collapsed version that conceals it.

When those who study action from the normative point of view use narrow or collapsed descriptions, the technological component is glossed over. What is missed is that particular movements of an agent's body could not have had their associated effects were it not for an artifact. Noting the artifacts involved in moral behavior is the first step in gaining a better understanding of the role of the artifactual platform in morality. Becoming aware of this platform allows us to see that a good deal of moral behavior is action *with* technology. In this respect, moral actions, agents, and patients are not sufficient for an ontology of morality; artifacts are also part of the moral world. The task of understanding the role of artifacts in morality is, then, a matter of recognizing the difference it makes for humans to live in a world with the particular artifacts that currently exist or might exist in the future.

Nevertheless, realizing that moral action takes place *with technology*, and on or from an artifactual platform, does not go far enough. As indicated, technological artifacts with their particular features are matters of human choice. Just as humans deliberate about and choose their actions, some humans (artisans and engineers) deliberate about and create artifacts; other humans (consumers and users) choose and employ artifacts that enable and constrain moral action. Human agency is significantly affected by technological artifacts. It may be augmented, constrained, or merely altered. The design, availability, and employment of technology shapes what humans can do, and what they end up doing.

What, then, is the significance of technology? Technology expands and constrains the range of human moral behavior, and changes the character of that behavior. Technology is far from neutral in its combination with human behavior. Can one say that it has moral agency? This question can be pursued by considering relations between human moral agency and technology.

The Moral Agency of Technology

The question of the moral agency of technology can be used as an entry point for exploring the role of technology in morality. Grounding it in philosophical concepts, the analysis starts with the traditions of ethical theory and action theory and the accounts of human moral agency they provide. In ethical theory, the standard account of the responsibility of moral persons (acting without technology) says that individuals are primarily responsible for their voluntary, intended behaviors. In action theory, there is a broader account of intentionality, in which intentional states ("intendings" as well as desires, beliefs, plans, etc.) are the causes of action. The intentionality of these states is a property that relates them to states of affairs and objects in the actual world and in possible worlds. Intentionality, then, is "aboutness" or directedness. On this view, voluntary action or intended behavior is understood to be outward behavior caused by a complex of internal mental states. By stipulating the specific kind of intending, desiring, and believing that causes a particular action, philosophers have distinguished moral action from nonmoral behavior. Because the outward behavior in moral action is the result of these internal mental states, it is amenable both to a causal explanation and to a "reason explanation" (see Davidson 2001). That is, when we ask why someone acted in a particular way, he or she can offer antecedent intendings, beliefs, desires, and other intentional states as reasons for the action.

The standard philosophical account is spelled out in contemporary work in ethical theory and action theory, but the roots of the account are much older. The subject matter of moral appraisal even as far back as Aristotle (384–322 B.C.E.) has been understood to be intended, voluntary behavior. This is action, conduct, or the commission of a deed, as opposed to "mere" reaction or nonvoluntary behavior. In contemporary action theory, Aristotle's basic view is elaborated upon, and this produces the following conditions for moral action. First, there is a potential agent with an internal state. The internal state consists of intentional mental states, one of which is, necessarily, an intending to act. Together, the intentional states (e.g., belief that X is possible, desire to X, plus an intending to X) constitute a reason for X-ing. Second, there is an outward, embodied event—the agent does something, moves his or her body in some way. Third, the internal state is the cause of the outward event; that is, the movement of the body is rationally directed and is an action insofar as it is caused by an internal state. Fourth, the outward action has an outward effect. Finally, the effect has to be on a patient—the recipient of an action that can be harmed or helped. Moral patients are typically human beings, but the class may include other beings or things as well. Some ethicists now include higher functioning animals, entire species, and even ecosystems in the class of moral patients, and clearly technology does seriously affect ecosystems and nonhuman animals.

The convergence of these parts of ethical theory and action theory has produced a plausible account of the connection between thought and action, and has helped locate the focal point of moral agency. We adopt this account as the framework in which to consider the moral agency of technology. In other words, whether or not or in what ways technology has moral agency can best be revealed by comparing features of technology with the standard account of moral action as derived from ethical theory and action theory.

Interesting work has been done in the late twentieth and early twenty-first centuries along these lines, as philosophers have turned to consider the possibility of nonhuman moral agents (Allen, Varner, and Zinser 2000, Floridi and Sanders 2001, Brooks 2002, Kurzweil 1999, Danielson 1992). Most attention has been given to artificially intelligent computers as the best candidates for agency. Computers have drawn attention in part because of the interest in the precise nature of intelligence. Some philosophers of artificial intelligence (AI) seem to think that intelligence can emerge out of the complex states of computers. This view implies that the ability of a computer to generate intentional states on its own would go a long way toward making it like a human moral agent. (Researchers in AI are primarily interested in engineering robotic computers to do things such as sense, recognize, navigate, and modify, and not, in the main, concerned with the deeper implications of AI for a philosophical account of intelligence.) A thrust of the account here is to draw attention away from the project of considering intelligence and computers, and instead to explore technological artifacts more broadly, as entities that have intentional states that are not mental states.

At the heart of our argument for the moral significance of technology is the claim that artifacts have intentionality, the property of "aboutness" or directedness toward the actual world and a future designed world. One of the reasons so little attention has been given to ethics and technology seems to be a failure to recognize the intentionality designed into technological artifacts. On the one hand, the only type of intentionality of interest to ethicists has been the type found in the mental states of *human* agents. With its focus on human agents, ethical theory has not recognized the importance and relevance of the design and use of technological artifacts by human agents. On the other hand, scholars in science and technology studies have introduced the idea of technology having a kind of agency (Law 1987, Callon 1986). However, they have not recognized the ethical implications of this move. Nor have they related technological agency to the broader philosophical literature on action. The argument in this essay brings ethical theory and action theory to bear on the moral agency of technology.

Because the program outlined here builds on our claim that artifacts have intentionality, it will be helpful to discuss the theoretical apparatus traditionally used to describe intentionality in moral action. In order for a human action to be both open to "reason explanation" and subject to moral appraisal, there must be in the agent some collection of intentional mental states connected to the action in some fairly specific ways. Agents are subject to moral appraisal in virtue of those intentional acts that have morally relevant effects on moral patients. Intentional acts are caused by a variety of intentional states and/or entities: beliefs, desires, intendings, maxims, plans, and the like. An agent is a being who acts, with the cause of the action originating in the agent's mind as the complex of intentional states. The cause of the action is the primary reason for the action, and the cause as a whole can be seen as a collection of intentional states that serve as a "reason explanation" of the action. Intentional entities are entities that are capable of having intentional states; intentional actions are those actions that are caused by intentional states.

Our extension of this view of agency does not entail that artifacts have mental states or the ability to intend. We claim only that artifacts have intentionality or directedness at users and environments, and that this intentionality is causally efficacious. In proposing that intentionality is designed into technological artifacts, we avail ourselves of a quite general definition of intentionality, according to which it is the property of something, such that it is directed at or represents an object or state of affairs. The term *intentionality* is broadly construed so that intentional entities can be states of mind, sentences, speech acts, maps, or any designed object. Though this view of intentionality is quite broad, we nonetheless agree with the traditional view that humans are intentionality-generating beings. Their states of mind are directed at or about objects and states of affairs, and it is this original power of mind as intentionality generating that accounts for the intentionality in nonmental entities.

Humans have the ability to externalize their intentional states in speaking and writing. Spoken and written declarative sentences are intentional, just as are the beliefs that they express. While sentences and signs originate in the processes of the mental realm, these entities come into being only when they are expressed outwardly. Clearly, some intentional entities remain *internal* to humans, such as mental states of belief, desire, and visual perception. Internal intentional states explain the actions of human moral agents in that the intentional entities cause the actions and count as reasons why the agent committed the act. As for the external intentional entities, once they come into being and are (by definition) physically separated from the human who generated them, they still rely on a community of intentionality-generating beings (interpreters) in order to be intentional—in order for their intentionality to be grasped. Examples are maps, chairs, sentences in a natural language, and works of art. External intentional entities, like their internal counterparts, can cause and explain action. For example, the stop sign causes drivers to step on the brakes and bring their vehicles to a stop; the speech act of commanding individuals to behave in a certain way may cause individuals to do what is commanded; and so on.

The internal/external distinction in intentional entities takes into consideration the kinds of intentionality in human minds, in tangible expressions such as sentences and speech acts, and in representational states that are found in designed artifacts. Internal intentional states are those that necessarily remain mental; external intentional states, by contrast, are expressed in the form of entities that exist outside of the mind. An internal intentional state such as a belief often leads to an external intentional entity by means of a process not yet fully understood, but still assumed to be causal in nature. We argue that designed artifacts such as maps, computer programs, cars, and the like are externalized expressions of internal intentional states. They are intentional entities that cause action with morally relevant effects.

The most difficult part of the account here is the claim that things other than mental states can be about, be directed at, or represent objects and states of affairs. This claim seems noncontroversial when applied to sentences, speech acts, and maps. For instance, John R. Searle (2001) describes maps and house blueprints as intentional entities. Thus it should not be controversial when it comes to technological artifacts. While we claim that technological artifacts are intentional entities, we acknowledge that in the standard account of agency and action, agents have a specific intentional state of intending to perform a particular action, plus some more basic intentional states such as beliefs and desires. Because we claim that artifacts are intentional entities, the obvious question is what kind of intentionality do they have? That is, do they have something akin to the basic intentional states of humans, such as beliefs and desires, or something like the specific states of intending?

The Functionality and Intentionality of Artifacts

Our argument for the intentionality of technological artifacts is based on a particular understanding of the intentional states that artifacts can have. These intentional states cannot be fully understood without reference to the functions of the artifact. Accordingly, our account of the functionality of artifacts will be developed by answering three questions. What are functions in an artifact? How do they get into the artifact? What do users do with functions?

WHAT ARE THE FUNCTIONS IN AN ARTIFACT? Typically artifacts are thought to have functions, and their functionality is framed in terms of purposive or teleological explanation. While we do not reject this approach, we want to suggest a different view—one that allows for the flexibility we find in the design and use of artifacts. We base our understanding of the functionality of artifacts on the model of mathematical functions. An artifact has a function when it takes some input from a domain of human behaviors and produces a result within a range—what we generically call the output. The behavior of the user with the artifact fits the mathematical model of functions in that it consists of a relational triple: input, rule of transformation, and output. In the case of both mathematical functions and artifacts, one of two things can happen in the functional transformation. Either an input maps onto exactly one output (in which case the relation is one-to-one), or many different inputs map onto one output (a many-to-one relation). The definition of a function precludes the possibility that a particular input will deliver varying outputs (except in the case of artifacts such as slot machines whose one output is to produce varying outputs). This is an important condition for mathematical functions as well as artifactual ones. An artifact ceases to be useful (or even sometimes safe) when its output is unpredictable (except, again, when unpredictability is the designed output), and this is exactly what happens when a user gets different outputs for the exact same input on different occasions.

Here is an example of a technological function. A designer of a braking system for cars would model input by considering reaction times, leg position, pedal pressure, and stopping force for drivers who wish to control a typical car by pressing on the brake pedal. This process of design begins to reveal how the artifact becomes intentional; the input model is "about" driver capabilities and driving conditions—what we can gloss as "input" and "environment" aspects of the model. The transformation rule for the function, which is embodied in the mechanical parts of the braking system, turns those anticipated inputs into a result: The car slows at an appropriate speed. This is how the intentional states are actually manifested in the artifact; they are "materialized" in the way the artifact transforms the input. A successful braking system will incorporate realistic reaction times and pressures for the vast majority of drivers, and will reliably transform those inputs into the safe braking of a car under most conditions. A proper braking system will not map the different outcomes "stop the car" and "accelerate the car" to the exact same driver behavior. Design functions, like mathematical functions, are not one-to-many relations.

When an artifact *appears* to function differently with the same inputs, either the artifact is broken or there is a mistake about the sameness of inputs. The input mode for many complex artifacts such as computers is context dependent. For example, when the input of "striking the return key" on the keyboard yields different results at different times, this is because the computer is in different states during the respective inputs. In some programs, a query can be answered affirmatively by striking the return key. In others—word processors, for example—striking the return key places a hard return in a document. The lesson is that inputs are always tied to context. The condition that the artifactual functions borrow from mathematical ones reveals that there will never be more than one output for an *input in a context*. We may get spaces in some word-processing documents when we push the return key, and affirmations to queries when running other programs, but we will never get spaces sometimes and affirmations other times, in the exact same input context.

HOW DO FUNCTIONS GET INTO AN ARTIFACT? Crucial to this account is the fact that transformation rules of functions cannot be built into artifacts without applying intentional models of users and the world in which they operate.

There are two immediate senses in which the intentionality that begins with design is connected to technological artifacts in use. The act of design always requires intentionality—the ability of a designer to represent, model, perceive, and the like. Similarly, the use of an artifact—grasping a tool, following the user's guide—requires typical forms of cognition that feature intentionality. But there are deeper ways intentionality connects to designed functions and uses, ways that go beyond the intentionality of designers and users. When designers design artifacts, they poise them to behave in certain ways. Those artifacts *remain* poised to behave in those ways. They are designed to produce unique outputs when they receive inputs. They are directed at states of affairs in the world and will produce other states of affairs in the world when used. The telephone is "about" typical human fingers and ears, auditory capacities, and the physics of sound—it is intentional with respect to certain organisms and their environments. In a complicated way, the intentionality of the telephone is required to make it work as a communication device. But the telephone is also directed at certain social facts; it is about a world in which individuals want to talk with others who are beyond the reach of (unassisted) human voices. The telephone also requires that users memorize or keep a record of numbers attached to persons. Otherwise, a potential caller will not be able to use the telephone. Long after the designer has poised the artifact, the functions still reside in it and make complex actions possible. The argument here receives support from an analysis by Fred Dretske (1989) of what he terms the "design problem," as exemplified by how to get a mechanical system to do something that its designers find important, such as how to get a temperature indicator to be a switch for turning on a furnace.

WHAT DO USERS DO WITH FUNCTIONS? Users do not merely comply with the behavioral requirements designed into artifacts; they do not merely "satisfy" the model of use. They can add to the functions of an artifact by envisioning an unanticipated input that yields a novel output. This envisioning itself begins as an intentional state in the user, but it is then manifest in outward ways. An example of this is when someone picks up a television and throws it at an attacker to stop the attack. Here the user sees that by providing a particular kind of input (lifting and throwing), the television can be used to produce an output that it was not originally designed to produce.

The intentional states of artifacts are the result of the work of the artifact designer; designers *mold* intentionality into artifacts by concretizing the intentional models so that they enable the transformations promised by the functions. Users then deploy these functions by supplying inputs to the artifacts, under the prescribed conditions. Our argument is thus more than that the intentionality of designers and users becomes operative when artifacts get put to use. Our claim is that artifacts are in some sense chunks of intentionality, externalized by artifact designers and deployed by users in particular contexts.

When the intentionality and functionality of artifacts are seen in this light, it becomes difficult to locate precisely the agency in human actions with technological artifacts. There is intentionality in the mind of the artifact user, in the intentional states and functions of the artifact, and in the designer who created the intentionality and functionality embodied in the artifact. What may begin as the intentional model of a designer gets molded into an artifact and then deployed by the user. Hence, there is a complex of agency with human and nonhuman components.

We thus acquire a picture of moral action with technology as a complex combination of the intentionality of artifact designer, the intentionality of the artifact, and the intentionality of the user. Does this mean that artifacts are moral agents? If we return to the standard account of moral agency, it is now clear that artifacts meet most but not all of the conditions. Remember that on the standard account, human moral agency includes the stipulation of a potential agent with internal mental states, and one of these states is an intending to act. The agent does something, moves his or her body in some way, such that the internal states are the cause of the movement. The internal, mental states are thus also the reason for the action. The movement or behavior has an effect on a moral patient, someone or something that can be harmed (or helped).

Our analysis of human-action-with-artifact overlaps significantly with standard (nontechnological) human action, even though it locates agency in the triad of designer, artifact, and user. We have found that intentional states are spread out over designers, artifacts, and users, so that the action of the human-agent-with-artifact is caused by intentional states in each member of the triad. A complete reason explanation must include an account of the intentional states and functions of the artifact, because these states and functions play a causal role in the eventual action. The causal role of the artifact is necessary, but not sufficient, for the effect on the moral patient. True, artifacts alone are not agents, nor are their intentional states in any way internal mental states. Likewise, artifacts alone do not intend. But the intentional states of artifacts shape and cause external or embodied movement, both in terms of functional inputs of users and in terms of artifactual output. And intentional, caused, embodied movement can have morally relevant effects on patients. Thus, the intentionality and functionality of artifacts are important components of a full picture of moral action.

This account has implications for the notion of moral responsibility. Because philosophers and others may resist the idea of any kind of agency or even intentionality being attributed to technology because it may appear to deflect responsibility from human actors, it is appropriate to consider the issue of responsibility in a case study. Can technological artifacts be said to bear moral responsibility, or even to be morally good or bad entities?

An Illustration: The Moral Evaluation of Computers

At first glance, the idea of artifacts bearing moral responsibility appears implausible. There is, however, a form of human moral responsibility that is applicable to certain kinds of computer systems that may have broader application to other technologies. We refer here to the responsibility of human surrogate agents to their clients. Human surrogate agents are those who act on behalf of others. For example, lawyers, tax accountants, estate executors, and managers of performers and entertainers pursue the interests of their clients. The behavior of these agents is evaluated in terms of how well they pursue their client's interest while staying within the constraints and expectations associated with their roles. Like surrogate agents, computer systems pursue interests of their users; hence, their behavior can be evaluated in terms of how well they pursue the interests of their users.

If computer systems can be understood as surrogate agents for their human users, it would seem that role morality can be extended to computer systems, and this is a reason for attributing moral responsibility to computer systems and for morally evaluating such systems. In essence, the suggestion here is that the concept of role morality can be understood as a set of constraints on behavior, based on the interests of others, and can be applied to the functionality of particular computer systems. Just as human surrogate agents are evaluated in terms of whether they adequately understand and represent the point of view of their clients, one can evaluate computer systems in terms of how they represent and pursue the user's interests. Such an evaluation would involve many aspects of the system, including what it allows as user input and how it goes about implementing the interests of the user.

Consider the search engine surrogate that pursues a user's interest in finding web sites on a particular topic. Whether the search engine lists web sites in an order that reflects highest use, or fails to list some sites, or gives priority to sites for which the owner has paid to be listed—all of this can have moral implications (Introna and Nissenbaum 2000). We might say, then, that the computer system takes on a third-person, interested perspective, either of the user or of someone else. Several important questions arise. Does the system act on the actual user's interests, or on a restricted conception of the user's interests? Does the system competently pursue the user's interests, without pursuing other, possibly illegitimate interests such as those of advertisers, computer hardware or software manufacturers, government spying agencies, and the like? Are faulty or buggy computer systems analogous to misbehaving human surrogate agents? Do they fail to do the tasks (or to adequately do the tasks) that users employ them to do?

The foregoing suggests the kind of moral evaluation that can be made when computer systems are seen as surrogate agents. Tax preparation programs perform like tax advisers; contract-writing programs perform some of the tasks of attorneys; Internet search engines seek and deliver information like information researchers or librarians. Other types of programs and computer systems serve the interests of users, but there are no corresponding human surrogate agents with whom to compare them. Spyware programs uncover breeches in computer security, but when they do so for the user, they do not replace the tasks of a private detective or security analyst. Increasingly, computer systems do more for us than human surrogates could do. This is why it is all the more important to have a framework for morally evaluating computer systems, especially a framework that acknowledges that computer systems can do an incompetent job of pursuing the interests of their users and can misbehave in their work on behalf of users.

To claim that computer systems (and possibly other technologies) have moral responsibility and can be morally evaluated is *not* to claim that the responsibility or blameworthiness of users or system designers is thereby diminished. We anticipate that the standard response to our argument will be that the attribution of responsibility to various agents is a zero-sum situation—that designers are "let off the hook" when we turn to the moral evaluation of computer systems. In response, we deny that moral evaluation is zero sum. Computer systems behave. Their behavior is intentional, and it can have effects on humans and can be morally appraised independently of an appraisal of their designers' behavior. What the designer does and what the computer does (in a particular context) are different, albeit closely related. To think that only human designers are subject to morality is to fail to recognize that technology has intentionality, and its intentionality plays a causal role in the effects that computer systems can have on moral patients.

So the point of emphasizing the moral responsibility and moral evaluation of computer systems is not to deflect responsibility away from system designers or users. Because a computer system is conceptually distinct from the computer system designer and user, all three should come in for moral scrutiny. Computer systems are an interesting case here because they are becoming increasingly sophisticated, in both technical and social dimensions. Though the first computer systems may have been simple utilities or "dumb" technologies designed to help humans connect phone calls, calculate bomb trajectories, and do arithmetic, computer systems are increasingly taking over roles once occupied by human surrogate agents. This continuous change would suggest that, somewhere along the way, computer systems changed from mere tool to component of a complex agent. Now, it can no longer be denied that computer systems have displaced humans—both in the manufacturing workforce, as has long been acknowledged, and more recently in the service industry. It would be peculiar, then, for users to recognize that computers have replaced human service workers who have always been supposed to have moral constraints on their behavior, but to avoid the ascription of similar moral constraints to computer systems.

We introduced this discussion of computer systems as a way of opening up the possibility of technology bearing moral responsibility and being subject to moral evaluation. The challenge of the program we propose is to explore this territory in relation to both smart as well as more mundane (less complicated) technologies. The larger program will have to come to grips with the triad involved in moral action and agency: designers, artifacts, and users.

Conclusion

The line of reasoning developed here sketches an account of the role of technology in moral action. We began with the distinction between natural and human-made objects and noted that moral philosophy has neglected the importance of the artifactual platform in which human action occurs. We argued that artifacts have intentionality and gave an account of this intentionality using the functionality of artifacts and their directedness at states of affairs in the world; in this way, artifacts are comparable to speech acts. Building on our account of the intentionality of artifacts, we considered whether artifacts have moral agency. Here we argued that there are three forms of intentionality at work in moral action with technology: the intentionality of the artifact designer, the intentionality of the artifact, and the intentionality of the artifact user. Allowing for the agency of artifacts does not diminish the responsibility of human actors. To address the issue of the responsibility and moral evaluation of artifacts, we examined computer systems as surrogate agents. We argued that the responsibility of human surrogate agents provides a good model for making sense of the responsibility of computer systems. Computer systems can be morally evaluated in terms of their roles in relation to users. We have long known that computer systems can err; our account suggests that they can also misbehave.

The set of issues discussed here constitute a program for future research. Technology has not been a significant focus in moral philosophy, and yet it shapes the human moral universe in significant ways. Attention to technology promises to open up a range of interesting, complex, and important philosophical issues.

BIBLIOGRAPHY

Allen, Colin; Gary Varner; and Jason Zinser. (2000). "Prolegomena to Any Future Artificial Moral Agent." *Journal of Experimental and Theoretical Artificial Intelligence* 12(3): 251–261.

Allenby, Braden R. (2004). "Engineering and Ethics for an Anthropogenic Planet." In *Emerging Technologies and Ethical Issues in Engineering*. Washington DC: National Academies Press.

Brooks, Rodney A. (2002). *Flesh and Machines: How Robots Will Change Us*. New York: Pantheon.

Callon, Michel. (1986). "Some Elements of a Sociology of Translation: Domestication of the Scallops and the Fishermen of St. Brieuc Bay." In *Power, Action, and Belief: A New Sociology of Knowledge?* ed. John Law. London: Routledge and Kegan Paul.

Danielson, Peter. (1992). *Artificial Morality: Virtuous Robots for Virtual Games*. London: Routledge.

Davidson, Donald. (2001). *Essays on Actions and Events*, 2nd edition. Oxford: Clarendon Press.

Dretske, Fred. (1989). "Reasons and Causes." *Philosophical Perspectives* 3: 1–15.

Feinberg, Joel. (1970). "Action and Responsibility." Chap. 6 in *Doing and Deserving: Essays in the Theory of Responsibility*. Princeton, NJ: Princeton University Press.

Floridi, Luciano, and J. W. Sanders. (2001). "Artificial Evil and the Foundation of Computer Ethics." *Ethics and Information Technology* 3(1): 55–66.

Introna, Lucas D., and Helen Nissenbaum. (2000). "Shaping the Web: Why the Politics of Search Engines Matters." *Information Society* 16(3): 169–185.

Johnson, Deborah G., and Thomas M. Powers. (2005). "Computers as Surrogate Agents." In *Information Technology and Moral Philosophy*, eds. Jeroen Van Den Hoven and John Wechert. Cambridge: Cambridge University Press.

Kurzweil, Ray. (1999). *The Age of Spiritual Machines: When Computers Exceed Human Intelligence*. New York: Viking.

Law, John. (1987). "Technology and Heterogeneous Engineering: The Case of Portuguese Expansion." In *The Social Construction of Technological Systems*, eds. Wiebe E. Bijker, Thomas P. Hughes, and Trevor J. Pinch. Cambridge, MA: MIT Press.

Searle, John R. (2001). *Rationality in Action*. Cambridge, MA: MIT Press.

RESEARCH ETHICS, ENGINEERING ETHICS, AND SCIENCE AND TECHNOLOGY STUDIES

RONALD R. KLINE

• • •

The fields of research ethics and engineering ethics, as well as programs in science, technology, and society, were established in the United States in the late 1960s and early 1970s amid concerns about fraud in science, engineering-management disasters such as the Ford Pinto gas tank explosions, the role of technologies such as Agent Orange in fighting an unpopular Vietnam War, and environmental degradation. Concerns about scientific scandals and engineering disasters thus shaped the fields of research ethics and engineering ethics. More recent approaches in science and technology studies can complement and supplement methods from moral philosophy to do research in, and teach courses on, social and ethical issues in engineering.

Issues in Research Ethics and Engineering Ethics

The disjunction between the fields of research ethics and engineering ethics is striking. The literature is divided along that amorphous but venerable boundary erected and maintained to separate science from engineering (Kline 1995). Of the dozen or so textbooks on engineering ethics published since the early 1980s, only one, by Caroline Whitbeck (1998), treats research issues in engineering, but sharply divides it from engineering practice. By "practice," Whitbeck means activities other than research, that is, the development, design, testing, and selling of structures and consumer products. The journal *Science and Engineering Ethics*, established in 1995, publishes articles that mainly discuss ethics in science *or* in engineering. Only a few are on matters relating to both science *and* engineering. The Committee on Science, Engineering, and Public Policy, a joint effort of the U.S. National Academy of Sciences, the National Academy of Engineering, and the Institute of Medicine, published a little booklet on research ethics entitled *On Being a Scientist: Responsible Conduct in Research*, the second edition of which appeared in 1995. It does not address the product development or design side of engineering, which is of so much concern to professional engineering societies such as the Institute of Electrical and Electronics Engineers (IEEE) and the American Institute of Chemical Engineers (AIChE).

Although writers work hard to maintain these boundaries, publications such as *On Being a Scientist* and engineering codes of ethics (Martin and Schinzinger 1996, appendix; Anderson et al. 1993) list very similar ethical issues, but with the order of importance inverted. This difference is also seen in the amount of attention given to cases involving these issues in research and engineering ethics:

Main Issues in Research Ethics

Integrity of research

Credit and authorship

Conflicts of interest

Welfare of subjects, experimenters, and the environment

Social implications of research

Main Issues in Engineering Ethics

Public's health, safety, and welfare, including the environment

Being a faithful agent of the employer

Conflicts of interest

Credit (e.g., intellectual property provisions)

Integrity of reports

How does one explain the reversal in priority given to these issues by scientific and engineering organizations? An older view in the history of technology held that science and technology, especially engineering, are mirror-image twins. Science values theory and ideas, whereas engineering values practice and the design of products (Layton 1971). This explanation helps one understand how leaders in science and engineering reproduce these stereotypes of the two fields. Yet it is unsatisfactory because it has been common for engineers to do theoretical research (Kline 1992) and for scientists to build instruments (Galison 1997).

Another way to investigate this difference in priority is to look at how the engineering disasters and the scandals in science of the 1970s and 1980s helped shape the issues in the two fields, and how scientific and engineering societies reacted to these threats to their authority and to the public image of science and engineering.

Scandals and Disasters

All of the above issues have been prevalent in science and engineering for a long time. Charles Babbage (1792–1871) spoke about "forging," "trimming," and "cooking" (serving up the best results) in a book on reforming science in England in the early nineteenth century (Babbage 1989 [1830], pp. 90–91). Scientists and engineers have questioned the social implications of science and technology since the United States dropped atomic bombs on Japan in 1945 (Boyer 1985). The American mathematician Norbert Wiener criticized the militarization and secrecy of science after the war, as well as the possible ill effects of cybernetics, the very field he created (Heims 1980). One of the most famous disputes about credit in science was that between Isaac Newton and Gottfried Leibniz in the eighteenth century over the innovation of the calculus (Westfall 1980). Engineers have long been concerned about public reactions to their work, concerns that intensified with the professionalization of their field in the late nineteenth century (Layton 1986 [1971]).

Why was there such a great interest in research and engineering ethics in the 1970s? It seems probable that public concerns were part of a broader critique of cultural authority at the time, which included a general criticism of science and technology, protest against the Vietnam War, the rise of the environmental and appropriate technology movements (Pursell 1993), and the national scandal of Watergate. In the 1970s, charges of misconduct in science and dangerous designs in engineering grew into public scandals about "fraud" in science and amoral calculation in engineering. Accounts of scientific scandals and engineering disasters filled newspapers, calling forth responses from the scientific and engineering communities, as well as from social scientists and philosophers. This public outcry also helped create the fields of research ethics and engineering ethics, as well as programs to study issues in science, technology, and society (Mitcham 2003a, 2003b).

Perhaps the book that did the most to publicize "fraud" in science was *Betrayers of the Truth* (1982), written by the science journalists William Broad and Nicholas Wade. That same year, a young congressman from Tennessee, Al Gore, held congressional hearings on fraud in biomedical research, drawing on many of the cases reported by Broad and Wade in the journal *Science* (Kevles 1998).

Despite its sensational and naive title, *Betrayers of the Truth* discusses subjects that have been of keen interest in science and technology studies, such as differences between ideology and practice in science, problems with replication, and trust relations. The authors severely criticized historians, philosophers and sociologists of science for upholding the myth of science as a rational, autonomous, verifiable producer of certain knowledge. Their main targets seem to be Karl Popper, Robert Merton, and internalist historians of science. They cite Thomas Kuhn appreciatively. In regard to the first issue in research ethics mentioned above, integrity of research, they questioned the objective ideology of science and the autonomy and effectiveness of its system of checks and balances—peer review, refereeing, and replication.

The more sensational part of the book described the prevalence of what they called "fraud" in science, under which they included the issues of integrity of research and credit and authorship. An appendix lists thirty-four cases of fraud, dating from the Greek astronomer Hipparchus in the second century B.C.E., who "published a star catalog taken from Babylonian sources as if it were the results of his observations" (p. 226), to three cases of falsification of data in biomedical research in 1981. Most of the then-recent cases occurred in biology, including that of Mark Spector, a graduate student with "golden hands" working at Cornell University under the biologist Efraim Racker. Racker and Spector announced a novel theory of cancer causation in 1981, only to find out later that Spector had forged experiments. Broad and Wade conclude that "Pride, ambition, excitement at a new theory, reluctance to listen to bad news, unwillingness to distrust a colleague" were the "ingredients that caused the kinase cascade theory to go so far.... Replication was the *last* step in the episode, undertaken when everything else had failed and only after plain evidence of forgery had come to light" (p. 63, their emphasis). In regard to the self-policing mechanism of science, they give a structural explanation. "The roots of fraud lie in the barrel, not in the bad apples that occasionally roll into public view" (p. 87).

The scientific community responded to the publicity surrounding these cases by conducting investigations, issuing reports, and publishing educational materials. The first edition of *On Being a Scientist* appeared in 1989. In 1992 the National Research Council defined misconduct as "fabrication, falsification, and plagiarism in proposing, conducting, and reporting research" (Whitbeck 1998, p. 201; Mitcham 2003a, p. 277). The cold fusion controversy in 1989 (Lewenstein 1992), the David Baltimore case in biomedicine in 1991 (Kevles 1998), and the early 2000s case of data fabrication at Bell Labs by the rising "star physicist" Jan Hendrik Schön in research on organic semiconductors and nanoscience (Levi 2002) have kept the topic in the news and before the scientific community.

The issues raised and discussed during these "scandals" have dominated thinking on research ethics by scientists and ethicists. The booklet *On Being a Scientist* and the journal *Science and Engineering Ethics* both devote much more space to questions of integrity of research and credit and authorship, than to conflicts of interest and social implications of research. This priority existed before the 1970s, but it seems that the charges of fraud and responses to it have reinforced the status of these issues in research ethics and lessened that of other issues, such as gender and other power relations.

The field of engineering ethics has a similar history. In the Progressive Era of the late nineteenth and early twentieth centuries in the United States, professional engineering societies developed codes of ethics in order to raise the status of the field, to make it look more like a learned profession such as medicine, which was considered socially responsible (Layton 1986 [1971], 1978). The codes played this role for a short time around World War I, but they became rather obscure documents thereafter.

How obscure the codes were is revealed in the Bay Area Rapid Transit (BART) case in San Francisco. In late 1971, three engineers working for the BART district brought concerns about the safety of an automated train project to the attention of a member of the board of directors, after getting no satisfaction from a supervisor. Their analysis predicted, for example, that doors would open before the train entered the station. Instead of investigating the engineers' allegations of a dangerous design, the board investigated who the anonymous engineers were and had them fired (Friedlander 1974). The IEEE came to their assistance in early 1975 by filing a "friend of the court" brief. The IEEE proposed the novel argument that BART had violated the employment contract of the fired engineers because, as professional engineers, they were obligated to abide by the code of ethics of their profession and "hold paramount" the public's safety. The IEEE referred to the code of ethics of the Engineers' Council for Professional Development, an umbrella group for all engineers and the predecessor to the current Accreditation Board for Engineering and Technology, because the IEEE did not know it had an existing code on the books, created in 1912. Still unaware of the earlier code, it wrote a new one in 1979 (Kline 2001/2002). The IEEE's argument in the BART case did not set a precedent. The two engineers settled out of court when they realized that some false statements they had made to management would probably hurt their case (Unger 1994).

The BART case is one of a litany of disasters and near disasters used in teaching engineering ethics in the United States. Among them are:

- Gas tank ruptures of rear-ended Ford Pintos that caused burn injuries and deaths in the 1970s. Dozens of lawsuits were subsequently brought against Ford Motor Company (Camps 1981; De George 1981).
- The crash of a Turkish Airlines DC-10 near Paris in 1974, killing all 346 people aboard, attributed to a poorly designed cargo latch system. A test facility in Long Beach, California, said it had completed design changes when it had not (Fielder and Birsch 1992).
- The Three Mile Island nuclear power plant accident of 1979, resulting in a partial meltdown of its core and a lengthy and costly cleanup (Ford 1982).
- The crash of another DC-10, this time upon takeoff from Chicago in 1979 when an engine sepa-

rated from the plane. All 271 people aboard were killed, as well as two persons on the ground. The airline used shortcuts in maintenance procedures (Fielder and Birsch 1992).

- The collapse of a fourth-floor walkway in the atrium of the Hyatt Regency Hotel in Kansas City, Missouri, in 1981, killing 114 partygoers (Rubin and Banick 1986).
- The Union Carbide Corporation Bhopal disaster in India in 1984 (Stix 1989).
- The space shuttle *Challenger* accident of 1986 (Vaughan 1996).
- The space shuttle *Columbia* accident of 2003.

In all of these cases, investigation showed that engineers had known about, and often raised issues about, what they considered to be risky and unsafe designs from an early stage in the design process.

The cases are usually taught as a conflict between engineers wanting to create a safe design and managers wanting to push the products out the door because of time and financial constraints. But as Diane Vaughan (1996) has argued in the case of the space shuttle *Challenger*—a favorite in engineering ethics courses and literature—assumptions of amoral calculation by managers and engineers should be reexamined. Vaughan focuses instead on the construction of acceptable risk in the work-group cultures of day-to-day engineering practices, which led up to the fateful decision to launch the *Challenger*.

These disasters have greatly shaped the field of engineering ethics. The code of ethics of the Engineers' Council for Professional Development (1978) had been rewritten in 1974 to contain the obligation that the engineer "shall hold paramount the safety, health, and welfare of the public." Other engineering professional societies followed suit. This revision aimed to assure the public that engineers, if not their managers, were socially responsible. (See Davis 2001 for an argument that the original codes stressed social responsibility). It was a move to protect the autonomy of the engineering profession as a self-policing group that did not need government oversight. Of course, the increased amount of damages awarded in lawsuits and the rise of strict product liability laws have resulted in another type of oversight.

Most textbooks rely on these large cases to discuss safety, risk, whistle-blowing, conflicts of interest, rights of engineers in corporations, and so forth (Martin and Schinzinger 1996; Whitbeck 1998; Harris, Pritchard, and Rabins 1995; Unger 1994. Herkert 2000 takes a broader approach by including articles on history and policy). They are a major avenue for students to consider the messy complexity of engineering practices in a world of multinational corporations, subcontractors, liability laws, government regulation and deregulation, consumer activities, and what Bryan Wynne has called "unruly technology" (1988).

But the cases have also helped shape the field such that some issues are marginalized. The relationship between gender and product design, lack of access to new technology, and the flexible interpretation of test results are not visible because they have been invisible in the way the disasters have been reported in the newspapers, investigated by government committees, and analyzed by scholars in engineering ethics. Vaughan's participation in the board appointed by the National Aeronautics and Space Administration (NASA) to investigate the *Columbia* space shuttle accident is a recent and much welcomed exception.

Science and Technology Studies

In the late 1990s a movement began aiming to bring science and technology studies (S&TS) to bear on research and teaching in engineering ethics (e.g., Herkert, 2000; Lynch and Kline 2000; Kline 2001/2002). Textbooks on the subject typically show students how to apply moral philosophy to ethical issues, especially to moral dilemmas (see, e.g., Martin and Schinzinger 1996). Consider the hypothetical case of an engineer asked by his supervisor to "do the math backwards" to come up with data to support a design recommendation that, based on engineering judgment, contradicts suspected test results (Kohn and Hughson 1980). Students are often asked to identify the rights, duties, and consequences in this case and weigh them to make a decision. Textbooks usually do not prescribe the correct (ethical) courses of action, but present methods for engineers to use to sort out and identify ethical issues, to understand the basis for their decision, and to consider innovative alternatives to escape the horns of the dilemma (see, e.g., Harris, Pritchard, and Rabins 1995).

Textbooks treat "large cases," the lengthy descriptions of engineering disasters, in much the same way. For example, the complexities of the *Challenger* case are often reduced to the mythic moment of the night before the launch when Jerry Mason, a senior vice president at Morton Thiokol, the maker of the rocket boosters, asked Robert Lund, the vice president of engineering, to take off his engineering hat and put on his management hat to make a decision. The case is presented as one of *amoral calculation* on the part of managers, pressured by time schedulers and political necessities to overturn a *sound* engineering recommendation (Lynch and Kline 2000).

One disadvantage of this approach is that it provides, even in the big cases, a very *thin description* of engineering practice. Work relations among engineers, technicians, and managers are flattened and described from the agent-centered perspective favored by these textbooks, engineering professional societies, accreditation agencies, and moral philosophers. Power relations are often reduced to engineers versus management, and gender relations are virtually ignored. The production of engineering knowledge is usually seen as unproblematic, as are conceptions of risk and safety. The textbook by Mike W. Martin, a philosopher, and Roland Schinzinger, an engineer, is better in some of these respects. It discusses different perceptions of risk and safety, as well as work relations in corporations—under the rubric of rights of engineers in the workplace—and proposes the idea that engineering is a social experiment (Martin and Schinzinger 1996). That idea resonates well with literature in the history and sociology of engineering and technology, but those fields are underutilized in engineering ethics literature.

William T. Lynch and Ronald R. Kline (2000) pointed to the work of Diane Vaughan—her "historical ethnography" of the *Challenger* case—as one approach to take to bring S&TS to bear on engineering ethics. Vaughan (1996) concluded that the acceptable risk of flying with solid rocket booster O-rings that did not seat as they were designed to was constructed by a process of "normalization of deviance" from original design specs in the "production of culture" within engineer-manager work groups. This construction was supported by the "culture of production" of the wider engineering community and the "structural secrecy" of passing information up through bureaucratic channels. Engineers thought they were gaining a better technical understanding of how O-rings behaved in this harsh, complex environment and thus considered the erosion of O-rings by hot gases to be "normal" and under their control. The proposed launching at a low temperature "outside their experience base" brought about the conflict with management during the famous teleconference on the eve of the launch. The engineers' perception that NASA and the managers involved had reversed the ground rules and now asked them to prove the shuttle was *unsafe* to fly brought about the charges of amoral calculation by managers.

Although Vaughan draws on some S&TS ideas, such as the concepts of unruly technology and the interpretative flexibility of test results, she does not cover the entire field of technology studies. In fact, her structuralist approach collides with social constructivists' accounts. Its chief merit is its detailed historical ethnography of engineering practice.

The history, philosophy, and sociology of engineering also provide a wealth of information about engineering practice. There are accounts of the professionalization of engineering, engineering education, the relationship between scientific and engineering research, and the production of engineering knowledge (Leslie 1993; Downey and Lucena 1995); the engendering of engineering as a masculine profession (Oldenziel 1999); and the processes of design and testing (Vincenti 1990; Kline 1992; Latour 1996; Alder 1997; Cooper 1998; Thompson 2002).

S&TS scholars can draw on many concepts to illuminate social and ethical issues in engineering. These include:

- Gender and technology: gender relationships built into buildings; masculinity and technical competence (Wajcman 1991).
- Trust in numbers: why quantitative arguments carry more weight than qualitative ones in a bureaucratic setting (Porter 1995).
- Tacit knowledge: for example, the phenomenon of "golden hands" in research (Collins 1985).
- Risk: construction and communication of risk (Herkert 2000).
- User studies: interpretable flexibility of consumer products (Oudshoorn and Pinch 2003).
- Trust relations: assumptions of trust in research, design, and testing (Shapin 1995).
- Boundary work: separation of science from engineering, experts from laypeople, technology from politics, and so on. (Gieryn 1995).
- Politics of artifacts: by choice of design, "nature" of the design (Winner 1986).

Thick Description and Moral Prescription

One criticism of bringing S&TS to bear on engineering ethics is that is provides a better description of engineering practice, but does not directly address normative concerns. This work is in its infancy, but there are at least three ways in which the theory-based "thick description" provided by history and sociology of science and technology can lead to moral prescriptions.

The first is by telling a moral tale, such as the account of Robert Moses designing low bridges on the Long Island Expressway that prevented buses from the inner city from going to Jones Beach (Winner 1986). Although Moses's motives in this story may not have been racially discriminatory, and African Americans may have found other ways to travel to Long Island (Joerges 1999), more accurate stories of this kind can warn engineers of the unintended political consequences of their designs.

A second way is that thick descriptions can open new avenues of moral inquiry. Although moral philosophers rightly question the amount of ethical reflection permitted by Vaughan's concept of the normalization of deviance (1996), it can, if used properly, alert engineers and teachers of engineering ethics to the moral implications of everyday decisions made in engineering practice.

Finally, thick descriptions can provide a basis, by analogy, for taking a normative position. Martin and Schinzinger's concept of engineering as a social experiment (1996), for example, shows that engineers cannot know the precise technical or social *outcome* of a technology in the design stage, no matter how many computer simulations they run. The normative implications from this description of engineering are that the engineering experiment should be conducted in a morally responsible way, which means—after learning the lessons of the horrors of the Nazi medical experiments of World War II—*monitoring* the experiment, providing a *safe exit*, and ensuring that there was *informed consent* on the part of those being experimented upon.

In these and other ways, S&TS scholars can find ways to collapse or problematize the boundaries between description, analysis, and normative conclusions, to ask how they can relate to or perhaps strengthen each other. By bringing an extensive body of research in the history and sociology of engineering to bear on engineering and research ethics, S&TS scholars can improve humankind's understanding of the complex social and moral issues in science and engineering, and perhaps influence the practice of these fields as well.

BIBLIOGRAPHY

Alder, Ken. (1997). *Engineering the Revolution: Arms and Enlightenment in France, 1763–1815*. Princeton, NJ: Princeton University Press.

Anderson, Ronald E.; Deborah G. Johnson; Donald Gotterbarn; and Judith A. Perrolle. (1993). "Using the New ACM Code of Ethics in Decision Making." *Communications of the ACM* 36(2): 98–107.

Babbage, Charles. 1989 (1830). *Reflections on the Decline of Science in England and on Some of Its Causes*. London: William Pickering.

Boyer, Paul. (1985). *By the Bomb's Early Light: American Thought and Culture at the Dawn of the Atomic Age*. New York: Pantheon.

Broad, William, and Nicholas Wade. (1982). *Betrayers of the Truth*. New York: Simon and Schuster.

Camps, Frank. (1981). "Warning an Auto Company about an Unsafe Design." In *Whistle Blowing! Loyalty and Dissent in the Corporation*, ed. Alan F. Westin. New York: McGraw-Hill.

Collins, H. M. (1985). *Changing Order: Replication and Induction in Scientific Practice*. London: Sage.

Cooper, Gail. (1998). *Air-Conditioning America: Engineers and the Controlled Environment, 1900–1960*. Baltimore: Johns Hopkins University Press.

Davis, Michael. (2001). "Three Myths about Codes of Engineering Ethics." *IEEE Technology and Society Magazine* 20(3): 8–14.

De George, Richard T. (1981). "Ethical Responsibilities of Engineers in Large Organizations: The Pinto Case." *Business and Professional Ethics Journal* 1(1): 1–14.

Downey, Gary Lee, and Juan C. Lucena. (1995). "Engineering Studies." In *Handbook of Science and Technology Studies*, ed. Sheila Jasanoff, Gerald E. Markle, James C. Petersen, and Trevor Pinch. Thousand Oaks, CA: Sage.

Engineers' Council for Professional Development. (1978). "Code of Ethics of Engineers. Approved, October 1, 1974." In *Ethical Problems in Engineering*, ed. Robert J. Baum and Albert Flores. Troy, NY: Rensselaer Polytechnic Institute, Center for the Study of the Human Dimensions of Science and Technology.

Fielder, John H., and Douglas Birsch, eds. (1992). *The DC-10 Case: A Study in Applied Ethics, Technology, and Society*. Albany: State University of New York Press.

Ford, Daniel F. (1982). *Three Mile Island: Thirty Minutes to Meltdown*. New York: Viking.

Friedlander, Gordon D. (1974). "The Case of the Three Engineers vs. BART." *IEEE Spectrum* 11(10): 69–76.

Galison, Peter. (1997). *Image and Logic: A Material Culture of Microphysics*. Chicago: University of Chicago Press.

Gieryn, Thomas F. (1995). "Boundaries of Science." In *Handbook of Science and Technology Studies*, ed. Sheila Jasanoff, Gerald E. Markle, James C. Petersen, and Trevor Pinch. Thousand Oaks, CA: Sage.

Harris, Charles E.; Michael S. Pritchard; and Michael J. Rabins. (1995). *Engineering Ethics: Concepts and Cases*. Belmont, CA: Wadsworth.

Heims, Steve J. (1980). *John von Neumann and Norbert Wiener: From Mathematics to the Technologies of Life and Death*. Cambridge, MA: MIT Press.

Herkert, Joseph R., ed. (2000). *Social, Ethical, and Policy Implications of Engineering: Selected Readings*. New York: IEEE Press.

Joerges, Bernward. (1999). "Do Politics Have Artefacts?" *Social Studies of Science* 29(3): 411–431.

Kevles, Daniel J. (1998). *The Baltimore Case: A Trial of Politics, Science, and Character*. New York: Norton.

Kline, Ronald R. (1992). *Steinmetz: Engineer and Socialist*. Baltimore: Johns Hopkins University Press.

Kline, Ronald R. (1995). "Construing 'Technology' as 'Applied Science': Public Rhetoric of Scientists and Engineers in the United States, 1880–1945." *Isis* 86(2): 194–221.

Kline, Ronald R. (2001/2002). "Using History and Sociology to Teach Engineering Ethics." *IEEE Technology and Society Magazine* 20(4): 13–20.

Kohn, Philip M., and Roy V. Hughson. (1980). "Perplexing Problems in Engineering Ethics." *Chemical Engineering* 87(9): 100–107.

Latour, Bruno. (1996). *Aramis; or, The Love of Technology*, trans. Catherine Porter. Cambridge, MA: Harvard University Press.

Layton, Edwin T., Jr. (1971). "Mirror-Image Twins: The Communities of Science and Technology in Nineteenth-Century America." *Technology and Culture* 12(4): 562–580.

Layton, Edwin T., Jr. (1978). "Engineering Ethics and the Public Interest: A Historical Review." In *Ethical Problems in Engineering*, ed. Robert J. Baum and Albert Flores. Troy, NY: Rensselaer Polytechnic Institute, Center for the Study of the Human Dimensions of Science and Technology.

Layton, Edwin T., Jr. 1986 (1971). *The Revolt of the Engineers: Social Responsibility and the American Engineering Profession*. Baltimore: Johns Hopkins University Press.

Leslie, Stuart W. (1993). *The Cold War and American Science: The Military-Industrial-Academic Complex at MIT and Stanford*. New York: Columbia University Press.

Levi, Barbara Goss. (2002). "Investigation Finds that One Lucent Physicist Engaged in Scientific Misconduct." *Physics Today* 55(11): 15–17.

Lewenstein, Bruce V. (1992). "Cold Fusion and Hot History." *Osiris*, 2nd series, 7: 135–163.

Lynch, William T., and Ronald R. Kline. (2000). "Engineering Practice and Engineering Ethics." *Science, Technology, and Human Values* 25(2): 195–225.

Martin, Mike W., and Roland Schinzinger. (1996). *Ethics in Engineering*, 3rd edition. New York: McGraw-Hill.

Mitcham, Carl. (2003a). "Co-responsibility for Research Integrity." *Science and Engineering Ethics* 9(2): 273–290.

Mitcham, Carl. (2003b). "Professional Idealism among Scientists and Engineers: A Neglected Tradition in STS Studies." *Technology in Society* 25(2): 249–262.

Oldenziel, Ruth. (1999). *Making Technology Masculine: Men, Women, and Modern Machines in America, 1870–1945*. Amsterdam: Amsterdam University Press.

Oudshoorn, Nellie, and Trevor Pinch, eds. (2003). *How Users Matter: The Co-construction of Users and Technologies*. Cambridge, MA: MIT Press.

Porter, Theodore M. (1995). *Trust in Numbers: The Pursuit of Objectivity in Science and Public Life*. Princeton, NJ: Princeton University Press.

Pursell, Carroll. (1993). "The Rise and Fall of the Appropriate Technology Movement in the United States, 1965–1985." *Technology and Culture* 34(3): 629–637.

Rubin, Robert A., and Lisa A. Banick. (1986). "The Hyatt Regency Decision—One View." *Construction Lawyer* 6(4): 20.

Shapin, Steven. (1995). "Trust, Honesty, and the Authority of Science." In *Society's Choices: Social and Ethical Decision Making in Biomedicine*, ed. Ruth Ellen Bulger, Elizabeth Meyer Bobby, and Harvey V. Fineberg. Washington, DC: National Academy Press.

Stix, Gary. (1989). "Bhopal: A Tragedy in Waiting." *IEEE Spectrum* 26(6): 47–50.

Thompson, Emily. (2002). *The Soundscape of Modernity: Architectural Acoustics and the Culture of Listening in America, 1900–1933*. Cambridge, MA: MIT Press.

Unger, Stephen H. (1994). *Controlling Technology: Ethics and the Responsible Engineer*, 2nd edition. New York: Wiley.

U.S. National Academy of Sciences; National Academy of Engineering; and Institute of Medicine. Committee on Science, Engineering, and Public Policy. (1995). *On Being a Scientist: Responsible Conduct in Research*, 2nd edition. Washington, DC: National Academy Press.

Vaughan, Diane. (1996). *The* Challenger *Launch Decision: Risky Technology, Culture, and Deviance at NASA*. Chicago: University of Chicago Press.

Vincenti, Walter G. (1990). *What Engineers Know and How They Know It: Analytical Studies from Aeronautical History*. Baltimore: Johns Hopkins University Press.

Wajcman, Judy. (1991). *Feminism Confronts Technology*. University Park: Pennsylvania State University Press.

Westfall, Richard S. (1980). *Never at Rest: A Biography of Isaac Newton*. Cambridge, UK: Cambridge University Press.

Whitbeck, Caroline. (1998). *Ethics in Engineering Practice and Research*. Cambridge, UK: Cambridge University Press.

Winner, Langdon. (1986). "Do Artifacts Have Politics?" Chap. 2 in his *The Whale and the Reactor: A Search for Limits in an Age of High Technology*. Chicago: University of Chicago Press.

Wynne, Brian. (1988). "Unruly Technology: Practical Rules, Impractical Discourses, and Public Understanding." *Social Studies of Science* 18(1): 147–167.

NANOSCIENCE, NANOTECHNOLOGY, AND ETHICS: PROMISE AND PERIL

RAY KURZWEIL

• • •

Our rapidly growing scientific and technological ability to manipulate matter and energy at ever smaller scales promises to transform virtually every sector of society, a phenomenon that presents manifest ethical responsibilities. There will be increasing overlap between nanotechnology and other technologies, such as biotechnology and artificial intelligence. And as with these previous scientific and technological transformations, we will be faced with deeply intertwined promise and peril.

The Nano-Frontier

Nanoscience and nanotechnology today have been expanded to include essentially any science or technology where the key features are measured in a modest number of nanometers (under 100 by some definitions). By this standard, contemporary electronics has already passed this threshold. Eric Drexler has further developed the concept of building molecule-scale devices using molecular assemblers that would precisely guide chemical reactions by means of information. Moreover, just as technologies related to information develop at an exponential pace, generally doubling in capability and price-performance every year, so the size of technology is itself inexorably shrinking, and most of technology will be "nanotechnology" by the 2020s.

This era will bring us the ability to essentially convert software, that is, information, directly into physical products. We will be able to produce virtually any product for pennies per pound. Computers will have greater computational capacity than the human brain, and we will be completing the reverse engineering of the human brain to reveal the software design of human intelligence. We are already placing devices with narrow intelligence in our bodies for diagnostic and therapeutic purposes. With the advent of nanotechnology, we will be able to keep our bodies and brains in a healthy, optimal state indefinitely. Nanotechnology and related advanced technologies will bring us the opportunity to overcome age-old problems, including pollution, poverty, disease, and aging.

Many object to the intermingling of the so-called natural world with the products of our technology. However, the increasing intimacy of our human lives with our technology is not a new story. Human life expectancy was thirty-seven years in 1800. Most humans at that time lived lives dominated by poverty, intense labor, disease, and misfortune. We are immeasurably better off as a result of technology, but there is still a lot of suffering in the world to overcome. We have a moral imperative, therefore, to continue the pursuit of knowledge and of advanced technologies that can continue to overcome human affliction. There is also an economic imperative to continue .

Nanotechnology is advancing on hundreds of fronts. We cannot relinquish its pursuit without essentially relinquishing all of technology, which would require acts of totalitarianism inconsistent with the values of our society. Technology has always been a double-edged sword, and that is certainly true of nanotechnology. However, we will have no choice but to confront the challenge of guiding nanotechnology in a constructive direction. Any broad attempt to relinquish nanotechnology will only push it underground, which would interfere with the benefits while actually making the dangers worse.

With the human genome project, three to five percent of the budgets were devoted to the ethical, legal, and social implications (ELSI) of the technology. A similar commitment for nanotechnology would be appropriate and constructive. Near-term applications of nanotechnology are more limited in their benefits and more benign in their potential dangers. We cannot say a priori that all nanoengineered particles are safe, nor would it be appropriate to deem them necessarily unsafe. Environmental tests thus far have not shown reasons for undue concern.

I believe that existing regulatory mechanisms are sufficient to handle near-term applications of nanotechnology. As for the long term, we need to appreciate that a myriad of nanoscale technologies are inevitable. The current examinations and dialogues on achieving the promise while ameliorating the peril are appropriate and will deserve increased attention as we get closer to realizing these revolutionary technologies.

The Nano-Background: Models of Technology Trends

Models of technology trends show that nanotechnology and related advanced technologies are inevitable. They are deeply integrated into our society and are advancing on many diverse fronts, comprised of hundreds of small steps, each benign in itself.

INTUITIVE LINEAR AND HISTORICAL EXPONENTIAL VIEWS. Although exponential trends did exist a thousand years ago, they were at that very early stage where it is so flat and so slow that it looks like no trend at all. Today, everyone expects continuous technological progress and the social repercussions that follow. But the future will nonetheless be far more surprising than most observers realize because few have internalized the fact that the rate of change itself is accelerating.

Most long-range forecasts of technical feasibility underestimate the power of future developments because they are based on the "intuitive linear" view of history rather than the "historical exponential" view. We will not experience a hundred years of progress in the twenty-first century; rather we will witness on the order of twenty thousand years of progress (at today's rate of progress). An unexamined intuition provides the impression that progress changes at the rate that we have recently experienced because an exponential curve approximates a straight line when viewed for a brief duration.

But an assessment of the history of technology shows that technological change is exponential. Indeed, we find "double" exponential growth, meaning that the rate of exponential growth is itself growing exponentially. These observations are based on a rich model of diverse technological processes.

THE LAW OF ACCELERATING RETURNS. The ongoing acceleration of technology is the inevitable result of the "law of accelerating returns," which describes the acceleration of the pace and the exponential growth of the products of an evolutionary process, including technology, particularly information technologies.

The law of accelerating returns has three key features. First, evolution applies positive feedback as the more capable methods resulting from one stage of evolutionary progress are used to create the next stage. As a result, the rate of progress of an evolutionary process increases exponentially over time, as the "returns" of that process (e.g., speed or cost-effectiveness) increase exponentially. As an evolutionary process becomes more effective, greater resources are invested in it, resulting in a second level of exponential growth (i.e., the rate of exponential growth itself grows exponentially).

A second feature is "technological paradigm shifts." A specific paradigm (a method or approach to solving a problem) provides exponential growth until the method exhausts its potential. When this happens, a paradigm shift (a fundamental change in the approach) occurs, which enables exponential growth to continue. Each paradigm follows an "S-curve," which consists of slow growth, followed by rapid growth, followed by a leveling off as the particular paradigm matures. During this third phase in the life cycle of a paradigm, pressure builds for the next paradigm shift. The acceleration of the overall evolutionary process proceeds as a sequence of S-curves, and the overall exponential growth consists of this cascade of S-curves.

A third key feature is that the resources underlying the exponential growth of an evolutionary process are relatively unbounded. One resource is the order of the evolutionary process itself. Each stage of evolution provides more powerful tools for the next. The other required resource is the "chaos" of the environment in which the evolutionary process takes place and which provides the options for further diversity. In technological evolution, human ingenuity and the ever-changing market sustain innovation.

The evolution of life forms and technologies constantly accelerates. With the advent of a technology-creating species, the exponential pace became too fast for evolution through DNA-guided protein synthesis and moved on to human-created technology. Technology goes beyond mere tool making; it is a process of creating ever more powerful technology using the tools from the previous round of innovation. The first technological steps took tens of thousands of years. For people living in this era, there was little noticeable technological change. By 1000 C.E., progress was much faster and a paradigm shift required only a century or two. The nineteenth century saw more technological change than in the nine centuries preceding it. Then in the first twenty years of the twentieth century, we saw more advancement than in all of the nineteenth century. Now, paradigm shifts occur in only a few years. The paradigm shift rate is currently doubling every decade. So the twenty-first century will see about a thousand times greater technological change than its predecessor.

MOORE'S LAW AND BEYOND. The exponential trend that has gained the greatest public recognition has become known as "Moore's Law." Gordon Moore, one of the inventors of integrated circuits, noted in the mid-1970s that we could squeeze twice as many transistors on an integrated circuit every twenty-four months. Given that the electrons have less distance to travel, the circuits also run twice as fast, providing an overall quadrupling of computational power.

However, the exponential growth of computing is much broader than Moore's Law. If we plot the speed per price of forty-nine famous calculators and computers spanning the twentieth century, we note that there were four paradigms that provided exponential growth in the price-performance of computing before integrated circuits. Therefore, Moore's Law was the fifth paradigm to exponentially grow the power of computation. When Moore's Law reaches the end of its S-Curve, the exponential growth will continue with three-dimensional molecular computing, constituting the sixth paradigm.

Moore's Law narrowly refers to the number of transistors on an integrated circuit of fixed size. But the most appropriate measure to track is computational speed per unit cost. This takes into account many levels of innovation in computer design. For example, there are many nascent technologies that build circuitry in three dimensions in a way that mimics the parallel organization of the human brain. One cubic inch of nanotube circuitry would be a million times more powerful than the human brain. There are more than enough new computing technologies now being researched to sustain the law of accelerating returns as applied to computation.

Specific paradigms do ultimately reach levels at which exponential growth is no longer feasible. That is why Moore's Law is an S-curve. But the growth of computation will continue exponentially. Paradigm shift, or innovation, turns the S-curve of any specific paradigm into a continuing exponential. A new paradigm takes over when the old paradigm approaches its natural limit.

OTHER TECHNOLOGIES. There are many examples of the exponential growth implied by the law of accelerating returns in technologies as varied as DNA sequencing, communication speeds, brain scanning, electronics of all kinds, and even in the rapidly shrinking size of technology. Exponential growth in communications technology has been even more explosive than in computation. Miniaturization is a trend that will have profound implications for the twenty-first century. The salient implementation sizes of technologies, both electronic and mechanical, are shrinking at a double-exponential rate.

The future nanotechnology age will result not from the exponential explosion of computation alone, but rather from the synergies that will result from intertwined technological revolutions. Every point on the exponential growth curves represents an intense human drama of innovation and competition. It is remarkable that these chaotic processes result in such smooth and predictable exponential trends.

Examples of True Nanoscience and Nanotechnology

Ubiquitous nanoscience and nanotechnology is two to three decades away. One forthcoming achievement will be "nanobots," small robots the size of human blood cells that can travel inside the human bloodstream. There have already been successful animal experiments using this concept.

In addition to human brain reverse engineering, these nanobots will be able to perform a broad variety of diagnostic and therapeutic functions inside the human body. Robert Freitas, for example, has designed robotic replacements for human blood cells that perform thousands of times more effectively than their biological counterparts. His "respirocytes" (robotic red blood cells) could allow one to sprint for fifteen minutes without taking a breath. His robotic macrophages will be far more effective than our white blood cells at combating pathogens. His DNA repair robot would be able to repair DNA transcription errors, and even implement needed DNA changes. Although Freitas' conceptual designs are two or three decades away, there has already been progress on bloodstream-based devices.

Nanobot technology has profound military applications, and any expectation that such uses will be relinquished is highly unrealistic. Already, the U.S. Department of Defense (DOD) is developing "smart dust," or tiny robots to be used for surveillance. Billions of invisible spies could monitor every square inch of enemy territory and carry out missions to destroy enemy targets. The only way for an enemy to counteract such a force is with their own nanotechnology. Nanotechnology-based weapons will obsolete weapons of larger size.

In addition, nanobots will be able to expand our experiences and our capabilities. Nanobot technology will provide fully immersive virtual reality by taking up positions in close proximity to every interneuronal connection related to the senses. If we want to enter virtual reality, the nanobots suppress all of the inputs coming from the real senses, and replace them with the signals that would be appropriate for the virtual environment.

Scientists at the Max Planck Institute have developed "neuron transistors" that can detect the firing of a nearby neuron, or alternatively, can cause a nearby neuron to fire, or suppress it from firing. This amounts to two-way communication between neurons and the electronic-based neuron transistors. The scientists demonstrated their invention by controlling the movement of a living leech from their computer.

The Internet will provide many virtual environments to explore. We will be able to "go" to these virtual environments and meet others there, both real and simulated people. Of course, ultimately there will not be a clear distinction between the two. By 2030, going to a web site will mean entering a full-immersion virtual-reality environment, encompassing all of the senses and triggering the neurological correlates of emotions and sexual experiences.

"Experience beamers" circa 2030 will beam a person's entire flow of sensory experiences and emotions. We'll be able to go to a web site and experience other people's lives. Full-immersion visual-auditory environments will be available by 2010, with images written directly onto our retinas by our eyeglasses and contact lenses. The electronics will be embedded in our glasses and woven into our clothing, so computers as distinct objects will disappear.

The most significant implication of nanotechnology and related advanced technologies of the twenty-first century will be the merger of biological and nonbiological intelligence. Nonbiological intelligence is growing at a double-exponential rate and will vastly exceed biological intelligence well before the middle of this century. However, in my view, this nonbiological intelligence should still be considered human, as it is fully derivative of the human-machine civilization.

Our brains are relatively fixed in design, but brain implants based on massively distributed intelligent nanobots will ultimately expand our memories a trillion fold and improve all of our cognitive abilities. Since the nanobots are communicating with each other over a wireless network, they can create any set of new neural connections, break existing connections, create new hybrid biological-nonbiological networks, and add new nonbiological networks.

Using nanobots as brain extenders is a significant improvement over surgically installed neural implants. Nanobots will be introduced without surgery and can be directed to leave, so the process is easily reversible. They can change their configuration and alter their software. Perhaps most importantly, they are massively distributed and can take up billions or trillions of positions throughout the brain, whereas a surgically introduced neural implant can only be placed in a few locations.

The Economic Imperatives of the Law of Accelerating Returns

The economic imperative of a competitive marketplace is driving science and technology forward and fueling the law of accelerating returns, which, in turn, is transforming economic relationships. We are moving toward nanoscale, more intelligent machines as the result of many small advances, each with their own particular economic justification.

There is a vital economic imperative to create smaller and more intelligent technology. Machines that can more precisely carry out their missions have enormous value. There are tens of thousands of projects that are advancing the various aspects of the law of accelerating returns in diverse incremental ways. Regardless of near-term business cycles, the support for "high tech" in the business community has grown enormously. We would have to repeal capitalism and every visage of economic competition to stop this progression.

The economy has been growing exponentially throughout this century. Even the Great Depression of the 1930s represented only a minor blip compared to the underlying pattern of growth. Recessions, including the Depression, represent only temporary deviations from the underlying curve. Statistics in fact greatly understate productivity growth (economic output per worker), which has also been exponential.

Inflationary factors are offset by the double-exponential trends in the price-performance of all information-based technologies, which deeply affect all industries. We are also undergoing massive disintermediation in the channels of distribution through the Internet and other new communication technologies and escalating efficiencies in operations and administration. Current economic policy is based on outdated theories that do not adequately model the size of technology, bandwidth, megabytes, intellectual property, knowledge, and other increasingly vital constituents that are driving the economy.

Cycles of recession will not disappear immediately. However, the rapid dissemination of information, sophisticated forms of online procurement, and increasingly transparent markets in all industries have diminished the impact of these cycles. The underlying long-term growth rate will continue at a double-exponential rate. The rate of paradigm shift is not noticeably affected by the minor deviations caused by economic cycles. The overall growth of the economy reflects completely new forms of wealth and value that did not previously exist: nanoparticle-based materials, genetic information, intellectual property, communication portals, web sites, bandwidth, software, data bases, and many other new technology-based categories.

Another implication of the law of accelerating returns is exponential growth in human knowledge, including intellectual property, education, and learning. Over the course of the long twentieth century we increased investment in K-12 education by a factor of ten. We have a one hundred fold increase in the number of college students. Automation has been eliminating jobs at the bottom of the skill ladder while creating new and better paying jobs at the top. So, the ladder has been moving up, and we have been exponentially increasing investments in education at all levels.

Promise and Peril

Science and technology have always been double-edged swords, bringing us longer and healthier life spans, freedom from physical and mental drudgery, and many new creative possibilities, while at the same time introducing new and salient dangers. We will need to adopt strategies to encourage the benefits while ameliorating the risks. Relinquishing broad areas of technology, as some critics have proposed, is not feasible, and attempts to do so will only drive technology development underground, which will exacerbate the dangers.

As technology accelerates toward the full realization of biotechnology, nanotechnology and "strong" AI (artificial intelligence at or above human levels), we will see the same intertwined potentials: a feast of creativity resulting from greater human intelligence combined with many new dangers. Nanobot technology requires billions or trillions of such intelligent devices to be useful. The most cost-effective way to scale up to such levels is through self-replication. A defect in the mechanism curtailing nanobot self-replication could be disastrous. There are steps available now to mitigate this risk, but we cannot have complete assurance in any strategy that we devise today.

Other primary concerns include "Who is controlling the nanobots?" and "Who are the nanobots talking to?" Organizations or individuals could put undetectable nanobots in water or food supplies. These "spies" could monitor and even control thoughts and actions. Existing nanobots could be influenced through software viruses and other software "hacking" techniques. My own expectation is that the creative and constructive applications of this technology will dominate, as they do today. But we need to invest more heavily in developing specific defensive technologies.

There are usually three stages in examining the impact of future technology: awe at its potential to overcome problems; then a sense of dread at a new set of dangers; followed by the realization that the only viable and responsible path is to set a careful course that can realize the promise while managing the peril.

Bill Joy, cofounder of Sun Microsystems, has warned of the impending dangers from the emergence of self-replicating technologies in the fields of genetics, nanotechnology, and robotics, or "GNR." His concerns include genetically altered designer pathogens, self-replicating entities created through nanotechnology, and robots whose intelligence will rival and ultimately exceed our own. Who's to say we will be able to count on such robots to remain friendly to humans? Although I am often cast as the technology optimist who counters Joy's pessimism, I do share his concerns regarding self-replicating technologies. Many people have interpreted Joy's article as an advocacy of broad relinquishment, not of all technology, but of the "dangerous ones" like nanotechnology. Joy, who is now working as a venture capitalist with the legendary silicon valley firm of Kleiner, Perkins, Caufield & Byers investing in technologies such as nanotechnology applied to renewable energy and other natural resources, says that broad relinquishment is a misinterpretation of his position and was never his intent. He has recently said that the emphasis should be to "limit development of the technologies that are too dangerous," not on complete prohibition. He suggests, for example, a prohibition against self-replicating nanotechnology, which is similar to the guidelines advocated by the Foresight Institute.

Others, such as Bill McKibben, the environmentalist who was one of the first to warn against global warming, have advocated relinquishment of broad areas such as biotechnology and nanotechnology, or even of all technology. However, relinquishing broad fields would be impossible to achieve without essentially relinquishing all technical development.

There are real dangers associated with new self-replicating technologies. But technological advances, such as antibiotics and improved sanitation, have freed us from the prevalence of such plagues in the past. We may romanticize the past, but until fairly recently, most of humanity lived extremely fragile lives. Many people still live in this precarious way, which is one reason to continue technological progress and the economic enhancement that accompanies it. Should we tell the millions of people afflicted with devastating conditions that we are canceling the development of all bioengineered treatments because there is a risk that these same technologies may someday be used for malevolent purposes? Most people would agree that such broad-based relinquishment is not the answer.

THE RELINQUISHMENT ISSUE. Relinquishment at the right level is part of a responsible and constructive response to these genuine perils. The issue, however, is: At what level are we to relinquish technology? Ted Kaczynski (the Unabomber) would have us renounce all of it. This is neither desirable nor feasible. McKibben takes the position that many people now have enough wealth and technological capability and should not pursue more. This ignores the suffering that remains in the human world, which continued technological progress could alleviate.

Another level would be to forego certain fields (such as nanotechnology) that might be regarded as too dangerous. But such sweeping strokes of relinquishment are untenable. Nanotechnology is the inevitable result of the persistent trend toward miniaturization that pervades all of technology. It is not a single centralized effort, but is being pursued by a myriad of projects with many goals.

Kaczynski argued that modern industrial society cannot be reformed because technology is a unified system in which all parts are dependent on one another. It is not possible to get rid of the "bad" parts of technology and retain only the "good" parts. He cited modern medicine as an example, arguing that progress depends on several scientific fields and advancements in high-tech equipment. Kaczynski was correct on the deeply entangled nature of the benefits and risks, but his overall assessment of the relative balance between the two was way off. Joy and I both believe that technology will and should progress, and that we need to be actively concerned with the dark side. Our dialogue concerns the granularity of relinquishment that is feasible and desirable. Abandonment of broad areas of technology will only push them underground where development would continue unimpeded by ethics and regulation. In such a situation, it would be the less-stable, less-responsible practitioners who would have all the expertise.

One example of relinquishment at the right level is the proposed ethical guideline by the Foresight Institute that nanotechnologists agree to relinquish the development of physical entities that can self-replicate in a natural environment. Another is a ban on self-replicating physical entities that contain their own codes for self-replication. Such entities should be designed to obtain codes from a centralized secure server, which would guard against undesirable replication. This "broadcast architecture" is impossible in the biological world, which represents one way in which nanotechnology can be made safer than biotechnology. Such "fine-grained" relinquishment should be linked to professional ethical guidelines, oversight by regulatory bodies, the development of technology-specific "immune" responses, as well as computer assisted surveillance by law enforcement agencies. Balancing privacy rights with security will be one of many challenges raised by some new nanotechnologies.

Computer viruses serve as a reassuring test case in our ability to regulate nonbiological self-replication. At first, concerns were voiced that as they became more sophisticated, software pathogens had the potential to destroy computer networks. Yet the "immune system" that has evolved in response to this challenge has been largely effective. Although self-replicating software entities do cause damage from time to time, no one would suggest we do away with computers and the Internet because of software viruses. This success is in a highly productive industry in which there is no regulation, and no certification for practitioners.

DEFENSIVE TECHNOLOGIES AND THE IMPACT OF REGULATION. Arguments such as McKibben's for relinquishment have been influential because they paint a picture of future dangers as if they were released into an unprepared world. But the sophistication and power of our defensive technologies and knowledge will grow along with the dangers. When we have "gray goo" (unrestrained nanobot replication), we will also have "blue goo" ("police" nanobots). We cannot say with assurance that we will successfully avoid all misuse. We have been able to largely control harmful software virus replication because the requisite knowledge is widely available to responsible practitioners. Attempts to restrict this knowledge would have created a far less stable situation.

The present challenge is self-replicating biotechnology. By reprogramming the information processes that lead to and encourage disease and aging, we will have the ability to overcome these afflictions. However, the same knowledge can also empower a terrorist to create a bioengineered pathogen.

Unlike biotechnology, the software industry is almost completely unregulated. Although bioterrorists do not need to put their "innovations" through the FDA, scientists developing defensive technologies are required to follow regulations that slow innovation. It is impossible under existing regulations and ethical standards to test defenses to bioterrorist agents on humans. Animal models and simulations will be necessary in lieu of infeasible human trials, but we will need to go beyond these steps to accelerate the development of defensive technologies.

We need to create ethical and legal standards and defensive technologies. It is quite clearly a race. In the software field the defensive technologies have remained ahead of the offensive ones. With extensive regulation in the medical field slowing down innovation, this may not happen with biotechnology.

There is a legitimate need to make biomedical research as safe as possible, but our balancing of risks is skewed. The millions of people who need biotechnology advances seem to carry little political weight against a few well-publicized casualties from the inevitable risks of progress. This equation will become even starker with the emerging dangers of bioengineered pathogens. We need a change in public attitude in terms of tolerance for necessary risk.

Hastening defensive technologies is vital to our security. We need to streamline regulatory procedures to achieve this. However, we also need to greatly increase our investment explicitly in defensive technologies. In the biotechnology field, this means the rapid development of antiviral medications.

The comparable situation will exist for nanotechnology once replication of nano-engineered entities has been achieved. We will soon need to invest in defensive technologies, including the creation of a nanotechnology-based immune system. Such an immune system may itself become a danger, but no one would argue that humans would be better off without an immune system because of the possibility of autoimmune diseases. The development of a technological immune system for nanotechnology will happen even without explicit efforts to create one.

It is premature to develop specific defensive nanotechnologies as long as we have only a general idea of the threat. However, there is a dialogue on this issue, and expanded investment in these efforts should be encouraged. The Foresight Institute, for example, has devised a set of ethical standards and strategies for assuring the development of safe nanotechnology. They are likely to be effective with regard to preventing accidental release of dangerous self-replicating nanotechnology entities. But the intentional design and release of such entities is more challenging.

Conclusion

Protection is not impossible, but we need to realize that any level of protection will only work to a certain level of sophistication. We will need to continue to advance the defensive technologies and keep them ahead of the destructive technologies. The challenge of self-replication in nanotechnology impels us to continue the type of study that the Foresight Institute has initiated. With the human genome project, three to five percent of the budget was devoted to the ethical, legal, and social implications (ELSI) of the technology. A similar commitment for nanotechnology would be appropriate and constructive. Science and technology will remain double-edged swords, and the story of the twenty first century has not yet been written. We have no choice but to work hard to apply these quickening technologies to advance our human values, despite what often appears to be a lack of consensus on what those values should be.

BIBLIOGRAPHY

Drexler, Erik K. (1986). *Engines of Creation*. Garden City, NY: Doubleday.

Drexler, Erik K. (1992). *Nanosystems: Molecular Machinery, Manufacturing, and Computation*. New York: Wiley.

Freitas, Robert, Jr. (1999). *Nanomedicine*. Vol. 1: *Basic Capabilities*. Georgetown, TX: Landes Bioscience.

Joy, Bill. (2000). "Why the Future Doesn't Need Us." *Wired*262.

Kurzweil, Ray. (1999). *The Age of Spiritual Machines: When Computers Exceed Human Intelligence*. New York: Viking.

Kurzweil, Ray. (2005). *The Singularity Is Near: When Humans Transcend Biology*. New York: Viking.

McKibben, Bill. (2003). *Enough: Staying Human in an Engineered Age*. New York: Henry Holt.

RECOGNIZING THE STRUCTURAL FEATURES OF NEW TECHNOLOGIES

HANS LENK

• • •

Any assessment of the ethical issues associated with new technologies must take into account their special structural features. Single-factor theories of technology, highlighting just one trait (such as the domination of nature), are insufficient for grasping the multiple levels and aspects of contemporary technologies or technological societies. This is all the more true in what I have analyzed since the 1970s as our information-and-systems technological era, with its ever more tightly coupled systems and relationships between systems, the linking of information in global networks, and the comprehensive management of technologies in organizational systems defined in terms of abstract procedures and formalized functions.

Traditional analyses have described technologies as human organ projections, sensorimotor skills, applied science, efficient action, the pursuit of power, the physical realization of ideas, and more. (Mitcham 1994 provides one review of such traditional definitions.) In each case the attempt was to identify something "essential." But such one-factor descriptions apply only to some limited aspect of any technology, and fail to appreciate the spectrum of diverse elements now involved. Although traditional analyses may continue to be useful, they are more and more embedded in new trends along with their social, intellectual, material, and ecological contexts. Analyses of the structural features of new technologies oriented toward an ethical assessment would thus do well to consider at least the following emerging and interrelated traits.

1. Operations, Procedures, and Processes

Technology is not comprised only of machines, instruments, and other technical products. Instead there is a growing orientation toward technological processes, operations, and procedures. Process control and managerial phenomena are key features of modern technological and industrial production. This extends an earlier trend in which energy-transforming machines and systems became widespread in assembly-line production. More recently, "the real is the process" has become a characteristic feature of technologies.

2. Systematic Methods and Methodologies

Not only methods but also methodologies are increasingly essential. This trend is found in all science-based technological developments as well as in administration. Such trends increasingly characterize fields that have been captured by operations technologies such as process controls, systems engineering, and operations research.

3. Informatization, Abstraction, Formalization

Computerization and informatization, along with the use of formal and functional operations technologies such as flowcharts and network analyses, create increasingly comprehensive processes, organizations, and interrelations. (One example: the manufacturing–inventory–sales chains characteristic of retail giants.)

4. Systems Engineering and Technology

Different technical realms, including engineering and economics, are increasingly related by means of systems engineering and technology. This creates a positive feedback loop in which initial interactions promote the development of further and more thoroughgoing interactions.

5. Options Identifications Precede Problem Formulations and Needs Generation

In research and development (R&D), systems characteristics have been apparent for some time. R&D work systematically inventories and then exploits potentials, possibilities, and options (see Klages 1967). Only after having identified several products or processes by means of systematic research will investigators formulate a problem to be addressed or discover a new "need" that can be met by the already achieved technological development. In such cases the technological solution or invention precedes the problem or need. (This reversal was already anticipated by Karl Marx in the nineteenth century.)

6. Interdisciplinarity

Interdisciplinarity is promoted by spillovers from one science to another science, and from there to technological invention, innovation, and application—in both the laboratory and society at large. Interdisciplinary interactions are increasingly embedded in developmental processes. Systems technologies require practical interdisciplinarity.

7. Artificiality

The human world is increasingly shaped by technogenic relationships, properties, and artifacts to such an extent that it itself takes on the character of the artificial. The "second nature" or "symbolic universe" described by the German philosophers Helmut Plessner (1892–1985), Arnold Gehlen (1904–1976), and Ernst Cassirer (1874–1945) has now become a *technological* second nature. Moreover, this technologically enacted second nature is characterized by information networks of ever-greater extent and impact. Media technicalize a kind of second-hand reality, which becomes the socially real reality.

8. Virtuality

Humans now experience the virtualization of the artificial and symbolic worlds through information technologies, as well as by means of images and models and the related interpretations they superimpose on real life.

9. Multimedia

Systematic accumulations of technomedia yield multimedia. The manifold technicalizations of the symbolic, of virtual representations and their respective interpretations, lead to a kind of *coaction* or *coevolution* of diverse information technologies and media. There is a progressive universality and commonality of impact as well as systems integration. Humans find themselves increasingly living in a multiple-mediated technogenic world impregnated by multimedia—in short, in a multimedia technoworld.

10. Simulations

Computer hardware, software, and other successful efforts to improve and optimize the relevant information models by way of programming and computer-graphic constructions provide rapid, efficient, and inexpensive simulated solutions to all kinds of design and construction tasks. This includes scientific modeling, as in molecular design, and the technical development and construction of new machines, processes, and systems in the narrow sense.

The computer has turned out to be a universal, easily employable, and representative "can-do-anything" instrument that can identify variable routes for technical and nontechnical action, each described in terms of alternative materials and energy costs. Trial-and-error learning and physical work is reduced to a minimum when models are simulated in advance without real risks.

11. Flexibility

Computerized models allow the virtually risk-free simulation and testing of all kinds of hypotheses, inventions, and constructions in advance. This is generally if not universally true for models in science, planning, and administration. Systems organizing and management are rendered more flexible and variable than in the past.

12. Modularity

In a movement that began with the standardization of interchangeable parts, technology is often structured around modules, functional building blocks, and functionally integrated units. One good example is integrated circuits or microprocessors, which can be inserted by way of open interfaces into larger modules or systems. Such structures promote not only replaceability of obsolete or failed parts but also technical progress as a new peripheral (such as a video display or printer) is purchased to replace an old one or software programs are themselves continuously enhanced with updates.

13. User-Friendliness

New technologies have gradually become more user-friendly, more anthropomorphic in their reactions, often displaying a self-explanatory design that minimizes or even eliminates the need for technical manuals or instruction. One example is the context-dependent help menu in a computer application package. Another is the automated external defibrillator, which when placed on a person's chest can identify sudden cardiac arrest and then voices instructions for use to a responder.

14. Remote Control and Intelligent Sensing

New electronic and multimedia technologies allow remote control and intelligent sensing at a distance or in inaccessible environments. Intelligent sensing involves systems that mimic human senses such as sight, smell, or taste. When coupled with remote control technologies, intelligent sensing allows robot manipulation in nuclear plants or outer space. These devices multiply manipulative and technological power in extension and scope. Intelligent sensing can also involve the creation of "smart technologies" such as buildings that monitor their own structural characteristics.

15. Robotization

Robotization is proliferating and becoming widely disseminated in all fields of technology-guided production.

16. Smart Technology and Systems Autonomy

Feedback control and "intelligent decision-making" techniques and procedures are being introduced not only in sensing and remote control instrumentation, but in a plethora of machines, creating a kind of flexible systems autonomy. (Such developments simply extend a trajectory that can be traced back to the replacement of meters and gauges with warning lights, sometimes coupled with automatic control mechanisms. In some airplanes if a human being tries to override an automatic pilot when it is not safe to do so, the automatic pilot will continue to exercise control.)

17. Meta-autonomy

In the designing, constructing, and monitoring of machines, programs, or technological and organizational systems, there is a tendency to eliminate human interference. Machines can be used to build other machines or to check lower-level machines. It is increasingly programs that control and check machines, and programs that check programs. In effect this involves a meta-level technicalization in terms of a higher-order self-applicability of overarching abstract procedures and programs. This may be described as a sort of "reflexive" or "self-referential" applicability leading to what might be termed "meta-feasibility" or "meta-functionality" with regard to models and metamodels.

18. Computerization and Multifunctionality

Universal machines such as the computer provide a kind of abstract, software-determined processing and control. Along with techno-organizational systems, these are progressively maximizing flexibility, speed, smart machine autonomy, modularity, and more.

19. Mega-information Systems

There is a tendency to conceive of the world as a technology-dominated, manipulatable organization shaped by technosystems. Ecosystems and social systems come to be conceived as subordinate to techno-ecosystems or eco-technosystems and sociotechnical systems, respectively. The trend is toward thinking in terms of mega-information systems or a mega-world machine dependent on the meta-functionality of technological and operational processing or the multiple applicability of machines, processes, and programs.

20. Globalization

The overwhelming global success of technology and the technicalization of almost everything leads to a new technogenic world unity—one that is integrated technologically and informationally and is interactive. Increasingly humans live in a media-electronic global village. Technology appears to take on the character of a fate or destiny, with human survival appearing to be increasingly dependent on technological, social, political, and ecological change. This change or progress thus exhibits its own inner orientations and momentum.

21. Telematization

Telematization, in which everything is ubiquitously present (24/7/365), gives rise to locally separated but functionally coordinated teams working on giant virtual projects, designs, or networks.

22. Information-Technological Historicity

Information technology development has a history. The history of information systems, expert systems, and computerized decision-making systems designed, developed, and controlled by diverse agents mirrors the development of the notion of system itself. *Quod non in systemis non in realitate* (What is not in the systems is not real).

23. Intermingling and Interdependence

The systematized, interdisciplinary, functional integration and interrelation of activities in all aspects of the human lifeworld are weaving together mutual dependencies. These dependencies are at the same time susceptible to informational and operational manipulations, including economic manipulations. (Manipulation, however, does not always equal control. Interdependencies often have their own characteristics that will be asserted as unintended consequences when they are not acknowledged or respected.)

24. Sociotechno-systems

Nature and nurture are interdependent. Systems orientation, systems engineering, and the establishment and maintenance of sociotechnical systems all point toward an inseparable, indissoluble social systems complex characterized by ever-growing, accelerating, and ever-more encompassing technologies. One might even talk of socio-eco-techno systems.

25. Systems-Technocratic Tendencies

Systems-technocratic tendencies will gain in significance. Contemporary political, cultural, and human problems are increasingly conceived in systems-technological terms. But within systems-technological approaches to problems there lurk systems-technocratic dangers. (See entry on *Technocracy*.)

26. Data Protection

With information technologies, social and legal problems of data protection and privacy acquire new urgencies. This urgency carries over as well to concerns for protection of the integrity and dignity of the human person, respect for human values, and even reflection on what it means to be human.

27. Unforeseeable Risks

Technological systems are susceptible to risks that are often in principle not able to be foreseen. Increasing complexities in technological systems and the variability of human responses make predictions difficult if not impossible over certain distances and time frames. The persistence of risk within well-designed systems is illustrated by such simple occurrences as repeatedly occurring electrical blackouts in large metropolitan areas. Some technologically engendered dangers such as radioactivity may even go unobserved by most people who are affected.

28. Miniaturization

The trajectory of technological miniaturization in both part and whole of processes, products, and systems produces another kind of achievement and challenge: the "chipification" of things and functions or, ironically, above almost everything. From microsystems to nanotechnology these trends bring about new levels of manipulability and new degrees of difficulty in understanding and management.

29. Impacts Multiplication

Systems and information technologies multiply both positive and negative impacts, successes and failures. With the nearly unimaginable explosion of human technological powers through the vast extension of energy technologies and information systems, direct and indirect consequences, both successes (domination and control) and failures (accidents, "normal" or otherwise), pose extraordinary problems. They appear to exceed the human grasp, in literal as well as figurative senses.

30. Distributed Responsibility

Who bears responsibilities within ever-extended technological systems? The enlarged powers of multiple-distributed technological systems—systems that in some instances such as the Internet have become global—pose challenging ethical questions. How is it possible to deal with, divide up, or share responsibility in or for such systems? Responsibilities for general systems phenomena, for the detailed consequences of technological entanglements, and even for individual decision-making at strategic points within system contexts are not properly borne by individuals within current legal and moral frameworks. Thus many sociotechnical activities appear to evade responsible decision-making, calling forth the need to develop new forms of distributed responsibility.

As will have been apparent, this nonsystematic review of a series of structural features associated with new technologies has increasingly emphasized ethical and political issues. Perhaps it will eventually be necessary to analyze possible combinations and conditional relationships among the many characteristics mentioned here, and to investigate their associations with particular types of technology or technological fields, as well as with sociotechnical contexts and ethical problems. Such analyses could help refine many ethical and policy debates, which too often attempt to transfer an assessment from one context to another—at times even from one context in which it may well be appropriate to another in which it fails to be as genuinely relevant.

BIBLIOGRAPHY

Klages, Helmut. (1967). *Rationalität und Spontaneität* [Rationality and spontaneity]. Gütersloh, Germany: Bertelsmann.

Lenk, Hans. (1971). *Philosophie im technologischen Zeitalter* [Philosophy in the technological age], 2nd edition. Stuttgart: Kohlhammer.

Lenk, Hans. (1982). *Zur Sozialphilosophie der Technik* [Toward a social philosophy of technology]. Frankfurt am Main: Suhrkamp.

Lenk, Hans. (2003). *Grasping Reality: An Interpretation-Realistic Epistemology*. River Edge, NJ: World Scientific.

Lenk, Hans, and Matthias Maring, eds. (2001). *Advances and Problems in the Philosophy of Technology*. Münster, Germany: Lit.

Lenk, Hans, and Günter Ropohl. (1976). *Technische Intelligenz im systemtechnologischen Zeitalter* [Technical intelligence in the systems-technological age]. Düsseldorf: VDI Verlag.

Mitcham, Carl. (1994). *Thinking through Technology: The Path between Engineering and Philosophy*. Chicago: University of Chicago Press.

THE ETHICS OF EVIDENCE: A CALL FOR SYNTHESIS

VALERIE MIKÉ

• • •

As a mathematical scientist engaged in medical research in the 1970s, I became increasingly aware of the poor quality of much clinical research and the need for better assessment of medical technology. Of relevance here was the Hippocratic maxim "help or at least do no harm," the basis of the ethical tradition of Western medicine. If a treatment lacked proper evaluation, then no one could know whether it helped or harmed the patient, and that raised the question of ethics and its connection to statistics. In 1977 I organized an international symposium exploring these issues, "Medical Research: Statistics and Ethics," and articles based on the presentations were published in the journal *Science* (Miké and Good 1977). One of the editors told me that they did not, as a rule, publish conference proceedings, but this was special: "Your theme is in the air." These and related issues were further developed—and later published in book form—at a 1981 weeklong conference, at which I chaired a panel discussion addressing ethical, legal, and psychological aspects of clinical trials (Miké 1982).

A major success achieved by medical technology was the survival of smaller and smaller newborn infants. But many were impaired, and what to do about them became the subject of national debate. The issue centered on whether treatment should be withheld from some of these babies to let them die, and who should decide. Scholars in the new field of bioethics were usually trained in philosophy or law, so that questions of scientific assessment tended to be absent from the discourse. The physiology of the infants' disabilities was often poorly understood, treatments being applied had not been evaluated, and there was no reliable information on prognosis. Opposing conclusions were likely to be based on differing philosophical views of the "sanctity of life" versus the "quality of life." Ideology was taking the place of evidence.

There was also the role of social factors in disease and the outcome of treatment. A stark example concerned a tiny, premature infant at Babies Hospital at Columbia University in New York. A team of neonatal experts provided high-technology intensive care to save the child's life, and after three months and enormous expense the baby was well enough to be sent home to a nearby Harlem apartment. Later the doctors learned that the little boy died during the night when a rat chewed off his nose (Silverman 1980).

I became convinced that the problem was so broad that a new term was needed for the spectrum of related issues involving science, technology, uncertainty, philosophy, and society. Deciding on *Ethics of Evidence*, I used it for the first time in 1987 in the title of a lecture, illustrating it with the treatment of impaired newborns (Miké 1989b).

Medicine places societal concerns in sharp relief, because it is at the interface of technology and the deepest questions of human existence: the meaning of life, of suffering, and death. The Ethics of Evidence, an approach for dealing with uncertainty, had its primary focus on medicine, but was then seen to be more widely applicable. It calls for using the best possible evidence for decision-making in human affairs, in a continuous integration of the emerging results of relevant disciplines, but with recognition of the ultimately irreducible nature of uncertainty. Being well- informed and aware should form the basis of responsible action. The Ethics of Evidence—symbolized by a lighthouse—serves to provide guidance (Miké 2003).

After some general comments on the concept of evidence, this essay focuses on the uncertainties of scientific evidence. It sets the stage with the loss of certainty in mathematics itself, affecting what since ancient times had been considered self-evidently true. It sketches the scope of probability theory and statistical inference, used in the evaluation of scientific evidence. It discusses two important examples. The first one concerns evidence in a contemporary context: risk assessment. The second pertains to evidence in a historical context: evolution. This is followed by a more detailed discussion of the Ethics of Evidence. The final section addresses a long-range goal, the call for a philosophical synthesis, and presents a possible blueprint.

The relationship between statistics and ethics goes back to the late nineteenth century, to the English scientist Francis Galton (1822–1911), founder of modern statistics. Galton's work was inspired by a vision he named *eugenics*—improving the human race through controlled breeding. He championed social Darwinism and the eugenics movement, which would spread to other nations, including the United States. Forced sterilization of those deemed "socially inadequate" became legal in more than 30 states and was declared constitutional by the U.S. Supreme Court in the landmark case of Buck v. Bell (1927). Some 60,000 Americans were subjected to eugenic sterilization over the years, sanctioned by laws based on ideology and deeply flawed science (Reilly 1991).

Another area involving statistics and ethics was experimentation on humans, such as the Tuskegee Syphilis Study and other shocking medical practices reported into the 1970s. There were no pertinent laws in the United States, but at the time of the Nazi atrocities Germany already had legally binding regulations on human experimentation, issued in 1931, and these were more stringent than the subsequent Nuremberg Code (Miké 1990). Professional responsibility, the ethics of research and therapy, informed consent, and quality of proposed research were addressed in detail.

In 1974 the U.S. Congress passed the National Research Act and created a commission to propose ethical principles and guidelines for the protection of human research subjects, to be used in the development of federal regulations. In what came to be known as *The Belmont Report*, the commission identified three basic ethical principles consonant with the major traditions of Western thought: respect for persons, beneficence, and justice (U.S. National Commission 1979).

Ongoing concerns include end-of-life issues, embryo research, cloning, and the fundamental question of what it means to be human. The twentieth century made dazzling advances in science and technology, but it also produced unspeakable horrors, and it discovered the limits of scientific knowledge. To counter the pervasive skepticism of contemporary philosophy, the twenty-first century must accept the challenge of a new intellectual synthesis.

Introductory Remarks on Evidence

Evidence is defined as the data on which a judgment or conclusion may be based. In a court of law, evidence comprises the material objects and the documentary or verbal statements admissible as testimony, to be used by the jury in its verdict to convict or acquit the accused. In criminal cases the prosecution is to prove guilt "beyond a reasonable doubt," whereas in civil court "a preponderance of evidence" produced by the plaintiff is sufficient.

Evidence is often highly technical, presented by expert witnesses, and statisticians may be called to testify concerning the interpretation of empirical evidence (Gastwirth 2000). Tort cases may deal with injury due to exposure to a toxic chemical or drug, with each side offering its own supporting testimony. DNA evidence, not always clear-cut, may be decisive in a criminal trial. But scientific evidence is important in other areas, such as economic, social, and medical affairs, and as a guide in the formulation of public policy. Evidence of safety and effectiveness is critical in the use of drugs to treat or prevent disease.

Because evidence is intended to persuade others to take some action or to convince them of some belief, it has an intrinsic ethical component. Assertions that the evidence proves a claim can mislead and manipulate the uninformed. Evidence is not fixed and permanent; it is whatever is accepted as support for a conclusion by a given community (scholars, jurors, members of society) at a given point in time, and is subject to change with new developments. Statistical DNA evidence, if judged to be of acceptable quality, may exonerate someone convicted of a serious crime, even when the conviction was based on the evidence of eyewitness testimony. Eyewitnesses may identify someone in a lineup who closely resembles the perpetrator actually observed. There is always a subjective element, an element of uncertainty.

Mathematics and Uncertainty

Mathematics can be remarkably effective in the exploration of physical, measurable phenomena. But it is a creation of the human mind. Long-held beliefs about its absolute and certain nature were destroyed by discoveries made in the nineteenth and early twentieth centuries. Albert Einstein (1879–1955) stated it clearly: "As far as the laws of mathematics refer to reality, they are not certain; and as far as they are certain, they do not refer to reality" (1983 [1923], p. 28). Euclidean geometry is no longer seen as a true description of space, nor does mathematical logic claim to grasp all reality. Concurrent with these discoveries was the emergence of the theories of probability and statistics, as a way to assess observed variability and uncertainty.

THE LOSS OF CERTAINTY: NON-EUCLIDEAN GEOMETRY. In the 1820s the Hungarian mathematician János Bolyai (1802–1860) and the Russian Nikolai Lobachevsky (1793–1856) showed independently that by changing a supposedly "self-evident" postulate of Euclidean geometry another logically consistent system of geometry could be developed. This discovery dealt a fatal blow to the notion of Immanuel Kant (1724–1804) that Euclidean geometry inheres in the human mind as a priori knowledge that is necessarily true, imposed by the mind on an unknown and unknowable reality. These geometries were now seen to be human constructs, not intrinsic to the mind, applied as different models to the universe that existed "out there" and was thus observable and real.

In the Bolyai-Lobachevsky system Euclid's fifth postulate, stating that through a point in a plane only a single line can be drawn parallel to a given line, was replaced by the assumption that an infinite number of lines can be drawn through a point parallel to the given line. A few decades later the German mathematician Bernhard Riemann (1826–1866) developed another consistent geometry with the axiom that *no* line can be drawn through a point parallel to a given line—in other words, that all lines intersect. This became the basis of Einstein's general theory of relativity.

Strictly speaking, Euclidean geometry is wrong in the real world; space is curved by gravity. But for practical purposes, because the curvature is very slight even for enormous distances, it is a very good approximation. The philosophical impact of the discovery, however, was radical: For any axiom considered to be self-evidently true in an earlier age, it is wiser to say that it may not be so.

LOSING MORE GROUND: THE INCOMPLETENESS THEOREM. But the twentieth century revealed an ultimate barrier to scientific knowledge of reality. In 1931 the Austrian logician Kurt Gödel (1906–1978) proved what is known as the *incompleteness theorem:* Any consistent mathematical system that includes even as little as the arithmetic of whole numbers contains statements that cannot be proved either true or false within the system. No mathematical system can encompass all truth; there will always be some truths that are beyond it. This result precludes a full grasp by logic of all reality.

ASSESSING UNCERTAINTY: PROBABILITY AND STATISTICS. The theory of probability, with its axiomatic foundation, is a vigorous branch of modern mathematics. Statistical inference, based on probability, reached maturity in the twentieth century and is central to much of modern technology. As induction, its methods of inference pertain to the philosophy of science.

Typically, there is interest in some characteristic of a *population* from which a *representative sample* is selected. It is assumed that the identical experiment of selecting the sample can in principle be *repeated indefinitely*, such as drawing ten balls from an urn containing red and black balls (replacing each after noting its color). Statistical inference provides methods for reaching conclusions about the population from the sample, such as the proportion of red balls in the urn, with predetermined limits of *sampling error*. It is often impossible to sample the actual *target population* of interest, and there remains the difficult step of going from the population sampled to the target population. For example, a drug may be tested on patients selected in a given hospital to assess its response rate, but the target population includes all patients with the disease now and in the future. But even within the hospital, no two patients are identical, so the study has to consider factors that may affect the outcome of the trial, such as age, sex, other medical conditions, and so on. There may also be relevant factors as yet unknown. The simple model of drawing balls from an urn may be assumed by the theory, but it is rarely found in practice.

In classical statistical inference the sample is used to test, and then reject or accept (the latter, strictly speaking, should be *not reject*), a null hypothesis of interest, and confidence intervals are constructed for point estimates. The American statistician Allan Birnbaum (1923–1976) undertook studies to develop principles of *statistical evidence* in this framework (1969). Statistical evidence as part of induction, based on different interpretations of probability, is the subject of ongoing research in epistemology and the philosophy of science (Taper and Lele 2004).

Evidence in a Contemporary Context: Risk Assessment

Risk is the probability that something bad may happen. Much effort is devoted to identifying hazards in the environment and the workplace that are harmful to health, with controversial claims of evidence seeking to affect government regulation. Another area concerns the control of risk, such as the use of drugs for the prevention of disease.

RISK ASSESSMENT: VAST UNCERTAINTY. The uncertainties in the risk assessment of chemical hazards to health were explored at an international workshop held in Italy in 1998, with extensive use of real-life examples (Bailar and Bailer 1999). Uncertainty results from inaccurate and incomplete data, incomplete understanding of natural processes, and the basic ways of viewing the questions. There is uncertainty in hazard identification, exposure assessment, dose–response modeling, and the characterization and communication of risk. There is also true variability in risk across space, time, and among individuals.

Assessments of risk of the same hazard frequently differ by factors of 1,000 or more. For example, four estimates of the added lifetime risk of kidney cancer from the chemical Tris, used as a flame retardant in children's sleepwear, ranged between 7 and 17,000 per million children exposed. Random deviations of a sample from a specified model, addressed by the methods of statistics, are but a small component of uncertainty in risk assessment. In any particular case, the public needs insight into the nature of the uncertainties involved, in order to participate in meaningful discourse.

MENOPAUSAL HORMONE THERAPY: A STARTLING REVERSAL. For decades millions of postmenopausal women were routinely prescribed hormone replacement therapy (HRT), first introduced for the alleviation of menopausal symptoms, then believed to offer protection against coronary heart disease, the leading cause of death for women in most developed countries. Observational studies of HRT, as well as meta-analyses (formally combined evaluations) of these studies, had suggested a 35 to 50 percent reduction in coronary events. But carefully designed randomized clinical trials began to report, culminating in results published in 2002, not prevention, but an increased risk of heart disease, heart attacks, and stroke in HRT users, in addition to the known increased risk of breast cancer. The results indicated that about 1 percent of healthy postmenopausal women on HRT for five years would experience an excess adverse event, a substantial number when applied to the estimated 10 million American women taking hormones. Much research remains to be done, but a subsequent meta-analysis of the earlier observational studies found that HRT users differed from nonusers in important characteristics. Adjusting the data for socioeconomic status, education, and major coronary risk factors eliminated the apparent cardiac protection of HRT, the evidence that had once so firmly convinced the medical community (Wenger 2003).

Evidence in a Historical Context: Evolution

The issue here is not the fact of evolution, but the mechanism of evolution. Are random variation and natural selection sufficient to explain the origin of life and the complexity of living systems, or are there other forces driving evolution? Is there purpose or design in what is observed? Many scientists hold that Charles Darwin's theory of evolution, or a more elaborate version of it, provides a natural explanation for the existence of all living systems. Others are challenging this view, and books published by either side may contain the word *evidence* in their titles.

CLAIMS OF EVIDENCE IN OPPOSING VIEWS. A popular book on the Darwinian view is *The Blind Watchmaker: Why the Evidence of Evolution Reveals a Universe without Design* (1986), by the British zoologist Richard Dawkins, who holds a chair in the public understanding of science at Oxford University. The title refers to the argument for design in the universe by the eighteenth-century English theologian William Paley (1743–1805), who used the analogy that finding a watch would lead one to conclude that it was made by someone, that there was a watchmaker. Dawkins aims to show that evolution took place entirely by chance variation and small changes, by natural forces without purpose, so that the watchmaker is blind. But he assumes that life was already on hand, that it came from entities so simple as to require no explanation. He leaves the details of their origin to physicists, although the latter have in fact encountered high specificity in the systems of modern cosmology.

Other scientists hold that a further evolutionary structure is needed beyond variation and natural selection. Advances in the fields of biochemistry and molecular biology, as well as the new information sciences, are being used to explore the question, with explanations sought in the natural order. A still different view is presented in *Science and Evidence for Design in the Universe* (2000), by the American researchers Michael J. Behe, William A. Dembski, and Stephen C. Meyer, trained in biochemistry, mathematics, and philosophy. Analyzing the latest scientific developments, they argue that the complex specified information encountered in the cosmos, including irreducibly complex biochemical systems, cannot be generated by a chance mechanism, that there is evidence of intelligent design. If patterns are broken down into a series of steps guided by what has gone before, as in evolutionary algorithms proposed by Dawkins and others, then there is built-in purpose or predetermined design.

It is helpful here to review the methodology claiming to provide evidence.

HISTORICAL SCIENCE: INFERENCE TO THE BEST EXPLANATION. The American philosopher Charles Sanders Peirce (1839–1914) distinguished three modes of inference (Peirce 1998 [1923]). These were deduction (reasoning from general to particular), induction (reasoning from particular to general), and what he called *abduction* or *hypothesis* (reasoning from effect to cause). An example of deductive inference is proving the theorems of Euclidean geometry from its axioms. Induction includes the customary use of probability in science, where results from the observed sample can be confirmed in further experiments to describe a natural process or mechanism of action. Abduction is not directly related to probability. The cause is not observed, and the question is which of any rival hypotheses gives the best explanation of the observed effect. As historical science, exploring the origin and evolution of the universe is in this category. It occurred once in the distant past, and the aim is to explain what may have caused it to happen. Probability enters only as the chance of realization of a particular path among all possibilities in assumed evolutionary mechanisms.

Contemporary philosophers of science speak of abduction in terms of *explanatory power* or *inference to the best explanation*, with three proposed criteria. Hypothesis A is the best explanation for observed outcome B if: (1) A is consonant (consistent, in harmony) with B, (2) A adds something to the understanding of B, and (3) A adds more to the understanding of B than its rival hypotheses. Scientific naturalists consider only material hypotheses to explain the visible universe and its living systems. But because the ultimate goal is to understand all of life, the full range of human experience, others argue that it is not rational to arbitrarily exclude any viable hypotheses, including that of intelligent design.

EVIDENCE AND THE LIMITS OF SCIENTIFIC KNOWLEDGE. Caution in making claims of evidence was advised by Ronald A. Fisher (1890–1962), British pioneer of the fields of statistics and genetics and the mathematical theory of evolution. Fisher showed Gregor Johann Mendel's laws of inheritance to be the essential mechanism for Darwin's theory of evolution (Fisher 1930), but as a Christian he saw no conflict between science and his own faith. In a 1955 radio address on the BBC he referred to his own work as "the study of the mode of inheritance of the heritable characteristics of animals, plants and men" (Fisher 1974, p. 351), and spoke of the evil of misleading the public to believe that science is the enemy of religion. He urged scientists to acknowledge the limits of their own discipline:

> In order to know, or understand, better, it is necessary to be clear about our ignorance. This is the research scientist's first important step, his *pons asinorum*, or bridge which the asses cannot cross. We must not fool ourselves into thinking that we know that of which we have no real evidence, and which, therefore, we do not know, but can at

> most accept, recognizing that still we do not actually know it. (pp. 351–352)

The recurring appearance of conflict between the exact sciences and the philosophical search for truth cannot be decided in favor of either side by careless or ignorant trivialization, by attributing to the other side a simplistic conceptual framework, as is especially often the case against believers. Theology has traditionally been defined as *fides quaerens intellectum* (faith seeking understanding), and this means being open to new insights of all human endeavors, including science. As stated, for example, by Pope John Paul II (1920–2005):

> Only a dynamic relationship between theology and science can reveal those limits which support the integrity of each discipline, so that theology does not profess pseudo-science and science does not become an unconscious theology. Our knowledge of each other can lead us to be more authentically ourselves. (1988, p. M14)

To attain consensus in a pluralistic culture, it is necessary to seek common ground, common principles to serve as guide to life in a world of uncertainty.

The Ethics of Evidence

The notion of an Ethics of Evidence, proposed initially for dealing with uncertainty in medicine (Miké 1991, 1999, 2003), applies equally to other difficult issues encountered in daily life.

TWO IMPERATIVES OF THE ETHICS OF EVIDENCE. The Ethics of Evidence can be expressed in two simple rules or imperatives. The first imperative calls for the creation, dissemination, and use of the best possible evidence for decision-making in human affairs. Complementing it, the second imperative focuses on the need to increase awareness of, and come to terms with, the extent and ultimately irreducible nature of uncertainty.

Evidence here means the information obtained and interpreted by the highest standards of scholarship in each relevant field, with the minimal requirement of internal logical consistency. It allows for diverging views within a field, as there is a range of uncertainty, but the points of divergence should be clear. It assumes a philosophy of realism, the conceptual framework of the scientist, who believes in an external world of order that is accessible to human inquiry. It differentiates between two kinds of uncertainty: *Scientific uncertainty*, essentially dynamic, constantly changing with progress in research, but never fully eliminated, because of intrinsic limitations of the scientific method; and *existential uncertainty*, also invariably present, because the question of ultimate meaning, the deepest mystery, is beyond the scope of science.

Evidence is complex and fragile. Proof by experiment covers little beyond the laws of physics. Mathematical models may not apply to reality, and even the logic of mathematics is limited in its scope. Standards for proof of causation vary by field, and it is the consonance of data from diverse sources that provides the strongest evidence.

INVOLVEMENT OF VARIOUS DISCIPLINES. Affirming the complexities of dealing with uncertainty, research in cognitive psychology has shown that intuitive judgments do not follow the laws of probability; people tend to be overconfident in their conclusions (Kahneman, Slovic, and Tversky 1982). Findings in cognitive neuroscience suggest that emotion is an integral part of the reasoning process (Damasio 1994).

Positivist views of objectivity in science were challenged by the physical chemist Michael Polanyi (1891–1976), who turned to philosophy to develop his concept of *personal knowledge*, the vast domain of tacit assumptions, perceptions, and commitments of the persons who hold it (Polanyi 1958). Science must be consistent with the evidence, but the ultimate commitment is that of personal judgment. Hungarian-born like Polanyi, the mathematician George Polya (1887–1985) gained recognition for his skill in sharing insight into the heuristics of plausible reasoning (Polya 1954).

Relevant to contemporary social upheavals is the thought of Viktor E. Frankl (1905–1997), founder of the so-called third Viennese school of psychotherapy (after those of Sigmund Freud and Alfred Adler). Frankl's approach, called logotherapy (after *logos*, the Greek concept of rational principle), was derived from his vast experience as a psychiatrist and as survivor of concentration camps in World War II (Frankl 1992). He held that the search for meaning is the basic motivation of human life. Frankl saw the *existential vacuum* of present times—a pervasive lack of purpose or meaning—as the major cause of the triple plague afflicting society, that of depression, aggression, and addiction. These insights, too, need to be considered in analyzing the troubling issues of the day.

Without a critical attitude to empirical data and insight into the nature of science and evidence, the public is vulnerable to manipulation by special interest groups and the market. The many conflicts of interest and misleading reports in the media, often with improper use of statistics, have been well documented by sociologists and others (Best 2001). Professionals with a poor understanding of statistical concepts may agitate with false charges (Miké 1989a).

An example of a complex problem in need of impartial discussion of the evidence from a variety of sources is that of abortion. Confrontational bandying of slogans for a generation has not resolved the national debate, a standard feature of political campaigns and perhaps the most divisive issue in American society.

The Ethics of Evidence urges focus on what is known about the subject or calls for further study, without the barrier of ideology. What does biomedical science know about the human embryo, from its origin as a single cell? Can direct visualization of the developing organism by contact embryoscopy be made widely available to the public? What are the demographics of the women having abortions? Why do women have abortions? Are many of them pressured into the decision by others? What are the economic issues involved for the women and the abortion industry? What is known about the long-term consequences of abortion? Scholarly research addressing these and related questions by the relevant disciplines could be reported by the mainstream media, including prime-time television, on a regular basis. Given that 45 million abortions have been performed in the United States since the procedure was legalized in 1973, a great deal of source material is available. Objective and ongoing presentation of the best available evidence, with emphasis on quality and completeness, would encourage open discussion and informed judgments by all concerned, especially the young who have not as yet taken sides in the debate.

OVERVIEW. The Ethics of Evidence is a means of consciousness-raising, of urging society to examine all aspects of vexing issues, to be wary of facile claims of evidence, to recognize conflicts of interest. It is consistent with the accepted norms of science that include intellectual integrity, objectivity, doubt of certitude, tolerance, and communal spirit (Miké 1999). More generally, the Ethics of Evidence is supported by the principles of honesty and literacy. No one would question the ideal of honesty, of telling the truth and being trustworthy. But a democratic society must also strive to be a literate, well-informed society, and this includes scientific literacy, with insight into the scope of science and its methods of inference. The Ethics of Evidence implies responsibilities for professionals as well as the public, and a central role for education. Looking to the future, it calls for the creation of a new philosophical synthesis as a central challenge of the twenty-first century.

Toward a Philosophical Synthesis

René Descartes (1596–1650) chose *thought* as the first principle of his philosophy. The discoverer of analytic geometry, he saw in the absolute certainty of mathematics a way to impose the certainty of rational knowledge on all reality. Descartes, a brilliant dreamer, did not know about non-Euclidean geometry (not discovered for another 200 years) or the incompleteness theorem of mathematics (not discovered for another 300). What crystallized in his mind as the first principle, his famous *Cogito, ergo sum* (I think, therefore I am), would lead to rationalism, and had already been analyzed by Saint Augustine of Hippo (354–430) in four of his books. Both used it to counter the skepticism of their age and to develop an ontological argument for the existence of God. But unlike Descartes, Augustine did not adopt the principle as the basis of a philosophical system.

A different perspective was proposed by the French philosopher and medieval scholar Étienne Gilson (1884–1978). In 1936 Harvard University marked the 300th anniversary of its founding, and as part of the celebration Gilson was invited to be a visiting professor. He accepted the lectureship named in memory of William James (1842–1910), the founder of American pragmatism, and his lectures were published in 1937 as *The Unity of Philosophical Experience*.

Gilson sees the unity of philosophical experience in the persistent search for a first principle, by a naturally transcendent human reason, to explain what is given in sense experience. He argues that the many previous attempts in the history of Western philosophy eventually failed, because philosophers took a part of the system for the first principle. He holds that the first principle of human knowledge is *being*, and it therefore has to be the first principle of metaphysics.

Gilson insists: "Man is not a mind that thinks, but a being who knows other beings as true, who loves them as good, and who enjoys them as beautiful" (1999 [1937], p. 255). In the search for philosophical synthesis, he is not suggesting some new system of tomorrow or the reviving of some old system of the past:

> The three greatest metaphysicians who ever existed—Plato, Aristotle and St. Thomas Aquinas—had no system in the idealistic sense of the word. Their ambition was not to achieve philosophy once and for all, but to maintain it and to serve it in their own times, as we have to maintain it and to serve it in our own. For us, as for them, the great thing is not to achieve a system of the world as if being could be deduced from

thought, but to relate reality, as we know it, to the permanent principles in whose light all the changing problems of science, of ethics and of art have to be solved. (p. 255)

This philosophy of realism is for Gilson a continuous process, a constant analysis of experience:

> A metaphysics of existence cannot be a system wherewith to get rid of philosophy; it is an always open inquiry, whose conclusions are both always the same and always new, because it is conducted under the guidance of immutable principles, which will never exhaust experience, or be themselves exhausted by it. For even though, as it is impossible, all that which exists were known to us, existence itself would still remain a mystery. (pp. 255–256)

BIBLIOGRAPHY

Bailar, John C., III, and A. John Bailer. (1999). "Risk Assessment—The Mother of All Uncertainties: Disciplinary Perspectives on Uncertainty in Risk Assessment." In *Uncertainty in the Risk Assessment of Environmental and Occupational Hazards: An International Workshop*, ed. A. John Bailer, Cesare Maltoni, John C. Bailar III, et al. New York: New York Academy of Sciences.

Behe, Michael J.; William A. Dembski; and Stephen C. Meyer. (2000). *Science and Evidence for Design in the Universe*. San Francisco: Ignatius Press.

Best, Joel. (2001). *Damned Lies and Statistics: Untangling Numbers from the Media, Politicians, and Activists*. Berkeley and Los Angeles: University of California Press. Incisive commentary by a sociologist.

Birnbaum, Allan. (1969). "Concepts of Statistical Evidence." In *Philosophy, Science, and Method: Essays in Honor of Ernest Nagel*, ed. Sidney Morgenbesser, Patrick Suppes, and Morton White. New York: St. Martin's Press.

Damasio, Antonio R. (1994). *Descartes' Error: Emotion, Reason, and the Human Brain*. New York: Putnam.

Dawkins, Richard. (1986). *The Blind Watchmaker: Why the Evidence of Evolution Reveals a Universe without Design*. New York: Norton.

Einstein, Albert. (1983 [1923]). "Geometry and Experience." In his *Sidelights on Relativity*. New York: Dover. Brief article, based on a talk addressed to a general audience.

Fisher, R. A. (1930). *The Genetical Theory of Natural Selection*. Oxford: Clarendon Press; 2nd edition, New York: Dover, 1958. A classic work integrating Mendelian genetics with Darwin's theory of evolution.

Fisher, R. A. (1974). "Science and Christianity: Faith Is Not Credulity." In *Collected Papers of R. A. Fisher*, Vol. 5 (1948–1962), ed. J. H. Bennett. Adelaide, Australia: University of Adelaide. A 1955 talk given in London on BBC radio.

Frankl, Viktor E. (1992). *Man's Search for Meaning: An Introduction to Logotherapy*, 4th edition. Boston: Beacon Press. Originally published in German as *Ein Psycholog erlebt das Konzentrationslager*, 1946.

Gastwirth, Joseph L., ed. (2000). *Statistical Science in the Courtroom*. New York: Springer. A collection of essays by statisticians experienced in giving court testimony.

Gilson, Étienne. (1999 [1937]). *The Unity of Philosophical Experience*. San Francisco: Ignatius Press. Analysis of the history of Western philosophy with a proposed new philosophical synthesis.

Goldstein, Rebecca. (2005). *Incompleteness: The Proof and Paradox of Kurt Gödel*. New York: Norton. A biography of Gödel, with explanation of his famous theorem for the general reader.

Hacking, Ian. (2001). *An Introduction to Probability and Inductive Logic*. Cambridge, UK: Cambridge University Press. Introductory textbook for students of philosophy, explaining different interpretations of probability, with examples from everyday life.

Hayes, Zachary. (2001). *The Gift of Being: A Theology of Creation*. Collegeville, MN: Liturgical Press. A concise introduction to the relationship between modern science and Christian theology, written for the general reader.

John Paul II, Pope. (1988). "Message of His Holiness Pope John Paul II to Conference Participants." In *Physics, Philosophy, and Theology: A Common Quest for Understanding*, ed. Robert J. Russell, William R. Stoeger, and George V. Coyne. Vatican City: Vatican Observatory. Delivered at a Vatican conference marking the 300th anniversary of the publication of Isaac Newton's *Philosophiae Naturalis Principia Mathematica*.

Jones, James H. (1981). *Bad Blood: The Tuskegee Syphilis Experiment*. New York: Free Press. Study sponsored by the U.S. Government, in which 400 black sharecroppers afflicted with syphilis, but not told they had the disease, were kept under observation without treatment, even after penicillin became widely available.

Kahneman, Daniel; Paul Slovic; and Amos Tversky, eds. (1982). *Judgment under Uncertainty: Heuristics and Biases*. Cambridge, UK: Cambridge University Press.

Kline, Morris. (1980). *Mathematics: The Loss of Certainty*. New York: Oxford University Press. Historical analysis of the fundamental changes in the understanding of the nature of mathematics.

Miké, Valerie. (1982). "Clinical Trials: Exploring Ethical, Legal, and Psychological Issues" (panel discussion). In *Statistics in Medical Research: Methods and Issues, with Applications in Cancer Research*, ed. Valerie Miké and Kenneth E. Stanley. New York: Wiley.

Miké, Valerie. (1987). "Saving Money and Improving Health by Evaluating Medical Technology." *Technology Review* 90(3): 22–25.

Miké, Valerie. (1989a). "Philosophers Assess Randomized Clinical Trials: The Need for Dialogue." *Controlled Clinical Trials* 10(3): 244–253.

Miké, Valerie. (1989b). "Toward an Ethics of Evidence—and Beyond: Observations on Technology and Illness." *Research in Philosophy and Technology* 9: 101–113.

Miké, Valerie. (1990). "Ethics, Evidence, and Uncertainty." *Controlled Clinical Trials* 11(3): 153–156.

Miké, Valerie. (1991). "Understanding Uncertainties in Medical Evidence: Professional and Public Responsibilities." In *Acceptable Evidence: Science and Values in Risk Management*, ed. Deborah G. Mayo and Rachelle D. Hollander. New York: Oxford University Press.

Miké, Valerie. (1992). "Quality of Life Research and the Ethics of Evidence." *Quality of Life Research* 1: 273–276.

Miké, Valerie. (1999). "Outcomes Research and the Quality of Health Care: The Beacon of an Ethics of Evidence." *Evaluation & the Health Professions* 22(1): 3–32. Commentary in the same issue by Edmund D. Pellegrino, "The Ethical Use of Evidence in Biomedicine," pp. 33–43.

Miké, Valerie. (2003). "Evidence and the Future of Medicine." *Evaluation & the Health Professions* 26(2): 127–152.

Miké, Valerie, and Robert A. Good. (1977). "Old Problems, New Challenges." *Science* 198(4318): 677–678. Introduction to the Birnbaum Memorial Symposium "Medical Research: Statistics and Ethics." Issue contains articles based on symposium lectures.

Miké, Valerie; Alfred N. Krauss; and Gail S. Ross. (1993). "Neonatal Extracorporeal Membrane Oxygenation (ECMO): Clinical Trials and the Ethics of Evidence." *Journal of Medical Ethics* 19(4): 212–218.

Nuremberg Code. (1949). *Trials of War Criminals before the Nuremberg Military Tribunals under Control Council Law* 10(2): 181–182. Washington, DC: U.S. Government Printing Office.

Paley, William. (1802). *Natural Theology; or, Evidences of the Existence and Attributes of the Deity, Collected from the Appearances of Nature*. London: Faulder.

Peirce, Charles Sanders. (1998 [1923]). "Deduction, Induction, and Hypothesis." In *Chance, Love, and Logic: Philosophical Essays*. Lincoln: University of Nebraska Press.

Polanyi, Michael. (1958). *Personal Knowledge: Towards a Post-Critical Philosophy*. Chicago: University of Chicago Press.

Polya, George. (1954). *Mathematics and Plausible Reasoning, Vol. 2: Patterns of Plausible Inference*. Princeton, NJ: Princeton University Press.

Silverman, William A. (1980). *Retrolental Fibroplasia: A Modern Parable*. New York: Grune and Stratton. Author was director of the neonatal intensive care unit at Columbia University in New York.

Reilly, Philip R. (1991). *The Surgical Solution: A History of Involuntary Sterilization in the United States*. Baltimore: Johns Hopkins University Press.

Taper, Mark L., and Subhash R. Lele, eds. (2004). *The Nature of Scientific Evidence: Statistical, Philosophical, and Empirical Considerations*. Chicago: University of Chicago Press. Collection of essays on scientific evidence based on different interpretations of probability, using ecology as the illustrative science.

U.S. National Commission for the Protection of Human Subjects of Biomedical and Behavioral Research. (1979). *The Belmont Report: Ethical Principles and Guidelines for the Protection of Human Subjects of Research*. Washington, DC: U.S. Department of Health, Education, and Welfare.

Wenger, Nanette K. (2003). "Menopausal Hormone Therapy and Cardiovascular Protection: State of the Data, 2003." *Journal of the American Medical Women's Association* 58(4): 236–239.

TOWARD AN ETHICS OF SCIENCE AND TECHNOLOGY AS KNOWLEDGE

CARL MITCHAM
ROBERT FRODEMAN

• • •

A central feature of science and technology is their character as knowledge. Not only is science commonly described as both cognitive activity and a body of knowledge, but technological power has become increasingly knowledge-dependent. Unlike power, knowledge is often judged an unqualified good. But in a world in which technoscientific knowledge offers along side its manifest benefits unparalleled opportunities for destructive utilization, and in which individuals are increasingly challenged to come to terms with scientific and technological perspectives on the natural world and themselves, the moral status of knowledge deserves substantive consideration.

Knowledge Questions

Knowledge has been defined since Plato as "justified true belief," that is, as true opinion with reason or *logos* (*Theatetus* 201d–210d). Epistemology or the theory of knowledge examines what counts as the reasoning that can convert true opinion (which may be quite accidental) into knowledge. Does epistemic rationality require reference to empirical data, systematic coherence, covering laws, or what?

Precisely because of its various possible justifications, knowledge comes in many forms. Bertrand Russell (1910), for instance, distinguished knowledge by description (scientific propositions) and knowledge by acquaintance (including technical know how). In relation especially to science and technology each type raises ethical as well as epistemological issues that have seldom been addressed in standard philosophical discussions. Is it not possible for certain types of propositional knowledge or their pursuit to distract human beings from more important activities and ends? Might not knowledge by acquaintance be ethically or politically problematic?

The first question has been broached on the margins of philosophy in information science and knowledge management. These contemporary disciplines have, for instance, examined the relations between data, information, knowledge, and wisdom—distinctions first suggested by the poet T. S. Eliot in "Chorus from *The Rock*" (1934). Economist and diplomat Harland Cleveland (1982) and operations research scientist Russell Ackoff (1989) have each proposed different versions of these distinctions that highlight how knowledge and understanding can be obscured by data or information.

The second question has been raised in relation to forms of knowledge as diverse as nuclear engineering and genetic screening. For the master inventor of the atomic bomb, J. Robert Oppenheimer (1947), "In some sort of crude sense which no vulgarity, no humor, no overstatement can quite extinguish, the physicists have known sin." For philosopher Ruth Chadwick (1997), information about genetic abnormalities constitutes a kind of knowledge that patients may have a "right not to know" in order to lead their lives without excessive worry. How did it come about that knowledge, which has so often been seen as a pristine virtue, is now manifest in the contemporary world as both benefit and burden?

Historical Emergence

Reflection on the role of knowledge in society goes back to the origins of European civilization. Pre-Socratic philosophers were largely concerned with the natural world, but by the mid-fifth century B.C.E. this had changed. According to Plato, Socrates suggested that he could learn little of human importance from nature (*Phaedrus* 230d), and in the *Republic* he set up a keen tension between knowledge and politics.

The *Republic* begins with an account of the various ways societies can be governed: through violence, religious authority, tradition, or discursive rationality. The first three play an inevitable role in society. Governments must possess a monopoly over violence, while religious authority and tradition provide the guidance needed to establish social norms. Plato is nevertheless often interpreted as launching the West on a 2500 year trajectory to progressively free rationality from the constraints imposed by these other approaches, a process of disengagement that reached apotheosis in the Enlightenment. In the dialogues, however, Plato repeatedly emphasizes the tension between philosophy and power. Socrates must be (playfully) coerced to reveal what he knows, and even then he carefully reminds his listeners that the philosophic knowledge of the few looks topsy turvy to the many.

The dialogue reaches a climax in the myth of the divided line and allegory of the cave, where Socrates once again yokes knowledge to politics. These images describe the difficulty of distinguishing truth from shimmering illusions, as well as the way that falsehoods can blind a person to the truth. The difficulties are multiplied, however, by Socrates's view that knowledge can also cripple. Inverting the Homeric story in which Odysseus visits the underworld in order to gain the knowledge needed for practical matters, Socrates describes how philosophers can become so dazzled by the brilliance of their insights as to lose any sense of how to relate them to everyday experience.

In his *Nicomachean Ethics* Aristotle also emphasizes the relation between knowledge and desire: "Both the reasoning must be true, and the desire right, if the choice is to be good" (VI, 2; 1139a25). On Aristotle's account, excellence in reasoning and right desire are cultivated through the moral and intellectual virtues. Moral virtues such as courage, generosity, and magnanimity are governed by a principle, the doctrine of the mean, that seeks out the midpoint between the extremes of excess and deficiency. The intellectual virtues—which Aristotle examines in order as *episteme* (science), *techne* (craft skill), *phronesis* (practical judgment), and *nous* (intuition)—identify the different ways human beings can acquire truth.

Crucially, however, there is no principle of the mean to govern these intellectual virtues. There is no discussion of the possibility that there could be an excess as well as a deficiency in any intellectual virtue after the manner of the moral virtues. Nor for that matter is there any account of how the moral and intellectual virtues relate to one another. When Aristotle turns to a fifth intellectual virtue, *sophia* (wisdom), he describes it as the combination of intuition and science—leaving out technical skill and practical judgment. Wisdom consists of theoretical knowledge lacking any clear relation to practical matters. For Aristotle, the highest form of knowledge appears to escape any Platonic problematic.

In the Platonic tradition, which became through Augustine a vehicle for Christian theological reflection, this problematic finds multiple expressions. Consider the story of Leontius (*Republic* IV, 439e ff.). Walking along the wall outside the Piraeus, Leontius spies a corpse from an execution, and desires to feast his sight on the repugnant image. Recognizing this as a degraded use of the most noble and cognitive of the senses, he struggles to resist temptation. Failing in moral stamina, he finally runs toward the rotting body and exclaimed to in sarcastic irony: "Look, you damned wretches, take your fill of the fair sight!"

The problem of the custody of the eyes becomes in fact a major moral issue in the Christian tradition. According to the biblical narrative, the knowledge of good and evil was associated with a tree in the midst of the Garden of Eden that was "a delight to the eyes" (Genesis 3:6), but from which Adam and Eve had been forbidden to eat. When they succumbed to the visual temptation, their eyes were opened in new ways that brought hardship upon them. During the medieval period this notion of dangerous knowledge was elaborated especially in the monastic tradition. In an extended commentary on chapter seven of the *Rule of St. Benedict*, Bernard of Clairvaux, in the *Steps of Humility* (1120), criticizes "curiositas" as a form of pride. Thomas Aquinas, working under the influence of Aristotle, sought to qualify such criticism, although even he admitted that "curiosity about intellective sciences may be sinful" (*Summa theologiae* II-II, Q.167, art.1). But with the coming of the modern age the restriction on knowing was set aside in favor of a view of knowledge as an unqualified good in an even stronger sense than found in Aristotle himself.

In the modern era, traditional boundaries on scientific pursuits began to drop away as interrogation undertook new active forms in dealing with both nature (the performing of autopsies and experimentation) and the sacred (subjecting the Bible to the same kinds of analysis as any other book). René Descartes represents a signal turning point. Offering a distinctively modern scientific sense of reason, he claimed that with his method "there is no need for the mind to be contained within any limits" (*Rules for the Direction of the Mind*, 1620s). For Descartes there were new rules to replace those of the monasteries, and new meditations to replace spiritual reading, through which human beings might become the "masters and possessors of nature" for which they had been divinely predestined. This project approaches fulfillment in the twenty-first century, as scientific and technological advances create possibilities that herald wholesale changes in nature, society, body, and mind.

E. F. Schumacher (1977) in a simple but insightful characterization, describes the transition introduced by Descartes and others as one from the pursuit of "science as understanding" to "science for manipulation." Whereas the former sought to integrate the knower with the known, to raise human beings out of their material state by means of insight into higher things, the latter began with a sense of the knower as separate from the known and sought to assert this separation by means of analysis. The overarching theme concerning knowledge since the 1500s has been the progressive application of the principle of analysis. Descartes provides the classic statement of the analytic method in his *Discourse on the Method for Rightly Conducting the Sciences* (1637). Items were to be understood by being broken into their constituent pieces. The goal was to arrive at the smallest possible elements. Once these "simples" were identified and completely examined knowledge would be reconstructed upon an unimpeachable foundation. Complemented by the empiricist methods developing from Francis Bacon, who also sought new forms of knowledge, there has flowed forth an ever widening stream of results, including but not limited to the growth of academic disciplines.

The New World of Knowledge and Its Production

In the epistemological world opened up by Descartes and Bacon new categories and forms of knowledge multiply without bounds. In the nineteenth century natural philosophy divided into physics, chemistry, and mathematics, while natural history morphed into biology with an experimental component that challenged the traditional emphasis upon description and taxonomy. The social sciences—sociology, psychology, economics, political science, and anthropology—arose to address the new social conditions, applying a scientific approach to the problems of industrialized experience.

The disciplines that become known as the humanities—philosophy, classical languages, modern languages, history, art, and music—formed a rump out of what was left over after the extraction of these other new specialties. The term itself was an adaptation from the Renaissance *studia humanitatus*, when humanist scholars looked to ancient thinkers such as Cicero for inspiration and guidance. A few of these latter day humanists protested the rise of specialization and disciplinarity and the new emphasis on research, but in general the humanities accommodated themselves to the novel paradigms of knowledge. Abandoning the traditional notion of expounding a perennial philosophy, fields such as literature and philosophy now trained specialists whose role was to develop new insights. Having given over the study of nature to the physical sciences, and the study of culture to the social sciences, the humanities were left with conducting meta-analyses or pursuing one or another version of *l'art pour l'art*.

Analytic assumptions concerning knowledge also promoted the concept of expertise. Expertise in the modern sense depends on phenomena being able to be understood in isolation from each other. In politics this makes democracy at once necessary and problematic—necessary to do the relating that can no longer be done by knowledge, and problematic to the degree that intelligent decision making requires specialized knowledge. Specialization and expertise lead to what can be called epistemological myopia, where a powerful understanding of the details of comes at the cost of appreciating the larger implications of a phenomenon. This in turn has led to calls for interdisciplinary approaches to knowledge.

While problematic even within the sciences, the analytic approach to knowledge has had its most destructive effects in the humanities. Even as the intellectual division of labor has become more and more fine-grained, there was no part of knowledge explicitly concerned with the development of and relation of knowledge between and across the disciplines. Philosophy, the traditional location of such knowledge, also embraced specialization and professionalization, and new claimants to interdisciplinarity such as the sociology of knowledge or science, technology, and society studies, have nevertheless in short order come under the gravitational attraction of their own disciplinary formations. Disciplinary myopia in turn has run parallel to and contributed to the progressive loss in public ability to rationally debate the ends of life, which has reached the point that to even speak of "the good life" often invites derisive commentary—or relegation to the private sphere of personal preference.

Disciplinary specialization and its corresponding cognitive productivity have thus been bought at the cost of ignoring the lateral connections between one subject and the rest of the universe of thought and action. The issue here is the dominance of the metaphor of the laboratory, which presumes that it is relatively unproblematic to separate a bench experiment from the world at large: creating conditions that can be replicated, by controlling the materials used and constraining the parameters of the experiment (Frodeman 2003). Even fields quite far from, and in some cases quite disdainful of, science have applied this presumption to their own work. To offer just one example, it is presumed by literary scholars that it is more central to the work of their field to further probe the depths of the *Prelude* than to see how William Wordsworth might illuminate the experience of employees of U.S. National Parks, and through them, the park-visiting public.

The Knowledge Explosion and Its Discontents

Despite the tremendous explosion of knowledge, there is no discipline that takes as its provenance understanding the relation between the disciplines. Knowledge and information workers multiply ever faster. Hundreds of thousands of bachelor degrees and tens of thousands of doctorates are awarded each year; the annual U.S. federal support of science approaches $150 billion (with twice as much more coming from private sources); and a sky-rocketing stream of publications floods the infosphere in hardcopy, electronic, and various other media. As more than one social commentator has repeated, we are increasingly the most information and knowledge-intensive society in history (see Machlup 1962, Rubin et al. 1986, Castells 1996, and Mokyr 2002). To adapt a prescient distinction from Albert Borgmann (1999), knowledge about reality (science) and knowledge for reality (engineering) have morphed into knowledge as reality. But the knowledge society appears to have little or no program for how to live in or with this information rich possibility space other than to affirm the personal construction of meaning, some automatic synthesis (perhaps by means of Adam Smith's "invisible hand" or G. W. F. Hegel's "cunning of reason"), or Vannevar Bush's linear hypothesis from *Science: The Endless Frontier* (1945): just fund basic science and good results will flow for national security, healthcare, and the economy.

In the area of science policy, selective voices have questioned the received view that all knowledge production is good knowledge production. According to Daniel Sarewitz (1996), David Guston (2000), and Philip Kitcher (2001) there are good reasons to doubt that simply giving more money to science is always the best social investment. A few isolated analyses point in rather more radical directions, with provocative studies on the theme of "forbidden knowledge" by Nicholas Rescher (1987), Roger Shattuck (1996), and Agnieszka Lekka-Kowalik and Daniel Schulthess (1996). Among others, Carl Mitcham and Robert Frodeman (2002) have sought to extend the argument for balance in science funding to a broader balance in knowledge production. Subsequent to September 11, 2001, new forms of knowledge restriction have been debated in the sciences themselves. All together, such efforts suggest that the traditional research philosophy in favor of unfettered scientific autonomy and unrestricted knowledge production is running up against both epistemological and political limits. The *epistemological* limits of knowledge production are evident in the increasingly complex nature of both knowledge and societal problems: our lives are becoming more interwoven on global scales, and many of the problems that are most easily isolated have already been addressed. The *political* limits are found in the increasingly public demand that publicly funded research and education clearly show their connections to community needs. Although the repeated call for interdisciplinarity in education and research is often an effort to respond to such problems, in many instances the interdisciplinarity that emerges does little to address such issues since it leads only to more and more refined disciplinarity.

What Is to Be Done?

Existing ethical assessments of science focus on methodological norms in knowledge production. In exceptional cases, critics have contested claims to scientific knowledge on ethical and religious grounds (as in the challenge to evolutionary theory), although they have not questioned the value of knowledge per se. Existing ethical assessments of engineering and technology focus largely on the active use of technical knowledge rather than the knowledge itself. By contrast, the argument here is that knowledge itself deserves ethical analysis and criticism.

What would this involve? To begin with, it will depend on some recognition, however provisional, of knowledge as an ethical issue beyond the belief in knowledge as an unqualified good. But such acknowledgment could also find support from one or more of five complementary approaches to the knowledge question.

First, is phenomenological work on the character of scientific knowledge by philosophers such as Hans Jonas (1966 and 1974) who has argued the inherently practical character of modern natural science. Such an argument poses obvious challenges for any classical defense of knowledge as inherently good or neutral.

Second, is the argument by scientists themselves from the 1970s on who considered the possible dangers in and limitations to scientific research, because of the complexities with which it had become involved. Although some of the early arguments to this effect (e.g., Holton and Morison 1978) were subsequently challenged, later studies in complexity theory (e.g., Pagels 1988) raise related issues that have yet to be fully appreciated.

Third, virtue epistemology makes a case for relating knowledge and virtue that also has implications for relating knowledge and vice. Virtue epistemology is concerned with identifying the virtues that could transform true belief into knowledge that make knowing possible (see, e.g., Zabzebski 1996). But here ethics is simply incorporated into an ethical epistemology, while what is equally called for is an epistemic ethics and metaphysics.

Fourth, information ethics in its two forms—the ethics of library science and the ethics of computer information generation and manipulation—both suggest the need for ethical assessments of knowledge in relation to issues of privacy and equity. How can all knowledge be inherently good when some of it is inherently invasive or promotes inequalities? Moving in the directions of moral psychology, there is also research that suggests certain types of propositional knowledge might limit the exercise of intuitive knowledge (Gladwell 2005). Extending such a notion, is it not possible that certain types of knowledge could distract human beings from more important goods? Is the acquaintance with some types of things on which know how depends never psychologically problematic?

Finally, science studies research on transformations in the social character of knowledge production have developed suggestive analyses that have implications for any ethics of knowledge. A useful reference here is the work of Michael Gibbons and others, *The New Production of Knowledge* (1994), which distinguishes what it terms "Mode 1" and "Mode 2" knowledge. Mode 1 is the standard form of modern knowledge generated in disciplinary and academic frameworks. Mode 2 knowledge is a new kind of knowledge originating outside academic research institutions. Mode 2 knowledge production

- is governed by practical, problem solving concerns (rather than by more academic or epistemic ones),
- is transdisciplinary in character,
- engenders linkages among subfields and heterogeneous sites,
- is subject to economic and social accountability, and
- incorporates social, economic, and political interests.

Although this analysis and a companion volume by Helga Nowotny and others (2001) suggests little more than adaptive strategies in response to such transformations, they open up space for more normative assessments. Deborah Johnson (1999), for instance, has argued that recognition of the new social constructive context of science offers opportunities for reframing the question of forbidden knowledge.

On the basis of these kinds of existing research one may propose the following overlapping questions for any future ethics of science and technology:

1. Historically and socially, what is the moral status of a kind of knowledge with inherently applied characteristics? Is the distinction between ancient, contemplative knowledge and modern, inherently manipulative knowledge defensible? Furthermore, has the character of technoscientific knowledge itself undergone morally relevant change of the types suggested by social studies of science?
2. Conceptually, what are the ethical dimensions of distinctions between the forms knowledge (in the general sense) as data, information, knowledge (in a strict sense), and wisdom?
3. From the political and policy perspectives, what is the proper balance between knowledge and knowledge production in the technosciences, the social sciences, the humanities, and the arts? How do different forms of cognition properly interact, not just to produce knowledge but to promote the good life?
4. Psychologically, what are the moral implications of the proliferation of technoscientific knowledge? Does more knowledge always promote better thinking or acting?
5. Ethically (in a narrow sense): What are the morally relevant consequences of knowledge and knowledge production? Are there no deontological limits on knowledge and knowledge production? With regard to virtue, are there no extremes to epistemological practice that deserve censure?

Although not exhaustive of any future ethics of science and technology as knowledge, responses to these kinds of questions might provide guidance for the co-creative interaction between knowing, making, and doing in the expansively human sense.

BIBLIOGRAPHY

Ackoff, Russell L. (1989). "From Data to Wisdom." *Journal of Applied Systems Analysis* 16: 3–9.

Borgmann, Albert. (1999). *Holding on to Reality: The Nature of Information at the Turn of the Millennium*. Chicago: University of Chicago Press.

Bush, Vannevar. (1945). *Science: The Endless Frontier*. Washington, DC: U.S. Government Printing Office.

Castells, Manuel. (2000). *The Information Age: Economy, Society and Culture*. Vol. 1: *The Rise of the Network Society*. 2nd edition. Oxford: Blackwell.

Chadwick, Ruth; Mairi Levitt; and Darren Shickle. (1997). *The Right to Know and the Right Not to Know*. Brookfield, VT: Aldershot.

Cleveland, Harland. (1982). "Information as Resource." *Futuris* December: 34–39.

Frodeman, Robert. (2003). *Geo-Logic: Breaking Ground between Philosophy and the Earth Sciences*. Albany: State University of New York Press.

Gibbons, Michael; Camille Limoges; Helga Nowotny; Peter Scott; Simon Schwartzman; and Martin Trow. (1994). *The New Production of Knowledge: The Dynamics of Science and Research in Contemporary Societies*. London: Sage.

Gladwell, Malcolm. (2005). *Blink: The Power of Thinking without Thinking*. Bostom: Little Brown.

Guston, David. (2000). *Between Politics and Science: Assuring the Integrity and Productivity of Research*. New York: Cambridge University Press.

Holton, Gerald, and Robert S. Morison, eds. (1979). *Limits of Scientific Inquiry*. New York: W.W. Norton.

Johnson, Deborah. (1999). "Reframing the Question of Forbidden Knowledge for Modern Science." *Science and Engineering Ethics* 5(4): 445–461.

Jonas, Hans. (1966). "The Practical Uses of Theory." In *The Phenomenon of Life: Toward a Philosophical Biology*. New York: Harper and Row.

Jonas, Hans. (1974). *Philosophical Essays: From Ancient Creed to Technological Man*. Englewood Cliffs, NJ: Prentice Hall.

Kitcher, Philip. (2001). *Science, Truth, and Democracy*. New York: Oxford University Press.

Lekka-Kowalik, Agnieszka, and Daniel Schulthess. (1996). *Forbidden Knowledge*. Theme issue, *Monist* 79(2) (April). Includes eight articles all questioning the autonomy of science.

Machlup, Fritz. (1962). *The Production and Distribution of Knowledge in the United States*. Princeton, NJ: Princeton University Press.

Mitcham, Carl, and Robert Frodeman. (2002). "The Plea for Balance in the Public Funding of Science." *Technology in Society* 24(2): 83–92.

Mokyr, Joel. (2002). *The Gifts of Athena: Historical Origins of the Knowledge Economy*. Princeton, NJ: Princeton University Press.

Nowotny, Helga; Peter Scott; and Michael Gibbons. (2001). *Re-Thinking Science: Knowledge and the Public in an Age of Uncertainty*. Cambridge, UK: Polity Press.

Oppenheimer, J. Robert. (1947). *Physics in the Contemporary World*. Cambridge, MA: MIT Press.

Pagels, Heinz. (1988). *The Dreams of Reason: The Computer and the Rise of the Sciences of Complexity*. New York: Simon and Schuster.

Rescher, Nicholas. (1987). *Forbidden Knowledge and Other Essays on the Philosophy of Cognition*. Boston: D. Reidel.

Rubin, Michael Rogers; Mary Taylor Huber; and Elizabeth Lloyd Taylor. (1986). *The Knowledge Industry in the United States*. Princteon, NJ: Princeton University Press.

Russell, Bertrand. (1910). "Knowledge by Acquaintance and Knowledge by Description." *Proceedings of the Aristotelian Society* 11: 108–128. Reprinted in *Mysticism and Logic* (1963), pp. 152–167. London: Allen and Unwin.

Sarewitz, Daniel. (1996). *Frontiers of Illusion: Science, Technology, and the Politics of Progress*. Philadelphia: Temple University Press.

Schumacher, E. F. (1977). *A Guide for the Perplexed*. New York: Harper and Row.

Shattuck, Roger. (1996). *Forbidden Knowledge: From Prometheus to Pornography*. New York: St. Martin's Press.

Zabzebski, Linda. (1996). *Virtues of the Mind*. Cambridge, UK: Cambridge University Press.

VALUES IN TECHNICAL DESIGN

HELEN NISSENBAUM

• • •

Although their precise parameters and significance are easily debated, it is generally recognized that values influence the design of scientific experiments. Because scientific research is designed to yield answers to specific questions, truth values in operational forms internal to science play prominent roles in the structuring of research activities. Moreover, when experimentation takes place with human subjects or dangerous reagents there are further values of respect for persons and public safety that readily take the stage. It is thus not difficult to argue that values regularly and properly are embodied in scientific activities—and that the practice of science can have value implications for the larger social contexts in which they are pursued.

The idea that values may also be embodied in engineered products, processes, and systems is perhaps more controversial, although the thesis is now commonly argued in a variety of disciplines relevant to questions of science, technology, and ethics (e.g., Winner 1986, MacKenzie and Wajcman 1999). Moreover, a practical turn from what has sometime been a largely descriptive posture sets forth values as a design aspiration, exhorting engineers, producers, and consumers to include values in the criteria by which technological excellence is judged (Mitcham 1995). For those committed to bringing selected values to bear in technical design, the ideal result is a world of artifacts that embody not only such instrumental values as effectiveness, efficiency, safety, reliability, and ease of use, but promote (or at least do not undermine) substantive values to which the surrounding societies or cultures subscribe. In liberal democracies, such values may include, among others, liberty, justice, privacy, security, friendship, comfort, trust, autonomy, and transparency.

But it is one thing to subscribe to such ideals and another to put them into practice. Putting values into practice is often dismissed as a form of political or moral activism irrelevant to the designing of technical systems such as software programs. Experienced software engineers will recall the not too distant past when interface and usability were also overlooked features of software system design (Adler and Winograd 1992). While these and other aspects of design have now entered the mainstream, we are still at the shaky beginnings of thinking systematically about the practice of technical design and values (Norman 2002). Even designers who support the principle of integrating values into systems are likely to have trouble applying standard design methodologies, honed for the purpose of meeting functional requirements, to the unfamiliar turf of values. There are at least two factors that contribute to the difficulty of embodying values in the design of technical systems and devices—one epistemological, the other practical.

Epistemological Challenges

One reason the study of human or social dimensions of technology is so demanding is that the areas of knowledge and relevant methodologies are far-flung and self-contained. This dispersion is reflected in the disciplinary organization of universities, in which science and technology are typically segregated from the social sciences and humanities. Yet the successful embodying of values in technical design demands simultaneous engagement with these distinct areas of knowledge and their respective methodologies. For technical design purposes, what is readily drawn from these fields is sufficient, whereas for others the puzzles raised push beyond standard boundaries. Either case, however, calls for more comprehensive interactions among diverse areas of knowledge than is customary—in the first instance requiring enough knowledge to identify existing, relevant insights; in the second, deliberate efforts to extend what is known in order to address the hard and sometimes novel questions that arise.

In practical terms, these active interdependencies may be understood through the metaphor of "balls in the air." Conscientious designers must juggle and keep in play the results of at least three modes of knowledge: foremost those from the relevant scientific and technical fields; beyond these, philosophical reflections on relevant values; and finally empirical findings regarding relations between values, individuals, and their societies. The balls in play metaphor reflects the need to direct attention to all three aspects simultaneously, keeping an eye not only on each factor but also on how the three factors shift in relation to each other.

TECHNICAL MODES. In the technical mode, a designer or design team brings to bear state-of-the-art scientific knowledge and technical know-how on particular design specifications that realize given values in an overarching design project. In a project to build a hospital patients record system, for example, designers might be charged with the task of building privacy protection into the software. In responding to this charge, they might aim for a design that enables access to particular fields of data only by specific, authorized members of the hospital staff. With this goal in mind, they set about designing system constraints, and selecting or creating mechanisms to attain them.

These steps, comprising the technical ball-in-play, are familiar to technical system designers. The sole departure in the present instance is that they are described as undertaken in the name of values and not, as is typically the case, in the name of technical functionality and efficiency.

PHILOSOPHICAL MODE. While designers and engineers seek and invent mechanisms to meet design specifications for promoting values, the philosophical perspective is generally overlooked. But values are more than simple givens. Values can themselves be examined in terms of their origins and scope of relevance, their meanings, and as the basis for normative influence—especially when it is necessary to resolve conflicts.

At the foundation of such philosophical reflection lies an account values that may be quite contentious. There are extensive debates about the precise character of values, for instance, whether they are subjective or objective. Nevertheless, within a broad construction of values as interests, purposes, or ends in view, those of greatest concern in the present context are values that can be construed as social, moral, or political. This still wide-ranging category includes abstractly conceived values such as freedom, autonomy, equality, justice, and privacy, as well as concrete values such as friendship, safety, sociality, and comfort.

The question of whether any such values are universal to all humans or are always locally defined by nations, societies, cultures, religions, communities, or families deserves to be appreciated for its moderating influence. Designers and developers of technology in the United States (and other technology producing liberal democracies) may confidently reach for constitutional values such as freedoms of speech, association, and religion; protections of property, equality, due process, and privacy; or cultural values such as individualism and creativity. But they should at the very least also consider whether such values are always appropriate to other countries where their products may be distributed. At the same time, taking the Universal Declaration of Human Rights as a guide, it is reasonable to postulate a few basic values as common to all humanity, with specific interpretations subject to local variation—a position that nevertheless remains subject to philosophical analysis and empirical assessment.

In seeking to promote the embodying of values in technologies it is a designer's understanding that will guide how they are "cashed out" as system features. In the case of the electronic patient records example, concerned developers seek specifications that will yield privacy and not something else, and a key factor will be defining privacy. Evaluating the proposal mentioned earlier to operationalize privacy by giving variable access to the different fields of information, a philosophical critic might argue that a different interpretation of privacy would support a system whose default is to give access to the patient only, as a way to embody privacy as control over information about oneself.

An ability to consider and discuss such alternatives is a significant component of what it takes to keep the philosophical ball in play. In some instances this means turning for insights to a long tradition of philosophical and political thought that guides the moral and political systems of the different technology producing liberal democracies. Because many of the most important and contested value concepts have evolved within these traditions, design teams might need to plumb them for sound, workable concepts. Failure to take these concepts seriously can lead to bungled interpretations in the specification of design features.

Two caveats: First, it is unrealistic to expect designers always to work from first principles and grapple directly with abstract conceptions of value. Yet over time, one can imagine an emerging database of analyses specifically developed for the context of technology design. Second, traditional analyses may not be sufficient when technology itself has brought about such radical change in the social and material world that certain values themselves demand reconsideration. In such cases, as with privacy in the wake of information technologies, keeping the philosophical ball-in-play means producing original research analyzing on the concepts at issue.

Finally, the philosophical mode engages with issue of normative force, providing rationale or justification for commitments to particular values in a given device or system. With the electronic patient record system, one might consider why privacy is relevant, important, or necessary. Frequently, the answers to such questions are to be found in surrounding moral and political theories that explain why and when certain values ought to be promoted. This is particularly needed when conflicts among values result from specific design choices. Normative theory can guide resolution or tradeoffs. In the patient records system, finding that access is slowed as a result of privacy constraints, designers might return to the underlying theory of a right to privacy to learn the circumstances under which privacy claims may justifiably be diminished or overridden.

EMPIRICAL MODE. Empirical investigation answers questions that are as important to the goal of embodying values in design as the philosophical and technical. Not only does it complement philosophical inquiry into what values are relevant to a given project, but it is the primary means for addressing, systematically, the question of whether a given attempt at embodying values "worked"—that is, whether the intentions of designers were fulfilled.

Philosophical inquiry can take us only so far in determining the values that ought to be considered in relation to given technological projects. Even if one holds to the existence of a basic set of universal human values, the people affected by these projects are likely to subscribe to a far richer set of values determined by their cultural, historical, national, ethnic, and religious affiliations. It may be even more crucial to attend to these commitments when engineers face choices among design alternatives. Despite the enormous attention philosophers, and others, have given to the problem of systematically resolving values (and rights) conflicts, this remains notoriously difficult. For such situations, ascertaining the preferences of affected parties is a sound practical response, using such methods as surveys, interviews, testing under controlled conditions, and observation in the field. In the conflict between efficient access to information and its confidentiality in a patient records system, for instance, designers should at least consult preferences among affected parties.

Empirical investigation is also necessary for ascertaining whether a particular design embodies intended values. Again in the case of the electronic patient records system, designers might learn from observing patterns of usage if security mechanisms for restricting access to the appropriately authorized personnel are so onerous that many users simply bypass them, thus leaving the records more vulnerable than ever. They might thus discover that their attempts to promote privacy are thwarted by a design that does not achieve its intended results—information crucial to any values in technical design analysis.

VALUES IN PLAY. The metaphor of balls-in-play includes not simply the need to incorporate three distinct modes of knowing into the design context but an effort to iteratively integrate these modes. Because findings from each of the areas affect or feed back into others, members of a design team cannot seek solutions in each area independently. Although the hardest cases might call for innovation within each of the three modes (and hence diverse expertise), many cases will be able to rely on what is already known in at least one or two.

Consider, for example, the task of building a system that provides fair access to information to diverse members of a community. Designers might quickly settle on accessibility to all mentally able individuals as the embodiment of the value of fairness, while it struggles with the technical questions of how to go about doing so and, later, testing empirically whether particular designs have succeeded. It is reasonable, furthermore, to hope that with greater attention to the study of values in technology a body of findings, experience, results, and definitions will develop that gradually will alleviate some of the epistemological burdens.

Practical Challenges

In addition to epistemological challenges, the practical challenge engineers face is the sparseness of methodologies for embodying values in system design, due in part to the newness of the endeavor. If we think of what we need to know constitutes the ingredients for a recipe, then what remains is the equally important method for combining them into a dish. Attempts to fill this methodological gap are new and evolving. Some that have been around longer are restricted to certain specialized areas of application.

One of the best known in the latter category is an approach known as "participatory design." Having evolved in Scandinavia, in the context of the workplace, the methodology is committed to democratic participation by those likely to be affected by new technologies as well as design outcomes that enhance not only efficiency of production and quality of product but the skill and well-being of workers. Emerging methods include value sensitive design, which recognizes the importance of technical, conceptual, and empirical investigations to the purpose of bringing values to bear in the design of information technologies generally. Another approach developed by Mary Flanagan, Daniel

Howe, and Helen Nissenbaum (2006) posits a methodology comprising four constitutive activities for embodying values in design—discovery, translation, resolution, and verification—which, in order to illustrate possibilities, can be considered here.

DISCOVERY. The activity of discovery involves identifying values that are relevant to or might inform a particular design project by looking to key sources of values in the context of technical design and asking what values they bring to the project in question. The specific list of values will vary considerably from project to project. But one promising heuristic is simply to ask "What values are involved here?" and then brainstorm possible answers. Sometimes values are expressed explicitly in the functional definition of a deliverable (as grasped through the technical mode of knowing). But all designs are underdetermined by explicit functional requirements, leaving designers and developers numerous alternatives as they proceed through an iterative design process.

Open-endedness calls forth the implicit values of designers themselves (and thus may be furthered by the philosophical mode of reflection). Sometimes designers unconsciously assume that they are the likely users of their work and act accordingly. But values reflection in technical design can almost always be deepened by efforts to critically identify implicit values in both designers and potential users (as accessed by means of the empirical mode of inquiry), and subsequent critical assessments of and dialogue between such values.

TRANSLATION. In the activity of translation, a design team operationalizes value concepts and implements them in design. The values discovered in the first moment of reflection are not only multiple but they tend to be abstract. To become concretely accessible in the design context they will need to be rendered into operational or functional forms. This translation activity will almost certainly involve some input from the philosophical mode of knowing. No matter how well value concepts are operationalized, the efforts of conscientious designers are easily undermined if the historical traditions and substantive characteristics of particular values are incorrectly interpreted. With values such as privacy, for example, clarity, good intentions, and technical competence can be misdirected when not adequately backed up with sensitive analyses of various philosophical approaches to privacy itself.

RESOLUTION. Translation is key to any implementation of discovered values. But implementation and the corresponding transfer of values into design specifications also calls for the resolution of any potential incompatibilities in a values possibility space. One of the major challenges of implementation is resolving conflicts that arise as a result of specific design choices.

Conflicts arise when designers who have committed to some set of discovered values, further discover that it is practically impossible to embody all of them equally well within some product, process, or system. Engineering is rife with such conflicts: whether to favor safety over cost, transparency over privacy, aesthetics over functionality, with many more appearing at layers of finer granularity. Resolving such conflicts is by no means a challenge for engineering alone, but is manifest as one of the enduring problems of practical ethics, politics, and law. But this means again that the resources of the philosophical mode of thinking may be of special benefit to this moment in practical values design work.

VERIFICATION. Finally, the activity of verification involves assessing whether values have been successfully embodied in design. Verifying the inclusion of values is likely to draw on both technical and empirical thinking. It can easily begin with internal testing by the design team but will not be complete without user testing in controlled environments.

It might be useful in this regard to consider the possibility of some approach analogous to that of clinical trials for pharmaceuticals. In phase one trials the basic question concerns whether a drug is safe. Phase one studies, which are short term, are done to gather preliminary data on chemical action and dosage using healthy volunteers, and there is no comparison with any control group. In phase two trials, which take longer, the basic question is whether the drug works to achieve a desired therapeutic end. Is it an effective treatment? Now the trials are done with patients who exhibit a target disease or illness, and there are control groups for comparison. Finally, phase three trials focus on the long-term effects in larger populations. Only after this phase is complete may a drug be widely marketed. In a like manner one might construct a series of alpha, beta, and gamma testings of new technologies to assess how values may have been embodied in technical designs, using initially small groups of technical volunteers, then non-technical users with the need that a new technology aims to address, and finally longer-term monitoring of larger populations of consumers and users.

Open Questions

It is too early to judge the long-term success of any method for embodying values in technical design, because few projects have proceeded through the various milestones characteristic of the lifespan of technologies—including, sometimes, unintended (often negative) consequences. The method nevertheless deserves serious consideration in any discussion of science, technology, and ethics—not only in relation to the kind of case referenced here (that is, software design) but across the technology spectrum, from machines and structures to systems and software. Moreover, critical consideration may also throw light on the roles of values in design of scientific experimentation.

Two other potentially critical stances are worth mentioning. Taking a social constructivist stance, critics might question the supposition that key social, ethical, and political aspects of technologies are attributable either to their blueprints or physical shape. What imbues technologies with values are not any of their objective functions but their meanings, generated by the interpretive forces of history, culture, politics, and a myriad other social contingencies. An ironically related stance holds that technologies are neutral. The extent to which systems or devices promote values is a function of the individual uses to which they are put; technologies are mere tools of human intention. Although the view of technology as neutral is currently out of favor in scholarly circles, it remains a common presumption with which those interested in values in technical design must contend.

BIBLIOGRAPHY

Adler, Paul S., and Terry Winograd. (1992). *Usability: Turning Technologies into Tools*. New York: Oxford University Press. Seven useful papers.

Agre, Philip E. (1997). "Toward a Critical Technical Practice: Lessons Learned in Trying to Reform AI." In *Social Science, Technical Systems, and Cooperative Work: Bridging the Great Divide*, eds. Geoffrey Bowker, Susan Leigh Star, William Turner, and Les Gasser. Mahwah, NJ: Erlbaum.

Bødker, S., and K. Grønbæk. (1991). "Design in Action: From Prototyping by Demonstration to Cooperative Prototyping." In *Design at Work: Cooperative Design of Computer Systems* eds. Joan Greenbaum and Morton Kyng. Mahwah, NJ: Erlbaum.

Ehn, Pelle. (1992). "Scandinavian Design: On Participation and Skill." In *Usability*, eds. Paul S. Adler and Terry Winograd. New York: Oxford University Press.

Flanagan, Mary; Daniel Howe; and Helen Nissenbaum. (2006). "Values in Design: Theory and Practice." In *Information Technology and Moral Philosophy*, eds. Jeroen van den Hoven and John Weckert. Cambridge, UK: Cambridge University Press.

Friedman, Batya. (1996). "Value-Sensitive Design." *Interactions* 3(6): 17–23.

Friedman, Batya; Peter H. Kahn Jr.; and Alan Borning. (2002). *Value Sensitive Design: Theory and Methods*. Technical Report 02-12-01, Department of Computer Science and Engineering, University of Washington, Seattle, WA. Available from http://www.urbansim.org/papers. Revised version in *Human Computer Interaction in Management Information Systems: Foundations*, eds. Ping Zhang and Dennis Galletta. New York: M. E. Sharpe, 2006.

Friedman, Batya, and Helen Nissenbaum. (1996). "Bias in Computer Systems." ACM *Transactions on Information Systems* 14(3): 330–347.

MacKenzie Donald, and Judy Wajcman, eds. (1999). *The Social Shaping of Technology*, 2nd edition. Milton Keynes: Open University Press.

Mitcham, Carl. (1995). "Ethics into Design." In *Discovering Design*, eds. Richard Buchanan and Victor Margolis. Chicago: University of Chicago Press.

Nissenbaum, Helen. (1998). "Values in the Design of Computer Systems." *Computers in Society*, March: 38–39.

Norman, Donald. (2002). *The Design of Everyday Things*. New York: Basic Books. First published as *The Psychology of Everyday Things* New York: Doubleday, 1990.

Pfaffenberger, Bryan. (1992). "Technological Dramas." *Science, Technology, and Human Values* 17(3, Autumn): 282–312.

Winner, Langdon. (1986). "Do Artifacts Have Politics?" In his *The Whale and the Reactor: A Search for Limits in an Age of High Technology*. Chicago: University of Chicago Press.

LIST OF ARTICLES

• • • B

• • • C

• • • D

• • • E

• • • F

• • • H

HABERMAS, JÜRGEN
Antti Kauppinen

HACKER ETHICS
Deborah G. Johnson

HALDANE, J. B. S.
Krishna R. Dronamraju

HARDIN, GARRETT
Craig A. Straub

HARDWARE AND SOFTWARE
Keith W. Miller

HAZARDS
Daniel Sarewitz

HEALTH AND DISEASE
Dominic A. Sisti

HEAVY METALS
Wil Lepkowski

HEGEL, GEORG WILHELM FRIEDRICH
Craig Condella

HEIDEGGER, MARTIN
Mark Blitz

HINDU PERSPECTIVES
Subhash Kak

HIROSHIMA AND NAGASAKI
Dennis W. Cheek

HIV/AIDS
Marilyn Averill

HOBBES, THOMAS
Adam Briggle
Carl Mitcham

HOLOCAUST
Darrell J. Fasching

HOMOSEXUALITY DEBATE
Robert Alan Brookey

HORMESIS
Adam Briggle

HUMAN CLONING
Glenn McGee

HUMAN GENOME ORGANIZATION
Tind Shepper Ryen

HUMAN NATURE
Larry Arnhart

HUMANISM
Fred Edwords

HUMANITARIAN SCIENCE AND TECHNOLOGY
Carl Mitcham
Juan Lucena
Suzanne Moon

HUMANIZATION AND DEHUMANIZATION
Carl Mitcham
Adam Briggle

HUMAN RIGHTS
Audrey R. Chapman

HUMAN SUBJECTS RESEARCH
Jeffrey P. Spike

HUME, DAVID
Shane Drefcinski

HUSSERL, EDMUND
Michael R. Kelly

HUXLEY, ALDOUS
Adam Briggle

HYPERTEXT
Antonio Rodríguez de las Heras
Carl Mitcham

• • • I

IBERO-AMERICAN PERSPECTIVES
José A. López Cerezo

ILLICH, IVAN
Carl Mitcham

IN VITRO FERTILIZATION AND GENETIC SCREENING
Andrea L. Bonnicksen

INCREMENTALISM
Marvin J. Croy

INDIAN PERSPECTIVES
Uday T. Turaga

INDIGENOUS PEOPLES' PERSPECTIVES
John A. Grim

INDUSTRIAL REVOLUTION
Robert Martello

INFORMATION
Luciano Floridi

INFORMATION ETHICS
Edwin M. Elrod
Martha M. Smith

INFORMATION OVERLOAD
A. Pablo Iannone
Adam Briggle

INFORMATION SOCIETY
Bruce Bimber

• • • M

• • • P

• • • S

• • • T

LIST OF CONTRIBUTORS

Evandro Agazzi, Prof., Philosophy
Univ. of Genoa, Italy
AXIOLOGY

Janaki R. R. Alavalapati, Assoc. Prof., Resource Economics and Policy
Univ. of Florida, Gainesville
MULTIPLE-USE MANAGEMENT

Braden R. Allenby, Prof., Civil and Environmental Engineering
Arizona State Univ., Tempe
EARTH SYSTEMS ENGINEERING AND MANAGEMENT

Andoni Alonso, Prof., Philosophy
Univ. of Extremadura, Cáceres, Spain
CYBERCULTURE; WITTGENSTEIN, LUDWIG

Hussein A. Amery, Assoc. Prof., Liberal Arts and Int'l. Studies
Colorado Sch. of Mines, Golden
WATER

Agustin Araya, Prof., Computer Science
San Jose State Univ., CA
CYBERSPACE; INTERNET

Larry Arnhart, Prof., Political Science
Northern Illinois Univ., DeKalb
ATLANTIS, OLD AND NEW; BACON, FRANCIS; BIOTECH ETHICS; BRAVE NEW WORLD; DESCARTES, RENÉ; EVOLUTION-CREATIONISM DEBATE; EVOLUTIONARY ETHICS;HUMAN NATURE; NATURE, PRESIDENT'S COUNCIL ON BIOETHICS

Christopher Arroyo, Teaching Fellow, Philosophy
Fordham Univ., Bronx, NY
LEIBNIZ, G. W.

Vicki Arroyo, Dir. of Policy Analysis
Pew Ctr. on Global Climate Change, Arlington, VA
PUBLIC POLICY CENERSS

Iñaki Arzoz Karasusan, Freelance writer and artist
Pamplona, Spain
CYBERCULTURE

Peter M. Asaro, Grad. Teaching Asst., Philosophy
Univ. of Illinois at Urbana-Champaign
PARTICIPATORY DESIGN

Mie-Sophia Augier, Post-doctoral Fellow
Stanford Univ., CA
SIMON, HERBERT A.

Marilyn Averill, Ph.D. candidate, Political Science
Univ. of Colorado, Boulder
HIV/AIDS; INTELLECTUAL PROPERTY; UNINTENDED CONSEQUENCES

Ignacio Ayestarán, Assoc. Prof., Philosophy
Univ. of the Basque Country, Donostia-San Sebastián, Spain
BIODIVERSITY COMMERCIALIZATION

Osman Bakar, Malaysia Chair of Islam in Southeast Asia
Georgetown Univ., Washington, DC
ISLAMIC PERSPECTIVES

David Baker, Prof., Politics and Int'l. Studies
Univ. of Warwick, Coventry, UK
FASCISM

Dominic Balestra, Prof., Philosophy; Dean of Arts and Sciences Faculty
Fordham Univ., Bronx, NY
POPPER, KARL

Patrick Ball, Dir., Human Rights Programs
The Benetech Initiative, Palo Alto, CA
FREE SOFTWARE

David P. Barash, Prof., Psychology
Univ. of Washington, Seattle
AGGRESSION; DOMINANCE; ETHOLOGY; SELFISH GENES

Lowell W. Barrington, Assoc. Prof., Political Science
Marquette Univ., Milwaukee, WI
AUTHORITARIANISM; TOTALITARIANISM

Mary Catherine Bateson, Pres.
Inst. for Intercultural Studies, Washington, DC
MEAD, MARGARET

Benjamin Bederson, Prof. Emeritus, Physics
New York Univ., New York
EINSTEIN, ALBERT; OPPENHEIMER, J. ROBERT; TELLER, EDWARD

Jeremy Beer, Ed.-in-Chief
ISI Books, Wilmington, DE
AGRARIANISM

Vicente Bellver Capella, Assoc. Prof., Philosophy of Law
Univ. of Valencia, Spain
ENVIRONMENTAL RIGHTS; ORTEGA Y GASSET, JOSÉ

Juan Bautista Bengoetxea, Visiting Scholar, Philosophy and Science, Technology, and Society
Colorado Sch. of Mines, Golden
CHEMISTRY

Françoise Baylis, Prof., Bioethics
Dalhousie Univ., Halifax, N.S., Canada
GENE THERAPY

Rosalyn W. Berne, Asst. Prof., Science, Technology, and Society
Sch. of Engineering, Univ. of Virginia, Charlottesville
NANOETHICS

Mary Besterfield-Sacre, Asst. Prof., Industrial Engineering
Univ. of Pittsburgh, PA
ETHICS ASSESSMENT RUBRICS

Bruce Bimber, Dir., Ctr. for Information Technology and Society
Univ. of California, Santa Barbara
INFORMATION SOCIETY

Matthew Biniek, Science instructor
Grand Army of the Republic Memorial High Sch., Wilkes-Barre, PA
TECHNOCOMICS

Stephanie J. Bird, Ed., *Science and Engineering Ethics*
Massachusetts Inst. of Technology, Cambridge
MISCONDUCT IN SCIENCE: OVERVIEW; NEUROETHICS; RESEARCH ETHICS; RESPONSIBLE CONDUCT OF RESEARCH

Douglas Birsch, Assoc. Prof., History, Philosophy
Shippensburg Univ., PA
FORD PINTO CASE

Kenneth C. Blanchard, Jr., Prof., Political Science
Northern State Univ., Aberdeen, SD
ARISTOTLE AND ARISTOTELIANISM; FREE WILL; LOCKE, JOHN; MACHIAVELLI, NICCOLÒ; NATURE VERSUS NURTURE; PLATO; STRAUSS, LEO

Robert H. Blank, Prof., Public Policy
Brunel Univ., Uxbridge, UK
FETAL RESEARCH; GENETHICS

Thomas Blau, Prof., National Security Executive Education
Nat. Defense Univ., Washington, DC
MISSILE DEFENSE SYSTEMS

Mark Blitz, Fletcher Jones Prof., Political Philosophy
Claremont McKenna Coll., Claremont, CA
HEIDEGGER, MARTIN

Maeve A. Boland, Assoc. Moderator, McBride Program in Public Affairs for Engineers
Colorado Sch. of Mines, Golden
GEOGRAPHIC INFORMATION SYSTEMS; GLOBAL POSITIONING SYSTEM; MINING; SOFT SYSTEMS METHODOLOGY

Albert L. Bonnicksen, Presidential Research Prof., Political Science
Northern Illinois Univ., DeKalb
EMBRYONIC STEM CELLS; IN VITRO FERTILIZATION AND GENETIC SCREENING

Ann Borgmann, Prof., Philosophy
Univ. of Montana, Missoula
MATERIAL CULTURE

Ann Bostrom, Assoc. Prof., Sch. of Public Policy
Georgia Inst. of Technology, Atlanta
RISK ASSESSMENT; RISK PERCEPTION

David R. Bonneville, P. E., S. E., Senior Principal
Degenkolb Engineers, San Francisco, CA
EARTHQUAKE ENGINEERING

Jeffrey Boutwell, Exec. Dir.
Pugwash Conferences on Science and World Affairs, Washington, DC
PUGWASH CONFERENCES; ROTBLAT, JOSEPH

Margaret M. Brace, Research Analyst
Vanderbilt Univ. Medical Ctr., Nashville, TN
PSYCHOLOGY: HUMANISTIC APPROACHES

Perry C. Bramlett, Rev.
C. S. Lewis for the Local Church Interstate Ministries, Louisville, KY
ANGLO-CATHOLIC CULTURAL CRITICISM; LEWIS, C. S.; TOLKIEN, J. R. R.

Philip Brey, Assoc. Prof. and Vice Chair, Philosophy
Univ. of Twente, The Netherlands
PROSTHETICS; VIRTUAL REALITY

Søren Brier, Prof., Management, Politics and Philosophy
Copenhagen Business Sch., Denmark
SEMIOTICS: NATURE AND MACHINE

Adam Briggle, Ph.D. candidate
Univ. of Colorado, Boulder
ADVERTISING, MARKETING, AND PUBLIC RELATIONS; ASILOMAR CONFERENCE; BETHE, HANS; BIOETHICS COMMITTEES AND COMMISSIONS; BIRTH CONTROL; BLACKETT, PATRICK; BUSH, VANNEVAR; CULTURAL LAG; DOUBLE EFFECT AND DUAL USE; ENVIRONMENTAL REGULATION; FAUST; FEDERATION OF AMERICAN SCIENTISTS; HOBBES, THOMAS; HORMESIS; HUMANIZATION AND DEHUMANIZATION; HUXLEY, ALDOUS; INFORMATION ETHICS; INFORMATION OVERLOAD; INSTITUTIONAL BIOSAFETY COMMITTEES; INSTITUTIONAL REVIEW BOARDS; LIFE; MANAGEMENT: OVERVIEW; MUMFORD, LEWIS; NORMAL ACCIDENTS; OPEN SOCIETY; POLICE; RELIABILITY; RESEARCH ETHICS: OVERVIEW; SCIENCE: OVERVIEW; SIMON, JULIAN; SPEED; SPENCER, HERBERT; TECHNOLOGY: OVERVIEW; TERRORISM; TERRORISM AND SCIENCE; TRADEOFFS; WATSON, JAMES; WORLD COMMISSION ON THE ETHICS OF SCIENTIFIC KNOWLEDGE AND TECHNOLOGY

Robert Alan Brookey, Assoc. Prof., Communication
Northern Illinois Univ., DeKalb, IL
HOMOSEXUALITY DEBATE

Taft H. Broome, Jr., Prof., Civil Engineering
Howard Univ., Washington, DC
PRAXIOLOGY

Ronald D. Brunner, Prof., Political Science
Univ. of Colorado, Boulder
LASSWELL, HAROLD D.

Richard Buchanan, Prof., Design
Carnegie Mellon Univ., Pittsburgh, PA
DESIGN ETHICS

George Bugliarello, Prof., Civil Engineering; Chancellor
Polytechnic Univ., Brooklyn, NY
BIOENGINEERING ETHICS

Mark A. Bullock, Principal Scientist, Dept. of Space Studies
Southwest Research Inst., Boulder, CO
ASTRONOMY; COSMOLOGY

Radford Byerly Jr., Visiting Fellow, Ctr. for Science and Technology Policy Research
Univ. of Colorado, Boulder
NATIONAL AERONAUTICS AND SPACE ADMINISTRATION; SPACE SHUTTLES *CHALLENGER AND COLUMBIA* ACCIDENTS

Edmund F. Byrne, Prof. Emeritus, Philosophy
Indiana Univ., Indianapolis
WORK

Emily Lynn Brumm, Engineering
Univ. of Virginia, Charlottesville
ENTREPRENEURISM

Gregory Caicco, Lincoln Asst. Prof. of Ethics in Architecture and Environmental Design
Arizona State Univ., Tempe, AZ
ARCHITECTURAL ETHICS

J. Baird Callicott, Prof., Philosophy and Religion
Univ. of North Texas, Denton
BIOPHILIA

Luis Camacho, Prof., Philosophy
Univ. of Costa Rica, San José
DEVELOPMENT ETHICS

Nancy D. Campbell, Asst. Prof., Science and Technology Studies
Rensselaer Polytechnic Inst., Troy, NY
DRUGS

Penelope Canan, Exec. Officer
Global Carbon Project, Tsukuba, Japan
MONTREAL PROTOCOL

Nicholas Capaldi, Legendre-Soulé Distinguished Chair in Business Ethics
Sch. of Business Administration, Loyola Univ., New Orleans, LA
MILL, JOHN STUART

Arthur L. Caplan, Emanuel & Robert Hart Prof. of Bioethics; Dir., Ctr. for Bioethics
Univ. of Pennsylvania, Philadelphia
BIOETHICS CENTERS

Joseph Carroll, Prof., English
Univ. of Missouri, St. Louis
DARWIN, CHARLES

William D. Casebeer, Assoc. Prof., Philosophy
U.S. Air Force Academy, CO
SCIENTIFIC ETHICS

Leah Ceccarelli, Assoc. Prof., Communication
Univ. of Washington, Seattle
RHETORIC OF SCIENCE AND TECHNOLOGY

Daniel P. Cérézuelle, Scientific Dir.
P.A.D.E.S, Paris, France
FRENCH PERSPECTIVES

Ruth Chadwick, Prof., Inst. of Environment, Philosophy, and Public Policy
Lancaster Univ., UK
APPLIED ETHICS; EUTHANASIA; RIGHT TO DIE; RIGHT TO LIFE; RIGHTS AND REPRODUCTION; RIGHTS THEORY

Audrey R. Chapman, Dir., Science and Human Rights Program
American Association for the Advancement of Science, Washington DC
HUMAN RIGHTS

Anne Chapman, Teaching Asst., Inst. of Environment, Philosophy, and Public Policy
Lancaster Univ., UK
ARENDT, HANNAH; GREEN IDEOLOGY

Dahlia Chazan, Co-Dir., Sustainability Indicators Program
Redefining Progress, Oakland, CA
ECOLOGICAL FOOTPRINT

Dennis Cheek, Sr. Principal Consultant
Science Applications Int'l. Corp., Exton, PA
ATOMIC BOMB; ENTERTAINMENT; HIROSHIMA AND NAGASAKI; RAMSEY, PAUL; ROADS AND HIGHWAYS

Fan Chen, Prof., Research Ctr. for Philosophy of Science and Technology
Sch. of Humanities and Law, Northeastern Univ., Shenyang, China
THREE GORGES DAM

Joshua L. Cherniss, Ph.D. candidate, Balliol Coll.
Univ. of Oxford, UK
BERLIN, ISAIAH; ETHICAL PLURALISM; LEVI, PRIMO

Jennifer Chesworth, Independent Scholar
Boalsburg, PA
SANGER, MARGARET

William Christian, Prof., Political Science
Univ. of Guelph, Ont., Canada
GRANT, GEORGE

Clifford Christians, Prof., Inst. of Communications Research
Univ. of Illinois at Urbana-Champaign
COMMUNICATION ETHICS

Daryl E. Chubin, Senior Vice Pres. for Policy and Research
Nat. Action Council for Minorities in Engineering, Inc., White Plains, NY
PEER REVIEW

Eric D. Claus, Research Asst., Psychology
Univ. of Colorado, Boulder
CHOICE BEHAVIOR; PSYCHOPARMACOLOGY

Vincent Colapietro, Prof., Philosophy
Pennsylvania State Univ., University Park
PEIRCE, CHARLES SANDERS

Carol Colatrella, Prof., Literature, Communication, and Culture; Co-Dir., Ctr. for the Study of Women, Science, and Technology
Georgia Inst. of Technology, Atlanta
SCIENCE, TECHNOLOGY, AND LITERATURE

Craig Condella, Teaching Fellow, Philosophy
Fordham Univ., Bronx, NY
HEGEL, GEORG WILHELM FRIEDRICH

Daniel Conway, Prof., Philosophy
Pennsylvania State Univ., University Park
KIERKEGAARD, SØREN; NIETZSCHE, FRIEDRICH W.

Martin L. Cook, Prof., Philosophy
United States Air Force Academy, CO
MILITARY ETHICS

Thomas D. Cornell, Assoc. Prof., History
Rochester Inst. of Technology, NY
FORD, HENRY

Ross B. Corotis, Prof., Civil, Environmental and Architectural Engineering
Univ. of Colorado, Boulder
RELIABILITY

Margaret Cosentino, Researcher
Ctr. for Strategic and Int'l. Studies, Washington, DC
BARUCH PLAN; LIMITED NUCLEAR TEST BAN TREATY; NUCLEAR NON-PROLIFERATION TREATY

Jessica L. Cox, Researcher
Ctr. for Strategic and Int'l. Studies, Washington, DC
BARUCH PLAN; LIMITED NUCLEAR TEST BAN TREATY; NUCLEAR NON-PROLIFERATION TREATY

Robert P. Crease, Prof., Philosophy
State Univ. of New York, Stony Brook
EXPERTISE

Kem D. Crimmins, Ph.D. candidate, Philosophy
Fordham Univ., Bronx, NY
LYOTARD, JEAN-FRANÇOIS

Jennifer L. Croissant, Assoc. Prof., Women's Studies
Univ. of Arizona, Tuscon
SOCIAL THEORY OF SCIENCE AND TECHNOLOGY

Marvin J. Croy, Assoc. Prof., Philosophy
Univ. of North Carolina, Charlotte
INCREMENTALISM; SOCIAL ENGINEERING

Ana Cuevas-Badallo, Asst. Prof., Philosophy
Univ. of Salamanca, Spain
MONDRAGÓN COOPERATIVE CORPORATION

Charles M. Culver, Prof., Medical Education
Barry Univ., Miami Shores, FL
THERAPY AND ENHANCEMENT

Stephen H. Cutcliffe, Prof., History; Prof. and Chair, Science and Technology Studies
Lehigh Univ., Bethlehem, PA
SCIENCE, TECHNOLOGY, AND SOCIETY STUDIES

Martin Daly, Prof., Psychology
McMaster Univ., Hamilton, Ont., Canada
CRIME

Marcel Danesi, Prof., Anthropology; Dir., Program in Semiotics and Communication Theory
Univ. of Toronto, Ont., Canada
POPULAR CULTURE; POSTMODERNISM

Peter Danielson, Prof., Dir., Ctr. for Applied Ethics
Univ. of British Columbia, Vancouver, Canada
ARTIFICIAL MORALITY; MONITORING AND SURVEILLANCE; ROBOT TOYS; TURING TESTS

Marcy Darnovsky, Assoc. Exec. Dir.
Ctr. for Genetics and Society, Oakland, CA
SEX SELECTION

Edward B. Davis, Prof., History of Science
Messiah Coll., Grantham, PA
BOYLE, ROBERT

Michael Davis, Prof., Philosophy
Illinois Inst. of Technology, Chicago
CODES OF ETHICS; CONFLICT OF INTEREST; PROFESSION AND PROFESSIONALISM

Juan de Castro, Full-time faculty, Eugene Lang Coll.
New School Univ., New York, NY
COLONIALISM AND POSTCOLONIALISM; JUANA INÉS DE LA CRUZ

José M. de Cózar-Escalante; Prof., History and Philosophy of Science, Education and Language
Univ. of La Laguna, Spain
EFFICIENCY

Inmaculada de Melo-Martín, Assoc. Prof., Philosophy
St. Mary's Univ., San Antonio, TX
ASSISTED REPRODUCTION TECHNOLOGY; BIRTH CONTROL; GENETIC RESEARCH AND TECHNOLOGY

Wesley Dean, Post-doctoral Research Assoc., Veterinary Integrative Biosciences & Rural Sociology
Texas A&M Univ., College Station
EMERGENT INFECTIOUS DISEASES

Thomas R. DeGregori, Prof., Economics
Univ. of Houston, TX
GENETICALLY MODIFIED FOODS

Willard Delavan, Visiting Asst. Prof., Economics
Rochester Inst. of Technology, NY
CAPITALISM; PROPERTY; VEBLEN, THORSTEIN

Christelle Didier, Ethics
Institut Catholique de Lille, France
ENGINEERING ETHICS: EUROPE

Shane Drefcinski, Philosophy
Univ. of Wisconsin, Platteville
HUME, DAVID

Krishna R. Dronamraju, Pres.
Foundation for Genetic Research, Houston, TX
HALDANE, J. B. S.

Ruth E. Duerr, Archive Services Mgr., Nat. Snow and Ice Data Ctr.
CIRES, Univ. of Colorado at Boulder
ANSCOMBE, G. E. M.; ROYAL SOCIETY

Paul T. Durbin, Prof. Emeritus, Philosophy
Univ. of Delaware, Newark
BELL, DANIEL, CONSERVATION AND PRESERVATION; DEFORESTATION AND DESERTIFICATION; PRAGMATISM; RAIN FOREST

Roger W. Dyson, Dir., Ctr. for the History of Political Thought
Univ. of Durham, UK
THOMAS AQUINAS

Frank H. W. Edler, Instructor, Philosophy
Metropolitan Community Coll., Omaha, NE
LUDDITES AND LUDDISM

Fred Edwords, Ed., *Humanist*; Editorial Dir.
American Humanist Association, Washington, DC
HUMANISM

Deni Elliott, Poynter Jamison Chair in Media Ethics and Press Policy
Univ. of South Florida, St. Petersburg
JOURNALISM ETHICS

Edwin M. Elrod, Adjunct Prof., Coll. of Information Science and Technology
Drexel Univ., Philadelphia, PA
INFORMATION ETHICS

Charles Ess, Distinguished Research Prof., Interdisciplinary Studies Ctr.
Drury Univ., Springfield, MO
DISCOURSE ETHICS

Amitai Etzioni, Prof., Sociology, *George Washington Univ.*
Founder and Dir., *The Communitarian Network Washington, DC*
COMMUNITARIANISM

Dylan Evans, Senior Lecturer, Computing, Engineering and Mathematical Sciences
Univ. of the West of England, Bristol, UK
EMOTION

John R. Fanchi, Prof., Petroleum Engineering
Colorado Sch. of Mines, Golden
OIL

Samar Farage, Sociology
Pennsylvania State Univ., University Park
GALENIC MEDICINE

Darryl L. Farber, Asst. Prof., Science, Technology, and Society
Pennsylvania State Univ., University Park
CHERNOBYL; NUCLEAR REGULATORY COMMISSION; NUCLEAR WASTE

Joshua C. Farley, Asst. Prof., Community Development and Applied Economics; Fellow, Gund Inst. for Ecological Economics
Univ. of Vermont, Burlington
ECOLOGICAL ECONOMICS

Darrell J. Fasching, Prof., Religious Studies
Univ. of South Florida, Tampa
HOLOCAUST; TECHNICIZATION

Marc R. Fellenz, Assoc. Prof., Philosophy
Suffolk County Community Coll., Brentwood, NY
ANIMAL EXPERIMENTATION; ANIMAL RIGHTS

John Fielder, Prof., Philosophy
Villanova Univ., PA
DC-10 CASE

Autumn Fiester, Senior Fellow, Ctr. for Bioethics
Dir. of Graduate Studies, Medical Ethics
Univ. of Pennsylvania, Philadelphia
BIOETHICS CENTERS

Robert Figueroa, Asst. Prof., University Studies
Colgate Univ., Hamilton, NY
ENVIRONMENTAL JUSTICE

Kevin Finneran, Ed.-in-Chief
Issues in Science & Technology, Washington, DC
NATIONAL ACADEMIES

Erik Fisher, Doctoral student, Environmental Studies
Univ. of Colorado, Boulder, CO
TWO CULTURES

Kenneth R. Fleischmann, Asst. Prof., Coll. of Information
Florida State Univ., Tallahassee
MODELS AND MODELING

Luciano Floridi, Assoc. Prof., Philosophy
Oxford Univ., UK and Bari Univ., Italy
INFORMATION

James R. Flynn, Prof., Political Studies
Univ. of Otago, Dunedin, New Zealand
IQ DEBATE

Franz Allen Foltz, Assoc. Prof., Science, Technology & Society
Rochester Inst. of Technology, NY
ENVIRONMENTAL IMPACT ASSESSMENT; POPULATION; SOCIAL CONTRACT THEORY

Rayvon Fouché, Asst. Prof., Science and Technology Studies
Rensselaer Polytechnic Inst., Troy, NY
RACE

Mark S. Frankel, Dir., Scientific Freedom, Responsibility and Law Program
American Association for the Advancement of Science, Washington, DC
AMERICAN ASSOCIATION FOR THE ADVANCEMENT OF SCIENCE

Maarten Franssen, Assoc. Prof., Philosophy
Delft Univ. of Technology, The Netherlands
TECHNICAL FUNCTIONS

Anthony Freeman, Managing Ed.
Journal of Consciousness Studies, Exeter, UK
CONSCIOUSNESS

Volker Friedrich, Prof., Creative Writing
Hochschule Konstanz, Germany
FAUST; RUSSELL, BERTRAND

Robert Frodeman, Assoc. Prof. and Chair, Philosophy and Religion Studies
Univ. of North Texas, Denton
ACID MINE DRAINAGE; DECISION SUPPORT SYSTEMS; GLOBAL CLIMATE CHANGE; KNOWLEDGE; PLACE; TOWARD AN ETHICS OF SCIENCE AND TECHNOLOGY AS KNOWLEDGE

Jun Fudano, Prof., History of Science and Technology; Dir., Office of Int'l. Programs
Kanazawa Inst. of Technology, Japan
JAPANESE PERSPECTIVES

Wilhelm E. Fudpucker, *Massenheim Inst.*
TECHNICISM

Lars Fuglsang, Assoc. Prof., Social Sciences
Roskilde Univ., Denmark
SCANDINAVIAN AND NORDIC PERSPECTIVES

Steve Fuller, Prof., Sociology
Univ. of Warwick, Coventry, UK
KUHN, THOMAS; SOCIAL CONSTRUCTION OF SCIENTIFIC KNOWLEDGE

Giorgio Gallo, Prof., Computer Science
Univ. of Pisa, Italy
OPERATIONS RESEARCH

Arun Gandhi, Founder and Pres.
M.K. Gandhi Inst. for Nonviolence, Memphis, TN
GANDHI, MOHANDAS

José Luís Garcia, Asst. Prof., Sociology
Univ. of Lisbon, Portugal
SPENGLER, OSWALD; THEODICY

Marcos Garcia de la Huerta, Prof., Philosophy
Univ. of Chile, Santiago
JÜNGER, ERNST

Jerry Gaston, Prof., Sociology
Texas A&M Univ., College Station
MERTON, ROBERT

Bernard Gert, Stone Prof. of Intellectual and Moral Philosophy
Dartmouth Coll., Hanover, NH
THERAPY AND ENHANCEMENT

Onkar Ghate, Senior Fellow
Ayn Rand Inst., Irvine, CA
RAND, AYN

Tina Gianquitto, Asst. Prof., Liberal Arts and Int'l. Studies
Colorado Sch. of Mines, Golden
LEOPOLD, ALDO; TREAT, MARY

David W. Gill, Freelance ethics writer, educator, and consultant
Berkeley, CA
BUSINESS ETHICS

Nicholas W. Gillham, James B. Duke Prof. Emeritus, Biology
Duke Univ., Durham, NC
GALTON, FRANCIS

Michael Gilmore, B.S., Fellow of the British Interplanetary Society
SPACE EXPLORATION

Hebert Gintis, Emeritus Prof., Economics
Univ. of Massachusetts, Amherst
DECISION THEORY; ECONOMICS AND ETHICS; GAME THEORY; PRISONER'S DILEMMA; RATIONAL CHOICE THEORY

Ranulph Glanville, Freelance Prof., Cybernetics and Design
CybernEthics Research, Southsea, UK
CYBERNETICS

Andrew J. Goddard, Tutor, Christian Ethics
Wycliffe Hall, Oxford Univ., UK
ELLUL, JACQUES

Sara Goering, Asst. Prof., Philosophy
Univ. of Washington, Seattle
AUTONOMY; COSMETICS

Steven L. Goldman, Andrew W. Mellon Distinguished Prof. in the Humanities
Lehigh Univ., Bethlehem, PA
PROGRESS; SCIENTIFIC REVOLUTION; TECHNOLOGICAL INNOVATION

David L. Goodstein, Vice Provost; Prof. of Physics and Applied Physics
California Inst. of Technology, Pasadena, CA
MISCONDUCT IN SCIENCE: PHYSICAL SCIENCES CASES

Gennady Gorelik, Research Fellow, Ctr. for Philosophy and History of Science
Boston Univ., MA
SAKHAROV, ANDREI

Michael E. Gorman, Prof., Science, Technology, and Society
Univ. of Virginia, Charlottesville
EARTH SYSTEMS ENGINEERING MANAGEMENT; ENTREPRENEURISM

Vitaly Gorokhov, Prof., Philosophy
Universität Karlsruhe, Germany
RUSSIAN PERSPECTIVES

Gabriella Gosman, M.D., Asst. Prof., Obstetrics, Gynecology, and Reproductive Sciences
Univ. of Pittsburgh Sch. of Medicine, Magee-Womens Hospital, PA
ABORTION

John A. Grim, Prof., Religion
Bucknell Univ., Lewisburg, PA
INDIGENOUS PEOPLES' PERSPECTIVES

Jim Grote, Freelance business writer
Louisville, KY
GATES, BILL; GIRARD, RENÉ

Alastair Gunn, Assoc. Prof., Philosophy
Univ. of Waikato, Hamilton, New Zealand
ENGINEERS FOR SOCIAL RESPONSIBILITY; INSTITUTE OF PROFESSIONAL ENGINEERS NEW ZEALAND

Claus Günzler, Prof. Emeritus, Philosophy
Univ. of Education, Karlsruhe, Germany
SCHWEITZER, ALBERT

David H. Guston, Assoc. Prof., Public Policy
Rutgers Univ., New Brunswick, NJ
RESEARCH INTEGRITY; SOCIAL CONTRACT FOR SCIENCE

Edward J. Hackett, Prof., Sociology
Arizona State Univ., Tempe, AZ
PEER REVIEW

Patrick W. Hamlett, Assoc. Prof., Science, Technology, and Society
North Carolina State Univ., Raleigh, NC
CONSENSUS CONFERENCES

Eric T. Hammer, Spiritual Care Coordinator and Chaplain
VistaCare Hospice, Temple, TX
DEATH AND DYING

J. Craig Hanks, Assoc. Prof., Philosophy
Texas State Univ., San Marcos
CRITICAL SOCIAL THEORY; DEWEY, JOHN

Sven Ove Hansson, Prof., Philosophy
Royal Inst. of Technology, Stockholm, Sweden
EXPOSURE LIMITS; RISK ETHICS; SAFETY ENGINEERING: PRACTICES; SAFETY FACTORS

Robin M. Harrap, Adjunct Prof., Geological Sciences and Geological Engineering; Assoc. Dir., Research, GIS Laboratory
Queen's Univ., Kingston, Ont., Canada
DECISION SUPPORT SYSTEMS

Poul Harremoës, Prof., Environment and Resources
DTU Technical Univ. of Denmark, Kongens Lyngby
PRECAUTIONARY PRINCIPLE

Charles E. Harris, Jr., Prof., Philosophy
Texas A&M Univ., College Station
ENGINEERING ETHICS: OVERVIEW

David M. Hart, Assoc. Prof., Sch. of Public Policy
George Mason Univ., Arlington, VA
POLITICAL ECONOMY

John Hart, Researcher, Chemical and Biological Warfare Project
Stockholm Int'l. Peace Research Inst., Sweden
CHEMICAL WEAPONS

Leonard Hayflick, Adjunct Prof., Anatomy
Univ. of California, San Francisco Sch. of Medicine
AGING AND REGENERATIVE MEDICINE

Maril Hazlett, Ph.D.
Independent Scholar
CARSON, RACHEL

Topi Heikkerö, Researcher, Systematic Theology
Univ. of Helsinki, Finland
MATERIALISM; VEGETARIANISM; VON WRIGHT, GEORG HENRIK

Susan Hendler, Assoc. Prof., Sch. of Urban and Regional Planning
Queen's Univ., Kingston, Ont., Canada
PLANNING ETHICS

Bertrand Hériard Dubreuil, S.J. Researcher, Centre de Recherche et d'Action Sociale, Paris
Centre Ethique Technique et Société, Inst. Catholique d'Arts et Metiérs de Lille, France
ENGINEERING ETHICS: EUROPE

Joseph R. Herkert, Assoc. Prof., Program Dir., Science, Technology, and Society
North Carolina State Univ., Raleigh
BHOPAL CASE; INSTITUTE OF ELECTRICAL AND ELECTRONICS ENGINEERS; PRODUCT SAFETY AND LIABILITY

Peter D. Hershock, Educational Specialist
East-West Ctr., Honolulu, HI
BUDDHIST PERSPECTIVES

David J. Hess, Prof., Science and Technology Studies
Rensselaer Polytechnic Univ., Troy, NY
CANCER; COMPLEMENTARY AND ALTERNATIVE MEDICINE

Eric Higgs, Dir., Assoc. Prof., Sch. of Environmental Studies
Univ. of Victoria, B.C., Canada
ECOLOGICAL RESTORATION

Thomas C. Hilde, Visiting Asst. Prof., Research Scholar, Inst. for Philosophy and Public Policy
Univ. of Maryland, Washington, DC
COMMUNITY; TOURISM

Kenneth Einar Himma, Assoc. Prof., Philosophy
Seattle Pacific Univ., WA
NATURAL LAW

Andrew P. Hodges, Lecturer, Mathematics
Wadham Coll., Univ. of Oxford, UK
TURING, ALAN

J. Britt Holbrook, Asst. Prof., Philosophy and Religion Studies
Univ. of North Texas, Denton
NATIONAL SCIENCE FOUNDATION SECOND MERIT CRITERION

Rachelle D. Hollander
Baltimore, MD
NATTIONAL SCIENCE FOUNDATION

Alexander E. Hooke, Prof., Philosophy
Villa Julie Coll., Stevenson, MD
FOUCAULT, MICHEL; VIRTUE ETHICS

James P. Horne, P.E., Design Engineer
KL&A, Inc., Golden, CO
EARTHQUAKE ENGINEERING

Gilbert Hottois, Prof., Philosophy
Université Libre de Bruxelles, Belgium
TECHNOSCIENCE

M. Ann Howard, Assoc. Prof., Science, Technology, Society
Rochester Inst. of Technology, NY
NONGOVERNMENTAL ORGANIZATIONS; URBANIZATION

Craig R. Humphrey, Sociology
Pennsylvania State Univ., University Park
SUSTAINABLILTY AND SUSTAINABLE DEVELOPMENT

Morgan J. Hurley, Technical Dir.
Society of Fire Protection Engineers, Bethesda, MD
FIRE

Héctor José Huyke, Prof., Humanities
Univ. of Puerto Rico at Mayagüez
DISTANCE

A. Pablo Iannone, Prof., Philosophy
Central Connecticut State Univ., New Britain, CT
INFORMATION OVERLOAD; POLLUTION; SPEED

Andoni Ibarra, Prof., Logic and Philosophy of Science
Univ. of the Basque Country, Donostia-San Sebastián, Spain
LOGICAL EMPIRICISM

Dolores Ibarreta, Biologist
Inst. for Prospective Technological Studies, Joint Research Centre, European Commission, Seville, Spain
DES (DIETHYLSTILBESTROL) CHILDREN

Don Ihde, Distinguished Prof. of Philosophy; Dir., Technoscience Research Group
State Univ. of New York, Stony Brook
BODY; INSTRUMENTATION

Lucas D. Introna, Prof., Organisation, Technology and Ethics
Lancaster Univ., UK
LEVINAS, EMMANUEL; PHENOMENOLOGY

Bernhard Irrgang, Prof., Technikphilosophie
Technical Univ. of Dresden, Germany
GERMAN PERSPECTIVES

Paul Israel, Dir., Thomas A. Edison Papers
Rutgers Univ., Piscataway, NJ
EDISON, THOMAS ALVA

Fatima L. C. Jackson, Prof., Anthropology
Univ. of Maryland, College Park
TUSKEGEE EXPERIMENT

Struan Jacobs, Senior Lecturer, Sociology
Deakin Univ., Geelong, Victoria, Australia
LIBERALISM

John E. Jalbert, Prof., Philosophy and Religious Studies
Sacred Heart Univ., Fairfield, CT
MONEY

Stephen Janos, Friar
Sts. Peter & Paul RO Rectory, Buffalo, NY
BERDYAEV, NIKOLAI

Lonny Jarrett, M.Ac., Acupuncturist
Stockbridge, MA
ACUPUNCTURE

Sheila Jasanoff, Pforzheimer Professor of Science and Technology Studies
Kennedy Sch., Harvard Univ., Cambridge, MA
TECHNOLOGIES OF HUMILITY: CITIZEN PARTICIPATION IN GOVERNING SCIENCE

Helena Mateus Jerónimo, Univ. Lecturer, Sociology, Inst. of Economics and Business
Administration of the Technical Univ. of Lisbon and Research Ctr. on Economic Sociology and the Sociology of Organizations
Lisbon, Portugal
RISK SOCIETY; WASTE

James V. Jesudason, Assoc. Prof., Liberal Arts and Int'l. Studies
Colorado Sch. of Mines, Golden
SINGAPORE

Sujatha Jesudason, Gender, Justice and Human Genetics Program Dir.
Ctr. for Genetics and Society, Oakland, CA
SEX SELECTION

Deborah G. Johnson, Anne Shirley Carter Olsson Prof. of Applied Ethics; Chair, Science, Technology, and Society
Univ. of Virginia, Charlottesville
COMPUTER ETHICS; ETHICS AND TECHNOLOGY: A PROGRAM FOR FUTURE RESEARCH; HACKER ETHICS; RESPONSIBILITY: ANGLO-AMERICAN PERSPECTIVES; SOCIAL CONSTRUCTION OF TECHNOLOGY

James Turner Johnson, Prof., Religion
Rutgers Univ., New Brunswick, NJ
JUST WAR

Michael E. Johnson-Cramer, Management
Bucknell Univ., Lewisburg, PA
STAKEHOLDERS

Stephen Johnston, Adjunct Prof., Engineering
Univ. of Technology Sydney, Australia
AUSTRALIAN AND NEW ZEALAND PERSPECTIVES

Alicia Juarrero, Prof,. Philosophy
Prince George's Community Coll., Largo, MD
COMPLEXITY AND CHAOS

Matthias Kaiser, Dir.
Nat. Committee for Research Ethics in Science and Technology, Oslo, Norway
INTERNATIONAL COUNCIL FOR SCIENCE

Subhash Kak, Delaune Distinguished Prof. of Electrical Engineering and Prof. of Asian Studies and Cognitive Science
Louisiana State Univ., Baton Rouge
HINDU PERSPECTIVES

Mark A. Kalthoff, Chairman, Assoc. Prof., History
Hillsdale Coll., MI
GALILEI, GALILEO

David M. Kaplan, Visiting Asst. Prof., Humanities and Social Sciences
Polytechnic Univ., Brooklyn, NY
MARCUSE, HERBERT; WORLD TRADE ORGANIZATION

James E. Katz, Prof., Communication
Rutgers Univ., New Brunswick, NJ
TELEPHONE

Antti Kauppinen, Philosophy
Univ. of Helsinki, Finland
HABERMAS, JÜRGEN

Christine Keiner, Asst. Prof., Science, Technology, and Society
Rochester Inst. of Technology, NY
DDT; DUBOS, RENÉ

Evelyn Fox Keller, Prof., Science, Technology, and Society
Massachusetts Inst. of Technology, Cambridge
MCCLINTOCK, BARBARA

Michael R. Kelly, Adjunct Asst. Prof., Philosophy
Boston Coll., MA
HUSSERL, EDMUND

Habibul H. Khondker, Assoc. Prof., Sociology
Nat. Univ. of Singapore
SINGAPORE

Gavin Kitching, Prof., Politics and Int'l. Relations
Univ. of New South Wales, Sydney, Australia
GLOBALISM AND GLOBALIZATION; NATIONALISM; NEOLIBERALISM

Chuck Klein, NRA and TDI certified firearms instructor
Cincinnati, OH
POLICE

Julie Thompson Klein, Prof., Humanities, Interdisciplinary Studies
Wayne State Univ., Detroit, MI
INTERDISCIPLINARITY

Edward H. Klevans, Prof. Emeritus, Nuclear Engineering
Pennsylvania State Univ., University Park
CHERNOBYL; NUCLEAR WASTE; THREE-MILE ISLAND

Ronald R. Kline, Bovay Prof. in History and Ethics of Engineering
Cornell Univ., Ithaca, NY
RESEARCH ETHICS, ENGINEERING ETHICS, AND SCIENCE AND TECHNOLOGY STUDIES; STEINMETZ, CHARLES; WIENER, NORBERT

Arnold Kling, Contributing Ed.
TechCentralStation.com
LIBERTARIANISM; MARKET THEORY

Billy V. Koen, Prof., Mechanical Engineering
Univ. of Texas, Austin
ENGINEERING METHOD

David Kolb, Charles A. Dana Prof. of Philosophy
Bates Coll., Lewiston, ME
EXISTENTIALISM

Katherine J. Komenda Poole, Prof., Religious Studies
Naropa Univ., Boulder, CO
SHINTO PERSPECTIVES

Maria Kostyszak, Ph.D., Adjunct
Inst. of Philosophy, Wroclaw, Poland
CENTRAL EUROPEAN PERSPECTIVES

Steven Kreis, Instructor, History and Humanities
Wake Technical Community Coll., Raleigh, NC
COMTE, AUGUSTE

Manfred Kroger, Prof. Emeritus, Food Science
Pennsylvania State Univ.,University Park
FOOD SCIENCE AND TECHNOLOGY

Wolfgang Krohn, Prof., Sociology
Univ. of Bielefeld, Germany
ENQUETE COMMISSIONS; EXPERIMENTATION

Robert Kugelmann, Prof., Psychology
Univ. of Dallas, Irving, TX
STRESS

Ray Kurzweil, inventor, entrepreneur and author; Pres.
Kurzweil Technologies, Inc., MA
NANOSCIENCE, NANOTECHNOLOGY, AND ETHICS: PROMISE AND PERIL

Gregory Clifford Kunkle, Adjunct Prof., Science, Technology & Society
Lehigh Univ., Bethlehem, PA
OFFICE OF TECHNOLOGY ASSESSMENT

Hugh Lacey, Scheuer Family Prof. Emeritus of Philosophy
Swarthmore Coll., PA
FACT/VALUE DICHOTOMY

Frank N. Laird, Assoc. Prof., Graduate Sch. of Int'l. Studies
Univ. of Denver
ALTERNATIVE ENERGY; DEMOCRACY

Rob Latham, Assoc. Prof., English, American Studies
Univ. of Iowa, Iowa City
SCIENCE FICTION

Peter Augustine Lawler, Dana Prof., Department Chair, Government and Int'l. Studies
Berry Coll., Mount Berry, GA
TOCQUEVILLE, ALEXIS DE

P. Patrick Leahy, Assoc. Dir. for Geology
U.S. Geological Survey, Reston, VA
NATIONAL GEOLOGICAL SURVEYS

Sarah A. Leavitt, Assoc. Historian, Office of NIH History
Nat. Institutes of Health, Bethesda, MD
NATIONAL INSTITUTES OF HEALTH

Hans Lenk, Prof., Philosophy
Univ. of Karlsruhe, Germany
NEW TECHNOLOGIES; RECOGNIZING THE STRUCTURAL FEATURES OF NEW TECHNOLOGIES; RESPONSIBILITY: GERMAN PERSPECTIVES; SCHWEITZER, ALBERT; TECHNOCRACY

Mary Lenzi, Asst. Prof., Philosophy
Univ. of Wisconsin, Platteville
MURDOCH, IRIS; PROMETHEUS; RAWLS, JOHN

Wil Lepkowski, Independent scholar, journalist
Reston, VA
ARSENIC; CORRUPTION; HEAVY METALS; SCIENCE AND ENGINEERING INDICATORS

Peter Lévesque, Deputy Dir., Knowledge Products and Mobilization Division
Social Sciences and Humanities Research Council of Canada, Ottawa
ROYAL COMMISSIONS; SCIENCE SHOPS

Norman Levitt, Prof., Mathematics
Rutgers Univ., New Brunswick, NJ
SOKAL AFFAIR

Tammy L. Lewis, Asst. Prof, Sociology and Anthropology
Muhlenberg Coll., Allentown, PA
SUSTAINABILITY AND SUSTAINABLE DEVELOPMENT

Jon A. Leydens, Writing Program Administrator, Liberal Arts and Int'l. Studies
Colorado Sch. of Mines, Golden
PLAGIARISM

Li Chenyang, Chair and Prof., Philosophy
Central Washington Univ., Ellensburg
CONFUCIAN PERSPECTIVES

Li Shixin , Asst. Prof., Philosophy of Science and Technology
Beijing Inst. of Technology, China
CHINESE PERSPECTIVES

Liu Zeyuan, Sch. of Humanities and Social Science
Dalian Univ. of Technology, China
CHINESE PERSPECTIVES: ENGINEERING ETHICS

Liu Xiaogan, Prof., Philosophy
Chinese Univ. of Hong Kong, Shatin, New Territories
DAOIST PERSPECTIVES

George M. Logan, James Cappon Prof. of English Language and Literature
Queen's Univ., Kingston, Ont., Canada
MORE, THOMAS

José A. López Cerezo, Prof., Philosophy of Science
Univ. of Oviedo, Spain
IBERO-AMERICAN PERSPECTIVES

Michael C. Loui, Prof., Electrical and Computer Engineering
Univ. of Illinois at Urbana-Champaign
ASSOCIATION FOR COMPUTING MACHINERY

Juan Lucena, Assoc. Prof., Liberal Arts and Int'l. Studies
Colorado Sch. of Mines, Golden
HUMANITARIAN SCIENCE AND TECHNOLOGY

Heinz C. Luegenbiehl, Prof. of Philosophy and Technology Studies
Rose-Hulman Inst. of Technology, Terre Haute, IN
JAPANESE PERSPECTIVES; WHISTLEBLOWING

José Luis Luján, Assoc. Prof., Philosophy of Science
Univ. of the Balearic Islands, Palma de Mallorca, Spain
ENVIRONMENTAL REGULATION

James A. Lynch, Doctoral candidate, Comparative Literature; Spanish and Portuguese
Indiana Univ., Bloomington
BENJAMIN, WALTER; BUTLER, SAMUEL; TECHNOCOMICS

Tibor R. Machan, Prof., Argyros Sch. of Business & Economics
Chapman Univ., Orange, CA
SKINNER, B. F.

Robert Mackey, Independent Scholar
Fairlawn, VA
TOOLS AND MACHINES

Janet Malek, Asst. Prof., Medical Humanities
Brody Sch. of Medicine at East Carolina Univ., Greenville, NC
INFORMED CONSENT; MISCONDUCT IN SCIENCE: BIOMEDICAL SCIENCE CASES

Gary E. Marchant, Prof., Dir., Ctr. for the Study of Law, Science & Technology
Arizona State Univ. Coll. of Law, Tempe, AZ
SCIENCE, TECHNOLOGY, AND LAW

Jean-Michel Margot, Pres.
North American Jules Verne Society
VERNE, JULES

Genevieve Maricle, Ph.D. candidate, Environmental Studies
Univ. of Colorado, Boulder
MILITARY-INDUSTRIAL COMPLEX

Allison C. Marsh, History of Science and Technology
Johns Hopkins Univ., Baltimore, MD
MUSEUMS OF SCIENCE AND TECHNOLOGY

Robert Martello, Asst. Prof., History of Science and Technology
Franklin W. Olin Coll. of Engineering, Needham, MA
INDUSTRIAL REVOLUTION

Carol Martin, Assoc. Prof., Drama
Tisch Sch. of the Arts, New York Univ., New York
BRECHT, BERTOLT

Glen A. Mazis, Prof., Philosophy and Humanities
Pennsylvania State Univ., Harrisburg
EARTH

Glenn McGee, John A. Balint Endowed Prof. of Medical Ethics
Alden March Bioethics Inst., Albany Medical Coll., NY
HUMAN CLONING

William C. McGrew, Prof., Zoology
Miami Univ., Oxford, OH
ANIMAL TOOLS

Ross McKitrick, Assoc. Prof., Economics
Univ. of Guelph, Ont., Canada
ECONOMICS: OVERVIEW

Eric McLuhan, Part-time lecturer
York Univ.; Univ. of Toronto; Harris Inst. for the Arts, Toronto, Ont. Canada
MCLUHAN, MARSHALL

Elizabeth C. McNie, Graduate Research Asst., Ctr. for Science and Technology Policy Research
Univ. of Colorado, Boulder
ARCHEOLOGICAL ETHICS; BRENT SPAR; GENOCIDE; WEAPONS OF MASS DESTRUCTION; ZOOS

Samuel C. McQuade III, Asst. Prof., Criminal Justice
Rochester Inst. of Technology, NY
BIOSECURITY; FORENSIC SCIENCE; MONITORING AND SURVEILLANCE

Andy Miah, Lecturer in Media, Bioethics & Cyberculture
Univ. of Paisley, Scotland, UK
SPORTS

Alex C. Michalos, Dir., Inst. for Social Research and Evaluation
Univ. of Northern British Columbia, Prince George, Canada
SOCIAL INDICATORS

Diane P. Michelfelder, Dean, Coll. of Arts and Sciences
Indiana State Univ., Terre Haute
PETS

Valerie Miké, Clinical Prof., Public Health
Weill Medical Coll. of Cornell Univ., Ithaca, NY
BIOSTATISTICS; EPIDEMIOLOGY; ETHICS OF EVIDENCE; NIGHTINGALE, FLORENCE; META-ANALYSIS; PASCAL, BLAISE; PROBABILITY; STATISTICS

Keith W. Miller, Prof., Computer Science
Univ. of Illinois at Springfield
COMPUTER VIRUSES/INFECTIONS; HARDWARE AND SOFTWARE

Fabio Minazzi, Prof., Information and Communication
Università degli Studi, Milan, Italy
ITALIAN PERSPECTIVES

Carl Mitcham, Prof., Liberal Arts and Int'l. Studies
Colorado Sch. of Mines, Golden
Fac. Affil., Ctr. for Science and Technology Policy Research, Univ. of Colorado, Boulder
BIODIVERSITY COMMERCIALIZATION; BIOETHICS COMMITTEES AND COMMISSIONS; BUSH, VANNEVAR; CAPITALISM; CULTURAL LAG; CYBERCULTURE; DEATH AND DYING; DEMATERIALIZATION AND IMMATERIALIZATION; DESSAUER, FRIEDRICH; DIRECT DEMOCRACY; ECONOMICS: OVERVIEW; ETHICS: OVERVIEW; HOBBES, THOMAS; HUMANITARIAN SCIENCE AND TECHNOLOGY; HUMANIZATION AND DEHUMANIZATION; HYPERTEXT; ILLICH, IVAN; LIFE; MISCONDUCT IN SCIENCE: SOCIAL SCIENCE CASES; MUMFORD, LEWIS; OPEN SOCIETY; OPPENHEIMER, FRANK; POLANYI, KARL; PROMETHEUS; RELIABILITY; RESPONSIBILITY: OVERVIEW; RISK AND SAFETY: OVERVIEW; RUSSELL, BERTRAND; SCIENCE: OVERVIEW; SEMIOTICS: OVERVIEW; SIMON, JULIAN; SOCIAL CONTRACT THEORY; SYSTEMS AND SYSTEMS THINKING; TERRORISM; TOOLS AND MACHINES; TECHNOETHICS; TECHNOLOGY: OVERVIEW; TOWARD AN ETHICS OF SCIENCE AND TECHNOLOGY AS KNOWLEDGE; UNCERTAINTY; VALUE AND VALUES

Kristen Renwick Monroe, Prof., Political Science
Univ. of California, Irvine
ALTRUISM

Jensen R. Montambault, Ph.D. candidate, Sch. of Natural Resources and Environment
Univ. of Florida, Gainesville
MULTIPLE-USE MANAGEMENT

Suzanne Moon, Asst. Prof., Liberal Arts and Int'l. Studies
Colorado Sch. of Mines, Golden
HUMANITARIAN SCIENCE AND TECHNOLOGY; GREEN REVOLUTION

Steven A. Moore, Assoc. Prof., Architecture and Planning
Univ. of Texas, Austin
BUILDING CODES

Thomas Mormann, Prof., Logic and Philosophy of Science
Univ. of the Basque Country, Donostia-San Sebastián, Spain
LOGICAL EMPIRICISM

Janice M. Morse, Prof., Nursing; Scientific Dir., Int'l Inst. for Qualitative Methodology
Univ. of Alberta, Edmonton, Canada
QUALITATIVE RESEARCH

Elizabeth J. Mulligan, Psychology
Univ. of Colorado, Boulder
CHOICE BEHAVIOR; PSYCHOLOGY: OVERVIEW; PSYCHOPHARMACOLOGY

Phil Mullins, Prof., Humanities
Missouri Western State Coll., St. Joseph
POLANYI, MICHAEL

Joshua Muravchik, Resident Scholar
American Enterprise Inst. for Public Policy Research, Washington, DC
COMMUNISM; SOCIALISM

Deena M. Murphy-Medley, Ph.D. candidate, Psychology
North Carolina State Univ., Raleigh
BHOPAL CASE

Anna Nagurney, John F. Smith Memorial Prof., Isenberg Sch. of Management
Univ. of Massachusetts, Amherst
NETWORKS

Indira Nair, Prof., Engineering and Public Policy
Carnegie Mellon Univ., Pittsburgh, PA
ETHICS OF CARE

Massimo Negrotti, Prof. of Methodology of Human Sciences
Univ. of Urbino, Italy
ARTIFICIALITY

Annemiek Nelis, Deputy Dir.
Ctr. for Society & Genomics, Nijmegen, The Netherlands
DUTCH PERSPECTIVES

Bryon Newberry, Assoc. Prof., Mechanical Engineering
Baylor Univ., Waco, TX
PROFESSIONAL ENGINEERING ORGANIZATIONS; TECHNOLOGICAL FIX

Francis R. Nicosia, Prof., History
Saint Michael's Coll., Colchester, VT
NAZI MEDICINE

Helen Nissenbaum, Assoc. Prof., Culture and Communication, Computer Science
New York University, New York
VALUES IN TECHNICAL DESIGN

Bryan G. Norton, Prof., Sch. of Public Policy
Georgia Inst. of Technology, Atlanta
BIODIVERSITY; WILDERNESS; WILDLIFE MANAGEMENT

Pelucy Ntambirweki, Advisor, Children Affected by HIV and AIDS
UNICEF, Swaziland
AFRICAN PERSPECTIVES: HIV/AIDS IN AFRICA

Barbara Olds, Prof., Liberal Arts and Int'l. Studies
Colorado Sch. of Mines, Golden
ETHICS ASSESSMENT RUBRICS

Richard Olson, Prof. of History and Willard W. Keith Jr. Fellow in Humanities
Harvey Mudd Coll., Claremont, CA
ENLIGHTENMENT SOCIAL THEORY; SECULARIZATION

Carleton A. Palmer, Asst. Prof. of Psychology
Inst. for the Psychological Sciences, Arlington, VA
DIAGNOSTIC AND STATISTICAL MANUAL OF MENTAL DISORDERS

Eul-Soo Pang, Prof., Liberal Arts and Int'l. Studies
Colorado Sch. of Mines, Golden
POLITICAL RISK ASSESSMENT

Robert J. Paradowski, Prof., Science, Technology, and Society
Rochester Inst. of Technology, NY
FOOD AND DRUG AGENCIES; NUTRITION AND SCIENCE; PAULING, LINUS; TAYLOR, FREDERICK W.

Govindan Parayil, Prof., Ctr. for Technology, Innovation and Culture
Univ. of Oslo, Norway
CHANGE AND DEVELOPMENT

John S. Partington, Ed., *The Wellsian*
The H. G. Wells Society, Reading, UK
WELLS, H. G.

Ernest Partridge, Independent Scholar
Gadfly Enterprises, Cedarpines Park, CA
FUTURE GENERATIONS

Robert T. Pennock, Assoc. Prof., Philosophy
Michigan State Univ., East Lansing
DETERMINISM

Ted Peters, Prof., Systematic Theology
Pacific Lutheran Theological Seminary and the Graduate Theological Union, Berkeley, CA
DIGNITY; PLAYING GOD

Joseph A. Petrick, Prof., Management; Dir., Inst. for Business Integrity
Wright State Univ., Dayton, OH
MODELS OF MANAGEMENT

Sarah K. A. Pfatteicher, Asst. Dean, Engineering Academic Affairs
Univ. of Wisconsin-Madison
BUILDING DESTRUCTION AND COLLAPSE

Robert Phillips, Assoc. Prof., Sch. of Business Administration
Univ. of San Diego, CA
STAKEHOLDERS

Peter W. B. Phillips, Prof., Political Studies; Dir., Coll. of Biotechnology
Univ. of Saskatchewan, Saskatoon, Canada
CANADIAN PERSPECTIVES

Roger A. Pielke, Jr., Prof., Environmental Studies
Dir; Ctr. For Science and Technology Policy Research
Univ. of Colorado, Boulder
SCIENCE POLICY; SENSITIVITY ANALYSES

Martin T. Pietrucha, Dir., Science, Technology, and Society Program; Assoc. Prof. of Civil Engineering
Penn State Univ., University Park
ERGONOMICS

Rosa L. Pinkus, Prof., Ctr. for Bioethics and Health Law
Univ. of Pittsburgh, PA
CICERO'S CREED

Joseph C. Pitt, Prof., Head, Philosophy; Adjunct Prof., Science and Technology Studies
Virginia Tech, Blacksburg
SPACE TELESCOPES

Louis P. Pojman, Visiting Fellow
Cambridge Univ., UK
DEATH PENALTY; EQUALITY; JUSTICE

Stephen G. Post, Prof. of Bioethics, Case Sch. of Medicine
Case Western Reserve Univ., Cleveland
POSTHUMANISM

William C. Potter, Inst. Prof. and Director, Ctr. for Nonproliferation Studies
Monterey Inst. of Int'l. Studies, CA
ATOMS FOR PEACE PROGRAM

Thomas M. Powers, Research Fellow, Science, Technology, and Society
Univ. of Virginia, Charlottesville
CONSEQUENTIALISM; DEONTOLOGY; ETHICS AND TECHNOLOGY: A PROGRAM FOR FUTURE RESEARCH

Andrew Price, Graduate Researcher
Univ. of Colorado, Boulder
ANTIBIOTICS

Michael S. Pritchard, Prof., Philosophy
Western Michigan Univ., Kalamazoo
ENGINEERING ETHICS: OVERVIEW

Qian Qian, Ed.
China Aerospace Engineering Consultancy Ctr.
POPULATION POLICY IN CHINA

Wang Qian, Prof., Philosophy
Sch. of Humanities and Social Science, Dalian Univ. of Technology, China
CHINESE PERSPECTIVES: RESEARCH ETHICS; POPULATION POLICY IN CHINA

Michael Rabins, Prof. Emeritus of Mechanical Engineering
Texas A&M Univ., College Station
ENGINEERING ETHICS: OVERVIEW

Eric S. Rabkin, Prof., English Language and Literature
Univ. of Michigan, Ann Arbor
FRANKENSTEIN; SHELLEY, MARY WOLLSTONECRAFT; ZAMYATIN, YEVGENY IVANOVICH

Majid Rahnema, Visiting Prof., Int'l and Intercultural Studies
Pitzer Coll., Claremont, CA
UNITED NATIONS EDUCATIONAL, SCIENTIFIC, AND CULTURAL ORGANIZATION

Kim Alaine Rathman, Development Consultant, Los Angeles Accelerated Schools Ctr.
California State Univ., Los Angeles
COMMON HERITAGE OF MANKIND PRINCIPLE

Thomas C. Reeves, Prof., Instructional Technology
Univ. of Georgia, Athens
DIGITAL LIBRARIES

Nancy Reichman, Assoc. Prof., Sociology
Univ. of Denver, CO
MONTREAL PROTOCOL

Jacob Dahl Rendtorff, Assoc. Prof., Social Sciences
Roskilde Univ., Denmark
SCANDINAVIAN AND NORDIC PERSPECTIVES

Matthias Rieger, Adjunct faculty, Sociology
Univ. of Hannover, Germany
MUSIC

Jean Robert, Prof., City Planning and Urban History, State Univ. of Morelos
Prof. of Architecture Theory and History of Ideas, La Salle Univ.

Cuernavaca, Mexico
ALTERNATIVE TECHNOLOGY; ENERGY; SPACE

Jason Scott Robert, Asst. Prof., Sch. of Life Sciences
Arizona State Univ., Tempe
GENE THERAPY; GENETICS AND BEHAVIOR

John B. Robinson, Prof., Sustainable Development Research Initiative
Univ. of British Columbia, Vancouver
GEORGIA BASIN FUTURES PROJECT

Sheila Rock, Senior Lecturer, Computer Science
Univ. of the Witwatersrand, Johannesburg
AFRICAN PERSPECTIVES: COMPUTERS IN SOUTH AFRICA

Antonio Rodríguez de las Heras, Prof., Humanities and Communication
Unversidad Carlos III de Madrid, Spain
HYPERTEXT

Sabine Roeser, Asst. Prof., Philosophy
Delft Univ. of Technology, The Netherlands
RISK AND EMOTION

Bernard E. Rollin, Univ. Distinguished Prof. of Philosophy
Colorado State Univ., Fort Collins
ANIMAL WELFARE

Holmes Rolston III., Prof., Philosophy
Colorado State Univ., Fort Collins
ECOLOGY

Günter Ropohl, Dr.-Ing., Prof. Emeritus of General Technology
Univ. of Frankfurt on Main, Germany
INVENTION

Sue V. Rosser, Ivan Allen Dean's Chair of Liberal Arts and Technology; Prof., History, Technology, and Society
Georgia Inst. of Technology, Atlanta
SEX AND GENDER

David Rothenberg, Prof., Philosophy
New Jersey Inst. of Technology, Newark
AIR

Christina Rudén, Senior Researcher, Philosophy
Royal Inst. of Technology, Stockholm, Sweden
REGULATORY TOXICOLOGY

Michael Ruse, Lucyle T. Werkmeister Prof. of Philosophy
Florida State Univ., Tallahassee
SOCIOBIOLOGY

Martin Ryder, Adjunct Prof., Information and Learning Technologies
Univ. of Colorado, Denver
DIGITAL DIVIDE; POLYGRAPH; SCIENTISM; SEMIOTICS: LANGUAGE AND CULTURE; TECHNOLOGY: OVERVIEW

Tind Shepper Ryen, Ctr. for Science and Technology Policy
Univ. of Colorado, Boulder
BRENT SPAR; COMMUNICATIONS REGULATORY AGENCIES; HUMAN GENOME ORGANIZATION

Wolfgang Sachs, Head of Globalization and Sustainablity Project
Wuppertal Inst., Berlin, Germany
LIMITS; POVERTY

Mark Sagoff, Sr. Research Scholar, Inst. for Philosophy and Public Policy
Univ. of Maryland, College Park
ENVIRONMENTAL ECONOMICS

Silja Samerski, Adjunct faculty, Sociology
Univ. of Hannover, Germany
GENETIC COUNSELING

Sajay Samuel, Instructor, Accounting
Pennsylvania State Univ., University Park
ACCOUNTING; ALTERNATIVE TECHNOLOGY; ENERGY

Stephen K. Sanderson, Prof., Sociology
Indiana Univ. of Pennsylvania, Indiana, PA
MARX, KARL

José Sanmartín, Prof., Philosophy of Science
Univ. of Valencia
Dir., The Queen Sofía Ctr. for the Study of Violence, Spain
VIOLENCE

Amy Santamaria, Psychology
Univ. of Colorado, Boulder
PSYCHOLOGY: OVERVIEW

Daniel Sarewitz, Prof.; Science and Society; Dir., Consortium for Science, Policy, and Outcomes
Arizona State Univ., Tempe
AFFLUENCE; GAIA; GOVERNANCE OF SCIENCE; HAZARDS; PREDICTION; UNCERTAINTY

Mónica Edwards Schachter, Science Education
Univ. of Valencia, Spain
ACTIVIST SCIENCE EDUCATION

A. George Schillinger, Prof. Emeritus, Management
Polytechnic Univ., Brooklyn, New York
AUTOMATION

Lawrence E. Schmidt, Prof., Dept. and Ctr. for the Study of Religion
Univ. of Toronto at Mississauga, Ont., Canada
WEIL, SIMONE

Richard Schmitt, Prof. Emeritus, Philosophy

Brown Univ., Providence, RI
ALIENATION

Johan Schot, Prof., History of Technology
Eindhoven Univ. of Technology, The Netherlands
CONSTRUCTIVE TECHNOLOGY ASSESSMENT

Ernst Schraube, Asst. Prof., Psychology
Roskilde Univ., Denmark
ANDERS, GÜNTHER

H. Morgan Scott, Asst. Prof., Veterinary Integrative Biosciences
Coll. of Veterinary Medicine, Texas A&M Univ., College Station
EMERGENT INFECTIOUS DISEASES

James Seaton, Prof., English
Michigan State Univ., East Lansing
CONSERVATISM

Jerome M. Segal, Senior Research Scholar, Inst. for Philosophy and Public Policy
Univ. of Maryland, College Park
SIMPLICITY/SIMPLE LIVING

Edward Seiler, Author
asimovonline.com
ASIMOV, ISAAC

Evan Selinger, Asst. Prof., Philosophy
Rochester Inst. of Technology, NY
EXPERTISE; PARTICIPATION; TELEVISION

Adil E. Shamoo, Prof.
Univ. of Maryland Sch. of Medicine, Baltimore
ACCOUNTABILITY IN RESEARCH

Suzanne Shaw, Dir. of Communications
Union of Concerned Scientists, Cambridge, MA
UNION OF CONCERNED SCIENTISTS

Helena Sheehan, Senior Lecturer, History of Ideas
Dublin City Univ., Ireland
BERNAL, J. D.; LYSENKO CASE; MARXISM

Garrett Ward Sheldon, The John Morton Beaty Prof. of Political and Social Sciences
The Univ. of Virginia Coll. at Wise
JEFFERSON, THOMAS

Robert Shelton, Lyman Briggs Sch. of Science
Michigan State Univ., East Lansing
UTOPIA AND DYSTOPIA

Richard Sherlock, Prof., Philosophy
Utah State Univ., Logan
BIOETHICS; MEDICAL ETHICS

Michael Shermer,Ed.-in-Chief, *Skeptic* magazine; Contrib. Ed., *Scientific American*
Pasadena, California
PSEUDOSCIENCE; SKEPTICISM

William M. Shields, Science and Technology in Society
Virginia Tech, Blacksburg
FREUD, SIGMUND; JUNG, CARL GUSTAV, SAFETY ENGINEERING: HISTORICAL EMERGENCE

William O. Shropshire, Emeritus Callaway Prof., Economics
Oglethorpe Univ., Atlanta, GA
SMITH, ADAM

Larry Shuman, Assoc. Dean for Academic Affairs; Prof. of Industrial Engineering
Sch. of Engineering, Univ. of Pittsburgh, PA
CICERO'S CREED; ETHICS ASSESSMENT RUBRICS

Alan Sica, Prof., Sociology
Pennsylvania State Univ., University Park
DURKHEIM, ÉMILE; PARETO, VILFREDO; PARSONS, TALCOTT; WEBER, MAX

Murli M. Sinha, Prof., Sociology
Rochester Inst. of Technology, NY
FAMILY; TECHNOLOGY TRANSFER

Dominic A. Sisti, Researcher, Ctr. for Bioethics
Univ. of Pennsylvania, Philadelphia
HEALTH AND DISEASE

Eugene B. Skolnikoff, Prof. Emeritus, Political Science
Massachusetts Inst. of Technology, Cambridge
INTERNATIONAL RELATIONS

Martha M. Smith, Assoc. Prof., Coll. of Information Science and Technology
Drexel Univ., Philadelphia, PA
INFORMATION ETHICS

Martijntje W. Smits, Asst. Prof., Technology Management
Technical Univ. Eindhoven, The Netherlands
CYBORGS; PLASTICS

Marilyn Berlin Snell, Writer/Ed.
Sierra Magazine, San Francisco, CA
SIERRA CLUB; NATIONAL PARKS; UNITED NATIONS ENVIRONMENTAL PROGRAM

Stephen D. Snobelen, Asst. Prof., History and Philosophy of Science
Univ. of King's Coll., Halifax, Canada
NEWTON, ISAAC

Marlene K. Sokolon, Asst. Prof., Political Science
Concordia Univ., Montreal
EMOTIONAL INTELLIGENCE

Wha-Chul Son, Doctoral candidate, Philosophy
Catholic Univ. of Leuven, Leuven, Belgium

AUTONOMOUS TECHNOLOGY

Leonard R. Sorenson, Prof., Politics
Assumption Coll., Worcester, MA
ROUSSEAU, JEAN-JACQUES

Robert J. Sparrow, Lecturer, Sch. of Philosophy and Bioethics
Monash Univ., Melbourne, Australia
ANDROIDS; ROBOTS AND ROBOTICS

Raymond E. Spier, Emeritus Prof. of Science and Engineering Ethics
Univ. of Surrey, Guildford, UK
JEWISH PERSPECTIVES; NUCLEAR ETHICS; REGULATION AND REGULATORY AGENCIES; VACCINES AND VACCINATION

Jeffrey P. Spike, Assoc. Prof., Medical Humanities and Social Sciences
Florida State Univ. Coll. of Medicine, Tallahassee
HUMAN SUBJECTS RESEARCH; PERSISTENT VEGETATIVE STATE

Robert L. Sprague, Prof. Emeritus, Kinesiology
Univ. of Illinois, Urbana-Champaign
MISCONDUCT IN SCIENCE: SOCIAL SCIENCE CASES

Peter Stansky, Frances and Charles Field Prof. of History, Emeritus,
Stanford Univ., CA
MORRIS, WILLIAM

Nicholas H. Steneck, Prof., History
Univ. of Michigan, Ann Arbor
OFFICE OF RESEARCH INTEGRITY

Sharon Stoerger, Human Subjects Research Specialist
Univ. of Illinois at Urbana-Champaign
SOCIOLOGICAL ETHICS

Craig A. Straub, Restoration Ecologist
Straub Ecological Services, Cincinnati, OH
HARDIN, GARRETT

Sytse Strijbos, Prof., Philosophy
North West Univ., Potchefstroom, South Africa
Vrije Univ., Amsterdam, The Netherlands
SYSTEMS THINKING

David Strong, Prof. of Philosophy and Environmental Studies
Rocky Mountain Coll., Billings, Montana
ENVIRONMENTAL ETHICS

John P. Sullins III, Asst. Prof., Philosophy
Sonoma State Univ., California
ARTIFICIAL INTELLIGENCE

Shanna H. Swan, Research Prof., Family and Community Medicine
Univ. of Missouri, Columbia
DES (DIETHYLSTILBESTROL) CHILDREN

James Tansey, James Martin Lecturer in Science and Technology Studies
James Martin Inst. for Science and Civilization, Univ. of Oxford, UK
GEORGIA BASIN FUTURES PROJECT

Herman T. Tavani, Prof., Chair, Philosophy
Rivier Coll., Nashua NH
TRANSACTION GENERATED INFORMATION AND DATA MINING

Bob Pepperman Taylor, Prof., Political Science; Dean, Honors Coll.
Univ. of Vermont, Burlington
ENVIRONMENTALISM; THOREAU, HENRY DAVID

Eugene TeSelle, Emeritus Prof. of Church History and Theology
Vanderbilt Divinity Sch., Nashville, TN
AUGUSTINE

J. Brian Thomas, Lecturer, Engineering
Baylor Univ., Waco, TX
RADIO

J. Mark Thomas, Sociology
Madison Area Technical Coll., Madison, WI
TILLICH, PAUL

Paul B. Thompson, W.K. Kellogg Prof. of Agricultural, Food and Community Ethics
Michigan State Univ., East Lansing
AGRICULTURAL ETHICS

Ole Thyssen, Prof., Management, Philosophy and Politics
Copenhagen Business Sch., Denmark
LUHMANN, NIKLAS

Pieter Tijmes, Prof. Emeritus
Univ. of Twente, Enschede, The Netherlands
EUTHANASIA IN THE NETHERLANDS

Jeffrey R. Timm, Prof., Religion
Wheaton Coll., Norton, MA
BHUTAN

Nancy Tuana, Prof., Philosophy; Dir., Rock Ethics Inst.
Pennsylvania State Univ., State College
FEMINIST ETHICS; FEMINIST PERSPECTIVES

Uday T. Turaga, Independent Scholar
INDIAN PERSPECTIVES

Christine Turner, Research Geologist
U.S. Geological Survey, CO
NAT. GEOLOGICAL SURVEYS

Stephen H. Unger, Prof., Computer Science

Columbia Univ., New York, NY
BAY AREA RAPID TRANSIT CASE

Nicanor Ursua, Prof., Philosophy
Univ. of the Basque Country, San Sebastián-Donostia, Spain
PUBLIC UNDERSTANDING OF SCIENCE

Ibo van de Poel, Asst. Prof., Philosophy
Delft Univ. of Technology, The Netherlands
ENGINEERING DESIGN ETHICS

Jeroen van den Hoven, Philosophy
Delft Univ. of Technology, The Netherlands
PRIVACY

Ton van der Valk, Asst. Prof., Philosophy
Free University, Amsterdam, The Netherlands
JONAS, HANS

Kenneth W. Van Treuren, Assoc. Prof., Engineering
Baylor Univ., Waco, TX
AIRPLANES; AVIATION REGULATORY AGENCIES

Willem H. Vanderburg, Dir., Ctr. for Technology and Social Development
Univ. of Toronto, Ont., Canada
PREVENTIVE ENGINEERING

Jason Venetoulis, Senior Research Assoc.
Redefining Progress, Oakland, CA
ECOLOGICAL FOOTPRINT

Peter-Paul Verbeek, Asst. Prof., Philosophy
Twente Univ., The Netherlands
JASPERS, KARL

Jason M. Vogel, Ph.D. candidate, Environmental Studies
Univ. of Colorado, Boulder
ORGANIC FOODS

Eugene Volokh, Prof. of Law
UCLA Sch. of Law
SLIPPERY SLOPE ARGUMENTS

Rudi Volti, Prof., Sociology
Pitzer Coll., Claremont, CA
AUTOMOBILES

Raban Graf von Westphalen, Prof.
Technische Fachhochschule Berlin, Germany
TECHNOLOGY ASSESSMENT IN GERMANY AND OTHER EUROPEAN COUNTRIES

Leonard J. Waks, Prof. of Educational Leadership and Policy Studies
Temple Univ., Philadelphia, PA
EDUCATION

Henry H. Walbesser, Provost and Chief Operating Officer
Henry Cogswell Coll., Everett, WA
FULLER, R. BUCKMINSTER; VON NEUMANN, JOHN

Jonathan Wallace, Publisher, *The Ethical Spectacle*
www.spectacle.org
ADVERTISING, MARKETING, AND PUBLIC RELATIONS; BRIDGES; DAMS; DEMATERIALIZATION AND IMMATERIALIZATION; MODERNIZATION; MOVIES; NORMAL ACCIDENTS; ORGANIZATION FOR ECONOMIC COOPERATION AND DEVELOPMENT; RAILROADS; SHIPS; WORLD BANK; WORLD HEALTH ORGANIZATION

William A. Wallace, Prof., Decision Sciences and Engineering Systems
Rensselaer Polytechnic Inst., Troy, NY
MODELS AND MODELING; THOMISM

Brent Waters, Dir., Stead Ctr. for Ethics and Values
Assoc. Prof. of Christian Social Ethics
Garrett-Evangelical Theological Seminary, Evanston, Illinois
CHRISTIAN PERSPECTIVES: CONTEMPORARY ASSESSMENTS OF SCIENCE; CHRISTIAN PERSPECTIVES: CONTEMPORARY ASSESSMENTS OF TECHNOLOGY; CHRISTIAN PERSPECTIVES: HISTORICAL TRADITIONS

Charles Weijer, Assoc. Prof., Bioethics, Medicine and Surgery; Adjunct Prof., Philosophy
Dalhousie Univ., Halifax, NS, Canada
CLINICAL TRIALS

Richard Weikart, Prof., History
California State Univ., Stanislaus
EUGENICS; SOCIAL DARWINISM

Vivian Weil, Prof. of Ethics; Dir., Ctr. for the Study of Ethics in the Professions
Illinois Inst. of Technology, Chicago
CONTRACTS

Daryl J. Wennemann, Assoc. Prof., Philosophy
Fontbonne Univ., St. Louis
FREEDOM; KANT, IMMANUEL

Laura Westra, Prof. Emeritus, Philosophy, Univ. of Windsor
Adjunct Prof. of Social Science, York Univ., Toronto Ontario, Canada
ECOLOGICAL INTEGRITY

Caroline Whitbeck, Elmer G. Beamer-Hubert H. Schneider Prof. in Ethics
Case Western Reserve Univ., Cleveland, OH
Dir., Online Ethics Ctr. for Engineering & Science, http://onlineethics.org
TRUST

Mark R. Wicclair, Prof., Philosophy and Adjunct Prof., Community Medicine, West Virginia Univ., Morgantown, WV

Adjunct Prof., Medicine and Faculty, Ctr. for Bioethics and Health Law, Univ. of Pittsburgh, Pittsburgh, PA
ABORTION

Per Wikman, MSc., Philosophy
Royal Inst. of Technology, Stockholm, Sweden
INTERNATIONAL COMMISSION ON RADIOLOGICAL PROTECTION; RADIATION

William H. Wilcox, Assoc. Prof., Philosophy
Utah State Univ., Logan
PETS

David R. Williams, Nat. Space Science Data Ctr.
NASA Goddard Space Flight Ctr., Greenbelt, MD
APOLLO PROGRAM

Glenn R. Willis, The Divinity Sch.
Vanderbilt Univ., Nashville, TN
PSYCHOLOGY; TOLSTOY, LEO

Margo Wilson, Psychology
McMaster Univ., Hamilton, Ont., Canada
CRIME

James J. Winebrake, Prof., Science Technology and Society/Public Policy
Rochester Inst. of Technology, NY
POWER SYSTEMS

Gregor A. Wolbring, Biochemist, Adjunct Asst. Professorhips, Ethics, Disability Studies, Community Health,
Univ. of Calgary; Univ. of Alberta, Edmonton, Canada
DISABILITY

Mark J. P. Wolf, Assoc. Prof., Communication
Concordia Univ. Wisconsin, Mequon
SPECIAL EFFECTS; VIDEO GAMES

Harvey Wolfe, Prof., Industrial Engineering
Univ. of Pittsburgh, PA
CICERO'S CREED

Marsha C. Woodbury, Lecturer, Computer Science
Univ. of Illinois, Urbana-Champaign
SECURITY

Edward J. Woodhouse, Assoc. Prof., Science and Technology Studies
Rensselaer Polytechnic Inst., Troy, NY
AFFLUENCE; CITIZENSHIP; CIVIL SOCIETY; CLASS; CONSUMERISM

Russell J. Woodruff, Asst. Prof., Philosophy
St. Bonaventure Univ., NY
NEUTRALITY IN SCIENCE AND TECHNOLOGY; PURE AND APPLIED

Bilge Yesil, Instructor, Communication Studies
New York Univ., New York
BIOMETRICS

Yin Dengxiang, Prof., Philosophy of Science and Technology; Dir., Research Ctr. for Science, Technology and Society
Chinese Academy of Social Sciences, Beijing, China
CHINESE PERSPECTIVES

Stuart J. Youngner, Susan E. Watson Prof., Bioethics
Case Western Reserve Univ., Cleveland
BRAIN DEATH

Jean Pascal Zanders, Dir.
BioWeapons Prevention Project, Geneva, Switzerland
BIOLOGICAL WEAPONS

Yinghuan Zhao, Assoc. Prof., Research Ctr. for Philosophy of Science and Technology
Northeastern Univ., Shenyang, P. R. China
THREE-GORGES DAM

Sheldon Zink, Dir., Program for Transplant Policy and Ethics
Univ. of Program for Transplant Policy and Ethics, Ctr. for Bioethics, Univ. of Pennsylvania, Philadelphia
ORGAN TRANSPLANTS

Hub Zwart, Prof., Philosophy, Science Studies; Dir., Ctr. for Society & Genomics
Radboud Univ. of Nijmegen, The Netherlands
DUTCH PERSPECTIVES

TOPICAL OUTLINE

The following classification of articles provides an analytic summary of the Encyclopedia contents. It is intended to assist the user, whether researcher or browser, in appreciating the scope of coverage and in locating articles broadly related to a given theme. Nevertheless, because the field of science, technology, and ethics is an emerging interdisciplinary effort, it is not as easily parsed as traditional scholarly disciplines. One alternative classification scheme, for instance, would list under each specialized introduction all related articles—an analysis that would, of course, have required extensive repetitions. In the present instance, despite the fact that topic headings are not always mutually exclusive, entries are not listed more than once. It is assumed that any user will supplement use of the topical outline with the list of related articles that follows each article, and with the index.

Introductory Essays

Eight synthetic essays to introduce the encyclopedia as a whole.

- Ethics and Technology: A Program for Future Research
- The Ethics of Evidence: A Call for Synthesis
- Nanoscience, Nanotechnology, and Ethics: Promise and Peril
- Recognizing the Structural Features of New Technologies
- Research Ethics, Engineering Ethics, and Science and Technology Studies
- Technologies of Humility: Citizen Participation in Governing Science
- Toward an Ethics of Science and Technology as Knowledge
- Values in Technical Design

Introductions and Overviews

SPECIALIZED INTRODUCTIONS

Specialized introductions provide entrances into issues of science, technology, and ethics from the perspectives of recognized fields of study, many in applied ethics, relevant to science, technology, and ethics.

- Agricultural Ethics
- Applied Ethics
- Archeological Ethics
- Architectural Ethics
- Bioengineering Ethics
- Bioethics
- Biotech Ethics
- Business Ethics
- Communication Ethics
- Computer Ethics
- Design Ethics
- Development Ethics
- Engineering Design Ethics
- Environmental Ethics
- Evolutionary Ethics
- Genethics
- Information Ethics
- Journalism Ethics
- Medical Ethics
- Military Ethics
- Nanoethics
- Neuroethics
- Nuclear Ethics: Industrial Perspectives
- Nuclear Ethics: Weapons Perspectives
- Planning Ethics
- Rhetoric of Science and Technology
- Science Fiction
- Science Policy
- Science, Technology, and Law
- Science, Technology, and Literature
- Science, Technology, and Society Studies
- Scientific Ethics
- Sociological Ethics

OVERVIEWS

Overview articles introduce specific themes that either are dealt with by more than one entry or have multiple branches out into other entries.

- Economics: Overview
- Engineering Ethics: Overview
- Ethics: Overview
- Management: Overview
- Misconduct in Science: Overview
- Psychology: Overview
- Research Ethics: Overview
- Responsibility: Overview
- Risk and Safety: Overview
- Science: Overview
- Semiotics: Overview
- Social Institutions: Overview
- Technology: Overview

Concepts, Case Studies, Issues, and Persons

KEY CONCEPTS

Analyses of special concepts that often play significant roles in discussions of science, technology, and ethics. Such concepts can often be distinguished in those that arise from an ethical-political or a scientific-technological base.

CONCEPTS, ETHICAL AND POLITICAL

- Aggression
- Alienation
- Altruism
- Animal Rights
- Animal Welfare
- Autonomy
- Change and Development
- Cicero's Creed
- Citizenship
- Civil Society
- Class
- Codes of Ethics
- Common Heritage of Mankind Principle
- Community

Conflict of Interest
Cultural Lag
Death and Dying
Determinism
Dignity
Disability
Dominance
Double Effect and Dual Use
Environmental Justice
Environmental Rights
Equality
Ethical Pluralism
Ethics Assessment Rubrics
Fact/Value Dichotomy
Free Will
Freedom
Future Generations
Genocide
Human Rights
Incrementalism
Informed Consent
Intellectual Property
Just War
Justice
Limits
Participation
Plagiarism
Playing God
Poverty
Precautionary Principle
Privacy
Profession and Professionalism
Property
Research Integrity
Responsibility: Anglo-American Perspectives
Responsibility: German Perspectives
Right to Die
Right to Life
Rights and Reproduction
Security
Sensitivity Analyses
Skepticism
Slippery Slope Arguments
Stakeholders
Tradeoffs
Trust
Two Cultures
Utopia and Dystopia
Values and Valuing
Whistleblowing
Work

CONCEPTS, SCIENTIFIC OR TECHNOLOGICAL

Animal Experimentation
Artificial Intelligence
Artificiality
Autonomous Technology
Biodiversity
Biophilia
Complexity and Chaos
Cyberspace
Dematerialization and Immaterialization
Ecological Footprint
Ecological Integrity
Ecology
Efficiency
Energy
Engineering Method
Ethology
Experimentation
Expertise
Hardware and Software
Health and Disease
Human Nature
Information
Invention
Models and Modeling
Multiple-Use Management
Nature
Networks
Normal Accidents
Participatory Design
Peer Review
Pollution
Population
Praxiology
Prediction
Preventive Engineering
Prisoner's Dilemma
Progress
Progress
Pseudoscience
Pure and Applied
Regulation and Regulatory Agencies
Reliability
Responsible Conduct of Research
Risk Assessment
Risk Perception
Safety Factors
Scientism
Selfish Genes
Social Engineering
Social Indicators
Stress
Sustainability and Sustainable Development
Systems Thinking
Technical Functions
Technicism
Technoethics
Technological Fix
Technological Innovation
Technology Transfer
Technoscience
Therapy and Enhancement
Tools and Machines
Turing Tests
Uncertainty
Unintended Consequences
Waste
Wilderness

CASE STUDIES

Presentations of a broad sample of influential, historical cases in science, technology, and ethics discussions.

Abortion
Accountability in Research
Accounting
Acid Mine Drainage
Aging and Regenerative Medicine
Apollo Program
Artificial Morality
Asilomar Conference
Atomic Bomb
Atoms for Peace Program
Baruch Plan
Bay Area Rapid Transit Case
Bhopal Case
Bhutan
Biosecurity
Birth Control
Brain Death
Brent Spar
Building Destruction and Collapse
Cancer
Chernobyl
DC-10 Case
DDT
DES (Diethylstilbestrol) Children
Digital Libraries
Ford Pinto Case
Georgia Basin Futures Project
Green Revolution
Hiroshima and Nagasaki
HIV/AIDS
Limited Nuclear Test Ban Treaty
Lysenko Case
Military-Industrial Complex
Misconduct in Science: Biomedical Science Cases
Misconduct in Science: Physical Sciences Cases
Misconduct in Science: Social Science Cases
Missile Defense Systems

Mondragón Cooperative Corporation
Montreal Protocol
Nuclear Non-Proliferation Treaty
Nuclear Waste
Pets
Rain Forest
Robot Toys
Robots and Robotics
Science Shops
Singapore
Sokal Affair
Space Shuttles *Challenger* and *Columbia* Accidents
Space Telescopes
Special Effects
Technocomics
Three Gorges Dam
Three-Mile Island
Tuskegee Experiment
Video Games

ISSUES

Issue articles overlap with Key Concepts and Case Studies, by reporting on topics that have received often contentious treatment because of substantial disagreements about their scientific, technological, or ethical aspects. Some issues selected to provide historical perspective. Like concepts, issues often arise from different bases, this time either historical and social or scientific-technological bases. With regard to issues, it is also possible to distinguish a third base in phenomena.

ISSUES, HISTORICAL AND SOCIAL

Activist Science Education
Advertising, Marketing, and Public Relations
Affluence
Automation
Biodiversity Commercialization
Consensus Conferences
Conservation and Preservation
Constructive Technology Assessment
Consumerism
Contracts
Corruption
Death Penalty
Deforestation and Desertification
Digital Divide
Direct Democracy
Entertainment
Entrepreneurism
Galenic Medicine
Green Ideology
Hacker Ethics
Hazards
Homosexuality Debate
Humanitarian Science and Technology
Humanization and Dehumanization
Industrial Revolution
Information Overload
Information Society
IQ Debate
Luddites and Luddism
Material Culture
Materialism
Modernization
Monitoring and Surveillance
Nazi Medicine
Political Risk Assessment
Posthumanism
Product Safety and Liability
Public Understanding of Science
Race
Risk and Emotion
Scientific Revolution
Secularization
Sex and Gender
Simplicity/Simple Living
Social Darwinism
Technicization
Technocracy
Terrorism
Terrorism and Science
Theodicy
Tourism
Urbanization
Vegetarianism
Violence

ISSUES, SCIENTIFIC OR TECHNOLOGICAL

Alternative Energy
Alternative Technology
Animal Tools
Arsenic
Building Codes
Choice Behavior
Clinical Trials
Complementary and Alternative Medicine
Computer Viruses/Infections
Consciousness
Decision Support Systems
Diagnostic and Statistical Manual of Mental Disorders
Ecological Restoration
Embryonic Stem Cells
Emergent Infectious Diseases
Emotional Intelligence
Environmental Impact Assessment
Eugenics
Euthanasia
Evolution-Creationism Debate
Exposure Limits
Fetal Research
Forensic Science
Free Software
Gene Therapy
Genetic Counseling
Genetic Research and Technology
Genetically Modified Foods
Genetics and Behavior
Heavy Metals
Human Cloning
Human Subjects Research
In Vitro Fertilization and Genetic Screening
Nature versus Nurture
Neutrality in Science and Technology
Nutrition and Science
Organ Transplants
Organic Foods
Persistent Vegetative State
Regulatory Toxicology
Sex Selection
Social Contract for Science
Soft Systems Methodology
Space Exploration
Transaction Generated Information and Data Mining
Virtual Reality

ISSUES, PHENOMENA

Air
Body
Cosmetics
Crime
Cyberculture
Distance
Earth
Emotion
Fire
Gaia
Global Climate Change
Holocaust
Hormesis
Hypertext
Instrumentation
Interdisciplinarity
Life
Money
Oil
Place
Plastics

Popular Culture
Prosthetics
Qualitative Research
Risk Society
Safety Engineering: Historical Emergence
Safety Engineering: Practices
Science and Engineering Indicators
Space
Speed
Water

PERSONS AND FIGURES

Biographical entries on persons and a few mythical figures of continuing contemporary relevance to science, technology, and ethics. Includes philosophers, scientists, engineers, and writers. The division into time periods provides a quite appreciation of proportions. Although there are a few exceptions, in the third period the general rule has been to avoid entries on living persons.

PERSONS AND FIGURES, PREMODERN

Aristotle and Aristotelianism
Augustine
Plato
Prometheus
Thomas Aquinas

PERSONS AND FIGURES, MODERN PERIOD TO WORLD WAR I

Bacon, Francis
Boyle, Robert
Butler, Samuel
Comte, Auguste
Darwin, Charles
Descartes, René
Durkheim, Émile
Faust
Frankenstein
Galilei, Galileo
Galton, Francis
Hegel, Georg Wilhelm Friedrich
Hobbes, Thomas
Hume, David
Jefferson, Thomas
Juana Inés de la Cruz
Kant, Immanuel
Kierkegaard, Søren
Leibniz, G. W.
Locke, John
Machiavelli, Niccolò
Marx, Karl
Mill, John Stuart
More, Thomas
Morris, William
Newton, Isaac
Nietzsche, Friedrich W.
Nightingale, Florence
Pareto, Vilfredo
Pascal, Blaise
Peirce, Charles Sanders
Rousseau, Jean-Jacques
Shelley, Mary Wollstonecraft
Smith, Adam
Spencer, Herbert
Steinmetz, Charles
Thoreau, Henry David
Tocqueville, Alexis de
Tolstoy, Leo
Treat, Mary
Verne, Jules
Weber, Max

PERSONS AND FIGURES, POST WORLD WAR I

Anders, Günther
Anscombe, G. E. M.
Arendt, Hannah
Asimov, Isaac
Bell, Daniel
Benjamin, Walter
Berdyaev, Nikolai
Berlin, Isaiah
Bernal, J. D.
Bethe, Hans
Blackett, Patrick
Brave New World
Brecht, Bertolt
Bush, Vannevar
Carson, Rachel
Dessauer, Friedrich
Dewey, John
Dubos, René
Edison, Thomas Alva
Einstein, Albert
Ellul, Jacques
Ford, Henry
Foucault, Michel
Freud, Sigmund
Fuller, R. Buckminster
Gandhi, Mohandas
Gates, Bill
Girard, René
Grant, George
Habermas, Jürgen
Haldane, J. B. S.
Hardin, Garrett
Heidegger, Martin
Husserl, Edmund
Huxley, Aldous
Illich, Ivan
Jaspers, Karl
Jonas, Hans
Jung, Carl Gustav
Jünger, Ernst
Kuhn, Thomas
Lasswell, Harold D.
Leopold, Aldo
Levi, Primo
Levinas, Emmanuel
Lewis, C. S.
Luhmann, Niklas
Lyotard, Jean-François
Marcuse, Herbert
McClintock, Barbara
McLuhan, Marshall
Mead, Margaret
Merton, Robert
Mumford, Lewis
Murdoch, Iris
Oppenheimer, Frank
Oppenheimer, J. Robert
Ortega y Gasset, José
Parsons, Talcott
Pauling, Linus
Polanyi, Karl
Polanyi, Michael
Popper, Karl
Ramsey, Paul
Rand, Ayn
Rawls, John
Rotblat, Joseph
Russell, Bertrand
Sakharov, Andrei
Sanger, Margaret
Schweitzer, Albert
Simon, Herbert A.
Simon, Julian
Skinner, B. F.
Spengler, Oswald
Strauss, Leo
Taylor, Frederick W.
Teller, Edward
Tillich, Paul
Tolkien, J. R. R.
Turing, Alan
Veblen, Thorstein
von Neumann, John
von Wright, Georg Henrik
Watson, James
Weil, Simone
Wells, H. G.
Wiener, Norbert
Wittgenstein, Ludwig
Zamyatin, Yevgeny Ivanovich

Sciences, Technologies, Institutions, and Agencies

PARTICULAR SCIENCES AND TECHNOLOGIES

A selective examination of ethics issues related to specific sciences and technologies.

- Acupuncture
- Airplanes
- Androids
- Antibiotics
- Assisted Reproduction Technology
- Astronomy
- Automobiles
- Biological Weapons
- Biometrics
- Biostatistics
- Bridges
- Chemical Weapons
- Chemistry
- Cosmology
- Cybernetics
- Cyborgs
- Dams
- Decision Theory
- Drugs
- Earth Systems Engineering and Management
- Earthquake Engineering
- Economics and Ethics
- Epidemiology
- Ergonomics
- Food Science and Technology
- Game Theory
- Geographic Information Systems
- Global Positioning System
- Internet
- Management: Models
- Meta-analysis
- Mining
- Movies
- Music
- Operations Research
- Polygraph
- Power Systems
- Probability: Basic Concepts of Mathematical Probability
- Probability: History Interpretation and Application
- Psychology: Humanistic Approaches
- Psychopharmacology
- Radiation
- Radio
- Railroads
- Rational Choice Theory
- Roads and Highways
- Semiotics: Language and Culture
- Semiotics: Nature and Machine
- Ships
- Sociobiology
- Statistics: Basic Concepts of Classical Inference
- Statistics: History, Interpretation, and Application
- Telephone
- Television
- Vaccines and Vaccination
- Weapons of Mass Destruction

SOCIAL INSTITUTIONS

How a few leading social institutions are influenced by and influence science and technology.

- Education
- Family
- International Relations
- Museums of Science and Technology
- Police
- Sports
- Zoos

ORGANIZATIONS AND AGENCIES

Includes government agencies and NGOs at the national and international levels, emphasizing how these institutions are related especially to the creation and management of science and technology.

- American Association for the Advancement of Science
- Association for Computing Machinery
- Aviation Regulatory Agencies
- Bioethics Centers
- Bioethics Committees and Commissions
- Communications Regulatory Agencies
- Engineers for Social Responsibility
- Enquete Commissions
- Environmental Protection Agency
- Environmental Regulation
- Federal Aviation Administration
- Federation of American Scientists
- Food and Drug Adminstration
- Food and Drug Agencies
- Human Genome Organization
- Institute of Electrical and Electronics Engineers
- Institute of Professional Engineers New Zealand
- Institutional Biosafety Committees
- Institutional Review Boards
- International Commission on Radiological Protection
- International Council for Science
- National Academies
- National Aeronautics and Space Administration
- National Geological Surveys
- National Institutes of Health
- National Parks
- National Science Foundation
- National Science Foundation Second Merit Criterion
- Nongovernmental Organizations
- Nuclear Regulatory Commission
- Office of Research Integrity
- Office of Technology Assessment
- Organization for Economic Cooperation and Development
- President's Council on Bioethics
- Professional Engineering Organizations
- Public Policy Centers
- Pugwash Conferences
- Royal Commissions
- Royal Society
- Sierra Club
- Union of Concerned Scientists
- United Nations Educational, Scientific, and Cultural Organization
- United Nations Environmental Program
- World Bank
- World Commission on the Ethics of Scientific Knowledge and Technology
- World Health Organization
- World Trade Organization

Philosophical, Religious, and Related Perspectives

PHILOSOPHICAL PERSPECTIVES

Articles highlighting how different philosophical traditions, schools of thought, or theories relate to science, technology, and ethics.

- Axiology
- Consequentialism
- Critical Social Theory
- Deontology
- Discourse Ethics
- Ethics of Care
- Existentialism
- Feminist Ethics
- Feminist Perspectives
- Logical Empiricism
- Natural Law
- Phenomenology
- Postmodernism
- Pragmatism
- Rights Theory

Risk Ethics
Social Contract Theory
Thomism
Virtue Ethics

RELIGIOUS PERSPECTIVES

Articles exploring the views of selected religious traditions about issues related to science, technology, and ethics issues.

Anglo-Catholic Cultural Criticism
Buddhist Perspectives
Christian Perspectives: Contemporary Assessments of Science
Christian Perspectives: Contemporary Assessments of Technology
Christian Perspectives: Historical Traditions
Confucian Perspectives
Daoist Perspectives
Hindu Perspectives
Islamic Perspectives
Jewish Perspectives
Shinto Perspectives

POLITICAL AND ECONOMIC PERSPECTIVES

Articles on how science and technology might be assessed by different political or economic movements, ideologies, or theories.

Agrarianism
Atlantis, Old and New
Authoritarianism
Capitalism
Colonialism and Postcolonialism
Communism
Communitarianism
Conservatism
Democracy
Ecological Economics
Enlightenment Social Theory
Environmental Economics
Environmentalism
Fascism
Globalism and Globalization
Governance of Science
Humanism
Liberalism
Libertarianism
Market Theory
Marxism
Nationalism
Neoliberalism
Open Society
Political Economy
Social Construction of Scientific Knowledge
Social Construction of Technology
Social Theory of Science and Technology
Socialism
Totalitarianism

CULTURAL AND LINGUISTIC PERSPECTIVES

Perspectives based in different cultural, linguistic, and/or national traditions on the ethical dimensions of science and technology.

African Perspectives: Computers in South Africa
African Perspectives: HIV/AIDS in Africa
Australian and New Zealand Perspectives
Canadian Perspectives
Central European Perspectives
Chinese Perspectives
Chinese Perspectives: Engineering Ethics
Chinese Perspectives: Research Ethics
Dutch Perspectives
Engineering Ethics: Europe
Euthanasia in the Netherlands
French Perspectives
German Perspectives
Ibero-American Perspectives
Indian Perspectives
Indigenous Peoples' Perspectives
Italian Perspectives
Japanese Perspectives
Population Policy in China
Russian Perspectives
Scandinavian and Nordic Perspectives
Technology Assessment in Germany and Other European Countries

ABORTION

In the United States and in some other countries, abortion is one of the most divisive moral and political issues. Developments in abortion techniques, such as medical abortion and intact dilation and evacuation ("partial-birth" abortion), have prompted responses in law, policy, and ethical scholarship, which in turn have influenced abortion technology and provision. The emphasis here will be on the definition of abortion, abortion techniques, ethical issues, and law and public policy, focusing primarily on the United States.

Abortion Definition and Techniques

Abortion is the termination of a pregnancy and the expulsion of pregnancy tissue, including embryo/fetus, placenta, and membranes. In principle, pregnancy begins with conception (in vivo fertilization of an ovum by a spermatozoon). The earliest that a pregnancy can be clinically recognized, however, is when a serum pregnancy test becomes positive (approximately one week to ten days after ovulation). In a spontaneous abortion, also called a miscarriage, the termination of pregnancy is not intentional. In popular usage, as in the present case, the term *abortion* refers solely to an intentionally induced termination of a clinically recognized pregnancy.

References to abortion techniques describing both medication and surgical measures appear in the records of ancient civilizations, including those of China, Greece, and Rome. The modern surgical technique, which was developed in the nineteenth century, involves dilation (opening the cervix) and sharp curettage (removing the uterine contents with a sharp instrument). This procedure had the potential to be safer and more effective than the pre-nineteenth-century alternative that involved the administration of various compounds presumed to have abortifacient properties. When performed with unsterile instruments or by unskilled practitioners, however, surgery involved high risks of infection and uterine damage. In the twentieth century, the introduction of vacuum aspiration curettage improved the safety of surgical abortion. This method for dilation and curettage (D&C) achieved widespread use in the United States in the 1960s and became the dominant method for first trimester abortion. Improvements in effective local anesthesia made it possible to perform the procedure in a medical clinic or office. By 2000, only 5 percent of all abortions were performed in hospitals. These developments in medical technology presented a serious challenge to the claim that abortion poses a significant risk to the health and safety of women.

In the United States, "medicinal" or "pharmacological" abortion using pharmacologic means, which is referred to as "medical abortion," became available as a safe and effective alternative to surgery for early abortions in the mid-1990s. The drugs used for medical abortion are methotrexate or mifepristone, followed by a dose of prostaglandin. Mifepristone (Mifeprex, or RU-486), developed in France in the 1980s, attained U.S. Food and Drug Administration (FDA) approval for this indication in September 2000, by which time more than 600,000 women in Europe had used the drug. In the United States, more than 200,000 women took mifepristone for this purpose during its first three years on the market. Medical abortion involves three doctor's office visits over a two-week span. Patients can expect to bleed and spot for nine to sixteen days. Approximately 1 percent of women will require a D&C for excessive bleeding. Approximately 2 to 5 percent of women will require a D&C because tissue is incompletely expelled from the uterus. In the first few years of mifepristone's use, approximately 2 to 8 percent of eligible women in the United States chose this medical regimen over surgical abortion. European experience with the drug suggests that this may increase gradually with time. Pro- and antiabortion forces alike had predicted that the introduction of mifepristone would increase the availability of abortion. In its first six months on the market, however, mifepristone was administered primarily by physicians who already provided abortions, suggesting that the drug does not dramatically increase abortion access.

Beyond the first trimester, medical abortion methods induce labor-like uterine contractions that result in the expulsion of the fetus and other pregnancy tissues from the uterus. The most common procedure for second trimester surgical abortion is dilation and evacuation (D&E). Surgery is the safer second trimester technique until about eighteen weeks of gestation. A variant of D&E, intact D&E (called by some "partial-birth" abortion or "dilation and extraction" [D&X]) differs with respect to how the fetus is removed from the uterus. In a D&E, the fetal parts are separated before removal. Intact D&E involves a procedure to decompress the fetal skull so that the fetus can be removed in its entirety. Intact D&E accounted for 0.17 percent of all abortions in 2000.

Ethical Issues

Under what conditions, if any, is having and performing an abortion ethically permissible? This deceptively simple question is the subject of often heated controversy and has generated a wide range of answers—from "never" or "only to prevent a pregnant woman from dying," at one extreme, to "whenever a women decides to have one," at the other. In between are a variety of views that distinguish between acceptable and unacceptable reasons and/or draw a line at a particular gestational stage, such as onset of brain activity or viability. That there are several points of contention adds to the complexity of the debate.

One point of contention concerns the moral status of human fetuses (the term *fetus* is used here as a generic term referring to a developing organism between conception and birth). Proponents of the view that abortion generally is ethically unacceptable often claim that human fetuses have full moral standing (i.e., moral status equivalent to that of adult humans) and a right to life beginning at conception. For example, John T. Noonan Jr. (1970) claims that possession of a "human genetic code" is a sufficient condition of full moral standing. Those who deny that abortion generally is unethical often reject the claim that fetuses have full moral standing and a right to life. For example, Mary Anne Warren (1973) argues that to be genetically human is neither a necessary nor a sufficient condition of full moral standing. Only *persons* are said to have full moral standing, and Warren identifies five criteria for personhood: consciousness, reasoning, communication, self-motivated activity, and self-concepts. With the possible exception of consciousness, human fetuses prior to birth fail to satisfy these criteria. As critics have observed, however, human infants also fail to satisfy Warren's criteria. Michael Tooley (1972) proposes a more demanding set of criteria for personhood, which requires complex cognitive capacities, including self-consciousness. Clearly, neither human fetuses nor infants satisfy these criteria, and Tooley presents arguments in support of both abortion and infanticide.

Opponents of abortion sometimes attempt to avoid the controversial issue of whether a living organism with a human genetic code *is* a person by claiming that human fetuses are *potential* persons. This strategy, however, simply shifts the debate's focus from whether fetuses *are* persons to whether *potential* persons have full moral standing and a right to life.

Don Marquis (1989) adopts an antiabortion strategy that does not rely on potentiality. He argues that killing human fetuses is seriously immoral for the same reason that it is seriously immoral to kill adult humans: Killing deprives them of *their futures* (i.e., the experiences, activities, projects, and the like that would have comprised their future personal lives if they were not killed). This line of argument, however, may be vulnerable to the objection that, unlike adult humans, fetuses do not have a *present* as an *experiencing subject*, and therefore fetuses cannot have a future as the same experiencing subject.

Some commentators have claimed that even if human fetuses do not have full moral standing, there still might be grounds for ethical constraints on abortion. For example, Jane English (1975) claims that insofar as fetuses in later stages of development are "person-like nonpersons" (e.g., they resemble babies), failing to ascribe any moral standing to them might undermine our moral commitments. Daniel Callahan (1970) claims that a human fetus has partial moral standing because it is a developing human life.

Other commentators have made the opposite claim, arguing that even if it is assumed that human fetuses have full moral standing and a right to life, it does not follow that abortion generally is unethical. Judith Jarvis Thomson (1971) presents an argument along these lines, claiming that the right to life does not entitle a fetus to use a pregnant woman's body without her permission.

People who believe that abortion is morally acceptable are unlikely to favor restrictive abortion laws and policies. A belief that abortion is unethical, however, is not necessarily linked to support for restrictive abortion laws and policies. For example, a person might believe that such restrictions would result in more harm than good or that the government should not take sides when there are persistent disagreements about fundamental values.

Law and Policy in the United States

U.S. law and public policy regarding abortion are constantly evolving. Because it concerns the practice of medicine, abortion legislation is often enacted on the state level. Through the early nineteenth century, in most states abortion was legal prior to quickening (the time at which the woman senses fetal movement), which occurs at approximately twenty weeks of gestation. Later in that century, however, most states enacted legislation that provided criminal penalties for women and/or practitioners for abortions performed at any time in gestation. Many physicians and the American Medical Association supported this transformation in the law, arguing that abortion endangers women and is immoral.

This approach continued through the early 1960s, when all fifty states had restrictive abortion laws, and many states permitted abortions only to protect the woman's life. During the late 1960s and early 1970s, however, more than ten states liberalized their statutes by permitting abortion not only to prevent a woman's death but also in cases of medical necessity, fetal defect, rape, or incest. During this period, several states passed laws that placed even fewer limits on early abortions. For example, New York allowed abortion on demand up to twenty-four weeks' gestation.

Two 1973 U.S. Supreme Court cases, *Roe* v. *Wade* and *Doe* v. *Bolton*, substantially curtailed the legal authority of states to prohibit abortion. These opinions declared that abortion decisions are protected by a Constitutional right to privacy, the same right that in a 1965 case, *Griswold* v. *Connecticut*, the Court applied to decisions about birth control. In *Roe*, the Court adopted a trimester analysis, ruling as follows: (a) Prior to third trimester viability (the point at which the fetus could survive outside the uterus), a woman's right to an abortion always trumps the state's interest in fetal life. It is only after viability that states *may* prohibit abortion, but such laws *must* include exceptions for cases in which an abortion is necessary to *protect* a woman's life or health. (b) During the first trimester, states *may not* impose any restrictions on abortion. (c) From the beginning of the second trimester, states *may* impose restrictions that are designed to protect maternal health.

In the years following *Roe*, the Supreme Court reviewed a number of state abortion statutes that set limits on legal abortion and struck down the provisions it considered to be incompatible with that decision. For example, the Court invalidated laws that required extensive physician disclosure and counseling procedures, spousal consent, limitations on the facilities where abortions could be performed, and limitations on the specific abortion technique used. Beginning in the late 1980s, however, the Supreme Court became more tolerant of abortion restrictions. State regulations that were upheld include bans on abortions in publicly funded facilities, bans on abortions by publicly paid physicians, and mandatory viability testing prior to abortions. In *Rust* v. *Sullivan*, the Court approved the "gag rule" policy issued by the U.S. Department of Health and Human Services regarding abortion counseling in family planning clinics funded by Title X of the Public Health Services Act. This 1988 policy prohibited clinic employees from providing counseling about, or referring patients to, abortion services. President Bill Clinton suspended the "gag rule" in 1993, and regulations instituted in 2000 revoked the rule.

In the 1992 case *Planned Parenthood* v. *Casey*, a Supreme Court sharply divided five to four affirmed *Roe* v. *Wade*. However, neither *Roe*'s trimester framework nor its reliance on privacy commanded a Court majority. A joint opinion by Justices Sandra Day O'Connor, Anthony Kennedy, and David Souter substituted an "undue burden" test for the trimester framework of *Roe* and cited liberty as the basis of a constitutionally protected right to abortion. As in *Roe*, *Casey* holds that after viability states may prohibit abortion except when it is necessary to protect the life or health of pregnant women. Prior to viability, state restrictions may not present a "substantial obstacle" to women who seek an abortion. In *Casey*, the Court reviewed five Pennsylvania requirements: informed consent, a twenty-four-hour waiting period, parental consent for minors (with a judicial bypass procedure), spousal notification, and a reporting requirement. Only spousal notification was determined to be an undue burden by a majority of justices. As the Court noted, advances in neonatal care subsequent to *Roe* pushed the onset of fetal viability earlier into gestation. With additional technological advances in neonatology and obstetrics, this trend will continue.

Federal legislative and domestic policy activity related to abortion has addressed access to abortion, antiabortion violence, and the late-term abortion procedure sometimes called partial-birth abortion. The Hyde Amendment, first enacted in 1976, withholds abortion coverage for beneficiaries of Medicaid and other federal programs, with the exception of procedures performed because pregnancy threatens a woman's life or resulted from rape or incest. Since enactment, this amendment has been maintained as a rider to federal appropriations bills. The Supreme Court upheld this law in 1980 in *Harris* v. *McRae*. Nevertheless, a number of states use their public funds to pay for abortions for poor women.

The U.S. Congress responded to escalating antiabortion force and violence, such as blockades, arsons, bombings, and murders, with the Freedom of Access to Clinic Entrances Act (FACE) of 1994. This law makes it a federal crime to use force or threat of force to impede abortion providers and/or potential patients, or to intentionally damage abortion facilities. Many states passed similar laws. Subsequent federal legislation has focused on outlawing the intact D&E or partial-birth abortion procedure. More than half of all states have passed laws banning the procedure. In 2000 the Supreme Court reviewed and rejected Nebraska's law for several reasons: The statute was vaguely worded and could have been interpreted to include a ban on standard abortion procedures; the law had no exception for the protection of a woman's health; and it posed an "undue burden" to women seeking abortions. The U.S. Congress has worked since the mid-1990s to pass similar legislation. President Clinton twice vetoed bills passed by Congress, but President George W. Bush signed the Partial Birth Abortion Ban Act of 2003. This legislation does not include an exception for a woman's health. Additional federal legislative and policy efforts include legislation and federal regulations to give fetuses legal status. Two such laws have been enacted: the Born-Alive Infants Protection Act of 2002 and the Unborn Victims of Violence Act of 2004. A federal regulation extends insurance coverage under the State Children's Health Insurance Program of the Centers for Medicare and Medicaid Services to fetuses.

The abortion controversy and resulting policies have had a far-reaching impact on medical care and research in the United States. Abortion opponents have supported restrictions on research using embryos and fetal tissue. These restrictions have affected care for patients with infertility and have hampered efforts to develop stem cell or fetal tissue transplant treatments for diseases such as spinal cord injury, juvenile diabetes, and Parkinson's.

Internationally, U.S. policy has focused on not subsidizing overseas abortion. The Helms Amendment passed in 1973 prohibited and continues to prohibit the use of U.S. foreign aid money to fund abortions abroad. Presidents Ronald Reagan, George H. W. Bush, and George W. Bush built upon this policy by instituting what opponents call the "global gag rule." Under this rule, international family planning organizations that receive U.S. aid cannot perform abortions (even if funded by other sources), refer patients to abortion services, offer abortion counseling, or advocate for proabortion policies in their country.

Law and Policy Outside the United States

A comprehensive 1999 United Nations report on abortion policies around the world revealed significant differences between abortion law and policy in more and less well developed regions (United Nations, World Abortion Policies 1999, available from http://www.un.org/esa/population/publications/abt/fabt.htm). Out of a total of 48 more developed countries, abortion on request was legally permitted in 31 (65%). By contrast, out of a total of 145 less developed countries, abortion on request was permitted in only 21 (17%). A similar disparity can be seen between more and less developed countries in relation to the legality of abortion in other situations: economic or social difficulty (75% vs. 19%); fetal impairment (81% vs. 26%); rape or incest (81% vs. 30%); to protect mental health (85% vs. 54%); and to protect physical health (88% vs. 55%). The only reason for which there was no significant difference is to prevent the death of the pregnant woman (96% vs. 99%). In many developing countries, maternal morbidity and mortality from unsafe abortions is a significant contributor to overall maternal morbidity and mortality. Policies associated with a decline in abortion morbidity and mortality include the following: increased access to safe abortions, increased contraception, increased abortion provider experience and/or the use of modern medications, and increased availability of life-saving care for women with abortion complications (World Health Organization 1997).

Legal restrictions against abortion in Europe were eliminated or reduced in the last half of the twentieth century, due in part to a concern about mortality and morbidity associated with unsafe illegal abortions. In 1999, out of forty-two European countries, abortion on request was legal in twenty-eight (United Nations World Abortion Policies 1999, available from http://www.un.org/esa/population/publications/abt/fabt.htm). However, most of these countries imposed a limit on gestational age, typically twelve weeks. A majority of the countries that limited abortion on request to a certain gestational age permitted later abortions under specified conditions, such as to protect the physical and/or mental health of the pregnant woman. Malta was the only European country in which abortion was illegal. In four countries (Ireland, Andorra, San Marino, and Monaco) abortion was legal only to prevent the death of the pregnant woman.

In 1999, out of forty-six Asian countries, abortion on request was legal in sixteen (United Nations 1999). All forty-six countries permitted abortion to prevent the death of the pregnant woman; and this was the only permitted reason in seventeen. China's abortion policy was among the most liberal, permitting abortion on request. The primacy of population control concerns in China trump political and ethical arguments against abortion.

In both the United States and internationally, it is to be expected that abortion will continue to provide a paradigm example of the interaction of technology, ethics, law, and public policy.

MARK R. WICCLAIR
GABRIELLA GOSMAN

SEE ALSO *Birth Control; Fetal Research; Medical Ethics; Right to Life.*

BIBLIOGRAPHY

Callahan, Daniel. (1970). *Abortion: Law, Choice, and Morality*. New York: Simon and Schuster. A defense of the view that human fetuses have only *partial* moral standing.

English, Jane. (1975). "Abortion and the Concept of a Person." *Canadian Journal of Philosophy* 5(2): 233–243. A defense of a moderate position on abortion.

Finer, Lawrence B., and Stanley K. Henshaw. (2003). "Abortion Incidence and Services in the United States in 2000." *Perspectives on Sexual and Reproductive Health* 35(1): 6–15.

Marquis, Don. (1989). "Why Abortion Is Immoral." *Journal of Philosophy* 86(4): 183–202. An anti-abortion argument based on the claim that killing fetuses deprives them of their futures.

Mohr, James C. (1978). *Abortion in America: The Origins and Evolution of National Policy, 1800–1900*. Oxford: Oxford University Press. The most comprehensive historical reference on abortion in the United States.

Noonan, John T., Jr. (1970). "An Almost Absolute Value in History." In *The Morality of Abortion: Legal and Historical Perspectives*, ed. John T. Noonan Jr. Cambridge, MA: Harvard University Press. A critique of abortion based on the claim that fetuses have full moral standing starting from the moment of conception.

Paul, Maureen; Steve E. Lichtenberg; Lynn Borgatta; et al. (1999). *A Clinician's Guide to Medical and Surgical Abortion*. New York: Churchill Livingstone. Medical textbook covering all issues relevant to the performance of abortions.

Rubin, Eva R., ed. (1994). *The Abortion Rights Controversy: A Documentary History*. Westport, CT: Greenwood. An annotated collection of excerpts from many of the primary sources relevant to the history of abortion.

Thomson, Judith Jarvis. (1971). "A Defense of Abortion." *Philosophy and Public Affairs* 1(1): 47–66. A defense of abortion based on the claim that a right to life does not entitle one person to use another person's body.

Tooley, Michael. (1972). "Abortion and Infanticide." *Philosophy and Public Affairs* 2(1): 37–65. A defense of abortion and infanticide.

Warren, Mary Anne. (1973). "On the Moral and Legal Status of Abortion." *Monist* 57(1): 43–61. A defense of abortion based on the claim that fetuses lack significant moral standing.

United Nations, Department of Economic and Social Affairs, Population Division. (1999). *World Abortion Policies*. Publication ST/ESA/SER.A/178. New York: United Nations, Department of Economic and Social Affairs, Population Division.

World Health Organization, Department of Reproductive Health and Research. (1997). *Unsafe Abortion: Global and Regional Estimates of Incidence of a Morality Due to Unsafe Abortion, With a Listing of Available Country Data*. 3rd ed. Publication WHO/RHT/MSM/97.16. Geneva: World Health Organization, Department of Reproductive Health and Research.

INTERNET RESOURCES

"Mifeprex: The Early Option Pill." Danco Laboratories. Available from http://www.earlyoptionpill.com. Mifeprex manufacturer's site, containing information about the procedure of medical abortion and the most up to date information about use of the drug in the United States.

"Mifepristone Information." U.S. Food and Drug Administration. Center for Drug Evaluation and Research.

Available from http://www.fda.gov/cder/drug/infopage/mifepristone/. FDA documents relevant to the drug's approval and use appear on this site.

ACCIDENTS

SEE *Normal Accidents; Unintended Consequences.*

ACCOUNTABILITY IN RESEARCH

• • •

Accountability is a central issue in ethics and politics, one closely related to other concepts such as responsibility, integrity, and authenticity. In ethics, individuals are held accountable for their actions. In a democracy, citizens of the state ultimately hold politicians accountable. In both instances, however, there are questions about how such accountability is to be practiced, and in reference to what standards. Similar questions arise with regard to accountability in scientific and engineering research. *Accountability in research* or *research accountability* as general terms may thus refer to a range of concerns and practices related to the philosophies, policies, systems, procedures, and standards for analyzing and promoting ethical conduct in research.

In the worlds of business, finance, and government, accountability also implies a more specific reference to accounting in the sense of bookkeeping methods that involve maintaining the financial records of monetary transactions and the regular preparation of statements concerning the assets, liabilities, and operating results of some activity. To assure the accuracy of such financial accounts, one well-developed dimension of the accounting profession is auditing. Audits review and examine accounts to determine whether they are reasonably accurate and presented in an understandable manner. The attempt to adapt such methods from the fields of business and finance to those of scientific research is called data auditing (DA), and constitutes a special effort to assure accountability in research.

Historical Background

In the early history of modern natural science the methodological requirement that experimental results be reported in such a way that they could be reproduced by others, and the practice of accepting into the body of scientific knowledge only those results that had been reproduced, effectively made auditing a standard part of research practice. Even so, William Broad and Nicholas Wade (1982) argue that what is now called *creative accounting* was sometimes practiced in scientific research. For example, there is evidence that physicist Isaac Newton (1642–1727) made experimental data fit his theories, and that chemist John Dalton (1766–1844) cleaned up his data to obtain whole numbers for ratios on chemical reactions. Biologist Louis Pasteur (1822–1895) is alleged to have announced an anthrax vaccine before completing his experiments, and Cyril Burt (1883–1971) may have fabricated intelligent quotient (IQ) test results. Even Nobel Prize winner Robert Millikan (1868–1953) may have fudged his data. Other examples include the fabrication of animal test data by Industrial Biotech Corporation for the Food and Drug Administration (FDA), and the conduct of unethical high-risk experiments on psychiatric patients. In some cases serious adverse events, including deaths, were not reported (Shamoo and Resnik 2003).

A number of surveys indicate that students and researchers suspect that questionable conduct in research is widespread, accounting for 0 percent to 50 percent of all research. The actual percentage of questionable research practices is probably much lower—in the single digits (Shamoo and Resnik 2003, LaFollette 2000). The scientific community was initially slow to call for reforms in dealing with scientific misconduct. In response to media coverage of some serious lapses, commissions were formed and congressional hearings held to discuss accountability in research. Then-senator Albert Gore chaired hearings to examine concerns and urge reforms (LaFollette 1994).

The modern explosion in attainment of knowledge has resulted in profound changes in the social character of science. In 2002, nearly 3 million individuals worked as researchers in the United States alone, with about 1 million holding post-graduate degrees and controlling a budget of more than $250 billion. Science in has become *mass science* in the pattern of mass production and mass culture. Traditional means of apprenticeship and social pressure are not effective ways to uphold high standards for scientific knowledge in the early-twenty-first century. More explicit approaches must be developed.

It was in this context that the term *data audit* first began to be used (for complete references on this topic see Loeb and Shamoo 1989). Following a 1988 conference on the subject, the inaugural issue of the journal *Accountability in Research* announced its intention to "serve as a catalyst for the development of specific procedures and standards for acquiring, analyzing, and auditing" (Shamoo 1989, p. i).

DA Theory and Practice

The concept of auditing has a long history, including efforts in early Egyptian, Greek, and Roman civilizations by governments to develop ways to expose cheating by accountants. Modern accounting and auditing procedures have their immediate origins in response to the enormous expansion in business enterprises since the nineteenth century.

There are several kinds of auditors. External auditors are independent auditors who work in public accounting firms for identified clients. However because third parties use the information in the financial statements generated by these auditors, external auditors can be said to work also in the interests of society. Internal auditors work as employees within organizations; public and private corporations and government agencies have internal auditors. Government auditors are employed by government agencies to audit outside entities and individuals. An example of a government auditor is the Internal Revenue Service (IRS).

In addition, there are various types of auditing. Financial auditing examines the accuracy of an entity's financial statements. The resultant report can be used inside or outside the entity. Operational or performance auditing examines performance, management, or value-added operations, including cost-economy, efficiency, and effectiveness. Compliance auditing examines whether an organization is in compliance with specific rules and regulations, whether issued internally or imposed on the entity by a third party. Attestation engagements are given to public accounting firms for the purpose of examining the representations of an entity other than those that are traditionally included in financial statements, for example, those regarding systems of internal accounting control or investment performance statistics (Loeb and Shamoo 1989).

Auditing is an independent activity that reviews accounting, but is separate and apart from it. Its methods rely on logic, not accounting principles, to evaluate concrete issues. DA, as proposed by Adil Shamoo in the late 1980s, is modeled after financial auditing. The purpose of DA is to check the accuracy of derived research data by comparing it to the original raw data. This method can be used either randomly for a small number of data determined by a statistical method or when the data are suspect. Several publications have outlined the method since its initial introduction (Shamoo and Annau 1987).

The Future of DA

Accountability in research requires reviewing institutional policies (for example, those of universities) and examining the attitudes and behavior of researchers. Institutional policies are key because they dictate the tone and culture of tolerance in research conduct and are major influences on how and why researchers work on particular issues (Shamoo and Dunigan 2000).

Society demands accountability from researchers. This is especially true when the results of particular research affect individuals and communities. In the early-twenty-first century, accountability in research is an important and expanding area of interest to both professionals and the general public.

ADIL E. SHAMOO

SEE ALSO *Misconduct in Science; Peer Review; Research Ethics.*

BIBLIOGRAPHY

Broad, William, and Nicholas Wade. (1993 [1982]). *Betrayers of the Truth: Fraud and Deceit in the Halls of Science*. New York: Simon and Schuster.

LaFollette, Marcel C. (1994). "The Pathology of Research Fraud: The History and Politics of the U.S. Experience." *Journal of Internal Medicine* 235: 129–135.

LaFollette, Marcel C. (2000). "The Evolution of the 'Scientific Misconduct' Issue: An Historical Overview." *Experimental Biology and Medicine* 224: 211–215.

Loeb, Stephen, and Adil E. Shamoo. (1989). "Data Audit: Its Place in Auditing." *Accountability in Research* 1: 23–32.

Shamoo, Adil E, ed. (1989). *Accountability in Research* 1(1). Inaugural issue of the journal; contains complete references.

Shamoo, Adil E., and Zoltan Annau. (1987). Correspondence (Letter to the Editor), June 18, 1987. *Nature* 327(6123): 550.

Shamoo, Adil E., and Cheryl D. Dunigan. (2000). "Ethics in Research." *Experimental Biology and Medicine* 224: 205–210.

Shamoo, Adil E., and David B. Resnik. (2003). *Responsible Conduct of Research*. Oxford and New York: Oxford University Press.

ACCOUNTING

• • •

Accounting comprises techniques to record, verify, report, plan, and analyze governmental, commercial, or personal financial transactions. As such, accounting is related to science, technology and ethics in two ways. First, particularly since the early-twentieth century, accounting is understood as a technology rationalized by scientific theories and practiced by professionals requiring ethical guidance (Whitley 1986). Second, both large-scale technological projects and big budget science are increasingly subjected to accounting-based evaluations. Accordingly, both scientific data and procedures, and fiscal accounts of technoscience, are being audited to verify the ethical behaviors of engineers and scientists.

Varieties of Bookkeeping

The emergence of modern accounting in the mid-nineteenth century is symbolically marked by a fire and an avalanche. In 1834 the British House of Commons was razed in a blaze fed by a pyre of wooden tallies, which had been used by the Exchequer since the thirteenth century. Around the same time, the social and natural worlds began to be blanketed by an "avalanche of printed numbers" that only gathered in force over the subsequent two centuries (Hacking 1982, p. 279). Accounting becomes recognizably modern when numbers are exclusively used to ascribe economic value not only to things and events but also to people.

Vernacular Accounting

On the far shore of modern accounting lie the myriad vernacular ways of counting wealth and recording transactions. It is perhaps anachronistic to speak of bookkeeping before books or of accounting before counting. Indeed recent archeological evidence of an *archaic bookkeeping* around 3000 B.C.E. suggests that it is reckoning that gave rise to alphabetic script and homogenous number (Schmandt-Besserat 1992). If bookkeeping refers to the ways of reckoning and recording trade and commerce, its history is overwhelmingly about how the unlettered and the innumerate kept count.

For most of human history, fingers, pebbles, abaci, and counting boards were used for calculating, while transactions were recorded as knots on strings, notches on sticks, inscriptions on tablets, pipe-rolls, and parchment. The diversity of vernacular accounting is exemplified by the tally stick on which peasants and princes, from China to Europe, recorded and verified commercial, tax, and even credit transactions through notches, incisions, and cuts that varied by region, by village, and even within villages and among products (Menninger 1992). Such heterogeneous measures of things and products were bound to place and purpose, and usually rooted in the human form, of which the *foot* remains a dim reminder. A bushel in Cracow was different in girth and height from that in Gdansk; Alpine peasants recorded the sale of sheep with different inscriptions than those for the sale of cheese because sheep were qualitatively distinct from cheese (Kula 1986).

Double Entry Bookkeeping

The homogenization of vernacular counting and recording is related to the emergence of double-entry bookkeeping (DEB), which is also the framework for modern accounting. It was popularized by Luca Pacioli (1445–1514), called the *father of accounting*, largely because his was the first book printed and published on the subject of DEB. For Pacioli, a friar and contemporary of Leonardo Da Vinci, the visual order and quantitative balance of DEB served to justify commerce by showing every transaction as the result of an equal and, therefore, fair exchange. Accounting in the DEB form lent credence to business as an ethical enterprise at a time when commercialism was viewed with some suspicion (Aho 1985).

The technique of DEB involves recording every transaction twice: once each as a debit and a credit in two distinct accounts. For example, purchasing a computer for cash would require recording the increase in the value of an asset by debiting the computer account through a debit and recognizing the reduction in cash by crediting the cash account. It is the sum of such equal and opposing effects of a transaction that produces the famed doubled balance of DEB. At a technical level, the genesis and diffusion of DEB presupposed the replacement of Roman numbers by Hindu-Arabic numerals, the loss of the symbolic power of numbers, and perhaps crucially, the emergence of the text that had to be seen to be read. For example, 0 had to be rethought as a mere numeral instead of evoking the horror of nothingness (Rotman 1987), and the inherently temporal events of giving and taking became reduced to a spatially arranged textual record of equal exchange (Clanchy 1999).

The popular belief that DEB stimulated profit seeking and, therefore, capitalism was first suggested by the sociologist Werner Sombart (1863–1941). However, though the distinction between profit and capital is necessary to regularly calculate the rate of profit, the distinction itself is not necessary for DEB, which emerged no later than the fourteenth century in the time of little, if any, capitalist activity (De Roover 1974). It was disseminated, though spottily, throughout Europe only after the Italian Renaissance. Indeed DEB was not instrumental to the pursuit of profit well into the eighteenth century (Yamey 1964). For the Fuggers of the fifteenth century, the Dutch East India Company of the seventeenth, and numerous factories of the eighteenth, DEB played no part in the quest for profitable trade and commerce. The bilateral, columnar ordering of debits and credits in tables of interconnected parts that balanced is therefore better understood as an instrument of visualization and legitimization rather than one of economic rationalism (Crosby 1997).

Modern Accounting

Only after the early-nineteenth century did accounting became a technique to calculate the economic productivity of all factors of production (Hoskins and Macve 2000). Modern accounting is not mere record keeping of materials used, wages paid, and profits made as it was in the eighteenth century and before. Rather accounting achieves its contemporary status as the *sine qua non* of economic rationalism, which implies the coordination and control of humans, materials, and machines, only when human actions are rendered into a calculable form and people therefore measured as economic resources.

The modern technique for a *system of accountability* was forged in the classrooms of the U.S. Military Academy at West Point. Since 1817, each cadet has been subjected to a regimen of written and graded examinations (Hoskins and Macve 1988). When employed as managers in such companies as the Springfield Armory and the Pennsylvania Railroad during the 1830s, some graduates of West Point used the technique of student grading as a template to measure and calculate human performance in general. For example, the quantity of widgets producible after eight hours of effort, under normal conditions, can be measured and then used as a benchmark to calculate the productivity of a particular worker. Modern accounting thus induces double vision: On one hand, it reduces human action to a countable economic resource, while on the other, it fosters the belief that such accountability is ethical.

This writing of objects, events, *and* persons in financial terms soon spread to both the emerging governmental bureaucracies and large scale corporations during the latter half of the nineteenth century (Hoskins and Macve 1986). Modern accounting is thus coeval with large-scale corporations—the visible hand in modern economies—that manage resources across space and time to harness productivity, reduce costs, and increase profits. Economic rationalism, rooted in management by the numbers, hence came to fruition only by the late-nineteenth century; it is not coincidental that the word capitalism flowers when the invisible hand of markets begins to wither (Braudel 1982).

By the mid-twentieth century, modern accounting as performance evaluation had become a pervasive, if almost unseen, technique for controlling human action and holding people accountable (Hoskins and Macve 2000). Through accounting, governments, schools, hospitals, and even countries, as well as bureaucrats, students, doctors, and elected officials were increasingly described as economic objects and stimulated to behave as economic resources (Miller 1992). One measure of the current ubiquity of accounting is the extent to which the behavior of scientists and engineers are motivated, monitored and controlled through accounting-based techniques. This has been pronounced since the postwar years when both engineering projects and scientific research began to absorb ever increasing sums of money from both public and private sources. Public projects such as highways and dams are routinely subjected to cost-benefit analysis; time and cost overruns and penalties are measured and charged against budgeted figures; laboratory notebooks are maintained and used as evidence of employee input and performance in a manner similar to time cards in factories. despite its many failings, such as using budgets to evaluate inherently unpredictable long term projects, accounting-based techniques seem necessary to manage large institutions, whether governments, corporations, or technoscentific practices.

Accounting Science, Profession, and Ethics

Since the 1970s, accounting techniques have also gained much in the way of scientific respectability. Economic, sociological, and psychological theories of human behavior have transformed the study of accounting into a social science based on mimicking the methods of the natural sciences: the use of mathematical models, experimental tests, and statistical results. However because people are not atoms, the predictive and explanatory power of accounting theories is necessarily far below that of physics. Moreover, because it is based on the fact-value distinction scientific accounting research cannot prescribe changes in accounting techniques to better modify behaviors and decisions. In the breach between low explanatory power and the even lower normative force of scientific accounting, the mass production and *ritual verification* of accounting numbers continues unabated (Power 1999).

The perceived objectivity of numbers is a fundamental vehicle by which accounting techniques spread as a bureaucratic method to manage people in a manner consistent with liberal government (Porter 1995). However, by now, most students of accounting agree that all valuation techniques—whether of things or persons—are the result of conventional rules and not laws of nature. Accordingly the claim to objectivity in accounting should be understood less as an unbiased reflection of natural processes and instead as the adherence to conventional standards of measurement and calculation.

During the twentieth century, the accounting profession used the notion of objectivity as a lever to promote the idea of accountants as disinterested professionals. As part of this attempt at professionalizing accounting practice the newly formed American Institute of Accountants established a code of professional conduct in 1917. Throughout the twentieth century, the code was to become both wider in scope and more specific in detail. For example, what started as a list of eight rules in 1917 had expanded to list of six principles and a series of five rules, each with a host of related "interpretations" (Preston et al. 1995), the elaboration of the code of conduct has been accompanied by a shift in the social status of the accounting professional: Increasingly the profession has disavowed its professionalism and embraced its function as service provider (Zeff 2003). Perhaps the strongest evidence of this shift away from professionalism is that accountants are no longer barred from advertising their services as they were until the 1970s.

In this context corporate bankruptcies and managerial misconduct can be understood. The much-publicized saga of the Enron Corporation reveals that greed and envy continue, with predictable frequency, to prompt fraud and duplicity by corporate chieftains, government officials, and accountants. The response, exemplified by the recently passed Sarbanes-Oxley Act (2002), has been equally predictable: Additional accounting techniques are instituted to engineer valued behaviors, including cost-benefit analyses, risk assessments, audits, and budgets. In the blind spot of this spiraling cycle, the foundational questions of whether it is ethical to reduce human action to a quantity, whether engineered behavior is akin to ethical action, and whether human failings can be eradicated by technical devices remain.

SAJAY SAMUEL

SEE ALSO *Economics and Ethics; Science, Technology, and Society Studies.*

BIBLIOGRAPHY

Aho, James. (1985). "Rhetoric and the Invention of Double Entry Bookkeeping." *Rhetorica* 3: 21–43. An insightful rhetorical analysis of DEB emphasizing its religious underpinnings.

Bailey, Charles N., James R. Hasselback, and Julia N. Karcher. (2001). "Research Misconduct in Accounting Literature: A Survey of the Most Prolific Researchers' Actions and Beliefs." *Abacus* 37: 26.

Braudel, Fernand. (1982). *The Wheels of Commerce*. New York: Harper & Row.

Clanchy, Michael T. (1993). *From Memory to Written Record*. London: Basil Blackwell. A seminal work on the transition in Europe from oral to literate culture, focusing on legal documents, wills and accounting records.

Cohen, Patricia. (1999). *A Calculating People*. New York: Routlege.

Crosby, Alfred. (1997). *The Measure of Reality*. Cambridge, UK: Cambridge University Press. A cultural history of the rise of quantification in different fields ranging from art and music to architecture and commerce. Convincingly argues the textual and therefore visual form of DEB.

De Roover, Raymond. (1974). *Business, Banking and Economic Thought*. Chicago: University of Chicago Press.

Hacking, Ian. (1982). "Biopower and the Avalanche of Printed Numbers." *Humanities in Society* 5: 279–295. Documents the rise of quantification in the newly emerging social sciences of the early nineteenth century.

Hoskins, Keith, and Richard Macve. (1986). "Accounting and the Examination: A Genealogy of Disciplinary Power." *Accounting, Organizations and Society* 11: 105–136.

Hoskins, Keith, and Richard Macve. (1988). "The Genesis of Accountability: The West Point Connection." *Accounting, Organizations and Society* 13: 37–73.

Hoskins, Keith, and Richard Macve. (2000). "Knowing More as Knowing Less? Alternative Histories of Cost and Management Accounting in the U.S. and the U.K." *Accounting Historians Journal* 27: 91–149. A deft and clear summary of the decade-long argument among accounting scholars on the origins of modern accounting. Places the original research presented in the papers listed below within a neatly formulated 'dialogue' with contesting interpretations.

Kula, Witold. (1986). *Measures and Men*, trans. Richard Szreter. Princeton. NJ: Princeton University Press. An indispensable reference for the social history of modern measurement.

Menninger, Karl. (1992). *Number Words and Number Symbols*. New York: Dover Publications. A delightful book on the world history of numbers that includes excellent pages on the prehistory of modern accounting.

Miller, Peter. (1992). "Accounting and Objectivity: The Invention of Calculating Selves and Calculable Spaces." *Annals of Scholarship* 9: 61–86.

Porter, Theodore M. (1995). *Trust in Numbers: The Pursuit of Objectivity in Science and Public Life*. Princeton, NJ: Princeton University Press. A historian's documentation of the uses and abuses of 'objectivity' in the social sciences.

Power, Michael. (1999). *The Audit Society: Rituals of Verification*. Oxford: Oxford University Press. A theoretically informed analyses of the contemporary proliferation of auditing in numerous arenas of life.

Preston, Alistair; David Cooper; Paul Scarbourgh; and Robert Chilton. (1995). "Changes in the Code of Ethics of the U.S. Accounting Profession, 1917 and 1988: The Continual Quest for Legitimacy." *Accounting, Organizations and Society* 20: 507–546.

Rotman, Brian. (1987). *Signifying Nothing: The Semiotics of Zero*. Stanford, CA: Stanford University Press.

Schmandt-Besserat, Denise. (1992). *Before Writing*. Austin: University of Texas Press. Original argument from an anthropologist of the Near East on 'accounting' as a precursor of alphabetic script and numbers.

Whitley, Richard. (1986). "The Transformation of Business Finance into Financial Economics: The Roles of Academic Expansion and Changes in U.S. Capital Markets." *Accounting, Organizations and Society* 11: 171–192.

Yamey, Basil S. (1964). "Accounting and the Rise of Capitalism: Some Further Notes on a Thesis by Sombart." *Journal of Accounting Research* 11: 117–136. Seminal paper arguing against the so-called Sombart thesis about the relation between DEB accounting and capitalism.

Zeff, Stephen. (2003). "How the U.S. Accounting Profession Got Where It Is Today: Part II." *Accounting Horizons* 17: 267–285.

ACID MINE DRAINAGE

Acid mine drainage (AMD), along with acid rock drainage (ARD), is a problem of water quality that is common to rivers and lakes that receive water draining from mine sites. Although not usually viewed as a first-tier environmental problem, AMD is a critical water-quality issue around the world, affecting nations from the Far East to Europe and the Americas. In the United States it occurs in wide areas in the East as a result of coal mining. In the American West several hundred thousand abandoned hard rock mines have contaminated thousands of miles of streams and thousands of lakes. Sites, streams, and lakes that require attention number in the thousands, according to the Mineral Policy Center (1997), which estimates that the cleanup in the United States alone will cost more than $10 billion. Acid mine drainage also provides an object lesson in the complex relationships among engineering, communities, and ethics and values and in the evolving nature of environmental debates.

The Problem

Apart from questions of causation and remediation, the production of acid drainage is a complex process that involves chemistry, geology, and biology. Exposing sulfur-rich rocks to air and water causes sulfide minerals such as pyrite, galena, and sphalerite to oxidize. An example is provided by pyrite (FeS2), also known as fool's gold. Rainwater, snowmelt, and air break this iron sulfate mineral into its constituent parts: ferrous iron and sulfur. The sulfate ions react with the water to produce sulfuric acid, and the iron passes into the water column. By itself this chemical reaction is not energetic enough to produce much acid drainage, but the reaction increases exponentially in the presence of sulfur-oxidizing bacteria (genus thiobacilli), which cause a great expansion of the amount of acid drainage produced.

The pH of a solution is a measure of its acidity, based upon a logarithmic scale; going down the scale, each number represents a tenfold increase in the amount of acidity. Thus the difference between a pH of 7 and one of 3 is four orders of magnitude, or 10,000 times more acidic. The pH in AMD-affected streams can drop as low as 2 and 3 (lower than the pH of vinegar, and about the same as that of a car battery). Trout, for instance, die at pH values below 5.4. Therefore, contaminated mine water passing into streams and lakes can lower the pH of that water to the point where it stunts the development of, or kills, fish and invertebrates. In addition, the lower pH allows heavy metals to stay dissolved in the water column. Those metals can have a variety of effects on the streams: Zinc and copper kill aquatic life through their toxicity, and aluminum and iron settle on stream bottoms and disrupt the physical habitat of bottom-dwelling creatures, such as stone flies and caddis flies, that various aquatic species depend on for sustenance. These damaged waters also can have a negative impact on other species and the human communities living within the watershed.

Scientific, Technical, and Political Challenges

It is important to note that acid mine drainage is the human-caused analog of the natural processes of acid rock drainage. Acid rock drainage results from natural weathering processes, biological activity, and local or regional geology. Distinguishing between AMD and ARD—that is, separating natural background conditions from human-caused acid drainage—can be difficult and contentious, often uniting scientific, political, and ethical perspectives in a single debate.

Restoring streams, lakes, and landscapes damaged by acid mine drainage thus presents a challenge that is simultaneously scientific, technical, political, and philosophic. The issues in this area include the following:

- Scientific: How bad are the conditions? Are they natural or human-caused? What effects do they have on natural and human systems?

- Technical: Can a river, lake, or landscape be restored, and if so, at what cost and with what chance of long-term success?
- Political and philosophic: Who bears the cost of cleanup: the current landowner, the mineral industry, or society at large? Should restoration involve only areas damaged by human activity? Does it even make sense to speak of areas "damaged" by naturally occurring drainage? (Frodeman 2003)

Although they seldom are recognized, philosophical assumptions often guide people's thinking about how and whether to restore damaged landscapes. For instance, the attempt by scientists to distinguish between natural and human-caused acid drainage relates to the unspoken belief that the difference between the two provides a solid criterion for determining which areas should be cleaned up.

Another political and philosophic conundrum arises when parties to an AMD conflict feel that the very idea of "restoring" nature is misconceived, for what results is a dishonest attempt to pass off an artificial landscape as something natural (Elliott 1997). In contrast, scientists and technicians in the field of ecological restoration often fail to see anything wrong with intervening in compromised landscapes, viewing the development of restoration science as a positive sign of increasing technological prowess. Other participants in the AMD debate emphasize the political dimension of restoration, seeing it as offering a chance for a community to build a more harmonious relationship among its members as well as with nature (Gobster and Hull 2000).

Acid mine drainage is emblematic of a new phase in environmental thinking, where scientific, technical, political, and normative questions are tightly interlinked. Moreover, it also highlights the ongoing shift in environmental thinking from the preservation of pristine lands to the restoration of landscapes damaged by human actions.

ROBERT FRODEMAN

SEE ALSO *Ecological Restoration; Environmental Ethics; Mining.*

BIBLIOGRAPHY

Allenby, Brad. (1999). "Earth Systems Engineering: The Role of Industrial Ecology in an Engineered World." *Journal of Industrial Ecology* 2(3): 73–93. An important article arguing that environmental systems should be treated as large scale management tasks.

Elliott, Robert. (1997). *Faking Nature: The Ethics of Environmental Restoration*. Lanham, MD: Roman & Littlefield. Expansion of a classic essay on the philosophic aspects of environmental restoration.

Frodeman, Robert. (2003). *Geo-Logic: Breaking Ground between Philosophy and the Earth Sciences*. Albany: State University of New York Press.

Gobster, Paul H., and R. Bruce Hull, eds. (2000). *Restoring Nature: Perspectives from the Social Sciences and the Humanities*. Washington, DC: Island Press. Contains essays that explore a wide range of political and philosophic questions tied to environmental restoration.

Marshall, John, and Zeke Zanoni. (1996). *Mining the Hard Rock in the San Juans*. Silverton, CO: Simpler Way Press. Offers the miner's perspective on the joys and hazards of mining.

INTERNET RESOURCES

"AMD and Art: Artfully Transforming Environmental Liabilities into Community Assets." Available from http://www.amdandart.org/. Details an example of innovative combination of science, art, and community involvement for addressing AMD.

Mineral Policy Center, 1997. *Golden Dreams, Poisoned Streams*. Washington, DC: Mineral Policy Center. Available from http://www.mineralpolicy.org.

ACID ROCK DRAINAGE

SEE *Acid Mine Drainage*.

ACTIVIST SCIENCE EDUCATION

• • •

To what extent should science education in primary and secondary schools promote learning about science, technology, and ethics? At the primary and secondary school levels, ethical theory and issues of professional ethics would be inappropriate. At these levels one of the most common ethical issues has to do with the environment, which may thus serve as a case study here. But it must be recognized that environmental pollution and global climate change are controversial in ways not always easy to examine with primary and secondary school learners. Indeed many environmental education teachers also sometimes fail to critically assess their own beliefs.

Arguments for Activist Education

There are two basic arguments for activist science education to address environmental issues. One is a scientific and public consensus about its importance, another is the importance of democracy.

During the last half of the twentieth century, many environmental and social problems that drew public concern (climatic change, ecosystems degradation, demographic inequalities, migration, and terrorism, among others) expanded from local to global spheres. The situation had become so perturbing that science teachers often adopted the language of planetary crisis (Bybee 1991). During the United Nations Conference on Environment and Development, held in Rio de Janeiro in 1992, educators of every subject were asked to contribute to public awareness and understanding of the problems and challenges relating to the planet's future in order to enable the participation of citizens in well-grounded decision making. At the World Summit on Sustainable Development (2002), the consensus was that education is critical for promoting sustainable development, involving all levels of education in all countries.

Advances in science and technologies, because of their social impact, also call for a democratic debate on knowledge production and use. No members of early-twenty-first-century society can participate intelligently in the community without being familiar with how science and technology affect their daily life and future. Thus science education is considered a fundamental prerequisite for democracy and for ensuring sustainable development. Meaningful science education is more necessary than ever in order to develop and expand scientific and technological literacy in all cultures and sectors of society and thus improve public participation in decision making.

Activist Education Practices

But thirteen years after the Rio Conference, in spite of increasing international recognition of the fact that the challenges associated with environmental degradation and sustainable development have important implications for education, science education continues to demonstrate little concern for the present and future state of the world. There are numerous reasons for this insufficient response.

First, although the attainment of scientific and technological literacy (STL) is the main goal of curricular reforms in most countries, its meaning is still unclear. While some advocate a broadening of the knowledge base of the science curriculum to include greater consideration of interactions among science, technology, and society (STS), with more or less emphasis on environmental issues, others argue that educators must prepare students to compete effectively in the global marketplace (Hodson 2003).

The authors of *Science For All Americans*, for instance, direct attention toward scientific literacy for a more environmentally responsible democracy, stating that science can provide knowledge "to develop effective solutions to its global and local problems" and can foster "the kind of intelligent respect for nature that should inform decisions on the uses of technology" (AAAS 1989, p. 12). The "Standards for Technological Literacy" of the International Technology Education Association (ITEA) also establish requirements for technological literacy for all students; enforcing these standards, according to ITEA, will allow students to develop an understanding of the cultural, social, economic, political, and environmental effects of technology and of the role of society in the development and use of technology. By contrast, the National Research Council does not include such issues in the scientific literacy goals set out in its "National Science Standards."

Second, even when some environmental problems are incorporated in curricula, science education research has uncovered marked differences between the goals of curriculum designers and actual classroom practice. Such differences reveal that changes and reforms are difficult to put into practice and require significant changes in the values and beliefs of teachers.

Third, despite the enthusiasm that initially accompanied the appearance and promotion of environmental education (EE) with its varied proposals and projects, it continues to be a marginal and isolated subject in most education systems. Research frequently cites inadequate teacher preparation as a key obstacle to incorporating EE into school curricula. The situation is typical in a majority of countries (Poitier 1997, Gough 2002). In the United States Rosalyn McKeown-Ice surveyed 715 teacher education institutions and concluded that preservice teacher education programs seldom include EE. She also found that when such programs do include EE, the quality of it varies considerably. Thus EE teacher education is largely inadequate (McKeown-Ice 2000).

Fourth, most EE texts focus exclusively on local problems without addressing the global situation, display a reductionist approach, and ignore the strong connections between natural, environment and social, cultural, political, and economic factors (Tilbury 1995). These perspectives are beginning to change with such new approaches as Environmental Education For Sustainability (EEFS) and Science-Technology-Society-Environment (STSE) teaching materials.

Assessment

But, possibly, one of the main reasons for the inappropriate treatment of the global crisis resides in the perceptions of teachers and researchers. Analysis of articles published in thirty-two journals of research in science education (from 1992 to 2000) reveals that work on this problem is almost nonexistent. There are few contributions (4.5%) on particular problems and references to sustainability reach a scarce 10 percent. Extending this analysis to the contributions made at international congresses and conferences, and in handbooks on research in science education, the results are similar. A study involving science teachers from Spain, Portugal, and Latin America revealed substantially the same results and exposed the perceptions of science teachers as, in general, fragmentary and superficial, displaying a serious lack of knowledge and commitment. Only 5.3 percent of 848 science teachers raised sustainability issues (Edwards 2003). Critics, of course, argue that such attitudes are themselves more realistic than activist advocates would admit.

Despite the evidence of spreading environmental and social problems, the importance of EE has made little headway in the majority of schools. As activist science educator David Orr wrote in 1994, "We still educate the young ... as if there were no planetary emergency" (p. 27). But this reveals the problem at the heart of any activist science education program: how to get the majority involved. Education is needed to make it happen, but education itself is part of what needs to happen.

MÓNICA EDWARDS SCHACHTER

SEE ALSO *Education; Science, Technology, and Society Studies.*

BIBLIOGRAPHY

American Association for the Advancement of Science (AAAS). (1989). *Science for All Americans: A Project 2061 Report on Literacy Goals in Science, Mathematics, and Technology*. Washington, DC: Author. Also available from http://www.project2061.org/tools/sfaaol/sfaatoc.htm.

Bybee, Rodger W. (1991). "Planet Earth in Crisis: How Should Science Educators Respond?" *American Biology Teacher* 53(3): 146–153.

Edwards Schachter, Mónica E. (2003). *La atención a la situación del mundo en la educación científica* [The attention to the state of the world in science education]. Ph.D. diss.: Servei de Publicacions de la Universitat de València, CD edition. Also available from http://www.tdx.cesca.es/TESIS_UV/AVAILABLE/TDX-0211104-122215//edwards.pdf]

Gough, Annette. (2002). "Mutualism: A Different Agenda for Environmental and Science Education." *International Journal of Science Education* 24(11): 1201–1215.

Hodson, Derek. (2003). "Time for Action: Science Education for an Alternative Future." *International Journal of Science Education* 25(6): 645–670.

Knapp, Doug. (2000). "The Thessaloniki Declaration: A Wake-up Call for Environmental Education?" *Journal of Environmental Education* 31(3): 32–39.

McKeown-Ice, Rosalyn. (2000). "Environmental Education in the United States: A Survey of Preservice Teacher Education Programs." *Journal of Environmental Education* 32(1): 4–11.

National Research Council. (1996). "National Science Education Standards." Washington, DC: National Academy Press. Also available from http://www.nsta.org/standards.

Orr, David W. (1994). *Earth in Mind. On Education, Environment, and the Human Prospect*. Washington, DC: Island Press.

Poitier, Michel. (1997). *Le développement durable: stratégies de l'OCDE pour le XXIè siècle* [Sustainable development: strategies of the OCDE for the twenty-first century]. Paris: OECD.

Tilbury, Daniella. (1995). "Environmental Education for Sustainability: Defining the New Focus of Environmental Education in the 1990s." *Environmental Education Research* 1(2): 195–212.

INTERNET RESOURCE

International Technology Education Association (ITEA). "Listing of *STL* Content Standards." Available from http://www.iteawww.org/TAA/Publications/STL/STLListingPage.htm.

ACUPUNCTURE

All science is based on assumptions that define, a priori, the relative weaknesses and strengths of their practical application. As western science and technology have run up against limits to their comprehension and effectiveness, other approaches to both knowledge and practice have emerged to complement them. Nowhere is this more an issue than in medicine: Acupuncture has become a popular alternative to the drugs and surgeries offered by the biomedical sciences. In its fundamentally holistic approach acupuncture also presents an implicit ethical challenge to western technoscience to see the human patient in his or her entirety and within the context of the patient's life circumstances.

Acupuncture is the practice of inserting thin needles into the body to influence physiological functioning. It is an integral part of Chinese medicine, which also includes herbal medicine, massage, nutrition, and exercise. Chinese medicine began to take form during the Shang dynasty (1766–1050 B.C.E.), and an early form of acupuncture might have been practiced then, with the oldest needles having been made of sharpened stone (Gwei-Djen and Needham 1980, Unschuld 1985). There are bronze needles dating from the Chou dynasty (approximately 600 B.C.E.). By the Warring States period (475–221 B.C.E.) the classic acupuncture text, the *Huang Di Nei Jing Su Wen* [Yellow emperor's classic of internal medicine], had appeared.

Nature and Origins

The practice of acupuncture is thought to have started when shamans used needles to kill evil spirits that were thought to cause illnesses (Unschuld 1985). Over thousands of years the properties of specific points were discovered empirically, and those observations were tied in to traditional theories. What originally began as a superstitious ritual gradually became a flourishing medical field. The practice has grown further since its introduction to the West in the 1970s; there are more than fifty accredited schools of Chinese medicine in the United States, and practitioners are licensed independently in over forty states.

Chinese medicine was first introduced to Europe in the 1600s by Jesuit priests returning from the Orient. By the 1950s major schools of acupuncture were established in England and France. Acupuncture lost state support in China by the late 1800s and languished until a decree by Chairman Mao in 1958 that Chinese medicine should be revived according to the principles of dialectical materialism. Despite the "scientization" of Chinese in China, older traditions more grounded in a spiritual world view have survived both in Europe and in other parts of Asia that were not suppressed by the Chinese totalitarian regime.

Philosophical Orientation

According to the *Shen Nong Ben Cao*, one of China's oldest medical texts (second century C.E.), the highest aspect of healing involves helping patients fulfill their destiny so that they can live out the years allotted to them by heaven. The next highest aspect is the nourishment of people's inborn nature. Finally, the lowest class of healing is to treat specific physical illnesses. In its highest form, then, Chinese medicine focuses on individuals' health in the overall context of their lives. Health is manifested when one lives in harmony with the laws of nature and represents a profound integration of function on all levels: spiritual, mental, and physical. The presence of illness represents a denial and loss of the true self.

As a holistic practitioner an acupuncturist uses several diagnostic methods to determine the overall functional balance of a patient. Diagnoses occur largely within the perspective of the Chinese models of the universal poles of *yin* and *yang* and the five-element system, both of which provide qualitative standards for interpreting a range of physiological phenomena. From the *yin/yang* perspective practitioners consider observations in terms of internal/external, soft/hard, deficient/excess, and cold/hot, all of which point toward understanding the particular thermodynamic state of individuals and the unique manifestations of their illnesses or imbalances. For example, a practitioner might note that cold in nature tends to have a slowing and contracting influence. If the patient's pulse is slow and his or her muscles are tight the practitioner might deduce the presence of cold.

The five-element system (*wuxing*) was elaborated fully around 350 B.C.E. by Zou Yen (Kaptchuk 1983). The term *wuxing* denotes five dynamic movements—water, wood, fire, earth, and metal—that continually transform into each other as the seasons do. The language used by the early Chinese to describe their world was one of simple poetic images rich in allusions. Water is the element associated with winter because of its tendency to freeze and become focused in that season. Wood is associated with spring because it grows rapidly at that time of the year. Fire is associated with summer because of the increased heat during those months as the sun reaches its zenith. Earth is associated with late summer when the fields are full of the earth's bounty. Minerals are a natural expression of the metal element because they lie hidden beneath the ground; they symbolize the essential, precious, and rarefied aspects of life. Metal is associated with the fall, when what is of value must be harvested by the farmer's knife and everything else must be left to wither in the fields (Connelly 2002, Jarrett 1999).

Over the course of thousands of years laws were discovered and codified that described the functional dynamics of natural change. The five-element model is one example of these laws. Relating physiological functions to these qualitative standards, an acupuncturist is able to generate a diagnosis that is unique to each individual. The goal of treatment is to harmonize individuals both internally and within the context of their natural environment. The internal health of the

FIGURE 1

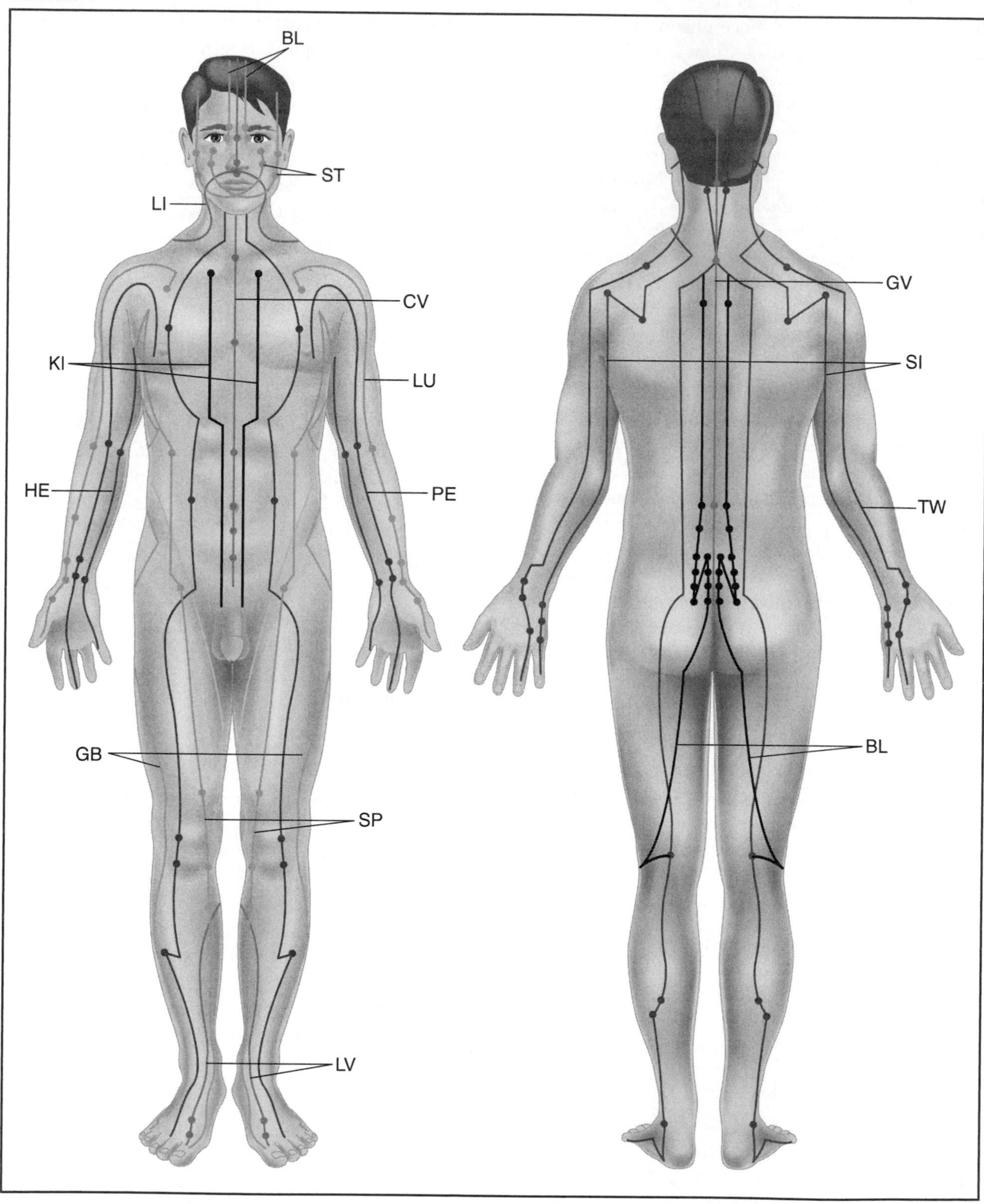

Diagram showing acupuncture meridians on a male body. A meridian is a group of acupuncture points all associated with the function of a particular internal organ system. *(Electronic Illustrators Group. Reproduced by permission of the Gale Group.)*

individual and the integrity of the natural environment are seen as mutually dependent, a worldcentric view that is especially relevant at a time when technoscience has achieved the power to destroy much of nature.

How Acupuncture Works

The attempts by Western scientific research to describe how acupuncture works rely on modern biomedical concepts. Popular theories include the notion that the mode of efficacy of acupuncture can be attributed to its influence on the structure and function of the body's different systems, including the nervous, circulatory, and immune systems (National Institutes of Health 1997, World Health Organization 2002). However, to appreciate acupuncture on its own terms one must understand the traditional explanations of how acupuncture works.

Western biomedicine focuses on the quantitative analysis of physical structure; it is mechanistic and reductionist in character. By contrast, Chinese medicine focuses on the qualitative analysis of function; it is holistic and synthetic in nature (Jarrett 1999, 2003). Over four millennia the Chinese have developed a rigorous language for discussing the subtleties of human physiological function. The central physiological concept is predicated on the notion of *qi* (*chi*), a universally present influence that maintains the functional integrity not only of the organism but of all natural processes (Porkert 1982, Jarrett 1999). The functions of *qi* are manifest in five forms: movement, transformation, protection, retention, and warming. Any dysfunction of these attributes in any aspect of being, whether physical, psychological, or spiritual, is said to be an imbalance of *qi*.

Acupuncture points are discrete locations on the external surface of the human body where the internal function of the organs can be influenced and the quality and directionality of their *qi* can be mediated. Points that are functionally related are said to constitute a specific meridian. Each meridian is associated with the function of an internal organ system or "official." Rather than naming specific organs anatomically, the ancient Chinese conceived of each organ as being an official with a specific duty to fill. When each official did his duty, health and harmony resulted. In the *Huang Di Nei Jing Su Wen* each organ is personified as being in charge of specific functions (Larre and de la Vallée Rochat 1987). For example, the fourteen points most closely associated with the function of the liver official constitute the liver meridian. The liver traditionally is likened to a military general in charge of planning and decision making. Its function is associated with growth, vision, and flexibility in all aspects of being. Hence, visual disturbances, poor planning, frustration, and tightness in the tendons that limits flexibility all can be treated through acupuncture points on the liver meridian.

Each point harmonizes an unbalanced aspect of function on a continuum ranging from deficient to excessive. For example, if a patient's heart rate is too slow or too fast, an acupuncture point such as Heart-7 (*shenmen*, or "Spirit Gate") can be used to increase or decrease the pulse to achieve the correct rate. Similarly, a point such as Liver-14 (*qimen*, or "Gate of Hope") can be used to help calm a belligerent person or enhance self-esteem in a timid person.

Acupuncture has evolved as a sophisticated science of human function for at least 2,500 years. As Chinese medicine is integrated into Western cultures, patients are afforded the benefits of both biomedical and functional medicine. The worldcentric and holistic view of Chinese medicine holds special promise for helping humanity face the unique challenges of the dawn of the twenty-first century.

LONNY S. JARRETT

SEE ALSO *Complementary and Alternative Medicine; Confucian Perspectives; Daoist Perspectives; Galenic Medicine; Medical Ethics.*

BIBLIOGRAPHY

Connelly, Dianne M. (2002). *Traditional Acupuncture: The Law of the Five Elements*. Columbia, MD: Traditional Acupuncture International.

Gwei-Djen, Lu, and Joseph Needham. (1980). *Celestial Lancets: A History and Rationale of Acupuncture and Moxa*. Cambridge, UK, and New York: Cambridge University Press.

Jarrett, Lonny S. (1999). *Nourishing Destiny: The Inner Tradition of Chinese Medicine*. Stockbridge, MA: Spirit Path Press.

Jarrett, Lonny S. (2003). *The Clinical Practice of Chinese Medicine*. Stockbridge, MA: Spirit Path Press.

Kaptchuk, Ted J. (1983). *The Web That Has No Weaver: Understanding Chinese Medicine*. Chicago: Congdon & Weaver.

Larre, Claude, and Elisabeth Rochat de la Vallée. (1987). *The Secret Treatise of the Spiritual Orchid*. East Grinstead, UK: International Register of Oriental Medicine.

Porkert, Manfred. (1982). *The Theoretical Foundations of Chinese Medicine*. Cambridge, MA: MIT Press.

Unschuld, Paul U. (1985). *Medicine in China: A History of Ideas*. Berkeley: University of California Press.

World Health Organization. (2002). *Acupuncture: Review and Analysis of Reports on Controlled Clinical Trials*. Geneva: Author.

INTERNET RESOURCE

National Institutes of Health. (1997). *Acupuncture*. NIH Consensus Statement 15(5): 1–34. Available from http://odp.od.nih.gov/consensus/cons/107/107_intro.htm

ADVERTISING, MARKETING, AND PUBLIC RELATIONS

The relationships between advertising, marketing, and public relations are not well defined. In general, however, advertising and public relations are considered components of marketing. Marketing is the craft of linking producers of a product, service, or idea with existing and potential consumers. Marketing techniques are most generally associated with transactions in capitalist economies, but they are also applied in religion, politics, and other aspects of public life. Advertising is part of an overall marketing strategy, and it involves the paid promotion of goods, services, ideas, and companies by an identified sponsor. Public relations connotes a broad spectrum of communication either within a group (e.g. company, political party, scientific community) or between that group and specific publics with the intent of informing and influencing their behavior and perceptions in ways that are favorable to that group.

Technology, Science, and Advertising

Advertising, like any transmission of information, requires a medium, and the biggest impact that technology has had on advertising is the expansion of media outlets. Initially vendors had to rely only on the spoken word and hand written signs. Then the printing press allowed for the first rudiments of mass media marketing, as advertisers could reach wider audiences through handbills and the inclusion of advertisements in books. Radio, television, and the Internet have further expanded media options for advertisers. In addition, logos printed on clothing and other products, billboards, and even skywriting ensure that our world is increasingly saturated by advertisements and brand names. In fact, it is estimated that the average North American child views roughly 40,000 television commercials per year (Strasburger 2001). As advertising becomes more sophisticated and the products more technologically complex, consumers today are less able to judge quality than they were even 100 years ago, when they themselves were involved in the production of simple crafts and thus more skilled in judging the quality of the things they bought. So as advertising becomes a more pronounced element of our cultural environment, the context of a global system of production causes our understanding of the goods being advertised to decline. This in turn means that we rely more heavily on regulatory agencies and advertising codes of ethics to ensure fairness and truth in advertising.

Technology has not only changed media and the societal dimensions of advertising but it has changed the nature of advertising as well. Handbills and other printed materials are relatively passive and static, whereas television commercials, and to an increasing extent internet advertisements, tend to be dynamic, employing rapidly changing images. The increasing pace of modern, technological societies and rising costs of marketing tend to condense both political and product advertisements into short clips. Improvements in information technology allow marketers to more quickly and flexibly respond to changes in consumer behavior. On the downside, however, increasingly complex technological tools and information systems can overload marketing managers and distract them from the creativity and judgment that remain central to successful advertising strategies.

The emergence of advertising on a large scale coincided with the rise of consumerism-fueled industrial capitalism. Although the development of new technologies for transmitting advertisements and managing marketing strategies is a key element of this process, so too is the continuing creation of marketing as a science. The traditional advertiser's dilemma was expressed in this way, "I know half my advertising is wasted, but I don't know which half!" In response to this inefficiency and the demand to create new markets to increase sales (or in politics, the demand to win over more voters), various social and behavioral sciences have been applied to advertising. Marketing research and motivation analysis are just two of the terms that signify the rise of a systematic science of advertising. Techniques include mathematical models, game theory, multivariate analyses, econometric analyses, psychometric approaches, and choice models (see Sutherland and Sylvester 2000). Several institutions carry out this research, including the Academy of Marketing Science, which publishes the journal *Academy of Marketing Science Review* (AMS).

Advertising is open to several interpretations, but one of the most influential remains Vance Packard's indictment of the advertising industry, *The Hidden Persuaders* (1957). Packard examined the use of psychoanalysis and other scientific techniques to understand human behavior and guide campaigns of persuasion and manipulation. These image-building campaigns are launched at both consumers and citizens; they are both about what to buy in the market and how to act in the polis. He labels these efforts "hidden," because they take place beneath our level of awareness. Packard claims that we are duped into believing that rather than buying lipstick, oranges, and automobiles we are acquiring hope, vitality, and prestige. Although sometimes constructive or amusing, most of these practices "represent regress rather than progress for man in his long struggle to become a rational and self-guiding being" (p. 6). This Orwellian interpretation is probably hyperbolic, but Packard is more convincing in his modest claim that "These depth manipulators are . . . starting to acquire a power of persuasion that is becoming a matter of justifiable public scrutiny and concern" (pp. 9–10). This power raises several ethical concerns about deception, the manipulation of behavior and self-image, and the exploitation of weaknesses and fears.

Ethical and Societal Issues of Marketing

Early advertising and marketing techniques were disreputable due in part to the lack of established laws and codes of conduct, which allowed deceptive advertising practices to flourish unchecked. In the United States, early development of the industry was largely driven by the marketing of patent medicines and "nostrums," and by spectacles such as P.T. Barnum's circus and museum. In later years, rather than traveling with his circus, Barnum concentrated on advertisement, creating a whole new species of marketing rhetoric that persists to this day. His colorful descriptions of sideshow mermaids and white elephants (the first a stuffed monkey sewed to a fish-tail, the latter a white-washed gray elephant) are classics in the psychology of marketing. Although Barnum commented that "the people like to be humbugged," he also said that "you may advertise a spurious article and induce many people to buy it once, but they will gradually denounce you as an impostor" (Ogilvy 1988, p. 156).

After the turn of the century, some members of the nascent advertising industry wished to distinguish themselves from their less reputable colleagues, and the first trade associations and codes of practice were established. Around 1900, the Curtis Code of magazine publishers stated: "We exclude all advertising that in any way tends to deceive, defraud or injure our readers." In 1910, the Association of Advertising Clubs of America adopted "Truth in Advertising" as its slogan. Four years later, the Audit Bureau of Circulations was formed, with the job of verifying the circulations reported by magazine publishers, on which ad space prices were based. In 1917, the American Association of Advertising Agencies issued a code that included a prohibition on copy "knocking" a competitor's product and on ads with "immoral or suggestive" content; banned the use of the word "free" unless the item offered was actually free; and declared that installment plans were inherently suspect.

These and other efforts by the marketing industry were attempts at self-regulation, partially motivated by the desire to avoid Congressional regulations. Nonetheless, Congress did become involved with the 1914 Federal Trade Commission Act, which empowered a commission to enforce rules designed to prevent deceptive and unfair practices in advertising. With the passage of the 1938 Wheeler-Lea Amendment the jurisdiction of the Federal Trade Commission (FTC) was broadened to include the advertisement of food, drugs, cosmetics, and therapeutic devices. The truth in advertising rules of the FTC not only require advertising to be fair and nondeceptive, but also hold advertisers responsible for producing evidence to substantiate their claims. FTC rules apply to all media, including the Internet.

Despite drastic political and technological changes through the history of modern advertising, ethical concerns about advertisements that misrepresent the capabilities of products and negative or "attack advertising" have remained constant (see *The Ethical Problems of Modern Advertising*, 1978). This suggests that in marketing, new technologies may exacerbate perennial ethical problems more than raise entirely novel ones. Additionally, some of the same industries have sustained a steady level of controversy pertaining to ethics in advertising. A good example is the tobacco industry, which caused conflict even during the 1930s. All does not stay the same, however, since new technologies give new form to old ethical problems. A good example is pornographic "pop-up" advertisements on computers linked to the Internet.

Ambiguity enters the ethical debates, because advertising need not be based on facts alone. Indeed a certain appeal to emotion is ethical and even necessary for successful marketing. Likewise, there is no formula for determining when omission constitutes deception. Thus the charge that a group owes the public "truthful" advertising requires significant acts of judgment as general rules must be interpreted within specific cases. There is clearly a spectrum of ethical severity involved, from advertising new car models that have only cosmetic but no functional improvements, to the increasing commercialization of public schools, to advertising a new drug without fully studying or disclosing possible harmful effects. Foregoing some practices, such as negative or attack ad campaigns, may be based more on marketing strategies than ethics, as managers (or politicians) attempt to gauge whether their target audience will be offended by aggressive attacks on the competition. However, even in these cases ethical concerns cannot be wholly avoided. One classic example from the 1930s was the ad campaign that enticed its audience to "reach for a Lucky instead of a sweet," which angered the candy industry because of the unfounded insinuation that smoking cigarettes is more healthy than eating candy.

A certain element of popular opinion views advertisements as socially invidious, leading to shallow, self-absorbed behavior, fostering negative body-image issues and poor self-esteem, and wrecking devastation on the natural environment and the larger social fabric via large-scale consumerism. Yet even among those who feel these concerns, behavior is seldom altered, as the experience of an individual purchase is difficult to link to these larger effects. Several academic analysts have attempted to confirm and articulate the corrupting influence of advertising on individuals and society. Many, like Packard, portray it as psychic manipulation, exploiting human insecurity to drive product sales. It is a truism in such writing, for example, that problems such as bad breath and body odor, considered normal and tolerable in the nineteenth century, were recast as unalloyed evils, sources of personal shame and social isolation, by twentieth century advertising in the service of product sales.

Richard Pollay (1986) provides a taxonomy of academic complaints about advertising. It can be simplified into two multifaceted claims that advertising is: (a) "intrusive, environmental, inescapable, and profound" and reinforces "materialism, cynicism, irrationality, selfishness, anxiety, social competitiveness, sexual preoccupation, powerlessness and/or a loss of self-respect" (Pollay, p. 18); and (b) "essentially concerned with exalting the materialistic virtues of consumption by exploiting achievement drives and emulative anxieties.... generally reducing men, women and children to the role of irrational consumer" (Pollay, p. 21). He cites a National Science Foundation study from 1978, which found that advertising encourages unsafe behavior, inappropriate standards for choice, and parent-child conflict; models hazardous behavior, such as malnutrition and drug abuse; and reinforces sex-role stereotypes, cynicism and selfishness. Pollay concludes that advertising in our age has become a ritualistic "social guide," promoting ideas about "style, morality, behavior."

Feminist analysts claim that some advertising causes harm by educating young girls to covet unnaturally thin bodies and driving anorexia and bulimia as unintended side-effects. They see advertising as a tool of social repression, keeping women subservient. "The female body is represented as the dream image that disguises her own exclusion.... But the ideals sold us are impossible to live, creating a hunger that keeps us unsatisfied and forever buying" (Schutzman 1999, p. 3). Mady Schutzman says that advertising makes women neurotic: "What advertising prescribes, women regurgitate in rage, histrionics, amnesia and paralysis" (p. 115).

Jean Kilbourne argues that advertisements create an image of women as "sophisticated and accomplished, yet also delicate and child-like" (1999, p. 137). Kilbourne collects print advertisements that share a common theme of encouraging young women to be silent and let their nail polish, clothes, perfume or make-up do their communicating, a message which she states has a "serious and harmful" impact. In their drive to sell products, ads communicate messages, which put young women in severe conflict, promising them "fulfillment through being thin and through eating rich foods" (p. 145), or through being virginal yet sexually wild. While she does not believe that ads directly cause anorexia, the "images certainly contribute to the body-hatred so many young women feel" (p. 135). She points out that in Fiji, well-fleshed women constituted the feminine ideal, and eating disorders were unknown until the introduction of television.

These critiques raise questions about how far the ethical obligation of advertisers should extend. But they also echo Packard's concerns about the degree to which our self-image and behavior are influenced by the environment of advertisements. They are made all the more important by the ability of modern technology to saturate our surroundings with advertisements, each not only promoting a product or idea but also transmitting cultural messages about what is appropriate and desirable. The technologically enhanced barrage of advertisements recalls Langdon Winner's (1986) insight that "technologies are not merely aids to human activity, but also powerful forces acting to reshape that activity and its meaning" (p. 6). Advertising shapes our shared world and thus to some extent it orients us within a web of meanings and influences our identity.

David Ogilvy, advertising executive, presents an optimistic take on the social benefits of advertising. He quotes Franklin Delano Roosevelt:

> If I were starting life over again, I am inclined to think that I would go into the advertising business in preference to almost any other. . . .The general raising of the standards of modern civilization among all groups of people during the past half century would have been impossible without the spreading of the knowledge of higher standards by means of advertising. (1988, p. 150)

He then quotes Winston Churchill: "Advertising nourishes the consuming power of men. . . . It spurs individual exertion and greater production" (p. 150). Ogilvy was a proponent of informative advertising and an extremely honest man, so his personal traits certainly provided a rose color to the advertising industry.

Yet he also admitted some of the negative aspects of advertising. For example, Ogilvy considered the economic effects of advertising and concluded that ads probably result in lower prices by driving sales volume. At the same time, they may contribute to monopolization by companies large enough to afford their costs. Ogilvy detested the trend of using Madison Avenue techniques to sell politicians. He addressed the criticism that ads influence the editorial content of magazines and newspapers, and argued that advertising serves as a force of social cohesion, building community and national identity.

James B. Twitchell (1996) argues that our culture is not just driven by advertising; it *is* advertising. Indeed he maintains that culture is just advertising's way of ensuring its own survival. He traces an unbroken line from religion and rituals to advertising: "[B]y adding value to material, by adding meaning to objects, by branding things, advertising performs a role historically associated with religion" (p. 11). "[A]dvertising is the gospel of redemption in the fallen world of capitalism" (p. 32). Advertising is "an ongoing conversation within a culture about the meaning of objects" (p. 13). Globally, "Adcult," the powerful, pervasive social, psychological, and cultural phenomenon of worldwide advertising, homogenizes cultures and exploits human doubt and insecurity and, accordingly, has become "the dominant meaning-making system of modern life because of our deep confusion about consumption, not only about what to consume, but how to consume" (p. 253).

The world described by Twitchell is very different than the future envisioned by Ogilvy. The information-filled ads championed by the latter are largely a thing of the past, with modern television ads relying largely on emotion and desire, based on numerous, almost subliminal rapid images and sounds of the lifestyle the audience is urged to associate with the product. The time is long gone when consumers care about the type of stitching or fabric used in a shirt; the sale today relies on the way the shirt will make you feel about yourself, the members of the opposite sex it will attract, and the access it will grant you to a better life.

Twitchell, like Packard, believes that the implications of modern advertising for human freedom, especially freedom of speech, are bleak. He argues that advertisers are the primary censors of media content in the United States. *Adbusters*, a monthly magazine, attempts to raise these issues to the consciousness of consumers by criticizing, deconstructing, and parodying ads. Twitchell asks if advertising is an inherently unethical medium and concludes that it is best conceived as amoral rather than immoral. Advertisers primarily want to sell products; their main goals are not reinforcement of stereotypes, or the exploitation of insecurities, which are often, however, secondary effects of what they do. If advertisers believe they can sell more products by portraying strong, independent women rather than childlike, dependent ones, they will do so; the ads for Charlie perfume were an early example, presenting a self-assured businesswoman, to whom the men in the ads were subservient. (However, she was also young, thin, and beautiful.)

Public Relations of Science

Since at least the mid-nineteenth century, scientists and scientific institutions have engaged in public relations activities in order to improve their social status, sway public policy with respect to science and technology, and promote greater public support of research and science in general. Although this attempt to improve the relationship between the public and science usually benefits science, it has also been couched in arguments that are less directly self-serving. These arguments are often grouped under the general labels of "public understanding of science" or "scientific literacy." Some of the more common justifications for enhanced public understanding of science are that it can bring benefits to national economies, boost national power and influence, improve individuals' chances in the job market, inspire greater intellectual, aesthetic, and moral achievements, and benefit democratic government and society in general. Jacob Bronowski (1974) voiced this last justification in terms of a "democracy of the intellect," in which the distance between power and the people can be closed only if scientific knowledge is dispersed broadly.

Though indirect, almost all of these reasons for enhanced public understanding of science will benefit science by leading to greater public support and investment. The opposite effect is possible, however. Greater understanding of science can lead to increased public scrutiny and skepticism or even control over research agendas and practices. Partly in response to just such a possibility, Steve Fuller argues that "science may be popular *precisely* because it is misunderstood. Thus, a movement genuinely devoted to 'public understanding of science' may have some rather unintended consequences for the future of science" (1997, p. 33). Dorothy Nelkin (1995) adds that "While scientists see public communication of scientific information as necessary and desirable, they are also aware that it extends their accountability beyond the scientific community" (p. 148).

Own a practical level, the public relations of science arose from the insight that peer review is not sufficient to maintain research support and favorable public policies. Thus information must be directed not just at peers, but also at corporations, policy makers, and the general public, highlighting the fact that science cannot survive as an autonomous enterprise. Nelkin traces the history of science public relations and argues that government science agencies, scientific journals, science-based corporations, and individual scientists have developed sophisticated ways to utilize and even manipulate the media to put a positive image on their work. These tactics span a spectrum from employing public relations officers to directly restricting journalists' access to information. The restrictions placed on reporters at the 1975 Asilomar Conference on recombinant DNA research are an example of the latter form of image control. Another problem that can arise from public relations efforts in the medical sciences is the improper inflation of hopes that a cure for the disease under research is immanent. This is exacerbated by the increasing pressure on journalists to be the first to report the most sensational claims, rather than well-researched and balanced news. In general, as fiscal and societal pressures mount on scientists to demonstrate the relevance, safety, and importance of their work, it becomes more difficult to see through tactics of self-promotion in order to gain a balanced understanding of the issues.

JONATHAN WALLACE
ADAM BRIGGLE

SEE ALSO *Business Ethics; Communication Ethics.*

BIBLIOGRAPHY

Bronowski, Jacob. (1974). *The Ascent of Man*. London: Little Brown. Companion to a television series, argues for the extension of scientific knowledge to the general public.

Fuller, Steve. (1997). *Science*. Minneapolis: University of Minnesota Press. Explores the normative and social aspects of science in order to undermine naïve assumptions about scientific realism.

Kilbourne, Jean. (1999). *Deadly Persuasion: Why Women and Girls Must Fight the Addictive Power of Advertising*. New York: Free Press. How advertising plays to the neurotic preoccupations of girls and women.

National Science Foundation. (1978). "Research on the Effects of Television Advertising on Children: A Review of the Literature and Recommendations for Future Research." Washington, DC: Author.

Nelkin, Dorothy. (1995). *Selling Science: How the Press Covers Science and Technology*. New York: W.H. Freeman. Analyzes the coverage of science and technology in the popular media. Contains a chapter on the public relations of science, pp. 124–143.

Ogilvy, David. (1988). *Confessions of an Advertising Man*. New York: Atheneum. First printed in 1962 and re-issued with minimal revisions. The autobiography of an ethical advertising man, with reflections on the culture of advertising.

Packard, Vance. (1957). *The Hidden Persuaders*. New York: David McKay. Argues that advertising science is manipulating human behavior and threatening to undermine our free will.

Pollay, Richard. (1986). "The Distorted Mirror: Reflections on the Unintended Consequences of Advertising" *Journal of Marketing*, vol. 50 (April), pp. 18–36. A review of the available literature on the social impact of advertising.

Schutzman, Mady. (1999). *The Real Thing: Performance, Hysteria and Advertising*. Hanover: Wesleyan University Press. How advertising caters to neurosis and insecurity in women.

Strasburger, Victor C. (2001). "Children and TV Advertising: Nowhere to Run, Nowhere to Hide," *Journal of Developmental and Behavioral Pediatrics*, vol. 22, no. 3, pp. 185–187.

Sutherland, Max, and Alice K. Sylvester. (2000). *Advertising and the Mind of the Consumer*. 2nd edition St. Leondards, Australia: Allen and Unwin. Aimed at both advertisers and consumers to explain which strategies work and why.

The Ethical Problems of Modern Advertising. 1978. New York: Arno Press. A reprint of an early twentieth century collection of essays on ethics in advertising.

Twitchell, James B. (1996). *Adcult USA: The Triumph of Advertising in American Culture*. New York: Columbia University Press. A cynical and influential book on the new language and psychological effectiveness of advertising.

Twitchell, James B. (2000). *Twenty Ads That Shook the World: The Century's Most Groundbreaking Advertising and How It Changed Us All*. New York: Crown Books. An overview of the most influential ads and why they were so effective.

Winner, Langdon. (1986). *The Whale and the Reactor: A Search for Limits in an Age of High Technology*. Chicago: University of Chicago Press. Presents a philosophy of technology as a form of life and argues that technologies have politics.

AFFLUENCE

If affluence is defined as an abundance of money and material goods, more humans than ever before are affluent beyond what could have been imagined a few generations ago. This growth and diffusion of affluence has been made possible in large part by advances in science and technology. Indeed, the political, social, and economic viability of contemporary market democracies has become linked to a considerable extent to the ability of scientific research and technological innovation to catalyze the growth of affluence. But technologically enabled material success brings with it substantial contradictions, in terms of distributional equity, environmental impacts, and the very notion of what "quality of life" means. These contradictions in turn challenge conventional thinking about the pursuit of affluence and the role of science and technology in society.

Economist Robert Solow (1957) estimated that technological innovation accounts for about half of all economic growth, and subsequent research has reinforced the idea that fields such as solid-state physics, computer science, material science, aeronautics, and genomics are the primary forces creating the new and diverse products and services associated with an affluent way of life (Nelson 1996). Government support of research and development tends to be justified on this basis.

Distributional Implications

The central role of science and technology as the engine of economic growth obscures other important outcomes. For example, the complex processes by which innovation translates into growing industrial productivity also can lead to the disruption or destruction of labor markets and social networks. Controversy in the United States over the outsourcing of high-tech jobs to developing nations is the most recent example of such disruption, whose devastating social consequences were portrayed compellingly in the nineteenth-century fiction of Charles Dickens (1812–1870). Economic theory views such conditions as an unfortunate consequence of a process of "creative destruction" that "incessantly revolutionizes the economic structure *from within*" (Schumpeter 1975, p. 83) to generate more jobs and more affluence. What is being destroyed in the process may be entire sectors of the economy and the livelihoods that depend on them.

Moreover, the distribution of benefits may be extremely uneven, inasmuch as wealth creation may be accompanied by increasing unemployment or underemployment, decreasing or stagnant real wages, enormous wage inequality, and increasing concentration of wealth both within nations and between nations (Noble 1995, Arocena and Senker 2003). Between 1960 and 1997 the income gap between the top and bottom 20 percent of the world population increased from 30:1 to 74:1, meaning that the poorest fifth of humanity now earns a little more than one percent of that earned by the wealthiest fifth (United Nations Development Programme 1999).

Although the exact causes of these trends can be debated, there is little question that they reflect the capacities of some individuals, sectors of society, and nations disproportionately to capture the benefits of scientific research and technological innovation. This asymmetry is now being reinforced by international rules governing intellectual property and other aspects of innovation policy (Commission on Intellectual Property Rights 2002).

Environmental Implications

Rising consumption has increased use of natural resources and generation of wastes. At least since the work of economist Thomas Malthus (1766–1834), many people have doubted that increasing production and consumption could be sustained indefinitely because of limited resources, and observers in our day have echoed the concern that ever-increasing material affluence is an unsustainable endeavor (Meadows, Randers, and Meadows 1972). To date, however, technologists have pushed back whatever limits may exist by improving resource extraction, by using less material inputs per unit of output, and by substituting artificial products for natural ones. These processes have permitted not only the exponential growth of human populations, but increasing material standards of living for many.

Technological optimists believe this can continue indefinitely as eco-efficiency improvements are enabled by ongoing innovation (Lomborg 2001). Less optimistic observers point to species extinctions, increasing production and proliferation of toxic materials, and other threats to long-term sustainability. If the technological optimists are ultimately proven wrong, and the environment does not sustain endlessly increasing material affluence, major shifts would be required in economic thought, in technological R&D, and perhaps even in the basic political rationale for contemporary market democracies, where worries about inequality have been swept aside by a focus on the pursuit of greater material affluence (Daly 1991).

Quality of Life Implications

A basic tenet of this rationale is that a growing gross domestic product per capita leads to a higher material standard of living, which in turn translates into a higher overall quality of life. All modern societies embrace this formula, though perhaps not to the same degree; this was captured memorably in the phrase that underlay the 1992 campaign strategy of presidential candidate Bill Clinton: "It's the economy, stupid." When economic growth slows or stops, political upheaval often follows.

The contribution of science and technology to the growth of affluence must be understood not just in terms of increased efficiency and diversity of production but also in terms of the willingness, even ardor, of people to consume the results of this productivity. As Rosenberg and Birdzell (1986, p. 264) note, "the long growth in scientific and technical knowledge could not have been transformed into continuing economic growth had Western society not enjoyed a social consensus that favored the everyday use of the products of innovation." This consensus feeds back into the economy to promote more innovation and growth but also feeds back into society, which is transformed continually in ways both expected and surprising by the introduction of new products and systems of technology. To remark that science and technology have resulted in a society that bears little resemblance to that of a century ago is a truism, but hidden beneath the obvious is the more subtle reality that commitment to this path of technological self-transformation is founded on a belief in the equivalence of affluence and quality of life.

But are they equivalent? Research on subjective well-being in countries throughout the industrialized world demonstrates that people's happiness and satisfaction with their lives have not increased during the historically unprecedented scientific, technological, and economic advancement of recent decades. Indeed, there has been a decline in some measures of life satisfaction (Lane 2000). Many people are richer and live longer, healthier lives; but most do not *feel* better off (Diener and Suh 1997).

These results should not be surprising, for moral traditions and common wisdom long have emphasized spiritual and social relationships over material ones as sources of satisfaction and meaning. Who would really suppose that marginal increases in affluence in already affluent societies would greatly enhance the quality of life? What luxury expenditures could add as much to people's comfort as did indoor plumbing, central heating, and related innovations of an earlier era?

What Goals for an Affluent Civilization?

If affluence raises both ethical and practical issues about how to use technical capacities wisely and fairly, what sorts of inquiries and deliberations might be warranted about the future relations of science, technology, and affluence? One source of inspiration for such queries can be found in John Kenneth Galbraith's *The Affluent Society*, first published in 1958, which posed fundamental questions about the "social balance" between private and public spending.

Galbraith argued that "the affluent society" was on the wrong track by continuing to behave as if it were living in an age of scarcity, rather than reshaping goals in accord with new priorities appropriate for an age of affluence. A preoccupation with unending increases in "the production of goods . . . (is) compelled by tradition and by myth," Galbraith said, not by thoughtfully chosen goals that "have a plausible relation to happiness" (Galbraith 1958, pp. 350–351). In effect, he argued that what economists call "diminishing marginal returns" had set in, such that additional increments of private affluence would not bring very much net gain in people's sense of well being. In contrast, he asserted, great gains in a society's overall quality of life could be obtained by aiding the poor, making work life more enjoyable, investing in scientific research, and generally shifting priorities away from private consumption and toward public purposes. For example, Galbraith recommended instituting larger sales taxes, both to reduce consumption and to assure that those who consume large quantities of private goods contribute commensurately to public services.

That the pursuit of technology-driven affluence remains the political raison d'être of the modern market economy may be less a reflection of "human nature" than one of enormously successful salesmanship by business executives, government officials and politicians, technologists, and economists. As Galbraith concluded, "To furnish a barren room is one thing. To continue to crowd in furniture until the foundation buckles is quite another. To have failed to solve the problem of producing goods would have been to continue man in his oldest and most grievous misfortune. But to fail to see that we have solved it and to fail to proceed thence to the next task, would be fully as tragic" (Galbraith 1958, pp. 355–356).

DANIEL SAREWITZ
EDWARD J. WOODHOUSE

SEE ALSO *Consumerism; Money.*

BIBLIOGRAPHY

Arocena, Rodrigo, and Peter Senker. (2003). "Technology, Inequality, and Underdevelopment: The Case of Latin America." *Science, Technology, & Human Values* 28(1): 15–33. Argues that technology-based wealth creation is not enough to reduce pervasive and persistent economic inequality, and that government interventions into economic life are required to alleviate poverty.

Daly, Herman E. (1991). *Steady-State Economics*, 2nd edition. Covelo, CA: Island Press. A systematic analysis of the possibility of operating an affluent economy sustainably, without net growth, based on the principle that "enough is best."

Diener, Ed, and Eunkook Suh. (1997). "Measuring Quality of Life: Economic, Social, and Subjective Indicators." *Social Indicators Research* 40: 189–216. Argues that measuring quality of life requires a combination of social indicators such as health and levels of crime, subjective well-being measures, and economic indices.

Galbraith, John Kenneth. (1958). *The Affluent Society*. Boston, MA: Houghton Mifflin. A classic mid-century analysis of problems and opportunities confronting affluent societies, focusing primarily on the U.S.

Lane, Robert E. (2000). *The Loss of Happiness in Market Democracies*. New Haven, CT: Yale University Press. A painstaking review of psychological research showing that increasing affluence does not bring much net improvement in life satisfaction, once a family is beyond poverty level.

Lomborg, Bjørn. (2001). *The Skeptical Environmentalist: Measuring the Real State of the World.* Cambridge, UK: Cambridge University Press. A careful, well documented, and thoroughly partisan argument that the global environment has been improving during recent decades.

Meadows, Donella H., Jorgen Randers, and Dennis L. Meadows. (1972). *Limits to Growth*. New York: Universe. One of the original warnings about environmental deterioration and the desirability of curtailing economic growth.

Nelson, Richard R. (1996). *The Sources of Economic Growth*. Cambridge, MA: Harvard University Press.

Noble, David F. (1995). *Progress without People: New Technology, Unemployment, and the Message of Resistance*. Toronto: Between the Lines.

Rosenberg, Nathan, and L. E. Birdzell, Jr. (1986). *How the West Grew Rich: The Economic Transformation of the Industrial World*. New York: Basic Books. The fundamental causes of the West's unprecedented affluence were not technological innovation and business entrepreneurship per se, but were the political pluralism and flexible institutions that encouraged scientific, cultural, and political freedoms.

Schumpeter, Joseph A. (1975 [1942]). *Capitalism, Socialism, and Democracy*. New York: Harper. A classic analysis of innovation, economic growth as creative destruction, and the interpenetration of market and democracy.

Solow, Robert M. (1957). "Technical Change and the Aggregate Production Function." *Review of Economics and Statistics*, August, pp. 214–231.

United Nations Development Programme. (1999). *World Development Report*. New York: Oxford University Press.

INTERNET RESOURCE

Commission on Intellectual Property Rights. (2002). *Integrating Intellectual Property Rights and Development Policy* London: Author. Available at http://www.iprcommission.org/graphic/documents/final_report.htm.

AFRICAN PERSPECTIVES

COMPUTERS IN SOUTH AFRICA

With its history of apartheid and its current mix of third- and first-world values, facilities, and services, the role of computer technology and the associated science in South Africa is different from most other countries, both in Africa and on other continents. Ethical considerations include such standard ones as employment, job losses, and social inclusion, but there are differences due to economic distortions caused by the legacy of apartheid.

Apartheid Legacy

Unemployment figures can be misleading in this economy. In the formal sector, unemployment seems to be at non-critical levels, but in the informal sector, joblessness is extremely high—2004 estimates are 40 percent—leading to crime and other social problems. South Africa nevertheless is very attractive to immigrants from other parts of sub-Saharan Africa, and there has been an influx of people seeking work.

Most students from disadvantaged backgrounds who are studying computer science are under some pressure to earn an income as soon as possible after they gain their first qualification, as they may have families to support, often families that have scrimped and saved to send a chosen member to university. Hence the desire to continue with any research or postgraduate study is disproportionately clustered in the privileged community, which historically is mainly white. In addition, it is comparatively easy in this field for individuals with degrees to obtain well-paid jobs, further lessening the incentive to contribute to research in computer science. This trend exacerbates the predominantly white presence of academics. A similar situation was faced by women in the 1980s. At the same time, at the beginning of the twenty-first century there is a great spirit of entrepreneurial activity among individuals while they are students, and small and medium-sized enterprises are being developed in response to a diverse range of needs, from computerizing legal records to computer control of traffic lights, all of which were largely ignored because the apartheid machinery had no need to optimize them.

Ethical Applications and Issues

South Africa's history of apartheid has left the country with an unusual technological infrastructure. During the apartheid years, it built an intensive war economy, supported by research and development in universities and in industry. Educators in liberal educational institutions faced the dilemma that their students would end up as engineers and computer scientists supporting this industry, engaged in activities of, at best, dubious morality. This dilemma no longer exists, but the legacy of the infrastructure still does, and so there is an imbalance in appropriate technology and expertise that is yet to be resolved. Furthermore, there are many areas, including high-density ones such as the historical townships, where electrical and telephonic infrastructure remains underdeveloped, impacting the ability to use technology—it is relatively easy to obtain old computers that are still usable, but there are no power sockets into which to plug them.

The most extensive legacy of apartheid is, of course, a huge gap between rich and poor, which actually continues to increase. Computer technology has a role to play, both in contributing to the gap and in lessening it. Because South Africa has traditionally had a labor-intensive economy, with labor being cheap and plentiful, the computerization of various work functions readily removes unskilled workers from the labor force, thus increasing unemployment and poverty. At the same time, the innovative use of computer technology and the development of local industry such as the excellent mobile phone network tend to bridge the rich-poor divide. Indeed, mobile telephony is especially appropriate in a country that is geographically large and whose fixed line telephonic network has been concentrated exclusively in wealthy urban areas. Mobile telephony has also empowered entrepreneurs by allowing them easy and efficient communications without the need to invest in anything more than a prepaid mobile phone.

For similar reasons, free and open source software is being embraced in South Africa, as in many other countries (especially in the developing world). Some of these motivations have an ethical or political component, such as the desire to promote the local software industry rather than enrich foreign corporations, while the free software movement has always claimed an ethical basis for shunning proprietary code. The collaborative maintenance model of open source software also seems to have opened up new possibilities—for example, translations to languages ignored by mainstream software manufacturers: In the South African context, the work of The Translate Project (http://translate.org.za) stands out. The appropriate application of computational linguistics techniques also has the potential for fostering social inclusion, by using machine translation to enable text in only English or Afrikaans (historically the two official languages) to be translated to the other nine official languages of South Africa, which include Zulu, Xhosa, Sotho, and other indigenous tongues.

One issue that might not otherwise be thought related to computers is that of the HIV/AIDS epidemic in South Africa: Between 20 percent and 40 percent of the population is directly affected by the disease, with a significant fallout effect on those indirectly connected. Some educational institutes have taken the stance that all subject areas have an ethical responsibility to educate about and mitigate the effect of the epidemic. While HIV might seem to have no direct impact on areas such as computer science, this is not actually the case. Research is currently underway in areas such as bioinformatics (http://www.sanbi.ac.za), including, for example, the modeling of the development of viral activity. Additionally, the epidemic affects educational institutions on a daily basis simply because it affects individuals on a daily basis. Many university students are already supporting extended families, on wages from part-time employment, and when a parent has the virus, the burden falls on the supportive child to look after younger children. In education this can often have the effect that completing practical assignments or studying for exams is relegated to the second tier of priority, once the caring for others has been done, resulting in poor performance from otherwise capable students.

In a country where a lot of dialogue about constitutional issues has been taking place since the 1990s, it is appropriate that the new South African constitution gives strong rights to individuals to access all information, including electronic information, held about themselves, especially by government bodies. In August 2002, the Electronic Communications and Transactions

Act became law. This is a wide piece of legislation pertaining to e-commerce and e-government, whose aim is to facilitate business, and it is descriptive rather than prescriptive. In contrast to the Data Protection Act of the UK, for example, there is no requirement for compliance. The chapter regarding personal information and privacy protection describes a voluntary regime that data collectors may subscribe to if they wish, so issues of personal privacy are still of concern.

SHEILA ROCK

SEE ALSO *Computer Ethics; Digital Divide.*

BIBLIOGRAPHY

Stallman, Richard M. (2002). *Free Software, Free Society: Selected Essays of Richard M. Stallman*, ed. Goshua Gay. Boston: GNU press. This is a collection of essays and speeches about open source by Richard Stallman—one of the primary pioneers of the free software movement.

INTERNET RESOURCE

Electronic Communications and Transactions Act (2002), South Africa. Available from (http://www.internet.org.za/ect_act.html). The ECT act describing appropriate use of computer technology in commerce and government.

HIV/AIDS IN AFRICA

From the perspective of Africa, HIV/AIDS is one of the most significant ethical and political issues involved with science and technology. The spread of HIV/AIDS in Africa has the potential to undermine almost any other positive benefits of, for example, scientific education and research or sustainable technological development. Of particular importance is the fact that increasing numbers of children are being orphaned and made vulnerable by HIV/AIDS, and the traditional extended family is being strained to the breaking point. To appreciate the extent of the challenge, it is necessary to have some appreciation of the origins, spread, and impact of HIV/AIDS in Africa, and of the debates regarding response and treatment.

African Origins and Impact

HIV is sexually transmitted, and can be passed on through direct blood contact (for example, blood transfusion). In addition to blood, semen, and vaginal fluids, there are sufficient amounts of HIV in breast milk to cause transmission from mother to child. The genesis of HIV is not clear; however, some postulate a link between the virus and oral polio vaccines distributed in the Democratic Republic of Congo in the late-1950s that may have been contaminated by the simian immunodeficiency virus (SIV). Though the theory is largely discredited, the possibility of a connection between the two viruses is still debated (Worobey et al. 2004).

FIGURE 1

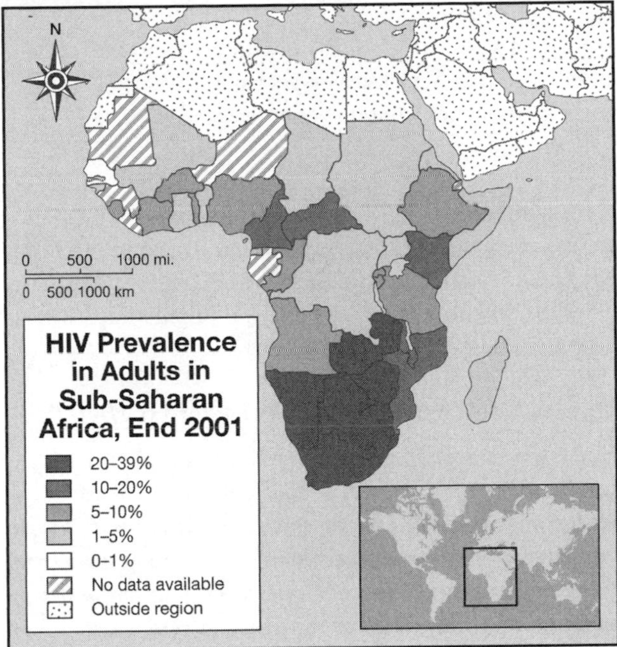

While more than 70 percent of HIV infection worldwide is through heterosexual sex, in sub-Saharan Africa the percentage is higher (Jackson 2002). The second most important route of transmission in the region is from an HIV-infected mother to her child. In Africa transmission via sex among men is far less common, and infection by drug users through sharing contaminated needles is relatively infrequent. Other means of transmission are through use of non-sterile needles and cutting implements in medical procedures, unscreened blood, and inadequate hygiene precautions in the care of AIDS patients. The map below shows the concentration levels over the continent.

Seventy-nine percent of AIDS deaths worldwide have occurred in sub-Saharan Africa. An estimated 71 percent of all adults and 87 percent of all children living with the disease in the early-twenty-first century reside in this region. Eighty-eight percent of all children who have been orphaned by AIDS live in sub-Saharan Africa (AIDS Epidemic Update 2002).

Researchers debate the reasons for the patterns of HIV/AIDS infection in different parts of Africa. Some believe that these patterns are influenced by whether the population is affected by HIV-1, HIV-2, or other strains of the virus, some of which are more virulent than others. Other observers focus on the social and cultural differences among countries. Researchers Jack Caldwell and Pat Caldwell, for example, see a coincidence between low infection rates and male circumcision, which improves personal hygiene and corresponds to low rates of sexually transmitted disease (STDs). Muslim countries in North Africa have relatively low rates of infection, as do Muslim populations within countries that are highly infected.

Factors Contributing to the Spread of HIV/AIDS

Since the sixteenth century, violence and disorder have upset the political and social culture of Africa. To understand the devastating spread of HIV/AIDS on the continent, one must consider events including war and desperate poverty that continue to be familiar and persistent conditions in many African nations.

MIGRATIONS. Massive migrations of displaced persons due to war, social unrest, and economic disadvantage are key contributors to the spread of the virus. In some cases, refugees flee their homelands to countries where the infection rate is already high. Upon resettlement, the refugees bring the disease home with them.

Due to economic depression, workers are forced to look for jobs far from home. For example, many from eastern and southern Africa went to work in the mines of South Africa, living in conditions of poverty and social unease. Poor hygiene, multiple sexual partners, and other social and economic factors that affect such workers promote infection at an accelerated rate.

WAR. Wars and other conflicts raged across Africa in the late-twentieth century and continued into the early-twenty-first century. Refugees help spread the epidemic. But the various armies involved in these conflicts are even more efficient sources of infection. Military personnel, both combatants and peacekeepers with regular pay, are more likely to contract HIV than civilians; in addition, they have higher rates of STDs, a factor known to correlate with easier transmission of the virus. Resolving these conflicts is key to a sustained, effective response to HIV/AIDS (Mills and Sidiropoulous 2004).

POVERTY. At the beginning of the twenty-first century, sub-Saharan Africa accounted for 32 of the 40 least developed UN member states. The region's total income is about the same as that of Belgium. (World Bank 2000).

Poverty leads to health conditions that promote spread of the disease, including chronic, severe malnutrition. In addition, people living in poverty have less access to basic education and health services. Extreme poverty is linked to an increase in commercial sex among women, who have the fastest growing infection rate.

SILENCE, STIGMA, AND DISCRIMINATION. Some African governments have denied the extent of the problem or that it exists at all. In addition, stigma attached to the infection has caused many people to refuse to become involved in finding solutions (Campbell 2003). For example, for several years in the 1980s former Kenyan president Daniel Arap Moi denied that HIV/AIDS infection existed in his country for fear of destroying the tourist industry, a key source of national income. As a result, there was little if any effort to promote precautions against transmission of the virus and the disease spread unabated (Singhal and Rogers 2003).

Social Impacts

HIV/AIDS will have enormous implications for the future of Africa. This entry will address just a few of the most pressing issues at the beginning of the twenty-first century.

ORPHANED CHILDREN. The main impact of the disease is felt through the loss of economically active people in their child rearing years, between the ages of fifteen and forty-five. UNICEF's *Africa's Orphaned Generations* (2003) puts the number of African children orphaned by AIDS at 11 million, with an estimate that the disease will ultimately rob 20 million children of their parents. Figure 2 shows the increasing numbers of children who will become orphans as a result of the epidemic.

IMPACT ON GOVERNMENT AND SERVICES. Many countries in eastern and southern Africa are already burdened by weak government infrastructures and inadequate human resources, compounded by the migration of skilled professionals due to economic reasons. The epidemic has exacerbated the situation with the attendant loss of workers in their most productive years. Staff attrition in key sectors such as education and agriculture outpaces replacement, causing a loss of institutional memory and low morale. Nongovernmental organizations (NGOs), which have been central to the struggle to control the disease, are focusing more energy on caring for the sick and less on education, prevention, and self-help initiatives in the community. Disintegration of national institutions such as the army and police threaten the security and political stability of many nations. Effects of the disruption of governance, such as displacement, food insecurity, and conflict, spur transmission of the disease, and contribute to the continent's downward cycle.

FIGURE 2

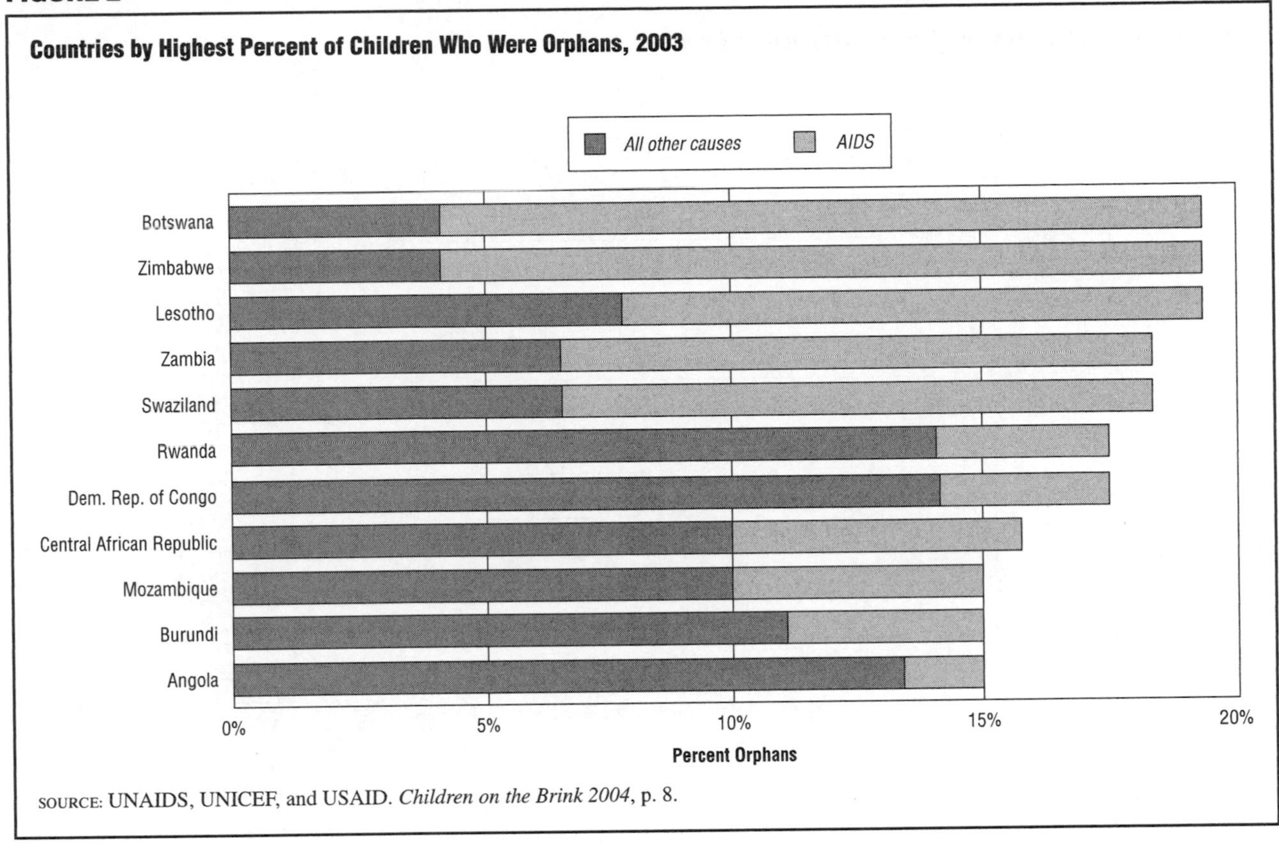

SOURCE: UNAIDS, UNICEF, and USAID. *Children on the Brink 2004*, p. 8.

IMPACT ON NATIONAL ECONOMIES. The World Bank (2001) estimates that per capita growth in half of Africa's countries is falling by 0.5 to 1.2 percent annually as a direct result of HIV/AIDS; by 2010, GDP in some of the countries most affected will drop as much as 8 percent. According to the Food and Agricultural Organization of the United Nations (FAO) (2004), two-person years of labor are lost for each AIDS death. In addition to the stark loss of life, HIV/AIDS deaths contribute to the loss of local knowledge of farming practices and forces communities to opt for less labor-intensive, less productive cropping patterns (FAO 2001).

WOMEN. According to the United Nations Programme on HIV/AIDS (UNAIDS) AIDS Epidemic Update for 2004, 76 percent of all young people (ages fifteen to twenty-four) in sub-Saharan Africa who are infected with HIV are female. Females are three times more likely to be infected than males in this age range. Gender inequality is the most important reason that HIV/AIDS infection has transformed into an epidemic that affects women and girls in disproportionate numbers. Women in Africa hold a lower socioeconomic position than do men. They are likely to be poorer and have less education and less access to social services than men do. Women faced with limited options to earn money sometimes turn to commercial sex; in some cases, for example in areas affected by sustained drought, women and girls resort to exchanging sex for food or other basic survival needs. Other factors related to the imbalance in power between men and women including sexual violence, early marriage, and poor access to information about transmission of the disease (even as relates to mother-child transmission) contribute to the infection rate. Adding to the problem is the fact that women are physiologically more vulnerable to being infected with the virus.

Response and Treatment Debates

Antiretroviral drugs (ARVs) are a great advance in the treatment of HIV/AIDS patients. Such drugs do not prevent infection or cure the virus. They do, however, disrupt the life cycle of the virus, preventing its reproduction. ARVs can reduce the patient's viral load tenfold within eight weeks, and lower it to undetectable levels

FIGURE 3

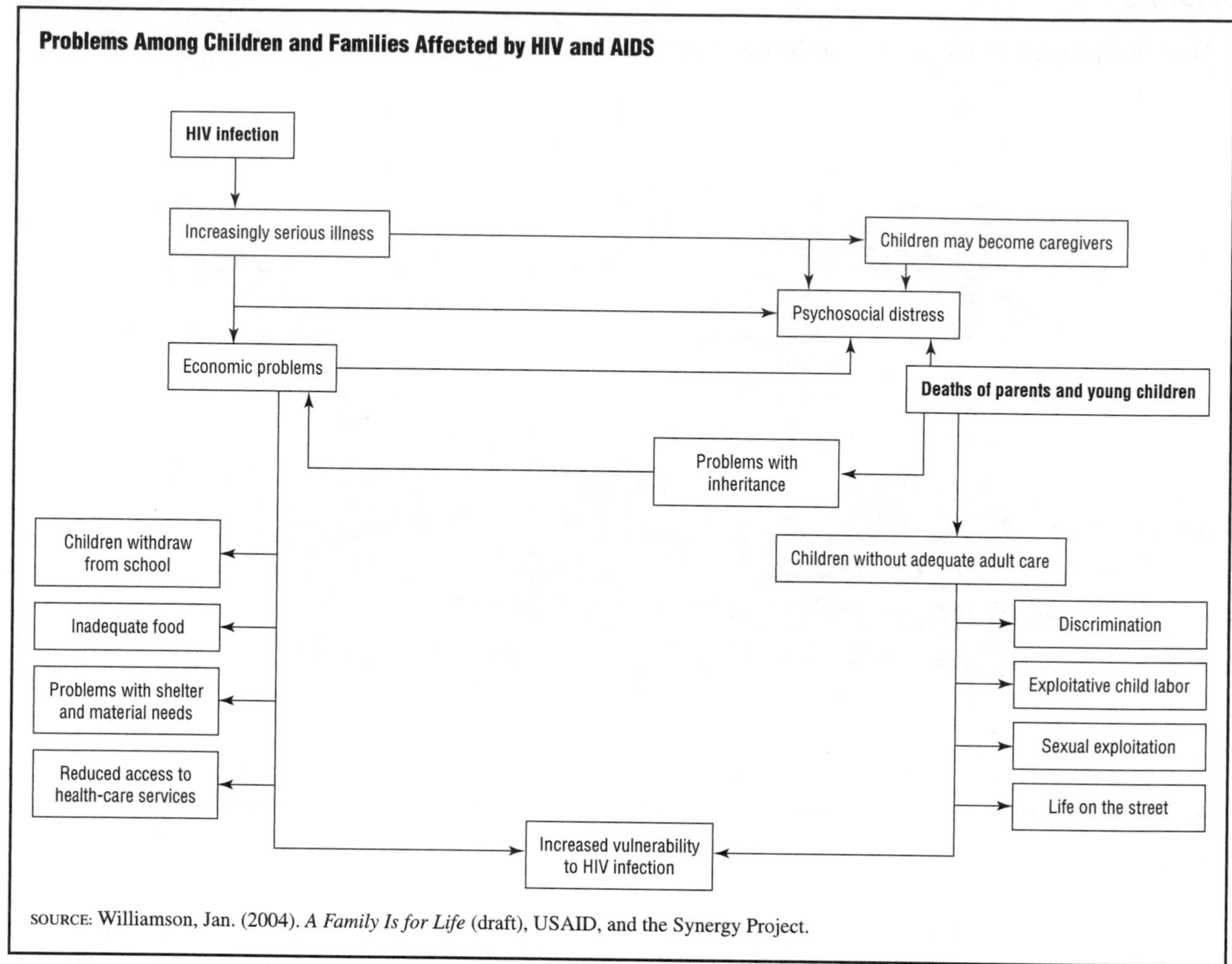

SOURCE: Williamson, Jan. (2004). *A Family Is for Life* (draft), USAID, and the Synergy Project.

within six months. For those infected with HIV, the onset of AIDS can be delayed indefinitely. Patients live longer, gain weight, and feel better.

ARVs were unaffordable in Africa until 2001 when an Indian drug company, Cipla, offered to provide a year's supply for $350, one-fortieth the cost in countries such as the United States. Although the price of ARVs has fallen dramatically, few Africans have access to the drugs. In addition, ARVs work most effectively when people are well nourished and have acceptable hygiene standards. In Africa the provision of ARVs is linked not only to challenges to improve the living conditions of sufferers, but to improving distribution of the drugs by strengthening public health systems.

The World Health Organization (WHO) plans to distribute ARVs to 3 million people in Africa by the end of 2005 through its "3 by 5" initiative. In addition to prolonging lives, this effort will slow the rate of orphanhood of the children of HIV/AIDS victims. Major drug companies, due to pressure from the global community, have recognized the need to reduce the cost of life-saving treatments. In an attempt to undo a public relations nightmare caused by the public perception of avarice, some companies provide the drugs free of charge; others have built medical clinics.

However there are those who argue that ARVs will not address HIV/AIDS in Africa due to the scope of the problem and the price of the therapy, and that an effective vaccine is necessary. Where to test such a vaccine, who to test it on, and what treatment should be provided to vaccine subjects who are already infected (where the vaccine is not a preventative but works to slow replication of the virus) are all questions with both medical and ethical importance.

Other efforts continue. Of particular note is the work of the Bill and Melinda Gates Foundation. The foundation's top global health priority is to stop transmission of the HIV virus and it has given more than 1.4

billion dollars toward that goal since 1994 (Gates Foundation).

Conclusion

In the early-twenty-first century, many African governments finally declared the HIV/AIDS epidemic national emergencies—a necessary first step to beginning HIV-prevention programs. Progress to control the epidemic has been made, but spread of the virus continues to outpace such efforts. Denial of the scope of the problem and stigmatization of victims continue. The most daunting task is to acquire the funds and means necessary to develop proven interventions, and provide them to sufferers. Promoting education, developing treatments, and providing relief to victims of the disease in Africa poses ethical challenges to scientists and technicians, not just in the field of medicine, but in host of other fields as well.

PELUCY NTAMBIRWEKI

SEE ALSO *Development Ethics; Equality; HIV/AIDS.*

BIBLIOGRAPHY

Barnett, Tony, and Alan Whiteside. (2002). *AIDS in the Twenty-First Century: Disease and Globalization*. Hampshire: Palgrave Macmillan.

Caldwell, Jack C., and Pat Caldwell. (1996). "The African AIDS Epidemic." *Scientific American* 274(3): 62–68.

Campbell, Catherine. (2003). *Letting Them Die: Why HIV/AIDS Prevention Programmes Fail*. Oxford: International African Institute.

Guest, Robert. (2004). *The Shackled Continent: Africa's Past, Present and Future*. Oxford: Macmillan.

Hunter, Susan, S. (2000). *Reshaping Societies HIV/AIDS and Social Change: A Resource Book for Planning, Programs and Policy Making*. New York: Hudson Run Press.

Hunter, Susan, S. (2003). *Who Cares?: AIDS in Africa*. New York: Palgrave Macmillan.

Jackson, Helen. (2002). *AIDS Africa: Continent in Crisis*. Zimbabwe: SAfAIDS.

Madraa, Elizabeth, and Major Ruranga-Rubamira. (1998). "Experience from Uganda." In *Partners in Prevention: International Case Studies of Effective Health Promotion Practise in HIV/AIDS*, eds. Anne Malcolm and Gary Dowsett. Geneva: UNAIDS.

Mills, Greg, and Elizabeth Sidirpoulos, eds. (2004). *New Tools for Reform and Stability: Sanctions, Conditionalities and Conflict Resolution*. Johannesburg: South African Institute of International Affairs.

Root-Bernstein, Robert S. (1993). *Rethinking AIDS: The Tragic Cost of Premature Consensus*. New York: Free Press.

Singhal, Arvind, and Everett M. Rogers. (2003). *Combating AIDS: Communication Strategies in Action*. New Delhi: Sage Publications.

Topouzis, Daphne. (1994). *Uganda: The Socio-Economic Impact of HIV/AIDS on Rural Families with an Emphasis on Youth*. Rome: Food and Agriculture Organization of the United Nations.

UNICEF. (2003). *Africa's Orphaned Generations*. New York: Author.

Wakhweya, Angela, et al. (2002). *Situation Analysis of Orphans in Uganda: Orphans and Their Households: Caring for the Future–Today*. Kampala: Uganda AIDS Commission.

Whiteside, Alan, and Clem Sunter. (2000). *AIDS: The Challenge for South Africa*. Cape Town: Human and Rousseau.

World Bank. (2000). *Can Africa Claim the 21st Century?* Washington, DC: Author.

World Bank. (2001). *Turning the Tide Against HIV/AIDS: Time for Global Action to Match a Global Threat*. Washington, DC: Author.

Worobey, Michael, et al. (2004). "Origin of AIDS: Contaminated Polio Vaccine Theory Refuted." *Nature* 428: 820.

INTERNET RESOURCES

Committee on World Food Security. "Food Security and HIV/AIDS: An Update." Food and Agricultural Organization of the United Nations (FAO). Available from http://www.fao.org/DOCREP/MEETING/006/Y9066E/Y9066E00.HTM. Report of the Committee's Twenty-Ninth Session, May 12–16, 2003, Rome.

Food and Agriculture Organization of the United Nations (FAO). "HIV/AIDS and Food Security." Available from http://www.fao.org/hivaids/.

Gates Foundation. Available fromwww.gatesfoundation.org.

Joint United Nations Programme on HIV/AIDS. "AIDS Epidemic Update: December 2002." UNAIDS. Available from http://www.unaids.org/html/pub/publications/irc-pub03/epiupdate2002_en_pdf/epiupdate2002_en_pdf_1.jpg.

Joint United Nations Programme on HIV/AIDS. "AIDS Epidemic Update: December 2003." UNAIDS. Available from http://www.unaids.org/html/pub/publications/irc-pub06/JC943-EpiUpdate2003_en_pdf/JC943-EpiUpdate2003_en_pdf_1.jpg.

Joint United Nations Programme on HIV/AIDS. "AIDS Epidemic Update: December 2004." UNAIDS. Available from http://www.unaids.org/wad2004/EPIupdate2004_html_en/epi04_00_en.htm.

Joint United Nations Programme on HIV/AIDS. "Children on the Brink." UNAIDS. Available from http://www.unaids.org/EN/resources/epidemiology/epi_recent_publications/childrenonthebrink.asp.

Joint United Nations Programme on HIV/AIDS. "Ethical Considerations in HIV Preventive Vaccine Research." UNAIDS. Available from http://www.unaids.org/html/pub/publications/irc-pub01/JC072-EthicalCons_en_pdf/JC072-EthicalCons_en_pdf_1.jpg.

AGGRESSION

The word aggression comes from the Latin roots *ag* (before) and *gred* (to walk or step). Hence to aggress is to step before or in front of someone, to initiate something, commonly an attack. Aggression—whether by a state or an individual—refers to an unprovoked, offensive action against another. It is useful to contrast aggression with violence, which derives from the Latin root *vio*, which refers to force. Dictionary definitions include "rough, unjust, unwarranted and injurious physical force or treatment," as well as "immoderate vehemence, typically causing injury, pain or gross distortion." It is possible to talk about a violent storm, or an earthquake of exceptional violence, but the term is most often applied to human actions, in which case it generally implies that pain or injury is intentionally inflicted on someone or something.

By contrast aggression is not necessarily hurtful: A person may promote a viewpoint aggressively, for example, which implies initiative, forcefulness and assertiveness, but without injury. It is admirable to conduct an aggressive campaign against cancer, poverty, or illiteracy. One may even seek to aggressively oppose violence. Nonetheless aggression as such is not highly regarded; it, like its frequent concomitant, violence, is typically considered undesirable, at least from the perspective of most ethicists.

Aggression Among Animals

Aggression is widespread among animals, especially those living in social groups. Although it sometimes takes the form of clear, outright violence, aggression is more often subtle, involving intimidation and the displacement of one individual by another, typically in the context of established dominance hierarchies. Early scientific studies of animal behavior emphasized that animal aggression very rarely results in serious injury or death, and that most living things with the capacity of inflicting serious harm on one another have evolved inhibitory mechanisms that prevent them from doing so. As ethological studies have gotten more sophisticated, however, it has become clear that these generalizations were idealized and exaggerated. In fact animals, even members of the same species, do kill one another. There is, however, some truth to the generalization that many living things have evolved behaviors that make lethal aggression less frequent than might otherwise be expected.

Increasingly sophisticated field studies of animal behavior show that animal aggression is not limited to inter-individual events; inter-group aggression has also been documented—for example, between lion prides or chimpanzee groups. Lethal aggression, in these cases, is most likely when the groups in question consist of genetically unrelated individuals, just as within-group aggression is also significantly modulated among close relatives, as predicted by *selfish gene theory*.

Aggression Among Human Beings

There has been considerable research into the causes of aggression, especially among human beings. Aggression is caused by many different factors; indeed, virtually every scientific specialty has its own *take* on which factors are especially important. For psychoanalysts, aggression derives largely from innate human destructiveness, what Freud called *thanatos*, or the *death instinct*. Although biologists are particularly unconvinced by this approach (it is difficult to imagine a situation for which a death instinct—especially when directed toward one's self—would be selected), there are parallels between this and another instinctivist approach, best articulated by the ethologist Konrad Lorenz (1903–1989). Lorenz hypothesized that aggression has evolved in a variety of circumstances, including spacing and population control, and provides an opportunity for competition within a species, as a result of which the most fit members will emerge to produce the next generation, and also establishes a means whereby the pair bond is strengthened, when, for example, a mated pair demonstrates shared aggression against competitors.

Sociobiology and evolutionary psychology provide updated biological explanations for human and animal aggression, emphasizing the degree to which aggression is adaptive rather than somehow mandated by the genome. This approach focuses on the way particular behavior patterns are maintained and promoted in a population because they contribute to the reproductive success of individuals (and their genes), as opposed to groups or species. For example, the *adaptionist* evolutionary view of aggression examines such phenomena as ecological competition, male-male competition, and the role of kinship patterns in directing aggressive behavior in particular ways. It also focuses on aggression as a *response* to circumstances rather than an innate need. Adaptionists do not argue that aggressiveness will emerge despite affirmative constraints. Rather, proponents maintain that living things have the capacity to behave aggressively when such behavior maximizes their fitness, and to behave pacifically when that response is in their best evolutionary interest.

It should be emphasized that predatory behavior—hunting—is different from aggressive behavior. The fact that certain Australopithecines and other prehuman species were evidently meat-eaters does not in itself mean that they were aggressive. Aggressive behavior is most prominent within a species, not between species. Lions, for example, often behave aggressively toward other lions, in which case they make themselves conspicuous and threatening; by contrast, when hunting zebras, lions employ very different behavior patterns, making themselves inconspicuous until the actual attack, and not relying on bluff or other means of aggressive intimidation.

The mainstream view among social scientists is that aggression is almost entirely a response to specific circumstances. So-called *frustration theory* has been especially influential; it posits that whenever aggressive behavior occurs, there must be frustration, and similarly, whenever frustration occurs, it always produces aggression of some kind. Other psychological approaches focus on the role of social learning, such as conditioning theory in which aggressiveness—by groups as well as individuals—is more likely when such behavior has been *positively reinforced*, and less likely when *negatively reinforced*. In short aggression is crucially modified by its consequences.

Social psychologists, by contrast, focus on the degree to which individuals can be socialized to aggressiveness, just as sociologists examine the role of social structures (religion, family, work ethos, mythic traditions) in predisposing toward aggression. Special consideration has been given to matters of ethnic, racial, and religious intolerance. Ironically, although most scientists agree that race has no genuine biological meaning, theories that focus on the importance of stereotyping and of *in-group amity*, *out-group enmity* have gained increasing attention.

For anthropologists interested in cross-cultural comparisons of human aggression, a paramount consideration is the extent to which aggression may be *functional* in acquiring land, access to mates, or status, as well as in regulating population, organizing social relationships within the group, and even influencing the pressure that tribal units place upon agricultural productivity and/or human population or the wild game on which they may depend. The prehistory of human aggressiveness remains shrouded in mystery, although most specialists agree that primitive human groups engaged in substantial violence as well as cannibalism.

For many political scientists, relevant considerations include the role of rational calculations of state benefit and national power. An important underlying assumption is that states behave aggressively when it is in their perceived interest to do so, perhaps because of the prospect of enhancing their influence and power (*realpolitik*), minimizing potential decrements to it, or enhancing the political viability of national leaders, among other reasons. Approaches run the gamut from mathematical models created by game theoreticians to analyses of historical cycles, matters of national prestige, and economic/resource based considerations.

Aggression and Ethics

Ethical analyses of aggression are nearly as diverse as efforts to explain its occurrence. Although aggression among animals is not susceptible to ethical judgments, human aggression certainly is. Indeed ethical assessments—often negative—may be especially directed toward cases of aggression. Such judgments may be absolute, on the order of philosopher Immanuel Kant's (1724–1804) *categorical imperative*, which maintains that any act or aggression is acceptable only if it could be reasonably seen to be based on general principles of behavior. However *situational ethics* typically emphasize that aggression should be evaluated with regard to the conditions in which it occurs. Thus self-defense—whether by an individual or a group—is enshrined in most legal and moral codes, whereas aggression is widely considered to be unacceptable when it occurs without adequate provocation, or preemptively.

The degree to which such ethical judgments are supported or undermined by scientific studies is open to debate. For instance, some believe that scientific knowledge of the biological mechanisms of aggressive behavior demonstrates that cultural moderation, in the form of moral sanctions, is a continuation of nature in nurture. Others argue that the widespread presence of aggression among animals legitimates its presence among humans. In the end, the tensions between these arguments point toward granting moral judgments or values some degree of independence in assessing human behavior, although such judgments will, by necessity, be refined as science advances additional theories to explain the complexities of aggression. Finally, the discussion of whether and to what extent science and technology can be characterized as aggressive activities, although again somewhat independent of scientific research, is furthered by reflection on the scientific study of the phenomena of aggression.

DAVID P. BARASH

SEE ALSO *Darwin, Charles; Ethology; Just War; Nature versus Nurture; Sociobiology; Violence.*

BIBLIOGRAPHY

Barash, David P. (2001). *Understanding Violence*. Needham Heights, MA: Allyn & Bacon. Contains many crucial pieces concerning violence, from a variety of disciplines.

Barash, David P., and Charles P. Webel. (2002). *Peace and Conflict Studies*. Thousand Oaks, CA: Sage Publications. A comprehensive textbook that introduces the field of peace and conflict studies.

Eibl-Eibesfeldt, Irenaus. (1979). *The Biology and Peace and War: Men, Animals and Aggression*. New York: Viking. Presents traditional ethological ideas concerning aggression, with special attention to human peace and war.

Lorenz, Konrad Z. (2002). *On Aggression*. New York: Routledge. A readable and only slightly outdated account of the classical ethological approach to animal and human aggression.

Moynihan, Martin H., and Michael H. Robinson. (1998). *The Social Regulation of Competition and Aggression in Animals*. Washington, DC: Smithsonian Institution Press. A technical compendium of source material concerning causes of, and constraints upon, aggressive behavior in animals.

AGING AND REGENERATIVE MEDICINE

Advances made in research on the biology of aging and on the repair, replacement, and regeneration of tissues and organs (regenerative medicine) have drawn attention to old and new ethical issues. The principal concern among those who anticipate intervention in the aging process is whether or not attaining the power to do so is a desirable goal. The issues for those who are concerned about using human cells or tissues for research or therapeutic purposes are as follows: (1) whether the donor is, or is not, capable of giving informed consent; (2) if not, whether it is ethical for others to make that decision; and (3) whether the taking of one or a few potential lives for the benefit of many is ethically sound. When the source of the cells or tissues is a fetus or embryo, debate centers on the ethics of using tissue from induced or spontaneous abortion and when human life begins.

Where some see only benefit in the ability to slow, stop, or even reverse the aging process, others see an array of unintended consequences. There have been efforts made to intervene in the aging process throughout recorded history, and also warnings given that doing so could lead to undesirable consequences (Hayflick 2000).

The only way humans have succeeded in extending their longevity is by eliminating or delaying the appearance of disease or pathology. The greatest success occurred during the twentieth century when actuaries recorded the largest increase in human life expectancy at birth in developed countries.

The thirty-year increase from about forty-seven years in 1900 to about seventy-seven years in 2000 resulted from the implementation of public health measures for the control of acute infectious diseases, the discovery of antibiotics and vaccines, and the great advances made in other medical and health care disciplines. The result has been an enormous reduction in mortality rates in early life and a concentration of deaths in later years. In the early-twenty-first century, in developed countries, infectious diseases are no longer a leading cause of death. They have been replaced by cardiovascular disease, stroke, and cancer.

The maximum number of additional years attainable by the elimination of deaths caused by disease or pathology is between fifteen and twenty (Hayflick 2000). Once the leading causes of death are resolved, immortality will not occur, but we will have revealed the underlying, inexorable aging process that leads to causes of death attributable to the loss of function in some vital organ.

The Aging Process and the Ethics of Intervention

The process of aging is the inevitable loss of molecular fidelity that occurs randomly in the molecules of most animals after reproductive maturation. The status of biomolecules before they undergo age changes determines potential longevity or degree of resistance to age changes. No intervention that increases the stability of biomolecules before they undergo the aging process is known, nor is any method that can slow, stop, or reverse the aging process in humans.

The fundamental aging process increases vulnerability to what is written on the death certificates of older people. There seem to be no ethical issues that would oppose the goal of eliminating all causes of death attributable to pathology, violence, or accidents. But ethical issues do arise when considering interventions in either the aging process or the determinants of longevity. However, at this time, these can only be considered in the abstract.

Those who favor intervention in either the aging process or the determinants of longevity see a benefit in increasing the chronological time during which life satisfaction and good health are at their maximum levels. Critics see an array of ethical issues. These include the determination of when to intervene, because in order to determine when life satisfaction is at its greatest one must experience that time of life. If subsequent events reduce life satisfaction, choosing to return to a former happier state will depend on methods known to occur only in science fiction. It is also science fiction to expect that the environment that contributed to such better conditions will remain unchanged (Hayflick 2004).

Of the many bizarre scenarios that can be imagined, one would not be surprised to find families in which adult children, who chose not to slow or stop the aging process, are themselves biologically older than their parents, who did. Finally people who are highly satisfied with the quality of their lives are those most likely to contemplate arresting the aging process. It is not likely to be an attractive option for a substantial part of the world population, that is, the poor, oppressed, and sick.

Hundreds of thousands of septagenarians, and even older people, who are in relatively good health say that their current age is the happiest time of their lives. They contend that arresting the aging process at an earlier age would have either denied or delayed for them the contentment of retirement, travel, freedom from child-rearing responsibilities, and time to pursue personal interests that do not demand income generation. Human interactions depend to a substantial degree on perceptions of relative age. The destruction of those relationships could have enormous negative personal and societal consequences.

Presumably any method for intervening in the aging process would first become available to those able to afford expensive treatment and would be unavailable to those who could not. The intervention would also become available to antisocial and asocial persons, as well as those who do not harm or who benefit human civilization. The effect of manipulating the aging process could be disastrous for many human institutions.

Proposals to circumvent aging by replacing all old parts with younger parts are unlikely to be an option. For example, replacement of the brain could not only compromise one's sense of self-identity but the attendant loss of memory would erase the most essential part of what makes one human. Absent unrealistic scenarios in which a computer might be used to first upload the contents of an aging brain, cleanse it of old thoughts, and then download it to a new erased brain, it is unlikely that replacement of one's brain would ever be an attractive option. Also the eventual replacement of all old parts with younger or new parts in both animate and inanimate objects would result in both the physical and philosophical dilemma of having lost the original entity.

If it is true that mental processes continue to change for the better with age, one might equate the goal of arresting the aging process with that of arresting developmental processes. Arrested mental development in childhood is viewed universally as a serious pathology. If it is undesirable to retard the physical and mental development of a seven-year-old for ten years in order to gain an equivalent increase in longevity, arresting one's aging processes in later life should not be attractive for the same reasons.

Perhaps the least imperfect scenario would have each person live to be 100, while remaining in good physical and mental health, and then quickly and painlessly die at the stroke of midnight (as in "The Deacon's Masterpiece or The Wonderful One Hoss Shay" by Oliver Wendell Holmes [1857–1858]).

Humankind will probably not face these ethical issues in the near future because it is unlikely that biogerontologists will find ways to intervene in the fundamental aging and longevity determining processes for several reasons. First, most research done under the rubric of aging research in humans is done on age-associated diseases, the resolution of which cannot extend human longevity more than fifteen years. This accomplishment will not provide any insight into the fundamental aging process. The resources devoted to research on the underlying aging process are, by comparison, infinitesimal. Second, there are no generally acceptable criteria for measuring whether or not an intervention in humans is affecting either the aging process or the determinants of longevity. Finally, although the determinants of longevity might be altered, the aging process, because it is a fundamental property of all matter, is unlikely to be changed.

Regenerative Medicine Research and Ethical Considerations

The ethical issues that derive from research in regenerative medicine are more immediate than those that might result from intervening in the aging or longevity determining processes. In the early twenty-first century, several major advances are close to, or have become, reality (Cibelli, Lanza, Campbell, and West 2002).

The central ethical issue in regenerative medicine is whether the taking of human cells or tissues for research or therapeutic purposes is acceptable when a donor is incapable of giving informed consent. If no informed consent is possible, does consent by others, for the purpose of promoting research that will benefit society, outweigh the taking of what some believe to be a potential life?

When the potential source of cells or tissues is a fetus or embryo, ethical considerations usually center on the pros and cons of induced abortion. The arguments are frequently based on some arbitrary time in embryonic or fetal development when it is thought that human life begins. Many biologists argue that human life does not have a beginning (except on an evolutionary timescale) because both sperm and egg cells must be alive from the start and fusion of the two is simply another of the many critical steps that, if successful, can lead to the development of a viable offspring. Others contend that the potential for human life only occurs at the moment of conception. This is another arbitrary point because equally critical events must occur both before and after fertilization to insure that the potential for human life is realized. This issue could become even more clouded if it is shown, as it has been in some animals, that a jolt of electricity or a needle prick can stimulate an unfertilized egg to develop—a process known as parthenogenesis.

However the vast majority of sperm and eggs produced never fuse to form a zygote and if they do, a substantial number of zygotes subsequently are lost naturally. Yet this enormous loss of potential human life that far exceeds the number of successful births is rarely deplored.

In order to circumvent some ethical objections, the use of somatic cell nuclear transfer (SCNT) has been shown to be a practical alternative. Here the nucleus from a body cell (other than a gamete or its precursors) is inserted into an egg whose nucleus has been removed. This is done in vitro with the resulting dividing cells used for research or for potential therapy in the nuclear donor where problems of immunological incompatibility are reduced. Like the fusion of a sperm and egg in vitro, it is not possible for this cluster of cells to become a viable embryo unless the zygote is implanted into a uterus. Despite the fact that the nucleus used in SCNT comes from a single donor, the cells that form the zygote, or later developmental stages, could be used therapeutically when compatibility problems are overcome.

What must be weighed in considering the taking of human fetal cells or tissue is if anyone has the right to make the decision and whether or not the benefit that might accrue to many potential recipients outweighs the loss of one or a few potential lives. One significant precedent for making this decision in the affirmative is the often overlooked fact that, in the last forty years, hundreds of millions of people throughout the world have benefited from the use of many common virus vaccines, all of which were produced (and are still produced) in cells obtained from one or two surgically aborted human fetuses on which research had been publicly supported (Hayflick 2001).

LEONARD HAYFLICK

SEE ALSO *Biotech Ethics; Medical Ethics; Posthumanism; President's Council on Bioethics.*

BIBLIOGRAPHY

Binstock, Robert H. (2003). "The War on 'Anti-aging Medicine.'" *The Gerontologist* 43(1): 4–14. A scholarly analysis of how mainstream biogerontologists are countering the practitioner's of anti-aging medicine who biogerontologists believe are exploiting the publics' ignorance of the science for financial gain.

Cibelli, Jose; Robert P. Lanza; Keith H. S. Campbell; and Michael West. (2002). *Principles of Cloning.* San Diego, CA: Academic Press.

Gruman, Gerald J. (1966). "A History of Ideas about the Prolongation of Life." *Transactions of the American Philosophical Society* 56, Part 9. Probably the best history ever written on the evolution of human thought about longevity from the first written records until 1800.

Hall, Stephan S. (2003). *Merchants of Immortality.* New York: Houghton Mifflin. Developments in the fields of research on aging, stem cells, regenerative medicine and cloning are considered in this well-written and carefully researched history of these fields. Biological advances, political positions, and ethical debates are covered in detail.

Hayflick, Leonard. (1994). *How and Why We Age.* New York: Ballantine Books. An overview of the aging process written for the lay reader with a chapter on the probability and desirability of intervention.

Hayflick, Leonard. (2000). "The Future of Aging." *Nature* 408: 37–39.

Hayflick, Leonard. (2001). "A Brief History of Cell Substrates Used for the Preparation of Human Biologicals." In *Developments in Biologicals* Vol. 106: *Evolving Scientific and Regulatory Perspectives on Cell Substrates for Vaccine Development*, ed. Fred Brown, Keith Peden, Philip Krause and Andrew M Lewis. Basel, Switzerland: Karger.

Hayflick, Leonard. (2004). "'Anti-aging' Is an Oxymoron." *Journal of Gerontology* 59A(6): 573–578. The failure to define common terms in the field of gerontology has contributed to the many conceptual misunderstandings that have hindered progress and contributed to serious communication failures.

Hayflick, Leonard; Stanley Plotkin; and Robert E. Stevenson. (1987). "History of the Acceptance of Human Diploid Cell Strains as Substrates for Human Virus Vaccine Manufacture." In *Developments in Biological Standardization* Vol. 68, eds. John C. Petricciani, and W. Hennessen. Basle, Switzerland: Karger Publishing. Also published in Bookchin, Debbie, and Jim Schumacher. (2004). *The Virus and the Vaccine.* New York: St. Martin's Press. How the struggle to have normal human fetal cells used for the production of many of the world's human virus vaccines was ultimately won. The early use of primary monkey kidney cells for this purpose revealed contamination with several dangerous viruses. Yet it took more than a decade to overcome the scientific, political, and economic resistance to have them replaced with the safe, normal, human fetal cell strain WI-38 discovered by Leonard Hayflick. Now many safe and efficacious vaccines made in Hayflick's human fetal cells have been administered to almost one billion people.

Olshansky, S. Jay; Leonard Hayflick; and Bruce A. Carnes. (2002). "No Truth to the Fountain of Youth." *Scientific American* June: 92–95. A declaration with much supporting evidence that, in the early-twenty-first century, there is no way of intervening in the fundamental aging process in humans. Signed by fifty-one of the world's leading researchers on aging.

AGRARIANISM

Agrarianism may be defined as the view that the practices of the agricultural life, and the types of technology on which that life has historically been based, are particularly effective in promoting various important personal, social, and political goods. The precise character of these goods—and the respective roles of science, technology, government, society, and individuals in procuring them—varies according to which thinker or stream of agrarian thought one wishes to consider. Two different sources of modern agrarian thinking will be considered here: (1) the agrarianism of the "Old Whig," anti-federalist American founders, itself a self-conscious effort to retrieve the agrarian and republican values of the classical world; and (2) the agrarianism promoted by antimodern thinkers of the twentieth and twenty-first centuries. A third stream of agrarian thought and practice may be found among dissenting religious groups such as the Amish and other Anabaptist sects. However, with the exception of their theologically-grounded suspicion of scientific inquiry, these religious groups' ethical critique of science and technology is more fully articulated by the antimodern agrarians. Indeed, the antimodern agrarians' political-ethical critique of modern science and technology, though not especially well known, is one of the more original to have emerged in the last century and is arguably becoming more influential.

From Old Whig to Antimodern Agrarianism

As every schoolchild knows, Thomas Jefferson contended for an agrarian vision of America. As unsystematic in his approach to this subject as he was to most others, Jefferson scattered his brief observations about the value of the agricultural life throughout his letters and other documents. Most famously, in query XIX of his *Notes on the State of Virginia* (1781–1782), Jefferson argued that agriculturalists were especially apt to be virtuous: "Those who labour in the earth are the chosen people of God, if ever he had a chosen people, whose breasts he has made his peculiar deposit for substantial and genuine virtue. . . . Corruption of morals in the mass of cultivators is a phenomenon of which no age nor nation has furnished an example." By virtue, Jefferson and most of the other anti-federalists had foremost in mind a certain spirit of self-reliance that made economic—and therefore genuine political—independence possible. Yeoman farming was the indispensable support of republican government.

Jefferson was a reliable spokesman for republican agrarianism, but its most penetrating theorist was probably John Taylor of Caroline (1753–1824), a leading Virginia planter whose agrarian treatise *Arator* was first published as a series of newspaper articles in 1803. Much of the book consists of Taylor's practical suggestions, based on his own analysis, observation, and experiments, for improving American agriculture (eight numbers alone are devoted to Taylor's thoughts on the topic of "manuring"), the condition of which he lamented ("Let us boldly face the fact. Our country is nearly ruined").

Taylor's defense of republican agrarianism rests on much the same ground as Jefferson's. Political independence, Taylor agrees with Jefferson, cannot be secured by "bankers and capitalists." But not only does he place more emphasis than does Jefferson on the role of agriculture as "the mother of wealth" as well as "the guardian of liberty," he goes further in articulating the personal benefits afforded by life on the land. Farming, he maintains, brings more pleasure than other modes of employment. It provides continual novelty and challenges to the mind. It meets the physical needs of the body. It promotes the virtue of liberality and rewards almost every other virtue. It is an aid in the quest for eternal life, for it feeds the hungry, clothes the naked, and gives drink to the thirsty. And because it is a vocation inevitably more concerned with practical affairs than abstract speculations, it is the "best architect of a complete man." Virtually every claim for the farming life to be made by agrarian thinkers in the following centuries is anticipated here.

As M. E. Bradford points out in his introduction to a 1977 reissue of *Arator*, Taylor, Jefferson, and their fellow Old Whigs, such as Edmund Ruffin (1794–1865), quite consciously saw themselves as retrieving the classical agrarian tradition represented by figures such as Hesiod (c. 600 B.C.E., in *Works and Days*), Marcus Porcius Cato (234–149 B.C.E., in *De Agri Cultura*), Marcus Terentius Varro (116–27 B.C.E., in *Re Rustica*), and Virgil (70–19 B.C.E., in *Georgics*). Such figures were, like the Old Whigs, concerned with the relationship between politics and farming, and they therefore also tended to celebrate the personal and civic virtues associated with farming—economic independence, willingness to engage in hard work, rural sturdiness, hatred of tyranny—that the Old Whig founders saw themselves as protecting through the American Revolution.

The celebration of the farmer's life in America at the time of the founding of the nation was not limited to southern Republicans. One must note, for instance, J. Hector St. John Crèvecoeur's *Letters from an American Farmer* (1782). But the approach of someone such as Crèvecoeur (or, in the nineteenth century, writers such as Donald Grant Mitchell [1822–1908]) is that of the pastoral—which is to say, the use of farming principally as a literary device or metaphor for the exploration of other themes. In Crèvecoeur's case, this would include the nature of Nature, with a capital and Rousseauean N. *Letters from an American Farmer* is thus a literary more than an agrarian classic, and philosophically Crèvecoeur is more nearly a forerunner of later environmentalists than he is of the agrarians, who typically display a more profound awareness than he of the imperfectability of the human and natural world.

Although republican agrarianism would continue to permeate American politics and literature for many years—and indeed, continues to find resonance in contemporary works such as Victor Davis Hanson's influential *The Other Greeks: The Family Farm and the Agrarian Roots of Western Civilization* (1995)—by the mid- to late-1800s defenses of agrarian ways had become entangled with populist politics and as such were less explicitly focused on the goods of the farming life per se than on the interests of farmers. But with the closing of the North American frontier at the end of the nineteenth century, and with the concomitant slow decline in the number of Americans living on farms (the U.S. farm population began to decrease as a proportion of the whole after 1917), a new generation of self-consciously agrarian thinkers began to emerge. These included the American economist Ralph Borsodi (1888–1977), the founder of the Country Life movement Liberty Hyde Bailey (1858–1954), and the Harvard sociologist Carle Zimmerman (1897–1983), all of whom—along with several others—are profiled in Allan Carlson's indispensable history, *The New Agrarian Mind: The Movement Toward Decentralist Thought in Twentieth-Century America* (2000).

As Carlson shows, this group heralds the advent of a new and distinct type of agrarianism. Although its proponents' political affiliations varied widely (some were radical progressives, some liberals, some conservatives, and at least one a self-described reactionary), they all shared a deep dissatisfaction with many aspects of modern economic, political, social, and religious structures. The urbanized, mass consumerism of industrial society had come into focus for them as a characteristic feature of modernity in a way that it could not have for the earlier republican agrarians. Some form of resistance to modernity, some alternative, was therefore needed.

The men and women associated with the so-called Southern Agrarians arguably constituted the most important group of antimodern agrarian thinkers. Their manifesto, *I'll Take My Stand,* was published in 1930. An oft-overlooked sequel, *Who Owns America?* (which also featured contributions from prominent English distributists like Hilaire Belloc [1870–1953] and Douglas Jerrold [1803–1857]), appeared six years later. The leaders of the Southern Agrarians—John Crowe Ransom (1888–1974), Donald Davidson (1893–1968), Allen Tate (1899–1979), and Andrew Nelson Lytle (1902–1995)—would continue to develop agrarian themes and arguments for some years, although Ransom bowed out of the struggle earlier than the others. While they shared the republican concerns of their southern forebears Jefferson and Taylor, they also charged modern industrialism with promoting irreligion, extinguishing great art and high culture, degrading the quality of human relations, and, not least, destroying the old rural, aristocratic southern culture they preferred to the industrial culture of the North.

The Southern Agrarians hoped to spark a "national agrarian movement," in Ransom's words. In this they failed spectacularly, but they did leave behind some successors, most notably the University of Chicago rhetoric professor Richard M. Weaver, the literary critic and American founding scholar M. E. Bradford, and the novelist, essayist, poet, and farmer Wendell Berry, unquestionably North America's leading contemporary agrarian writer.

Although Berry belongs to some extent to the Southern Agrarian tradition, his agrarianism has several other sources, as well. He represents the agrarianism associated with radical and progressive movements—the mid-century "Back to the Land" movement, the eco-agrarianism loosely associated with the postwar counterculture (Berry has been active, for instance, in antinuclear efforts), and the movement toward green or organic farming and against agribusiness and genetically modified foods. In Berry the common ground held by all of these sources of modern agrarian discontent becomes clear.

Agrarians, Science, and Technology

Agrarianism, in its republican version, was generally associated with a positive view of the ability of science and technology to aid agriculture in its effort to bring about a wealthier and more comfortable existence. "If this eulogy should succeed in awakening the attention of men of science to a skilful practice of agriculture," wrote Taylor of his *Arator*, "they will become models for individuals, and guardians for national happiness." Classical in inspiration, even the practical Taylor's republican agrarianism conformed to the rationalism of the Enlightenment.

Indeed, even the antimodern agrarianism of people such as Borsodi and Bailey, who were more concerned with the urbanization and centralization of modern life than they were with its secularization and the cultural ascendance and authority of science, represented a version of Enlightenment rationalism. But a few agrarians developed rather sophisticated and original critiques of scientific rationality and technological society. Most worthy of mention in this regard are Ransom, Tate, and Berry. For them, mass technological-industrial society was the consequence and analog of the scientific mode of thinking.

Some of Ransom's best work on this subject is included in his first two books, *God Without Thunder* (1930) and *The World's Body* (1938), in which he argued that reality does not inhere in the abstract, universal laws proposed by science as a way of "explaining" all phenomena, but rather in concrete, particular objects. These particular objects cannot be known as particulars via scientific reason, because science depends on the method of abstraction, which sees a particular only as an instance of a more universal category. A poetic or aesthetic approach, by contrast, does justice to the world by attempting to create a vision of the whole of reality with all its messy and mysterious particularity. In Ransom's historiography, the world had moved first from the perceptual (or premodern) "moment," thence to the "conceptual/scientific," and finally must now progress to the "aesthetic."

Tate makes a similar argument in "Remarks on the Southern Religion," first published in *I'll Take My Stand*. Where Ransom posits poetry or the aesthetic mind as conserving the "whole" object (or, in his vocabulary, "the world's body") for consideration, Tate posits a religious approach as the antithesis of abstraction. Modern science, writes Tate, reduces objects to those qualities they share with other objects of the same type, and to what they can do or how they work—"the American religion" to which the southern religion of his title is opposed. For Tate, an obsessively quantitative way of seeing the world had become characteristic of the modern Western mind.

Among agrarians, Berry has articulated the most radical critique of scientific rationality and technological progress. Like Ransom and Tate, he defends the validity of a particularist epistemology and maintains that only a limited portion of the truth of experience can be known by the reductionist methods of science. His ethical critique of modern society rests, like his epistemological critique, on the argument that mass technological industrialism collaborates with science to enshrine a view of human beings and the natural world that treats objects and people as essentially interchangeable. Such arguments can be found throughout Berry's corpus, but they are brought together most systematically in his *Life Is a Miracle: An Essay Against Modern Superstition* (2000), which attacks the scientism promoted by E. O. Wilson in *Consilience: The Unity of Knowledge* (1998). The skepticism displayed by Ransom, Tate, and Berry with regard to the truth claims of science has obvious resonances with postmodern thought while resisting temptations to indulge in relativism.

Agrarianism Outside America

It is difficult to generalize about the relationship between agrarianism and science and technology as that relationship has taken shape outside the North American context. Often, the so-called agrarian social movements of Latin America, Asia, and Eastern Europe have been allied with, or inspired by, anarchist or Marxist revolutionary ideologies (for example, most repulsively, Mao Zedong's Cultural Revolution and Pol Pot's Khmer Rouge). In the case of Marxist or neo-Marxist agrarians, their accompanying attitudes toward scientific rationality have hardly been similar to those of the antimodern agrarians.

Yet few non-American thinkers or activists commonly associated with agrarianism seem especially worthy of mention. Prince Pyotr Alekseyevich Kropotkin (1842–1921) was beloved by many on the American left in the late nineteenth and early twentieth centuries, including radical agrarian-oriented writers such as Dorothy Day (1897–1980), who saw Kropotkin and Leo Tolstoy as promoting essentially the same sociopolitical vision espoused by the English distributists. Both groups advocated the decentralization of economic power and associated agrarian ideals of one kind or other. But by and large these Russian and British thinkers did not share deeply in nor anticipate the kind of antimodern critiques of science and technology discussed above. Kropotkin was in fact a scientist, an accomplished geographer whose anarchism was at least in part the consequence of his scientific view that the natural, animal, and social worlds were not inevitably grounded in the law of competition, as the Darwinists (social and otherwise) taught, but cooperation and mutual aid. And neither G. K. Chesterton (1874–1936) nor Belloc, though certainly not philosophical "modernists," was as skeptical of the epistemological power of discursive rationality as were antimodern American agrarians such as Ransom or Berry. Frankly, the social philosophies of the English distributists and Russian anarchists were too broad and diffuse to be called properly "agrarian," although they certainly had agrarian components. The same could probably be said of Mohandas Gandhi (whose agrarian views were much inspired by Tolstoy).

In his specifically agricultural writings, there may be no one whom Berry cites more frequently than Sir Albert Howard, the English scientist whose *An Agricultural Testament* (1940), which was chiefly concerned with the rehabilitation of soil fertility, played an important role in creating the organic farming movement. The new agricultural science—and, hence, agriculture—promoted by Howard and those he influenced, including the American organic gardening/farming pioneer J. I. Rodale (1898–1971) and The Land Institute founder and director Wes Jackson, may possibly be considered as constituting yet another stream of agrarian thought.

But note that this tradition of agricultural thought does not concern itself so much with the larger philosophical question of how agrarian practices and culture contribute to the good life, as with attempting to deepen our understanding of what kind of farming techniques are truly sound, arguing on a scientific basis that, for instance, small, family-owned and -operated organic farms are more practical in the long run. There tends to be a confluence between it and antimodern agrarianism because it tends to reject the scientific specialization characteristic of the modern West, and especially its close relationship to industrialism. Although Berry, for one, has clearly been heavily influenced by this tradition, he is typically much more skeptical of the epistemological sufficiency of science and the social beneficence of technology than are the representatives of scientific agrarianism.

Conclusion

At the beginning of the twenty-first century, in North America and Europe at least, political radicals and progressives are most usually identified with resistance to the large-scale agriculture embodied by contemporary agribusiness and the technological triumph it symbolizes: think, for instance, of Theodor Shanin's work in peasant studies or José Bové, the southern French farmer and activist famous for his attacks on McDonald's and the globalization of the food market generally. However, the fact that Berry's work has registered appeal across the political spectrum indicates that concern for the fate of the independent farmer and the land of which he is the steward continues to draw on popular agrarian ideals. Thus, to the modern republican agrarian, agribusiness represents the application of commercial and industrial techniques to farming. And to the antimodern agrarian, genetic engineering represents a misplaced faith in the beneficence of technological experimentation. It seems likely that the intellectual future of agrarianism lies in the success with which it is able to put forth a political and ethical philosophy that grounds such arguments convincingly, a task all the more difficult in a profoundly non-agrarian culture.

JEREMY BEER

SEE ALSO *Agricultural Ethics; Jefferson, Thomas.*

BIBLIOGRAPHY

Berry, Wendell. (1977). *The Unsettling of America: Culture and Agriculture*. San Francisco: Sierra Club Books. Without question, the most important agrarian work to be published in America during the last half of the twentieth century.

Berry, Wendell. (2002). *The Art of the Commonplace: The Agrarian Essays of Wendell Berry*, ed. Norman Wirzba. Washington, DC: Counterpoint. A well-selected collection of Berry's most important essays on agricultural themes.

Carlson, Allan. (2000). *The New Agrarian Mind: The Movement Toward Decentralist Thought in Twentieth-Century America*. New Brunswick, NJ: Transaction. Essential volume includes profiles of largely forgotten antimodern agrarians such as Father Luigi Ligutti, Frank Owsley, Troy Cauley, and Herbert Agar.

Freyfogle, Eric T., ed. (2001). *The New Agrarianism: Land, Culture, and the Community of Life*. Washington, DC: Island Press. Fine collection of essays by contemporary agrarian writers, including Gene Logsdon, Scott Russell Sanders, and David Kline, made all the more valuable by Freyfogle's learned and informative introduction.

Malvasi, Mark G. (1997). *The Unregenerate South: The Agrarian Thought of John Crowe Ransom, Allen Tate, and Donald Davidson*. Baton Rouge: Louisiana State University Press. Best book-length explication of the ideas of these three most important Southern Agrarians.

Shi, David E. (1985). *The Simple Life: Plain Living and High Thinking in American Culture*. New York: Oxford University Press. Highly regarded intellectual history provides a broader context for the consideration of American agrarian thought.

Smith, Kimberly K. (2003). *Wendell Berry and the Agrarian Tradition: A Common Grace*. Lawrence: University Press of Kansas. Only monograph to date to explore Berry's social and political thought; contains an excellent sketch of the intertwined histories of environmentalism and agrarianism in America.

Taylor, John. (1977). *Arator: Being a Series of Agricultural Essays, Practical and Political: In Sixty-Four Numbers*, ed. M. E. Bradford. Indianapolis, IN: Liberty Fund. Taylor, more than Jefferson, is the representative republican agrarian of the post-revolutionary period.

Twelve Southerners. (1930). *I'll Take My Stand: The South and the Agrarian Tradition*. New York: Harper & Brothers. The antimodern agrarian touchstone, suffused with intelligence, still proves inspiring to agrarians and irritating to critics.

AGRICULTURAL ETHICS

Agriculture is among the earliest, most enduring, and most fundamental domains of technology. Although associated primarily with the cultivation of food crops such as wheat, maize, and rice, the term *agriculture* covers a wide variety of activities, including animal husbandry, dairy production, fiber production (for example cotton, flax), fruit and wine production, and aquaculture, as well as the harvesting, storage, processing, and distribution of food and fiber commodities. Agriculture frequently is understood to include all forms of food, fiber, and subsistence production, including forestry and fishing, especially with respect to the organization of scientific research institutes and government regulatory agencies. For example, government ministries such as the United States Department of Agriculture (USDA), the United Kingdom Ministry of Agriculture, Food and Fisheries (MAFF), and the United Nations Food and Agricultural Organization (FAO) have a responsibility for forestry and fisheries in their mandates. In all cases agriculture is both deeply involved with technology and science and subject to technical reflection.

Technology and Science in Agriculture

What is the relationship between agriculture and technology? That question reflects the way agriculture has faded into the cultural background in contemporary life, as if foods naturally appeared on supermarket shelves without technological intervention. It also reflects the way technology is associated strictly with machinery, manufacture, and engineering. Yet even in this narrow view agriculture has been influenced deeply by mechanization and chemical technology for 250 years.

It is more informative to see the crop varieties that farmers plant as technological artifacts, along with the systems they develop for cultivating soil, applying water, controlling weeds and other pests, harvesting, and storing and distributing agricultural products. In any broad interpretation of technology, agriculture is fundamentally a technological activity, and a technically sophisticated approach to the production, harvesting, and distribution of food is a hallmark of all civilizations.

Technical innovation in agricultural practice has been continuous throughout human history. The simple act of cultivating plants and domesticating animals, as distinct from scavenging, marks a fundamental technical advance. Prehistoric innovations in agricultural technology include achievements such as the domestication of animals, the construction of complex systems for irrigation and water management, and the development of tools for turning and maintaining the soil. Farmers also developed sophisticated techniques for maintaining desirable traits in their crops long before the underlying genetic basis of those methods was understood. Recent research (see Richards 1985, Brush 1992, Bellon and Brush 1994) on traditional farming systems has documented the sophistication farmers have applied in adapting cultivation methods and the genetic stock of their crops and animals to local conditions. Seeing traditional agricultural methods as "pretechnological" is unwarranted in light of this research. Indeed, the "agricultural revolution" equals and may exceed the industrial revolution with respect to its impact on environment and subsequent human history.

Traditional agricultural systems take a wide variety of forms. Improving or maintaining soil fertility, for example, has a number of possible technical solutions, including the composting and application of human, animal or vegetable wastes. Alternatively, pastoralists can develop symbiotic relationships with settled cultivators, who allow animals in their fields to graze (especially on stubble) and derive the benefit of the animals' manure in exchange. Swidden or "slash and burn" agriculture involves the use of fire to release nutrients from indigenous vegetation followed by cultivation at the site until fertility created by this technique has been exhausted. Other key technical elements involve water, soil loss, and genetic diversity. Much traditional agriculture is rain fed, but massive irrigation systems were developed in ancient Egypt and China. Construction of terraces provided an ancient solution to erosion. Genetic diversity was traditionally enhanced by farmer observation of unique types (or sports) and subsequent experimentation with small plots until new traits were understood and could be integrated into the main crop (see Wilken 1987).

Technical innovation in agriculture continued in the modern era and has been continuous with the development of modern science. The link between science and agricultural improvements was mentioned prominently by the philosopher Francis Bacon (1561–1626). The agriculturist Jethro Tull (1674–1741) published a scientific treatise on tillage in 1733. Thomas Jefferson (1743–1826) made improvements to the moldboard plow and advocated the inclusion of agriculture in university curricula. Cyrus McCormick (1809–1884) developed a mechanical reaper that is regarded as one the signature technologies of the nineteenth century. The German chemist Justus von Leibig (1803–1873) often is identified as the founder of modern agricultural science. Von Leibig pioneered the use of controlled experimental approaches in soil chemistry and crop improvement.

In the early twenty-first century, many traditional agricultural practices coexist with highly industrialized production methods. Commercial fertilizers and insecticides are synthetic, petroleum based products that were developed in junction with military technologies (Russell 2001). Modern crop varieties (discussed below) provide the genetic basis for large scale monocultures. In contrast to plants from traditional crop varieties, which may vary greatly in size, shape, color and response to climatic conditions, plants from modern varieties are uniform in size. They germinate, flower and produce grain or fruit at the same time. As such they are well suited to mechanical cultivation and harvesting, as well as to large-scale management and marketing practices. They also require intensive management of factors (such as water, nutrients, diseases and insect pests) that would be highly variable under traditional conditions. All these characteristics of industrial agriculture tie it closely to an extensive science and technological support system.

Agricultural science became institutionalized in industrialized countries in the late nineteenth century with the establishment of government stations dedicated to agricultural research. The system in the United States combined the federally based Agricultural Research Service with existing state-based land grant universities that were chartered in 1862 as institutions dedicated to agriculture and engineering. In addition to offering education in agronomy and animal husbandry, land grant universities conducted research on local soil, climate, and crop interactions. Their findings were made available to farmers through state-operated extension services whose agents conducted demonstrations of new crop varieties, machinery, and management systems. That system was responsible for a number of technical advances of regional importance in the first half of the twentieth century, including new methods for testing soil chemistry and recommendations for the efficient application of fertilizer.

The historian Charles Rosenberg's *No Other Gods* (1976) argues that the early success of agricultural research conducted and disseminated through this three-way partnership of experiment stations, universities, and extension was responsible for the rising status of science in the United States during the early twentieth century. The example of agricultural technology also encouraged Americans to support the provision of public funds for science and engineering. The U.S. system of partnership between agricultural universities, experiment stations, and local extension services to develop technology for the benefit of citizen farmers continues to serve as a model for publicly funded and publicly managed approaches to the development and dissemination of technology.

Main Problems in Agricultural Ethics

The potential range of ethical issues in agricultural technology is extraordinary. Those issues can be conceptualized in three categories: (1) issues relating to human health and security; (2) issues relating to the broader environment; and (3) issues relating to the cultural, historical, and social significance of agriculture as a way of life and a system of connected institutions. The first category includes the availability of basic foods, diet, nutrition, and questions concerning food safety. The second category includes the philosophical status of agricultural ecosystems and their relationship to nature, along with questions about the standing of animals and human obligations to them. The third category concerns the social organization of agriculture and has focused on questions associated with the industrialization of farming. These categories clearly overlap, and the three-way division should be understood as a heuristic device rather than a philosophical classification scheme with ontological or ethical significance.

Hunger and food security usually are thought of as particularly compelling cases of the ethics of distributive justice: What constitutes a fair, just, or morally acceptable pattern of access to wealth and resources? Key problems include the ethical basis for framing moral obligations relating to food access: Is there a basic human right to food, as the International Declaration of Human Rights (1948) alleges, or do utilitarian models of human welfare provide a better approach to understanding the ethics of hunger? How should moral entitlements to food security be operationalized now and in the future? This question ties the discussion of food security to broad issues in economic development and especially to the challenge of population growth.

Problems related to nutrition are closely interwoven with the development of scientific nutrition in animal science departments at the end of the nineteenth century. Methodological issues figure importantly in ethical discussions of appropriate nutritional advice. Other issues involving food system risk and safety are closely tied to science and technology in two ways. First, risks frequently are associated with agricultural technologies such as chemical pesticides, food irradiation, and biotechnology. Second, scientific risk analysis is central to the debate over the appropriate response to those risks. Risk optimization, informed consent, and the precautionary principle represent three philosophical approaches to the way in which risk analysis should be applied in determining the acceptability of food system risks.

Similar risk issues are associated with the environmental consequences of agricultural technology, and transgenic crops and animals have been important case studies for risks to nonhuman organisms and ecosystem integrity. With respect to environmental impact, ethical analysis draws on debates in environmental ethics about the moral standing of nonhuman animals, wild nature, and the structure of ecosystems as well as duties to future generations. Sustainability has been proposed as a way to frame the ecologically desirable features of any agricultural system, and disputes over the appropriate specifications for a sustainable agriculture have been a major focus in agricultural ethics.

In the United States and Canada discussion of the sociocultural aspects of agricultural production systems often has been framed in terms of "saving the family farm." In Europe the debate has been framed in regard to the need to preserve traditional agriculture, and internationally the issues have been framed in terms of the industrialization and intensification of farming methods that continue to rely on a great deal of human and animal labor. These questions can be looked at strictly in terms of environmental and human well-being, but the structure of agriculture and the centuries-long transition that has seen fewer and fewer people employed in agriculture highlights an important dimension of the sociocultural aspects of agriculture as well as a significant link to the philosophy of technology.

Ethical Issues in Agricultural Science and Technology

The influence of publicly organized research conducted at experiment stations in industrialized countries and the organized attempt to extend those results throughout the world provide the basis for viewing agricultural science and technology as an applied science with explicit value commitments. Those values derive from the importance of food and fiber in meeting human subsistence needs, the vulnerability of virtually all people to food-borne risk, and the dependence of the rural countryside on agriculture as its key industry and dominant cultural force.

Although farming practice sometimes has adopted the stance of maintaining traditions and social institutions, modern agricultural science more typically has been guided by the maxim of increasing yield: Make two plants grow where one grew before. Thus, the underlying ethic of agricultural technology has been one of increasing efficiency. This ethic is can be interpreted most readily as a fairly straightforward application of utilitarianism: Research and technology development should aim to produce "the greatest good for the greatest number," primarily by increasing the efficiency of agricultural production.

This general orientation to science and technology has been challenged by the view that agricultural science should serve the specific interests of farmers and that researchers should be mindful of this constraint. The development of high-yielding varieties of hybrid maize is a case in point. In the 1950s Paul Mangelsdorf (1899–1989) of Harvard and Donald Jones (1890–1963) of the Connecticut Agricultural Experiment Station discovered and patented cytoplasmic male sterility as a method for producing hybrid varieties. Many technical advances of the early twentieth century had been distributed to farmers free of charge through state extension services, but hybrid seeds had to be produced anew for each growing season. Jones was censured publicly by his colleagues for seeking to patent his discovery despite the fact that, or perhaps because, its chief value was to the commercial seed industry. Mangelsdorf's affiliation with a private university shielded him from his colleagues' censure. Contrary to medicine and engineering, in which publicly funded research has been commercialized routinely through the use of patents, publicly sponsored agricultural research has been seen by some as a public good for the express benefit of farmers (see MacKenzie 1991).

The economist Willard Cochrane (b. 1914) developed an analysis of efficiently increasing agricultural technology that extended the scope of this concern. In referring to "the technology treadmill," Cochrane showed that because the market for food is limited in size, more efficient production always will lead to a reduction in prices. Farmers who adopt technology quickly can earn profits before prices adjust, but as prices come down, they will have "run harder just to stay in place" (produce more to earn the same level of income they had at the higher commodity price). Cochrane's analysis suggests that agricultural research typically does not benefit farmers; instead, the benefit goes almost exclusively to consumers in the form of lower food prices. It also implies that there is an underlying economic necessity to the trend for fewer and ever larger farms (see Browne et. al. 1992).

The technology treadmill argument places the utilitarian argument for efficiency against the idea that agricultural scientists have special moral duties and loyalties to rural communities. One still might argue for yield-enhancing technological improvements on the grounds that they provide small but universally shared (and hence additively large) benefits to food consumers. Those benefits almost certainly will outweigh the losses in the form of farm bankruptcies and depopulation of the rural countryside. However, this argument undercuts the populist ethical rationale for agricultural research as benefiting rural communities and preserving the family farm.

Cochrane's interest was in American farmers, but the economic logic of the technology treadmill plays out in developing countries as well. Perhaps the most controversial application of agricultural science in the twentieth century was the Green Revolution, an initiative sponsored by the Rockefeller Foundation in the 1950s and 1960s to make high-yielding crops available in depressed regions of developing countries. The program was rationalized in part as a response of the capitalist world to the growing influence of Soviet bloc socialism after World War II.

As a technical program the Green Revolution was a mixed success, with early efforts at improved crops foundering over local resistance to new methods and aesthetic differences in taste and cooking quality. Over time, however, improved varieties won out in most parts of the world, especially in India. Green Revolution rice and wheat varieties lie at the basis of a decade of surplus in India's total food production and one of best-fed populations outside the industrial West.

However, these increases in food availability came at a price. The use of Green Revolution varieties led to more food at lower prices, but the farmers with the smallest farms could not survive on lower profit margins. Furthermore, Green Revolution varieties were developed to be used with fertilizers and sometimes chemical pesticides as well. Poor farmers could not afford to purchase those inputs, and their use also created environmental problems in rural areas. The growing scale of farming in the developed world put farmers on a path toward the use of technology for weed control and harvest, whereas in the past those tasks had been performed by very poor landless laborers. Although one could argue that in the end the benefits of the Green Revolution have outweighed the costs, those costs were borne primarily by the poorest people in developing societies. The Green Revolution thus ran directly counter to the "difference principle" of justice elaborated by the philosopher John Rawls (1921–2002), which holds that social policies are justified to the extent that they tend to improve the lives of the group that is worst off. Vandana Shiva (1993; 1997) has been particularly influential in criticizing the Green Revolution on grounds of environmental damage and social inequality.

The environmental critique of Green Revolution technology addresses the utilitarian orientation to agricultural research in a different way. In treating the decision to develop new technology as an optimization problem, the utilitarian approach has a tendency to ignore impacts that are difficult to quantify. Environmental impacts are often externalities that do not figure in the costs a producer considers when deciding whether to use a particular technology. Furthermore, there are often no markets or forums available for those who bear environmental costs most directly to register their complaints. This is the case for future generations, for example, but also for animals, which can be placed in intolerable conditions in modern confined animal feeding operations. Thus, to be truly justified as producing the greatest good for the greatest number, agricultural technologies must not be plagued with externalities, and those who develop, evaluate, and utilize such technologies face a philosophical challenge in reflecting externalities in their decision making.

Since 1985 many of these issues have been revisited and revised in connection with the use of recombinant DNA techniques for transforming the genetic basis of agricultural plants and animals. Disputes over the patenting and ownership of genetic resources and intellectual property have been an especially prominent feature of this debate.

History of Agricultural Ethics

In one sense agricultural ethics is among the oldest philosophical topics. Classical figures such as Xenophon (444–375 B.C.E.) and Aristotle (384–322 B.C.E.) wrote lengthy discussions of agriculture and its relationship to the values and social institutions of Greek society. There is little doubt that those classical authors saw agriculture as a systematic human adaptation and modification of the natural environment rather than a natural system lacking a significant technological component. Furthermore, they saw the material basis of their society as playing a significant role in both shaping the *ethos* of Greek life and shaping the opportunities and requirements for political institutions. Brief and less systematic discussions of agriculture occur throughout the history of philosophy, though those discussions frequently involve technological changes in agricultural production methods. A typical example is the philosopher John Locke's (1632–1704) rationale for the enclosure of common lands as a strategy for increasing agricultural production through intensive farming in the *Second Treatise of Government* (1689).

The Baron de Montesquieu (1689–1755) made agriculture a main theme of his *Spirit of the Laws* (1748), arguing that climate and agricultural methods form the basis for population patterns, social institutions, and national identity. The philosopher Georg Wilhelm Friedrich Hegel (1770–1831) also offered extensive discussions of agriculture as a clue to the manifestation of Spirit. Hegel's account of the Greek food system, for example, notes that it was marked by rocky hills and mountains alternating with lowlands suitable for crop farming. Hegel noted that unlike China or India, the Greek landscape lacks a major inland waterway conducive to large-scale irrigation projects or the transport of harvested grain. In the place of centrally managed systems for irrigating and moving foodstuffs the Greeks developed a complex farming system that included a mix of tree and vine crops and did not depend on large pools of human labor for planting and harvesting. Hegel argued that this system favors democracy and the development of individuals who can see themselves as authors of moral judgment. This work in the Greek and European traditions of philosophy anticipates contemporary debates over the character of rural areas and the preservation of the family farm.

Ethical debate over hunger and food availability was comparatively rare until the eighteenth century, when important studies appeared in the work of the economists François Quesnay (1694–1774) and Adam Smith (1723–1790). The topic of hunger was of central importance for Thomas Malthus (1766–1834) and was discussed by Jeremy Bentham (1748–1832) and John Stuart Mill (1806–1873), all of whom were occupied at one time by the problem of "surplus population" and reform of England's corn laws. Malthus argued that the race between agricultural improvement and population growth would make hunger a continuing ethical issue.

In the twentieth century philosophers such as Peter Singer, Peter Unger, Onora O'Neill, and Amartya Sen were among the many who wrote about the ethics of hunger, questioning the moral basis of the obligation to address hunger and examining the moral implications of various economic regimes in light of hunger. Other recent work has been contributed by scientists such as Garrett Hardin (1915–2003) and Norman Borlaug, who have extended the Malthusian tradition of stressing the tension between the technical capacity for food production and population growth. With the exception of Sen, twentieth-century philosophical work on hunger seldom was attentive to science and technology.

Although philosophers writing before 1900 did not organize their work in terms of scientific or technological ethics, there is little doubt that they understood agriculture as a form of technology and were interested in the normative problems and implications of agricultural practice. For the most part the agricultural writings of past philosophers have been neglected. Singer's seminal article on world hunger in 1972 has virtually no discussion of agriculture and typically is not read as an exercise in either scientific or technological ethics. Recent work on hunger, as well as even more recent studies of agricultural biotechnology, makes virtually no reference to the philosophical-agricultural writings of the past. There is thus a large hiatus in the philosophical history of agricultural ethics as it relates to technology.

A few agricultural specialists contributed ethical studies on agriculture during the period from roughly 1900 to 1975. Liberty Hyde Bailey (1858–1954) was a leading American agricultural scientist who was known especially for his contributions to plant taxonomy. He chaired the Country Life Commission under President Theodore Roosevelt and was the main author of its report, which was an argument for egalitarian improvement of rural America through technological advance and social reform. Sir Albert Howard (1873–1947) was an English agronomist who conducted research on soil fertility. His books *An Agricultural Testament* (1940) and *Soil and Health* (1956) anticipated many contemporary ethical critiques of industrial agriculture and served as an inspiration for figures such as J. I. Robert Rodale, founder of the Rodale Press, and Wes Jackson, founder of the Land Institute. The anthropologist Walter Goldschmidt conducted a critical study of the social consequences associated with large-scale farming in California for the USDA in 1947, but many of his results were suppressed until they were published under the title *As You Sow: The Social Consequences of Agribusiness* in 1978. Rachel Carson (1907–1964) was the author of *Silent Spring* (1962), a polemical critique of agricultural pesticides that sometimes is credited with creating a popular environmental movement in the United States. The turn toward concern about the social and environmental effects of industrial agriculture paved the way for a rebirth of philosophical attention to agriculture as a form of technology in the last quarter of the twentieth century.

Aside from work by philosophers such as Singer, Unger, and O'Neill, who did not think of themselves as working in agricultural ethics, philosophical studies in agricultural ethics began anew around 1975 when Glenn L. Johnson (1918–2003), an agricultural economist, produced a series of articles on positivist influences in the agricultural sciences and called for renewed attention to normative issues. Agricultural issues came to the attention of philosophers largely through the work of Wendell Berry, a poet and novelist whose *The Unsettling of America* (1977) offered an extended philosophical critique of industrial agriculture, land grant universities, and modern agricultural science while putting forth an impassioned defense of the family farm. For a decade Johnson was known only to specialists in the agricultural science establishment, whereas Berry was regarded there as a meddling outsider with little credibility.

Johnson's call for normative reflection in the agricultural sciences was answered by Lawrence Busch, William Lacy, and Frederick Buttel, three sociologists who separately and in collaboration published many studies on the political economy of agricultural science during the last quarter of the twentieth century and also called for a philosophical and ethical critique of agricultural science and technology. They mentored a generation of sociologists who have examined normative issues, including Carolyn Sachs, who produced one of the first feminist studies of agriculture, and Jack Kloppenburg, Jr., author of *First the Seed* (1989), a normative history of plant breeding. Busch and Lacy brought the philosopher Jeffrey Burkhardt into their research group at the University of Kentucky in 1980. Paul B. Thompson was the first philosopher with an appointment in an agricultural research institution at Texas A&M in 1982. Thompson as well as a group at California Polytechnic University, including the philosopher Stanislaus Dunden, the agronomist Thomas Ruehr, and the economist Alan Rosenfeld, began to offer regular coursework in agricultural ethics in the early 1980s.

Institutional growth of agricultural ethics was stimulated by the W.K. Kellogg Foundation, which made many grants in that field in the 1980s and supported Richard Haynes in founding the journal *Agriculture and Human Values* and forming the Agriculture, Food and Human Values Society in 1988. In the 1990s Gary Comstock conducted a series of workshops on agricultural ethics at Iowa State University that brought the field to a larger audience. European interest in agricultural ethics lagged by about ten years. Led by Ben Mepham the agricultural research group at the University of Nottingham sponsored a seminal meeting on agricultural ethics in 1992. The European Society for Agricultural and Food Ethics was founded in 1998, and *The Journal of Agricultural and Environmental Ethics* became its official outlet in 2000. The first indication of interest in agricultural ethics beyond the West occurred with the launch of a series of papers on ethics at the FAO in 2000. Virtually all this work is focused closely on the ethical and policy implications of technological innovation and science-based decision making. The public debate over agricultural biotechnology has stimulated even more widespread interest in agricultural technology, and many individuals are conducting ongoing research.

PAUL B. THOMPSON

SEE ALSO *Agrarianism; Animal Rights; Animal Welfare; DDT; Deforestation and Desertification; Food Science and Technology; Genetically Modified Foods; Green Revolution; Environmental Ethics.*

BIBLIOGRAPHY

Bailey, Liberty Hyde. (1911). *Report of the Commission on Country Life*. With an introduction by Theodore Roosevelt. New York: Sturgis and Walton Co.

Bellon Mauricio R., and Stephen B. Brush. (1994). "Keepers of Maize in Chiapas, Mexico." *Economic Botany* 48(2): 196–209.

Berry, Wendell. (1977). *The Unsettling of America: Culture and Agriculture*. San Francisco: Sierra Club Books.

Browne, William; Jerry Skees; Louis Swanson; Paul Thompson; and Laurian Unneveher. (1992). *Sacred Cows and Hot Potatoes: Agrarian Myths and Policy Realities*. Boulder, CO: Westview Press.

Brush, Stephen B. (1992). "Ethnoecology, Biodiversity, and Modernization in Andean Potato Agriculture." *Journal of Ethnobiology* 12(2): 161–185.

Busch, Lawrence, and William B. Lacy. (1983). *Science, Agriculture, and the Politics of Research*. Boulder, CO: Westview Press.

Carson, Rachel. (1962). *Silent Spring*. Boston: Houghton, Mifflin.

Goldschmidt, Walter. (1978). *As You Sow: Three Studies in the Social Consequences of Agribusiness*. Montclair, NJ: Allenheld, Osmun.

Howard, Sir Albert. (1940). *An Agricultural Testament*. New York: Oxford University Press.

Howard, Sir Albert. (1956). *The Soil and Health: Farming and Gardening for Health or Disease*. New York: Devin-Adair.

Johnson, Glenn L. (1976). "Philosophic Foundations: Problems, Knowledge and Solutions." *European Review of Agricultural Economics* (July): 1–10.

Kloppenburg, Jack Ralph, Jr. (1988). *First the Seed: The Political Economy of Plant Biotechnology, 1492–2000*. New York: Cambridge University Press.

MacKenzie, David. (1991). "Agro-Ethics and Agricultural Research." In *Beyond the Large Farm: Ethics and Research Goals for Agriculture*, ed. Paul B. Thompson and Bill A. Stout. Boulder, CO: Westview Press.

Mepham, Ben., ed. (1996). *Food Ethics*. New York: Routledge.

O'Neill, Onora. (1986). *Faces of Hunger: An Essay on Poverty, Justice and Development*. London: Allen & Unwin.

Richards, Paul. (1985). *Indigenous Agricultural Revolution: Ecology and Food Production in West Africa*. Boulder, CO: Westview Press.

Rosenberg, Charles E. (1976). *No Other Gods: On Science and American Social Thought*. Baltimore: Johns Hopkins University Press.

Russell, Edmund. (2001). *War and Nature: Fighting Humans and Insects from World War I to Silent Spring*. New York: Cambridge University Press.

Sen, Amartya. (1988). *Poverty and Famine: An Essay on Entitlement and Deprivation*. Oxford: Oxford University Press.

Sen, Amartya. (1999). *Development as Freedom*. New York: Knopf.

Shiva, Vandana. (1993). *Monocultures of the Mind: Perspectives on Biodiversity and Biotechnology*. London: Zed Books.

Shiva, Vandana. (1997). *Biopiracy: The Plunder of Nature and Knowledge*. Boston: South End Press.

Singer, Peter. (1972). "Famine, Affluence and Morality." *Philosophy and Public Affairs* 1(Spring): 229–243.

Thompson, Paul B. (1995). *The Spirit of the Soil: Agriculture and Environmental Ethics*. New York: Routledge.

Unger, Peter. (1996). *Living High and Letting Die: Our Illusion of Innocence*. New York: Oxford University Press.

Wilken, Gene C. (1987). *Good Farmers : Traditional Agricultural Resource Management in Mexico and Central America*. Berkeley: University of California Press.

AIR

Air (and its variant spellings *eir, eyr, aier, ayre, eyir, eire, eyer, ayer, aire, ayere*, and *ayr*) all stem from the Latin *aer*. It is the most transparent but immediately necessary of all the classical Greek elements. It surrounds the Earth as atmosphere and was considered a mediating element, somewhere between fire and water, both warm and moist, the driving force behind the birth of the cosmos. As a spiritual element it pushed along the soul—the Greek work for spirit, *pneuma*, also means breath—and spread messages and ideas across the world in its guise as wind. In the early twenty first century, as gas, air represents one of the fundamental states of matter (the others being solid and liquid), while its pollution by technological activities constitutes a fundamental ethical challenge.

Air in Science

Air figures prominently in both physics and chemistry, and as atmosphere is subject to its own special science. Indeed among the achievements of early modern natural science was the distinction between air and atmosphere. In 1644 Evangelista Torricelli, a student of Galileo Galilei, invented the barometer and thereby discovered the phenomenon of atmospheric pressure. Later in the century it was shown that air/atmosphere is a mechanical mixture of at least two gases, and in the period from 1773 to 1774, Carl Wilhelm Scheele and Joseph Priestly are credited with identifying oxygen as one such element.

In 1784 Henry Cavendish published the first accurate information about the composition of naturally occurring air in the atmosphere, which is approximately 78 percent nitrogen and 21 percent oxygen. The remaining 1 percent is mostly argon (.9%) and carbon dioxide (.03%), with even smaller trace amounts of hydrogen, water, ozone, neon, helium, krypton, and xenon. Atmospheric air extends to approximately 350 miles above the Earth, is divided into a number of different layers (from the troposphere to the stratosphere and beyond), and undergoes tidal motions like the oceans. The study of those motions and other atmospheric phenomena, especially the weather, is known as meteorology. Of increasing importance as well is atmospheric chemistry and the study of air pollution.

Technologies of the Air

Even before the advent of humans the air served as a medium of communication for animals, a possibility that has been progressively developed by humans through speech and music. From early periods of human history the motion of air in the form of wind was been harnessed to power ships for transportation. During the late Middle Ages wind became a source of mechanical motion in windmills. And in the late-eighteenth and early-twentieth centuries it became a medium of transportation with the invention of balloons and the airplane, which has led to the science of aerodynamics and the technology of aeronautical engineering.

Air in the from of wind has also been a design problem, especially in the construction of tall buildings.

Since the late-twentieth century wind has again been exploited as a source for the creation of electrical power. From the earliest periods of human history, the heating of air has been a major technological issue, and as such air is closely associated with fire. With the advent of the Industrial Revolution the circulation and eventually the cooling of air became further technological design issues.

Toward an Ethics of the Air

The human ability to inhabit the world in a fashion that is sensitive toward the environment is reflected in the air people breathe. Throughout the course of the day each person consumes between 3,000 and 5,000 liters of air. But especially in the industrialized world, the air is full of notoriously harmful pollutants such as benzene, toluene, and xylenes, which are found in gasoline; perchlorethylene, which is used by the dry cleaning industry; and methylene chloride, which is used as a solvent by a number of industries. Examples of air toxics typically associated with particulate matter include heavy metals such as cadmium, mercury, chromium, and lead compounds; and semivolatile organic compounds such as polycyclic aromatic hydrocarbons (PAHs), which are generally emitted from the combustion of wastes and fossil fuels.

The latter (aromatic hydrocarbons) have to do with the formation of ground-level ozone. This is different from the stratospheric ozone that protects the Earth from ultraviolet radiation. Ozone is the same molecule regardless of where it is found, but its significance varies. Ozone (the name is derived from a Greek word meaning to smell) is a highly reactive, unstable molecule formed by reacting with nitrogen oxides from burning automobile fuel and other petroleum-based products in the presence of sunlight. It is also produced during lightning storms, which is why the air has that peculiar electrical odor during a storm. This type of ozone, however, is very short lasting and does not represent a significant risk to health. The real problem stems from certain volatile organic compounds such as those produced by the shellac of furniture finishing plants, cleaning solvents used by dry cleaners and computer manufacturers, and terpenes from trees; these atmospheric chemicals linger in the air and prevent the break up of the ozone molecule back into oxygen.

High concentrations of ground-level ozone may cause inflammation and irritation of the respiratory tract, particularly during heavy physical activity. The resulting symptoms may include coughing, throat irritation, and breathing difficulty. It can damage lung tissue, aggravate respiratory disease, and cause people to be more susceptible to respiratory infection. Children and senior citizens are particularly vulnerable. Inhaling ozone can affect lung function and worsen asthma attacks. Ozone also increases the susceptibility of the lungs to infections, allergies, and other air pollutants.

The greatest ethical issues concerning air involve the collective reluctance of humankind to take responsibility for the negative effects its way of life has upon the air, this essential element that has been recognized and harnessed for thousands of human years. Since the 1800s industry has been slow to admit that its technologies have seriously compromised the health of the air. In 1948 a *killer fog* caused the death of twenty and sickened 6,000 residents of the industrial town of Donora, Pennsylvania. For years local steel and zinc plants refused to admit that their effluents could have had anything to do with this *Act of God*. Thousands more died over the following decade. Even in the early twenty-first century industries tend to avoid taking responsibility for air pollution fatalities and illnesses caused by their routine operations.

This tendency to shirk responsibility extends to human obligations regarding the atmosphere as a whole, especially where the United States is concerned. Global climate change is one of the greatest harmful consequences of human industrial activity on Earth, and can only be controlled by managing air pollution.

DAVID ROTHENBERG

SEE ALSO *Earth; Environmental Rights; Fire; Water.*

BIBLIOGRAPHY

Abram, David. (1996). *The Spell of the Sensuous: Perception and Language in a More-Than-Human World.* New York: Vintage. Chapter 7 presents a phenomenological description of "The Forgetting and Remembering of the Air."

Davis, Devra. (2002). *When Smoke Ran Like Water: Tales of Environmental Deception and the Battle Against Pollution.* New York: Basic Books.

Olson, John. (2003). "Inebriate of Air." In *Writing on Air*, ed. David Rothenberg and Wandee J. Pryor. Cambridge, MA: MIT Press.

AIRPLANES

December 17, 2003, marked the 100th anniversary of the first heavier-than-air flight, or as pilot Orville Wright put it, "the first in the history of the world in which a machine carrying a man had raised itself by its own power into the air in full flight, had sailed forward without reduction of speed, and had finally landed at a point as high as that from which it started" (Anderson 1985, p. 2). Although their absolute priority has been contested—many were on the verge of heavier-than-air flight in the early 1900s—with their invention Orville and Wilbur Wright clearly helped change the world. Yet just as there are benefits of this technology, there are negative consequences. This article defines the airplane, examines historical developments, and introduces some of the ethical, political, and legal issues surrounding its future.

Definition and Developments, Military and Civilian

An airplane (or aircraft) is defined as a heavier-than-air machine that produces an upward thrust (lift) by passing air over its fixed wings and is powered by propellers or jet propulsion (thrust). As with any new technology, inventors immediately wanted to improve on the original design of the Wright Brothers and to develop versions of the airplane that would go higher, faster, and farther. During World War I, military planners used the airplane for war, first as reconnaissance platforms, but soon after as weapons. The two decades after the war saw the airplane become a more efficient military machine, as well as a commercial passenger carrier.

Charles Lindbergh, on May 20–21, 1927, captured public imagination by being the first person to fly alone non-stop across the Atlantic Ocean. Soon passenger airplanes were regularly making this transatlantic trip. World War II saw the golden age of piston airplanes. One of the most famous aircraft of the era, the Supermarine Spitfire, had a distinctive elliptical wing design and a remarkable maneuverability, which defended England during the Battle of Britain. Other fighter airplanes, such as the North American P-51 Mustang, designed and built in 140 days, were the technological pinnacles of the aircraft industry with speeds that topped 450 miles per hour.

Bomber airplanes showed it was possible to carry large payloads for long distances. These technological achievements directly led to the post-war development of commercial aviation. In the early 1950s, commercial piston-engine airplanes, such as the Lockheed Constellation, made travel by air practical and affordable for many people in developed countries.

One of the most significant advances in airplane technology occurred during World War II: the beginning of the jet age. Jet power greatly improved speed, by more than one hundred miles per hour. In 1952, the British DeHavilland Comet became the first jet powered commercial airliner but two fatal Comet crashes caused the public and commercial airlines to loose confidence in jet travel. Yet scarcely three years later, with the introduction of the Boeing 707 and the Douglas DC-8, commercial passenger jet service quickly revived. Since then a wide variety of commercial airplanes has satisfied the need to travel.

The DeHavilland Comet. In 1952, it became the world's first jet-powered aircraft. (© *Bettmann/Corbis.*)

Military airplanes also improved after World War II and on October 14, 1947, Capt. Charles Yeager flew the Bell X-1 faster than the speed of sound. The quest for improved performance airplanes that flew higher, faster, and farther would continue into space flight.

One of the more unique commercial aircraft was the Concorde, which was developed by a French-British consortium and flew from 1976 until 2003. It could carry approximately 128 passengers at more than twice the speed of sound. Funding for development of a similar supersonic transport (SST) was rejected by the U.S. Senate in 1973 because of concern for environmental damages from noise and stratospheric pollution. The SST was also very expensive and available only to the wealthy.

In the early twenty-first century, costs of manufacture, design, and operation, more than performance, influence many new designs, both civilian and military. It takes approximately five years to design and build an airplane from conception to rollout. Building a new gas turbine propulsion system takes longer, approximately ten years. As an example of current technology, a Boeing 747 cost approximately $160,000,000 in 2003 and

A British Airways Concorde taking off from London's Heathrow Airport. The Concorde was in use from 1976 to 2003 and flew at more than twice the speed of sound. (*© Kieran Doherty/Reuters/Corbis.*)

burned roughly 7,500 gallons of fuel on a typical 1,500 mile flight. Newer designs pose distinct commercial and technological risks.

From Propulsion to Application

Airplanes are often classified not only by their propulsion systems—propellers or jets—but also by application. Military airplanes can be classified by function: fighter, bomber, and reconnaissance; while civilian airplanes fall into two general categories: private and commercial. Private airplanes range from piston engine airplanes used for pleasure flying to private business jets carrying four to six passengers.

The future of commercial aircraft faces several issues highly influenced by technology. Cost often decides which technology will be incorporated into new or existing airplanes. In 2001, Airbus Industries announced a radical new design, the Airbus 380, expected to carry 550 passengers. To achieve this, the airplane features a twin-deck passenger compartment. More passengers should result in lower operating costs for the airlines and lower fares for customers.

The same questions that faced the Boeing 747 in 1968 must again be asked: Should an airplane carrying so many passengers be built? The A380 is technologically feasible, but is it safe? What if the A380 crashes? Is the public willing to accept loss of life on this scale? The collision of two Boeing 747s on the runway in Santa Cruz de Tenerife, Canary Islands, on March 27, 1977, resulted in 582 deaths, and is just one example of the large loss of life possible. Safety is a crucial issue in airplane design, but finances often influence decisions. As long as airlines see the need for these large airplanes to lower costs, they will be built.

Another controversial question is whether pilots or a computer should have the ultimate control authority over a commercial jetliner as the plane approaches its design limits in an emergency. The fly-by-wire flight control system of the Airbus A380 does not allow the pilot to override the computer, whereas a similar system on the Boeing 747 does allow for aviator override. Some forms of this technology provide "cues" that tell the pilot when the plane is approaching certain speed, load or attitude limits but allow the pilot to exceed these limits. For example, much more force is needed to pull back on the control column as an aircraft reaches its stall speed.

Economics, Safety, and the Environment

The need for new commercial airplanes is also the result of problems associated with an aging commercial airplane fleet. Fatigue and corrosion take their toll. Costs associated with replacement overshadow the timetable to replace these aircraft. Should old airplanes be repaired or new ones purchased? It often depends on the financial stability of an airline. New techniques need to be developed to detect structural problems before they become life-threatening. Development costs money. Should the government be responsible for such development? Are the airlines financially able to develop techniques and ethically equipped to enforce standards without government supervision?

On January 31, 2000, an MD-83 plunged into the Pacific Ocean, killing all eighty-eight people on board. Accident investigations pointed to substandard maintenance procedures causing the horizontal stabilizer to jam. Subsequent inspections of similar airplanes found twenty-three more with the same problems. Operators must make safety inspections regardless of cost. They are ethically bound to accomplish the proper repairs. The public deserves no less, but there are always temptations to cut corners and reduce cost.

The quest for more economical airplanes has driven the airplane industry to reduce airplane weight. Reducing weight improves fuel efficiency. To achieve this reduction, the industry is using materials such as composites, with which the military has some experience that is migrating to the civilian aircraft industry. But the crash of an Airbus 300 airliner on November 12, 2001, in a residential section of Queens, New York, has been blamed in part on the failure of composite material in the vertical tail. When is a technology sufficiently mature and when should it be applied in commercial airplanes? Often the answer is left to a private company or government regulatory agency that may not fully understand the technology. Airplane manufacturers and their suppliers also have an ethical obligation to ensure quality parts. Oversight and enforcement are difficult but necessary.

The future of large, commercial airplanes clearly shows two companies dominant: Boeing from the United States and Airbus Industries from Europe. Boeing traditionally had the majority of commercial passenger airplane sales in the world, but in 2003, Airbus superseded Boeing in the number of airplanes sold. Some believe the shift is due to a subsidy of Airbus by its parent countries. Boeing, not having direct subsidies, has protested that Airbus is able to undercut prices to attract business. At the same time, Boeing is directly supported by U.S. military contracts in ways Airbus is not.

Airplanes have been blamed for a number of environmental problems. The first is noise. Technological improvements have satisfied noise restrictions imposed by regulatory agencies. Compliance is mandatory, and older airplanes are either refitted with newer, quieter engines or "hush kits" are retrofitted to older engines. Engine emissions are also thought to impact the ozone layer and contribute to global climate change. Particulate emissions, such as carbon, can cause residues. While it might be possible to reduce the problems of pollution, it is often not economically feasible to fix older airplanes. The possibility of different propulsion systems, such as nuclear power, could solve some environmental problems, but would create others. Nuclear powered airplanes flying over populated areas would certainly cause public alarm.

Military Applications and New Civilian Options

Airplanes will continue to be used in military applications. The development of military airplanes and engines generally supports technological progress, which then finds commercial application. Fuel efficiencies and performance standards of contemporary commercial engines are a direct result of this technology transfer.

Other military technologies are maturing rapidly. Stealth technologies have given the United States an advantage in air warfare. In the Iraq conflicts (1992 and 2003), stealth airplanes were able to destroy command and control networks and anti-aircraft batteries prior to ground conflict. However, drug runners and other undesirable individuals could also use stealth to evade capture.

In 2003 another new airplane technology for the military was the remotely piloted Unmanned Aerial Vehicle (UAV). The military has successfully used these airplanes, such as the Predator or Global Hawk, to gather information and even launch attacks. The technology is reliable and may lead to UAV operation in U.S. airspace along with other airplanes. Automatic collision avoidance on UAVs and other airplanes would undoubtedly be part of such a development.

Is the public ready for the next step: Unpiloted Aerial Commercial Vehicles? It is possible to operate airplanes without pilots, because most of the systems on commercial airplanes are already fully automated. To eliminate the pilots would save the cost of their large salaries. Will the technology cost more?

Precedents for replacing crewmembers with technology already exist. Airplane manufacturers designed and built airplanes with advanced cockpits for two crewmembers in the 1980s. The traditional third crewmember, the flight engineer, was eliminated by improvements in system automation. This increased the workload for the two-person crew but the workload proved manageable. Perhaps the next step is a single pilot crew. However, what if this one pilot fell ill or died in flight? A totally automated system is feasible, but would face some acceptance issues.

Terrorism

Another aspect of aviation safety concerns terrorism. In the aftermath of September 11, 2001, efforts have been made to enhance the security of the commercial air travel system, including airplanes, against terrorism. Cockpit door reinforcements have been the most visible and immediate development, but some see this modification as ineffective. Allowing pilots to carry weapons is controversial. What else can be done to protect against terrorism?

One technology being discussed is the addition of infrared countermeasures to deter possible ground launched missiles. Israel's El Al airline has flare detection equipment installed on its aircraft but no active countermeasures. The estimated cost to equip the U.S. commercial aircraft fleet with countermeasures is $10 billion (Israel High Tech and Investment Report 2003). In 2003, the Bush administration committed $100 million to the first phase development of such a system in the United States. Another serious weakness is the absence of commercial cargo inspection on passenger airplanes. Clearly, more needs to be done, but economics will strongly influence the outcome.

Airplanes have proved indispensable to the contemporary world in ways the original inventors could not have predicted. Nor could they have predicted some of the negative consequences. Increased air travel has revolutionized how people think about the world and been a major contributor to globalization, which has created problems for both cultures and the environment. Safety, through responsible design, must be the main emphasis in the aviation industry, from the design and construction of new airplanes by the industry to the operation of the airplanes by the airlines. Professional engineers, as part of their ethical responsibility, must make sure that designs are safe. Industry also has a responsibility to the public to provide a quality product and to use that product responsibly. Trusting aircraft and aircraft related industries to accomplish this task without supervision would be naïve, but government regulation alone will not insure the desired outcome. Public awareness of and action on these issues may yet prove to be the most important factor in deciding the future of aviation.

KENNETH W. VAN TREUREN

SEE ALSO *Aviation Regulatory Agencies; DC-10 Case; Military Ethics.*

BIBLIOGRAPHY

Anderson, John. (2002). *Introduction to Flight*, 4th edition. New York: McGraw-Hill.

INTERNET RESOURCE

Israel High Tech and Investment Report. (2003). "Is $10 Billion for Protection Worth It?" From the October issue. Available from http://www.ishitech.co.il/1003ar2.htm.

ALIENATION

The word *alienation* has a checkered history. Drawn originally from the vocabulary of the law, the word later appeared in connection with the treatment of persons who were, as ordinary people say, "not themselves." In the eighteenth and nineteenth centuries, property given away or sold was said to have been "alienated." This usage survives in the expression "inalienable rights"—rights that cannot be taken away, given away, or traded. The physician who treated the mentally ill was formerly called an "alienist." In contemporary usage, one speaks of being alienated from a former friend for whom one's affection has cooled or from a group in which one feels no longer comfortable. Alienation, in everyday English, refers to a specific loosening of ties to another person or a sense of estrangement from a group.

Philosophies of Alienation

In philosophy, by contrast, the word alienation has been used in a different sense to refer to estrangement from *oneself*, a profound disturbance within persons, their selves, and their lives. There is conflict or disconnection at the very heart of the alienated person's existence. Alienated lives do not form an intelligible whole; the alienated cannot tell a coherent story about their lives. Their lives lack meaning.

Philosophers have always said that human lives are more than a series of unconnected episodes, that they should form an intelligible whole. Hence people might ask whether human life as a whole, and especially their own, makes sense and whether it is a good life that serves a purpose and is meaningful. At the beginning of Plato's *Republic*, Socrates raises those questions in conversation with an old man nearing the end of his life. Plato's answer is that a good life is a just one. A just life, he also thinks, can be lived only in a just society, and thus the conversation about one's life, seen as a whole, leads to a long investigation into the just society. Aristotle gives a different answer to the question: A good life is dedicated to acquiring a set of moral virtues such as courage, temperance (self-discipline), and wisdom.

Beginning in the eighteenth century, new answers surfaced about what makes a life good. What matters in human life, according to the French philosopher Jean-Jacques Rousseau (1712–1778), is not only its moral character, but whether a person manages to be an individual rather than a conformist—dominated by the beliefs, values, and practices of everyone else. Most persons, Rousseau complained, craved acceptance by their fellows and were willing, for the sake of this, to sacrifice any independent identity.

Since Rousseau, philosophic views of the good life have become divided. Many Continental European philosophers have demanded that human lives be not only morally good but also coherent and meaningful. The majority of Anglo-American philosophers, by contrast, have continued to think only about the moral rectitude of human lives, ignoring the question of alienation. The utilitarians, beginning with the Englishmen Jeremy Bentham (1748–1832) and John Stuart Mill (1806–1873), explicitly reject the possibility of alienation. They insist that a life is a good one if it contains more pleasant episodes than unpleasant ones; the connection between these different episodes is of no interest. Thus there are disagreements among philosophers—rarely articulated and more rarely debated—about the importance of the concept of alienation.

But many thinkers have taken the idea of alienation very seriously (even though not all used the word alienation to name the condition). Georg Wilhelm Friedrich Hegel (1770–1831), in *The Phenomenology of*

Spirit, described some forms that alienation takes in human lives. The alienated suffer from inner conflict and self-hatred. As a consequence, they are unhappy.

The Danish philosopher Søren Kierkegaard (1813–1855) agrees with Hegel on the self-hatred of the alienated and develops the idea further. Alienated lives are not disorganized by accident or because persons do not try to unify their lives, but because at the heart of alienation lies the unwillingness to be oneself. It is difficult for the alienated to accept themselves for who they are; it is more pleasant for one who is alienated to escape into fantasy and imagine oneself different: richer, more powerful, more intelligent, or more beautiful than one is. It is also difficult to accept responsibility for one's life. Alienation, Kierkegaard believes, cannot be overcome, but can be mitigated if one is fully in one's life by dedicating oneself to a single project in such a way that every part of one's everyday existence is affected by it. Kierkegaard also believed that this needed to be a Christian project—to live so as to manifest God's presence in even the smallest details of one's life, such as taking a walk in the park and thinking about what there will be for Sunday dinner.

The German philosopher and social critic Karl Marx (1818–1883) focused on alienation in a different aspect of life—namely at work. For Marx, working for wages was inevitably alienating. Wageworkers under the command of employers have no control over their work or even whether there is work for them at all. Employers are—for the majority of wage earners—able to hire and fire them at will. It is impossible to have a meaningful life if such a large part of it is under the control of another whose goals are at odds with one's own. The employer's goal is to make as much money as possible; workers want to earn as much as they can. But they also want their work to be clean, pleasant, and interesting. Employers care nothing about this as long as the money keeps rolling in. Spending a significant portion of one's life pursuing goals that are not one's own, alienates. It makes it impossible to be one's own person—one who pursues goals of one's own choosing.

The German philosopher Friedrich Nietzsche (1844–1900) elaborates on the theme of conformism in his discussion of the "last man." Such persons want above all to be comfortable; they eschew all effort and anything that is even faintly unpleasant. Hence it is important for them to get along. In order not to stir up controversy by disagreeing with others, they have no ideas of their own. They do not think for themselves. There is nothing they believe in fervently and nothing they are willing to stand up and fight for. They want life to be easy and pleasant. Avoiding all challenges is the only challenge that remains. Although Nietzsche did not use the term, his "last man" is clearly suffering from alienation.

Writing after World War I, the Hungarian philosopher György Lukács (1885–1971) returned to Marx and elaborated on the claim that alienation is intimately connected with capitalism. Persons who sell their ability to work in the labor market treat themselves or at least important aspects of their persons as commodities—things meant to be exchanged for money. The skills and talents of persons thus become commodities that can be bought and sold—"alienated" in the old, legal sense. One's person and how it develops is no longer one's proper project, but is governed by the impersonal forces of the labor market. People are not able to study what most interests them because expertise in Egyptology, for example, does not promise to bring in a lot of money. Instead they go to business school and prepare themselves for the life of a junior executive and will, if they are lucky and sufficiently pliable (that is, a "team player"), end up as senior managers with a good income. They are forced to live where the work takes them. They dress the part of the executive. If they happen to have unpopular opinions, they will be wise to keep those to themselves. After a few years they may well forget they ever held them.

Martin Heidegger (1889–1976), destined by his Bavarian family for the Catholic priesthood, became a secular philosopher instead. He rediscovered alienation when reading Kierkegaard and Nietzsche. In *Being and Time*, Heidegger argued that most people are not themselves. Their opinions ape everyone else's; they are addicted to all things new. There is nothing they stand for unless they manage to overcome the pressures toward alienation and win through to being "authentically" themselves.

The French existentialist Jean-Paul Sartre (1905–1980) argued in his early work, *Being and Nothingness*, that alienation is not merely commonly chosen—a view he ascribes to Heidegger—but is inherent in the structure of human beings. People do not only think and act but are observers and critics of themselves. They can never be fully engaged in any activity or relationship because a part of them always stands aside to observe and judge. Being split against oneself is essential to being human.

In the years after World War II, numbed by a new, hitherto unknown level of prosperity paired with insistent demands for political conformity, writers in the United States produced a sizable literature concerned with alienation. Philosophically inclined writers, such as Erich Fromm (1955) and Paul Tillich (1952), brought the previously unknown ideas of Continental existentialism to the English-speaking world. Poetry, novels, and popular works in social science deplored conformism. Variants on Marxist themes attracted considerable interest and discussion in the 1970s and 1980s when a number of authors, including Bertell Ollman (1976), István Mészáros (1975), and Richard Schmitt (1983) published studies on alienation that were clearly anchored in the Marxist tradition.

Origins of Alienation

Sometimes technology is named as the source of contemporary alienation. Because technology is always a means to some end, the dominance of technology in society assures that all attention is given to means while ends remain unexamined. In such a situation, human lives lack goals and purposes because, absorbed in technological efforts, humans are unable or unwilling to reflect about the purposes of their activities (Ellul 1967). This thesis, however, portrays human beings as the impotent playthings of technology and overlooks that technology is not only used by humans but is also our creation.

The question about the origins of alienation have occasioned other controversies. For many years, philosophers have debated whether alienation is intrinsic to human nature or the effect of specific social conditions. Kierkegaard, Heidegger, and Sartre place the origin of alienation in the structure of human existence. Marx and, in different ways, Nietzsche blame the existence of alienation on social and economic conditions. Existing social conditions produce alienation, but in a happier future alienation may well disappear. Neither side to this debate seems to have understood that the two alternatives—alienation as intrinsic to human nature or alienation as the product of social conditions—are not exclusive: Alienation is anchored in human nature, but it exists more acutely in some social settings than in others. Alienation is always possible. But in some societies it is well-nigh unavoidable, whereas in others it is only a remote possibility.

Alienation springs from human nature insofar as it is characteristic of human beings to reflect about their lives as a whole. They ask whether their lives have a purpose, or whether their identity is well integrated. Gifted with certain capacities for reflection and the need to be able to tell a coherent story about their lives, they are, therefore, susceptible to alienation. But these general human characteristics do not inevitably produce actual alienation. Alienation arises when societies, as does America's, make conflicting and irreconcilable demands, for instance, when it asks one to love one's neighbor as oneself at the same time as it exhorts people to be aggressive competitors who give no quarter in the great contest for wealth and power. American society asks its citizens to be free and autonomous beings after hours but, during the day, to work in hierarchical organizations, in which one must be subservient and obedient to employers and supervisors. Surrounding daily life is a chorus of voices telling people to buy this, to buy that, to look like this model, or to have their house look like some dream house. Americans are told how their children must appear and how they themselves must spend their days and enjoy their leisure. Throughout one's waking life, these voices are never silent. Consequently, it is no wonder that there is a pervasive sense among Americans that their lives are not their own (Schmitt 2003). A variety of aspects of American society make it extremely difficult for Americans to live lives that are coherent and to be persons who pursue goals of their own choosing. American society fosters alienation.

Questioning Alienation

However interesting, such historical discussions of alienation remain extremely general. A number of original thinkers have provided a range of insights into alienation, but professional philosophers have mostly been content to repeat and embroider these original insights instead of developing them in greater detail. As a consequence, many important questions about alienation remain unanswered.

The concept of alienation refers to important characteristics of the modern social world. But it also directly refers to each person separately. If alienation is pervasive in modern society, as many authors have alleged, people must reflect, each with respect to their own person, whether they are conformists and therefore alienated or whether they lead lives of their own. But such questions about one's own conformism or independence are not easily answered. Humans are social beings, learning from others and sharing ideas with them. As a social process, is that participation in thinking a sign of conformism and hence of alienation? For example, Western people share the belief that freedom is important and that democracy is preferable to tyranny. Does that make Westerners conformists and manifest their alienation? Surely, there is an important distinction between sharing the ideas of one's fellow citizens and being conformist. But that difference remains unclear, and the discussions of philosophers do not provide much help. The idea of conformism, as one finds it in the literature about alienation, is not sufficiently specific to be useful to the individual's self-examination with respect to conformism and alienation.

Conformism is only one of several constituent concepts of alienation that have not been sufficiently developed. The alienated are often described as not being themselves with lives lacking unity and identities fragmented. But postmodern thought has provided an important reminder: that selves are multiple and complex (Flax 1987). Most people have more capabilities than they are able to develop; in different contexts—as their parents' child or as the boss at work—their personalities differ. People change over a lifetime and are rather different persons at seventy than at seventeen. Are all these diversities within one person signs of alienation? Are there not important differences between the alienated personality, which is vague and poorly delineated, and the complexities of the multiple aspects that well-constituted persons display in the different contexts of their lives and over an entire lifetime?

Traditional discussions of alienation have concealed the complexity of alienation in another respect. Human beings are very different from one another; they lead different kinds of lives because they are born into different conditions, have different abilities and defects, think in different ways, and have different character structures. The general symptoms of alienation mentioned in the literature will manifest themselves differently in different lives. Aimlessness leads to complete idleness in some lives, whereas in others it takes the form of frantic busyness—all of it trivial. The self-hatred of the alienated appears in some persons as constant self-deprecation and jokes at one's own expense, and in others as pompous self-importance. One does not really understand alienation until one is able to tell many concrete stories about the alienation of different persons, differently situated and therefore manifesting alienation in very different, sometimes, flatly contradictory ways.

The possibility of alienation flows from the human need to reflect about one's life (to ask whether it is coherent and has a purpose) and about one's person (whether one is autonomous or conformist). It is tempting to evade these reflections because their results are often confusing or discouraging when one finds that one's life is aimless or one's person ill delineated. As Kierkegaard pointed out forcefully, one can evade the pain of reflection about one's life by discoursing abstractly about alienation while refusing to try to apply this abstract philosophical discourse to one's own person and one's own life. The refusal to take one's own life and person sufficiently seriously to reflect about their meaning and coherence is one form of being alienated, of being a fractured person leading a haphazard life. Philosophical discussions of alienation foster this form of alienation because the very generality and lack of precision of many philosophical discussions of alienation make it difficult to engage in serious self-reflection.

RICHARD SCHMITT

SEE ALSO *Existentialism; Freedom; Hegel, Georg Wilhelm Friedrich; Kierkegaard, Søren; Marx, Karl.*

BIBLIOGRAPHY

Aristotle. (1999). *Nichomachean Ethics*, 2nd edition, trans. Terence Irwin. Indianapolis, IN: Hackett.

Ellul, Jacques. *The Technological Society*. New York: Vintage Books, 1967.

Flax, Jane. (1987). "Postmodernism and Gender Relations in Feminist Theory." *Signs* 12(4): 621–638.

Fromm, Erich. (1955). *The Sane Society*. New York: Rinehart.

Gilman, Charlotte Perkins. (1979). *Herland*. New York: Pantheon.

Hegel, Georg Wilhelm Friedrich. (1952). *Phänomenologie des Geistes* [Phenomenology of spirit]. Hamburg, Germany: Felix Meiner.

Heidegger, Martin. (1953). *Sein und Zeit* [Being and time]. Tübingen, Germany: Max Niemeyer.

Kierkegaard, Søren. (1941). *Sickness unto Death*, trans. Walter Lowrie. Princeton, NJ: Princeton University Press.

Kierkegaard, Søren. (1954). *Fear and Trembling*, trans. Walter Lowrie. Garden City, NY: Doubleday.

Lukács, György. (1971). *History and Class Consciousness: Studies in Marxist Dialectics*, trans. Rodney Livingstone. Cambridge, MA: MIT Press.

Marx, Karl. (1964). *Early Writings*, ed. and trans. T. B. Bottomore. New York: McGraw-Hill.

Mészáros, István. (1975). *Marx's Theory of Alienation*, 4th edition. London: Merlin Press.

Mill, John Stuart. (1949 [1863]). *Utilitarianism*. In *The English Utilitarians*, ed. John Plamenatz. Oxford: Blackwell.

Nietzsche, Friedrich. (1954). *Thus Spoke Zarathustra*. In *The Portable Nietzsche*, ed. and trans. Walter Kaufmann. New York: Viking.

Ollman, Bertell. (1976). *Alienation: Marx's Conception of Man in Capitalist Society*, 2nd edition. Cambridge, UK: Cambridge University Press.

Plato. (1920). *Plato's Republic*. In *The Dialogues of Plato*, 2 vols., trans. Benjamin Jowett. New York: Random House.

Rousseau, Jean-Jacques. (1950). *Discourse on the Origin and Foundation of the Inequality of Mankind*. In *The Social Contract, and Discourses*, trans. G. D. H. Cole. New York: Dutton.

Sartre, Jean-Paul. (1956). *Being and Nothingness*, trans. Hazel E. Barnes. New York: Philosophical Library.

Schmitt, Richard. (1983). *Alienation and Class*. Cambridge, MA: Schenkman.

Schmitt, Richard. (2003). *Alienation and Freedom*. Boulder, CO: Westview.

Tillich, Paul. (1952). *The Courage to Be*. New Haven, CT: Yale University Press.

ALTERNATIVE ENERGY

The very notion that some sources of energy make up *alternative energy* demonstrates the way people impute normative values to technologies. For decades, proponents of alternative energy have done more than advocate particular technologies: They maintain that their proposed technologies are socially and morally better. These social and moral claims show that advocates regard alternative energy technologies as different in profound ways from existing conventional energy technologies.

Social Contexts

Alternative energy must be understood against a background of conventional energy. Conventional energy is not conventional just because it is in wide use. It is conventional in that it underlies the functioning and embodies the values of the conventional society. Thus coal, oil, and natural gas are conventional both because they dominate energy production in industrialized countries and, even more, because they make possible a high-consumption society and require large-scale industrial systems to extract, convert, and distribute the energy.

Dam with power lines in the background. Hydropower as a source of alternative energy is considered problematic by enviromental groups, as dams often have adverse effects on their surrounding ecosystems. (*© Royalty-Free/Corbis.*)

Advocates of alternative energy seek more than simply technological replacements for fossil fuels. They seek technological systems that will reinforce and embody alternative values, such as avoiding the exploitation of nonrenewable resources and people, favoring smaller scale production, and, most importantly, living in a manner more in concert with natural systems, in the early twenty-first century often termed *living sustainably*.

This normative orientation sets alternative energy advocates apart from people who simply advocate new technologies but have no interest in an alternative society. For example, consider the case of nuclear power. From World War II on, many scientists and others advocated the use of nuclear power plants to replace fossil fuels. They sought a new technology, one that was not then in widespread use. However their purpose was to reinforce, maintain, and enhance the existing social and economic system, with all the values that went with it, including a reliance on large-scale resource extraction and production. They simply thought that nuclear technology would do the job better, more cheaply, and for a longer period of time than fossil fuels. Since the 1990s, proponents of nuclear energy have also argued that it will meet what has become another more or less conventional goal, reduction in carbon emissions.

The alternative label does not necessarily apply to all people and groups who advocate renewable energy technologies, such as solar, wind, and biomass. Since the 1950s, many of those advocating such technologies have simply seen them as ways to preserve the social status quo and its values. For such advocates, photovoltaic panels (often called solar cells) are just another way of producing electricity, and biomass-derived alcohol is just another way of producing liquid fuels for internal combustion automobiles. In contrast, for others photovoltaic panels offer the means to live *off the grid*.

Alternative energy advocates thus make up a subset of advocates for particular energy technologies. These are advocates who seek not only different technologies but also to promote different social values to go along with them.

The Ethical Dimension

The driving ethical concern that motivates most alternative energy advocates is a particular type of environmental ethics. These people feel that the relationship of industrial society to nature is fundamentally flawed. They come out of the more radical wing of the environmental movement and the broader alternative (or appropriate or intermediate) technology movements. To understand this alternative it is therefore necessary to consider the conventional societal attitudes toward nature.

For an industrial society, nature offers a set of resources to be exploited. Material consumption and the use of natural resources that go with it are good things, ethically desirable, as well as pragmatically important. Even so, people committed to this industrial ethos recognize that resource exploitation causes certain problems. From the late 1960s onward, conventional political groups accepted the need to curb pollution from industrial production, and governments around the world passed numerous environmental laws and established new agencies to carry them out. The oil embargo of 1973 and the resulting shortages and price increases demonstrated clearly the financial and security risks of U.S. dependence on imported oil. But for conventional society, and most political elites, environmental pollution and security risks were no more than manageable problems to be solved. They did not cast doubt on the basic normative commitments to exploiting nature and maximizing material growth.

Alternative energy advocates, however, see the society-nature relationship quite differently. They believe that human beings must understand themselves as parts of ecosystems and that, therefore, human well-being depends on the health of those ecosystems. They want societal values to be more consonant with the way that ecosystems work and to regard ecosystems as things of inherent value, not just resources to be exploited. For these advocates it is not enough to put scrubbers on the smokestacks of coal-fired power plants or to reduce the emissions coming out of automobile exhausts. They seek instead a society that puts much less emphasis on high levels of material consumption, epitomized by the use of individual automobiles. Such a society would be organized very differently, with different values guiding both individual behavior and social and political institutions. These normative commitments lead them to advocate different energy technologies, ones that use renewable resources that could provide the foundation for a different type of society. However these commitments also lead them to make fine distinctions among these technologies, rejecting some, and to carry on vigorous debates about the merits of particular technologies and energy sources.

Alternative Energy Options

Against this background, it is thus possible to consider at least three proposed alternative energies: solar, hydro, and wind.

SOLAR ENERGY. Numerous technologies use sunlight directly to produce either heat or electricity. During the 1970s, ecologically oriented alternative energy advocates pushed for certain of these technologies and opposed others. In general, the more high-tech and large-scale the technology, the less such advocates liked them. They favored solar panels that use sunlight to heat air or water. Such panels consist of little more than a black metal plate, which absorbs sunlight, encased in a box with a glass cover. Air or water flows over or through the plate, heating it up, and then enters the building to supply heat or hot water.

The principles of such technologies are not complicated, although it is not easy to make panels that last a long time and function well. The fact that they are easy to understand, small, and seemingly unrelated to large industrial systems and produce no pollution in their operation appeals to the ecological ethic of alternative energy advocates.

At the other extreme are proposals for solar power satellites (SPS). The idea is to launch a satellite into a stationary earth orbit and to attach to it many acres of photovoltaic panels, semiconductor solar cells that convert sunlight directly into electricity. The satellite could produce electricity almost twenty-four hours per day and beam it back to a receiving station on earth. This is the ultimate high-tech solar technology. Alternative energy advocates are hostile to the SPS system because it both requires and supports the conventional industrial system. As a system that could produce large quantities of electricity around the clock, SPS could substitute in a straightforward way for conventional power plants, making it just another conventional technology, albeit a solar, nonpolluting one. Due mostly to cost considerations, no one has yet put such a satellite into orbit.

HYDROPOWER. Controversies over hydropower again demonstrate conflicts over values. Many environmental groups opposed the hydropower dams the federal government sponsored in the 1950s and 1960s. While their operation produced no emissions, as would a coal plant, the dams flooded large areas and dramatically changed the ecosystems in which they were located. Besides the scientifically measurable damage they did, for many environmental advocates the dams represented a problematic relationship, of dominance and exploitation, to nature.

Therefore alternative energy advocates in the 1970s talked favorably about hydropower only when referring to low-head hydro (very small dams) or what was called *run-of-the-river* hydro. This latter technology consists of power-generating turbines that are put directly into rivers, without any dam at all. These technologies have the virtue of being smaller in size, more modest in environmental disruption, and less like large-scale industrial production.

Assessing Values

In the 1970s advocates of alternative energy did so in the hopes of moving toward a different society. They sought energy-producing technologies that were smaller in scale and simpler to understand, promoted local self-reliance instead of global dependence, and embraced an ecocentric environmental ethic. They thought that such technologies would provide the means to live in a society that was not only environmentally more sustainable but also more socially harmonious and cooperative, with less domination, hierarchy, and inequality. The ecocentric environmental ethic was particularly important to this view. Advocates thought that human domination of nature got reproduced in the domination of people. The energy crisis of the 1970s raised public awareness of the importance of energy to every social and economic function. For this reason, alternative energy advocates regarded changes in energy technologies as central to realizing their social vision. A final argument often made for alternative energy is that it supported projects in the developing world.

Were they correct? For the most part, no. The alternative energy advocate's vision of a new society based on a new energy source embraces the notion of technological determinism: Build the right technology, and one can get the desired society. Numerous studies show that this theory is false. Society does not simply evolve from technological choices. Many different societies can come out of similar technological choices.

However one should not entirely discount the advocates' ideas about energy. Technological choices do have profound effects on society, which in turn affects future technological choices. Moreover those choices are often not easy to change. If a society invests trillions of dollars in an energy system, as the industrial countries have done, they are reluctant to make rapid changes, a phenomenon historians call path dependence or technological momentum. So energy choices are heavily value-laden, long-term choices. It is difficult, however, to know just how those choices will interact with complex societies.

The case of wind energy illustrates this. Alternative energy advocates embraced wind energy in the 1970s, believing that wind turbines could produce electricity on a small scale and enable homes or communities to be less dependent on central-station power plants and the massive electrical grid that distributes the electricity. Those advocates were critical of federal research programs on wind turbines because such programs sought to build large wind turbines that the utility industry could use instead of smaller, off-the-grid turbines. These large turbines eventually achieved economies of scale that reduced the price of wind-generated electricity toward price-competitiveness. In the early-twenty-first century the wind industry is growing rapidly, with ever-larger turbines coming online as part of the large-scale electric utility industry. This technology is certainly cleaner than coal-fired power plants, but other than that, it bears no resemblance to the social vision held by alternative energy advocates of the mid-1970s.

The history of wind energy emphasizes another point about normative values and energy. Alternative energy advocates in the 1970s thought that society was in deep crisis and that its core values were debatable. The signs seemed to be everywhere. The economy was in a long decline during the 1970s after dramatic growth and prosperity in the 1950s and 1960s. Along with economic stagnation came social problems such as rising crime rates and declines in urban fiscal health, symbolized by the fiscal crisis in New York City. The oil embargo, along with the end of the Vietnam War and other problems abroad, seemed to indicate a loss of international influence for the United States. Faced with these realities, alternative energy advocates thought they were in a position to push for a society based on radically different values.

But they clearly miscalculated. In particular, the value of economic efficiency, an important ethical norm for conventional society, one that valorizes markets, has been an important, though not the only, driver of energy technology. In the early-twenty-first century virtually all advocates of renewable energy seek ways in which such technologies can succeed in competitive markets. Alternative energy advocates of the 1970s pushed a social vision that was greatly divergent from existing society. They never produced a narrative compelling enough to lead to widespread acceptance of their normative values and consequently to their technological system. Their values rather than their technologies kept them marginalized.

FRANK N. LAIRD

SEE ALSO *Alternative Technology; Automobiles; Environmental Ethics; Sierra Club.*

BIBLIOGRAPHY

Laird, Frank N. (2001). *Solar Energy, Technology Policy, and Institutional Values*. New York: Cambridge University Press. Studies federal government policy for renewable energy from 1945–1981, with an emphasis on the ways in which values drove policy debates.

Laird, Frank N. (2003). "Constructing the Future by Advocating Energy Technologies." *Technology and Culture* 44 (1): 27–49. Examines the values that alternative energy advocates used in their works from World War II through the energy crisis, with a heavy emphasis on the 1970s.

ALTERNATIVE MEDICINE

SEE *Complementary and Alternative Medicine*.

ALTERNATIVE TECHNOLOGY

Any reflection on *alternative technology* (AT) prompts the question, Alternative in what sense? According to one AT theorist, there are three dimensions to this question (Illich 1997). The alternatives can be technical, ethical, or political. In the first case the divide is between hard (oversized machines) and soft (smaller, local tools), in the second between heteronomy and autonomy in technology, and in the third between centralized (right) and decentralized (left) technological systems.

Technical Alternatives

In 1917 D'Arcy Wentworth Thompson published *On Growth and Form*, a study of the relation of *shape* and *size* in living beings and artifacts. His *law of similitude* states that every natural and technical shape is *scale-variant*, that is, shape or form is strongly influenced by size. According to J. B. S. Haldane (1956), for instance, the form of all natural organisms is covariant with their scale: A cow the size of an elephant would need legs as strong as columns and could hardly support its horns. The Austrian economist Leopold Kohr (1967) applied these ideas to economics and the study of societies and is therefore the pioneer of *social morphology*. For Kohr, the size of a political unit entails a certain kind of polity, that is, a correspondence between the form of government and the scale to be governed. He was a major influence on, and a friend of, the German-born British economist Ernst Fritz Schumacher (1911–1977), whose phrase *small is beautiful* has become a world-famous lemma.

Schumacher is deservingly considered the father of the AT movement. In 1961 he took a trip to India that changed his vision. Impressed by the inherent viability of Indian agriculture, he firmly opposed replacing the traditional ox-drawn cart by tractors (Dogra 1983). Instead he imagined the carts equipped with ball bearings and rubber tires. On his return to England, he founded the journal *Intermediate Technology*, which would popularize the concepts of *appropriate technology* and later AT. Though superficially similar, the word appropriate points to something the other terms do not: the fitness of shape and size; the balance of power between autonomous action and what is done for one; and the importance of subjecting the relation between means and ends to political deliberation.

During the 1970s and 1980s, the AT movement gathered strength through numerous journals, publications, and associations. The *Whole Earth Catalog* in the United States and *Resurgence* in the United Kingdom became leading periodicals. Informative and influential books and articles appeared on alternative or appropriate technologies in general (Darrow and Pam 1976), on improvements to traditional rural practices (Devender 1978), on ecological houses (Farallones 1979), and on alternatives to energy-intensive industrial technology (Lovins 1977). As individuals and small groups of citizens retooled their homes and villages, nongovernmental organizations (NGOs) began to proliferate and spread the good news that there were better means to meet ends than energy-intensive industrial technologies. Yet insofar as the AT movement restricted attention to the technical choice between *hard* and *soft*, it was often dubbed the *soft technology* movement—and had little more than decorative influence over the technological world.

Ethical Alternatives

In the twenty-first century distributive justice often takes the industrial system for granted and strives to allot its outputs according to some equalitarian scheme. The alternative to this justice by arithmetic is *equity*, sometimes inaptly called *participative justice*. An equitable society is founded on an architecture of *civil liberties* that protects everyone's freedom to act. In an equitable society, each contributes threads to the weave of the social fabric rather than passively claims *outputs* from society. The enhancement of productive liberties does not mean a blind refusal to all claims of consumption. Rather it implies the recognition of a hierarchy: Just as autonomy is higher than heteronomy so also civil liberties are superior to social rights.

Many activists of the AT movement have argued that this hierarchy demands some limits on tools. In contrast to the automobile, the bicycle is an example of an industrial product that fosters the autonomy of its users: It increases access without driving others off the road. Just as the automobile enchains drivers to highways, the flush toilet, once the glory of industrial hygiene, turns its users into compulsive elements of the sewer system. Clean, cheap, and often ingenious alternatives to the costly industrialization of waste removal suggest the possibility of freedom from other heteronomous systems insofar as they can be intelligently worked out. Starting with Dr. Duc Nguyen's Vietnamese latrines in the 1960s, there have been a great variety of high quality dry toilets that unplug their users from the sewage pipes, reduce the destruction of land and waters, and cut a home water bill by more than half (Nguyen 1981, Lehmann 1983, Anorve 1999).

Political Alternatives

Proponents of alternatives to the service industry have emphasized that civil liberties can only be perverted by bureaucratic and professional government for the people. For example, from 1955 on, a group of Peruvian activists, builders, and lawmakers were joined nonconformist architects and sociologists from Europe and the United States to collectively give shape and credibility to an alternative understanding of poor neighborhoods (Turner 1968). They suggested that there were two ways of looking at a neighborhood. One is to evaluate the neighborhood in terms of its material characteristics as a bundle of *goods and services* that satisfy people's *housing needs*. This will, almost inevitably, identify what people lack and petrify corrective measures into scientifically established and bureaucratically managed standards. It is associated with centralism, authoritarianism, professionally diagnosed needs, and institutional services.

But a neighborhood can also be understood as a set of productive relationships among its inhabitants. Such a commonsense view of people is sensitive to what people can do—their abilities rather than their deficits—and will generate flexible rules that protect free people acting to fulfill their self-defined ends. The British architect John Turner became the most articulate voice of *housing by people* (rather than for them) as the paradigmatic example of an activity that is not a need, and proved the feasibility of subordinating heteronomous tools to autonomous initiatives (Turner 1978).

Assessment

AT has had technical, ethical, and political defenders. Contrary to what might be expected, ethical commitments based on faith have supported many of the more sustained AT efforts. Schumacher's essay on "Buddhist Economics" and Servants in Faith and Technology (SIFAT), a Christian evangelical NGO founded in 1979 in Tennessee, are two cases in point.

During the late 1980s, however, AT began to be envisioned as a means rather than an end—as a cheap alternative to high cost services rather than a replacement for such services. Governments started to support the NGOs that promoted AT when they presented themselves as development professionals who could diffuse AT to the third world as underdeveloped versions of high-tech educational, medical, transportation, or sanitary packages. Advocates of distributive justice fought for the right of the poor to an equal share of industrial outputs. Though it had inspired the pioneers of the AT movement, equity, conceived as the civil liberty to decide what to do and how, was progressively neglected. ATs were not only conceived as alternative ways to satisfy needs, but increasingly as first steps toward the *real thing:* Communal literacy was simply the first step toward schooling, barefoot doctors were unshod versions of those in white coats, bicycles were cheap imitations of cars, dry commodes were training tools for flush toilets, and muscles were painful alternatives to fuels.

In the high Middle Ages, Hugh of Saint Victor defined tools as appropriate remedies for the natural imperfections of human beings. In this sense, appropriateness, Latin *convenentia,* refers to the *proportional* relationship between the radius of action circumscribed by a person's innate powers and the power deposited in hands or under buttocks by tools. Appropriate technology is the search for the fitting and proper relationship between means and ends. Accordingly it has become more urgent to distinguish the alternative from the appropriate. Often the alternative is neither appropriate nor intermediate.

JEAN ROBERT
SAJAY SAMUEL

SEE ALSO *Alternative Energy, Engineering Design Ethics; Engineering Ethics.*

BIBLIOGRAPHY

Añorve, Cesar. (1999). *Remar contra corriente: Conversaciones para animar a los espíritus del tiempo*. Cuernavaca, Mexico: Centro de Innovaciòn Tecnólogica.

Borgstrom, Georg. (1971). "The Green Revolution." In *Focal Points*. New York: Macmillan.

Borremans, Valentina. (1979). *Guide to Convivial Tools*, Library Journal Special Report No. 13. New York: Bowker.

Darrow, Ken, and Rick Pam. (1976). *Appropriate Technology Sourcebook: For Tools and Techniques that Use Local Skills, Local Resources, and Renewable Sources of Energy*. Stanford, CA: Volunteers in Asia.

Devender, Kumar. (1978). "Further Reading: A Select and Relevant Bibliography on the Rural Challenge." *Seminar* 227: 44–48.

Dogra, Bahrat. (1983). "Traditional Agriculture in India: High Yields and No Waste." *Ecologist* 13: 2–3.

Farallones Institute. (1979). *The Integral Urban House*. San Francisco: Sierra Club Books.

Haldane, J. B. S. (1985). *On Being the Right Size and Other Essays*, ed. John Maynard Smith. Oxford: Oxford University Press. Essay originally published in 1928.

Illich, Ivan. (1973). *Tools for Conviviality*. London: Calder and Boyars.

Illich, Ivan. (1977). "Disabling Professions." In *Disabling Professions*, eds. Ivan Illich et al. London: Marion Boyars.

Kohr, Leopold. (1967). "Critical Size." *Resurgence* 1(8): 8–11.

Lehmann, Pierre. (1983). *Essais avec des toilettes à compostage: Gestion de l'eau et de "déchets" organiques dans l'habitat* [Essay on compost toilets]. Vevey, Switzerland: SEDE.

Lovins, Amory. (1977). "Soft Energy Paths: Towards a Durable Peace." In *Disabling Professions*, eds. Ivan Illich et al. London: Marion Boyars.

McKnight, John. (1977). "Professionalized Service and Disabling Help." In *Disabling Professions*, eds. Ivan Illich et al. London: Marion Boyars.

Nguyen, Dang Duc, et al. (1981). *Human Feces, Urine, and Their Utilization*. Bangkok: Environmental Sanitation Information Center (ENSIC).

Sachs, Wolfgang, ed. (1992). *The Development Dictionary: A Guide to Knowledge as Power*. London: Zed.

Schumacher, Ernst Fritz. (1973). *Small is Beautiful: Economics as if People Mattered*. New York: Harper Torchbooks. Contains the essay "Buddhist Economics."

Theophilus, Prysbter. (1961). *The Various Arts*, trans. Charles R. Dodwell. New York: T. Nelson.

Thompson, D'Arcy Wentworth. (1971). *On Growth and Form*. Cambridge, England: Cambridge University Press.

Turner, John. (1968). "Housing Priorities, Settlement Patterns, and Urban Development in Modernizing Countries." *Journal of the American Institute of Planners*. November: 354–363.

Turner, John. (1978). "Housing in Three Dimensions: Terms of Reference for the Housing Question Redefined." In *World Development*, ed. Ron Burguess. London: Development Planning Unit, University College.

ALTRUISM

Altruism often is defined as an action intended to benefit another person even when that action could lead to sacrifices to the welfare of the actor. Altruism thus presents an issue for ethical reflection and a thorny problem for many scientific models of human behavior. It does not fit easily into the dominant theoretical paradigms of most behavioral sciences, which assume that self-interest is the drive that underlies human behavior. When presented with examples of altruism, analysts often dismiss them as too rare to be of practical significance or as representing self-interest in disguise. Scientific frameworks that continue to struggle with the theoretical challenge presented by altruism include evolutionary biology, whose paradigm suggests that altruistic behavior should be driven out by behavior guaranteed to produce greater evolutionary fitness; economics, which assumes that actors, whether they are people, firms, or countries, pursue perceived self-interest subject to information and opportunity costs; and rational choice theory, which was derived from economic theory but has become prevalent throughout social science and decision-making theory in the form of the cost-benefit model.

Explaining Human Altruism

Because altruism should not exist according to the basic premises of these theoretical models, much early work on altruism attempted to explain it away as a disguised form of self-interest. Economists minimized altruism by explaining it as behavior that is engaged in to provide psychic gratification, deferred material gain, or group welfare (group altruism). Using similar concepts, often under slightly different names, biologists dismissed altruism as acts designed to encourage similar behavior in the future (reciprocal altruism) or further the transmission of genetic material (kin selection). Some work on animal behavior (DeWaal 1996) suggesting that animals demonstrate strong evidence of cooperation and altruism and that human altruism may be part of people's makeup as primates has been ignored by most theorists in evolutionary biology.

Among scientists who have taken human altruism seriously as an empirical reality, not merely an aberration, much of the best work has been based on experimental laboratory experiments such as that by Daniel Batson (1991) on empathic altruism. However, experimental work cannot simulate fully the more complex interactions in the sociopolitical world. This is where political analyses, even those based on small samples, provide rich insight.

Nonlaboratory analyses of human altruism include work on why people give blood (Titmuss 1997) and extensive work on philanthropists and heroes who save others (Latané and Darley 1980, Monroe 1996). Some of the most interesting studies focus on rescuers of Jews, a group of individuals who have intrigued scientists both because of the extremity of their potential sacrifice—their families also were doomed to execution if the altruists were caught—and because they represent altruism in a situation in which their immediate society as a whole condemned their acts.

Altruism Personified

Much of the early work on rescuers is autobiographical, written by rescuers (Gies 1987) or survivors (Wiesel 1986 [1960]), and consists of anecdotal portraits designed to document rescue activity. Little early work was focused on rescuers' motivations until Perry London's 1970 book. Early social science works on altruism were correlational and inquired about a wide variety of sociocultural factors, such as religion (Hunecke 1981), social class (Klingemann and Falter 1993), and gender (Fogelman 1994). Analysts slowly zeroed in on the psychological underpinnings of rescue behavior, focusing first on general psychological factors such as the thrill of adventure involved in rescuing or a sense of social marginality in which the rescuer felt an empathic bond with the persecuted because of the rescuer's own feeling of being an outsider.

A focus on the self began with Nechama Tec (1986), whose work highlighted personality factors, arguing that rescuers had a strong sense of individuality or separateness. Tec concluded that rescuers were motivated by moral values that did not depend on the support or approval of other people as much as it did on their own self-approval. The first important systematic analysis of rescuers established personality as the critical explanation. Samuel and Pearl Oliner's *The Altruistic Personality: Rescuers of Jews in Nazi Europe* (1988) located the drive for altruism in habitual behavior, encouraged by parents or other significant role models, that led to habits of caring that effectively became structured as an altruistic personality. In the same year a filmed documentary in which survivors as well as rescuers were interviewed argued that rescuers "had to do it because that's the kind of people they were" (Immanuel Tanay in *The Courage to Care*, a 1988 Academy Award–nominated documentary by Rittener and Myers [1986]).

Later analysts (Fogelman 1994, Monroe 1996) also noted the psychological importance of reinforcing empathic and humane behavior and stressed critical psychological factors related to the sense of self in relation to others. The values associated with altruism always included tolerance for differences among people and a worldview characterized as "extensivity" (Reykowski 2001).

Altruism, Cognition, and Categorization

The critical variable in explaining altruism seems to be the actor's internal psychology, and analysts interested in human altruism focus on the internal cognitive forces that drive altruism, asking how the altruistic personality or an altruistic worldview can influence altruistic acts. The psychological process seems to be as follows: People use categories to organize experience. The vast literature on social identity theory makes it clear that people categorize themselves in relation to others and then compare themselves with those critical others. However, there are many ways in which people may make that comparison. This means that analysts must ask not just how people construct categories but how they accord moral salience to them. Rescuers of Jews, for example, did draw distinctions between Jews and Nazis, but those categories were not relevant for the rescuers. They did not accord moral salience to those categories; both Jews and Nazis were supposed to be treated as human beings. Instead, rescuers constructed a broader or alternative category that was deemed morally salient. For rescuers the morally salient category was the human race, not ethnicity, religion, or political affiliation.

This raises an important question and gives altruism importance for more general ethical concerns: Is it the recognition of common membership in a category that is necessarily relevant for people's treatment of others? Or is it merely that shared membership in a category makes it more likely that one will treat other members of the same category well? The cognitive recognition of a shared category may tend to accord moral salience, but that is not necessarily the case. The empirical evidence from altruists suggests that it is not enough to say that people divide the world into divisions of in-group and out-group. One must ask how the categories are constructed and then how they are invested with moral salience.

The rescuers' categorization schema, for example, seemed to be one in which all people could exhibit individual and group differences but still be placed in the common category of human being. That category took on a superordinate moral status in which all people deserved to be treated with respect and dignity. The cognitive process by which rescuers viewed others—their categorization and classification of others and their perspective on themselves in relation to those others—had a critical influence on rescuers' moral actions. The cognitive process included an affective component that served as a powerful emotional reaction to another person's need. It created a feeling, possibly arising from heightened hormonal activity akin to the biochemical changes in the amygdala during fear or flight situations, that made altruists feel connected to people in need. That reaction provided the motive to work to effect change.

Is there a "scientific" process through which the psychology of altruism affects the ethical treatment of others? A critical part of the process appears to involve identity. Something in the external situation triggers a perception by the altruist that there is a shared bond: Perhaps the person in need is a helpless child or reminds the altruist of someone she or he once knew and liked. Perhaps someone with the potential altruist indicates a sense of concern for the needy person. This perception causes the altruist to place the needy person in the category of someone who needs help and whose situation of neediness is relevant for altruism. The categorization and perspective on the needy person in relation to the actor cause the altruist to feel a moral imperative to act, to move beyond feeling sympathy and become involved in an active sense.

Altruism thus is related to the manner in which the external environment taps into the altruist's core self-concept, which is distinguished by the altruist's self-image as a person who cares for others. As a general rule it is this perspective that links the altruist's self-image to the circumstances of others by highlighting the situation of the person in need in a way that accords a moral imperative to the plight of others. When one taps into this self-concept, the suffering of others becomes morally salient for altruists in the way the plight of one's child or parent would be salient for most people.

Because the values of caring for others are so deeply integrated into altruists' self-concepts, these values form a self-image that constitutes the underlying structure of their identities. This means that the needs of others frequently are deemed morally salient for altruists. This self-concept transforms altruists' knowledge of another person's need into a moral imperative that requires them to take action. Their self-concepts are so closely linked to what is considered acceptable behavior that altruists do not merely note the suffering of others; that suffering takes on a moral salience, a feeling that they must do something to help. Even in the extreme situation of the Holocaust the suffering of Jews was felt as something that was relevant for the rescuers. It established a moral imperative that necessitated action.

Although hard data are difficult to obtain, the fact that those rescuers felt a moral imperative to help is evident in statements that reveal their implicit assumptions about what ordinary decent people should do. The unspoken expectations are embedded deep in a rescuers' psyche and are revealed in rescuers' descriptions of what was and what was not in their repertoire of behavior. For rescuers all people within the boundaries of their community of concern were to be treated the same, and their circle of concern included all human beings. That perception of a shared humanity triggered a sense of relationship to the other that made the suffering of another person a concern for the rescuers. Significantly, this extensivity included Nazis, with the rescuers demonstrating an extraordinary forgiveness of Nazis. It is the role of perspective to classify and categorize people and then to work through a cognitive process of salience that provides the link between the lack of choice and identity and the variation in a person's treatment of others.

The scientific literature thus suggests that the empirical evidence linking identity to altruism follows these critical links: (1) the innate human desire for self-esteem and the need for continuity of self-image; (2) core values stressing the sanctity of life and human well-being that are integrated into altruists' underlying concept of who they are; and (3) external stimuli that trigger critical aspects of altruists' multifaceted and complex identity in a way that compels them to notice and accord moral salience to the suffering of others.

KRISTEN RENWICK MONROE

SEE ALSO *Evolutionary Ethics; Game Theory; Selfish Genes.*

BIBLIOGRAPHY

Batson, Daniel C. (1991). *The Altruism Question: Toward a Social Psychological Answer.* Hillsdale, NJ: Erlbaum. A thorough empirical examination that focuses on determining whether altruism exists and emphasizes empathy as a critical predictor.

DeWaal, Frans. (1996). *Good Natured: The Origins of Right and Wrong in Humans and Other Animals.* Cambridge, MA: Harvard University Press. A noted zoologist and ethologist specializing in the study of primate behavior argues that ethical behavior is rooted in people's nature as primates.

Fogelman, Eva. (1994). *Conscience and Courage.* New York: Anchor. A survey of data on rescuers of Jews during the Holocaust emphasizes demographic correlates.

Gies, Miep. (1987). *Anne Frank Remembered: The Story of the Woman who Helped to Hide the Frank Family.* New York: Simon & Schuster. A memoir that describes how the Frank rescue was attempted.

Hunecke, Douglas K. (1981–1982). "A Study of Christians Who Rescued Jews During the Nazi Era." *Humboldt Journal of Social Relations* 9(1): 144–149. Emphasizes the demographic characteristics of rescuers, such as religion.

Klingemann, Ute, and Jurgen Falter (1993). *Hilfe fuer Juden waehrend des Holocaust.* Rheinland-Verlag GmbH Koeln. A comprehensive survey of rescuers that emphasizes demographic correlates.

Latané, Bibb, and John M. Darley. (1980). *The Unresponsive Bystander: Why Doesn't He Help?* New York: Appleton-Century-Crofts. A classic social psychological work on the importance of others for bystander/rescue behavior.

London, Perry. (1970). "The Rescuers: Motivational Hypotheses about Christians Who Saved Jews from the Nazis." In *Altruism and Helping Behavior,* ed. Jacqueline Macauley and Leonard Berkowitz. New York: Academic Press. The first important empirical examination of rescuers.

Monroe, Kristen Renwick. (1996). *The Heart of Altruism: Perceptions of a Common Humanity.* Princeton, NJ: Princeton University Press. Links empirical examination of altruism to social theories based on self-interest to argue that such theories offer only a partial account of human conduct.

Oliner, Samuel P. (2003). *Do Unto Others: Extraordinary Acts of Ordinary People.* Boulder, CO: Westview Press. Expands earlier work on rescuers into everyday life to reveal altruists at work.

Oliner, Samuel P., and Pearl Oliner. (1988). *The Altruistic Personality: Rescuers of Jews in Nazi Europe.* New York: Free Press. A comprehensive survey of rescuers during the Holocaust that emphasizes the ingrained habits of caring that form an altruistic personality.

Reykowski, Janusz. (2001). "The Justice Motive and Altruistic Helping: Rescuers of Jews in Nazi-Occupied Europe." In *The Justice Movement in Everyday Life: Essays in Honor of Melvin J. Lerner,* ed. Michael Ross and Dale T. Miller. New York: Cambridge University Press. Explains and advances extensivity as the psychological phenomenon that critically influences altruism.

Rittner, Carol, and Sondra Myers. (1986). *The Courage to Care.* New York: New York University Press. A moving documentary and photo book on rescuers during the Holocaust.

Tec, Nechama. (1986). *When Light Pierced the Darkness: Christian Rescue of Jews in Nazi-Occupied Poland.* New York: Oxford University Press. Builds on the author's experiences to explain the importance of the self-concept for altruism.

Titmuss, Richard M. (1997). *The Gift Relationship: From Human Blood to Social Policy.* New York: New Press. One of first serious social science explanations of altruism.

Wiesel, Elie. (1986 [1960]). *Night.* New York: Bantam. One of many books by a Nobel laureate on the Holocaust.

AMERICAN ASSOCIATION FOR THE ADVANCEMENT OF SCIENCE

• • •

The American Association for the Advancement of Science, or AAAS (triple A-S), founded in September 1848, began as an organization to establish a national identity and forum for U.S. scientists. It has become the largest federation of scientific societies in the world, with more than 250 affiliated institutions and 130,000 individual members. AAAS publishes the peer reviewed journal *Science,* and sponsors programs that include collaborations with organizations representing scientists and non-scientists throughout the world.

Science in Service of Society

Throughout its history, AAAS has addressed issues at the intersection of science and society. During World War I, as advances in science and technology created public expectations for progress, AAAS committed itself to "the use of science for public good" (Benson and Maienschein 1999, p. 3). In 1946, AAAS affirmed a commitment to bridging science and society by revising its Constitution to include objectives "to improve the effectiveness of science in the promotion of human welfare, and increase public understanding and appreciation of the importance and promise of the methods of science in human progress" (AAAS Constitution 1946).

The 1950s brought concerns due to increasing government secrecy restrictions, growing controversies over nuclear weapons, and anti-communist suppression of dissenting views. In 1958 the AAAS Board created the

Committee on Science in the Promotion of Human Welfare to recommend responses to the issues that concerned society. The Committee urged AAAS and the scientific community to fulfill "an obligation to call to public attention those issues of public policy which relate to science, and to provide for the general public the facts and estimates of alternative policies which the citizen must have . . . to participate intelligently in the solution of these problems" (AAAS Committee on Science in the Promotion of Human Welfare 1960, p. 71).

Scientists' Rights and Responsibilities

Social unrest in the 1960s and 1970s, fueled by antinuclear, environmental, and anti-Vietnam War movements, which argued that science was complicit in creating national problems rather than in solving them, led to public demands for greater accountability by scientists. In response AAAS created an ad hoc committee in 1970 to report on the "conditions required for scientific freedom and responsibility" (Edsall 1975, p. v). In its report the committee recommended that AAAS establish a more permanent committee to reassess boundaries of scientific freedom and responsibility in a world where science is increasingly "inextricably intertwined with major political, social, and economic problems" (Edsall 1975, p. ix).

As a result, the Association created a new standing Committee on Scientific Freedom and Responsibility in 1976 to "encourage and assist the AAAS . . . and other scientific groups to develop statements of principles governing professional conduct, and to . . . encourage scientists to accept their professional responsibilities both with regard to safeguarding the integrity of science and with regard to the application of science in the promotion of human rights and general welfare" (AAAS Committee on Scientific Freedom and Responsibility Internet site). In 1977 AAAS amended its Constitution to include "to foster scientific freedom and responsibility" in its mission and, in 1981, established the Scientific Freedom and Responsibility Award to "honor scientists and engineers whose exemplary actions have served to foster scientific freedom and responsibility."

Since the founding of the Committee on Scientific Freedom and Responsibility, AAAS ethics activities have focused on human rights and on the ethics associated with scientific research and the impacts of science and technology. The science and human rights activities of AAAS were initially influential in the 1970s and 1980s in defense of scientists, engineers, and health care professionals whose rights were violated by their governments. Collaborating with human rights groups, AAAS has helped to secure the freedom of scientists in the former Soviet Union as well as in Asia, Africa, Latin America, and the Middle East. These efforts have not been without risk, or setbacks. Committee members and staff have been harassed, even in one case arrested, while working on behalf of scientists in their home countries and accused of meddling in countries' sovereign political affairs.

In 1990 the Association established a Science and Human Rights Program that directed resources and expertise to use science to help bring notorious abusers of human rights to justice. AAAS pioneered the application of forensic science, genetics, and statistics to human rights investigations. Its work helped to unite families in Argentina, and identify victims of mass executions in Guatemala; in 2002 results of Program investigations were presented as evidence in the international war crimes trial of former Yugoslavian president, Slobodan Milosevic. The Program's work has made it a frequent technical consultant to *truth commissions* in many countries, including Haiti, Peru, and South Africa.

In 1991 AAAS reorganized its other ethics activities into a Program on Scientific Freedom, Responsibility and Law, which focuses on the ethics associated with the conduct of science as well as on the uses and impacts of advances in science and technology. AAAS has been in the vanguard of scientific societies in developing "a knowledge base to deal intelligently with misconduct" (Johnson 1999, p. 51) in science, in providing educational resources for scientists and administrators responsible for preserving the integrity of research, and in advocating a prominent role for scientific societies in promoting research integrity. Through a series of practicums begun in 1992, AAAS has helped prepare institutional officials for investigating allegations of research misconduct under federal regulations. A set of videos, produced by AAAS in 1996 and used to educate students and researchers in the ethics of conducting and reporting research, is a popular resource in hundreds of colleges and universities.

Engaging the Larger Public

To complement its work in human rights and ethics, in 1995 AAAS established the program of dialogue on science, ethics, and religion to promote scholarship on the religious implications of advances in science and technology and to facilitate communication between the scientific and religious communities. Through its programs, AAAS has recognized that the consequences of science and technology often challenge public and expert sensibilities about what is ethically acceptable,

Forensic anthropologist Clyde Snow on assignment for AAAS in Argentina, excavating a mass grave. AAAS pioneered the application of forensic science to investigations of human rights abuses. *(From the records of the Science and Human Rights Program, AAAS Archives.)*

and has highlighted the issues that may cause tension between the freedom of scientists and their social responsibilities. AAAS works to provide timely, credible, and balanced information to policy debates by bringing multidisciplinary analysis to bear on complex issues, and by brokering among a wide range of stakeholders to promote broad public dialogue on such matters as stem cell research, genetic modification, and human cloning. AAAS has used the knowledge and insights gained through these studies to brief the media, to provide testimony at legislative and administrative hearings, and to develop educational materials. It has also taken public positions on highly controversial issues, including the use of animals in research, the conduct of stem cell research, the prospects of human cloning, and post-9/11 debates over the impact of national security policies on the freedom of scientific inquiry. Although it is difficult to trace the precise influence that these efforts have had, it is testimony to AAAS's credibility that other scientific organizations, public interest groups, and government officials call on the organization for assistance (Teich 2002).

In 2002 under new executive leadership, AAAS revisited its historic mission and reinforced its commitment to "advance science and innovation throughout the world for the benefit of all people," and the priority to be accorded to the "responsible conduct and use of science and technology" (AAAS Mission 2002). As ethical issues associated with scientific research and technology continue to challenge public beliefs and attitudes, the professional responsibilities of scientists, and the capacity of public and private institutions to anticipate and respond effectively, AAAS has repositioned itself to be a more visible voice in science policy and reaffirmed its commitment to advancing science and serving society.

MARK S. FRANKEL

SEE ALSO *Federation of American Scientists; Nongovernmental Organizations; Profession and Professionalism; Royal Society.*

BIBLIOGRAPHY

American Association for the Advancement of Science. Committee on Science in the Promotion of Human Welfare. (1960). "Science and Human Welfare." *Science* 132: 68–73.

Benson, Keith, and Jane Maienschein. (1999). "Introduction: AAAS Narrative History." In *The Establishment of Science in America*, eds. Sally Gregory Kohlstedt; Michael M. Sokol; and Bruce V. Lewenstein. New Brunswick, NJ: Rutgers University Press. This well-documented history covers AAAS's first 150 years, and in the process connects the actions of the association with significant events related to the role of scientists in the United States and the changing relationship between science and American society.

Chalk, Rosemary, ed. (1988). *Science, Technology, and Society: Emerging Relationships*. Washington, DC: American Association for the Advancement of Science. This anthology includes eighty-five articles, editorials, and letters published in the journal *Science*, from 1949 until early 1988. Among the topics covered are science and responsibility, science and freedom, scientists and citizens, science and risk, and science and national security.

Edsall, John, ed. (1975). *Scientific Freedom and Responsibility*. (A Report of the AAAS Committee on Scientific Freedom and Responsibility.) Washington, DC: American Association for the Advancement of Science. This foundational document led AAAS to establish the committee on scientific freedom and responsibility. It maps the boundaries of scientific freedom and responsibility, prescribes criteria and procedures for the study of those boundaries, and recommends a role for AAAS in preserving the freedom of scientists and in encouraging scientists to address the broad social implications of their work.

Johnson, David. (1999). "From Denial to Action: Academic and Scientific Societies Grapple with Misconduct." In *Perspectives on Scholarly Misconduct in the Sciences*, ed. John M. Braxton. Columbus: Ohio State University Press.

Teich, Albert H. (2002). "AAAS and Public Policy: Speaking Softly and Carrying a Medium-Sized Stick." *Technology in Society* 24: 167–178.

INTERNET RESOURCES

American Association for the Advancement of Science. (1946). *AAAS Constitution*. Available from http://archives.aaas.org/docs/documents.php?doc_id=406.

American Association for the Advancement of Science. (2002). *Mission*. Available from http://www.aaas.org/about.

American Association for the Advancement of Science Archives. Available from http://archives.aaas.org.

American Association for the Advancement of Science. Committee on Scientific Freedom and Responsibility. Available from http://www.aaas.org/spp/sfrl/committees/csfr.

American Association for the Advancement of Science. Program of Dialogue on Science, Ethics, and Religion. Available from http://www.aaas.org/spp/dser.

American Association for the Advancement of Science. Science and Human Rights Program. Available from http://shr.aaas.org.

American Association for the Advancement of Science. Scientific Freedom and Responsibility Award. Available from http://www.aaas.org/about/awards/Freedom.shtml.

American Association for the Advancement of Science. Scientific Freedom, Responsibility and Law Program. Available from http://www.aaas.org/spp/sfrl.

ANDERS, GÜNTHER

Philosopher of technology Günther Anders (1902–1992), who was born in the city of Breslau (then a part of Germany) on July 12, developed a unique moral critique of modern technology. He studied psychology, history of art, and philosophy at the universities of Hamburg and Berlin, and, as a student of Edmund Husserl, received his Ph.D. from the university of Freiburg in 1923. Anders's escape from Nazi Germany in 1933, his exile in North America, and, most importantly, the events of Auschwitz and Hiroshima, formed the experiential background to his thought. He returned to Europe in 1950 and lived in Vienna until his death on December 17.

Anders's philosophy exemplifies that tradition of critical and enlightened thought that engages with the world and the concrete problems of its time, seeking to ground human actions and the necessity of morality and ethics from within actual historical conditions. Anders's extensive body of work analyzes the changes to which human beings, both individually and collectively, are subject in a technological world. In the early period of his development, he undertook socio-political analyses of human practice (e.g., studies on fascism and unemployment), while writing poems, philosophical novels, and other books on philosophy, literature, and art.

Concern with the world is such a strong feature of Anders's philosophical identity that, for him, theoretical analysis and practical engagement are inextricably linked. He was one of the first intellectuals who warned against the Nazis and he took part in the resistance against Hitler and fascism. Later he was an active anti-Vietnam War protester, and an initiator of the anti-nuclear and environmental movements. But as much as he was a political activist, he nonetheless recognized the vital role of theory in an increasingly scientific and technological world, and, in reversing Karl Marx's famous formulation, he emphasized: "It is not enough to change the world, we do this anyway. And it mostly happens without our efforts, regardless. What we have to do is to interpret these changes so we in turn can change the changes, so that the world doesn't go on changing without us—and does not ultimately become a world without us" (Anders 2002b [1980], p. 5).

Anders regarded the destruction of Hiroshima as year one of a new era, and as the event that crystallized a newly acquired human capacity for self-destruction. This step into a future continually threatened with its own finality represented for him a radically new context for human action, demanding a new ethics. Anders confronted this changed global reality, and from this point on concentrated his efforts on thinking through the new moral situation and elucidating the relationship between human beings and technology.

Human activity, through its development of technology, had begun to overreach itself in a fatal way. Because human faculties such as emotion, perception, or even the ability to assume responsibility, are relatively circumscribed when compared to the capacity to create new things, human beings are now faced, he says, with a *Promethean discrepancy* between the world of technology and human abilities to visualize it. The divide is primarily attributable both to the accelerated pace of technological development, and to the enormous complexity of the created things and their effects.

In this paradoxical situation, whereby *humans are smaller than themselves*, Anders sees the basic dilemma of the twenty-first century, a dilemma that can only be resolved by a *moral imagination* reconnecting production and visualization, creation and representation.

In his two-volume major work *Die Antiquiertheit des Menschen* (The Obsolescence of Human Beings) (2002a, 2002b), Anders develops the project moral imagination using a specific thing-cognizant approach. Because he realizes that acting has shifted (of course through human action) from the province of humans to the sphere of work and products, and that the created things are not simply neutral means to an end, but in fact represent *incarnated* or *reified actions*, he places the question of morality primarily in the realm of the things themselves. Therefore he is less concerned with listening to the voice of the heart (or examining the social processes of making or use), than with articulating the mute principles of work and the secret maxims of products, and trying to imagine how these embedded precepts are changing human beings and the fabric of daily life. Anders's work constitutes a new form of practical reason that attempts to reconnect modern technology to its human origins. "Have only those things," he formulates as a new categorical imperative, "whose inherent action maxims could become maxims for your own actions" (Anders 2002a [1956], p. 298).

ERNST SCHRAUBE

SEE ALSO *Arendt, Hannah; Atomic Bomb; German Perspectives; Weapons of Mass Destruction.*

BIBLIOGRAPHY

Anders, Günther. (1960). *Franz Kafka*, trans. A. Steer, and A. K. Thorlby. New York: Hillary House. A critical philosophical interpretation of the writings of Franz Kafka; the original German version appeared 1951 under the title "Kafka, pro und contra: die Prozessunterlagen." Munich: Beck.

Anders, Günther. (1962). "Theses for the Atomic Age." *Massachusetts Review* 3: 493–505.

Anders, Günther. (1989 [1962–]). *Burning Conscience: The Case of the Hiroshima Pilot, Claude Eatherly, Told in his Letters to Günther Anders*. New York: Paragon House. An influential dialogue on the moral situation in the atomic age; translated into more than 50 languages. The chapter "Commandments in the Atomic Age" appeared 1972 in *Philosophy and Technology: Readings in the Philosophical Problems of Technology*, eds. Carl Mitchum and Robert Mackey. New York: Free Press.

Anders, Günther. (2002a [1956]). *Die Antiquiertheit des Menschen, Band 1: Über die Seele im Zeitalter der zweiten industriellen Revolution* [The obsolescence of human beings, volume 1: The soul in the age of the second industrial revolution]. Munich: Beck. First volume of Anders's major work. A precise and thoughtful philosophical critique on how modern technologies are changing individual and social life; unfortunately, still not available in English.

Anders, Günther. (2002b [1980]). *Die Antiquiertheit des Menschen, Band 2: Über die Zerstörung des Lebens im Zeitalter der dritten industriellen Revolution* [The obsolescence of human beings, volume 2: The destruction of life in the age of the third industrial revolution]. Munich: Beck. Second volume of Anders's major work; unfortunately, still not available in English.

ANDROIDS

Androids are mechanical, or otherwise artificial, creations in the shape of humans. They have long been a staple of science fiction. From the clockwork persons of myth to Isaac Asimov's humanoid robots, to *Star Wars's* C-3PO, and to Steven Spielberg's *A.I. Artificial Intelligence*, imagined mechanical persons have enabled people to reflect upon what it means to be human.

The real world of androids is substantially more mundane than their appearance in science fiction. Although there exists a long history of clockwork automata and other mechanical imitations of persons, these have never been more than theatrical curiosities. The creation of more ambitious androids has had to await advances in robotics. Until the 1990s, the problems involved in creating a robot that could walk on two legs prevented robots from taking humanoid form. Yet if robotics technology continues to improve, then it seems likely that robots shaped like and perhaps even behaving like human beings will be manufactured within the twenty-first century.

For the purpose of considering the ethical issues they may raise, androids can be divided into three classes: Those that are merely clever imitations of human beings, hypothetical fully-fledged "artificial persons," and—in between—intelligent artifacts whose capacities are insufficient to qualify them as moral persons.

Existing androids are at most clever imitations of people, incapable of thought or independent behavior, and consequently raise a limited range of ethical questions. The use of animatronics in educational and recreational contexts raises questions about the ethics of representation and communication akin to those treated in media ethics. A more interesting set of questions concerns the ethics of human/android relations. Even clever imitations of human beings may be capable of a sufficient range of

Android SDR-3X. Sensory cells beneath its "feet," a camera, and a microphone enable it to walk, talk, and dance. (*© Hashimoto Noboru/Corbis Sygma.*)

responses for people to form relationships with them, which may then be subject to ethical evaluation. That is, people's behavior and attitudes towards such androids may say something important about them. Moreover, the replacement of genuine ethical relations with ersatz relations may be considered ethically problematic. This suggests that some uses of androids—for instance, as substitute friends, caregivers, or lovers—are probably unethical.

Any discussion of the ethical issues surrounding "intelligent" androids is necessarily speculative, as the technology is so far from realization. Yet obvious issues would arise should androids come to possess any degree of sentience. The questions about the ethics of android/human relationships outlined above arise with renewed urgency, because the fact of intelligence on the part of the android widens the scope for these relationships. If androids are capable of suffering, then the question of the moral significance of their pain must be addressed. Once one admits that androids have internal states that are properly described as pain, then it would seem that one should accord this pain the same moral significance as one does the pain of other sentient creatures.

There is also a set of important questions concerning the design and manufacture of such entities. What capacities should they be designed with? What inhibitions should be placed on their behavior? What social and economic roles should they be allowed to play? If androids were to move out of the research laboratory, a set of legal issues would also need to be addressed. Who should be liable for damage caused by an android? What rights, if any, should be possessed by androids? What penalties should be imposed for cruelty to, or for "killing," an android? Ideally, these questions would need to be resolved before such entities are created.

However, the major ethical issue posed by sentient androids concerns the point at which they move from being intelligent artifacts to "artificial persons." That is, when they become worthy of the same moral regard that individuals extend to other (human) people around them. If it is possible to manufacture self-conscious and intelligent androids, then presumably at some point it will be possible to make them as intelligent, or indeed more intelligent, than humans are. It would seem morally arbitrary to deny such entities the same legal and political rights granted human beings.

Importantly, any claim that this point has been reached necessitates a particular set of answers to the questions outlined above. If androids become moral persons then it is not only morally appropriate but required that humans should respond to the death of an android with the same set of moral responses as they do a human person; for instance, with horror, grief, and remorse. This observation alone is enough to suggest that the creation of artificial persons is likely to be more difficult than is sometimes supposed.

ROBERT SPARROW

SEE ALSO *Cyborgs; Robots and Robotics.*

BIBLIOGRAPHY

Brooks, Rodney Allen. (2003). *Robot: The Future of Flesh and Machines*. London: Allen Lane. A popular account of the history and probable future of robots and robotics by a leading robotics researcher.

Menzel, Peter, and Faith D'Aluisio. (2000). *Robo Sapiens: Evolution of a New Species*. Cambridge, MA: The MIT Press. A survey of the state of robotics research at the beginning of the twenty first century in the form of a photoessay and set of interviews with researchers around the world.

ANGLO-CATHOLIC CULTURAL CRITICISM

The terms *Anglo-Catholic* and *Anglo-Catholicism* are broad descriptions of people, groups, ideas, and practices that emphasize those dogmatic and sacramental aspects of the Church of England that promote continuity with Catholic tradition. Anglo-Catholicism formally began in 1833 with the *Oxford Movement* reaction to extreme liberal and conservative innovations of the Church of England, as argued most prominently in *Tracts for the Times*, eighty-eight pamphlets issued in five bound volumes (1834–1840), written by John Henry Newman, Edward B. Pusey, John Keble, and several others. Following is a brief discussion of several selected forerunners and heirs of Anglo-Catholicism, all who were and are important critics and interpreters of the culture of science in their time.

Jonathan Swift (1667–1745) was one of the keenest satirists and greatest masters of prose style that English literature has produced. His most famous work, *Gulliver's Travels* (1726), was a bitter satire of the politics and social attitudes of his day, and in Part One, "A Voyage to Lilliput," he satirized abstract science or technology. He was not opposed to science and scientific experimentation if it was benevolent, but he warned about putting too much faith in science, as he lived in an age when much that passed for science was pseudoscience, perpetrated by impostors. He was before his time in realizing that science could be put to evil as well as good use. Swift often painted science in a good light in *Gulliver's Travels*, as when Gulliver studies "Physick" at a renowned medical school, when he enthusiastically reports the scientific discoveries he encounters, and when he gives word of the discovery of the two moons of Mars by Laputan observers, 150 years before they were actually discovered in 1877. His attitude was in contrast to many critics of his day, who saw science as promoting intellectual arrogance which could lead a person away from God, and as a philosophy which would likely end in pure materialism.

John Henry Newman (1801–1890) was the Anglican, later Roman Catholic, theologian and churchman who was one of the chief founders of the Oxford Movement. Newman's views about the science of his day were decidedly pessimistic. He avoided the meeting of the British Academy for the Advancement of Science in 1832 because of its interests in theology, and also shunned later meetings of the British Association. He suggested that a person with simple faith had an advantage over an academic or scientist, particularly if the latter did not temper their empirical observations with proper moral quality and regard for faith. In *An Essay in Aid of a Grammar of Assent* (1870), Newman called attention to the faulty psychological presumptions of many scientific claims, with a specific reference to the search for extra-terrestrial intelligence. In *Letters and Diaries* (published posthumously in 1961), he voiced his indignation toward scientists who gave public talks on subjects other than their own.

Gilbert Keith Chesterton (1874–1936) was a convert to Roman Catholicism, social critic, Christian apologist, novelist, and popular speaker. As an apologist for the Catholic Church, Chesterton believed the Church to be a living institution, a meeting place for all truth, including science. But he was opposed to *scientism*, naturalistic science that left no room for metaphysical truth. The popularizers of science in his day (Thomas H. Huxley, H. G. Wells, and others) attacked religion openly, and statements about science as a new religion had become common in intellectual circles. Chesterton pointed out in such works as *All Things Considered* (1908) that scientists, in claiming to have no room for ultimate authority, violated their own rational-empirical methods by making dogmatic pronouncements about religion and God based solely on their own authority. He was critical of evolutionary theory in works like *Orthodoxy* (1908), and *The Everlasting Man* (1925), and reserved some of his harshest words for eugenics (*What's Wrong With the World* [1910]), declaring it would primarily be used to oppress the poor.

Dorothy L. Sayers (1893–1957) was a noted Christian apologist, Dantean scholar, playwright, and detective novelist. Her most original work was *The Mind of the Maker* (1941), in which she examined the creative instinct in human beings and speculated that the capacity to create was a human quality that mirrored the character of God. In that work and in *Begin Here* (1940), Sayers used Trinitarian analogy in describing the human soul. Theology interprets God in nature, humanity, and Christ; philosophy strives to understand humanity and its place in the universe; and science attempts to understand nature and how it should function. She saw science primarily as the study of means and instruments, and believed it could not deal with ultimate values. For Sayers, a Christian humanist, science was one part of the human soul, and it was God who created its possibility. Her creative thought was a synthesis of empiricism, reason, and revelation, all placed in the human spirit by God.

E. F. Schumacher (1911–1977) was born in Germany and was a Rhodes scholar at Oxford in the 1930s. From 1950 to 1970 he was an advisor to the British Coal Board, and his foresighted planning (he predicted the rise of the Organization of the Petroleum Exporting Countries [OPEC] and the problems of nuclear power) assisted Britain in its economic recovery from the war. A Roman Catholic convert, Schumacher's most famous work was *Small is Beautiful: Economics as if People Mattered* (1973), a blending of Christian principles and eastern belief systems (including those of Gandhi and Buddhism) that suggested for him an alternative to rampant accumulation and technology. He had the rare gift of being able to combine sound thinking with pragmatic common sense, and recognized that commitment to technology needed ethics to help give it balance in human affairs, as it had no natural controls or self-limitations. He understood the problem of expensive technology for underdeveloped nations, and proposed for them *intermediate* technology that was less efficient but employed more people and could be incorporated more easily into a poor culture. *A Guide for the Perplexed* (1977) extended his argument. Schumacher spent most of the latter part of his life teaching intermediacy and urging wealthy nations to share scientific advances and new technologies with less fortunate countries. His vision of intermediate technology and economics influenced the alternative technology movement in the developed countries and flourishes in the early twenty-first century in several countries in Africa and Asia.

E. L. Mascall (1905–1993) was a mathematically trained Anglican priest and for many years Lecturer at Christ Church, Oxford, and Professor of Historical Theology at King's College, London. Mascall argued in his *The Openness of Being* (1971) that the natural world reveals the presence of God, who is creator and sustainer. In this and other works such as *Christian Theology and Natural Science* (1956), he contended that the scientist should consider the idea that one does not start with the world and end up with God, but that God and the world can be perceived together in reality. In *The Secularization of Christianity* (1966), he praised those who argued that Christianity and science are compatible, and that scientific achievement only made sense when combined with a study of Christian doctrine. In *The Christian Universe* (1966), he deplored the decay of belief in God in his time, and urged his readers to see their vast world in light of the great creeds of Christendom.

John Polkinghorne (b. 1930) was Professor of Mathematical Physics at Cambridge and President of Queen's College, Cambridge until his retirement in 1997. A significant contributor in the dialogue between science and religion, his autobiography, *The Faith of a Physicist* (1994), was a best-seller. Polkinghorne is a rare combination of a working scientist and Christian apologist. In several of his works, including *The Way the World Is* (1983), and *Belief in God in an Age of Science* (1998), he initiates a place for natural theology (knowing God through reason and experience alone) in apologetics and theology. For Polkinghorne, natural theology is perhaps the crucial connection between the world of science and religion, and he asserts that one of the most important achievements of modern science has been its demonstration of a natural balance and ordering of the world. This leads him to ask in several of his works, where the balance and ordering of the world comes from.

PERRY C. BRAMLETT

SEE ALSO *Christian Perspectives: Historical Traditions; Lewis, C. S.; Tolkien, J. R. R.*

BIBLIOGRAPHY

Allitt, Patrick. (1997). *Catholic Converts: British and American Intellectuals Turn to Rome*. Ithaca, NY: Cornell University Press.

Barbour, Ian G. (2000). *When Science Meets Religion*. San Francisco: HarperSanFrancisco.

Chesterton, G. K. (1908). *All Things Considered*. London: Meuthen.

Chesterton, G. K. (1908). *Orthodoxy*. London: John Lane, the Bodley Head.

Chesterton, G. K. (1910). *What's Wrong with the World*. London: Cassell.

Chesterton, G. K. (1925). *The Everlasting Man*. London: Hodder & Stoughton.

Ferngren, Gary B., ed. (2002). *Science and Religion: A Historical Perspective*. Baltimore, MD: Johns Hopkins University Press.

Fox, Nichols. (2002). *Against the Machine: The Hidden Luddite Tradition in Literature, Art, and Individual Lives*. Washington, DC: Island Press.

Jaki, Stanley L. (1986). *Chesterton, A Seer of Science*. Urbana: University of Illinois Press.

Jaki, Stanley L. (2000). *Newman's Challenge*. Grand Rapids, MI: Eerdmans.

Lindberg, David C., and Ronald L. Numbers, eds. (1986). *God & Nature: Historical Essays on the Encounter between Christianity and Science*. Berkeley: University of California Press.

Mascall, E. L. (1956). *Christian Theology and Natural Science*. London: Longmans, Green & Co.

Mascall, E. L. (1966). *The Christian Universe*. New York: Morehouse-Barlow.

Mascall, E. L. (1966). *The Secularization of Christianity*. New York: Holt, Rinehart & Winston.

Mascall, E. L. (1971). *The Openness of Being: Natural Theology Today*. Philadelphia: Westminster Press.

McGrath, Alister E. (1999). *Science & Religion: An Introduction*. Oxford, England: Blackwell.

Newman, John Henry. (1870). *An Essay in Aid of a Grammar of Assent*. London: Burns, Oates.

Newman, John Henry. (1975). *The Letters and Diaries of John Henry Newman, Volume XXXVIII*. Oxford: Oxford University Press.

Newman, John Henry, et al. (1840). *Tracts for the Times*. Oxford: J. H. Parker.

Pearce, Joseph. (1999). *Literary Converts: Spiritual Inspiration in an Age of Unbelief*. San Francisco: Ignatius Press.

Polkinghorne, John. (1983). *The Way the World Is*. Grand Rapids, MI: Eerdmans.

Polkinghorne, John. (1998). *Belief in God in an Age of Science*. New Haven: Yale University Press.

Sayers, Dorothy L. (1940). *Begin Here*. London: Victor Gollancz.

Sayers, Dorothy L. (1941). *The Mind of the Maker*. New York: Harcourt, Brace, and Co.

Schumacher, E. F. (1973). *Small Is Beautiful: Economics As If People Mattered*. New York: Harper & Row.

Schumacher, E. F. (1977). *A Guide for the Perplexed*. New York: Harper & Row.

Swift, Jonathan. (1984). *Gulliver's Travels and Other Writings*. New York: Bantam.

Westfall, Richard S. (1973). *Science and Religion in Seventeenth-Century England*. Ann Arbor: The University Of Michigan Press.

ANIMAL EXPERIMENTATION

The use of animals in medical and other research has been a staple of modern scientific progress. In the early twenty-first century, biomedical research in the United States involves the use of several million animal subjects (mostly rodents) each year. With the rise of biotechnology and the techniques of genetic modification, the scientific use of animals will continue in novel forms. There are questions, however, about the reliability of information gained from animal experimentation, and whether it is morally defensible to exploit animals for the sake of scientific knowledge.

History

While animal experimentation might be thought of as a thoroughly modern practice, humans have been learning from animals since prehistory. Early human hunters' knowledge of the natural world was likely formed by their awareness of the life cycles and migration patterns of prey species. Prehistoric understanding of anatomy and physiology was no doubt the by-product of butchering animals for food. In classical antiquity, scientifically sophisticated knowledge of animal physiology emerged, indicating that the dissection of animals for the purpose of gaining such knowledge had begun. By the Roman era, dissection and vivisection (the dissection of live animals) were established scientific practices. Like much empirical science, these practices were squelched during the Middle Ages, only to reappear during the Renaissance.

By the seventeenth century, when William Harvey (1578–1657) revolutionized physiology, he and his colleagues relied almost exclusively on knowledge gathered from experiments on animals. Throughout the modern era, each subsequent advance in medical knowledge—the germ theory of disease, vaccinations, nutritional chemistry, surgery performed with anesthesia—was made possible by using animal subjects. In the early twenty-first century, virtually all medical therapies—drugs, vaccines, surgical techniques, prosthetics—are developed with the aid of animal subjects, and animal models play a significant role in psychological research. In the United States, the Food and Drug Administration (FDA) requires that all new medicines undergo animal testing to demonstrate safety before they are tested on humans. Other governmental agencies require that the safety and environmental impact of various consumer products be assayed, and, while not a legal requirement, manufacturers frequently rely on animal subjects to do so.

Given the omnipresence of medical and other technological goods to which animal experimentation has contributed, it is questionable whether moral objections to the practice can be consistently maintained in the modern world. For example, the animal rights theorist Tom Regan, in a paper delivered in May 2005, has raised the issue of whether respect for animals requires that one refuse all medical treatments that have been tested on animals, and thus whether animal advocates who continue to avail themselves of modern medicine are guilty of hypocrisy. Nonetheless, modern animal experimentation has been dogged by moral opposition throughout its history. Beginning in 1824, when the Royal Society for the Prevention of Cruelty to Animals was formed in England, many organizations rose to resist vivisection and other practices that inflict pain and take animal lives. This type of animal advocacy is continued in the early twenty-first century by People for the Ethical Treatment of Animals (PETA) and other international associations. Founded in 1980, PETA's early efforts in the United States led to the first successful criminal prosecution (later reversed on appeal) of a medical researcher on charges of animal cruelty.

The moral core of the opposition to animal experimentation is often overshadowed by the aggressive actions of extremist groups such as the Animal Liberation Front (formed during the 1970s in England by hunt saboteurs), whose members have been responsible for vandalizing animal research facilities and threatening violence against researchers who use animals. Nevertheless, moral concern for animals has also inspired the body of law under which animal experimentation is currently conducted. In the United States, the legal control of animal experimentation began in 1966 with the Animal Welfare Act. Animal research is regulated by the U.S. Department of Agriculture and the Department of Health and Human Services. These agencies require that research facilities establish institutional animal care and use committees (IACUCs) to evaluate the merits of research involving animals and monitor the treatment of experimental subjects.

Challenges

While most opposition to animal experimentation is based on moral considerations, some have also raised epistemological objections. Chief among these is the problem of species extrapolation. Because the relationship between an organism's higher functions and their underlying biology is very complex, it is impossible to predict with certainty how an agent will affect humans based on experiments done with other species. Detractors need only point to headlines from the early 2000s for examples of medicines that fared well during animal studies, but then produced problematic results when used widely on human patients. Proponents of animal testing acknowledge that identifying the animal species whose biology is most appropriate to a specific experiment is a daunting task, but it is not impossible. The number of instances in which failed species extrapolation led to significant harm to human patients is small when compared to the successes, proving that many biological analogies between humans and animals are sound.

This defense of the epistemological foundations of animal research has nevertheless provided the theoretical foundation for much of its moral criticism: If animals are sufficiently similar to humans to justify experimenting on them, it is likely that they also possess a degree of morally relevant attributes sufficient to render the experiments problematic. The point is especially significant for research involving primates. Opponents argue that if primates or other animals possess pain perception, emotional complexity, intelligence, or subjectivity comparable to that of humans, then at a minimum researchers are morally obligated to limit the impact their experiments have on animal subjects. Those who advocate the strong animal rights position argue for the abolition of animal research, even when the pains experienced by the subjects might reasonably be outweighed by gains in human well-being. Others stop short of rejecting all animal experiments, but rather draw attention to research that is redundant, poorly designed, or of dubious merit, or that inflicts a great deal of suffering.

In addition to the treatment that individual animals receive during the course of research, some have raised concerns about the commodification of life-forms that the acquisition of experimental subjects entails. Almost all laboratory animals are now "purpose bred" to make them compliant with the experimental conditions to which they will be subjected, and to ensure consistent data; thus, these living beings are essentially technological products, brought into existence for the purpose of their scientific use. The point is inarguable in the case of experimental subjects produced by means of genetic modification. In the most famous example, researchers at Harvard University developed through genetic modification a breed of mouse (dubbed the "OncoMouse™") with a disposition to develop cancer. Not only did the case raise the question of whether it is ethical to intentionally bring such genetically defective beings into existence, fundamental moral and legal issues were also raised by the researchers' efforts to patent the mice produced through their technique.

While the traditional defense against moral objections to animal research was to deny that animals possess the capacity for morally relevant experiences, that is a position seldom heard anymore. Indeed, many researchers speak in solemn terms about the sacrifices their animal subjects are forced to make; some Western research facilities have adopted a custom developed by Japanese scientists, who hold memorial observances for the animals they have used. Others admit to struggling with their natural inclination to empathize with the creatures they use (a fact that makes distancing techniques—such as limiting personal contact with animal subjects and assigning them numbers rather than names—part of standard laboratory practice). Nonetheless, some proponents make the argument that it is simply a misnomer to apply humankind's strongest moral categories (such as rights) to animals, which lack the capacities of rational self-awareness and moral autonomy that make human life so valuable. This point is buttressed by the clear benefits animal experimentation has brought: It is difficult to appreciate how much progress has been made in the treatment of human disease and the alleviation of human suffering, and how necessary the use of animals has been to this rate of progress. While opponents cite the availability of alternatives to animal research—such as tissue tests, computer models, epidemiological studies, and research involving human volunteers—proponents respond that they are not viable for all research situations, and that relying on them might lead to significant delays in gaining valuable medical knowledge. Given the health crises humankind still faces and the potentially great benefits to human well-being, many proponents argue that animal experimentation is not only defensible, but morally obligatory.

Despite the often heated controversy, a consensus ethic for animal research (the 3Rs approach) is beginning to emerge, with support among both animal advocates and proponents of scientific progress. It holds that researchers have a duty to *refine* experiments that use animals to ensure that the impact on them is proportionate to the potential benefits of the research; to *reduce* the number of animals sacrificed to the minimum that is statistically necessary to obtain the desired data; and, when possible, to *replace* research that uses mammals with nonmammalian or nonanimal alternatives.

MARC R. FELLENZ

SEE ALSO *Animal Rights; Animal Welfare.*

BIBLIOGRAPHY

Haraway, Donna J. (1997). *Modest_Witness@Second_Millennium.FemaleMan©Meets_OncoMouse™*. New York: Routledge. Explores the broad ramifications of the melding of technology and science embodied in the genetically modified laboratory animal.

LaFollette, Hugh, and Niall Shanks. (1996). *Brute Science: Dilemmas of Animal Experimentation*. London: Routledge. An analysis of the epistemological difficulties of animal experimentation, and their link to the moral case against using animals.

Langley, Gill, ed. (1989). *Animal Experimentation: The Consensus Changes*. New York: Chapman and Hall. Collection of essays by scientists and philosophers exploring the epistemological and moral objections to animal experimentation.

Paul, Elizabeth S. (1995). "Us and Them: Scientists' and Animal Rights Campaigners' Views of the Animal Experimentation Debate." *Society and Animals: Journal of Human–Animal Studies* 3(1): 1–21. An analysis of the human psychology of animal experimentation, involving a survey of attitudes on the part of both animal rights advocates and researchers who experiment on animals.

Paul, Ellen Frankel, and Jeffrey Paul. (2001). *Why Animal Experimentation Matters: The Use of Animals in Medical Research*. New Brunswick, NJ: Transaction Publishers. Collection of essays by biomedical researchers, social scientists, and philosophers responding to the strong animal rights objection to animal experimentation.

Regan, Tom. (2004). *The Case for Animal Rights*, 2nd edition. Berkeley and Los Angeles: University of California Press. The clearest articulation of the strong animal rights position.

ANIMAL RIGHTS

It is only recently, and in response to their perceived mistreatment by humans, especially in processes of industrial agricultural production and scientific research, that rights have been ascribed to animals. The concept remains contentious, especially insofar as in radical forms it would severely restrict the use of animals in scientific research and elsewhere, but has been defended on a number of grounds.

Historical Developments

The debate over whether animals possess rights must be viewed against the background of the ubiquitous use of animals to meet human needs and desires throughout history. Although interpreted in various ways, the status of animals is a significant economic and cultural category in every human society. Because the human connection to animals runs so deep, our shared history may amount to a form of coevolution: The selective breeding of domestic species has rendered them substantially different from their wild counterparts, and the effects of domestication on human social evolution have been profound, perhaps defining. At a minimum, because the benefits of this relationship are mutual (although rarely equal), domestication invites comparison to symbiosis.

However, because of the uniquely powerful effect of this symbiotic relationship, technological models contribute to the understanding of domestication. The environmental ethicist J. Baird Callicott (1980) argues that domesticated animals are essentially human inventions and should be viewed as technologies in their own right, to be evaluated in terms of their environmental impact. To a very different effect, the critic Donna J. Haraway (2003) uses the image of the cyborg to capture the complex layers of culture, nature, and technology that define both human and animal reality. This complexity is not limited to the special cases of genetically modified lab mice and artificial heart recipients: Haraway argues that humans and the "companion species" they have bred to work and live with them are equally significant others in an ecosystem that straddles the technological and the biological.

This multilayered, ambiguous relationship between humans and animals has both insulated animal exploitation from moral assessment, and made the assessment maddeningly complex. Any complacency over the possible rights of animals has been shaken over the last three centuries in light of some of the troubling effects of industrialization, including the physical and psychological pressures placed on domesticated animals in technologically intensive economies, and threats to the very survival of wild animal species. It is no coincidence that arguments on behalf of the moral claims of animals have risen in proportion to the distance that industrialization has placed between humans and the natural world.

The idea that animals deserve moral attention is not exclusively modern, however, but has been explored throughout European intellectual history; Pythagoras and Porphyry provided early philosophical arguments that using animals for food is morally problematic. Nonetheless, much of the tradition followed Aristotle in rejecting such arguments. His contention that nonhuman animals categorically lack reason and intellect was used for centuries to justify a moral divide between humans and animals: Irrational animals are natural slaves, and no positive human moral or political categories can govern humankind's relations with them.

Because it harmonized with the Judeo-Christian contention that God gave humans dominion over animals, this model of human/animal relations held sway through much of medieval Christendom. Despite the force of this tradition, a vocal minority argued that Western monotheism can and should accommodate moral concern for animals. (A contemporary example is Andrew Linzey (1994), who argues that animals possess *theos-rights*, and are owed justice simply in virtue of being creatures of the Creator.) This is noteworthy because the roots of the modern analysis of animal rights precede the Industrial Revolution, beginning in England with a sixteenth-century theological debate over whether animals are restored through the Incarnation. This debate expanded over the centuries that followed, inducing various English theologians, literary figures, political scholars, and philosophers to offer new analyses of the moral status of animals.

The result of these efforts was a sustained attempt to rethink the traditional Aristotelian position, and an intellectual climate ripe for the concept of animal rights. By the nineteenth century, the first animal advocacy groups were formed to speak out against the abuse of draft animals and to oppose vivisection, and the first modern legal protections of animals were established.

Basic Theories

The philosophical development in this period that had the greatest influence on subsequent discussion of animal rights is the advent of utilitarianism. Unlike other ethical theories that argue moral goods are the exclusive products of humans' rational nature, the early utilitarians held that the highest moral good is the happiness that results from maximizing pleasure and minimizing pain. Given the legacy of Aristotle, the claim that nonhumans possess anything comparable to the higher cognitive faculties of humans is unavoidably controversial; in comparison, the claim that animals seek comfort and shun suffering is an easy sell. Thus animal advocates found in utilitarianism a fitting ethical theory to make their case. As Jeremy Bentham (1748–1832), the father of utilitarianism, famously asserted, "The day *may* come, when the rest of the animal creation may acquire those rights which never could have been withholden from them but by the hand of tyranny.... [T]he question is not, Can they *reason?* nor, Can they *talk?* but, Can they *suffer?*"(Principles of Morals and Legislation, Chapter 17 1789).

Despite these bold words, Bentham was unopposed to using animals in science and agriculture. It would fall to later thinkers to argue that utilitarianism should force us to rethink these institutions. The most important figure to do so is the Australian philosopher Peter Singer, whose *Animal Liberation*, originally published in 1975, inspired much of the subsequent attention the issue has received. Making use of graphic depictions of how livestock are treated in intensive feeding operations, and the painful effects of product testing and medical and psychological research on primates and other mammals,

Singer argues that the equal consideration of a sentient animal's interest in avoiding suffering renders these common practices seriously immoral. To defend this conclusion, he offers the following analogy: Racism and sexism are immoral positions because they give undue importance to the morally irrelevant properties of race and gender; likewise, those who fail to extend moral consideration to other animals simply because of their species membership are guilty of a heretofore unrecognized offense: *speciesism*. Because modern science and industry routinely exploit animals in ways we would be loath to treat humans of comparable sentience (such as those with severe mental impairment), there are few citizens of modern industrialized societies whose lives are unaffected by speciesist practices.

While Singer's argument is the most famous in the contemporary debates, he makes clear that his conclusions do not hinge on the concept of animal rights per se, of which he is dubious. Those who try to make the explicit case for rights have generally followed Singer's lead by attempting to extend moral concepts traditionally reserved for humans to cover our treatment of animals as well. Callicott has termed this general approach *extensionism*. For example, Aristotelian ethics holds that the moral good for humans (virtue) is related to our *final cause*, the natural end or function that defines us (rationality). Bernard E. Rollin (1992) argues that this model can be extended to provide the basis for a theory of animal rights: He claims that moral concepts apply to our treatment of animals not simply because they can experience pleasure and pain, but because they, like us, have natural ends or functions that they have an interest in fulfilling. He concludes the most effective way of solidifying this concern is the establishment of legal and political rights for animals. Mark Rowlands (1998) forms an analogous argument to those of Singer and Rollin by extending social contract theory to articulate the rights of animals.

Some extensionists and many laypersons use the term *animal rights* as shorthand for the moral consideration humans owe animals, but they do not all envision the moral claims of animals as fully comparable to the natural rights that modern liberalism has ascribed to humans. Such a vision has been articulated by Tom Regan (2004). Rejecting the utilitarianism of Singer, Regan's argument extends the deontological theory of Immanuel Kant (1724–1804), which holds that all humans have an inherent right to moral respect in virtue of their rational nature. Regan argues that it is arbitrary to limit such respect to those who possess rationality; many humans cannot be described as fully rational, yet we do not therefore subject them to painful experiments or use them for food. Regan argues that all animals to which we can ascribe preferences qualify as *subjects of a life*; he claims this will include most mature mammals. All such beings, he concludes, have an *inherent* value that grounds natural rights to life and autonomy comparable to those of humans.

Critical Assessment

If any of these extensionist arguments are sound, it will require serious reappraisal of the place of domestic animals in society, and of human behavior toward wild animals. In its strongest forms, the claim that animals have rights implies that all forms of animal exploitation are seriously immoral: Vegetarianism is morally obligatory, all animal testing should be proscribed, and wild animals have a right to be left free of all human interference. At a minimum, granting that animals have some claim to direct moral attention would not only allow us to condemn overt acts of animal cruelty, but also raise serious doubts about the use of intensive industrial techniques in animal husbandry (factory farming), the use of animals to test medical technologies from which they will not benefit, and the genetic modification of animals to enhance their usefulness to humans. Although it is not clear where the moral limits to animal exploitation lie, there is a growing consensus that such limits do exist and that it is important that they be clarified.

But is the case for animal rights sound? Critics fall into two camps. First, those who uphold the traditional position argue that extending rights theory to animals goes too far. The category of rights emerged for the kinds of beings that only humans are: free, rational, autonomous agents who can form agreements, respect each other's interests, and operate politically. These critics argue that to apply the concept of rights to animals that do not have these attributes is to extend it beyond coherence. For some in this camp, the ground of their objections is metaethical: Their concern is the nature of moral language and whether it can have any meaning when extended to nonhumans. Others dispute the empirical bases of extensionist arguments, namely that animals possess psychological attributes—consciousness, capacity to suffer, subjectivity, personhood—that are morally relevant. Because the sciences of ethology, animal psychology, and animal welfare are relatively young, there is at present no consensus on which animals possess such attributes, or whether any nonhumans possess them to a degree that is morally significant. Thus, the status of debate on this point is ambiguous: Extensionists can muster enough empirical evidence to give their conclusions some rational support, but not enough to prove that the traditional position is unsustainable.

Second, other critics have argued that the concept of animal rights does not go far enough in expressing the value that animals possess and the challenge that it poses to humans. Some environmental ethicists (including deep ecologists and ecofeminists) have argued that because our moral categories are purely human creations, products of the same cultural tradition that sanctioned animal exploitation for millennia, they cannot simply be *extended* to cover animals, but must be radically rethought. Extensionism implies that animals are valuable to the extent that they can be assimilated to human moral reasoning, but there is another, more radical possibility: that animals should be valued for their differences from humans, and for those aspects of animal reality that lie beyond the reach of the traditional moral and political categories of humankind. Perhaps respecting animals is not simply a matter of protecting them from the effects of humankind's dependence on technology; by inviting us to appreciate the type of reality they occupy, a space where both technological and moral devices are unnecessary, animals may help us develop a critical perspective on the ends of human civilization.

MARC R. FELLENZ

SEE ALSO *Agricultural Ethics; Animal Experimentation; Animal Welfare; Consequentialism; Deontology; Pets.*

BIBLIOGRAPHY

Callicott, J. Baird. (1980). "Animal Liberation: A Triangular Affair." *Environmental Ethics* 2(4). Offers a criticism of animal rights arguments from the perspective of environmental ethics.

Cohen, Carl, and Tom Regan. (2001). *The Animal Rights Debate*. Lanham, MD: Rowman and Littlefield. A clear synopsis of the core arguments on both sides of the animal rights debate.

Fellenz, Marc R. (2006). *The Moral Menagerie: Philosophy and Animal Rights*. Urbana: University of Illinois Press. Surveys and assesses major arguments in the contemporary debate, and analyzes the importance of the debate for philosophy.

Haraway, Donna J. (2003). *The Companion Species Manifesto: Dogs, People and Significant Otherness*. Chicago: Prickly Paradigm Press.

Linzey, Andrew. (1994). *Animal Theology*. Champaign: University of Illinois Press. Develops the case for animal rights in the context of Christian theology.

Linzey, Andrew, and Dorothy Yamamoto, eds. (1998). *Animals on the Agenda: Questions about Animals for Theology and Ethics*. Urbana: University of Illinois Press. Essays by various authors exploring the implications of animal rights for ethics, theology, and human culture.

Miller, Harlan B., and William H. Williams, eds. (1983). *Ethics and Animals*. Clifton, NJ: Humana Press. A valuable selection of philosophically sophisticated essays on the possible rights of animals, the meaning of human/animal relations, and various forms of animal exploitation.

Regan, Tom. (2004). *The Case for Animal Rights*, 2nd edition. Berkeley and Los Angeles: University of California Press. The clearest articulation of the strong animal rights position.

Rollin, Bernard E. (1992). *Animal Rights and Human Morality*, rev. edition. Buffalo, NY: Prometheus Books. Argues that the telic nature of animals requires any adequate moral theory to address their interests, and makes the case for the legal rights of animals.

Rowland, Mark. (1998). *Animal Rights: A Philosophical Defence*. St. Martin's Press. Uses contract theory as the basis for understanding the rights of animals.

Singer, Peter. (1990). *Animal Liberation*, 2nd edition. New York: Random House. The foundational work of the recent debate (originally published in 1975), provides details on the nature of modern animal exploitation, and argues that the failure to give equal consideration to the interests of animals is as morally suspect as racism and sexism.

ANIMAL TOOLS

"Tools maketh man," so said Kenneth Oakley, the prehistorian. He meant that only human beings make tools of flaked stone. More generally, many species of animals make and use tools, both in nature and in natural captivity, from wasps to finches to apes, but many more do not. Few species have *tool kits* (repertoires of different types of tool for different purposes) or *tool sets* (two or more kinds of tools used in series to perform a task). Making sense of such behavioral variation is a challenge to scientists.

Tools

Definitions of tools vary (Beck 1980). In this entry, the following is used: a detached inanimate object used by a living creature to achieve a goal, typically to alter the state or position of another object. This includes constructing a nest, but not reclining on a bough, and cracking a snail with a stone but not with the teeth. It excludes glaciers moving stones across landscapes, but includes sea otters retrieving stones from the seabed. If these actions entail modifying the object so that is it more effective, then tool using becomes toolmaking. Tools can also be classed by function: subsistence (digging stick), social life (weapon), or self maintenance (napkin), or by mode of action: percussion (nut cracking), probe (termite fishing), barrier (leaf umbrella), and more.

For as long as scientists have paid attention to animal tool use, two vertebrate classes, birds and mammals, have predominated. Some examples are classic. California sea otters crack open mollusks on anvil stones balanced on their chests as they float on their backs (but their Alaskan cousins do not). Beavers fell trees and shrubs to construct dams and lodges that transform landscapes and watersheds. Woodpecker finches of the Galapagos Islands detach twigs or spines and use them to probe and to extract insects from cavities in woody vegetation. More magnificently, bowerbirds in Australia and New Guinea build and decorate complex structures and arenas. These edifices, which range from walls to spires, are not nests for residence or rearing young, but instead serve as advertisements by males to court females.

All of these examples of tool use vary across populations within a species or across individuals within populations. In many cases, they are *one-trick ponies*, that is, single, specialized adaptations: Sea otters in Monterey Bay that use anvil stones do not engage in any other type of tool use. The prize toolmaker among birds is probably the Caledonian crow of the south Pacific island of New Caledonia, which uses three types of tool in extractions foraging. By comparing twenty-one populations, scientists determined that tools have diversified over time and across space.

Some creatures with large brains (and so presumed high intelligence) even manage tool use without grasping appendages. The bottle-nosed dolphins of Shark Bay in western Australia carry sponges on their noses. Apparently they use these to root out prey from the sea floor, with the sponge serving as a *glove* to protect the rostra from abrasion.

Of the orders of mammals, the primates are the main tool users, especially the great apes (McGrew 1992). Apart from them, it is the capuchin monkeys of Central and South America that are best known for their tool behavior. Their use of wood or stone anvils to smash open hard-shelled fruits is widespread in rainforests. In harsher habitats, capuchin monkeys are even more enterprising: In the dry open country scrublands of Brazil, they use stones as hammers to crack nuts on anvils, and even as trowels to dig up tubers.

Of the four species of great apes (bonobo, chimpanzee, gorilla, and orangutan), there is surprising variety in nature despite the fact that in captivity, all show similar levels of intelligence. Wild gorillas, whether in lowland forest or on alpine slopes, exhibit no tool use. Similarly bonobos show little, apart from occasional use of leaves as rain shelters or felled saplings in branch-dragging displays; notably absent in these apes from the Democratic Republic of Congo is any tool use in foraging. Orangutans, in some high-density wild populations in Sumatra, are accomplished arboreal tool users, but their special feature is oral tool use, presumably because their hands are needed for support in the forest canopy. Using skillful movements of lips and teeth, tools of vegetation are used to process fruits with stinging hairs and to extract insects from rotten wood.

The champion tool user and maker of the animal kingdom is the chimpanzee, seen in captivity for more than eighty years from the experiments of Wolfgang Köhler and for more than forty years from the field observations of Jane Goodall. More than fifty populations of these wild apes across eastern, central, and western Africa are known to use tools (McGrew 2004). These include flexible probes made of vegetation to fish out termites from underground nests or ants from the cavities in trees, hammers of stone or wood to crack open nuts on anvils of root or stone, pestles of palm frond to smash the mortared heart of palm, crumpled leaves to sponge out water from tree holes, and leaves to wipe off bodily fluids in personal hygiene. Tools are transported from worksite to worksite, and sometimes are made in advance of use or kept to be used again. Termite fishing has been followed through four generations of chimpanzees at Gombe National Park in Tanzania. There are limits, however: No wild chimpanzee has yet been shown to purposefully modify stone for use as a tool, nor to use one tool to make another.

Technology

When the use of tools increases efficiency or convenience, or reduces risk, or opens up new ways to exploit resources, old or new, this knowledge is termed technology. As such, when time or energy is saved, or tasks are made easier or more comfortable, or danger to life or limb is lessened, or innovations yield new payoffs, however elementary, these may be thought of as the basis of material culture. When such techniques are invented and passed on by socially mediated processes of transmission, they come to approximate what in humans is called culture. Transmission within a generation is called horizontal; transmission passed down from one generation to the next is called vertical. The latter is termed tradition. This requires some form of social exposure or interaction between knowledgeable and naïve individuals, which may range from passive observational learning to active teaching. It takes careful experimentation to establish which mechanisms of transmission of knowledge are present, but in the end, what matters most is what technological transfer occurs, not how it gets done.

All known examples of technology in animals, as defined here, come from great apes. Often the first clue comes from observed behavioral diversity in wild populations (Whiten et al. 1999). The chimpanzees of Mahale ignore the fruits of the oil palm; those at Gombe eat the outer husk only and without tools; those at Tai crack open the nuts to extract the kernel; and those at Bossou sometimes modify the orientation of the anvil to make their nut cracking more efficient. The predator (ape) and prey (nut) are the same in all four places; what differs is technical knowledge. Similar cross-cultural differences have been reported for orangutans in Borneo and Sumatra (van Shaik et al. 2003), and bonobos in the Democratic Republic of the Congo (Hohmann and Fruth 2003). Recently studies of technology in animals have extended into the past, with archaeological excavations of chimpanzee nut cracking sites in Ivory Coast. These have yielded fragments of stone, and so give enduring time-depth to nonhuman technology (Mercader et al. 2002).

Are the differences between the elementary technology of nonhuman species and the more complex technology of human ones of degree or kind? This depends on the feature chosen for comparison: Some textbooks state that a key difference is that only humans depend on technology, while for other animals it is somehow optional. The logic is that because all human societies show technology, there must be dependence, but all known wild chimpanzee populations studied in the long term also show technology, so by the same yardstick they too depend on it. On other grounds, there seem to be differences: No known animal technology seems to be imbued with religious or supernatural significance, though it is hard to infer meaning from behavior.

These findings have not only scientific implications for the understanding of humans but ethical implications for the treatment of animals. Animals kept in captivity, but deprived of appropriate objects to manipulate (explore, play, and construct), may lead incomplete or distorted lives. Impoverished of raw materials, they may fail to show species-typical behavior, such as shelter making, or worse, develop abnormal patterns, such as coprophagy. Ecologically valid environmental enrichment means using the findings of field research to provide species-specific contexts for tool use and social settings for technology if animals are confined. This can be done through emulation (seeking to recreate nature, e.g., bamboo plantings) or simulation (seeking to mimic key features of nature, e.g., artificial termite mounds).

WILLIAM C. MCGREW

SEE ALSO *Evolutionary Ethics; Ethology; Sociobiology.*

BIBLIOGRAPHY

Beck, Benjamin B. (1980). *Animal Tool Behavior*. New York: Garland STPM Press. Twenty-five years after publication, this remains the "bible" on animal technology, as a comprehensive review of the literature and theoretical treatise on tool use from insects to apes.

Hohmann, Gottfried, and Barbara Fruth. (2003). "Culture in Bonobos?: Between-Species and Within-Species Variation in Behavior." *Current Anthropology* 44: 563–571.

McGrew, William C. (1992). *Chimpanzee Material Culture: Implications for Human Evolution*. Cambridge, UK: Cambridge University Press. Chimpanzee tool use reviewed and explained in terms of cultural processes as these might shed light on the evolutionary origins of human technology.

McGrew, William C. (2004). *The Cultured Chimpanzee: Reflections on Cultural Primatology*. Updates *Chimpanzee Material Culture* (1992) and expands discussion into related aspects of non-material culture; written in opinionated and accessible style. Cambridge, UK: Cambridge University Press.

Mercader, Julio; Panger, Melissa; and Christophe Boesch. (2002). "Excavation of a Chimpanzee Stone Tool Site in the African Rainforest." *Science* 296: 1452–1455. First attempt to practice archaeology on apes, by excavating stone artifacts from West African chimpanzees' nut-cracking sites.

van Schaik, Carel P.; Ancrenaz, Marc; Borgen, Gwendolyn; et al. (2003). "Orangutan Cultures and the Evolution of Material Culture." *Science* 299: 102–105. Following Whiten et al. (1999), researchers on wild orangutans find behavioral variation across populations of these apes in Borneo and Sumatra.

Whiten, Andrew; Goodall, Jane; McGrew, William C., et al. (1999). "Cultures in Chimpanzees." *Nature* 399: 682–685. Trail-blazing synthesis comparing behavioral diversity in seven populations of wild chimpanzees across Africa.

ANIMAL WELFARE

• • •

The concept of animal welfare was essentially unexamined until the 1970s. This is the case because, historically, the major use of animals in society was agriculture—that is, for food, fiber, locomotion, and power. The key to success in animal agriculture, in turn, was good husbandry (Rollin 1995).

The Husbandry Ideal

Husbandry involved putting animals into the best possible environment fitting their biological natures and needs, and then augmenting that environment with the provision by the agriculturalist of food during famine, water during drought, protection from predation, help in birthing, and medical attention. The resulting symbiotic relationships between farmers and their animals represented what has been called "a fair and ancient contract," with both animals and humans better off in the relationship than they would be outside it. Animals benefited from the care provided by humans; humans benefited from the animals' toil, products, and sometimes their lives. Proper animal treatment was assured by human self-interest; if the animals were made to suffer, their productivity was diminished. The only social ethic regarding animal treatment for most of human history was the prohibition of deliberate, sadistic, overt, willful, intentional cruelty, as encoded in anti-cruelty laws, to sanction sadists, psychopaths, and others not motivated by self-interest and likely to abuse humans as well as animals.

Thus animal welfare was not a conceptually problematic notion occasioning much reflection. If the animal was growing, reproducing, giving milk or eggs, or pulling the plow, it was surely enjoying good welfare. So powerful was the husbandry notion, in fact, that when the Psalmist looks for a metaphor for God's ideal relationship to humans, he chooses the shepherd in the Twenty-third Psalm: "The Lord is my shepherd, I shall not want. He leadeth me to green pastures; he maketh me to lie down beside still water; he restoreth my soul." Humans want no more from God than what the good husbandman provides to animals.

From Husbandry to Industrialized Agriculture

Beginning in the 1940s, changes in animal use were catastrophic for animal husbandry. In agriculture, this period saw the rise of the application of industrial methods to the production of animals to greatly increase efficiency and productivity, and academic departments of Animal Husbandry symbolically changed their names to Animal Science. In the industrialized confinement of "factory farming," technoscientific developments such as antibiotics, vaccines, hormones, and air-handling systems allowed human beings to force animals into environments not fitting their natures; these animals continued to be economically productive while their well-being was impaired. Animals thus suffered in four major ways (Rollin 2004).

First, probably the major new source of suffering in confinement agriculture resulted from physical and psychological deprivation for animals in confinement: lack of space, lack of companionship for social animals, inability to move freely, boredom, austerity of environments. Breeding sows, for example, spend their entire productive lives in stalls measuring seven feet by two feet by three feet, so small that the animals cannot turn around or sometimes even stretch out. Because the animals evolved for adaptation to extensive environments but are now placed in truncated environments, such deprivation is inevitably abusive.

Second, in confinement systems, workers may not be "animal smart"; the "intelligence," such as it is, is the mechanized system. Instead of husbandmen, workers in swine factories are minimum wage, often illegal immigrant labor. These workers often have no empathy with, or concern for, the animals. The Biblical shepherds have become detached (and often themselves oppressed) factory assembly-line workers.

Third, the huge scale of industrialized agricultural operations—and the small profit margin per animal—militate against the sort of individual attention that typified much of traditional agriculture. In traditional dairies as late as 1950, one could make a living with a herd of fifty cows. By 2000, one needed literally thousands. In the United States, dairies may have 6,000 cows. In swine operations, sick piglets are sometimes killed, not treated. Agricultural veterinary medicine is far more concerned with "herd health" than with treating sick individuals.

Finally, "production diseases" arise from the new ways animals are produced. For example, liver abscesses in cattle are a function of certain animals' responses to the high-concentrate, low-roughage diet that characterizes feedlot production. Although a certain percentage of the animals get sick and die, the overall economic efficiency of feedlots is maximized by the provision of such a diet. The idea of a method of production creating diseases that were "acceptable" would be anathema to a husbandry agriculturalist.

Thus, in industrialized agriculture, the tie between productivity and welfare was broken. The agriculture community nevertheless continued to insist that if animals were productive, they were well off, despite the fact that welfare applies to individual animals and productivity is an economic measure of an operation as a whole.

The same historical moment also saw the rise of large amounts of animal research and animal testing. This again differed from husbandry in that the animals did not benefit from being in research. Indeed, research deliberately hurt animals, gave them diseases, burns, fractures, and so on, with no compensatory benefit to the animals—although there was undeniable benefit to humans and other animals from the knowledge and therapies produced.

Criticizing Animal Treatment

Since the 1960s, beginning in Great Britain, Western society has become increasingly concerned about animal treatment in agriculture that is industrial, not husbandry-based, and in research and testing. Initially, such uses were seen as "cruel." Yet, as mentioned, the anti-cruelty ethic and laws were designed for deviant behavior, not common uses. In order to rationally capture concern about animal treatment that results from putatively decent motives, such as increasing productivity or studying disease, new conceptual tools were needed. First of all, a new ethic for animal treatment was needed to address suffering not resulting from intentional cruelty. Second, some notion of animal welfare or well-being was needed, given that productivity no longer assured welfare. In both cases, preserving or restoring the fairness inherent in husbandry served as an implicit standard.

Animal-using industries, however, continued to define animal welfare in terms of human goals for the animal. For example, the official agricultural industry response to burgeoning social concern for animal treatment, the Council for Agricultural Science and Technology (CAST) Report of 1981, defined farm animal welfare as follows: "The principle [sic] criteria used thus far as indexes of the welfare of animals in production systems have been rate of growth or production, efficiency of feed use, efficiency of reproduction, mortality and morbidity" (Council for Agricultural Science and Technology 1981).

When dealing with adults and ethics, one does better to remind than teach. New ethical challenges are likely to be answered only by appeal to unnoticed implications of extant ethical principles, rather than by creation of a new ethic *ex nihilo*. Thus the civil rights movement did not invent a new ethic; it rather reminded society that segregation violated basic ethical principles American society took as axiomatic. In the same way, society has looked to the ethic for the treatment of humans to derive an ethic for animals (Rollin 1981).

Specifically, every society faces a conflict between the good of the group and the good of individuals, as when a wealthy person is taxed to support social welfare. In totalitarian societies, the good of the individual is subordinated to the group. Democratic societies, however, build "protective fences" around individuals to protect basic aspects of human nature from being submerged for the general good. These fences protect freedom of speech, freedom of religion, property ownership, privacy, and so on. These are called *rights*, and are a morally-based legal notion. Society has reasoned that if animal use for human benefit is no longer naturally constrained by the need for good husbandry, such proper treatment must be legally imposed. This concept is well-illustrated by the proliferation of laws in Western society to protect animal welfare in research, agriculture, zoos, shows, and elsewhere.

Thus the notion that animals should have rights or legal protections for basic elements of their natures—a notion embraced by more than 80 percent of the U.S. public (*Parents Magazine* 1989)—represents a rational ethical response to the end of husbandry as well as to other factors that have focused social concern on animal treatment. These factors include the urbanization of society and correlative shrinkage in numbers of people making a living from animals; the emergence of companion animals as a paradigm for all animals; the mass media focusing on animal issues as a way of garnering audiences; the shining of a moral searchlight on the traditionally-disenfranchised—minorities, women, the handicapped—out of which movements many of the leaders of animal activism emerged.

Thus animal rights as a mainstream phenomenon captures the social demand for legal codified animal protection and assurance of welfare. In this sense, *animal rights is simply the form concern for animal welfare has taken when animal use is no longer constrained by husbandry*. This sense should not be confused with the vernacular use of "animal rights" as referring to the view of some activists that no animals should ever be used by humans, a view better termed "animal liberation." The two views are clearly distinguished by the fact that most people in society wish to see animals protected while used for human benefit, but do not wish such uses eliminated.

The Good of Animals

Any attempt to protect animals and their interests depends on some socially accepted view of animal welfare, some account of the good of animals themselves and what they are owed by humans to reach an acceptable quality of life. Providing an account of welfare, therefore, is going to involve both factual and value judgments. The factual part involves empirical studies of animal natures—what has been called their *telos*—nutritional needs, social needs, health needs, psychological needs, exercise needs, and needs arising from species-specific behavior (Fraser and Broom 1990). This is the purview of an emerging field known as *animal welfare science*. The value judgment component in addressing animal welfare comes from the moral decision entailed by deciding which of these multiple needs will be met, and to what extent. For example, in zoos during the 1970s, tigers were typically kept in austere cell-like cages and fed horse meat. At the beginning of the twenty-first century, they may have ten acres to prowl. But the natural tiger range is miles, and tigers kill their food. Clearly the situation now is better than the previous one, but major needs are still unmet, because the tigers are not allowed predation and their range has been truncated. Similarly, health is obviously fundamental to welfare, but analysis reveals that the concept of health includes significant value judgments (Rollin 1979). Indeed, the CAST Report definition of welfare as equating to productivity bespeaks a set of quite controversial value judgments.

One additional crucial component is essential to understanding animal welfare. In the early 1980s, a number of philosophers and scientists (Rollin 1981, Duncan 1981, Dawkins 1980) pointed out that, ultimately, animal welfare is most crucially a matter of the animal's subjective experience—how the animal feels, whether it is in pain or suffering in any way, a point that is obvious to ordinary people but which conflicted with the scientific ideology that dominated twentieth century science (Rollin 1998). This ideology affirmed that all legitimate scientific judgments had to be empirically testable. Value judgments and statements about human or animal subjective awareness, thoughts, or feelings were ruled out by fiat. Because most scientists were indoctrinated with this ideology, the scientific community was ill-equipped to deal with ethical issues occasioned in the public mind by scientific activity, the first historically being the ethics of animal research. In any case, the failure to recognize the need for value judgments in general and ethical judgments in particular, as well as judgments about animal feelings, helps explain why the scientific community has not been a major contributor to public understanding of animal welfare.

Assessment

There is no reason to believe that animal welfare issues will not continue to dominate the public imagination. Public fascination with animals, animal treatment, animal thought and feeling, is manifest in the many television programs, newspaper and magazine articles, books, and films devoted to these issues. Every area of human-animal interaction, be it agriculture, research, hunting, trapping, circuses, rodeos, zoos, horse and dog racing, product extraction, and even companion animals, is fraught with ethical and welfare issues. (Currently, a major social concern is elevating the monetary value of companion animals above mere market value.) As these issues are engaged, it is likely that human understanding of animal welfare will be deepened, as it must be to provide rational legislated protection for these fellow creatures.

BERNARD E. ROLLIN

SEE ALSO *Agricultural Ethics; Animal Experimentation; Animal Rights; Pets.*

BIBLIOGRAPHY

Benson, G. John, and Bernard E. Rollin. (2004). *The Well-Being of Farm Animals: Problems and Solutions*. Ames, IA: Blackwell. An anthology with major experts addressing animal welfare issues in agriculture.

Council for Agricultural Science and Technology. (1981). *Scientific Aspects of the Welfare of Food Animals*. Ames, IA: Council for Agricultural Science and Technology.

Dawkins, Marian Stamp. (1980). *Animal Suffering: The Science of Animal Welfare*. London: Chapman and Hall. Classic work defending the notion that animals can suffer and how this can be determined.

Duncan, Ian. J. H. (1981). "Animal Rights—Animal Welfare: A Scientist's Assessment." *Poultry Science* 60: 489–499. Classic paper defending the view that welfare is a function of animal consciousness.

Fraser, Andrew Ferguson, and Donald Broom. (1990). *Farm Animal Behaviour and Welfare*, 3rd edition. London: Bailliere Tindall. Standard textbook of welfare and behavior.

"Parents Poll on Animal Rights, Attractiveness, Television, and Abortion." (1989). *Parents Magazine* (Sept–Oct).

Rollin, Bernard E. (1979). "On the Nature of Illness." *Man and Medicine* 4 (3): 157–172.

Rollin, Bernard E. (1981). *Animal Rights and Human Morality*. Buffalo, NY: Prometheus Books. Account of the emerging social ethic for animals.

Rollin, Bernard E. (1983). "The Concept of Illness in Veterinary Medicine." *Journal of the American Veterinary Medical Association* 182: 122–125.

Rollin, Bernard E. (1995). *Farm Animal Welfare*. Ames: Iowa State University Press. A survey of issues in farm animal welfare.

Rollin, Bernard E. (1998). *The Unheeded Cry: Animal Consciousness, Animal Pain, and Science*, expanded edition. Ames: Iowa State University Press. Exhaustive defense of the legitimacy of talking about animal consciousness and animal pain.

Rollin, Bernard E. (2004). "Animal Agriculture and Emerging Social Ethics for Animals." *Journal of Animal Science* 82: 955–964.

ANSCOMBE, G. E. M.

Gertrude Elizabeth Margaret (G. E. M.) Anscombe (1919–2001), arguably England's greatest female philosopher and one of the great philosophers of the twentieth century, was born on March 18 in South London and died on January 5 in Cambridge, England. Trained at Cambridge and Oxford universities in the classics, ancient history, and philosophy, Anscombe converted to Catholicism while at college. She married Peter Geach, also a philosopher and converted Catholic, with whom she had seven children.

A student and friend of Ludwig Wittgenstein, Anscombe was one of his three literary executors (along with Georg von Wright and Rush Rhees) and was tasked with translating much of Wittgenstein's work. Her *An Introduction to Wittgenstein's Tractactus* (1959) is considered the basic analysis of that work. The recipient of many honors and awards, Anscombe eventually succeeded to Wittgenstein's chair of philosophy at Cambridge. A renowned debater, she was reputably responsible for C. S. Lewis's decision to give up theology and turn to writing children's literature.

While steeped in all aspects of philosophy, Anscombe was well aware of progress in the sciences and humanities, discussing the implications of modern physics on causality (referencing works by Erwin Schrödinger, Albert Einstein, Richard Feynman, and Max Born), noting that there was no point in continuing to work on moral philosophy until psychology was better understood, as well as delving deeply into medical ethics in areas such as abortion, euthanasia, and contraception. A moderately prolific writer, Anscombe wrote for two distinct audiences, her professional colleagues and the Catholic community. Throughout her life she showed no hesitation in publicly acting on her beliefs.

In 1939, while still an undergraduate, she and Norman Daniel coauthored a pamphlet examining British participation in World War II. They concluded that, despite the injustices perpetrated by Nazi Germany, the role of the United Kingdom in the war was immoral. Anscombe argued that U.K. intentions in terms of means, ends, and net probable effects were unjust. In particular, Anscombe predicted, correctly as it turns out, that attacks on civilian targets were likely (blockades were already in effect) and that such actions would constitute murder.

Years later, Anscombe opposed an Oxford University plan to confer an honorary degree on U.S. President Harry S. Truman on similar grounds. The basis of her objection was that Truman was ultimately responsible for what she considered to be the murder of thousands of civilians during the bombings of Hiroshima and Nagasaki. This principle, the immunity of innocents, carried forward in the early-twenty-first century in the international law of war, is the basis for discussions of *collateral damage* and is one driver for the development of more precise munitions.

When the birth control pill and other devices became generally accessible, Anscombe supported Pope Paul VI's pronouncement that contraceptive measures other than the rhythm method were immoral. She wrote a series of articles aimed at the Catholic laity logically justifying the pope's conclusion. Catholics who support liberalization of the Church's policy on contraception have not successfully countered Anscombe's arguments. Interestingly non-Catholics contend that once the religious precepts of Catholicism are removed from Anscombe's arguments, she makes a persuasive case that nearly any sexual act or form of relationship should be permissible.

To Anscombe, abortion also represented an unjust killing of the innocent. In typical fashion, this motivated her in later years to participate in the British pro-life movement, eventually causing Anscombe and her daughters to be arrested for blocking an abortion clinic. In her life and work, Anscombe represents the possibility of an analytic philosopher taking substantive positions on a variety of issues related to science, technology, and ethics.

RUTH DUERR

SEE ALSO *Consequentialism; Just War.*

BIBLIOGRAPHY

Anscombe, Gertrude Elizabeth Margaret. (1959). *An Introduction to Wittgenstein's Tractactus*. London: Hutchinson.

Anscombe, Gertrude Elizabeth Margaret. (1981). *Collected Papers: 1. Parmenides to Wittgenstein. 2. Metaphysics and the*

Philosophy of Mind. 3. Ethics, Religion and Politics. Minneapolis: University of Minnesota Press.

Teichmann, Roger, ed. (2000). *Logic, Cause & Action: Essays in Honour of Elizabeth Anscombe*. Cambridge, UK: Cambridge University Press.

ANTIBIOTICS

The search for antibiotics began with general acceptance of the germ (bacteria) theory of disease. The first antibiotics were developed in the late 1800s, with Louis Pasteur (1822–1895) commonly given credit for discovering that the bacterial disease anthrax could be cured in animals with an injection of soil bacteria. But it was not until Alexander Fleming (1881–1955) discovered penicillin in 1928 that the great potential of antibiotics was recognized. Especially during World War II penicillin revolutionized medical practice, but the subsequent heavy reliance on penicillin and other antibiotic agents as general technological fixes for numerous diseases has led to problems that have distinctly ethical aspects.

Historical Development

Fleming's serendipitous discovery of penicillin came when he examined an old gelatin plate he had forgotten to submerge in detergent solution. Staphylococci, common skin bacteria, were growing on the plate, along with a mold. A product of the mold had seemingly killed some bacteria. Fleming was not the first person to observe the phenomenon of bacterial destruction by mold, but he had the foresight to recognize its potential medical importance. He named the mold product penicillin after the *penicillium* mold that had produced it. By extracting this substance from a culture of the mold, he was able to directly show its antibacterial properties.

An event in the 1930s also helped establish that chemicals taken internally can cure infectious diseases without harming the host. This was the discovery, made by Gerhard Domagk (1895–1964), that a newly patented chemical dye, Prontosil, could cure disease caused by streptococcus bacteria when injected into diseased mice. Interestingly Prontosil only worked when used internally and could not inhibit bacterial growth in a test tube. It was later shown that it was not the dye but a chemical attached to it, the sulfonamide portion, that was responsible for killing the bacteria. The sulfonamide portion was released during metabolism and was free to fight bacterial infections. The discovery of sulfonamides and penicillin as potent antibacterial agents created a strong motivation for developing other antibiotic agents.

The twenty-five years following the introduction of penicillin in 1942 was the heyday of antibiotic development. Developed antibiotics were either natural substances isolated from an organism, or synthetic agents, exemplified by penicillin and the sulfonamides respectively. Antibiotics also typically have a limited scope of effectiveness, often restricted to either *gram-positive* or *gram-negative* bacteria. This distinction in bacteria is named after Hans Christian Gram (1853–1938) who discovered that some bacteria stained with specific dyes kept their color following washing whereas other bacteria lost their color. Those that keep their color are gram-positive and those that lose color are gram-negative. Gram-positive and gram-negative bacteria differ in the composition of their cell walls, the outermost structure of bacteria. So-called broad-spectrum agents are effective against both gram-negative and gram-positive bacteria and include the antibiotics chloramphenicol and tetracycline, first isolated from soil bacteria in the late 1940s. Cephalosporins, first introduced in 1964, were other natural, broad-spectrum agents similar to penicillin. Modification of the cephalosporins and penicillin led to a number of semisynthetic agents with properties varying in adsorption, residence time in the body, spectrum of activity, and insensitivity to degradation by bacterial enzymes. A number of synthetic antibiotics were also introduced, mainly in the 1970s, following the introduction of natural ones. While some antibiotics have been introduced since the 1990s, the pace of discovery and introduction of new antibiotics has slowed markedly from its heyday.

Antibiotic Resistance

Initially seen as *miracle drugs*, antibiotics, once they became widely available, were used not only for bacterial infections, but for everything from the common cold to headaches. Indeed antibiotics were a godsend, drastically improving medicine and contributing significantly to the increase in life expectancy achieved during the twentieth century. Like many technological fixes, along with the positive benefits of antibiotics came negative side effects. Antibiotics can kill the many beneficial bacteria in the human body, for instance those that promote digestion, along with invasive bacteria. Another, unexpected, consequence is the ability of bacteria to overcome the mechanisms that give antibiotics their efficacy, rendering them useless. Antibiotic resistance, first a curiosity seen in the laboratory, became common among populations of bacteria exposed to antibiotics. In a matter of years following the introduction of penicillin, penicillin-destroying staphylococci appeared in hospitals where much of the early use of penicillin had taken place.

A similar response has occurred in various strains of bacteria in response to vastly different antibiotics. Resistance traits exist for every antibiotic available in the marketplace. In addition, bacteria are often resistant to multiple antibiotic agents, leaving only expensive and potentially toxic antibiotics to fight bacterial infection, assuming a patient is fortunate enough to have access to such medicines.

Mechanisms of antibiotic resistance vary markedly but have the same effect of increasing tolerance until the bacteria are resistant. These mechanisms first appear in a few bacteria as a result of random mutations that naturally occur in the DNA that defines the genetic makeup of the bacterium. In the presence of antibiotics the bacteria having these mutations are selected for survival over those that are susceptible. With increased exposure to antibiotics, eventually only those bacteria with the resistance trait will survive. Furthering the propagation of resistances is the presence of transferable elements that readily exchange genetic material between bacteria. These elements exist either as plasmids, circular rings of DNA outside the core genetic material (chromosomes) of the bacterium, or as transposons, regions of DNA that can *jump* between chromosomes. Transferable elements allow susceptible bacteria to acquire resistances from other bacteria, either alive or dead. In order to limit the rise and spread of resistant bacterial strains, measures have been developed to encourage the proper use of antibiotics.

Ethical Use of Antibiotics

Ironically antibiotics have become a victim of their own success. The ability of antibiotics to effectively kill bacteria has also created an environment that selects for resistant strains and allows them to propagate. Antibiotics stand alone as the only therapeutic that is detrimental to society through their usage by an individual. Aside from the individual risks of side effects and allergies, widespread use of antibiotics has a much greater societal effect. Any antibiotic use, regardless of need, will hasten the selection for and propagation of resistant bacteria. Despite this drawback, antibiotics continue to play an invaluable role in healthcare. For them to remain efficacious, the misuse and overuse of antibiotics must be curbed.

In most industrialized countries antibiotics are obtained only through prescriptions. Despite this control on availability many people acquire antibiotics by coercing doctors or hoarding leftover medicine. In some instances people will use antibiotics obtainable from pet stores without prescription. These actions may seem frivolous but in the quick-fix world of medicine many patients demand some form of treatment for every ailment. Additionally many still hold the outdated view of antibiotics as a panacea. Not only does improper use of antibiotics have the danger of side effects, anything short of a full treatment will not rid the patient's system of the entire infection. Because the surviving bacteria are often the ones with a greater tolerance to the antibiotic, the potential exists for the reemergence of an infection resistant to the antibiotic. Though potentially dire outcomes resulting from resistances occur in industrialized nations, such as the emergence of *staphylococcus aureus*, which is resistant to almost all antibiotics, developing countries face even greater hazards.

The overuse and misuse of antibiotics in the developing world far eclipses the abuses present in developed countries. The frequency of infections in the developing world is greater due to poor public sanitation. Infections normally treatable for patients in developed countries often prove fatal when acquired in less developed nations. The uneven distribution of wealth does not allow poorer countries to afford newer antibiotics to overcome infections resistant to the ones readily available. Even if proper medicines are available, they are often misused, encouraging the propagation of drug resistant bacteria. Where one day of treatment can equal the daily wage, many are forced to choose the savings over a full treatment. Medical usage of antibiotics is a huge concern to both developing and developed nations but is not the only use that results in antibiotic resistances.

Use of antibiotics in agriculture, aquaculture, and food animals has been a tenacious issue. Humans are not the only species affected by infectious diseases. Antibiotics can protect the food supply by limiting loss to disease and have frequently been administered as a preventive measure, though use on crops has been banned in many countries. Antibiotics have also been found to promote growth in food animals when given in low doses. The mechanism responsible for this action is not known, but it is speculated that low dose antibiotics reduce competition for nutrients from bacteria living in the guts of these animals. Antibiotics used for treating animals and crops have the same ability to select for resistance traits in bacteria. Even antibiotics not used in human medicine can help to create bacteria resistant to medicinally important antibiotics. Clearly measures for the proper use of antibiotics in food production and medicine need to be advocated.

The Future of Antibiotics

The introduction of antibiotics into medicine has improved the quality and longevity of people's lives. Infections that were once a death sentence are easily controllable in the early twenty-first century. But the misuse and overuse of antibiotics has threatened their ability to control disease. With few new antibiotics being introduced and little incentive for pharmaceutical companies to invest in their research and development, measures are being taken to protect the efficacy of already existing antibiotics. To address this problem more efforts at the local level are needed to ensure their proper use. To this end, an international group, the Alliance for the Prudent Use of Antibiotics (APUA), was established in 1981. The organization, with a presence in more than 100 countries, aims to promote the proper use of antibiotics and to protect their long-term efficacy through communication and education. Although APUA is a start, doctors, pharmaceutical companies, governments, and individual users must continue efforts to improve current usage of antibiotics in order to ensure that such drugs remain effective for future generations.

ANDREW PRICE

SEE ALSO *Bioethics; Clinical Trials; Emergent Infectious Diseases; Medical Ethics; Vaccines and Vaccination.*

BIBLIOGRAPHY

Ferber, Dan. (2000). "Superbugs on the Hoof?" *Science* 5: 792–794. Discusses findings connecting antibiotics used on livestock with outbreaks of disease caused by antibiotic-resistant bacteria.

Hadley, Caroline. (2004). "Overcoming Resistance." *EMBO Reports* 5(6): 550–552.

Levy, Stuart B. (2002). *The Antibiotic Paradox: How the Misuse of Antibiotics Destroys Their Curative Powers*. Cambridge, MA: Perseus Publishing. Classic primer on the misuse of antibiotics.

INTERNET RESOURCES

Alliance for the Prudent Use of Antibiotics (APUA). Available from http://www.antibiotic.org. Internet site of the organization.

U.S. Food and Drug Administration. Antibiotic Resistance page. Available from http://www.fda.gov/oc/opacom/hottopics/anti_resist.html. A comprehensive resource containing background articles and current news.

APOLLO PROGRAM

In the early twenty-first century, the Apollo program still is invoked as the ultimate technological achievement. In terms of percentage of the national budget, that effort to land astronauts on the moon was the largest single scientific program ever undertaken by the United States. Six successful lunar landings were accomplished from 1969 to 1972. The twelve astronauts who walked on the surface of the moon collected samples, set up equipment, and conducted scientific experiments. The scientific return from those missions revolutionized people's understanding not only of the moon, but of the earth and the rest of the Solar System. The program also raised many ethical concerns, notably its motivation, the safety of the astronauts, and its cost at the possible expense of other national needs.

The Origins of Apollo

In a speech to Congress on May 25, 1961, President John F. Kennedy stated, "I believe that this nation should commit itself to achieving the goal, before this decade is out, of landing a man on the moon and returning him safely to the earth." This marked the official genesis of the Apollo program, although the rationale had been building steadily since October 4, 1957, when the Soviet Union launched the first satellite, *Sputnik,* into space. A series of successful Soviet space missions followed, culminating with Yuri Gagarin becoming the first human in space during the voyage of *Vostok 1* on April 12, 1961. The United States countered with Alan Shepard's suborbital flight on May 5, 1961, but it was clear that the Soviet Union was the preeminent spacefaring nation and that the United States was losing international prestige.

Many people saw the *space race* as another front in the long-standing rivalry between capitalism and communism. Politicians and the general public also feared that the Soviet Union might use a dominant position in space to gain military advantage. In that climate, Kennedy decided that nothing short of becoming the first nation to put an astronaut on the moon would allow the United States to *win* the space race and regain its technological leadership in the eyes of the world. The National Aeronautics and Space Administration (NASA) was charged with developing a program to achieve that task before 1970. Clearly, the primary goals of the program were political rather than scientific.

Early Apollo Program

The Apollo program proceeded through a series of tests, each building on the one before it. The lunar missions were designed to launch on a *Saturn V* rocket. The first two stages of the *Saturn V* would boost the craft into space, and the third stage would put Apollo into an earth parking orbit and then fire a second time to send Apollo toward the moon. The Apollo spacecraft consisted of a command module that carried the three astronauts; a service module that held much of the water, oxygen, and fuel; and the lunar module, which was designed to bring two astronauts to the surface of the moon. The first Saturn rocket, a *Saturn I*, was launched on October 27, 1961. Through 1966 over a dozen uncrewed orbital and suborbital flights were completed. The components of Apollo were tested and determined to be ready to fly with a human crew.

APOLLO 1. The first crewed Apollo test flight was scheduled for early 1967 to carry the astronauts Virgil Grissom, Ed White, and Roger Chaffee. However, in a preflight test on January 27, 1967, fire broke out in the sealed command module. It grew explosively in the pure oxygen atmosphere and killed all three men. Intense public scrutiny was focused on the first U.S. spacecraft casualties, and a reexamination of NASA procedures resulted in new safety protocols. The public had been awakened to the dangers of space travel and to questions regarding the wisdom of using astronaut versus robots in space exploration.

APOLLO 11. Much testing and three more uncrewed flights followed the *Apollo 1* tragedy. *Apollo 4*, the first launch of a full *Saturn V*, took place on November 9, 1967. Confidence in the Saturn rocket and the Apollo spacecraft was so high that the first astronaut flight, *Apollo 7*, was launched on October 11, 1968. That was an earth-orbiting mission during which the Apollo command and service modules were tested thoroughly. On December 24, 1968, *Apollo 8* became the first crewed mission to reach and orbit the moon. *Apollo 9* and *Apollo 10* followed in early 1969, completing the testing of all the aspects of a lunar landing mission.

Apollo 11 launched on July 16, 1969, carrying the astronauts Neil Armstrong, Edwin "Buzz" Aldrin, and Michael Collins. It reached lunar orbit on July 19, and on July 20 Armstrong and Aldrin landed on the moon in the lunar module. Armstrong stepped onto the lunar surface at 10:56 P.M. Eastern Daylight Time, stating, "That's one small step for man, one giant leap for mankind," to an audience estimated to include half the world's population. The astronauts spent just over two hours on the lunar surface, collecting samples, taking pictures, and setting up experiments. They returned to earth on July 24, completing Kennedy's challenge. *Apollo 12*, launched on November 14, 1969, demonstrated the ability of Apollo to make a targeted landing on the moon and recovered pieces of the 1967 *Surveyor 3* lunar lander.

View of the earth from space. Thanks to the accomplishments of the Apollo program, images like this have a permanent place in the public consciousness. *(U. S. National Aeronautics and Space Administration [NASA].)*

APOLLO 13. The *Apollo 13* mission was the only Apollo mission failure. The explosion of an oxygen tank on April 14, 1970, on the way to the moon, forced the mission to be aborted. The spacecraft circled the moon and headed directly back to earth, overcoming a number of life-threatening problems through the coordinated work of the ground crew and the astronauts. The crew made it back to earth safely, but as had happened after the *Apollo 1* tragedy, the wisdom of risking astronauts' lives was questioned.

Later that year the Soviet Union launched the robotic probes *Luna 16* and *Luna 17* to the moon. *Luna 16* brought back a small sample from the moon, and *Luna 17* carried a rover, *Lunokhod 1*, that traveled across the lunar surface, remotely controlled from the earth, and sent back television images. Over the next six years the Soviets launched two more successful sample return missions and another lunar rover. Those missions demonstrated the capacity of uncrewed vehicles to do scientific work on the moon at a far lower cost and without the risk of astronaut missions. The Apollo missions had a far

Neil Armstrong on the surface of the moon. Armstrong, one of the three members of the crew of Apollo 11, became the first human to set foot on the surface of the moon on July 20, 1969. *(U. S. National Aeronautics and Space Administration [NASA].)*

greater scientific return, but as technology improves, the abilities of robotic probes will come closer to those of astronaut missions. Meanwhile, the dangers inherent in the astronaut program became even more apparent after the space shuttle *Challenger* and *Columbia* accidents.

End of the Apollo Program

The four missions that followed *Apollo 13* were increasingly ambitious, with each spending more time on the moon, setting up more scientific experiments, and returning with more samples, culminating in the *Apollo 17* mission. Three more missions originally had been planned. After *Apollo 11*, the prime motivation for the program had been achieved, and public and political support began to wane. Additionally, the argument was made that money going to Apollo could be spent better elsewhere. The total cost of the Apollo program was over $20 billion and accounted for more than 2 percent of U.S. budget appropriations in the middle to late 1960s. The country was still fighting an expensive war in Vietnam, and it was pointed out that many social programs were underfunded. The final three missions were canceled as a cost-cutting measure. Apollo spacecraft were used in 1973 to launch and bring three crews to Skylab and in 1975 for the Apollo-Soyuz earth orbiting mission, in which the United States and the Soviet Union cooperated in a joint rendezvous mission.

Was the Expense of the Apollo Program Justified?

One of the arguments routinely used to defend the cost of the Apollo program is the value of *spin-offs*, technological developments made in the course of building the spacecraft. Although this would be hard to quantify, many technological advances were made during the Apollo program that later had commercial applications. However, it also can be argued that the economic return would have been even greater if the Apollo budget had been spent directly on technological innovation.

The scientific return from Apollo is unquestioned, but the economic value of those achievements is difficult to quantify. Much current knowledge of the moon, the earth, and the solar system is a direct result of the data returned from the Apollo missions.

Another unmeasurable aspect of the Apollo program is the effect on the public of the moon landings and pictures of earth from space. Apollo represented a cultural as well as a scientific milestone. The pictures of earth and of the astronauts on the moon are among the most famous photographs ever taken.

Arguably, Apollo also gave an impetus to science programs in schools and inspired many young people to go into science and engineering. Although science was not the primary motivation behind the Apollo program, the scientific benefits derived from it are of inestimable value and could not have been garnered during that period in any other way.

DAVID R. WILLIAMS

SEE ALSO *National Aeronautics and Space Administration; Space Exploration; Space Shuttle* Challenger *and* Columbia *Accidents.*

BIBLIOGRAPHY

Chaikin, Andrew. (1994). *A Man on the Moon.* New York: Penguin. An account based on interviews with the Apollo astronauts.

Cortright, Edgar M., ed. (1975). *Apollo Expeditions to the Moon.* Washington, DC: National Aeronautics and Space Administration. A collection of chapters written by astronauts and Apollo managers and flight directors on all aspects of the Apollo program.

Holman, Mary A. (1974). *The Political Economy of the Space Program*. Palo Alto, CA: Pacific Press. A survey of the economic impacts of the Apollo program.

McDougal, Walter. (1985). *The Heaven and the Earth: A Political History of the Space Age*. New York: Basic Books.

Siddiqi, Asif A. (2000). *Challenge to Apollo: The Soviet Union and the Space Race, 1945–1974*. Washington, DC: National Aeronautics and Space Administration. A history of the Soviet Union's part in the space race.

Spudis, Paul. (1996). *The Once and Future Moon*. Washington, DC: Smithsonian Institution Press. A scientific view of the moon largely based on samples and data returned by the Apollo program.

Van Nimmen, Jane; Leonard C. Bruno; and Robert Rosholt. (1988). *NASA Historical Data Book*. Vol. 1: *NASA Resources 1958–1968*. Washington, DC: National Aeronautics and Space Administration. A collection of information and statistics on NASA programs.

APPLIED ETHICS

While applied ethics may appear to be a relatively recent development, serious philosophy has always had its applications. Since the time of Plato (fourth century B.C.E.), philosophers have been concerned with problems of living in the real world. Plato's *Republic*, for instance, concerned as it was with the nature of justice, discussed inescapable questions relevant to how one should live.

What is now known as applied ethics, however, came to prominence in the last third of the twentieth century, after a period in which the prevailing view, among Anglo-American philosophers at least, was that philosophy could not usefully be applied to practical problems. Instead, ethics had often been rejected as emotive and noncognitive in character or, in an effort to contribute to progressive clarity in moral discussions, philosophy devoted itself to metaethics or the analysis of ethical language. Applied ethics initially came to the fore in a medical context, where expanding commitments to human rights and developments in technology gave rise to challenging ethical issues related, for example, to the allocation of scarce resources such as kidney dialysis machines, the use of heart–lung devices, and organ transplantation protocols. Questions such as the extent to which health care professionals should intervene to extend life, along with the definitions of life and death themselves, became extensively debated in a new field called bioethics, defined as the study of the ethical, legal, social, and philosophical issues arising from advances in medicine and the life sciences.

Scope

Applied ethics, is, however, by no means confined to bioethics. Indeed, in its many iterations since the mid-1970s applied ethics has included the discussion of such diverse non-biomedical issues as capital punishment, economic development, free speech, human rights, pornography, poverty, social discrimination, and war. Applied ethical issues arise in any area of life in which the interests of individual or groups conflict, including not just national groups but even different species. Prominent branches of applied ethics include business ethics, environmental ethics, biomedical ethics, legal ethics, military ethics, and professional ethics.

Some of these branches are more directly involved with science and technology than others. Bioethics, which is obviously influenced by biomedical science and technology, has already been mentioned. Nuclear ethics, which deals with nuclear weapons and deterrence strategies, is also closely tied to developments in nuclear science and engineering. Another example is environmental ethics, which has acquired increasing importance as a reflection on the moral limits of industrial development and pollution. Environmental ethics is also pertinent to research on animals, the crossing of species boundaries by means of genetic engineering, and the impact of genetically modified crops on nature.

Agricultural ethics, computer ethics, and media ethics might be cited as still other examples. Agricultural and food ethics are two expanding fields concerned with the production and distribution of food as well as its genetic modification (thus overlapping with environmental ethics while at the same time opening up new perspectives on the issues). Ethics in relation to computing and information technology (IT) has raised the issue of whether there are new ethical questions to be answered, or just new versions of old questions. Arguably the creation of new entities such as web sites, along with new forms of human interaction, give rise to a unique set of issues, although there are also issues of scale relating for example to the power of IT to transform social institutions. Media ethics, as various forms of communications media technologies become digitalized, further overlaps with and extends computer ethics questions and concerns.

Professional ethics is also pertinent with respect to scientists and engineers. Questions arise about the professional responsibilities of scientists with regard to the setting of research agendas, the conduct of research, the use of results, and communication with the public and potential users. The move from programs of promoting the public understanding of science toward enhancing public engagement with science and technology has led to debates about how upstream in the research and development process such engagement should be. Is there a role for public involvement in deciding what research is carried out, or should the role of the public be limited to discussing the impact of research on society? The increasing commercialization of science and the changing social context in which scientists operate are areas that overlap with business ethics, which concerns itself with questions about conflicts of interest, the pressures of commercialization on the setting of research priorities, the sharing of the benefits of the outcomes of research, and whether there are some things (e.g., living organisms) that should not be commercialized and that should therefore be outside the patenting system.

Even more than scientists, however, professional engineers have developed explicit codes of ethics to guide their technical conduct. These now generally emphasize responsibilities to protect public safety, health, and welfare, as well as to promote the profession, protect confidentiality, and avoid conflicts of interest. Engineers may be confronted with situations of conflict, for example, in which one safety concern has to be traded off against another, or concern for public safety is in tension with protection of confidentiality or corporate interests. There may also be international engineering projects in which different standards are applicable in different countries.

Models

There are different models concerning what is involved in applied ethics. In addition to those areas in which particular issues arise, it is essential to reflect on what if anything is being applied. It is tempting to think that in order for ethics to be applied, there must be something such as a theory to apply, which is indeed one possible model. According to James M. Brown (1987), conceiving applied ethics as the application of theory may be described as a "fruits of theory" approach. Although it depends on the view that in applied ethics *some* theory is applied, it admits the application of a variety of possible theories. This is to be distinguished from what might be termed an "engineering approach" (cf. Caplan 1983), which holds that there is one particular theory that is to be drawn upon to address practical problems as and when they occur, and that will produce answers as a result. Because agreement is lacking on any *one* theory, the engineering approach has relatively few adherents, but the fruits of theory approach—that applied ethics must involve the application of *some* ethical theory—remains a popular conception of applied ethics.

Contemporary applied ethics, insofar as it *is* an application of theory, relies to a large extent on ethical theories that date from the eighteenth and nineteenth centuries: deontology and utilitarianism. Deontological ethics draws on the thought of Immanuel Kant (1724–1804) in a tradition that stresses respect for persons and notions of human rights and dignity, without necessarily being a strict application of Kant's own philosophy. Similarly, utilitarian ethics as it is employed today rarely attempts to reproduce the thought of its original authors, Jeremy Bentham (1748–1832) and John Stuart Mill (1806–1873).

An alternative to applying high-level theory is the appeal to mid-level principles as found in Tom L. Beauchamp and James F. Childress's influential text, *Principles of Biomedical Ethics* (2001). Mid-level principles are said both to be in accordance with the "common morality" and to be reconcilable with different underlying theories. This in part explains their appeal. The notion of the common morality on which the approach depends has nevertheless been questioned: Common to whom? The "four principles" in Beauchamp and Childress include autonomy, beneficence, nonmaleficence, and justice. Thus autonomy, for example, can be supported both from a Kantian and a utilitarian point of view, although the interpretation of autonomy will be different in either case. Utilitarian ethics portrays the agent as choosing to maximize his or her utility, while the Kantian moral agent's exercise of autonomy is in accordance with what is right, rather than a pursuit of the good.

The four principles have been regarded by some of their advocates as forming the basis of a "global bioethics" in that they represent values that can be supported by anyone, although they may be so for different reasons. Thus people from very different cultures might support autonomy and justice, even when they disagree about their meanings.

The transferability of the four principles to different cultural contexts has nevertheless been challenged, as has the priority commonly accorded to the principle of autonomy. Furthermore, it is important to note that the application of the four principles does not represent the application of a theory as such. The principles simply represent a useful framework for highlighting the moral dimensions of a situation, but a great deal of work is required in thinking about prioritizing, balancing, and specifying them.

Because the fruits of theory approach involves appeal to a level of abstraction in either theories or principles, other models for the practice of applied ethics have attempted a more contextual and relational approach (Alderson 1991). Feminist ethics, for example, critically examines issues of power, assessing them from the perspective of the more vulnerable party. In discussions of the abortion issue or of reproductive technology, feminist ethics will not in an abstract way discuss the status of the fetus or the right to life, nor does it operate with the ideal of the abstract autonomous individual (which might be regarded as prominent in several other approaches); rather it will look at the position of the woman who has to carry the fetus or who has to undergo assisted reproductive techniques, and at the ways in which power relations in society have an impact on options and decision-making.

Feminist ethics has some characteristics in common with virtue ethics, which, rather than trying to apply principles, asks what traits of character should be developed, and what a person who has the virtues would do in particular situations. The virtuous person is one who, because he or she has the virtues, can *see* what is appropriate in particular cases (cf. Statman 1997).

A problem with the fruits of theory approach, over and above the fact that there is considerable and apparently irresolvable disagreement about the theories themselves and the issue of abstraction, is that the model presupposes there is a clear understanding or agreed-upon description of what the theory in question should be applied *to*. Arguably a prior task of applied ethics is to elucidate what the ethical issues *are*—and there is concern, especially in ethics as applied to the professions, that those working in the field uncritically accept problems defined in a particular way (see, e.g., O'Neill 1986). Contemporary debates about ethical aspects of developments in science and technology frequently focus on issues such as informed consent, safety and risk, privacy and security, conflict of interest, and professional responsibility. It is important to ask if significant matters of ethical concern are overlooked, such as the factors that influence the choice of areas of research.

In the light of such various considerations, antitheorists argue the desirability of doing applied ethics without theory. One way this finds expression is in judgment about particular cases. Specific developments and particular cases may affect the development of appropriate theory, and some argue that there is room for a bottom-up rather than a top-down approach. The approach of casuistry, for instance, starts from cases (analogous to case law) and has principles emerge from these, rather than being developed in the abstract and applied from above (Jonsen and Toulmin 1988).

One may thus distinguish at least five general models for doing applied ethics: theory application, mid-level principle application, feminist contextualism, virtue contextualism, and case-based casuistry. The first two apply some form of theory and may be described as top-down models. The second two are more concerned to apply traditions of reflection that emphasize context. The last is a very bottom-up model that applies one case to another. In regard to issues related to science and technology, top-down models are perhaps more common, with much of the literature in biomedical or computer ethics tending to illustrate such an approach. Context models exercise a stronger role in discussions of the professional responsibilities of scientists and engineers. Casuistry is no doubt the least-common approach to doing applied ethics in science and technology, in part because many of the ethical problems associated with science and technology are so unprecedented that argument by case analogy is often a stretch.

Challenges

Against all models of applied ethics certain challenges remain. One focus of concern is the notion of the ethical "expert." What might be meant by ethical expertise is problematic, and this issue has become a high-profile one as applied ethics has become increasingly involved or even institutionalized in public policy. There is skepticism regarding whether any one group of people has privileged access to the truth about what ought to be done—although insofar as applied ethics admits a plurality of legitimate approaches this criticism can be moderated.

This issue is not, therefore, unconnected with that of the models of applied ethics being practiced. On the fruits of theory model, one concern is that principles developed in one field of expertise, such as philosophy, are applied to another area of activity, such as the health care professions (e.g., MacIntyre 1984). There are questions here about whether it is possible or desirable for principles to be developed externally rather than internally to the profession in question.

Are there alternative notions of expertise that might be available (Parker 1994)? One possibility is that expertise in ethics involves familiarity with a range of views, skills in reasoning and argumentation, and an ability to facilitate debate. Insofar as this is the case, applied ethics expertise could be committed to a kind of ethical pluralism. In applying ethics to particular issues, discussions from more than one perspective are to be preferred to discussions from only one perspective. For some, however, this liberal approach constitutes a kind of relativism.

There remain questions about the identification of the ethical problems for which such reasoning is required. Is this a matter for particular professional groups, or can they be identified from outside by ethical experts? It may be the case that this is not a situation in which an either/or approach is desirable, but that it should be a collaborative venture. Thus policymaking on science needs to include the perspectives of both science and ethics so that greater insight can be achieved through dialogue. It is essential that ethics in this area be scientifically informed, but it is also the task of ethics to question assumptions about aspects of science that may have been overlooked because they appear so unproblematic within the scientific community.

A more radical objection to the notion of expertise comes from those who see applied ethics, and in particular bioethics, as an assertion of power on the part of a certain group. Bioethicists themselves, from this perspective, arguably form a powerful professional group. Bioethics then becomes not a field of study, but a site of struggle between different groups, where philosophers, for example, claim to have a special role. In addition to these challenges to applied ethics in general, however, there are particular issues about the relationship between ethics, on the one hand, and developments in science and technology, on the other.

Science and Technology

The assessment of science and technologies is made more problematic by the ways they extend the reach of human power across ever-wider spatial and temporal scales (Jonas 1982). Nuclear weapons systems are the most dramatic example. Because science and technology were traditionally limited in the extent to which they could know the world and transform it, issues of scientific and technological ethics seemed marginal in relation to ethical reflection on politics and economics, in which contexts human behavior could have much larger impacts on other human beings. But in the contemporary world politics and economics have themselves been transformed by science and technology—while science and technology themselves directly challenge ethics as well. These considerations lend weight to the view that over and above the assessment of individual technologies, there is a need for attention to the overall impact of technology on the human condition. This is more apparent in Continental philosophy than in Anglo-American applied ethics (Mitcham and Nissenbaum 1998).

Even within the Anglo-American tradition, however, applied ethics is called to respond both to rapid developments in science and technology and expanding opportunities and potential for use. The speed of change requires a similarly swift response on the part of society in terms of ethics, policy, and legislation. It is frequently argued that ethical deliberation comes too late—although in the case of the Human Genome Project ethical research was funded alongside the science. The difficulties posed by the speed of change are further complicated by perceptions that in some instances the development of technologies may pose challenges to traditional ethical frameworks themselves. In other words, humankind can no longer continue to think in ways that were once comfortable.

This is not just a point about how attitudes *do* change: Certain ways of thinking turn out to be no longer *thinkable*. As Albert Einstein remarked with regard to how nuclear weapons had altered warfare, "a new type of thinking is essential if mankind is to survive and move toward higher levels" (Einstein 1960, p. 376). New technologies sometimes push ethical frameworks, such as just war, to their limits of applicability. Insofar as this is the case, even those who subscribe to a fruits of theory approach may find it necessary to rethink theories and concepts. Ethical theories emerge in particular social and historical contexts, so why should they be presumed to apply in all other contexts?

To cite one other example, there has been discussion about "genetic exceptionalism" or the extent to which genetics requires rethinking of ethical doctrines such as the importance of confidentiality, because blood relatives have an interest in genetic information about those to whom they are related. Should the principle of medical information privacy always apply? Is it to be broken only in the case of life-threatening communicable diseases? The thesis of genetic exceptionalism is, however, hotly contested by arguments that genetic information is no different in kind, only in degree, from other kinds of information. Whether and to what extent this implies a need to rethink the principles of information privacy in general becomes an issue for any applied ethical engagement with the information explosion that is associated with new scientific and technological transformations.

Whatever model of applied ethics is preferred, science and technology thus appear to give rise to basic questions for applied ethics. One of the most general concerns how to address the presentation of new possibilities for human action, such as whether or not the normal human lifespan should be extended by, say, fifty years. Should the burden of proof be on those who want to make the extension or on those who oppose it? That is, should new technological possibilities be guilty until proven innocent or innocent until proven guilty?

As new developments occur, even among those in favor, they easily give rise to anxieties about possible consequences, and these anxieties find expression in some commonly used arguments that are not tied to any particular theory. In part, such anxieties may arise from previous experiences of things going badly wrong. But anxiety may also arise precisely because there is no experience on which to draw. With regard to certain developments, the worry of crossing limits or boundaries that should not be crossed is one expression of such an anxiety. The related objections to "playing God" or going "against nature" are others. Advocates of caution sometimes deploy the precautionary principle, which has been used by a number of policymaking bodies. Slippery slope arguments are also frequently invoked. It is in the effort to think through such arguments that applied ethics in the Anglo-American analytic tradition may be called upon to make its most general contributions to assessing science and technology.

Tools

In light of the multiplicity of approaches to applied ethics (see Chadwick and Schroeder 2002), some of those working in the field have tried to identify ethical "tools" to assist in identifying the ethical dimensions of a variety of situations. One example is the ethical matrix developed by Ben Mepham (1996) in the context of food ethics. The matrix does not apply a theory as such, although it borrows from the Beauchamp and Childress principles of biomedical ethics. In so doing it provides a structured way of identifying interest groups affected by a given new development and assesses the ways in which they will be affected across a number of dimensions: autonomy and rights, well-being, and justice or fairness. It does not purport to be a decision procedure that will produce answers (as in the engineering model), but a useful tool to assist deliberation.

Although the debates about the relative merits of theory and antitheory continue, along with arguments about the nature of expertise, if such exists, what cannot be doubted is that there are questions to be addressed, and they are not ones that can be settled by opinion polls. Even when the majority agree that x ought to be done, it does not follow that x is right. At the same time ethical reflection cannot be undertaken independent of some empirical input from the social sciences. Insofar as applied ethics involves interactions among science, technology, ethics, and the social sciences it may thus also be described as a new form of interdisciplinarity. Applied ethics requires collaboration, not only between philosophers and professionals but also between different academic disciplines.

RUTH CHADWICK

SEE ALSO *Consequentialism; Deontology; Dutch Perspectives.*

BIBLIOGRAPHY

Alderson, Priscilla. (1991). "Abstract Bioethics Ignores Human Emotions." *Bulletin of Medical Ethics*, no. 68: 13–21.

Almond, Brenda, ed. (1995). *Introducing Applied Ethics*. Oxford: Blackwell. An introduction including more than twenty texts on family life, professional ethics, law, economics, and international relations.

Beauchamp, Tom L., and James F. Childress. (2001). *Principles of Biomedical Ethics*, 5th edition. New York: Oxford University Press. First edition published in 1979.

Brown, James M. (1987). "On Applying Ethics." In "Moral Philosophy and Contemporary Problems," ed. J. D. G. Evans. *Philosophy: The Journal of the Royal Institute of Philosophy*. Suppl., vol. 22: 81–93.

Caplan, Arthur L. (1983). "Can Applied Ethics Be Effective in Health Care and Should It Strive to Be?" *Ethics* 93(2): 311–319.

Chadwick, Ruth, ed. (1998). *Encyclopedia of Applied Ethics*. 4 vols. San Diego: Academic Press. A standard work in the field.

Chadwick, Ruth, and Doris Schroeder, eds. (2002). *Applied Ethics: Critical Concepts in Philosophy*. 6 vols. London: Routledge.

Davis, Michael. (1991). "Thinking Like an Engineer: The Place of a Code of Ethics in the Practice of a Profession." *Philosophy and Public Affairs* 20(2): 150–167.

Edel, Abraham; Elizabeth Flower; and Finbarr W. O'Connor. (1994). *Critique of Applied Ethics: Reflections and Recommendations*. Philadelphia: Temple University Press.

Einstein, Albert. (1960). *Einstein on Peace*, ed. Otto Nathan and Heinz Norden. New York: Simon and Schuster.

Jonas, Hans. (1982). "Technology as a Subject for Ethics." *Social Research* 49(4): 891–898.

Jonsen, Albert R., and Stephen Toulmin. (1988). *The Abuse of Casuistry: A History of Moral Reasoning*. Berkeley: University of California Press.

MacIntyre, Alasdair. (1984). "Does Applied Ethics Rest on a Mistake?" *Monist* 67(4): 498–513.

Mepham, Ben. (1996). *Food Ethics*. London: Routledge.

Mitcham, Carl, and Helen Nissenbaum. (1998). "Technology and Ethics." In *Routledge Encyclopedia of Philosophy*, ed. Edward Craig. London: Routledge.

O'Neill, Onora. (1986). "The Power of Example." *Philosophy* 61(235): 5–29.

Parker, Jenneth. (1994). "Moral Philosophy—Another 'Disabling Profession'?" In *Ethics and the Professions*, ed. Ruth Chadwick. Aldershot, UK: Avebury.

Statman, Daniel, ed. (1997). *Virtue Ethics*. Edinburgh: Edinburgh University Press.

Wilcox, Diane Michelfelder, and William H. Wilcox, eds. (1997). *Applied Ethics in American Society*. Fort Worth, TX: Harcourt Brace. More than sixty texts on abortion, euthanasia, the death penalty, welfare, discrimination, animals, freedom of expression, sexual rights, and privacy.

AQUINAS, THOMAS

SEE *Thomas Aquinas*.

ARCHAEOLOGICAL ETHICS

When the public thinks of archaeology, it may have mental images of the fictional character Indiana Jones, who travels to exotic places, overcomes numerous challenges to capture precious antiquities, and brings them back to the United States for display. Life as an archaeologist must be full of adventure. Although images such as these are based loosely on some events in archaeological history, archaeologists more typically "seek knowledge rather than objects that are intrinsically valuable ... to help us understand vanished peoples and cultures" (Stiebing 1993, p. 22).

Anthropology, history, and other fields all attempt to understand the past, but what sets archaeology apart from the other disciplines is the way it achieves understanding, particularly through discovering the physical objects and human remains left behind by ancient and not so ancient peoples. The emergence of archaeology as a science has enhanced the understanding of human history but in the process has given rise to important ethical questions relating to ownership of artifacts and the disturbance of gravesites, among other issues.

History and Development

Archaeological activity of one type or another has existed for millennia, whether in the form of treasure hunting, looting, or appreciating and seeking understanding of the past. The sixth-century B.C.E. kings of Babylon Nebuchadrezzar and Nabonidus excavated and even restored parts of the ancient city of Ur, and local antiquities were collected by a Babylonian princess (Daniel 1981). Many of the tombs of Egyptian pharaohs were looted by treasure hunters despite the elaborate methods employed by the tomb builders to thwart such breaches.

Some of the earlier accounts of archaeological exploration as it is understood in the early twenty-first century began in Europe during the sixteenth century when Henry VIII appointed the King's Antiquary, whose duties were to travel the land "describing things of antiquarian interest" (Daniel 1981, p. 25). Sweden led the rest of Europe in the study, teaching, and collecting of antiquities with an Antiquities College and Museum and an official proclamation protecting "ancient monuments ... and portable antiquities" (Daniel 1981, p. 32). During that time archaeological scholars carried on robust debates about the age of the world; some held to the biblical age of the earth (dating back to about 4000 B.C.E.), and others claimed that it had to be older in light of the types of artifacts being discovered throughout Europe, such as stone axes and knives.

The notion of the technological stages of human cultural evolution—the age of stone, characterized by weapons and tools constructed of wood and stone; the age of bronze, in which tools and weapons were constructed of copper and later bronze; and the age of iron, in which tools and weapons that had been constructed of bronze were replaced by those made of iron—was proposed as early as 1738. The Danish National Museum curator Christian Jurgensen Thomsen (1788–1865), however, is credited with systematizing the three technological stages in archaeology (Daniel 1981).

The ancient Roman cities of Herculaneum and Pompeii, which were destroyed in 79 B.C.E. by the eruption of Mount Vesuvius, were the subject of the first large-scale excavations in the modern era. The suddenness of the eruption, coupled with rapid burial from ash, mud flows, and lava, preserved both cities until their discovery sixteen centuries later. Initially the purpose of the excavations was not to understand the past but to extract valuables from the ruins, resulting in haphazard and destructive extraction methods. It was not uncommon for small and seemingly worthless artifacts to be destroyed, and systematic identification of the location and position in which the artifacts were found was not practiced.

Partly as a result of the discoveries of Herculaneum and Pompeii European interest in classical antiquity exploded. However, much of the activity was centered on the acquisition of antiquities for collectors and museums (Lynott 2003), not on the production of historical knowledge. To satisfy the desires of collectors, most antiquities were collected hastily without proper cataloging and recording of the context in which they were found, causing the loss of valuable historical information forever. Even though many of those antiquities have been preserved in European museums, the debate over the ownership of the antiquities and the unscientific methods of excavation continues, constituting one of the earliest ethical conflicts in the field (Lynott 2003).

Archaeologists' interests grew spatially during the late eighteenth and nineteenth centuries, with excavations occurring in Asia, India, the Near East, and the Americas. After Napoleon Bonaparte arrived in Egypt in 1798, his scholars conducted excavations and recorded a substantial amount of information. Most impressive was the 1799 discovery of the Rosetta Stone, which, after it was deciphered in 1822, provided the key to understanding Egyptian hieroglyphics. Other activities in that century included the founding of the American Antiquarian Society in 1812, extensive explorations and recording of Central American civilizations in the 1830s and 1840s, the first excavations of Mesopotamia in 1843 at Nineveh, and excavations in India throughout the first half of the nineteenth century (Daniel 1981).

Emergence as a Science

Early archaeological method was mostly descriptive, based on the objects that were found. Basic mapping and drawing of the artifacts was the common practice. Thomas Jefferson, who excavated burial mounds in Virginia, "became the first person . . . to have used the principles of stratigraphy to interpret archaeological finds" (Stiebing 1993, p. 173). Stratigraphy, or the study of sedimentary distribution, age, and strata, enables archaeologists to estimate the ages of artifacts. In 1860 Giuseppe Fiorelli (1823–1896) took over the excavations in Pompeii and developed several new methodological approaches. Fiorelli pioneered the approach of using plaster to cast the remains of humans and animals, initiated a top-down approach to excavating buildings to reduce the frequency of their collapse, and left large objects "in situ" (Stiebing 1993, Daniel 1981). General Augustus Pitt Rivers (1827–1900) is credited with systematizing modern excavation methods, including the careful recording of the site and location of all objects found, the reproduction of all notes and drawings in publications, and the practice of recording even small and seemingly worthless artifacts (Stiebing 1993, Daniel 1981).

One of the most important contributions to the field of archaeology was the discovery of carbon-14 by Willard Libby (1908–1980) in 1949. Carbon-14, a radioactive isotope of carbon, is used to date living and formerly living things (Stiebing 1993, Daniel 1981). Progress in the field of geology and dating rocks through a similar process also expanded methods to establish the archaeological record. Other ways to date artifacts include dendrochronology (counting tree rings) and paleomagnetic dating, which compares the magnetic orientation of earthenware with the past orientation of the magnetic poles. Other technologies in use to locate, describe, and record artifacts include x-ray technology, aerial photography, geographical information systems (GISs), computer software programs, ground-penetrating radar, and miniature cameras (Stiebing 1993, Daniel 1981).

The invention of the Aqua-Lung and scuba (self-contained underwater breathing apparatus) technology revolutionized maritime archaeology and allowed the exploration of thousands of previously untouched archaeological sites around the world. More recently the development of deep-sea submersibles, both manned and unmanned, extended exploratory reach further. In 1985 one of the most famous shipwrecks was discovered through the use of such technology: the SS *Titanic*, which sank in 12,500 feet of water on its maiden voyage in 1912, killing about 1,500 people (Ballard and McConnell 1995).

The contemporary archaeological process includes more than just anthropologists and archaeologists. The study of ancient peoples and cultures requires scientists from diverse fields such as botany, geology, medicine, computer science, and art, among others.

Legal Activities

The first national law in the modern era to address concerns about preserving archaeological sites was the Antiquities Act of 1906, which protected sites on government lands (Messenger 1999). The National Historical Preservation Act of 1966 established various institutions for dealing with historical preservation. Although those laws provided needed protection to valuable archaeological sites, they did not address the concerns of the Native Americans whose ancestors and their gravesites were the focus of research and excavation. In 1990 Congress passed the Native American Graves Protection and Repatriation Act (NAGPRA). That law clearly delegates ownership of artifacts to the Native American tribes that descend from the ancient people who are the subject of archaeological studies (Messenger

1999). Some archaeologists were surprised by passage of NAGPRA and "viewed the new law as antiscience and a threat to their access to the archaeological record" (Lynott 2003, p. 23).

The debate over "Kennewick Man" illustrates the ongoing ethical issues with regard to the ownership of artifacts and remains. In 1996 skeletal remains were discovered near Kennewick, Washington, and through the use of radio carbon dating were estimated to be about 9,000 years old (Smith and Burke 2003). Five local Native American Indian tribes claimed the remains under the provisions of NAGPRA, seeking to rebury the artifacts after proving their "cultural affiliation" with the remains, thus removing Kennewick Man from scientific investigation.

A group of scientists challenged the claim on two grounds. First, they argued that the characteristics of Kennewick Man's skull indicated that he may have been white and not Native American. Second, they argued that it was unlikely that the present-day Native Americans actually were descended from Kennewick Man in light of the passage of 9,000 years and the likelihood that there was much movement of the tribes in the intervening years. In 2002 a U.S. district court ruled in favor of the scientists, although the tribal coalition appealed the ruling. The findings of the court raise important questions about Native American connections to ancient remains and the conflict between Native American values and the desire to conduct scientific research (Smith and Burke 2003).

Archaeological discoveries also spur debates centered on economic issues, as in the case of Ötzi, also known as the Iceman, who was discovered by a hiker in the Alps in 1991. Ötzi's body, clothing, and tools were particularly well preserved after having been encased in ice for almost 5,300 years. Both Austria and Italy claimed ownership of Ötzi in a bitter custody battled until it was determined that Ötzi had been found in Italian territory. With the expectation that tourists would flock to see Ötzi, Italy constructed a museum to display him and expected to earn millions of dollars in museum entrance fees. The hiker who discovered Ötzi also demanded compensation, but it took twelve years before he was legally declared Ötzi's discoverer. The hiker is entitled to 25 percent of Ötzi's value, but determining that value is a difficult endeavor.

One of the more famous cases of ownership disputes centered on the Elgin Marbles, so called because Thomas Bruce, the seventh earl of Elgin, was responsible for transporting the marbles from Greece to England in 1806. Also called the Parthenon Marbles, the collection includes much of the surviving frieze and sculptures from the Parthenon and other Greek sites. Bruce later sold the marbles to the British government, which put them on display. Many people and organizations, particularly the Greek government, have called for the return of the marbles to Greece, but as of 2005 none has been returned.

Ethical Issues

Ethical standards in archaeology developed simultaneously with the maturation of the field. With the exception of the seventeenth-century decree to protect antiquities in Sweden, little was done with regard to ethics until the second half of the nineteenth century. During that period many of those who called themselves archaeologists and conducted excavations were not formally trained in the field. Poor excavation practices damaged and occasionally destroyed artifacts. According to Lynott (2003), ethical concerns in archaeology originally were focused on the need to preserve sites from destruction through vandalism, looting, and poor excavation practices. In the early twenty-first century many archaeologists view ruins as nonrenewable resources that should be protected accordingly (Warren 1999).

Professionalization of the field began in earnest in 1879 with the creation of the Archaeological Institute of America (AIA), followed by the Society for American Archaeology (SAA) in 1934. Concerns about professionalism and technique continued, resulting in the creation in 1976 of the Society of Professional Archaeologists (SOPA), which established a professional registry.

The first major effort to codify professional practices occurred in 1960 with the SAA's "Four Statements for Archaeology" (Lynott 2003), which defined the field, established guidelines for record keeping, suggested standards for training, and established ethical standards that focused primarily on professional practices related to the larger archaeological community. SOPA also established a grievance procedure and enforced its ethical standards (Lynott 2003).

Attitudes toward cultural artifacts changed during the 1980s, when indigenous people worldwide developed greater concern over the treatment and ownership of their ancestors' remains and artifacts (Lynott 2003). Ethical codes changed in response to those concerns, but there still is no single set of ethical standards that defines the field of archaeology. For example, the World Archaeological Congress (WAC) developed "eight principles to abide by and seven rules to adhere to"

(Lynott 2003, p. 23). The AIA established its "Code of Ethics" in 1990, and the SAA developed its new "Eight Principles of Archaeological Ethics," which it approved in 1996 (Messenger 1999). The SAA's principles address archaeologists' responsibility to affected peoples, stewardship and accountability to society, rejection of the commercialization of archaeology, public education and outreach, intellectual property, public reporting and publication standards, records and preservation of collections and artifacts, and training standards for archaeological professionals (Messenger 1999, Society for American Archaeology 2004).

Other ethical concerns in archaeology relate to occasional incidents of fraud or unscientific analyses. In 2000 Shinichi Fujimura, one of Japan's most respected archaeologists, was photographed planting stone tools at a site he claimed to be 600,000 years old. He later admitted to having planted dozens of items at several sites, raising questions of legitimacy with most of his work. Both the Tohoku Institute and the Japanese Archaeological Association expelled Fujimura, although the institute's reputation was "irreparably damaged" by the event (Romey 2001).

As in any field, establishing codes of ethics and practicing them are two different issues. However, the archaeological community seems to understand the important responsibility it has not only to further the understanding of the past but to do so in cooperation with and with respect for people who have vested cultural and ancestral interests in archaeological research. Not only is there a healthy and lively discussion within the community regarding ethics, modern students of archaeology are likely to take a course on ethics as part of their preparation to become professional archaeologists.

ELIZABETH C. MCNIE

SEE ALSO *Misconduct in Science; Museums of Science and Technology.*

BIBLIOGRAPHY

Ballard, Robert D., with Malcolm McConnell. (1995). *Explorations: My Quest for Adventure and Discovery under the Sea*. New York: Hyperion.

Daniel, Glyn. (1981). *A Short History of Archaeology*. London: Thames and Hudson. This book, although somewhat dated, provides an excellent overview of the history of archaeology and the people behind the story.

Delgado, James. (2001). *Lost Warships*. Vancouver: Douglas & McIntyre.

Lynott, Mark. (2003). "The Development of Ethics in Archaeology." In *Ethical Issues in Archaeology*, ed. Larry J. Zimmerman, Karen D. Vitelli, and Julie Hollowell-Zimmer. Walnut Creek, CA: AltaMira.

Messenger, Phyllis Mautch, ed. (1999). *The Ethics of Collecting Cultural Property*, 2nd edition. Albuquerque: University of New Mexico Press.

Smith, Clarie, and Heather Burke. (2003). "In the Spirit of the Code." In *Ethical Issues in Archaeology*, ed. Larry J. Zimmerman, Karen D. Vitelli, and Julie Hollowell-Zimmer. Walnut Creek, CA: AltaMira.

Stiebing Jr., William H. (1993). *Uncovering the Past: A History of Archaeology*. Buffalo NY: Prometheus Press. An excellent historical account of archeology.

Vitelli, Karen D., ed. (1996). *Archaeological Ethics*. Walnut Creek, CA: AltaMira.

Warren, Karen J. (1999). "A Philosophical Perspective on the Ethics and Resolution of Cultural Property Issues." In *The Ethics of Collecting Cultural Property*, 2nd edition, ed. Phyllis Mautch Messenger. Albuquerque: University of New Mexico Press.

Zimmerman, Larry J.; Karen D. Vitelli; and Julie Hollowell-Zimmer, eds. (2003). *Ethical Issues in Archaeology*. Walnut Creek, CA: AltaMira. Excellent resource on current ethical issues in archaeology.

INTERNET RESOURCES

Romey, Kristin M. (2001)." 'God's Hands' Did the Devil's Work." *Archaeology: A Publication of the Archaeological Institute of America*, vol. 54, no. 1. Available at http://www.archaeology.org/0101/newsbriefs/godshands.html.

Society for American Archaeology. (2004). Available at http://www.saa.org. Provides a complete list of the ethical standards required of all members.

ARCHITECTURAL ETHICS

It is estimated that 90 percent of contemporary human existence takes place within built environments. It is also well known that the onset of illness and death is more rapid and often more prevalent as a result of inadequate shelter than of inadequate food supply. As economies shift to urban centers throughout the world with little or no civic infrastructure to receive their bulging populations, homelessness has become a global pandemic—and yet buildings alone are now considered responsible for at least 50 percent of all environmental waste. It is therefore surprising that a comprehensive ethical discourse, compared to other disciplines or professions, is relatively nonexistent within contemporary architectural, graphic, interior, industrial, landscape, urban, and regional design practices. This, according to scholars, was not always the case. In most premodern societies, and in many traditional or non-Western societies in the early twenty-first century, making and ethics were, and are, intertwined if not inseparable. Whenever eighteenth-century Enlightenment principles were uncritically adopted or imposed by force around the world, architects and designers—often in tandem with their clients and communities of users—rapidly abandoned their traditional discourse and practice of ethics, bowing to the demands of utilitarian market forces.

The Central Issues

The recovery of an architecture and design ethics within this postindustrial context begins with four key questions: What is (and is not) architecture and environmental design? Who is ethically responsible for the built environment? What are they ethically responsible for? And, how is ethics manifest through architecture and environmental design?

WHAT IS (AND IS NOT) ARCHITECTURE AND ENVIRONMENTAL DESIGN? This question attempts to define the boundaries and scope of the terms within which an ethics can be discussed. The way in which these terms are defined, however, is an ethical task of the first order. Without clarity in language, slippages in moral reasoning follow. While some scholars believe the terminological division between built and natural environments is largely self-evident, upon closer examination the boundary becomes less clear. If the built environment includes all that is made by humans, what of those places or objects found by humans and inhabited or used in an unaltered state, such as a cave for dwelling or a stick for digging? Is the cave or stick no longer "natural" once a human perceives it as useful? Furthermore, what "natural" environment or object has not already been altered by pollution, acid rain, global warming, or, say, overharvesting in neighboring environments—all effects caused by humans—long before any human "discovers" it? Alternatively, many nonhuman sentient beings—from bacteria through to mammals—may be said to design and/or build their habitats with a care and complexity that often rivals human ability. Could these not be considered built environments? If one considers the effects of human-initiated training, husbandry, breeding, or genetic engineering to generate places or products more useful to humans, would these effects be considered "natural"? Conversely, if a human-built artifact is abandoned and thus deteriorates until it is entirely reinhabited, reshaped, and subsumed by flora and fauna, is this still considered a designed environment?

One response to such questions is to shift the focus from built products to human intentionality. The degree to which human interest and imagination has shaped a given place or thing over time is the degree to which it could be considered "designed." The inherent problem with this, however, is the equality with which imaginary works—from the very influential futuristic cities of "paper architects" to the use of architectural metaphor in poetic verse—may be considered an essential part of the human-built environment and thus answerable to an ethics. If one adds in the inevitable misunderstandings between languages and cultures in an ethical discourse that hopes to be anything but local, then careful attention to terminology must be an essential responsibility of all participants.

WHO IS ETHICALLY RESPONSIBLE FOR THE BUILT ENVIRONMENT? It is estimated that more than 95 percent of the built environment is vernacular, that is, "not designed by professionals" but designed "by the people for the people." The shapes of these places and objects are determined as much by needs, available materials, and traditional building techniques as by regional or local production codes. Ethical responsibility may be considered shared among the owner who determines the need; the builders and craftspeople involved in the project; the communal representatives who determine site selection, safety considerations, zoning, water supply, and local material production; and the users of the building or object, for their involvement in future renovations and maintenance. In many traditional societies this responsibility extends to the ancestors, gods, or spirits who may be seen as the main inspiration for, producers of, or maintainers of the artifact, as long as the community performs the proper rituals. In some societies, responsibility may be laid upon the building or object itself for its good or bad actions. In these cases a tool, building, or city wall may be ritually fed or killed depending on its perceived benefit to the community.

In modern economies, where an architect or designer is involved in a project, this professional would often collaborate with or oversee an enormous diversity of professionals such as engineers, lawyers, design professionals, consultants, researchers, sociologists, archaeologists, technicians, contractors, realtors, manufacturers, restorers, and artists, as well as clients, user groups, neighbors, and/or political representatives. Designers themselves are typically answerable to their peers and society for obtaining their educational requirements and upholding ethical guidelines, technical codes, and bylaws. The problem with accepting, let alone determining, precise ethical responsibility for a particular decision is thus often complex. The matter is further complicated by an often nonexistent or faulty ethical education among most of the participants in a given project, the absence or ineffective presence of professional disciplinary bodies, and the enormous costs of initiating fair legal proceedings or protecting whistleblowers. As a result of this unique and extraordinarily complex network of relationships compared to most professions, ethical responsibility or blame in the design world is often more difficult to designate.

WHAT ARE THEY ETHICALLY RESPONSIBLE FOR? Designers, unlike scientists or technicians, are essentially midwives for a "total artifact" in search of its status—at its highest vocation—as a living being. As such, the designer is responsible for the same development a parent would most want for a child: a life of health, truth, beauty, and meaning. In terms of health, the designer seeks to ensure that the artifact poses no safety risks such as dangerous misuse, collapse, toxicity, or disorientation. It also needs to be secure from intentional criminal activity such as vandalism, theft of its contents, or easy transformation into a weapon. Typically it must perform the tasks it was designed for with relative efficiency, longevity, flexibility, and low maintenance. But in the wake of the 1948 Universal Declaration of Human Rights, the 1992 Rio Declaration on Environment and Development (along with Agenda 21, also from 1992), and the 2000 Earth Charter, professional design bodies have been asked to go well beyond this prescriptive minimum. As a result, the International Union of Architects and the American Institute of Architects now encourage all their members to observe "the rights and well being of the Earth and its peoples, the integrity and diversity of the cultural heritage, monuments and sites, and the biodiversity, integrity and sustainability of the global ecosystem" (World Congress, Principle 9). In practice this involves a holistic approach to the life of any conceived artifact in the ecosystem—from lowering energy use and toxic emissions to using reusable/recyclable materials. These declarations demand the integration of rigorous research science, citizen participation, and interdisciplinary cooperation into the building process, with legislative and legal protection accompanying these efforts. They also state that women, youth, indigenous peoples, and other voiceless groups must be heard and addressed throughout the entire planning and implementation process.

Although health aspects are desirable, many designers claim that their primary drive is to make a beautiful object. Here the use of narrative or poetic reference to history aims to create emotional resonance among the artifact, its context, and human experience. For many of these designers and the communities for which they design, to create a kitsch object or ugly city is a profound breach of ethical practice. In a similar way, quite a few designers see their creations as vehicles for communicating if not bringing about the context for an experience of truth. Here the idea of health at the expense of meaning, the idea of safety or security at the expense of liberty or free expression, or the idea of biodiversity at the expense of fostering traditional craft techniques is critically addressed. As such, the artifact demonstrates its vocation as a rational being seeking understanding, balance, equality, and logical harmony. Within the upper echelons of the design world, it is often on this basis that architectural or design critics evaluate certain works as primarily ethical or unethical. Finally, some objects or sites have a uniquely spiritual, mystical, or imaginative characteristic that the architect or designer seeks to respect if not prioritize over other considerations. In this case the architect becomes less a fabricator or technician than someone in relationship with a special object or site to whom the object or site reveals its living self and true spirit. Ethical interventions, therefore, must be consonant with the needs and character of the spirit, god, or mystical religious tradition present in that place.

These are but some of the possible ethical priorities with which designers approach their commissions. As these priorities come into conflict, so begins the need for ethical discernment.

HOW IS ETHICS MANIFEST THROUGH ARCHITECTURE AND ENVIRONMENTAL DESIGN? There are many ways to understand architectural ethics and how it should be brought about. Within this diverse space, philosophers have identified three of the most common approaches operating in post-Enlightenment societies and influencing their built environments: (1) outcome ethics (otherwise known as consequentialism or utilitarianism), (2) principle ethics (otherwise known as deontology or Kantianism), and (3) character ethics (otherwise known as virtue ethics or Aristolelianism).

Outcome ethics aims to create a state of affairs utilizing any actions necessary to bring about maximum happiness, or the "good." Outcome-directed designers may focus their efforts entirely on bringing about the "good" product by the most efficient means necessary: The best modern tools for research, development, implementation, and maintenance of a product are employed to engineer the longest lasting ease, comfort, and social health. The belief here is that general happiness in society, or the "good," is proportional to the abundance of "good" products circulating in that society. This approach is clearly the most dominant within market-driven economies of the early twenty-first century.

Indeed, one cannot ignore the plethora of excellent tools, appliances, buildings, cities, and ecosystems that have truly made the world an easier if not happier place in which to live. Criticism of this approach, however, is twofold. First, because the means is subordinated to the end, an enormous amount of damage to the environment and/or human rights might be perpetrated in order to bring about the "good" product. Second, despite using the best research methods or real-world modeling available at the outset of a project, the guarantee of producing a lasting good, or any good at all, through this particular product always remains a conjecture.

A principle ethics approach to design focuses less on the "good" product, and more on "right" actions. The process must have logical, rational consistency with universal moral precepts or imperatives, unswayed by inordinate desires or "false promises" of happy outcomes. Principle-based designers are conspicuous for their production and upholding of the laws, codes, and guidelines within which architects and designers have traditionally operated. Their hope is that by training the will to follow reason based on moral duty, a calm, rational civility will then pervade society, regardless of its products, because acting right itself is the ultimate good. Criticism of this approach centers on its tendency toward rigidity in the face of changing ethical situations, as well as a devaluing of human experience, memories, and imagination.

Finally, character ethics steps outside the means/ends debate to focus on developing the best habits or character for the architect or designer. Proponents of this approach hope that through a humanities-based education with history and the arts at its core, designers will be better able to respond with compassion, virtue, and reason to the often unprecedented moral dilemmas the future world will surely present. Detractors question what would compel a designer who follows character ethics to consider the real facts of an ethical dilemma, rational operating procedures to solve it, or solutions to bring about the good if their analysis is primarily historical/poetic, their solutions experimental/creative, and their outcomes primarily evaluated on the presence of beauty or deep interpersonal harmony.

As with the need for clear terminology, determining responsibility, and clarifying design priorities, so is it critical that an ethical methodology is carefully negotiated among all involved in a conflict of values.

The Relation and Impact of Science and Technology on Ethics in the Built Environment

Because architecture and design have both technological and poetic components, any development in science or technology could become a physical element or methodology adopted by a built or fabricated work, as well as a potential subject about which the work might "speak." Thus, no ethical issue arising within science and technology can be completely outside the making and discourse of architecture and environmental design. For instance, a skyscraper adopting the braided form of a DNA molecule as it reaches the sky might be seen to take an outcome ethics stance on the wonderful benefits of genetic science. An urban garden in the adjoining lot designed using principle ethics, meanwhile, might be filled only with non–genetically modified plants. Advances in computing, engineering, environmental, and material research along with the ethical issues they raise concerning security, health, safety, and just distribution of resources would be likely to have an obvious and immediate impact on the physical shape, use, and placement in society of newly designed goods. Of course this does not mean that pure sciences could not have a similar impact on design; such an impact would depend on the ethical dimensions of a problem that are given new shape by a finding in one of its fields.

A holistic critique raised by many post-Enlightenment philosophers, including Friedrich Nietzsche (1844–1900), Edmund Husserl (1859–1938), Martin Heidegger (1889–1976), Michel Foucault (1926–1984), and Jacques Derrida (b. 1930), is the alienation or "loss of meaning" in society brought about by each new technology introduced into the built environment. According to this argument, modern technology and science begin with a daringly original transformation: the reduction of the mysterious complexity of the given world to distinct quantifiables, categories, or simple binary digits. Human community and activity are likewise reduced by technology to distinct quantifiable tasks and ever-smaller specializations. Once reduced, these units can be traded, discarded, calculated, or multiplied with ever-greater speed, acceleration, and automation. The degree to which this self-generation mimics natural growth is the degree to which an uncritical enthusiasm for its technology is assured. Once the domain of the ancient magician, technology self-generates its own awe, propaganda, and docile adherents awaiting the promise of a better and better world. Whereas humans were once communally and ecologically integrated, modern technology demands isolated consumers, globalized uniformity, communication as monologue, and being without death. Perhaps the most disturbing aspect of unchecked technology is its inherent irreversibility; once the automobile, the nuclear bomb, clear-cut forestry, or human cloning become possible, they then become necessary.

Technology, according to these thinkers, is the primary cause of the dominant characteristics of the modern city: ugliness, alienation, toxicity, danger, waste, and constant expansion. While in the current geopolitical environment, this harmful growth is unlikely to stop anytime soon, warnings based on the results of research science of an imminent worldwide ecological crisis through ozone depletion and global climate change are beginning to be heard. As well, a number of contemporary academics and policymakers are advocating a less polarized position. They contend that technology, although inherently unsafe, dehumanizing, and accelerating, is still controllable and able to be harmonized with the biosphere through the promotion of slower, appropriate, or "medium" technologies (the latter in contrast to high technologies), as well as lifestyle change, political action, poverty eradication, demilitarization, and worldwide consensus on tough global policies representing a diversity of voices.

History of Ethics in the Built Environment

Myth and origin cycles, guidelines, or commentaries on what constitutes right action concerning building, boundary determination, and ritual object or place making can be found throughout the earliest known examples of writing in almost every culture. According to archaeologists, writing developed independently in Egypt, Mesopotamia, and Harappa between 3500 and 3100 B.C.E. But the human ancestor *Homo erectus* had campsites, fire, and tools, conducted burials, and began erecting megaliths and dolmens (a type of monument) as early as 3,000,000 B.C.E.; the earliest known shelters date from 2,000,000 B.C.E.; and the first cities came into existence around 7500 B.C.E. in the Indus Valley (present-day Pakistan). While the configuration, orientation, material selection, and care or destruction of early objects, buildings, and settlements might in themselves communicate proper ethical action to its community, only in the relatively late appearance of writing can one find specific ethical statements relating to building, orientation, calendars, ritual, and myth that could be used to guide appropriate procedures of making in harmony with that of the gods. For instance, a Sumerian inscription from Lagash, circa 2500 B.C.E., lists the actions of a corrupt ruler, Urlumma, that should not be imitated because he "drained the boundary canal of Ningirsu, the boundary canal of Nina; those steles he threw into the fire, he broke [them] in pieces; he destroyed the sanctuaries, the dwellings of the gods, the protecting shrines, the buildings that had been made. He was as puffed up as the mountains" (Barton 1929, p. 63) The Egyptian *Proverbs of Ptahhotep* of circa 2400 B.C.E. suggest the best mind-set for establishing a dwelling: "When a man has established his just equilibrium and walks in this path, there where he makes his dwelling, there is no room for bad humor" (Horne 1917, p. 62). And the Indian *Rig Veda* of circa 1500 B.C.E. records how making and orientation itself must be attributed, and thus be in alignment with the goddess Aditi because "The earth was born from her who crouched with legs spread, and from the earth the quarters of the sky were born" (10.72.3-4).

Eventually entire texts emerged whose subject matter was building practice alone—none of which, until the nineteenth century C.E., separated ethics or poetics from making and technique. The Indian *Manasara* of circa 800 C.E., for instance, integrates ritual activity at every step of its guidelines for building in order to ensure the most auspicious blessings upon the construction. Not only are lotus, water lily, and corn offerings essential for constructing foundations, so must the architect be bathed, clothed, and purified in order to perform the rituals and meditate on the creator-god such that the building will stay strong. Deviation from these prescriptions constitutes the most serious ethical breach (Manasara 1994, 109–129). In the classical West, *De architectura* (translated as *The Ten Books on Architecture*), written by Vitruvius circa 25 B.C.E., details how architectural making seeks to preserve the traditional symbolic order handed down through the Greeks in order to set the conditions for virtuous, civic, and ethical behavior of inhabitants and visitors (Vitruvius 1999). Much the same can be said for the writings of Abbot Suger (1081–1151), Guillaume Durandus (c. 1230–1296), Leon Battista Alberti (1404–1472), Giacomo da Vignola (1507–1573), and Andrea Palladio (1508–1580)—all of whom, in their given context, sought to preserve the civic, religious, and ethical order of the dominant classes they served through architectural making (Suger 1979, Durandus 1843, Alberti 1988, and Palladio 1997).

There is, however, an equally long and eloquent tradition of anti-architectural writing in which the techniques and products of craftsmen are said to deeply offend the gods, disgrace the ancestors, and corrupt the people. This, for instance, is one of the most important themes from the Hebrew Bible through to the Christian New Testament. In Genesis, Cain, the city builder, slays out of jealousy his brother Abel, the wandering pastoralist, because God told Cain he prefers the nomadic life over a settled existence for his chosen people (*Gen.* 4:1–16). Moses was prohibited by god the use of tools in building altar stones because instrumental manipulation of holy objects profanes them (*Ex.* 20:25). According to the prophet Isaiah, even though cities were constructed out of human goodwill, all are cursed by God. The city is the agent of war, financial greed, sexual abandon, idols, and injustice, where humans become merchandise. Once built, Isaiah says, they can never be reformed but can only self-destruct, Like Sodom, Gomorrah, Nineveh, and Jericho, as well as Jerusalem and its temple mount (*Is.* 13:19, 22:1–4, 66:1). Isaiah's call for a return to desert simplicity that would permit an undistracted contemplation on the mysteries if not the architecture of heaven, is cited by Saint Stephen before his death in the New Testament's *Acts of the Apostles* (7:44–50), and was carried out by tens of thousands of Christian desert monks and wilderness hermits from Egypt to Italy since the second century C.E. In this tradition, one of the most notable critiques of dominant building and craft practices comes from the thirteenth-century poet, saint, and builder Francis of Assisi. In the rules he wrote for his order and in his final testaments, Francis insists that his followers refuse the ownership, size, and expense of the neighboring cathedrals and more powerful monasteries, preferring that they live instead "as pilgrims and strangers" renovating small, abandoned, and dilapidated churches and dwelling in mud and stick huts surrounded by walls made of hedges (Francis 1999, p. 126).

Architectural writings produced by the dominant world powers after this time eventually reduced and eliminated ethical precepts or discourse in favor of describing practical techniques to bring about the most efficient, cost-effective, and comfortable cities. Claude Perrault (1613–1688) was one of the first to promote architecture as a vehicle for the principles of modern science, declaring that "man has no proportion or relation with the heavenly bodies" (1692–96: Vol. 4, pp. 46–59), thus severing the traditional natural and religious orders from architectural making. By the late eighteenth century, architecture students at the École Polytechique studied Gaspard Monge's (1746–1818) *Géométrie descriptive* (1795; Descriptive geometry), which applied to the totality of human action a synthetic system of mathematics, measurement, and geometry, stripped of all previous symbolic content. One of the most influential nineteenth-century textbooks on architecture, the *Précis des Leçons d'Architecture* (1819; Précis of the lectures on architecture), was composed by Monge's follower, Jean-Nicolas-Louis Durand (1760–1834). Durand's philosophical foundation was triumphantly materialistic. Humans, he declared, exist for two reasons only: to increase their well-being and to avert pain. Such a harsh positivist viewpoint accrued wide acceptance. The only sustained critique of this reduction of architecture to engineering came from Charles-François Viel (1745–1819). Reminding his readers that the two foundational principles of architecture, according to the ancients, were proportion and eurythmy (or "rhythmic pattern"), Viel strove to bring nature, human experience, and the traditional symbolic order back into harmony with making. To Viel, applied geometry masquerading as architecture without care for character, beauty, or metaphysics was harmful and decadent if not evil.

Viel's critique, the first of its kind in architecture, did little to stem the tide of new civic works such as bridges, railway stations, factories, and city plans that were now problems best resolved by engineers. Ornament, once the existential infrastructure of making, was now reduced to mere decoration (Viel 1812, pp. 51–52). As a result, an ethical debate raged in Germany and England concerning which "style" would be most appropriate to decorate certain building types. The point quickly became moot once twentieth-century modernists such as Walter Gropius (1883–1969), Ludwig Mies van der Rohe (1886–1969), and the early Le Corbusier (1887–1965) entirely abandoned ornament for the power, height, and awe available through the bold "expression" of modern materiality: iron, steel, glass, and ferroconcrete. This ideology, now intricately tied to corporate-driven market economies, continued to dominate architecture and design throughout the world into the early twenty-first century.

Following in the footsteps of other professional fields, architecture and design are beginning to develop their own "ethical culture" appropriate to their unique problems and challenges. Only now are the champions of environmental sustainability in the construction and manufacturing sectors beginning to see the deeper implications necessary to have it take hold: slower, reusable, "medium" technologies; community-based participation; global–local integration; historical/poetic awareness; and the fostering of a diverse, intergenerational culture of care. Many of these same conclusions have already been reached by social and environmental scientists who were among the first to critique, along with Werner Heisenberg (1901–1976) and Thomas S. Kuhn (1922–1996), their own historical roots and research agendas. Scientists and designers each have a lot to gain from widening their present specializations, exchanging research independent of corporate sponsorship or private gain, and coming to the table as global citizens with the responsibility to speak for the voiceless: the dead, the yet to be born, the poor, the marginalized, and nature itself.

GREGORY CAICCO

SEE ALSO *Building Codes; Building Destruction and Collapse; Design Ethics; Engineering Design Ethics; Engineering Ethics.*

BIBLIOGRAPHY

Alberti, Leon Battista. (1988 [1450]). *On the Art of Building in Ten Books*, trans. Joseph Rykwert, Neil Leach, and Robert Tavernor. Cambridge, MA: MIT Press.

Barton, George A. (1929). "Inscription of Entemena #7." In his *The Royal Inscriptions of Sumer and Akkad*. New Haven, CT: Yale University Press.

Casey, Edward S. (1997). *The Fate of Place: A Philosophical History*. Berkeley and Los Angeles: University of California Press.

Durand, Jean-Nicolas-Louis. (1809). *Précis des leçons d'architecture données à l'École polytechnique*. 2 vols. Paris: Author.

Durandus, Gulielmus. (1973 [1843]). "Rationale Divinorum Officiorum." In *The Symbolism of Churches and Church Ornaments*, trans. John Mason Neale and Benjamin Webb. Leeds, UK: T.W. Green. Original work published c. 1286.

Fox, Warwick, ed. (2000). *Ethics and the Built Environment*. London: Routledge.

Francis of Assisi, Saint. (1999). "A Rule for the Hermitages, the earlier Rule, the Later Rule and Testament." In *Francis of Assisi: Early Documents*, vol. 1, *The Saint*, ed. Regis J. Armstrong, J. A. Wayne Hellmann, and William J. Short. Hyde Park, NY: New City Press.

Glowacka, Dorota, and Stephen Boos, eds. (2002). *Between Ethics and Aesthetics: Crossing the Boundaries*. Albany: State University of New York Press.

Harries, Karsten. (1997). *The Ethical Function of Architecture*. Cambridge, MA: MIT Press.

Horne, Charles F., trans. (1917). *The Sacred Books and Early Literature of the East*, Vol. II.: *Egypt*. New York: Parke, Austin, and Lipscomb.

Manasara. (1994). *Architecture of Manasara, Translated from Original Sanskrit*, trans. Prasanna Kumar Acharya. Manasara Series, Vol. 4. New Delhi: Low Price Publications. Original work published c. 800 C.E.

McDonough, William, and Michael Braungart. (2002). *Cradle to Cradle: Remaking the Way We Make Things*. New York: North Point Press.

Monge, Gaspard. (1811–1812). *Géométrie Descriptive*. Paris: J. Klostermann fils.

Palladio, Andrea. (1997 [1570]). *The Four Books on Architecture*, trans. Robert Tavernor and Richard Schofield. Cambridge, MA: MIT Press.

Pelletier, Louise, and Alberto Pérez-Gómez, eds. (1994). *Architecture, Ethics, and Technology*. Montreal: McGill-Queen's University Press.

Perrault, Claude (1964 [1693]). *Parallèle des anciens et des modernes en ce qui regarde les arts et les science*. Munich: Eidos.

Rig Veda. (1981). In *The Rig Veda: An Anthology*, trans. Wendy Doniger O'Flaherty. London: Penguin Books. Original work first appeared c. 1750–900 B.C.E.

Suger, Abbot of Saint Denis. (1979 [1144–1148]). "On What Was Done under His Administration." In *Abbot Suger on the Abbey Church of St-Denis and Its Art Treasures*, ed. and trans. Erwin Panofsky. Princeton, NJ: Princeton University Press.

University of Pennsylvania, Graduate School of Fine Arts Staff, eds. (1990). "Architecture and Ethics." *Via* no. 10 (special issue).

Viel de Saint-Marx, Charles-François. (1812). *Principes De L'ordonnance Et De La Construction Des Bâtiments*. 4 Vols. Paris: Author.

Vitruvius Pollio. (1999 [c. 25 B.C.E.]). "De Architectura." In *The Ten Books on Architecture*, trans. Ingrid D. Rowland; commentary and illustrations by Thomas Noble Howe. New York: Cambridge University Press.

World Congress of the International Union of Architects and the American Institute of Architects. (1993). *Declaration of Interdependence for a Sustainable Future*. Paris: International Union of Architects; Chicago: American Institute of Architects.

ARENDT, HANNAH

Hannah Arendt (1906–1975) was born in Hannover, Germany on October 14 to a Jewish family of Königsberg, East Prussia, Germany. She studied philosophy at Marburg, Freiberg, and Heidelberg. At Marburg she was a pupil of the philosopher Martin Heidegger (1889–1976), with whom she had an affair, and at Heidelberg she did her doctoral dissertation on love in Saint Augustine with the philosopher Karl Jaspers (1883–1969). When Hitler came to power in 1933, Arendt left Germany and for eighteen years was a "stateless person," first in Paris, where she worked with Jewish refugee groups, and then, after the outbreak of war, in the United States. From 1929 to 1937 she was married to Günther Anders (1902–1992), a journalist, philosopher, and essayist. Arendt became an American citizen in 1951 and for the rest of her life lived in New York with her second husband, the historian Heinrich Blücher (1899–1970). She died on December 14.

The Human Condition

Arendt's major work with implications for science and technology was *The Human Condition* (1958). It is an inquiry into the *vita activa*, that is, "human life in so far as it is actively engaged in doing something" (p. 22). Within the *vita activa* Arendt distinguishes between three fundamental human activities, labor, work and action, each of which corresponds to a different condition of human existence.

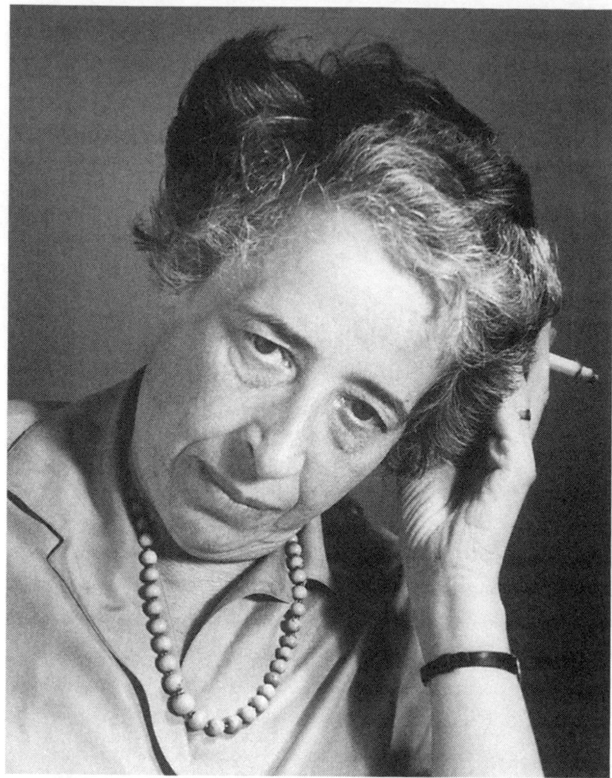

Hannah Arendt, 1906–1975. Arendt was a historian and philosopher of Jewish descent whose scholarly work is devoted to the study of the origins of totalitarianism and anti-semitism. (© *Bettmann/Corbis.*)

Labor includes all the repeated tasks of daily life—growing food, cooking, washing up, cleaning—to which there is no beginning or end. If labor "produces" anything at all, it is something, such as food, that is consumed almost as soon as it is produced. People labor because they are living, embodied beings; thus, life is the condition of labor.

Work is the activity through which people produce durable things—tables, chairs, buildings, but also institutions—that together form the world they inhabit. Humans may use the things of the world made by work and that use may wear those things out, but unlike the food that people consume, this destruction is incidental; it is not an inherent feature of that use. The durability of what work produces means that work has a definite end in the thing made as well as a clear beginning. People work to build a world, and so the world, or "worldliness," is the condition of work.

Action is the capacity to do something new, something that could not have been expected from what has happened before, that reveals who the actor is, and that cannot be undone once it has been accomplished. It derives from the fact of a person's uniqueness as an individual. Action is beginning a boundless unpredictable process of action and reaction. The condition of action is human plurality: a person can labor or work alone as well as with others, but action always requires the presence of others, who, like the actor, are unique human beings. Politics arises out of people acting together, so action constitutes the political realm.

Since the industrial revolution, new technology has transformed work in two ways. First, "automatic" machines and the assembly line transformed work into labor by transforming it into a process without beginning or end, done merely to "earn a living." This means that it is done for the sake of life rather than to build a world. Second, technologies such as nuclear power, synthetic chemicals, and genetic engineering all start new, unprecedented processes that would not exist on earth in the absence of those technologies. Because they are starting something new, the human capacity they make use of must be that of action. In the sphere of human affairs the boundlessness and unpredictability of action can be limited by promising and forgiveness, options that are not available with actions into nature. The inability to limit boundlessness and unpredictability has resulted in uncertainty becoming the defining characteristic of the human situation.

Arendt stresses that humans are "conditioned beings," although the conditions of human existence—the earth, birth and death as well as life, the world and plurality—never condition people absolutely. The earth is the natural environment in which people live, as other animals do, and is characterized by constant cyclical movement: Each new generation replaces the previous one in a process that is indifferent to the uniqueness of individuals.

The world is the condition of human existence that people have made themselves. Biological life is sustained by the earth, but life as a unique, human individual can be lived only in a durable, stable world in which that individual has a place—an identity. The world is always to some extent public in that unlike private thoughts and sensations, it can be perceived by others as well as by oneself. The presence of these others with different perspectives on a world that retains its identity when seen from different locations is what assures the individual of the reality of the world and of themselves (*Human Condition*, p. 50).

The world is related to action in that action always takes place in the world and is often about the world. Political action attempts to change the world. The deeds and words of action constitute an intangible but still real in-between, the web of human relationships that overlies the tangible objective reality of the world. Because it overlies the world, the forms that can be taken by the web of human relationships must depend on, although they are not determined by, what the world is like. To be a home for men and women during their lives on earth the world "must be a place fit for action and speech" (*Human Condition*, p. 173).

Implications

Arendt's most important work, *The Human Condition*, was only part of a lifelong effort to understand what happened to her world during the first half of the twentieth century. For instance, her first major work, *The Origins of Totalitarianism* (1951), analyzed the political systems of Nazi Germany and Stalinist Russia, with their historical roots in anti-Semitism and imperialism. Totalitarianism, Arendt concluded, required atomized, individualized masses: people who had lost any sense of living in a common world for which they shared responsibility. Totalitarianism made life subject to "inevitable" natural or historical processes and thus destroyed the possibility for human action.

Arendt's most controversial work was a report on the trail of the Nazi war criminal Adolf Eichmann (1906–1962). Her conclusion contained the phrase "the banality of evil," which encapsulated her view that the evil done by Eichmann was not a result of base motives, but of his inability to think. The evil resulting from modern technology could also be described as banal. It is not the result of extraordinary actions by people of ill intent, but of unthinking "normal" behaviour, using the technology that has become integral to everyday life in the western world.

In the posthumously published *The Life of the Mind* (1981) Arendt attempted to complement her interpretation of the *vita activa* with one of the *vita contemplativa*. This contains an account of thought that has important implications for thought and knowledge in science and for the relationship between science and technology.

In an approach clearly influenced by Arendt, Langdon Winner has suggested that the most important question to ask of technology is, "What kind of world are we making?" (Winner 1986). The clear implication of Arendt's argument is that questions concerning the nature of the world, and therefore of technology, are political questions. They cannot be decided simply by reference to science, or by technical decision procedures, but only through political debate: the exchange of opinions among people who share, but have different perspectives on, a common world. This position continues to animate many discussions of science, technology, and ethics in ways that can be deepened by dialogue with Arendt's thought.

ANNE CHAPMAN

SEE ALSO *Anders, Günther; Fascism; Heidegger, Martin; Holocaust; Socialism; Totalitarianism.*

BIBLIOGRAPHY

Arendt, Hannah. (1958). *The Human Condition*. Chicago: University of Chicago Press.

Arendt, Hannah. (1964). *Eichmann in Jerusalem: A Report on the Banality of Evil*. New York: Viking.

Arendt, Hannah. (1968). *Between Past and Future*. New York: Viking Compass Edition with additional text. This collection of essays contains further discussion of some of the concepts relevant to science and technology developed in the human condition.

Arendt, Hannah. (1973). *The Origins of Totalitarianism*. New York: Harcourt Brace Jovanovich.

Arendt, Hannah. (1981). *The Life of the Mind*. New York: Harcourt Brace Jovanovich.

Canovan, Margaret. (1992). *Hannah Arendt: A Reinterpretation of Her Political Thought*. Cambridge, UK, and New York: Cambridge University Press. A good general introduction to Arendt's work that emphasizes its origins in her thoughts on totalitarianism.

Young-Bruehl, E. (1982). *Hannah Arendt: For the Love of the World*. New Haven, CT, and London: Yale University Press. The main biography of Arendt.

Winner, Langdon. (1986). *The Whale and the Reactor: A Search for Limits in an Age of High Technology*. Cambridge, MA: MIT Press.

ARISTOTLE AND ARISTOTELIANISM

Aristotle (384–322 B.C.E.), born at Stagira, in northern Greece, was a philosopher and scientist, and a student of Plato (c. 428–c. 348 B.C.E.). The range and depth of Aristotle's thought is unsurpassed. He wrote on logic, physics and metaphysics, astronomy, politics and ethics, and literary criticism. His work formed the backbone of much Islamic and late medieval philosophy. In the early 2000s he is taken seriously as a social scientist and philosopher of biology. On a number of levels his thought is significant for science, technology, and ethics.

Basic Concepts

The root of Aristotle's thought lies in his response to the central puzzle of ancient Greek philosophy. For something to come to be, it must come either from what it is or from what it is not. But it cannot come from what it is, for what already exists cannot come to be; nor can it come from what it is not, because there would

Aristotle, 384–322 B.C.E. This Greek philosopher and scientist organized all knowledge of his time into a coherent whole which served as the basis for much of the science and philosophy of Hellenistic and Roman times and even affected medieval science and philosophy. *(NYPL Picture Collection.)*

not be anything for it to come to be from. Aristotle offered a commonsense solution to this conundrum: A kettle comes to be from and remains what it is, iron; but at the same time it comes to be from what it is not, not yet a kettle.

The confusion arises, Aristotle observes, because such concepts as "being" and "generation" are ambiguous; and this is because the objects to which they apply are not simple, but are compounds of *hule*, or matter, and *morphe*, or form. This view is known as *hulomorphism*. Matter is *potentiality*. The deer eats the corn and the hunter eats the deer: Thus the same materials are potentially herb, herbivore, and carnivore. The form is the *actuality*: This collection of matter at my feet is actually a dog.

Aristotle's primary interest was in the development of living things. He observed that an individual organism, say, Socrates, can change in a variety of formal ways: It can grow old or blush, but it remains the same thing—a human being and Socrates—all the while. Aristotle thus distinguished between accidents, such as Socrates' complexion, and his *substance*, which persists through many changes. Living organisms are clearly substances.

Aristotle's emphasis on substance reflects the general Greek view that what is most real is what persists through changes. But by this standard the species is more real than the individual, which dies. He thus felt compelled to distinguish between the primary substance, species, and secondary substances, individual organisms. This has its parallel in contemporary Darwinian thought: Some hold that it is the species form contained in the genes that is persistent and primary, and so is the real thing in evolution; others insist that the individual is the primary target of selection.

Perhaps Aristotle's most famous conceptual apparatus was his doctrine of causation, which he sometimes employed in analyzing technology but more often applied to living phenomena. The term for *cause* in Greek, *aitia*, indicates whatever is responsible for something being as it is or doing what it does. Aristotle distinguished four basic causes. Consider the growth of an animal from birth to adulthood. In part this happens because the organism is composed of certain materials, and these it may add to or subtract from itself (material causation). Second, it grows as it does because it is one kind of thing: Kittens become cats and never catfish (formal causation). Third, the form of the animal is more than a static arrangement; it is a complex dynamic process by means of which it is constantly recreating itself (efficient causation). Finally, this process does not proceed randomly but aims at some goal or *telos*, in this case the adult form (final causation).

Aristotle's insistence on teleological explanations (explaining something by explaining *what it is for*) became controversial in the modern period. But it amounts to two claims: First, many structural and behavioral features of organisms are clearly functional in design. Teeth are *for* biting and chewing. Second, organic processes are clearly self-correcting toward certain ends. The acorn grows toward an oak, its roots reaching down for water and minerals. The wolf weaves this way and that *in order to* bring down the fawn. Both of these claims are largely confirmed by modern biology.

From Biology to Politics

Aristotle's biology includes a distinctly nondualistic account of *psyche*, or soul, which in Greek refers to the principle of life. Rather than some separable substance, "soul" comprises all the processes by which the organism maintains itself and responds to its environment. In *On the Soul*, Aristotle distinguishes three types. Nutritive soul includes the capacity for self-nourishment, and so the possibility of growth and decay. All organisms possess this type of soul. Plants possess nutritive soul alone, but animal soul also includes perception and mobility. Finally, whereas animals are capable of pain and pleasure, human beings are capable of distinguishing what is really good and bad from what is merely attractive or unattractive, and so what is just from what is unjust. In addition to nutritive and animal soul, human beings possess *logos*, the power responsible for reason and speech. Aristotle's biology thus proceeds in a way similar to modern evolutionary accounts: Complex organisms are built by adding new levels of organization on top of existing ones.

Although Aristotle flirts only briefly with evolutionary explanations in biology, such an explanation is conspicuous at the beginning of his political science. Starting from political life as he knew it, he observed that the most elementary human partnerships were male and female (for the sake of procreation) and master and slave (for the sake of leisure). These comprise the household, which serves everyday needs. A union of households into a village serves more occasional needs, such as a barn raising. In turn, a union of villages makes up a polis, the independent city that was the foundation of classical political life. The polis is comprehensive: It incorporates all the elementary associations into a new, functional whole. Moreover the polis is self-sufficient, needing nothing more to complete it; and while it evolves for the sake of survival and comfort, it exists for the sake of the good life.

Aristotle's political science preserves the standard Greek classification of governments according to whether there is one ruler, a few, or many, as well as the argument that the primary tension in politics is between the few rich and the many poor. But it is not reductionist. What drives politics most of the time is not economic necessity but the desire for honor and wealth. Moreover, he recognizes a broad spectrum of regimes in place of simple kinds: Some monarchies are closer to aristocracies than others. In the body of the *Politics* Aristotle considers, from the point of view of various regimes, which institutions will tend to preserve that form and which will destabilize it. Because he includes even tyrants in this analysis, some have seen his approach as an example of a value-free social science. He also insists, however, that the authority of the ruling part of any partnership—father, king, or congress—is justified only to the degree that it serves the common good rather than the interest of the rulers. Aristotle's advice for more extreme regimes is also to move them in the direction of moderation, by broadening the base of citizens who benefit from the regime's rule. The goal of political action is the common good; authority should therefore be apportioned according to the contribution that each person or group can make toward that goal.

It is interesting to consider what Aristotle would have thought of modern technological and scientific expertise as a claim to rule. Unlike Plato, he does not explicitly consider the possibility of rule by trained elites. He does observe, however, that the best judge of a house is not the architect but the occupant, and similarly that the people collectively are better judges of policy outcomes than the best trained policymakers. Rule by experts would be safest in a regime with a substantial democratic element.

Aristotle's Ethics

The *Ethics*, like the *Politics*, begins with the observation that all human actions aim at some apparent good. But Aristotle distinguishes goods that are merely instrumental from those that are good in themselves. A person swallows a bitter medicine only for the sake of something else, health; but people seek out simple pleasures for their own sake. Aristotle argued that all the various good things can contribute to or be part of one comprehensive good, which he called *eudaemonia*, or blessedness. This term signifies a life that is complete and satisfying as a whole.

Eudaemonia requires certain basic conditions—such as freedom, economic self-sufficiency, and security—and it can be destroyed by personal tragedies. It is to this degree dependent on good fortune. Most important, however, are those goods of the soul that are largely resistant to fortune. The body of the *Ethics* is accordingly devoted to a treatment of virtues such as bravery, temperance, generosity, and justice. Perhaps Aristotle's greatest achievement was to have reconciled the concept of a virtuous action with that of a virtuous human being. Aristotle usually defines a virtuous action as a mean between two extremes. For example, a brave action is a mean between doing what is cowardly and what is foolhardy, in a given set of circumstances. But it is not enough merely to perform the appropriate action; virtue is also a matter of the appropriate emotional reactions, neither excessively fearful nor insensitive to genuine dangers. A virtue, then, is the power of acting and reacting in a measured way.

Virtues are different, however, from those powers that come directly from nature. In the case of sight, for example, one must first possess the power before one can begin to use it. By contrast, it is only by first doing brave things that one then becomes brave. Thus, a virtue requires cultivation. A virtuous person is someone who is habituated to acting properly in each situation, without hesitation, and who does so because it is the virtuous thing to do. The most important requirement of *eudaemonia* is the possession of a complete set of virtues.

Aristotle draws a clear distinction between moral and intellectual virtues. The former are acquired by habituation and produce right action in changing circumstances. The latter are acquired by learning and are oriented toward an understanding of the nature of things. Modern scientific and technological expertise certainly involves intellectual virtues as Aristotle understood them. But the one sort of virtue does not imply the other: A good ruler might be illiterate, or a scientist greedy and a coward. This is another Aristotelian reason why expertise alone cannot be a sufficient title to rule.

Aristotelianism

For well more than a thousand years after his death, and across several great traditions, Aristotle's works guided research in natural science, logic, and ethics. In Greek philosophy his own school, the Peripatos or Lyceum, long survived him; the first of many revivals of Aristotelianism occurred in the first century B.C.E., when Andronicus of Rhodes edited and published his major works. Aristotelianism thrived in centers of Hellenistic civilization and was revived again as part of a Byzantine scholarly renaissance in the ninth century C.E. By that time Aristotle's works had been translated into Syriac and Arabic, and in these languages became available both to Islamic and Jewish scholars. During the twelfth and thirteenth centuries the Aristotelian corpus was gradually translated into Latin and introduced to Western Christendom.

In all these traditions, his work served as a stimulus to scientific, ethical, and even technological progress. His natural science inspired his successor at the Lyceum, Theophrastus (c. 372–c. 287 B.C.E.), who produced an impressive botany. His logic, his empiricism, and his interest in nature inspired the stoics. Aristotle's work was instrumental to the medical researches of Galen (129–c. 199 C.E.) and the optics of Alhazen (965–1039 C.E.). Perhaps most importantly, the Jewish thinker Moses Maimonides (1135–1204) and the Christian Thomas Aquinas (1225–1274) wedded a modified Aristotelianism to existing theologies in attempts to create comprehensive systems of thought. Even his early modern critics such as Francis Bacon (1561–1626) and Thomas Hobbes (1588–1679) employed methods and concepts that were Aristotelian in origin.

Aristotle's reputation went into decline with the rise of early modern science and has only recently recovered. This is sometimes attributed to his scientific errors, which were many. He believed for example in spontaneous generation, the view that organisms can be produced by the action of heat and moisture on natural materials. He believed that in sexually reproducing species, the male provides all the form while the female provides only the matter. He believed that the function of the brain is to cool the blood. But such mistakes, amusing as they are, were due to the poverty of his experimental technologies and not to errors in his basic theories.

Nor do the flaws in his methods of investigation explain the modern decline of Aristotelianism. His logic was sound and is mostly preserved in contemporary philosophy. Moreover, contrary to a common prejudice, he and his students aimed at a rigorous empiricism. They gathered as much data as possible given the available technologies. It is true that Aristotle lacked a modern scientific method by which a hypothesis might be built and tested. But such a method could have as easily been employed to build on the Aristotelian foundation of premodern thought as to undermine it.

The reason for Aristotle's dismissal had more to do with the status of physics as the paradigmatic science. Confining itself to the mechanics of matter and energy, modern physics achieved a rigor previously matched only by abstract mathematics. On the topic of physics, Aristotle is embarrassingly weak, in part because he tried to extend biological reasoning to inorganic nature. Modern biologists, who might have defended him, suffered from their own inferiority complex. They were particularly embarrassed by the occasional flirtation of biologists with occult concepts, such as a mysterious "vital force" in living things. They accordingly pursued a rigorously reductionist view of organisms and tried to avoid any hint of purpose in their descriptive language. They could not afford to be seen in public with Aristotle, who was famous for teleological explanations.

Since the mid-twentieth century, the center of gravity in modern science has begun to shift from physics toward biology. This is marked by the quite literal drift of talented physicists into the laboratories of the biologists. One reason for this shift is the recognition that biology is in some ways a broader science than physics. No biologist is much surprised by the findings of chemists; but no physical scientist could remotely expect the existence of a cell from the principles of chemistry. As biology has become increasingly confident, it finds itself speaking in a language that is reminiscent of Aristotle. It is now safe to recognize him, in the words of the American zoologist Ernst Mayr

(b. 1904), as the greatest contributor to current knowledge of life before Charles Darwin (1809–1882).

In recent decades a number of thinkers have taken Aristotelian approaches to the philosophy of biology, bioethics, and political philosophy. The philosopher Hans Jonas (1903–1993) adopted a hulomorphism, teleology, and concept of life derived largely from Aristotle's *On the Soul*. Jonas (1966) argued that the greatest error of modern thought was dualism, in particular the isolation of the concept of mind from that of the living body. For Jonas, mind, and perhaps even some germ of consciousness, is present even in the simplest organisms. As in Aristotle, the natural history of mind and that of organic life are in fact the same study.

This rejection of dualism has important ethical as well as philosophical consequences. Modern ecological thought has largely discredited the early modern view of nature as a storehouse of materials to be manipulated by human will. If humans are as much a part of nature as any organic or inorganic process, then nature should be approached with respect, and cultivated rather than merely manipulated. Deeply influenced by Jonas, the philosopher Leon Kass (1985) puts special emphasis on the dignity of *human* life. As Aristotle argued, human beings share the capacities of soul that demarcate plants and animals but enjoy other capacities (such as speech and intelligence) that are found nowhere else in nature. Precisely if human nature is the result of an unrepeatable evolutionary process, we ought to take a cautiously ecological approach to biotechnology.

KENNETH C. BLANCHARD JR.

SEE ALSO *Natural Law; Plato; Scientific Ethics; Thomas Aquinas; Virtue Ethics.*

BIBLIOGRAPHY

Arnhart, Larry. (1998). *Darwinian Natural Right: The Biological Ethics of Human Nature*. Albany: State University of New York Press. An attempt at reconciling Aristotle with contemporary Darwinian social and political thought.

Barnes, Jonathan, ed. (1984).*The Complete Works of Aristotle*, 2 vols. Princeton, NJ: Princeton University Press. A good collection of Aristotle's major works.

Charlton, William, ed. (1984). *Aristotle's "Physics," Book One and Two*. Oxford: Clarendon Press. An excellent translation with notes of one of the most important passages in Aristotle's work. Includes discussions of hulomorphism, the doctrine of four causes, and teleology.

Gotthelf, Allan, and James G. Lennox, eds. (1987). *Philosophical Issues in Aristotle's Biology*. Cambridge, UK: Cambridge University Press. A collection of important essays by different scholars concerning Aristotle's philosophy of biology. All take Aristotle seriously from the point of view of modern biology.

Jaffa, Harry V. (1963). "Aristotle." In *History of Political Philosophy*, ed. Leo Strauss and Joseph Cropsey. Chicago: University of Chicago Press. Not included in later editions of the Strauss-Cropsey collection, this is possibly the best essay on Aristotle's politics in English.

Jonas, Hans. (1966). *The Phenomenon of Life: Toward a Philosophical Biology*. New York: Delta. A sustained attempt to reconstruct an Aristotelian philosophy of life out of the insights and evidence provided by modern biology.

Kass, Leon. (1985). *Toward a More Natural Science: Biology and Human Affairs*. New York: Free Press. A series of essays taking a mostly Aristotelian approach to contemporary bioethical questions.

Lennox, James G. (2001). *Aristotle's Philosophy of Biology*. Cambridge, UK: Cambridge University Press. A collection of essays by one of the foremost authorities on Aristotelian biology.

Nussbaum, Martha C., and Amélie Oksenberg Rorty, eds. (1992). *Essays on Aristotle's "De anima."* Oxford: Clarendon Press. A number of diverse essays organized around the question whether Aristotle's biological thought can be commensurate with modern biology.

ARSENIC

Arsenic has long been regarded as a dangerous poison and an environmental contaminant. But in the 1980s the focus on arsenic changed dramatically when approximately 3 million tube wells in Bangladesh and West Bengal, India, were found to be contaminated with that highly reactive chemical agent. By 2003 public health authorities estimated that as much as 40 million persons were being exposed to varying concentrations of the chemical in Bangladesh, plus another 3 million in West Bengal. The source of the arsenic came as a surprise to the toxic substance community in that the contamination was so widespread and came not from any industrial source but from rocks and sediment in the region's natural geological formations.

Arsenic is one of the most ubiquitous and paradoxical substances on Earth. In very small amounts, it is essential to life. In large amounts it is poisonous. While its inorganic forms are toxic, its organic forms are benign. Industrial arsenic is used for leather tanning, in pigments, glassmaking, fireworks, and medicinals, and as an additive that gives strength to metals. It is also a poison gas agent.

Arsenic's toxic effects vary according to exposure. Moderate levels (roughly 100 parts per billion and higher) can cause nausea and vomiting, decreased production of red and white blood cells, abnormal heart rhythm, and tingling in the hands and feet. Chronic exposure over time causes dark sores on hands, feet, and torso plus overall debilitation from damage to the cardiovascular, immune, neurological, and endocrine systems. Cancer can also occur after years of arsenic exposure at moderate to high levels.

After years of controversy over compliance costs, the U.S. Environmental Protection Agency in 2001 established a drinking water standard of 10 parts per billion, that was scheduled to go into effect in January 2006. The new rule supplanted the 50 ppb standard that had been in effect since 1975. The World Health Organization's has likewise adopted a 10ppb guideline. Arsenic readings in the Bangladesh/West Bengal groundwater frequently run from 200 ppb to 1,000 ppb. Deep wells, however, are not believed to be a problem.

Arsenic as both an industrial and natural pollutant is hardly a new phenomenon worldwide. High arsenic levels in air and water from mining and manufacturing operations from China to Peru have been well recognized though sporadically regulated for decades. Moreover, arsenic leached into waterways and aquifers from naturally occurring geological formations has been recorded in several regions. But because most of those areas are geographically remote, only the environmental toxicology community has taken much notice.

The ethics of arsenic control are vastly complex. The moment an environmental problem rises to crisis proportions in the industrial democracies of Europe and North America, the response is to assemble all possible mitigation techniques and human resources to attack the problem quickly. Nothing of the sort had happened in response to the disaster in South Asia, owing mainly to political graft, bureaucratic bloat, and the conflicting and poorly coordinated maze of national and international institutions whose involvement is required. The World Bank in 1998 issued a $32.4 million loan for the planning and execution of mitigation projects, but not until 2004 were the funds released for the project to begin.

As of 2005, the problem remained so widespread and Bangladesh was so lacking in resources that villagers themselves had to be taught to self-police and improve their water supplies by marking contaminated wells and using cheap and simple filtration techniques. For that to happen, inexpensive, mobile testing kits were needed and alternative sources of water had to be developed.

WIL LEPKOWSKI

SEE ALSO *Heavy Metals.*

BIBLIOGRAPHY

National Academy of Sciences. (2001). *Arsenic in Drinking Water: 2001 Update*. Washington, DC: National Academy of Sciences Press.

Patel, Priya Harish. (2001). "The Crisis of Capacity: A Case Study of Arsenic Contamination in Bangladesh and West Bengal." Cambridge, MA: Harvard University, Department of Environmental Science and Public Policy. Honors thesis.

INTERNET RESOURCES

Arsenic Crisis Information Center. Available at http://bicn.-com/acic/.

Harvard University Environmental Science Group. Available at http://phys4.harvard.edu/~wilson/arsenic. Provides valuable information on the Bangladesh crisis.

Jadavpur University Department of Environmental Studies. Available at http://www.sos-arsenic.net.

ARTIFICIAL INTELLIGENCE

Artificial Intelligence (AI) is the science and technology that seeks to create intelligent computational systems. Researchers in AI use advanced techniques in computer science, logic, and mathematics to build computers and robots that can mimic or duplicate the intelligent behavior found in humans and other thinking things. The desire to construct thinking artifacts is very old and is reflected in myths and legends as well as in the creation of lifelike art and clockwork automatons during the Renaissance. But it was not until the invention of programmable computers in the mid-twentieth century that serious work in this field could begin.

AI Research Programs

The computer scientist John McCarthy organized a conference at Dartmouth College in 1956 where the field of AI was first defined as a research program. Since that time a large number of successful AI programs and robots have been built. Robots routinely explore the depths of the ocean and distant planets, and the AI program built by International Business Machines (IBM) called Deep Blue was able to defeat the grand master chess champion Garry Kasparov after a series of highly publicized matches. As impressive as these accomplishments are, critics still maintain that AI has yet to achieve the goal of creating a program or robot that can truly operate on its own (autonomously) for any significant length of time.

AI programs and autonomous robots are not yet advanced enough to survive on their own, or interact with the world in the same way that a natural creature might. So far AI programs have not been able to succeed in solving problems outside of narrowly defined domains. For instance, Deep Blue can play chess with the greatest players on the planet but it cannot do anything else. The dream of AI is to create programs that not only play world-class chess but also hold conversations with people, interact with the outside world, plan and coordinate goals and projects, have independent personalities, and perhaps exhibit some form of consciousness.

Critics claim that AI will not achieve these latter goals. One major criticism is that traditional AI focused too much on intelligence as a process that can be completely replicated in software, and ignored the role played by the lived body that all natural intelligent beings possess (Dreyfus 1994). Alternative fields such as Embodied Cognition and Dynamic Systems Theory have been formed as a reply to this criticism (Winograd and Flores 1987). Yet researchers in traditional AI maintain that the only thing needed for traditional AI to succeed is simply more time and increased computing power.

While AI researchers have not yet created machines with human intelligence, there are many lesser AI applications in daily use in industry, the military, and even in home electronics. In this entry, the use of AI to replicate human intelligence in a machine will be called *strong AI*, and any other use of AI will be referred to as *weak AI*.

Ethical Issues of Strong AI

AI has and will continue to pose a number of ethical issues that researchers in the field and society at large must confront. The word *computer* predates computer technology and originally referred to a person employed to do routine mathematical calculations. People no longer do these jobs because computing technology is so much better at routine calculations both in speed and accuracy (Moravec 1999). Over time this trend continued and automation by robotic and AI technologies has caused more and more jobs to disappear. One might argue, however, that many other important jobs have been created by AI technology, and that those jobs lost were not fulfilling to the workers who had them.

This is true enough, but assuming strong AI is possible, not only would manufacturing and assembly line jobs become fully automated, but upper management and strategic planning positions may be computerized as well. Just as the greatest human chess masters cannot compete with AI, so too might it become impossible for human CEOs to compete with their AI counterparts. If AI becomes sufficiently advanced, it might then radically alter the kinds of jobs available, with the potential to permanently remove a large segment of the population from the job market. In a fully automated world people would have to make decisions about the elimination of entire categories of human work and find ways of supporting the people who were employed in those industries.

Other ethical implications of AI technology also exist. From the beginning AI raised questions about what it means to be human. In 1950 the mathematician and cryptographer Alan Turing (1912–1954) proposed a test to determine whether an intelligent machine had indeed been created. If a person can have a normal conversation with a machine, without the person being able to identify the interlocutor as a machine, according to the Turing test the machine is intelligent (Boden 1990). In the early twenty-first century people regularly communicate with machines over the phone, and Turing tests are regularly held with successful results—as long as the topic of discussion is limited. In the past special status as expert thinkers has been proposed as the quality that distinguishes humans from other creatures, but with robust AI that would no longer be the case. One positive effect might be that this technology could help to better explain the place of humans in nature and what it means for something to be considered a person (Foerst 1999).

The ethical responsibility that people have toward any strong AI application is a matter that must be taken into consideration. It does not seem moral to create thinking minds and then force them to do work humans do not want to do themselves.

Finally because AI technology deals directly with human operators, people must make decisions regarding what kind of ethics and morality are going to be programmed into these thinking machines. The scientist and fiction writer Isaac Asimov proposed in his writings three moral imperatives that should be programmed into robots and other AI creations:

- A robot may not injure a human being or, through inaction, allow a human being to come to harm.
- A robot must obey the orders given it by human beings except where such orders conflict with the first law.
- A robot must protect its own existence as long as such protection does not conflict with the first or second law.

These imperatives make for good reading but are sadly lacking as a solution to the problems presented by fully autonomous robotic technologies. Asimov wrote many stories and novels (The *Robot Series* [1940–1976] and *I,*

Robot [1950]) that used the unforeseen loopholes in the logic of these laws, which occasionally allowed for fatal encounters between humans and robots. For instance, what should a robot do if, in order to protect a large number of people, it must harm one human who is threatening others? It can also be argued that AI technologies have already begun to harm people in various ways and that these laws are hopelessly naïve (Idier 2000). Other researchers in the field nevertheless argue that Asimov's laws are actually relevant and at least suggest a direction to explore while designing a computational morality (Thompson 1999).

This problem is more pressing than it may seem, because many industrial countries are working to create autonomous fighting vehicles to augment the capabilities of their armed forces. Such machines will have to be programmed so that they make appropriate life and death choices. More subtle and nuanced solutions are needed, and this topic remains wide-open—widely discussed in fiction but not adequately addressed by AI and robotics researchers.

Ethical Issues of Weak AI

Even if robust AI is not possible, or the technology turns out to be a long way off, there remain a number of vexing ethical problems to be confronted by researchers and technologists in the weak AI field. Instead of trying to create a machine that mimics or replicates exactly human-like intelligence, scientists may instead try to imbed smaller, subtler levels of intelligence and automation into all day-to-day technologies. In 1991 Mark Weiser (1952–1999) coined the term *ubiquitous computing* to refer to this form of AI, but it is also sometimes called the *digital revolution* (Gershenfeld 1999).

Ubiquitous computing and the digital revolution involve adding computational power to everyday objects that, when working together with other semismart objects, help to automate human surroundings and hopefully make life easier (Gershenfeld 1999). For instance, scientists could imbed very small computers into the packaging of food items that would network with a computer in the house and, through the Internet perhaps, remind people that they need to restock the refrigerator even when they are away from home. The system could be further automated so that it might even order the items so that one was never without them. In this way the world would be literally at the service of human beings, and the everyday items with which they interact would react intelligently to assist in their endeavors. Some form of this more modest style of AI is very likely to come into existence. Technologies are already moving in these directions through the merger of such things as mobile phones and personal data assistants.

Again this trend is not without ethical implications. In order for everyday technologies to operate in an intelligent manner they must take note of the behaviors, wants, and desires of their owners. This means they will collect a large amount of data about each individual who interacts with them. This data might include sensitive or embarrassing information about the user that could become known to anyone with the skill to access such information. Additionally these smart technologies will help increase the trend in direct marketing that is already taking over much of the bandwidth of the Internet. Aggressive advertisement software, spying software, and computer viruses would almost certainly find their way to this new network. These issues must be thoroughly considered and public policy enacted before such technology becomes widespread.

In addition, Weiser (1999) argues that in the design of ubiquitous computing people should work with a sense of humility and reverence to make sure these devices enhance the humanness of the world, advancing fundamental values and even spirituality, rather then just focusing on efficiency. Simply put, people should make their machines more human rather then letting the technology transform human beings into something more machine-like.

A last ethical consideration is the possibility that AI may strengthen some forms of gender bias. Women in general, and women's ways of knowing in particular, have not played a large role in the development of AI technology, and it has been argued that AI is the fruit of a number of social and philosophical movements that have not been friendly to the interests of women (Adam 1998). Women are not equally represented as researchers in the field of AI, and finding a way to reverse this trend is a pressing concern. The claim that AI advances the interests of males over those of females is a more radical, yet intriguing claim that deserves further study.

AI continues to grow in importance. Even though researchers have not yet been able to create a mechanical intelligence rivaling or exceeding that of human beings, AI has provided an impressive array of technologies in the fields of robotics and automation. Computers are becoming more powerful in both the speed and number of operations they can achieve in any given amount of time. If humans can solve the problem of how to program machines and other devices to display advanced levels of intelligence, as well as address the many ethical issues raised by this technology, then AI may yet expand in astonishing new directions.

JOHN P. SULLINS III

SEE ALSO *Artificial Morality; Artificiality; Asimov, Isaac; Automation; Computer Ethics; Robots and Robotics; Turing, Alan; Turing Tests.*

BIBLIOGRAPHY

Adam, Alison. (1998). *Artificial Knowing: Gender and the Thinking Machine*. New York: Routledge. Adam is an AI researcher who writes from experience about the difficulties encountered by women working in AI. She also argues that if women were more involved in AI it would significantly alter and improve the results of this science. Accessible, informative, and includes a select bibliography.

Asimov, Isaac. (2004). *I, Robot*. New York: Bantam Books. Classic science fiction account of the social impact of robotics. The stories are very entertaining while at the same time exploring deeply philosophical and ethical questions.

Boden, Margret A., ed. (1990). *The Philosophy of Artificial Intelligence*. New York: Oxford University Press. A collection of the best papers written on the philosophical debate regarding the possibility of creating AI. The research included stops at the early 1990s but this is a good place to start one's study of the subject. Includes a select bibliography.

Dreyfus, Hubert L. (1994). *What Computers Still Can't Do*. Cambridge MA: MIT Press. This is a famous counter argument to the very possibility of achieving any results in strong AI. Largely dismissed by the AI community, it is still useful as an example of how one might argue that AI misses the important insights into human intelligence and consciousness that phenomenological psychology and Heideggerian philosophy might add.

Dreyfus, Hubert L., and Stuart E Dreyfus. (1986). *Mind over Machine: The Power of Human Intuition and Expertise in the Era of the Computer*. New York: Free Press. This book was written with the author's brother who worked with the AI technology questioned in the book. They claim that human intelligence is not completely captured by AI technology and give numerous examples to illustrate their claims. Many of their claims, such as the inability of AI to program a computer to beat a chess master, seem to have been incorrect but this book is an important first step in Dreyfus's critique of AI.

Foerst, Anne. (1999). "Artificial Sociability: from Embodied AI toward New Understandings of Personhood." *Technology in Society* 21(4): 373–386. Good discussion of the impact AI and robotics has on complicating the notion of *personhood*, or what are the conditions under which something is understood to be a person.

Gershenfeld, Neil. (1999). *When Things Start to Think*. New York: Henry Holt. Good general overview of the attempt to add intelligence to everyday technology. Gershenfeld is involved in this research and speaks with authority on the subject without being overly technical.

Idier, Dominic. (2000). "Science Fiction and Technology Scenarios: Comparing Asimov's Robots and Gibson's Cyberspace." *Technology in Society* 22: 255–272. A more up-to-date analysis of Asimov's prognostications about robotics in light of the more dystopian ideas of the science fiction author William Gibson.

Luger, George F., ed. (1995). *Computation and Intelligence*. Cambridge, MA: AAAI Press/ MIT Press. A collection of many of the best early papers written on AI by the pioneers of the field. These papers have shaped the field of AI and continue to be important. Includes an extensive bibliography.

Moravec, Hans. (1999). *Robot: Mere Machine to Transcendent Mind*. New York: Oxford University Press. One of the most successful AI researchers unabashedly predicts the future of humanity and the emergence of its robotic "mind children." Deftly presents very technical arguments for the general reader.

Nilsson, Nils J. (1986). *Principles of Artificial Intelligence*. Los Altos, CA: Morgan Kaufmann. Good introduction to the more technical issues of AI as a science.

Searle, John R. (1992). *The Rediscovery of the Mind*. Cambridge, MA: MIT Press. The American philosopher John Searle argues for the inability of AI to capture semantics (the meaning of words), with machines that can only manipulate syntax (the grammar and symbols of a language). If there is no mechanical way to create semantic understanding then there is no way to create thinking machines. This argument has not gone unchallenged by AI researchers and many of the entries in this bibliography attempt to refute Searle.

Thompson, Henry S. (1999). "Computational Systems, Responsibility, and Moral Sensibility." *Technology in Society* 21(4): 409–415. A discussion of how humans should be more sensitive to the moral implications of the computational systems they build.

Weiser, Mark. (1993). "Hot Topics: Ubiquitous Computing." *IEEE Computer* 26(10): 71–72. Also available from http://www.ubiq.com/hypertext/weiser/UbiCompHotTopics.html. Weiser introduces the concept of ubiquitous computing. This remains a very important defining moment for the field.

Weiser, Mark. (1999). "The Spirit of the Engineering Quest." *Technology in Society* 21: 355–361. Weiser argues that those who engineer devices that mediate the human-to-world interface should be especially careful that their work enhances human life rather than detracts from it.

Winograd, Terry, and Fernando Flores. (1987). *Understanding Computers and Cognition: A New Foundation for Design*. Boston: Addison-Wesley Publishing. Winograd and Flores argue for a form of AI that takes into account the important counterarguments to traditional AI, and instead of ignoring them, incorporates them into a better science that takes the embodiment of cognitive agents seriously.

ARTIFICIALITY

In the European and North American tradition, a thing is natural insofar as its existence does not depend on human intervention, while something is artificial if its existence depends on human activity. From this per-

English conjuror John Neville Maskelyne with two "musical automata" playing a trumpet and a tuba, c. 1890. These lifelike objects are an example of early forms of artifciality. *(Hulton Archive/ Getty Images.)*

spective, artificiality extends not just to some physical objects but also to intellectual phenomena such as science, art, and technology—to the extent that they are characteristic of human life. With regard to strictly physical artifice, Aristotle, in *Physics*, further notes that unlike natural objects, artifacts do not have internal sources of motion and rest. If a bed were to sprout, what would come up would not be another bed, but an oak tree (Book 2.1). In relation to both these extrinsic and intrinsic features, it has also been common to assess artifice, in comparison with nature, as a diminished level of reality, and sometimes as less valuable. The ethics of artifacts has usually been to argue their lesser intrinsic value but their greater extrinsic or instrumental value insofar as they benefit humans and moderate a sometimes harsh experience of nature.

From Nature to Technology and Back

Nevertheless it is necessary to distinguish at least two types of artificiality. For instance, Aristotle again distinguishes those technics that help nature do more effectively or abundantly what it already does to some extent on its own and those that construct objects that would not be found at all in nature if it were not for human ingenuity. The former or what might be called type A artifacts are associated with the techniques of agriculture, medicine, and education. The latter or type B artifacts are associated with architecture and more modern technologies. But type B artifacts need not always create things not found in nature such as right-angle buildings. Using technology it is also possible to create replacements or substitutes for natural objects in the form, for example, of artificial grass, artificial kidneys, and even artificial intelligence. The term *synthetics* is also sometimes applied to this class of artifacts, as with synthetic oil or synthetic wood. It is thus necessary to distinguish type B(1) and type B(2) artifice, and because of the special features of type B(2) artifacts it is useful to coin the term *naturoids*. Naturoids may include a variety of artifacts, from automatons, robots, and androids to humanoids, bionic humans, and more.

The field of naturoids is greatly advanced in the early twenty-first century thanks to developments in physics, chemistry, biology, materials science and technology, electronics, and computer science. Nevertheless its roots are quite ancient because, as Derek de Solla Price emphasizes, human history "begins with the deep-rooted urge of man to simulate the world about him through the graphic and plastic arts" (de Solla Price 1964, p. 8). Well-known are the efforts of eighteenth-century mechanics to build machines that would often mimic certain living systems, as in the cases of Jacques de Vaucanson, Julien Offray de Lamettrie, and Pierre Jacquet-Droz, as well as Karel Capek's image of a robot in the early twentieth century. Twenty-first century naturoids cover a wide range of machines, including artificial body parts and organs, advanced robots, and reproductions of other physical objects or processes—such as stone, grass, smell, and speech—and, on a software level, artificial intelligence or life.

Genetic engineering offers even more dramatic prospects, but in a different direction from naturoids' tradition. In fact, humans are able to change the architecture of DNA, but the final result is a quite natural system, though possibly unusual. At the contrary, a naturoid, even if built by means of nanotechnology, comes always from an analyical design within which all the components are replacements of the corresponding natural parts. Nevertheless, a new reality could come from mixed systems such as bionic ones, where natural subsystems are put at work along with artificial devices giving birth to fascinating and unexperienced problems even from an ethical point of view.

Embodied Ethics

Artificiality has often been criticized as opposed to the natural and the naturally human, and also for its unintended social, legal, and ethical consequences. Such attitude, which recalls the suspicion of sorcery directed at the mechanicians of the Renaissance, takes on a new form in the present. As Edward Tenner (1996) has argued, artifacts have a tendency, not unlike ill-behaved pets, to *bite back* through what he calls "the revenge of unintended consequences" (Tenner 1996).

However in discussions of unintended consequences—which is often taken as a fundamental ethical problem of artifacts—little effort has been made to distinguish among the types of artificiality already mentioned. In fact, while type I artifacts (such as pencils, rifles, cars, and cathode-ray tubes) return to human beings responsibility for their uses, type B artifacts, especially type B(2) artifacts or naturoids, as forms of objects and processes in nature, tend to embody ethical models in their own architecture.

The famous *Three Laws of Robotics*, proposed by Isaac Asimov, illustrate this phenomenon. Yet in fact every naturoid includes at its core not only some image of the natural exemplar it aims to recreate, but also its ideal function. For instance, an artificial organ embodies both the current knowledge of the natural organ and the views regarding its correct functioning in human physiology and even within human society. The same may be said for artificial intelligence programs, artificial life simulations, *virtual reality* devices, and other attempts to give birth to the entities of posthumanism.

Once some implicit or explicit ethical model is assigned to a naturoid, it will appear to be an actor itself, and people will interact with it as if they were interacting with something natural or social. This explains why some scholars such as Latour have begun to think that machines "challenge our morality" (quoted in Margolin 2002, p. 117) while others predict that they will soon be considered responsible actors.

The Third Reality

Unlike technologies that do not aim to produce anything immediately present in nature—that is, type B(1) artifacts—naturoid or type B(2) technologies emerge from a design process that begins with an idea not only of what a machine has to be and to do, but also of what the natural exemplar actually is and does. Nevertheless constructing a model of a natural exemplar requires some reduction in its complexity. This reductive process includes: (a) the selection or the construction of an observation level; (b) the simplification of the exemplar structure according to the selected observation level; (c) its isolation from the context in which it exists; and (d) the selection or the attribution of some performance function that designers judge essential in its behavior.

The adoption of materials that differ from those used by nature—and their interplay in a machine—makes the naturoid an alternative realization (Rosen 1993) when compared to the natural exemplar. All this, in turn, implies that the appearance and behavior of a naturoid will unavoidably overlap with only a limited set of properties from the natural exemplar, and thus importantly, give them a transfigured character in many respects and to various degrees (power, sensitivity, flexibility, side-effects, and so forth).

As a consequence, even the ethical models implicit in all naturoids will tend to work according to styles that are rather unusual in human behavior. This explains why, for example, automatic or artificial devices often appear too rigid in applying their rules. The same may be said for so-called *enhanced reality* devices—for example, deliberate transfigurations of some natural exemplar through its artificialization—because it is not possible to resort to any known or sufficiently established *artificial morality* model.

What must be emphasized is that naturoids are not simply devices humans *use*; rather, humans expect them to be self-adaptive and transparent *replacements* of natural objects. Therefore their way of being and acting is intrinsically presumed to be compatible with human ethics. Nevertheless naturoids are setting up a third reality, part natural and part artificial, whose ethical significance remains to be determined.

MASSIMO NEGROTTI

SEE ALSO *Artificial Intelligence; Artificial Morality; Plastics; Posthumanism.*

BIBLIOGRAPHY

Cordeschi, Roberto. (2002). *The Discovery of the Artificial: Behavior, Mind and Machines Before and Beyond Cybernetics.* Dordrecht, The Netherlands: Kluwer Academic Publishers.

de Solla Price, Derek. (1964). "Automata and the Origin of Mechanism." *Technology and Culture* 5(1): 24–42.

Latour, Bruno. (1992). "Where Are the Missing Masses: The Sociology of a Few Mundane Artifacts." In *Shaping Technology/Building Society: Studies in Sociotechnical Change,* ed. Wiebe E. Bijker and John Law. Cambridge, MA: MIT Press.

Margolin, Victor. (2002). "Human And Machine: Issues Of Collaborative Action." In *Yearbook of the Artificial, Nature, Culture and Technology,* ed. Massimo Negrotti. Bern, Switzerland: Peter Lang.

Negrotti, Massimo. (2002). *Naturoids: On The Nature of the Artificial*. River Edge, NJ: World Scientific Publishing.

Rosen, Robert. (1993). "Bionics Revisited." In *The Machine as Metaphor and Tool*, eds. Hermann Haken, Anders Karlqvist, and Svedin Uno. Berlin: Springer Verlag.

Simon, Herbert A. (1996). *The Sciences of the Artificial*, 3rd edition. Cambridge, MA: MIT Press. Earlier editions published in 1969 and 1981.

Tenner, Edward. (1996). *Why Things Bite Back: Technology and the Revenge of Unintended Consequences*. New York: Knopf.

ARTIFICIAL MORALITY

Artificial morality is a research program for the construction of moral machines that is intended to advance the study of computational ethical mechanisms. The name is an intentional analogy to artificial intelligence (AI). Cognitive science has benefited from the attempt to implement intelligence in computational systems; it is hoped that moral science can be informed by building computational models of ethical mechanisms, agents, and environments. As in the case of AI, project goals range from the theoretical aim of using computer models to understand morality mechanistically to the practical aim of building better programs. Also in parallel with AI, artificial morality can adopt either an engineering or a scientific approach.

History

Modern philosophical speculation about moral mechanisms has roots in the work of the philosopher Thomas Hobbes (1588–1679). More recently, speculation about ways to implement moral behavior in *computers* extends back to Isaac Asimov's influential three laws of robotics (1950) and pioneer cyberneticist Warren McCulloch's 1965 sketch of levels of motivation in games. On the lighter side, Michael Frayn's *The Tin Men* (1965) is a parody of artificial morality that features an experimental test of altruism involving robots in life rafts. Although there has been fairly extensive work in this field broadly considered, it is an immature research area; a recent article calls itself a "Prolegomena" (Allen, Varner, and Zinser 2000). The following survey will help explain some of the goals and methods in this young field.

Ethics in the Abstract

Consider first the easiest goal: to understand ethics in the abstract context provided by computer programs. Robert Axelrod (1984) made a breakthrough in the field when he organized tournaments by asking experts in decision and game theory to submit programmed agents to play a well-known game: the iterated prisoner's dilemma. That challenge entailed the basic computational assumption that everything relevant to such a player could be specified in a computer program. Although games-playing programs figured in the early history of artificial intelligence (for example, A. L. Samuel's [1959] checkers program), the prisoner's dilemma is a mixed motive game that models morally significant social dilemmas such as the tragedy of the commons. In such situations one alternative—overfishing or creating more greenhouse gas—is rational yet morally defective because it is worse for all.

These models have generated considerable interest in the question of the ways rational choice relates to ethics. By focusing on an abstract game Axelrod was able to avoid trying to model full human moral decision making. Nonetheless, the iterated prisoner's dilemma is a hard problem. There is a large strategy set, and good strategies must take account of the other players' strategies. Thus, unlike AI, which for much of its first generation focused on single agents, artificial morality began by focusing on a plurality of agents.

Ethics and Game Theory

One result of Axelrod's initiative was to unite ethics and game theory. On the one hand, game theory provides simple models of hard problems for ethics, such as the prisoner's dilemma. First, game theory forces expectations for ethics to be made explicit. Early work in this field (Danielson 1992) expected ethics to solve problems—such as cooperation in a one-play prisoner's dilemma—that game theory considers impossible. More recent work (Binmore 1994, Skyrms 1996) lower the expectations for ethics. Consider Axelrod's recommendation of the strategy tit-for-tat as a result of its relative success in his tournament. Because the game is iterated, tit-for-tat is not irrationally cooperative. However, its success shows only that tit-for-tat is an equilibrium for this game; it is rational to play tit-for-tat if enough others do. But game theory specifies that many—indeed infinitely many—strategies are equilibria for the iterated prisoner's dilemma. Thus game theory shifts the ground of ethical discussion, from a search for the best principle or strategy, to the more difficult task of selecting among many strategies, each of which is an equilibrium, that is to say, a feasible moral norm.

Artificial Evolution

Another result of Axelrod's work was to link ethics and the evolutionary branch of game theory and modeling.

Axelrod established equilibriums by means of an evolutionary simulation (a form of the standard replicator dynamics) of the initial results. His later work introduced agents whose strategies could be modified by mutation. Classic game theory and modern ethics share many assumptions that focus on a normative question: What should hyperrational, fully informed agents do, taking their own or everyone's interests into account, respectively? However, it sometimes is easier to discover which of many simpler, less well-informed agents will be selected for solving a problem, and generally evolution selects what rationality prescribes (Skyrms 1996). This change from attempting to discover the perfect agent to experimenting with a variety of agents is especially helpful for ethics, which for a long time has been divided among partisans of different ethical paradigms. Evolutionary artificial morality promises to make it possible to test some of these differences. One benefit of combining evolution and simple programmed agents is that one can construct, for example, all possible agents as finite state machines of a given complexity, and use evolutionary techniques to test them (Binmore 1994). Another example is provided by Skyrms (1996), who ran evolutionary simulations where agents bargain in different ways, characteristic of different approaches to ethics.

A third effect of this research program is more directly ethical. A common result of experiments and simulations in artificial morality is to heighten the role of reciprocity and fairness at the expense of altruism. This shift is supported by human experiments as well as by theory. Experiments show that most subjects will carry out irrational threats to punish unfair actions. The theory that supports these results shows that altruism alone will not solve common social dilemmas.

Moral Engineering

The previous examples illustrate the simplest cases of what more properly might be called artificial moral engineering. In this area theorists are happy to study simple agents in simple games that model social settings to establish proofs of the basic concepts of the field: that moral behavior can be programmed and that ethically interesting situations can be modeled computationally.

At the other end of the engineering spectrum are those who try to build moral agents to act in more realistic situations of real artificial agents on the Internet and in programs more generally (Coleman 2001). This highlights the most immediate importance of artificial morality: "The risks posed by autonomous machines ignorantly or deliberately harming people and other sentient beings are great. The development of machines with enough intelligence to assess the effects of their actions on sentient beings and act accordingly may ultimately be the most important task faced by the designers of artificially intelligent automata" (Allen, Varner, and Zinser 2000, p. 251).

However, this survey of artificial moral engineering would be misleading if it did not note that there is a well-developed sub-field of AI—multiagent systems—that includes aims that fall just short of this. In a successful multiagent system computational agents without a common controller coordinate activity and cooperate rather than conflict. No current multiagent system is ethically sophisticated enough to understand harm to humans, but the aims of these fields clearly are convergent.

Moral Science

All this is engineering, not science. Artificial moral science adds the goal of realism. An effective ethical program might work in ways that shed no light on human ethics. (Consider the analogy between cognitive engineering and science, in which the Deep Blue chess program would be the analogous example of cognitive engineering. The clearest cases of artificial moral science are computational social scientists who test their models of social interaction with human experiments. For example, Peter Kollock (1998) tests a model in which moral agents achieve cooperation by perceiving social dilemmas in the more benign form of assurance games by running experiments on human subjects.

Finally, one benefit of the computational turn in ethics is the ability to embed theories in programs that provide other researchers with the tools needed to do further work. Again there is an analogy with artificial intelligence, many early discoveries in which have been built into standard programming languages. In the case of artificial morality academic computational tools such as Ascape and RePast allow researchers to construct experiments in "artificial societies" (Epstein and Axtell 1996). A related benefit of the computational approach to ethics is the development of a common language for problems and techniques that encourage researchers from a range of disciplines, including philosophy, biology, computing science, and the social sciences, to share their results.

Computer Games

While the work discussed so far is academic research some of the issues of artificial morality have already come up in the real world. Consider computer games. First, some of the most popular games are closely related to the artificial society research platforms discussed above. The bestselling SimCity computer game series is a popularized urban planning simulator. The user can select policies favoring cars or transit, high or low taxes, police or education expenditures, but, crucially, cannot control directly what the simulated citizens do. Their response is uncontrolled by the player, determined by the user's policies and values and dynamics programmed into the simulation. This serves as a reminder that artificial morality is subject to the main methodological criticism of all simulation: Assumptions are imbedded in a form that can make their identification and criticism difficult (Turkle 1995, Chapter 2).

Second, as computer games make use of AI to control opponents and other agents not controlled by humans, so too they raise issues of artificial morality. Consider the controversial case of the popular grand theft auto series of games, in which the player can run over pedestrians or attack and kill prostitutes. The victims and bystanders barely react to these horrible acts. These games illustrate what one might call "artificial amorality" and connect to criticisms that video and computer games "create a decontextualized microworld" (Provenzo 1991, p. 124) where harmful acts do not have their normal social consequences.

Third, games and programmed agents on the internet raise questions about what features of artificial characters lead to their classification in morally relevant ways. Turkle (1995) shows how people adjust their category schemes to make a place for artificial agents they encounter that are "alive" or "real" in some but not all respects.

PETER DANIELSON

SEE ALSO *Altruism; Artificial Intelligence; Artificiality; Game Theory; Robots and Robotics.*

BIBLIOGRAPHY

Allen, Colin; Gary Varner; and Jason Zinser. (2000). "Prolegomena to Any Future Artificial Moral Agent." *Journal of Experimental and Theoretical Artificial Intelligence* 12: 251–261. A survey of problems to which artificial morality might apply.

Asimov, Isaac. (1950). *I, Robot*. New York: Fawcett. Classic science fiction source of the three laws of robotics.

Axelrod, Robert. (1984). *The Evolution of Cooperation*. New York: Basic Books. Accessible introduction to some game theory and evolution methods for studying moral problem of cooperation.

Binmore, Ken. (1994). *Game Theory and the Social Contract: Playing Fair*. Cambridge, MA: MIT Press. A sophisticated use of game theory to criticize and extend artificial morality

Coleman, Kari. (2001). "Android Arete: Toward a Virtue Ethic for Computational Agents." *Ethics and Information Technology* 3(4): 247–265.

Danielson, Peter. (1992). *Artificial Morality: Virtuous Robots for Virtual Games*. London: Routledge.

Danielson, Peter. (1998). *Modeling Rationality, Morality, and Evolution*. New York: Oxford University Press.

Danielson, Peter. (2002). "Competition among Cooperators: Altruism and Reciprocity." *Proceedings of the National Academy of Sciences* 99: 7237–7242.

Epstein, Joshua M., and Robert Axtell. (1996). *Growing Artificial Societies: Social Science from the Bottom Up*. Cambridge, MA: MIT Press. Introduction to agent based modeling in the social sciences.

Frayn, Michael. (1965). *The Tin Men*. New York: Ace. A parody of artificial intelligence and artificial morality.

Kollock, Peter. (1998). "Transforming Social Dilemmas: Group Identity and Co-operation." In *Modeling Rationality, Morality, and Evolution*, ed. Peter Danielson. New York: Oxford University Press. Illustrates how experiments and game theory can be used together.

McCulloch, Warren Sturgis. (1965). "Toward Some Circuitry of Ethical Robots or an Observational Science of the Genesis of Social Evaluation in the Mind-Like Behavior of Artifacts." In his *Embodiments of Mind*. Cambridge, MA: MIT Press. Early speculation of the ethical potential of simple machines and game-like situations.

Provenzo, Eugene F. (1991). *Video Kids: Making Sense of Nintendo*. Cambridge, MA: Harvard University Press.

Samuel, A. L. (1959). "Some Studies in Machine Learning Using the Game of Checkers." *IBM Journal of Research and Development* 3: 210–229. Classic article on machine chess.

Skyrms, Brian. (1996). *Evolution of the Social Contract*. Cambridge, UK, and New York: Cambridge University Press. Accessible introduction to evolutionary game theory and ethics.

Turkle, Sherry. (1995). *Life on the Screen: Identity in the Age of the Internet*. New York: Simon & Schuster.

ASILOMAR CONFERENCE

In February 1975 an international group of scientists met at the Asilomar Conference Grounds in Pacific Grove, California, to discuss the potential biohazards posed by recombinant DNA (rDNA) technology. The official title of the meeting was the International Conference on Recombinant DNA Molecules, but it is remembered simply as the Asilomar Conference. It established guidelines concerning the physical and biological containment of rDNA organisms that served as the model for the current guidelines used by the National Institutes of Health (NIH). Although the Asilomar Conference marks a watershed moment in the regulation of rDNA technology, its broader implications remain controversial (Barinaga 2000, Davatelis 2000). Some claim that it was an example of self-promotion by a small but powerful interest group. Others argue that the process was too alarmist and generated unfounded fears in the public. Still others contend that it was an instance of scientists successfully regulating their own work.

The Events Preceding Asilomar

The first successful trial of rDNA technology (a type of genetic engineering that involves splicing genes into organisms) was performed by Paul Berg and other researchers at Stanford University in the early 1970s. It quickly raised concerns about "playing God" and the potential biohazards posed by recombinant microorganisms. In an unprecedented call for self-restraint, prominent scientists sent letters to the journal *Science* calling for a temporary moratorium on rDNA research (Singer and Söll 1973, Berg et al. 1974). Concerns included the development of biological weapons and the potential for genetically engineered organisms to develop resistance to antibiotics or to escape control.

Singer and Söll's letter was the result of a June 1973 meeting of the Gordon Conference on Nucleic Acids. In response, Berg led a committee of the National Academy of Sciences (NAS) in April 1974 to formulate policy recommendations for the use of rDNA technologies. The Berg committee, which met at the Massachusetts Institute of Technology, was composed of leading molecular biologists and biochemists involved in the emerging rDNA field.

This committee produced three recommendations addressed at the scientific community and the NIH: (1) instituting a temporary moratorium on the most dangerous experiments; (2) establishing an NIH advisory committee to develop procedures for minimizing hazards and to draft guidelines for research (which became the Recombinant DNA Advisory Committee [RAC]); and (3) convening the Asilomar Conference. All three recommendations were implemented.

The Conference Itself

Participation in the NIH-sponsored Asilomar Conference was by invitation only. It was attended by 153 participants. Outside of sixteen members of the press and four lawyers, it was composed entirely of scientists, mostly molecular biologists from the United States. There were no representatives from ethics, social science, ecology, epidemiology, or public-interest organizations.

The formal task of the conference was to identify the potential biohazard risks involved with rDNA technology and design measures to minimize them. Yet there was also a more important informal task faced by the participants. The emerging rDNA technology presented novel problems of regulation characterized by vast uncertainties concerning potential environmental and public health threats. The conference was set within a cultural and political context marked by a growing awareness of these threats and an increasing suspicion of new technologies. Therefore, the informal task faced by the scientists was to regulate rDNA technology in such a way that satisfied the public and, most importantly, allowed the science to be self-governing.

A comment made by Berg, cochair of the conference, illustrates this mind-set:

> If our recommendations look self-serving, we will run the risk of having standards imposed. We must start high and work down. We can't say that 150 scientists spent four days at Asilomar and all of them agreed that there was a hazard—and they still couldn't come up with a single suggestion. That's telling the government to do it for us. (Wright 2001 [Internet source])

In order to achieve the goal of self-governance, the participating scientists narrowed the agenda such that the issue was defined as a technical problem. The organizers decided not to address ethical concerns but to focus on biohazard issues (Wright 1994). Defining the problem in technical terms legitimated the model of self-government by scientists, because they were the only group that could solve such problems.

The conference organizers shaped a consensus around this technical problem definition. There were, however, threats to consensus from both sides. Some participants were opposed to any type of regulation, because it would compromise their freedom of inquiry. Others wanted a broader agenda that included public input on ethical considerations and explicit bans on the development of biological weapons.

In the end, guidelines with respect to physical and biological containment of rDNA organisms were drafted that allowed the scientific community to police itself under the auspices of the NIH-RAC mechanism. The guidelines involved working with disabled bacteria that could not survive outside the lab and classifying experiments according to the level of containment necessary. They also called for an end to experimentation using known carcinogens, genes that produce toxins, and genes that determine antibiotic resistance. The Asilomar Conference established a general sense that the burden of proof rested with scientists and that they must proceed cautiously until they can show that their research is safe.

The laboratory guidelines also became the international standard for rDNA research (see Löw 1985, Wright 1994). In the United States, the system of self-policing avoided both the chaotic patchwork of local legislation established by community decision-making forums and the legal rigidity, yet political changeability, of federal legislation.

Conference Legacy

There have been changes to the guidelines drafted at Asilomar. The membership and role of the NIH-RAC have been expanded, and containment levels have been lowered for many experiments. More public involvement has been incorporated into decision-making processes, and subsequent Food and Drug Administration (FDA) and Environmental Protection Agency (EPA) rules have ensured that the private sector complies with rDNA guidelines as biotechnology has experienced an increasing corporatization. Despite such developments, the Asilomar Conference established the fundamental institutional mechanisms for decisions about rDNA technologies in the United States. It also heavily influenced rDNA research guidelines developed by other countries. In this sense, the legacy of Asilomar is unequivocal.

Yet in another sense its legacy remains controversial. Participants at Asilomar wrestled with two basic questions: How should the protection of scientific freedom of inquiry be balanced with the protection of the public good? How should decisions about scientific research and its technological applications in society be made, especially in climates of uncertainty? Evaluating the legacy in light of these questions points toward three possible conclusions.

First, it has been argued that the conference represented the use of covert power by special interest groups (Oei 1997). According to this claim, scientists marginalized social and ethical questions in order to legitimize the new rDNA technology and persuade the public that control of this technology is best left to scientists (Wright 1994). The Asilomar Conference is portrayed by external critics as an elitist process with a narrow agenda designed to justify the self-government of science.

Second, there is an internal criticism voiced by some within the scientific community. According to this conclusion, the process of the Asilomar Conference and the controversies over regulation were too alarmist. The conference set the precedent for debates that focus on worst-case scenarios and largely ignore a growing scientific consensus about the safety of many rDNA applications. Increasing public opposition to many types of genetic engineering may prevent beneficial uses of these technologies in agriculture and medicine.

The third conclusion is that, despite its shortcomings, the Asilomar Conference represents an unprecedented exercise of the social conscience of science. For the first time, scientists voluntarily halted their own work until the potential hazards could be assessed (Mitcham 1987). This made it one of the first instances of what came to be known as the "precautionary principle." The Asilomar Conference was a novel attempt to balance scientific self-interest with self-restraint. It has left a legacy that transcends rDNA technology by taking an important step in the process of integrating scientific progress into its environmental and social contexts.

ADAM BRIGGLE

SEE ALSO *Genetic Research and Technology; Governance of Science.*

BIBLIOGRAPHY

Barinaga, Marcia. (2000). "Asilomar Revisited: Lessons for Today?" *Science* 287: 1584–1585. Covers a meeting that examined the Asilomar Process twenty-five years later and its implications for contemporary societal issues surrounding rDNA and other technologies.

Berg, Paul; David Baltimore; Herbert Boyer, et al. (1974). "Potential Biohazards of Recombinant DNA Molecules." *Science* 185: 303. One of the letters that sparked the temporary moratorium.

Davatelis, George N. (2000). "The Asilomar Process: Is It Valid?" *Scientist* 14(7): 51. Examines a meeting held on the twenty-fifth anniversary of the Asilomar Conference and interprets the usefulness of the Asilomar process for the social regulation of current and future scientific research.

Grobstein, Clifford. (1979). *A Double Image of the Double Helix: The Recombinant DNA Debate*. San Francisco: Freeman. Surveys the main political and social issues surrounding rDNA science in its infancy.

Löw, Reinhard. (1985). *Leben aus dem Labor: Gentechnologie und Verantwortung—Biologie und Moral* [Life from the laboratory: Gene technology and responsibility, biology and morality]. Munich: C. Bertelsmann. An introduction to the German discussion on recombinant DNA technology.

Mitcham, Carl. (1987). "Responsibility and Technology: The Expanding Relationship." In *Technology and Responsibility*, ed. Paul T. Durbin. Dordrecht, Netherlands: Reidel Publishing. A broad treatment of the ethics of science and technology, which covers rDNA and Asilomar on pp. 11–13.

Oei, Hong Lim. (1997). *Genes and Politics: The Recombinant DNA Debate*. Burke, VA: Chatelaine Press. Explores policy-making processes in the early years of rDNA debates. Outlines three alternative policy processes: technocratic, democratic, and legislative. Offers recommendations for emerging controversies.

Richards, John, ed. (1978). *Recombinant DNA: Science, Ethics, and Politics*. New York: Academic Press. Contains a good annotated bibliography of primary, first-hand accounts of the Asilomar Conference.

Rogers, Michael. (1977). *Biohazard*. New York: Knopf. Explores the moral and ethical confrontations over rDNA research. Includes three full sections on the Asilomar Conference with detailed quotations from many top participants.

Singer, Maxine, and Dieter Söll. (1973). "Guidelines for DNA Hybrid Molecules." *Science* 181: 1114. Outlines the potential benefits and hazards of rDNA technology and propose an NAS committee to study the issue and "recommend specific actions or guidelines should that seem appropriate."

Wade, Nicholas. (1975). "Genetics: Conference Sets Strict Controls to Replace Moratorium." *Science* 187: 931–935. An eyewitness account of the Asilomar Conference.

Wright, Susan. (1994). *Molecular Politics: Developing American and British Regulatory Policy for Genetic Engineering, 1972–1982*. Chicago: University of Chicago Press. Chronicles the response of various groups in the United States and Britain to the emerging genetic engineering technologies, traces the evolution of policies and discourses, and includes a section on the Asilomar Conference and its legacy.

INTERNET RESOURCE

Wright, Susan. (2001). "Legitimating Genetic Engineering," *Foundation for the Study of Independent Social Ideas*. Available from http://www.biotech-info.net/legitimating.html. Examines the problem definition and the consensus forged at Asilomar and the process of persuading the public. A somewhat cynical interpretation of the self-governance of science.

ASIMOV, ISAAC

Author of more than 500 books on a multitude of subjects, Isaac Asimov (1920–1992) was born in Petrovichi, Russia on January 2. He emigrated to the United States in 1923, sold his first science fiction story at the age of eighteen, and went on to become one of the most prolific and well-known popularizers of science for the public in the post-Sputnik era. He died in New York City on April 6.

Asimov was a child prodigy who graduated from high school at the age of fifteen and earned his bachelor's degree at nineteen. His studies were delayed by World War II, after which he received a Ph.D. in Chemistry from Columbia University in 1948. He became an assistant professor of biochemistry at Boston University's School of Medicine in 1951. Asimov left the School of Medicine in 1958, but retained the title of associate professor, and was promoted to professor of biochemistry in 1979.

Asimov sold his first science fiction story at the age of eighteen. By 1950 he had become a well-known science fiction writer and by the end of that decade, published fifteen novels.

Asimov's best known science fiction includes his *Foundation* series of stories, which dealt with the decline and rebirth of a future galactic empire, and his positronic robot stories, in which he formulated the Three Laws of Robotics:

1. A robot may not injure a human being, or through inaction, allow a human to come to harm.
2. A robot must obey the orders given it by human beings except where those orders would conflict with the First Law.
3. A robot must protect its own existence except where such protection would conflict with the First or Second Law.

The Three Laws were designed as safeguards so that robots could be treated sympathetically, rather than be objects of fear as they were in many earlier science fiction stories. Asimov coined the word *robotics*, which later came to be the standard term used for the technology of robots. Many robotics researchers acknowledged that Asimov influenced their interest in their field of study, and almost universally have tried to design robots with the equivalent of his three laws, which required them to be safe, effective, and durable.

Asimov exploited ambiguities in the Three Laws to explore a variety of ethical issues associated with technology. His robot characters often faced difficult decisions in predicaments where they had to choose between alternatives in order to do the least harm to humans. Asimov's later robot novels featured self-aware robots that considered the consequences of obeying the Three Laws, and then formulated a Zeroth Law that applied not merely to individuals, but to all of humankind, which stated that a robot may not harm humanity, or through inaction, allow humanity to come to harm. The Zeroth Law considered humanity as a single entity, where the needs of the many outweighed the needs of the individual.

During the 1950s, Asimov had two careers, as an author and a biochemist. His scientific career was rather unremarkable, and he published only a small number of papers in scientific journals. However in one of them, he pointed out that the breakdown of carbon-14 in human genes always resulted in a mutation. Nobel Prize winning chemist Linus Pauling (1901–1994) later acknowledged that Asimov's notion of the dangers of carbon-14 was in his mind when he successfully campaigned for an end to atmospheric testing of nuclear weapons.

Asimov's career took a major turn after the launch of Sputnik I in October 1957. At that time he had published twenty-three books, most of them science fiction, but he immediately turned to concentrating on writing about science for the general public. In addition he began lecturing on the significance of space exploration and other science matters.

Asimov prided himself on his ability to write clearly rather than poetically, in both his fiction and nonfiction. He felt it was important to educate the public about science, so that people could make informed decisions in a world both dependent upon and vulnerable to advances in technology, mindful of the fact that poor decisions could potentially have catastrophic consequences.

Asimov wrote often about the dangers of overpopulation, and the importance of changing attitudes so that population could be held in check by a decrease in the birth rate rather than an increase in the death rate. He routinely spoke out against the dangers of the nuclear arms race, and believed that the exploration of space provided an opportunity for nations to put aside their differences and cooperate to achieve a common goal. Asimov argued that the most serious problems threatening humanity—such as overpopulation, nuclear war, the destruction of the environment, and shortages of resources—do not recognize international boundaries. Consequently he called for the establishment of a unified world government as the most sensible way to solve such global problems.

Asimov was a crusader against irrationality and superstition, and he believed strongly that the problems caused by science and technology could only be solved by further advances in science and technology.

EDWARD J. SEILER

SEE ALSO *Artificial Intelligence; Robots and Robotics; Science Fiction; Science, Technology, and Literature.*

Isaac Asimov, 1920–1992. As one of the world's leading and most prolific science writers, explaining everything from nuclear fusion to the theory of numbers, Asimov illuminated for many the mysteries of science and technology. *(AP/Wide World Photos.)*

BIBLIOGRAPHY

Asimov, Isaac. (1950). *I, Robot*. New York: Gnome Press. Asimov's earliest positronic robot stories, in which he introduced the three laws of robotics.

Asimov, Isaac. (1951, 1952, 1953). *Foundation; Foundation and Empire; Second Foundation*. New York: Gnome Press. The "foundation trilogy" of science fiction novels about the fall of a galactic civilization, and the efforts of two factions to minimize the duration of the dark age that follows.

Asimov, Isaac. (1962). *Fact and Fancy*. New York: Doubleday. The first collection of Asimov's science articles from the *Magazine of Fantasy and Science Fiction*.

Asimov, Isaac. (1979). *In Memory Yet Green*. New York: Doubleday. The first volume of Asimov's autobiography, detailing his life and career from 1920 to 1954.

Asimov, Isaac. (1980). *In Joy Still Felt*. New York: Doubleday. The second volume of Asimov's autobiography, covering the years 1954 to 1978.

Asimov, Isaac. (1985). *Robots and Empire*. New York: Doubleday. A science fiction novel in which Asimov introduced the Zeroth Law of robotics, and unified the fictional universes of his positronic robot and foundation stories.

Asimov, Isaac. (1991). *The Secret of the Universe*. New York: Doubleday. The twenty-third collection of Asimov's science articles from the *Magazine of Fantasy and Science Fiction*.

Asimov, Isaac. (1994). *I. Asimov: A Memoir*. New York: Doubleday. Asimov's third autobiographical volume, in which he uses recollections from throughout his life to present his thoughts and philosophy.

Gunn, James E. (1996). *Isaac Asimov: The Foundations of Science Fiction*. Lanham, MD: Scarecrow Press. A critical analysis of Asimov's science fiction. The author is a retired professor of English at the University of Kansas and has written twelve science fiction novels and five short story collections.

Olander, Joseph D., and Martin Harry Greenberg, eds. (1977). *Isaac Asimov*. New York: Taplinger. Part of the book series *Writers of the 21st Century*. Nine authors provide their literary perspectives of Asimov's science fiction.

ASSISTED REPRODUCTION TECHNOLOGY

On July 25, 1978, the work of Robert Edwards and Patrick Steptoe led to the birth of the first "test-tube baby," Louise Brown, in England. Since then, thousands of babies throughout the world have been born with the help of assisted reproduction technologies (ARTs). ARTs such as artificial insemination have been in use since the nineteenth century and, as with the technology that helped bring Louise Brown into existence, they still raise ethical concerns. Although ARTs are a common therapy to treat infertility, such treatments continue to provoke questions about safety and efficacy. Many of the ethical issues that appeared with the advent of these technologies continue to be relevant in the early twenty-first century.

Technologies

ARTs refer to a group of procedures, often used in combination, that are designed to establish a viable pregnancy for individuals diagnosed with infertility. The degree of sophistication of these techniques is highly variable. Artificial insemination (AI) requires the least technological complexity and is the oldest of such technologies. It consists of the mechanical introduction of sperm, from the husband or a donor, into a woman's reproductive tract. AI with the husband's sperm is indicated in cases where there are anatomical abnormalities of the penis, psychological or organic conditions that prevent normal erection and ejaculation, or female or male psychosexual problems that prevent normal intercourse. AI by donor is employed in cases of low sperm count or abnormal sperm function. It is also used by single women and by lesbian couples.

In vitro fertilization (IVF) is the quintessential type of ART. Approximately 1 million babies have been born worldwide through this procedure. In its most basic form (that is, the woman undergoing IVF provides her own eggs, and her husband or partner supplies the sperm), IVF consists of several stages. First doctors stimulate the woman's ovaries with different hormones to produce multiple oocytes. Next they remove the eggs from her ovaries through procedures such as laparoscopy or ultrasound-guided oocyte retrieval. After preparation of semen, specialists fertilize the mature eggs in a laboratory dish with the partner's sperm. If one or more normal looking embryos result, specialists place them (normally between three and five) in the woman's womb to enable implantation and possible pregnancy. The sperm and the eggs can also come from donors. Also the embryos might be cryopreserved for later use and transferred into the woman who supplied the eggs or into a surrogate. Similarly examination of sperm, eggs, and embryos for chromosomal and genetic abnormalities can be performed through preimplantation diagnosis. Although IVF was originally developed to use in cases of infertility when the woman's fallopian tubes were damaged, it soon became common treatment for other reproductive problems such as inability to produce eggs, poor sperm quality, endometriosis, or unexplained infertility.

Several modifications and variations from the basic IVF procedure exist. In the gamete intrafallopian transfer (GIFT), the specialists transfer both eggs and sperm to the woman's tubes. Thus conception occurs inside the woman's body. With the zygote intrafallopian transfer (ZIFT), fertilization, as with IVF, occurs in a petri dish. The difference here is that the fertilized egg is transferred to the fallopian tube eighteen hours after fertilization occurs. The newest of these procedures is intracytoplasmatic sperm injection (ICSI), which consists of the direct injection of one sperm into a harvested egg. Since the birth of Dolly the sheep in 1997, somatic cell nuclear transfer (SCNT) has been cited as another possible ART. In SCNT or reproductive cloning, the nucleus of a somatic cell is transferred into an egg cell from which the nucleus has been removed. Most countries have implemented bans or moratoriums on research directed to cloning human beings.

Although there are some ethical questions that are specific to particular reproductive technologies (for instance, manipulation of human embryos), many concerns are common to all. This entry will focus on ethical issues shared by all ARTs.

Procreation, Families, and Children's Well Being

Many of those who support the use and development of ARTs argue that people have a fundamental right to procreate. Thus the state should not interfere with the rights of infertile married couples to have offspring, unless compelling evidence of tangible harms is presented. Proponents claim that critics of ARTs have not offered such evidence (Robertson 1994). An emphasis on individual rights, however, might neglect the fact that reproduction is an act that clearly involves the community by bringing new persons into the world and by using societal resources.

From some religious perspectives, ARTs sever the natural link between sexual intercourse and procreation and, therefore, are impermissible. Many Christian theologians call the use of ARTs immoral because these technologies allow for the separation of procreation and sexual love between married partners (Ramsey 1970). Others contend that, within limits, ARTs can help infertile couples to reproduce and thus should not be completely rejected.

Some also argue that the use of ARTs challenges the traditional conception of the family by separating genetic, gestational, and rearing components of parenthood. Such criticisms assume that by *family* one can only mean a nuclear family composed of a male, a female, and their genetic offspring. They also ignore historical and anthropological evidence according to which humans have successfully adopted many kinds of family arrangements. Moreover such criticisms often fail to offer any compelling normative arguments that show that societies built of nuclear families, as generally understood, are better off than societies with other kinds of family arrangements (Coontz 1992).

The physical well being of children born through these technologies is another concern common to all the ARTs. Although initial assessments indicated that children born as a result of the use of ARTs did not suffer from more problems than did children born through conventional intercourse, such assessments are being questioned. Studies indicate that such children, especially those born through IVF and related techniques, are at increased risk of being premature, having low or extreme low birth weight, and suffering congenital malformations. It is still unclear, however, whether these risks are linked to the technologies themselves or to parental factors (Ludwig and Diedrich 2002).

Women's Well Being

Feminist criticisms have tended to focus on the effect of these technologies on the lives of women. They emphasize the risks that ARTs pose to women's health as well as their impacts on women's status in society. Some feminist groups argue that the new procedures are not designed to give women more choices but are based on the capitalist and patriarchal ideology of abusing, exploiting, and failing to respect women. They call attention to the dismemberment of women's bodies, the medicalization of the reproductive experience that puts pregnancy and birth in the hands of the medical profession, the commercialization of motherhood, and the eugenic and racist biases that the new technologies promote (Arditti et al. 1984.).

Other feminist authors have been less eager to completely reject ARTs. They maintain that assisted-conception techniques could be used to the advantage of women. Although they recognize that no technology is neutral, they reject the social and technological determinism that permeated initial feminist objections. These feminist critics acknowledge that the social policies surrounding ARTs harmed women's interests. However they oppose the image of women as brainwashed individuals, immersed in a world of constructed needs and unable to decide by themselves. They urge widespread public discussion and eventual political and legislative action to improve women's reproductive autonomy instead of a complete rejection of the new procedures (Callahan 1995).

Conception of Infertility

Another criticism common to all ARTs is that they reinforce a particular understanding of infertility as an individual medical failure to have children who are genetically related. Whether one views infertility mainly as a medical condition or also as a social one has important implications. Defining infertility as an individual medical difficulty suggests that a technological treatment is the appropriate response. Thus one might ignore that the causes of reproductive difficulties and the reasons that make infertility a serious concern are, in part, socially rooted. Analyzing infertility also as a socially generated problem indicates that social, ethical, and political solutions to reproductive difficulties should be considered. In this case there may be an emphasis on solutions such as preventive measures or social changes that might be more effective and less costly. This is especially noteworthy because sexual, contraceptive, and medical practices, occupational health hazards, environmental pollution, inadequate nutrition, and poor health are some of the main causes of infertility. Attention to these issues would require consideration of preventive measures rather than only curative treatments as solutions to the infertility problem.

Similarly the use of ARTs emphasizes the importance of genetic relationships in parenthood. One of the main goals of these technologies is to guarantee that at least one of the members of the couple would have genetically related offspring. Although a genetic link to one's offspring may be important, an emphasis on such a relationship might prevent social policies directed to facilitate and encourage adoption or other forms of parenting.

INMACULADA DE MELO-MARTÍN

SEE ALSO *Bioethics; Feminist Ethics; In Vitro Fertilization and Genetic Screening; Medical Ethics; Rights and Reproduction.*

BIBLIOGRAPHY

Arditti, Rita; Renate Duelli-Klein; and Shelley Minden, eds. (1984). *Test-Tube Woman: What Future for Motherhood?* London: Pandora Press. A collection of articles from a feminist perspective that address the ethical and medical implications of reproductive technologies for women's lives. Most of the authors are skeptical about the ability of these technologies to improve the lives of women.

Callahan, Joan. (1995). *Reproduction, Ethics, and the Law: Feminist Perspectives.* Bloomington: Indiana University Press. A collection of essays that consider moral and legal quandaries related to human reproduction. Without completely rejecting all ARTs, the authors call attention to new complexities brought about by these new technologies, such as the legal definition of parenthood, the status of frozen embryos, and the use of fetal tissue.

Coontz, Stephanie. (1992). *The Way We Never Were: American Families and the Nostalgia Trap.* New York: Basic Books. An analysis of some of the myths related to the so-called traditional two-parent nuclear family.

Ludwig, Michael, and Klaus Diedrich. (2002). "Follow-Up of Children Born after Assisted Reproductive Technologies." *Reproductive Biomedicine Online* 5(3): 317–322. Discusses the possibility that risks to the health of children born through ARTs might be related to the techniques themselves and not only to infertility related problems.

Ramsey, Paul. (1970). *Fabricated Man: The Ethics of Genetic Control.* New Haven, CT: Yale University Press. An evaluation of the implications of ARTs from a Christian perspective. The author rejects the use of these technologies for severing the sacred tie between procreation and married love.

Robertson, John. (1994). *Children of Choice: Freedom and the New Reproductive Technologies.* Princeton, NJ: Princeton University Press. A defense that the principle of procreative liberty should be used to resolve the ethical and legal controversies surrounding ARTs.

ASSOCIATION FOR COMPUTING MACHINERY

Founded in 1947, the Association for Computing Machinery (ACM) is a nonprofit scientific and educational organization devoted to advancing knowledge and practice in computing and information technology. The ACM comprises professionals, students, practitioners, academics, and researchers—a total of 75,000 members around the world. The ACM sponsors more than one hundred annual conferences and publishes magazines and journals in both print and electronic form. It provides expertise on social concerns and public policies related to computing and information technology, including ethical issues such as privacy, security, intellectual property, and equitable access to computing resources.

Major Activities

Within the ACM are several special interest groups. The Special Interest Group on Computers and Society (SIGCAS) sponsors activities in ethics. SIGCAS manages the quarterly online magazine *Computers and Society*, which publishes articles, book reviews, educational materials, and news reports related to the ethical and social impacts of computers. SIGCAS organizes occasional conferences and presents the annual Making a Difference Award to an individual who has contributed to understanding the ethical and social impacts of computers. The award has honored Deborah G. Johnson and James H. Moor for scholarly work on the philosophical foundations of computer ethics, and Ben Shneiderman for championing universal access to computing resources.

For many years, the ACM has promoted education in social and ethical issues in computing. The Special Interest Group on Computer Science Education (SIGCSE) usually schedules sessions on teaching computer ethics at the annual Technical Symposium on Computer Science Education. Two of the ACM's series of self-assessments focused on ethics in computing and information science (Weiss 1982, Weiss 1990). In 2001 a joint task force of the ACM and the Computer Society of the Institute of Electrical and Electronics Engineers (IEEE) produced recommendations for undergraduate curricula in computer science that require instruction in ethics in the context of professional practice. Unlike accreditation standards, these curricular recommendations are not mandatory, but they have influenced the development of undergraduate curricula.

The ACM Office of Public Policy and the U.S. Public Policy Committee of the ACM assist policymakers and the public in understanding social issues in information technology, with particular attention to legislation and regulations. For example, since publishing the report *Codes, Keys, and Conflicts: Issues in U.S.*

Crypto Policy in 1994, the ACM has advocated effectively against restrictions on the use of strong encryption. Although these restrictions were intended to thwart criminals and terrorists, they might instead reduce information security and harm electronic commerce. Recognizing ACM's concerns, the U.S. federal government relaxed export controls on encryption products. Since 1999, the ACM has criticized deficiencies in the Uniform Computer Information Transactions Act (UCITA), a proposed uniform state law that creates new rules for computerized transactions. The ACM believes that UCITA would threaten public safety and product quality, because the act would prevent software users from publicizing information about insecure products, and it would allow vendors to disable software remotely. Initially enacted by two states, UCITA has not been adopted by other states because of ACM's efforts.

Codes of Ethics

Like many professional organizations, the ACM has developed its own codes of ethics and professional conduct. In 1966 the ACM adopted its first codes, Guidelines for Professional Conduct in Information Processing (Parker 1968). These guidelines were expanded in 1972 into the ACM Code of Professional Conduct. In 1992 the ACM adopted the current Code of Ethics and Professional Conduct (Anderson et al. 1993).

The 1992 ACM code strives to educate computing professionals about professional responsibilities, rather than to regulate ACM members. In contrast with other professional codes of ethics, the ACM code has three notable features. First, each statement in the ACM code is supplemented by interpretive guidelines. For example, the guideline for the statement on confidentiality indicates that other ethical imperatives may take precedence:

> **1.8 Honor confidentiality** The principle of honesty extends to issues of confidentiality of information whenever one has made an explicit promise to honor confidentiality or, implicitly, when private information not directly related to the performance of one's duties becomes available. The ethical concern is to respect all obligations of confidentiality to employers, clients, and users unless discharged from such obligations by requirements of the law or other principles of this Code.

Second, a large section of the ACM code applies specifically to "organizational leaders"—typically technical managers. According to the code, organizational leaders must encourage subordinates to accept professional responsibilities, provide opportunities for subordinates to pursue continuing education, support policies that mandate appropriate uses of computing resources, and ensure that computing systems are designed to enhance the quality of life and protect the dignity of users. Third, the ACM code obligates members to "improve public understanding of computing and its consequences." It is unclear, however, whether this obligation applies to each member individually or to the computing community collectively.

Beginning in 1994 the ACM collaborated with the Computer Society of the IEEE to create the Software Engineering Code of Ethics and Professional Practices, drafted in 1997 and finalized in 1999 (Gotterbarn, Miller, and Rogerson 1999). Like the 1992 ACM code, the Software Engineering Code includes a section on the obligations of technical managers. Although the ACM participated in the development of the Software Engineering Code, the ACM opposes the licensing of software engineers (White and Simons 2002). (Both the 1992 ACM code and the Software Engineering Code appear in the appendix of this encyclopedia.)

Throughout its history, the ACM has dedicated attention to ethical issues in computing and information technology, both the impacts of computers on society and the responsibilities of individuals as professionals. The ACM will continue to emphasize these issues through conferences and publications, codes of professional conduct, educational activities, and public advocacy, particularly in the United States.

MICHAEL C. LOUI

SEE ALSO *Institution of Electrical and Electronics Engineers; Professional Engineering Organizations.*

BIBLIOGRAPHY

Anderson, Ronald E.; Deborah G. Johnson; Donald Gotterbarn; and Judith Perrolle. (1993). "Using the New ACM Code of Ethics in Decision Making." *Communications of the* ACM 36(2): 98–107. The current (1992) ACM code of ethics, with short fictional cases that illustrate uses of the code.

Gotterbarn, Don; Keith Miller; and Simon Rogerson. (1999). "Software Engineering Code of Ethics Is Approved." *Communications of the* ACM 42(10): 102–107. The full text of the software engineering code of ethics, developed by the ACN and IEEE Computer Society.

Landau, Susan; Stephen Kent; Clint Brooks, et al. (1994). *Codes, Keys, and Conflicts: Issues in U.S. Crypto Policy.* New York: Association for Computing Machinery. Raises questions and concerns about standards, laws, and policies for cryptography in the United States.

Parker, Donn B. (1968). "Rules of Ethics in Information Processing." *Communications of the ACM* 11(3): 198–201. Details the background and explains the motivation for the first ACM Code of Ethics.

Weiss, Eric A., ed. (1982). "Self-Assessment Procedure IX." *Communications of the ACM* 23(3): 181–195.

Weiss, Eric A., ed. (1990). "Self-Assessment Procedure XXII." *Communications of the ACM* 33(11): 110–132. Two self-tests on computer ethics; readers analyze several scenarios and compare their opinions with experts' opinions.

White, John, and Barbara Simons. (2002). "ACM's Position on the Licensing of Software Engineers." *Communications of the ACM* 45(11): 91. Justifies ACM's opposition to the licensing of software engineers.

INTERNET RESOURCES

Association for Computing Machinery. 2004. Available from http://www.acm.org.

Association for Computing Machinery. 2004. Special Interest Group on Computers and Society. *Computers and Society* online magazine. Available from http://www.computersandsociety.org.

Computer Society of the Institute for Electrical and Electronic Engineers and the Association for Computing Machinery. "Computing Curricula." Available from http://www.computer.org/education/cc2001/. Recommendations for undergraduate curricula in computer science, computer engineering, software engineering, and information systems.

ASTRONOMY

Astronomy, from the Greek *astron*, star, plus *nomos*, law—thus the laws or regular patterns of the stars—is now defined as the science of objects beyond the Earth's atmosphere, including their physical and chemical properties. This science of what is beyond the Earth paradoxically served as the model for the early modern effort to create a science of terrestrial phenomena. Because of their apparently more simple and necessary order, astral phenomena were the first to be subject to explanations in the form of "laws," the methods of which were then extended in modern physics to explain the dynamics of falling bodies at or near the Earth. Yet just as modern physics emerged to give human beings greater powers over material affairs than ever before, and thus pose a challenge to ethics, so subsequent developments in astronomy deprived humans of an order that could be perceived as a transcendent and normative guide for human conduct. Immanuel Kant (1724–1804) could still wonder at the correspondence between the "starry heavens above and the moral law within" (*Critique of Practical Reason*, p. 288), but the achievements of modern astronomy have left the moral law within to fend for itself.

Pre-Modern Astronomy

Astronomy has been called the world's second oldest profession. Notations found on artifacts scattered over Africa, Asia, and Europe dating from 30,000 B.C.E. appear to be rudimentary calendars based on the phases of the moon (Hartmann and Impey 1994). The transition from hunter-gatherers to life in stable villages, occurring around 10,000 B.C.E. with the rise of agriculture, required a refined estimation of the timing of seasonal changes. The sky, although no doubt deeply mysterious to these ancient cultures, was also reassuringly deterministic. By 4000 B.C.E., for instance, Egyptian astronomers knew that the first appearance of the brightest star in the dawn sky, Sirius, marked the beginning of the Nile's annual flooding. Many, probably most, cultures timed their agricultural activities based on similar annual celestial events.

The stars of course were also used for navigation. The Minoans of the island of Crete employed the stars to navigate the Mediterranean and to forge relationships with the Greeks as long ago as 2600 B.C.E. In developing this technology, they grouped the stars into pictures that gave rise to some of the constellations that we still know today (Hartmann and Impey 1994). The navigational prowess of the Polynesians is legend. The courage and faith these seafarers had in the heavens' ability to guide their way is astonishing. Crossing vast expanses of the Pacific, Polynesians discovered that if they sailed north until the Southern Cross dropped to a hand's length above the horizon, they would be at the latitude of Hawaii. To return, they would point their outriggers south until two stars, Sirius and Pollux, set together.

The megalithic monument Stonehenge on the Salisbury Plain in Great Britain had a utilitarian as well as spiritual design. On the longest day of summer, at solstice, the sun rose over a huge, notched boulder, the "Heel Stone," as seen from the center of concentric rings of massive boulders. Some weighed thirty to fifty tons (Hawkins and White 1965). The accompanying midsummer ritual 4000 years ago would have been an annual part of the cultural weaving of astronomy, beliefs, and values for the participants. Enormously demanding achievements such as the construction of Stonehenge and of the Egyptian pyramids are testament to the power the heavens exerted on the societies that built them.

Possibly the most extraordinary early example of institutional astronomy was that of the Mayans. The priest-astronomers that observed the heavens and performed the calculations to produce their calendars were publicly supported for at least 200 years around 400 C.E. The Mayan calendar did not only chart the seasons for agriculture. It also predicted eclipses, experienced by the Mayans as traumatic and darkly mysterious. Mayan astronomers computed the complex motions of Venus, believing it to be one god in the evening, and another when it reappeared in the morning. Venus's quasi-periodic disappearance and reemergence on the other side of the world was seen as a journey and transformation in the underworld (Aveni and Hotaling 1994). It appears that in all early cultures, astronomy and religion were deeply interconnected. Astronomy, by giving an accurate description of the motions of heavenly bodies, was at the same time a very powerful tool for sustaining civilization and exploring the world.

However it goes about it, religion seeks to provide guidance for living in harmony with the Earth, with other people, and with the universe. But peace, it could be argued, is only possible for human beings if they have in some way accepted what their lives mean. Religion addressed the human question of meaning, by defining our relationship with the cosmos. So astronomical questions, such as what brought forth the universe, how old it is, and what our place in it is, were religious questions. It has been suggested that the starkly hierarchical medieval (Aristotelian) cosmology, with the universe consisting of ten concentric spheres around the Earth (the outermost being heaven), was reflected in the rigidly hierarchical society that oppressed the vast majority of people (Abrams and Primack 2001).

The astronomical observations of Galileo Galilei (1564–1642), using the new technology of the telescope, began the fracture of science and religion that is today a deep chasm. As is well known, Galileo kept his head because he recanted his conclusions that the sun was at the center of the solar system and that the celestial bodies were not flawless. With improving technologies and the bold modern project begun by René Descartes (1596–1650), Francis Bacon (1561–1626), and John Locke (1632–1704), however, science and religion diverged under the auspices of an uneasy truce. As the quest for truth in the universe became a scientific endeavor, it was no longer part of the institution that spoke directly to meaning in human lives, to guidance for living in harmony, and for rules that guide human behavior.

Modern Astronomy and the Rise of Scientific Cosmology

Modern astronomy can be described in terms of its institutional structures, its intellectual debates, and its scientific discoveries.

NATIONAL, PRIVATE, AND UNIVERSITY OBSERVATORIES. Astronomy may have grown from a fundamental desire to understand the universe, but the use of heavenly motions as a powerful *technology* for navigation grew with it. Systematic observations of the heavens for centuries allowed us to chart the limits of our world, and to navigate confidently within it.

By the end of the nineteenth century, large national observatories existed in the United Kingdom, France, the United States, and Russia. Although originally designed to survey the heavens for applications in geodetics and navigation, these institutions also began to branch out and address more fundamental questions (Struve and Zebergs 1962). Especially as instrumentation improved, astronomers were increasingly making observations in attempts to understand the structure, history, and origin of the universe. Larger and larger telescopes would enable astronomers to see further into the universe and with ever greater sharpness. The excitement of this quest was felt keenly by a number of American philanthropists, and the late nineteenth century saw the rise of large, privately funded observatories such as the Lick (1888), the Lowell (1894), and the Yerkes (1897). Following these, construction of the last of the giant, privately funded observatories was completed with the McDonald Observatory in 1939 and the Palomar Observatory in 1947. The flagship of Palomar is the 200"-diameter Hale telescope, which reigned supreme as the largest and most capable telescope in the world until the launch of the Hubble Space Telescope into Earth orbit in 1990.

Hubble was born of the dreams of astronomer Lyman Spitzer (1914–1997), who, in the heady days of the postwar technology boom, first advocated a telescope in space to explore the universe with unprecedented clarity. Above the veil of obscuring atmosphere and luminous clamor of the Earth, a moderate telescope in space would see the universe 100 times clearer than the behemoths on Earth. This meant that it could see 100 times further away and 100 times further back in time. This it has done, and the images of the universe that it has returned have astonished us and enriched our lives.

Light is the only form of electromagnetic energy that is directly perceived by human beings. Electromag-

View of space from the Hubble Space Telescope. *(Courtesy NASA STScl.)*

netic waves are produced by a vast array of physical phenomena in the universe, including stars, planets, galaxies, supernovae remnants, black holes, and almost everything in between. Many of these emissions have wavelengths that are much longer than those of light; these are radio waves. Because they are absorbed by dust and gas less readily than is light, radio waves traveling through space allow a glimpse of parts of the Milky Way that cannot be seen by optical telescopes. In addition, radio waves are produced by different processes than those that create light, giving scientists insights into the physical processes and compositions of many objects in space.

Primitive radio receivers were first pointed at the sky in the early 1930s. It became clear soon thereafter that radio waves can be detected from all parts of the sky, but most especially from the center of the Milky Way. The rapid advances in electronics due to the technological efforts in waging World War II paved the way for vast improvements in radio telescope sensitivity. Serious construction of large astronomical radio telescopes began in 1947. Some are steerable, such as the 250-foot wire-mesh dish at Jodrell Bank in Great Britain. The largest is Arecibo, the immovable 1,000-foot dish carved into a limestone sinkhole in Puerto Rico. Today, enormous arrays of radio dishes are icons of modern astronomy, probing the universe's mysteries and listening for signs of alien minds.

THE ISLAND UNIVERSE DEBATE. On a clear night away from city lights, a ghostly swath cuts through the sky. It is thickest in the constellations of Sagittarius and Scorpio, and thins as its path is traced northeast through Cassiopeia or southwest through the summer constellations of Cygnus and Aquila. One of the great conceptual leaps of humanity was the realization that this apparition was our view of a great island universe, a galaxy, from the inside. The peculiar smudgy swirls seen in early telescopes, such as Galileo's, were vast communities of stars, comparable to ours but unimaginably far away. The close ones, such as Andromeda, can be seen to be in the shape of a pinwheel with a bright central bulge. As we look to Sagittarius, we look into the core of our galaxy from the inside of the disk. On the other side of the sky where the Milky Way is more diffuse, we can see dark lanes of dust obscuring stars, and the outline of spiral arms. Our sun is one dot in the multitudes that blend together with such promiscuity that they form the milk of the Milky Way.

By the end of the nineteenth century, astronomers knew that the Milky Way was a vast field of stars in which the sun and solar system were embedded. Systematic star counts led to estimates of the size and shape of our galaxy, but also to the erroneous conclusion that the sun was at the center of it. In spite of the Copernican revolution, subtle assumptions on the centrality and primacy of humans in the universe remained, skewing scientific interpretations of the observational data.

Our view of the Milky Way galaxy from within was sharpened considerably by the observations of Harlow Shapley (1885–1972). Shapley noticed that globular clusters—beautiful, tightly packed spherical aggregates of stars—tended to form a vast spherical halo around the nucleus of the Milky Way. His observations successfully set the stage for the twentieth-century view: that the sun exists in an enormous, flattened disk of stars, about two-thirds of the way from the center to edge. This final dethroning of the role of humans in the cosmos played out during the 1910s and 1920s and was one of the great classic scientific debates of the century. The new picture did little at first to illuminate what the universe was, or its extent. Was our disk, 100,000 light years wide and 10,000 light years thick, with a central bulge and 100 billion stars, *the* universe? What was outside of it, and how did it come to be? These questions could only be answered with improvements in telescope and photographic technology, which followed rapidly.

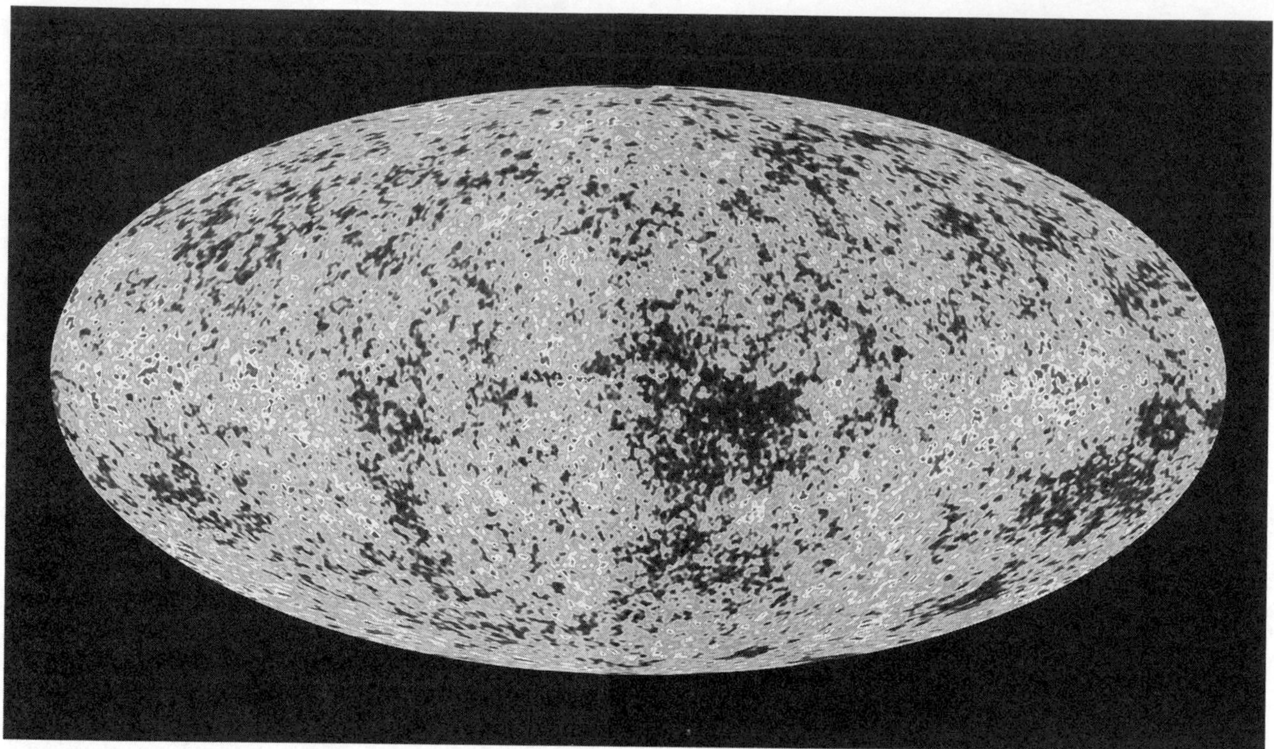

The afterglow of the Big Bang. This image is a map of the very edge of the Universe—looking so far back in time and space that all we see is the heat from the creation cataclysm. *(NASA/WMAP Science Team.)*

Kant proposed, in the eighteenth century, that the Milky Way we are inside of was a disk-shaped spiral, similar to the far-away spiral nebulae seen in telescopes at the time. He called these spirals "island universes." Kant's famous intuition turned out to be largely correct, although the scientific path to this conclusion did not end until the middle 1920s. During that decade, the shape of our galaxy's spiral arms came into focus, and the correspondence to the shapes of the far-off spiral nebulae became scientifically accepted. Until then, it was generally thought that the Milky Way was all that there was, and the large variety of spiral nebulae were smaller aggregates of stars within or just outside of it. As telescopic and photographic technology progressed in the twentieth century, and ever more detailed images of the deep heavens were acquired, this view began to change.

It was Edwin Powell Hubble (1889–1953) who eventually solved the mystery of the celestial spirals. It had long been known that a special class of variable stars, known as Cepheid variables, exhibited a well-determined relationship between periodicity and intrinsic brightness. Distance determinations to celestial objects were bootstrapped to ever more distant objects by noting the parallax shifts of nearby stars (including Cepheids) due to the earth's orbit around the sun. This technique was used to calibrate Cepheid variables at far more distant locales. Using the 100" telescope at Mt. Wilson observatory above Pasadena, then the largest instrument in existence, Hubble was able to resolve individual Cepheid variables in the Andromeda galaxy. Extrapolating from the period-luminosity relation for these variables in our own galaxy, in 1923 Hubble conclusively showed that the Andromeda galaxy was far, far away, about ten times further than the diameter of our own galaxy. So spiral galaxies are indeed island universes, vast collections of stars very much like our Milky Way, many with 100 billion stars or more. The press for larger, more powerful instruments in the early part of the twentieth century was on, driven almost entirely by a thirst for understanding the depth and breadth of all existence. This thirst was very much felt by society in general, and was part of the great scientific excitement of the time, which included the development of quantum mechanics and the deeper understanding of space and time worked out by Albert Einstein (1879–1955).

We now know that the Andromeda galaxy is only one of more than 100 billion such whirlpools of stars, making the observable universe an inconceivably large place, containing 100 billion times 100 billion stars, and perhaps almost as many solar systems. On a cloudless night in autumn, the Andromeda galaxy is clearly visible to the unaided eye. It is the farthest thing we humans can perceive directly. Light reaching us today left the galaxy 2.2 million years ago, traveling 10,000,000,000,000,000,000 miles before leaving its impression on our retinas and minds.

In his famous book *The Realm of the Nebula*, Hubble classified the vast diversity of extragalactic forms into a more-or-less coherent taxonomy (1926). The realization that spiral nebulae and their brethren, giant elliptical galaxies, were island universes, coequal with our own vast Milky Way, paved the way for one of the most extraordinary scientific discoveries of all time and gave birth to modern cosmology. In 1929, Hubble announced his discovery that the recessional velocities of galaxies were proportional to how far away they were. The furthest galaxies were receding the fastest, as measured by the Doppler shifts of their emitted light. The constant of proportionality became known as the Hubble constant. The implications of this relationship are profound. The simplest way to explain it is that at some time in the very distant past, all the galaxies were packed together. If we reverse the movie of the universe, all the galaxies speed in toward each other until—what? Georges-Henri Lemaitre (1894–1966) hypothesized that the movie takes us back to the primeval egg, a cosmology that poetically phrased the juxtaposition of myth and science. But how far one can extend the movie and continue to rely on the laws of physics as we know them is at the heart of modern cosmology. At the beginning of time and space, the galaxies or their precursors were propelled somehow from the egg. In this picture, the reciprocal of the Hubble constant is the age of the universe, and its extent is approximately the distance that light travels in this time. This theory became known as the Big Bang. Science has thus looked directly at *the* question: What is the origin of everything? We cannot go back: The countless and varied myths, societies' identification with the infinite, have been supplanted by the power of scientific truth.

THE MORALITY OF SUPERNOVAE. One of the great natural wonders of the universe is the supernova. In schoolchildren, descriptions of the great power of these exploding stars excite a keen intellectual wonder in the natural world. Stars are a great balance between gravity trying to squeeze them small, and nuclear-generated heat trying to pull them apart. The story of the supernova is awesome and kinetic, its wonders easily readable in the faces of children who listen to it. A single, supergiant star approaches the end of its life. As its final generation of fuel is exhausted, the giant radiation engine that supports the star shuts down. Massive collapse ensues, on a scale that is well beyond human comprehension. The implosion rebounds ferociously, spewing the alchemy of the old star into the cosmos. The transmuted elements are made nowhere else but here, the hellish belly of the most powerful beast of the universe. And these elements disperse through the cosmos—and become us.

Supernovae are so rare that one occurs in our galaxy, with 100 billion stars, only about once a century. For about a month, though, the maelstrom from that single, dying star is brighter than all of its 100 billion siblings combined. Overall, in the 100 billion galaxies that we can see from our vantage point, that means we have seen and measured and analyzed many hundreds of supernovae.

It isn't hard to see how a driving scientific curiosity could be drawn to trying to understand this thing. Indeed, supercomputer models of unimaginable explosions are quite refined, and scientific models of how stars explode have been highly successful. What is curious is that they are aided by a rather keen interest in an entirely different field: the nature and yield of human-made nuclear explosions. As declassification of the fundamental nuclear science of the 1940s and 1950s proceeded during the last decades of the twentieth century, there was a highly successful synergy between the study of the most fantastic, wondrous, violent explosions in our universe and the efficiency and effectiveness of nuclear weapons.

Conclusion

For 200,000 years, human beings have had an intense, powerful relationship with the skies above them. We all evolved within societies for which the sky was a pervasive source of magic, awe, religion, and art. For *every* human being, for 99.9 percent of the history of humankind, there was a personal relationship with the sky. For 10,000 generations, the sky had personal meaning to people, figuring in much of what they did and how they behaved, how they moralized, and how they loved. We were born with humanity's relationship to the sky in our genes. The scientific study of astronomy doesn't change this, although it has changed the feelings we have about our place in the universe. As humanity explores and understands the natural world, the ever-growing power it wields over nature demands clarity and wisdom. Shortly before his death in 1695, the eminent Danish astronomer Christiaan Huygens (1625–1695) wrote in *Kosmotheoros*, for his time and ours:

This shows us how vast those Orbs must be, and how inconsiderable the Earth, the Theater upon which all our mighty Designs, all our Navigations, and all our Wars are transacted, is when compared to them. A very fit Consideration, and matter of Reflection, for those Kings and Princes who sacrifice the Lives of so many People, only to flatter their Ambition in being Masters of some pitiful corner of this small Spot.

MARK A. BULLOCK

SEE ALSO *Cosmology.*

BIBLIOGRAPHY

Abrams, N. E., and J.R. Primack. (2001). "Cosmology and 21st-Century Culture." *Science* 293: 1769–1770. One of the few works that investigates the impact cosmology has had on culture.

Aveni, A.F., and L.D. Hotaling. (1994). "Monumental inscriptions and the observational basis of Maya planetary astronomy." *Archaeoastronomy* 19: S21–S54. The definitive popular work on Mayan astronomy by a leading expert in the field of archeoastronomy.

Hartmann, William K., and Chris Impey. (1994). *Astronomy: The Cosmic Journey*, 5th edition. Belmont, CA: Wadsworth. A standard undergraduate astronomy text book.

Hawkins, Gerald S., with John B. White. (1965). *Stonehenge Decoded.* Garden City, NY: Doubleday. A fascinating popular book on the mysteries of Stonehenge.

Hubble, Edwin P. (1936). *The Realm of the Nebula.* New Haven, CT: Yale University Press. The classic work on the taxonomy of galaxies.

Huygens, Christiaan. (1698). *Kosmotheoros.* London: Timothy Childe. A philosophical treatise addressed to Huygen's brother Constantijn on the construction of the universe and the habitability of planets.

Kant, Immanuel. (1956). *Critique of Practical Reason*, translated by Louis White Beck. New York: Liberal Arts Press. The second of Immanuel Kant's three critiques, dealing with moral philosophy.

Struve, Otto, and Velda Zebergs. (1962). *Astronomy of the 20th Century.* New York: Macmillan. A comprehensive text on the history of modern astronomy up to the early 1960s.

ATLANTIS, OLD AND NEW

The story of Atlantis was invented by Plato in an unfinished sequel to the *Republic* constituted by the *Timaeus* and the *Critias.* These two dialogues attempt to relate the political philosophy of the *Republic*, the argument of which is reviewed at the beginning to the *Timaeus*, to natural philosophy. The *Timaeus* describes a prehistorically virtuous Athens, embodying the natural harmonies argued in the *Republic*, that defeats attack from the unlawful empire of Atlantis, once located in the Atlantic Ocean beyond the Iberian peninsula and the North African coast. In defeat it sinks into the ocean. As Critias describes Atlantis, it was rich in both natural resources and technical developments—indeed, its technological works are described as "incredible" (*Critias* 118c) canals, fortifications, and palaces—but lacking in wisdom. With this story Plato raises questions about relationships between science and technology as well as technological and other forms of power.

Plato himself describes Atlantis as being recovered from the Egyptians, and the imagined island empire has exercised a continuing fascination in European literature. In the classical period, Aristotle, Herodotus, Proclus, Plutarch, Pliny, and others mention it. During the Middle Ages, interest languishes. With Francis Bacon's *New Atlantis* (1627), however, the story is critically revived to address precisely the issues raised by Plato but in a distinctly non-Platonic manner.

The New Atlantis: Salomon House

Bacon's imaginary story is of a society ruled by scientists dedicated to the technological conquest of nature. For those who share Bacon's vision of scientific progress, it is an inspiring vision of how modern science and technology could promote a good society. For those who disagree with Bacon, it is a disturbing depiction of how a scientific elite could use manipulation and secrecy to rule over a docile people.

The story is about European sailors who discover an island in the South Pacific inhabited by the people of Bensalem. These people live by laws and customs that secure a life that is free, healthy, and peaceful. They are Christians, although Jews and other religious believers are free to live there without persecution. Marital unions and family life are regulated to promote fertility, monogamous fidelity, and respect for the authority of fathers. Economic life is prosperous; political life is organized around a structure of offices with a king at the top, although the king's rule is cloaked in secrecy.

The most important institution in Bensalem is Salomon's House. Bacon's description of Salomon's House is remarkable, because it is the first account of a modern scientific research institution supported by public authority to promote progress in science and technology to conquer nature for human benefit. Salomon's House is said to have two purposes—"the knowledge of causes, and secret motions of things; and the enlarging of the bounds of human empire, to the effecting of all things possible" (Bacon 1989, p. 71). The first purpose is knowledge for its own sake. The second purpose is power over the world. The aim is to unite human knowledge and human power.

Salomon's House has facilities and tools for studying every realm of nature, including soil, minerals, air, light, wind, water, plants, animals, and human beings. Scientists work to produce new kinds of drugs, foods, and machines. They produce flying machines, boats that move under water, robotic devices that move like animals and human beings, powerful military weapons, and artificially created plants and animals. The scientists search for ways to preserve human health and prolong human life.

The scientists in Salomon's House are assigned various duties. Some travel throughout the world secretly gathering whatever experimental knowledge human beings have developed. Others draw out general conclusions from these experiments. Others apply these experiments to develop new inventions. Still others build on this knowledge to develop a comprehensive knowledge of nature. The scientists consult together to decide which inventions and experiments should be made public and which should be kept secret. They all take an oath of secrecy to conceal whatever should not be publicized. Inventions are particularly important in Salomon's House, and for every new invention, the inventor is honored with the erection of a statue. The scientists visit the major cities of Bensalem to announce useful inventions and to help people explain and protect themselves against natural dangers such as diseases, threatening animals, earthquakes, floods, comets, and scarcity of resources. Salomon's House conducts daily religious ceremonies to praise God for his works and to ask his aid in applying knowledge of his works to good and holy uses.

Heritage

Throughout his life, Bacon had tried unsuccessfully to persuade the British monarch to sponsor scientific research just as Bensalem supports the work of Salomon's House. After his death, many people were inspired by *New Atlantis* to devise plans to set up publicly supported scientific institutions for promoting experimental studies of nature and useful inventions. The establishment of the Royal Society of London in 1682, with a royal charter from Charles II, was one of the most successful outcomes. Contemporary institutions for collaborative scientific research dedicated to new discoveries and inventions such as the U.S. National Science Foundation also seem to follow the model first depicted in *New Atlantis*.

The careful reader of *New Atlantis* may wonder about the ethical problems that arise from possible conflicts between science, politics, and religion. The religious faith of Bensalem depends on a belief in a biblical God who performs miracles, and yet the scientists in Salomon's House are responsible for judging whether apparent miracles are true or fraudulent, which implies the rule of scientific reason over religious faith. Indeed it seems that the scientists rule Bensalem through a new religion of scientific technology that secures earthly life, which replaces the old religion of pious hope in heavenly redemption. The scientific research on prolonging life suggests that the new religion might even provide immortality through the scientific conquest of death. But one must wonder whether the abolition of death through scientific technology is possible or desirable.

The oath of secrecy in Salomon's House suggests that Bensalem cannot be a completely free and open society based on universal enlightenment. The scientific philosophers must hide from the general public those experiments, inventions, and discoveries that would be harmful if they were open to full public view. This implies that scientific and technological innovation can be dangerous for society, and therefore it needs to be regulated by those with the wisdom to understand the ethical problems of such innovation. The critics of Baconian science see this as confirming their fear that modern science and technology shape social life without the free and informed consent of ordinary citizens.

Yet defenders of Baconian science point out the theoretical understanding and practical usefulness that this science has produced. By executing Bacon's project, human beings have both a greater knowledge of nature and a greater power over nature than ever before. Some economic historians argue that economic growth in the Western world since the eighteenth century has been driven largely by a Baconian view of knowledge that connects science, technology, and industrial production. Since the late-twentieth century, Baconian principles are evident in biotechnological research for enhancing physical and mental health and perhaps prolonging life. People are moving toward "the enlarging of the bounds of human empire, to the effecting of all things possible" (Bacon 1989, p. 71). In many respects, human beings are now living in Bensalem.

Shadow

Indeed the effectiveness of Bacon's vision may even be reflected in the way the whole discussion of Atlantis, old and new, has turned away from philosophy and toward fiction and science. Ever since Captain Nemo's visit to Atlantis in Jules Verne's *Twenty-Thousand Leagues Under the Sea* (1870), the lost continent has been a persistent theme in contemporary entertainments. From the time Ignatius Donnelly, a congressman from Minnesota, published *Atlantis: The Antediluvian World* (1882), persistent interest has also focused on such historical and geographical issues such as whether Atlantis might have really existed and where. The journal *New Atlantis* (founded 2003) nevertheless seeks to return to that cluster of issues regarding science, technology, and philosophy that were at the heart of both the Platonic and the Baconian uses of the story of Atlantis.

LARRY ARNHART

SEE ALSO *Bacon, Francis; Governance of Science; Plato; Utopia and Dystopia.*

BIBLIOGRAPHY

Bacon, Francis. (1989). *New Atlantis and The Great Instauration*, ed. Jerry Weinberger. Wheeling, IL: Harlan Davidson.

Jones, Richard Foster. (1982). *Ancients and Moderns: A Study of the Rise of the Scientific Movement in Seventeenth-Century England*. New York: Dover Publications. History of Bacon's influence on early modern science.

Lampert, Laurence. (1993). *Nietzsche and Modern Times: A Study of Bacon, Descartes, and Nietzsche*. New Haven, CT: Yale University Press. Account of Bacon's philosophic vision for modern world.

Leiss, William. (1994). *The Domination of Nature*. Montreal: McGill-Queen's University Press. Critical history of the Baconian idea of mastering nature.

Merchant, Carolyn. (1980). *The Death of Nature: Women, Ecology, and the Scientific Revolution*. San Francisco: Harper & Row. Feminist historian's critical account of Baconian science.

ATOMIC BOMB

The mushroom-shaped cloud associated with the aboveground detonation of an atomic bomb is one of the most defining images and represents one of the most challenging moral imperatives to arise from the mid-twentieth century. The scientific, technological, political, sociological, psychological, religious, and ethical ramifications of humankind's ability to harness and release in a fraction of a second fundamental forces of nature make the atomic bomb one of the preeminent issues of modern society and human existence.

Bomb Engineering

An atomic bomb is a weapon that derives its energy from a nuclear reaction in which a heavy nucleus of an atom such as uranium or plutonium splits into two parts and subsequently releases two or three neutrons along with a vast quantity of energy. These nuclear reactions, if they can be induced rapidly and in quick succession across a critical mass of material, produce a cataclysmic release of energy of prodigious dimensions from a very small quantity of initial material.

Advances in the design of these weapons have focused on efficiency and effectiveness, including ways to produce purer initial materials, induce and sustain more rapid reactions, and produce similar effects with smaller amounts of material. As a result, nuclear devices now available to the armed forces can yield effects from a small warhead on a missile that compare favorably to those generated in the 1950s by weapons so large that bombers had to be specially adapted to haul and drop them. Advances in weapons construction techniques further allow experts to assemble even relatively impure materials into "dirty" bombs with limited yield but severe environmental effects.

Developments since the mid-1980s have posed new threats to world security as an ever-expanding set of nations gained access to suitable raw materials for constructing these devices. Global monitoring of these materials has become increasingly more difficult and nongovernmental organizations have sought, and probably have obtained, previously unavailable raw materials to construct small-scale nuclear devices to advance sinister purposes.

The technology behind atomic bombs dates to work in physics including the theoretical work of Albert Einstein at the beginning of the twentieth century and experimental work by Otto Hahn, Fritz Strassmann, Lise Meitner, Otto Robert Frisch, and others in Germany and Sweden in the late 1930s. Scientists in Germany, France, the United Kingdom, the Soviet Union, Japan, and the United States all realized that it might be possible to produce weapons of mass destruction as an extension of the work of the experimental physicists, but it was only in the United States that these efforts were organized and funded to achieve success.

State Construction

The Hungarian refugee physicist Leo Szilard organized his physics colleagues in the United States to petition President Franklin Delano Roosevelt to sponsor work to build an atomic bomb out of fear that the Germans were already well advanced in their efforts. (This claim was

"Fat Man" (left) and "Little Boy" (right), the only two nuclear weapons that have ever been used in warfare. The Little Boy was dropped on Hiroshima, Japan on August 6, 1945. The Fat Man was detonated over Nagasaki three days later. *(The Library of Congress.)*

later shown to be completely erroneous.) He enlisted the aid of Einstein in this cause, and Roosevelt responded in the fall of 1939 by devoting $6,000 in discretionary money to preliminary investigations by scientists. This sum had grown to $300,000 per year by 1941 with funds channeled through the National Bureau of Standards to hide the scientists' true intent. By 1941 Vannevar Bush, president of the Carnegie Institute of Washington, DC, had formed and chaired an Office of Scientific Research and Development to better harness the abilities of scientists in the United States to contribute substantially to the war effort. A series of experiments at the University of California at Berkeley, the University of Chicago, and a remote location in Oak Ridge, Tennessee, during the period of 1940 to 1941 established that a fission reaction could be created and controlled, that new elements were created in such reactions that could also be useful as sources for bomb materials, and that uranium-235 could be separated from the much more abundant but non-useful for bombs uranium-238 via a number of different means. Several of these separation techniques involved the use of highly reactive and corrosive materials, especially uranium-hexafluoride, in addition to a whole series of radioactive and dangerous by-products from the various processes associated with production of the basic materials needed for atomic bombs—by-products that continue to create problems of waste disposal and health impacts to this day.

Bush appointed a secret National Academy of Sciences (NAS) committee in 1941 to recommend whether it was feasible to build an atomic bomb. The committee, chaired by the Nobel Prize–winning physicist Arthur Holly Compton of the University of Chicago, concluded in May 1941 that an expanded six months of intensive research was needed before a decision could be rendered. Bush was dissatisfied with this report and responded by appointing more engineers to the committee and asking them to reconsider and produce a new report. This report, delivered on July 18, reached the same general conclusions as the prior one. By this point, Bush had a secret report from British scientists concluding that an atomic bomb could conceivably be built within the next few years.

Bush used this report and his own persuasive powers to convince President Roosevelt to give his full backing to proceeding with a large-scale effort to build the bomb. Roosevelt decreed that only four other people were to know: James B. Conant (Bush's deputy and president of Harvard University), Vice President Henry Wallace, Secretary of War Henry Stimson, and U.S. Army Chief of Staff George Marshall. Members of Congress were explicitly excluded from knowledge of the project and remained so throughout the war. The third and final NAS committee report completed in November 1941 provided a cost estimate of $133 million (in 1940 dollars)—a vast underestimate for a project whose final cost of $2 billion was about two-fifths of the entire military cost of World War II to the United States.

The U.S. Army Corps of Engineers (ACE) became the vehicle by which this massive endeavor would be hidden in the federal war budget because construction contracts were large and difficult to understand. The project was turned over to ACE in June 1942 and code-named the Manhattan Engineer District (MED) for its proposed base of administrative operations in New York City. MED became known colloquially as the "Manhattan Project," even though building the atomic bomb had little to do with the city of New York. Colonel Leslie Groves, the civil engineer who supervised the building of the Pentagon in record time, was promoted to brigadier general and given command of the Manhattan Project.

General Groves swiftly commandeered equipment, supplies, human resources, and the best scientists who could be assembled, and created a series of centers in remote locations in Hanford, Washington; Oak Ridge; and Los Alamos, New Mexico in addition to maintaining work at many universities and over 200 corporations including Stone and Webster, Dupont, Eastman Kodak, and Union Carbide. At its peak in 1944 there were more than 160,000 employees working on the project. This workforce overcame tremendous scientific and technical problems in the push to build "the device," and the first atomic bomb performed superbly at Alamogordo, New Mexico, on July 16, 1945. Three weeks later the first atomic bomb was used in war as the *Enola Gay* bomber dropped a single 90-kilogram device over Hiroshima, Japan, on August 6, 1945. Two days later the Soviets declared war on Japan and invaded Manchuria, and on August 9 a second atomic bomb weighing only 6.1 kilograms fell from the sky over Nagasaki, Japan, which created equally widespread destruction (because of its smaller size, the second bomb was considerably more powerful per kilogram). The emperor of Japan announced his intent to accept the Potsdam Proclamation and surrender to the Allied forces on August 14, 1945, with a formal surrender occurring on the 2nd of September.

Assessments

These first atomic bombs affected earth, water, air, and all living organisms in the targeted area. The Hiroshima bomb delivered the equivalent energy of 13.5 kilotons of TNT, while the much smaller but technically superior Nagasaki device yielded 22 kilotons of TNT. The fireball radius was 150 yards with a peak heat close to that of the center of the sun. These bombs leveled the core of these cities with a huge shock wave moving at the speed of sound and heat radiation moving at the speed of light that, while sustained for only a few seconds, vaporized entire structures and human beings, seriously burned thousands of others, and sowed radiation poisoning in human and animal tissue, water supplies, building remains, and the very earth itself, which would affect generations to come. J. Robert Oppenheimer, the scientific leader of the Manhattan Project, when viewing the test site explosion at Alamogordo was reminded of the words of Shiva from the Bhagavad Gita, a Vedic text of India, "I am become death, the destroyer of worlds."

Many scientists associated with the Manhattan Project went on to take leading roles in organizations such as the American Nuclear Society, Federation of Atomic (later American) Scientists, Union of Concerned Scientists, and International Pugwash that sought to stop the spread of nuclear weapons and better educate the public about the brave new world humanity entered with the creation and use of these devices. Einstein expressed deep regret at his own key role in getting the ear of President Roosevelt for Szilard. Einstein would later write, "the unleashed power of the atom has changed everything save our modes of thinking, and thus we drift toward unparalleled catastrophe ... [A] new type of thinking is essential if mankind is to survive and move toward higher levels." Szilard was appalled to learn that America had used the atomic bomb against Hiroshima and devoted himself to the post-war effort to restrict and control the development and use of nuclear weapons. Most nuclear scientists, however, went on to further government contract work on the construction of thermonuclear weapons that were more than one thousand times more powerful than those developed during the project or to work on peaceful uses of nuclear energy. Many scientists, joined by other scholars such as Pitirim Sorokin, Ruth Sivard, Alex Roland, Bruce Mazlish, Kenneth Waltz, and John Mearsheimer, agreed with the assessment of the nuclear scientist Donald York that providing these types of implements rendered war on a large scale too horrific to contemplate and consequently saved hundreds of millions of lives in the standoff between the United States and the Soviet Union known as the Cold War (1945–1989).

Karl Jaspers, a noted German philosopher, argued in *Atombombe und die Zunkunft des Menschen* (1958), that an entirely new way of thinking was required after the creation of the atomic bomb. The philosopher and mathematician, Bertrand Russell, argued in 1946 in "The Atomic Bomb and the Prevention of War" (*Bulletin of the Atomic Scientists* 2(5): p. 19), that the only way to prevent war was through an international government that possessed atomic weapons and was prepared to use them if nations would not heed its directives and settle their disputes amicably with one another.

In the years following the development and deployment of the atomic bomb, the United States and other nations went on to develop more powerful weapons and to repeatedly test them above and below ground. Tens of thousands of civilians and military personnel were exposed to increased amounts of radiation, many unwittingly and unknowingly. The balance of evidence and the opinion of the majority of scientists with expertise who have studied this issue, suggest that for the most part the effects were quite minimal, although whether these low levels of exposure have long-term detrimental health effects can neither be demonstrated nor conclusively denied. The government of the United States, throughout this period, consistently assured the American public that there were *no* risks, despite voluminous information from scientists and classified studies they had commissioned that showed such a claim to be preposterous.

Various ethical arguments have been advanced against nuclear weapons. For example, some have argued that atomic weapons are "unnatural" and on this basis alone should be banned. But all armaments beyond sticks and stones fall under the same charge. Massive fire bombings in World War II of British, German, and Japanese cities killed far more civilians and in ways every bit as horrendous. While an atomic weapon is more than the "beautiful physics" that Enrico Fermi declared when asked about any moral qualms he had about working on the bomb, it must be viewed on a long continuum of the technological evolution of warfare. Whether nations holding nuclear technologies can, and should be able to, prohibit others from acquiring such devices remains an open question to be decided in sociopolitical processes that will include but not be wholly determined by ethical criticism. There is little question that human thought as expressed in writings across a wide range of other subject areas has also been profoundly influenced by the genesis and spread of nuclear weapons. The future of the world is literally increasingly in the hands of a very small number of individuals.

DENNIS W. CHEEK

SEE ALSO *Baruch Plan; Einstein, Albert; Hiroshima and Nagasaki; International Relations; Limited Nuclear Test Ban Treaty; Oppenheimer, J. Robert; Rotblat, Joseph; Weapons of Mass Destruction.*

BIBLIOGRAPHY

Badash, Lawrence. (1995). *Scientists and the Development of Nuclear Weapons: From Fission to the Limited Test Ban Treaty, 1939–1963*. Atlantic Highlands, NJ: Humanities Press. This study portrays the scientific work, ethical and human dimensions, and societal interactions of scientists who worked on nuclear weapons over this period of time.

Bauckham, Richard J., and R. John Elford, eds. (1989). *The Nuclear Weapons Debate: Theological and Ethical Issues*. London: SCM Press. Series of essays by theologians and philosophers who take up central issues concerning nuclear weapons.

Bergeron, Kenneth D. (2002). *Tritium on Ice: The Dangerous New Alliance of Nuclear Weapons and Nuclear Power*. Cambridge, MA: MIT Press. Examines ramifications of a U.S. decision in 2003 to allow conventional nuclear power plants to produce tritium, a radioactive form of hydrogen used to convert atomic bombs into hydrogen bombs.

Canaday, John. (2000). *The Nuclear Muse: Literature, Physics and the First Atomic Bomb*. Madison, WI: University of Wisconsin Press. A detailed analysis of a variety of texts produced by physicists before, during and after World War II.

Einstein, Albert. (1960). *Einstein on Peace*, ed. Otto Nathan and Heinz Norden. New York: Simon and Schuster. A series of writings on peace by one of the most famous physicists of all time, including many reflections on the nuclear age.

Gusterson, Hugh. (2004). *People of the Bomb: Portraits of America's Nuclear Complex*. Minneapolis: University of Minnesota Press. A portrait based on fifteen years of research at weapons laboratories that shows how the military-industrial complex built consent for its programs and transformed public culture and personal psychology since the beginning of the nuclear age.

Hacker, Barton C. (1994). *Elements of Controversy: The Atomic Energy Commission and Radiation Safety in Nuclear Weapons Testing, 1947–1974*. Berkeley: University of California Press. A balanced portrait of weapons testing programs in the United States and their effects, including documentary evidence, clinical, and epidemiological studies. Contains an extensive bibliography.

Hashmi, Sohail, and Steven Lee, eds. (2004). *Ethics and Weapons of Mass Destruction: Religious and Secular Perspectives*. New York: Cambridge University Press. Structured dialogues among representatives of various religious and secular traditions along with essays on weapons of mass destruction and an analysis of existing agreements among nations.

Palmer-Fernandez, Gabriel. (1996). *Deterrence and the Crisis in Moral Theory: An Analysis of the Moral Literature on the Nuclear Arms Debate*. New York: Peter Lang. A systematic comparative study of the dominant views on nuclear arms. It suggests the weapons and the plans that accompany them challenge traditional moral reasoning and understandings of the relationship between intentions and actions.

Rhodes, Richard. (1986). *The Making of the Atomic Bomb*. New York: Simon and Schuster. The author won a Pulitzer Prize for this massive study that is rich in the human, political, and scientific dimensions of the making and use of the atomic bomb.

Smith, Alice Kimball. (1970). *A Peril and a Hope: The Scientists' Movement in America, 1945–47*. Cambridge, MA: MIT Press. Explores the political influence of scientists on arms control in the period just after the dropping of the atomic bomb.

ATOMS FOR PEACE PROGRAM

The Atoms for Peace program, announced by President Dwight Eisenhower at the United Nations in December 1953, constituted a new international effort to regulate the uses of nuclear energy. With its ethical and political justifications, it thus provides an important case study in the control of one specific form of science and technology.

Background

Following the Soviet Union's rejection of the 1945 Baruch Plan for the international control of atomic energy, passage of the Atomic Energy Act of 1946 established a U.S. policy to prevent the spread of nuclear technology by secrecy and denial. Even exchanges of information with U.S. allies who had cooperated in the development of the atomic bomb were prohibited.

By the end of 1953, however, it was apparent that the policy of restriction had failed. The Soviet Union had joined the United States as an atomic weapons state, and both the United States and the USSR had tested hydrogen bombs. In addition to the development of more sophisticated nuclear weapons, research also had progressed on the peaceful uses of nuclear power, especially in commercial applications. As Secretary of State John Foster Dulles noted during testimony before the Joint Committee on Atomic Energy, knowledge about atomic energy was growing in so much of the world that it was impossible for the United States to "effectively dam ... the flow of information." If the United States continued to try to do so, he observed, "we [would] only dam our influence and others [would] move into the field with the bargaining power that that involves" (Guhin 1976, p. 10).

The transition from a policy of secrecy and denial to active promotion of the peaceful applications of atomic energy was first clearly articulated in President Eisenhower's famous "Atoms for Peace" speech before the United Nations. There, Eisenhower acknowledged that the secret of the atom eventually would be acquired by other states, and he emphasized the need to exploit those properties in the atom that were good rather than evil. More specifically, he proposed that the governments principally involved in nuclear research and development make joint contributions from their stockpiles of fissionable materials to an International Atomic Energy Agency (IAEA).

The IAEA was to be set up under the jurisdiction of the United Nations and would be responsible for the storage and protection of contributed fissionable materials. It also was to have the important task of devising methods to distribute nuclear material for peaceful purposes, especially the production of electrical power. Eisenhower hoped that the contribution of fissionable products to the IAEA would assist arms control by diverting the stockpile of nuclear material from military to peaceful purposes. The contributing powers would, in Eisenhower's words, "be dedicating some of their strength to serve the needs rather than the fears of mankind" (*Papers of the Presidents of the United States: Dwight D. Eisenhower 1953*, pp. 813–822).

Implementation

It was not until 1957 that Eisenhower's Atoms for Peace proposals found fruition in the establishment of the IAEA. Not only did the Soviet Union's initial opposition need to be overcome, but substantial revisions had to be made in the very restrictive U.S. Atomic Energy Act of 1946. These changes, incorporated in the Atomic Energy Act of 1954, included removing most controls on the classifications of information regarding nuclear research, approving ownership of nuclear facilities and fissionable material by private industry, and authorizing the government to enter into agreements for cooperation with other nations on the peaceful uses of nuclear energy.

President Eisenhower's Atoms for Peace program ushered in a period of relaxed control over nuclear information, which, ironically, facilitated the development of a race between the United States and the Soviet Union for peaceful nuclear energy and prestige, in tandem with the superpower arms race. One aspect of the former competition was the rush by both the United States and the Soviet Union to declassify and disseminate a large volume of technical information. By 1958 this competition resulted in the adoption of new guidelines for information declassification in the United States that made it possible for any nation to gain access to almost all basic scientific information on the research, development, and operation of plants and equipment in the field of nuclear fission.

More than fifty years after president Eisenhower's "Atoms for Peace" speech, it is apparent that his initiative was a double-edged sword. Predicated on the belief—or at least the hope—that peaceful nuclear energy might be as beneficial to humanity as nuclear weapons were destructive, one indeed can observe many benefits derived from nuclear activities in the realms of medicine, agriculture, and industry. In addition, Eisenhower's initiative gave rise to a number of the most important components of the contemporary nonproliferation regime, including the IAEA and its international system of safeguards. However, one cannot ignore the fact that the Atoms for Peace program also accelerated nuclear proliferation by making it easier for some states to pursue their nuclear weapons ambitions. Although it may be more obvious today than in 1953, the fundamental dilemma remains unchanged—how can a policy prevent the proliferation of nuclear weapons capabilities while at the same time promoting the benefits of nuclear energy if the basic raw materials and technology for both are essentially the same?

WILLIAM C. POTTER

SEE ALSO *Baruch Plan; International Relations.*

BIBLIOGRAPHY

Guhin, Michael A. (1976). *Nuclear Paradox: Security Risks of the Peaceful Atom*. Washington, DC: American Enterprise Institute.

Lavoy, Peter R. (2003). "The Enduring Effects of Atoms for Peace." *Arms Control Today* 33(10): 26–30.

Papers of the Presidents of the United States: Dwight D. Eisenhower, 1953. (1960). Washington, DC: U.S. Government Printing Office.

Sokolski, Henry D. (2001). *Best of Intentions: America's Campaign against Strategic Weapons Proliferation*. Westport, CT: Praeger.

Weiss, Leonard. (2003). "Atoms for Peace." *Bulletin of the Atomic Scientists* 59(6): 34–41, 44.

INTERNET RESOURCE

Center for Global Security Research, Lawrence Livermore National Laboratory. (2003). "Atoms for Peace after 50 Years: The New Challenges and Opportunities" Available from http://www.eisenhowerinstitute.org/programs/globalpartnerships/safeguarding/AtomsPeace_Dec2003_whole.pdf.

AUGUSTINE

Augustine (354–430), born in Thagaste, North Africa, November 13, synthesized Platonism with Christian theology, and is considered a doctor of the European church. He taught rhetoric in Carthage, Rome, and Milan, until his conversion (386) and entry into monastic life; he became a presbyter (391) and bishop of Hippo (396), now Annaba, Algeria. Representative of the implications of his thought for science, technology, and ethics is the fact that in his early years he took an interest in one of the sciences of his day, astrology, and may even have practiced it himself; later he argued decisively against it. Augustine died in Hippo on August 28.

For Augustine, the chief concern of human beings ought to be God and the soul. This did not imply indifference to the material world and its events. When human beings perceive order in nature, he said, it points toward the realm of true happiness, the *intelligible realm* of divine ideas, which not only gives the world its form but enables the mind to discover both regularities in the world and rules for ethical behavior (*De ver. rel.* 29,52–36,67; 39,72–45,83). His general principle was that the mind judges things that are *inferior* to it, according to norms that are *above* it (*De ver. rel.* 31,58; 52,101). In the world presented by modern natural science, in which the order of the physical world appears to be the result of impersonal forces if not chance, the decisive question becomes to what extent the human mind can connect with realities superior to it.

In this journey from the outward to the inward and then upward, his most impressive venture was an analysis of music. In the sixth book of *De musica* (389), he traces the crucial role of proportions or *numbers*, starting with the physical sounds and moving inward to hearing, memory, speech, the spontaneous judgments that arouse *delight* at these proportions, and finally to the intelligible principles by which such judgments are made. His approach foreshadows modern interests in acoustics, the psychological effects of music, and the importance of music to the human spirit (for example, Arthur Schopenhauer).

Similarly he was aware of optics. When viewing a structure or a painting, humans spontaneously make judgments of harmony, he stated (*De lib. arb.* 30,54; 32,59). But there are complexities. An oar in water appears bent, but the light waves are not being *deceptive*; they act according to their nature as they are propagated through media of different densities, and what is fallacious is the premature judgment that the oar is really bent (*De ver. rel.* 33,62; 36,67).

Truth, he said, is God's wherever it is found; just as the Israelites were justified in appropriating the Egyptian's gold and silver because it belonged to God (*Ex.* 3:22, 11:2, 12:35), so Christians can appropriate all truth. The glory of the Gentiles, he said, is their science and philosophy (*Conf.* VII,9,15), though it must be transformed by the insights gained from revelation, which is the tradition of Israel. This early Christian attitude is continued by many modern Christians in dealing with secular science.

One of the major scientific disputes in which Augustine took part concerned the *antipodes*: Are there people living on the other side of a round earth, standing upside down? He regarded it as a matter of scientific conjecture rather than direct experience, but on the basis of Scripture he decided against it; he even thought that, if there should be people there, they could not be descendants of Adam and Eve (*De civ.*

Augustine, 354–430. Augustine was a Christian bishop whose vast literary output is an indispensable source for the religious and secular history of the twilight years of the Roman Empire in the West. (© *Bettmann /Corbis.*)

Dei XVI, 9). The eighth-century Irish monk Fergal or Vergilius in Salzburg was notorious for taking the opposite position. Gradually the question was seen as one for scientific inquiry rather than revelation, and Augustine's position was cited by Johannes Kepler, René Descartes, and the Encyclopedists as evidence of theological obscurantism.

Augustine's contributions relevant to science, technology, and ethics may be summarized in three ways. First his *last word,* at the end of *The City of God* (413–426), is an appreciation of human culture—the liberal arts (geometry, grammar, logic, and music); the fine arts, which use material things to convey thoughts and feelings (poetry, theater, painting, and architecture); and, perhaps most basic, the practical arts (domestication of plants and animals, the crafts, architecture and civil engineering, and navigation). These are indispensable, he said, to the life of the earthly city, even though the latter is not the highest end to be sought.

Second, in dealing with the issue of *natural evil,* Augustine acknowledged that humans live in a dangerous world, but saw this as an invitation to scientific inquiry and technological mastery. He argued that people are like visitors to a forge, surrounded by unknown implements; they resent falling against a furnace or a sharp tool, but the smith knows how to use each of these objects to accomplish his work (*De Gen. c. Man.* I,16,25–26). The venom of scorpions is poisonous, but it can also be put to medicinal use (*De mor.* II,8,11–12). The most personal kind of intervention is medicine, in which he finds many metaphors for the healing activity of God through Christ. In the early-twenty-first century, industry and government support both scientific inquiry and technological intervention.

Third, beyond these kinds of intervention in the world, Augustine suggests that human beings should not think solely in terms of their own discomfort or inconvenience; rather they should appreciate the intricate structure of all living forms, knowing that God created them though humans may not know why (*De civ. Dei* XII,4; XXII,24). In this respect he encouraged the later Christian Platonism of the Chartres school and of Kepler, which sought order in nature precisely because of the conviction that God rules intelligently and intelligibly.

EUGENE TESELLE

SEE ALSO *Christian Perspectives: Historical Traditions; Just War.*

BIBLIOGRAPHY

Delhaye, Phillipe. (1951). "La theacytelorie des antipodes et ses incidences théologiques." In *Le Microcosmus de Godefroy de St-Victor: Étude théologique,* ed. Phillipe Delhaye. Lille, France: Facultés Catholiques. History of the "round earth" question.

Ferrari, Leo C. (1996). "Augustine's Cosmography." *Augustinian Studies* 27: 129–177. The most complete account of Augustine's picture of the world.

Fortin, Ernest. (1984). "Augustine, the Arts, and Human Progress." In *Theology and Technology: Essays in Christian Analysis and Exegesis,* eds. Carl Mitcham and Jim Grote. Lanham, MD: University Press of America. Reprinted in *Ernest Fortin: Collected Essays,* Vol. III: *Human Rights, Virtue, and the Common Good: Untimely Meditations on Religion and Politics,* ed. J. Brian Benestad. (Lanham, MD: Rowman and Littlefield [1996]). Survey of Augustine's appreciation of culture and technology.

Martin, Thomas F., O.S.A. (2001). "Paul the Patient: *Christus Medicus* and the *Stimulus Carnis* (2 *Cor.* 12:7): A Consideration of Augustine's Medicinal Christology." *Augustinian Studies*: 219–256. Thorough discussion of the medical metaphor.

Nowak, Adolf. (1975). "Die 'numeri judicales' des Augustinus." *Archiv für Musikwissenschaft* 32: 196–207. A close analysis of Augustine's view of aesthetic judgment.

O'Daly, Gerard. (1987). *Augustine's Philosophy of Mind*. London: Duckworth. Comprehensive survey of Augustine's theory of knowledge.

Pickstock, Catherine. (1998). "Ascending Numbers: Augustine's *De Musica* and the Western Tradition." In *Christian Origins: Theology, Rhetoric and Community*, eds. Lewis Ayres and Gareth Jones. London and New York: Routledge. Defense of Augustinian aesthetics and cosmology by comparison with Indian, modern, and postmodern approaches.

AUSTRALIAN AND NEW ZEALAND PERSPECTIVES

Although they maintain their distinct identities, Australia and New Zealand are linked closely and together are often referred to as Australasia. Both countries claim to be "knowledge societies" and to value science and technology highly (if somewhat uncritically). Yet an apparent lack of understanding in government of the long-term character of scientific and technological development contributes to a perception among scientific and technical professionals that they have little political influence. Extensive corporatization and privatization of publicly-owned infrastructure during the 1990s was aimed largely at retiring government debt, while governments in both countries failed to provide effective support for the development of information- and sustainability-based technologies. However, early in the new century there were encouraging indications in New Zealand of government engagement with sustainability issues (Geddes and Stonyer 2001, Laugesten 2002). In Australia commitment to environmental/sustainability issues varies among and across the three tiers of government (federal, state/territory and local).

Historical Background

Australia and New Zealand had very different traditional cultures. Australian aborigines can demonstrate a continuous hunter-gatherer occupation of 40,000 years; in contrast, the Maori reached New Zealand as recently as 1000 to 1200 C.E., bringing with them a distinctive Polynesian cultural tradition. Australia became a British colony in the late eighteenth century, and New Zealand in the mid-nineteenth century. Invasion and settlement brought European religious and moral doctrines and European technologies designed to dominate the indigenous populations and exploit the natural environment.

Unfortunately, the colonists of both countries disdained indigenous knowledge and technologies. Only in the last quarter of the twentieth century did political activism lead to a broader appreciation of the depth of indigenous cultural and spiritual links with the land. There is increasing recognition that these values enrich the societies as a whole and in particular suggest important approaches to the search for sustainability. However, unresolved questions of reconciliation and compensation still constitute a major fault line in both societies and pose fundamental ethical dilemmas for their governments. This inevitably colors other ethics discussions on a host of issues related to ownership and custodianship of the land, including the use of natural resources and environmental degradation. As its population reached 20 million early in the twenty-first century, Australia became multi-cultural. New Zealand, with a population of 4 million, remains bicultural, with distinct Maori/Anglo (Pakeha) polarization.

As colonies and later as dominions within the British empire, Australia and New Zealand were until about 1950 major suppliers of food and raw materials to Great Britain and were captive markets for British manufactured goods. The colonial governments supplied essential infrastructure and took responsibility for funding science and technology, which tended to be applied and utilitarian, focusing initially on primary industries, particularly agriculture and mining. When multinational corporations set up substantial local operations after World War II, those operations were commonly "branch plants" with minimal research and development capability.

Although Australia and New Zealand have produced individual scientists and technologists who earned international acclaim, the technical culture in both countries was until relatively recently essentially derivative. Despite homegrown inventions and innovations, both countries were largely the recipients of technology transfer. While this tended to encourage a client-state mentality, valued local resources and technologies have been strongly defended, for example through resistance to the introduction of genetically modified crops.

While achieving rigorous academic standards, for many years the universities failed to provide an effective forum for broad ethical debate in science and technology. Higher education was based on British models, and into the second half of the twentieth century universities in Australia and New Zealand commonly looked to Britain for academic leadership. As in the rest of the "Western" world, scientific and technological advances were equated with social progress and the ethical focus was on gaining peer support, maintaining professional standards, and ensuring competent technical performance.

New Voices

Although the science and technology professions in Australia and New Zealand are well integrated into the global community, since the 1970s a distinctively Australasian voice has emerged, asserting that those professions must take a much broader approach to issues of ethical practice. There is growing awareness that science and technology involve social as well as technical practices (Johnston, Gostelow, and Jones 1999). Framing problems and choosing decision-making criteria increasingly are recognized as areas for professional judgment in which ethical choices are deeply embedded. For instance, in New Zealand Roy Geddes and Heather Stonyer (2001) highlight the ethical implications of setting national priorities and of deciding how far professionals should go in challenging government failure to provide adequate education and training in science and technology.

This groundswell of broader ethical awareness draws on worldwide developments in the scientific and technological communities, making the identification of distinctive local inputs and key national figures problematic. One person who stands out in this area is the Melbourne-born utilitarian moral philosopher Peter Singer, recognized for his courageous and challenging work on globalization, medical ethics and bioethics, and human relationships with the rest of the animal kingdom (Singer 2003).

The international partnership between New Zealand ethicist Alastair S. Gunn at the University of Waikato and U.S. civil and environmental engineer P. Aarne Vesilind also needs to be mentioned here. Their first book (Vesilind and Gunn 1986) was an important and timely contribution, not least because it argued that environmental ethics were relevant to the whole profession, and not only to environmental engineers. Two of their three books have been translated into Japanese and one into Chinese. Gunn has also been working with colleagues at the University of Malaya on an Internet site to provide ethics resources for technology professionals in Asia.

In Australia, Sharon Beder at the University of Wollongong is another public champion of ethical concerns, particularly within engineering. She has led a move away from paternalistic views of the public and toward greater transparency of professional action (Beder 1998). Until the 1980s government agencies in Australia that supplied major services and public utilities, including energy, communications, and water, were staffed mainly by engineers who prided themselves on doing the best they could with the resources allocated by the political process. Criticism of either the process or its outcomes was seen as bringing the profession into disrepute, and the profession's code of ethics was used to suppress internal dissent. Beder successfully challenged that limited approach to professional responsibility. By the 1990s the engineering profession in Australasia was looking outward and moving toward a clearly formulated emphasis on sustainability as a key ethical value.

In 1992 the Institution of Professional Engineers New Zealand (IPENZ) decided to revise its code of ethics. Gerry Coates argued that the new code should be values- rather than rules-based, provide high rather than low levels of guidance, and offer real ethical leadership for the profession. A key question was the extent to which technical and scientific professionals should be involved in political decision making. The change process took ten years and included extensive debate on the community-oriented values of *sustainable management* and *care of the environment*. However, respect for nonhuman life forms was considered too radical for inclusion at that time, and the revision did not provide guidance on the hierarchy of the values that were asserted (Coates 2000).

In Australia and New Zealand medical research became an important area of scientific and technical activity during the twentieth century. Since World War II there has been a worldwide strengthening of ethical guidelines and controls for research involving humans and animals and increasing awareness of environmental issues. The Australian National Health and Medical Research Council, a major channel for government funding, has exercised significant ethical leadership (National Health and Medical Research Council 2001). The Royal Society of New Zealand is also important in coordinating scientific and technical activity; its Code of Professional Standards and Ethics underscores legal and other constraints on professional behavior.

In both countries there is a powerful network of broadly based ethics committees in universities and research establishments that have a general commitment to ethical practice. There is frank debate in areas such as human stem-cell research, and ethics committees veto projects that do not satisfy their guidelines. However, globalization of research and pressures for economic returns promote increasing commercialization and public-private collaboration, and the traditional ideal of openness is under challenge.

Ethical issues have been highlighted in Australia since the late 1980s by dramatic business failures. Broad concerns have emerged about accountability and about the inward focus of much of the ethical debate in the professions, and the authority and influence of professional bodies have declined. Statutory anticorruption bodies and mechanisms such as commissions of inquiry appointed to look into specific problems or disasters now provide more effective sanctions against unethical behavior. In the public sector reliance on legislation and regulation remains fundamental.

Advancing Practice

By the 1990s ethics-focused research and guidance centers were emerging. With a focus on leadership rather than enforcement, Sydney's St. James Ethics Centre has an international reputation. Its executive director, Simon Longstaff, presents ethical practice in terms of building relationships, developing a well-informed conscience, being true to oneself, having the courage to explore difficult questions, and accepting the costs of ethical behavior. The center provides a framework for discussions that emphasize the recognition of the interests of stakeholders and the impacts of decisions. Developing involvement and avoiding polarization in ethical decision making require structure, space, and time (Taylor 1998). One facility provided by the center that is believed to be unique is a confidential ethics counseling help line for individuals.

There continue to be problems involving business ethics. In 2003 a royal commissioner reporting on the corporate culture that led to the multi-billion-dollar collapse of a major Australian insurance group, HIH, wondered if anyone had asked the simple question "Is this right?" The HIH demise highlighted problems with professional indemnity insurance. Some Australian states, in association with professional standards councils, have provided methods for limiting indemnity claims for professional groups that take specified steps to improve professional standards and protect consumers. Participating groups develop and adopt acceptable codes of ethics that are based on a model document that explains the nature and role of codes, describes their generic content, and outlines the development processes (Miller 2002). This approach encourages professional groups to acknowledge the non-technical aspects of problems; cross-disciplinary approaches are used to develop socially relevant project design criteria and address broad ethical issues.

One of the most promising developments has been a move toward exploration of the ways practitioners develop their own ethical frameworks. This work has led to programs that encourage and support students in recognizing, reflecting on, and dealing effectively with the ethical issues they encounter in practice (Johnston, McGregor, and Taylor 2000).

Ethical professional practice requires a broad awareness of social context, but this in itself is not sufficient. As Peter Singer pointed out, it is "clarity and consistency in our moral thinking [that] is likely, in the long run, to lead us to hold better views on ethical issues" (Singer 2003).

STEPHEN F. JOHNSTON

SEE ALSO *Engineers for Social Responsibility; Institute of Professional Engineers New Zealand.*

BIBLIOGRAPHY

Beder, Sharon. (1998). *The New Engineer: Management and Professional Responsibility in a Changing World*. South Yarra, Victoria, Australia: Macmillan Education Australia. A widely used text and a good introduction to her work, much of which is available at: http://www.uow.edu.au/arts/sts/sbeder/index.html.

Coates, Gerry. (2000). "Developing a Values-Based Code of Engineering Ethics." *IPENZ Transactions* 27: 11–16. Available from http://www.ipenz.org.nz; includes the IPENZ Code of Professional Ethics.

Johnston, Stephen F.; J. Paul Gostelow; and Evan Jones. (1999). *Engineering and Society: An Australian Perspective*. 2nd edition. Melbourne: Longman. A comprehensive exploration of the history and the social, political and economic context of engineering practice in Australia. A U.S./global version was published by Prentice Hall (2000).

Johnston, Stephen; Helen McGregor; and Elizabeth Taylor. (2000). "Practice-Focussed Ethics in Australian Engineering Education." *European Journal of Engineering Education* 25: 315–324.

Longstaff, Simon. (2003). *Hard Cases: Tough Choices*. Sydney: St James Ethics Centre. Available, with much other material, from http://www.ethics.org.au.

Miller, Seumas. (2002). *Model Code of Ethics Principles*. Sydney: Professional Standards Council.

National Health and Medical Research Council. (2001). *National Statement on Ethical Conduct in Research Involving Humans*. Canberra: Commonwealth of Australia.

Taylor, Elizabeth. (1998). "Peeling the Onion: A Technoprudence for Engineering." *Proceedings of the Tenth Annual Conference and Congress, Australasian Association for Engineering Education*. Gladstone: Central Queensland University: 311–314. Elizabeth Taylor is currently dean of the James Goldston Faculty of Engineering and Physical Systems at the University of Central Queensland.

Vesilind, P. Aarne; and Alastair S. Gunn. (1986). *Environmental Ethics for Engineers*. Chelsea, MI: Lewis. The latest collaboration by these authors, *Hold Paramount: The Engineer's Responsibility to Society* (Stamford, CT: Brooks Cole/Thomson, 2002), written as a novel, explores ethical dilemmas faced by Chris (gender undefined), a structural engineer, in his/her work.

INTERNET RESOURCES

Abrams, Louise. (2003). "Controlling the Direction of Research in Australia: An Ethical Debate. " Research Office, University of Technology, Sydney. Available from http://www.uts.edu.au/

Geddes, Roy, and Heather Stonyer. (2001). "Science and Technology: Participating in the 'Knowledge' Culture." *Proceedings of the Sixth Annual Conference Australian Institute of Health Law and Ethics*. Melbourne: June/July. Available from http://www.law.unimelb.edu.

Laugesen, Pauline, ed. (2002). "Creating Our Future: Sustainable Development for New Zealand." Wellington, New Zealand: Parliamentary Commissioner for the Environment. Available from http://www.pce.govt.nz/.

Singer, Peter. (2003). "Interview: Peter Singer." *Heilpädagogik Online* 1/03: 53. Available from http://www.petersingerlinks.com.

AUTHORITARIANISM

Along with totalitarianism and democracy, authoritarianism is one of the main types of political regimes or systems. Though different variants exist, all authoritarian systems share certain basic features that have significant implications for science, technology, and ethics. For instance, the easy flow of information that facilitates science and is promoted by communications technology creates both opportunities for and burdens on authoritarian leaders seeking to maintain their control over the political realm.

Prominent scholars of authoritarianism include Juan J. Linz and Guillermo A. O'Donnell. Linz (2000) highlights the differences between authoritarianism and totalitarianism, while also pointing out the possibility for authoritarianism to combine with the other two types of government in a hybrid form of political regime. O'Donnell (1973) emphasizes the importance of a bureaucratic form of authoritarianism, distinct from cases of traditional military regimes or authoritarian systems managed by a dominant political party, while highlighting differences among authoritarian systems based on the degree of modernization in particular countries.

TABLE 1

Comparison of Democracy, Authoritarianism, and Totalitarianism

Democracy	Authoritarianism	Totalitarianism
Political mobilization promoted	Political mobilization generally discouraged	Political mobilization promoted
Competing pro-democratic ideologies	No state ideology	State ideology
Legitimacy based on ideology, rule of law, and performance	Legitimacy based on performance	Legitimacy based on ideology and performance
Official and unofficial limits on government	No official limits on government	No official or unofficial limits on government

SOURCE: Courtesy of Carl Mitcham and Lowell W. Barrington.

Features of Authoritarianism

In its ideal form, authoritarianism exhibits four defining features: a depoliticization or demobilization of the general population, the lack of a central governing ideology, legitimacy based on performance, and the general absence of official limitations on government action. These features distinguish authoritarian systems from democratic and totalitarian ones (see Table 1).

Because authoritarian governments do not seek to remake society in the way totalitarian systems do, there are fewer reasons to mobilize the masses compared to the other types of political systems. When it occurs, mobilization is generally designed to enhance the legitimacy of the system (the belief by the general public in the right to rule of the governmental institutions and individual leaders). As Samuel P. Huntington (1968) has argued, instability in any political system is often the result of political participation that is not channeled into regime-supportive activities. Thus, although authoritarian political systems may hold elections, the campaigns for such elections are devoid of significant discussions of issues in a form critical of the government, and the outcome is not in doubt. If necessary, ballot boxes will be stuffed or results falsified. Likewise, political parties may exist, but they are not used to organize the masses as in a totalitarian system nor to aggregate and articulate issue positions to allow the masses to choose in free and fair elections as in democracies. Opposition political organizations are either tightly controlled or not tolerated at all.

Partly because authoritarian systems do not seek the remaking of society, ideology is less important than in either totalitarian states or democracies. This is not to imply that authoritarian systems lack goals or a vision for change; they tend to focus on a particular vision, what Linz (1975) has called the "mentality" of authoritarian systems. In cases in which an authoritarian system is established through the overthrow of a democracy, the authoritarian leaders may concentrate on the need to institute policy changes to bring economic stability or otherwise restore order to a chaotic situation. This is often welcomed by the masses, who will, in many cases, prefer order to freedom. Thus, an important part of the legitimacy for an authoritarian system is based on its performance. As long as it achieves its goals, the general population may be quite willing to tolerate the absence of freedoms and the lack of a check on government power.

The final central feature of authoritarian political systems—the lack of official limitations on government action—is one that these systems share with totalitarian regimes. As Mark Hagopian (1984) has argued, the lack of legal restraints helps define both totalitarian and authoritarian systems as dictatorships and allows one easily to distinguish them from constitutional democracies. There are differences, however, between authoritarian and totalitarian systems in this regard. One could argue that authoritarian systems have even fewer institutional constraints than do their totalitarian counterparts (because of the comparatively limited role of a ruling political party in most authoritarian regimes). On the other hand, totalitarian regimes lack the informal—or, to use Hagopian's (1984, p. 118) term, "extralegal"—limits on power found in most authoritarian systems. The lack of official constraints does not imply the absence of ruling institutions or an official constitution, nor does it mean that society is completely controlled or powerless. Instead, the official rules of the game are subordinate to the will of the authoritarian ruler or rulers. Checks and balances (including judicial review) and the rule of law, both of which are familiar to citizens of many democratic countries, are unusual in authoritarian states. To the extent that constraints exist, they tend to be informal or based on connections between the government and powerful figures in society such as the wealthy. Such figures, or social institutions such as the church, can have a degree of autonomy from the state—and in some cases even a degree of influence over it.

Types of Authoritarianism

There are as many variations of authoritarianism as there are of democracy. The three main forms, however, are: military, bureaucratic, and party. A military authoritarian system (such as Pinochet's Chile) is one in which the military actually controls the policymaking institutions. Military authoritarian systems can arise for several reasons: an external threat to the security of the country, instability within the country, or threats to the autonomy of the military and/or the degree of military spending by the government. A bureaucratic authoritarian system (for example, Brazil following the military coup in 1964, Argentina in 1966–1974) usually involves an uneasy relationship between the military and the bureaucracy. Experts in their fields hold important political positions, and the bureaucracy becomes a central actor in the creation and implementation of policy. This policy is designed to facilitate internal stability, foster economic development, and maintain a modern society (O'Donnell 1973). The goal of modernization helps justify the power of "technocrats" in this form of authoritarianism.

A party authoritarian system (such as Mexico during much of the twentieth century) uses an existing or newly created political party to organize political activity and enhance the legitimacy of the system. The party is less important than in totalitarian systems, though it can play a role in facilitating elite–mass linkages. During the long period of dominance of the Institutional Revolutionary Party (PRI) in Mexico in the twentieth century, connections between government officials and interests within society were maintained through the party rather than the state. Even the party authoritarian type can be dissected. Huntington (1970), for example, lists three forms of party authoritarianism. If control through a political party is combined with a broader effort to remake society, the result is a hybrid form of government bridging authoritarianism and totalitarianism.

A hybrid between authoritarianism and democracy is also possible. Some call this semi-democracy, while others have termed it semi-authoritarianism (Ottaway 2003). In these systems, certain aspects of democracy exist, though others—most commonly freedoms such as of speech and the press—are curtailed by government control and/or intimidation. Thus, elections may exist without significant fraud, but the range of opinions expressed during the campaign is limited; media coverage of the government leadership is uniformly favorable. Since the election (and reelection) of Vladimir Putin, Russia has moved more and more in this direction. In some countries, this semi-authoritarianism can act as a bridge to democracy. In others, as is arguably the case in Putin's Russia, it may signal a move away from liberal democracy and toward a more classic authoritarian system. But semi-authoritarianism can also be quite persistent and need not be a transition to something else.

Science, Technology, and Ethics

The impact of authoritarianism on science, technology, and ethics is significant. For authoritarian leaders, ethical considerations are usually secondary to the goals of maintaining power, fostering stability, and facilitating economic performance. The concept of the rule of law has no place in the ideal authoritarian system. Human rights violations are common, as those whom the government perceives to be potential political threats are harassed, arrested, or killed. As in totalitarian systems, scientists in authoritarian states face ethical dilemmas working with such governments. On the one hand, cooperation with the state may provide an essential opportunity to conduct research. On the other, such cooperation both sanctions the actions of the government and opens the door to government use of the research in ways scientists may find morally objectionable.

Likewise, science and technology in general are double-edged swords for authoritarian officials. Authoritarian leaders who emphasize economic development as a central goal must foster technological advancements. In addition, science and technology may be put to use in assisting the maintenance of authoritarian power. Though less so than in totalitarian systems, authoritarian governments monitor the actions of individuals who might threaten their political power. In China, leaders have sought to harness the power of new technology to spread regime-supportive propaganda.

But technology can also threaten authoritarian rule. Those leaders who emphasize as their defining goal the protection of national culture rather than economic development often see technology as a transmission belt for "foreign" (especially Western) values. Those leaders who seek to use technology to monitor the actions of individuals also find that the technology allows those individuals to hide from this monitoring. The information-enhancing capacity of the Internet can be harnessed by opponents as well as government officials. The Chinese government works diligently to shut down Internet sites of regime opponents. But as quickly as these sites are removed, others spring up. Simply put, the more advanced and complex the society, the more difficult it is to keep it under surveillance. Thus, some authoritarian leaders may actively discourage certain types of technological advancements in their country.

LOWELL W. BARRINGTON

SEE ALSO *Expertise; Fascism; Technocracy; Totalitarianism.*

BIBLIOGRAPHY

Hagopian, Mark N. (1984). *Regimes, Movements, And Ideologies: A Comparative Introduction To Political Science*, 2nd edition. New York: Longman.

Huntington, Samuel P. (1968). *Political Order in Changing Societies*. New Haven, CT: Yale University Press.

Huntington, Samuel P. (1970). "Social and Institutional Dynamics of One-Party Systems." In *Authoritarian Politics in Modern Society: The Dynamics of Established One-Party Systems*, ed. Samuel P. Huntington and Clement H. Moore. New York: Basic.

Linz, Juan J. (1975). "Totalitarian and Authoritarian Regimes." In *Handbook of Political Science*, Vol. 3: *Macropolitical Theory*, ed. Fred I. Greenstein and Nelson W. Polsby. Reading, MA: Addison-Wesley.

Linz, Juan J. (2000). *Totalitarian and Authoritarian Regimes*. Boulder, CO: Lynne Rienner.

O'Donnell, Guillermo A. (1973). *Modernization and Bureaucratic-Authoritarianism: Studies in South American Politics*. Berkeley: University of California, Institute of International Studies.

Ottaway, Marina. (2003). *Democracy Challenged: The Rise of Semi-authoritarianism*. Washington, DC: Carnegie Endowment for International Peace.

AUTOMATION

The term *automation* was coined by John Diebold (born 1926), a pioneering management consultant and entrepreneur, when he shortened the more correct term *automatization*. His classic book *Automation* (1952) was the first to advocate this process, and to consider the general implications of this process for manufacturing and society. Using what were at that time only emerging concepts of control, communication, and computers, he described the coming industrial world of automated production and predicted the incipient information revolution. In pioneering the automation of production systems, Diebold extended the concepts of materials handling to information handling, to analyzing information flows, and to studying ways to automate office processes.

The Automation Process

In general the term *automation* describes the employment of automatic devices as a substitute for human physical or mental labor. An automatic device is one that performs a specified function without human intervention. Critical in one form of this process is "feedback." For example, when an autopilot is set to fly an airplane along a given course, it will by itself correct any deviation from that course caused by air turbulence. It

Automobile assembly line with welding robots. The automobile industry relies heavily on automation in many parts of its manufacturing process. *(AP/Wide World Photos.)*

does this by sensing the degree of deviation from the course and then actuating the appropriate control surfaces, such as ailerons, elevators, and rudder of the plane in a way that will restore the desired heading.

The sensors of the autopilot measure the amount of deviation from the course and "feed back" the right amount of electrical signal or hydraulic pressure to restore the intended course. A system with these characteristics also is referred to as a servomechanism or control system. Norbert Wiener (1894–1964), a pioneer in developing the theory of control systems, used the word *cybernetics* to describe the science of control and communication in both machines and living organisms.

In automating a manufacturing process, functions once performed by humans are replaced by automatic devices that replicate those functions. Computers are widely used for process control. Manufacturers invest in automation because it increases productivity per worker employed, creates greater uniformity of product, and lowers cost per unit of output.

Economic Implications

The substitution of human and animal effort by machines has been pursued throughout history. Some substitutions have been beneficial in their effect, relieving humans from the need to do heavy physical labor. Automation is a relatively recent development in the long history of technological change and a new issue in long-standing debates about technological unemployment. These debates were particularly fierce during the Industrial Revolution in England, when new machines displaced workers and left many unemployable.

Many economists argue that automation, along with technological change in general, does not add to total unemployment. Total unemployment is not affected by technological innovation because although workers in one industrial sector lose jobs, others gain employment through the creation of new jobs. Frequently cited evidence includes the widespread anxiety in the 1950s that automation would lead to mass unemployment, which never materialized. This anxiety can be chalked up to the "lump of labor fallacy," which holds that there is a constant amount of work to be performed in the world, and therefore any increase in the productivity of workers reduces the number of available jobs.

Ethical Issues

Ethical issues arise when public policies or the strategies of industrial management lead to unemployment and other consequences that harm the life of the individual or lead to social dislocation. Diebold was much concerned with the social effects of automation and predicted that the "age of automation" would transform society as radically as did the Industrial Revolution but that that change would be more profound because the rate of change had become so much more rapid. He acknowledged that automation created some employment problems but stated that the social effects of communications and computer systems will be more insidious because information, its communication, and its use will change people's approach to work, society, and life.

It was Ben B. Seligman who in *Most Notorious Victory* (1966) cataloged the harmful consequences of automation. With a social scientist's broad interest in the human condition he systematically examined the economic, social, psychological, and philosophical implications of automation. His main indictment is implied in the title: The successful diffusion of wave upon wave of new technology threatens to destroy essential human qualities. New technologies render traditional work patterns obsolete, and the mechanization of labor may undermine the significance of work as a source of meaning for many people. Seligman also was concerned that complex technological issues that require the judgment of experts will weaken the democratic process and lead to a situation in which technocrats will chart the future of society.

Ethical issues that derive from automation will continue to confront society. There appears to be no end to technological innovation in the foreseeable future and to the application of automation to new areas. The new frontier for automation is no longer production but the service industries, prominent among them health care, financial services, telecommunications, retail, and transportation.

A. GEORGE SCHILLINGER

SEE ALSO *Artificial Intelligence; Autonomous Technology; Cybernetics; Wiener, Norbert; Work.*

BIBLIOGRAPHY

Diebold, John. (1952). *Automation: Its Impact on Business and Labor*. New York: Van Nostrand. The first comprehensive book on the subject. It is prophetic in its projections of how automation and computers will change production, management, labor and the nature of work in the office.

Quinn, Francis X., ed. (1962). *The Ethical Aftermath of Automation*. Westminster, MD: Newman Press.

Seligman, Ben B. (1966). *Most Notorious Victory: Man in an Age of Automation*. New York: Free Press. A systematic and sensitive analysis of the harm massive rapid technological development and automation in particular can inflict on the individual and society.

Zuboff, Shoshana. (1988). *In the Age of the Smart Machine*. New York: Basic Books.

AUTOMOBILES

One of the distinguishing characteristics of human beings is that they have always been mobile. From its origins on the African continent, the human species has traversed the earth and populated every continent but Antarctica. For most of human existence, land travel was entirely dependent on human and animal muscle power. Radical changes came in the nineteenth century with the invention of steam-powered locomotives, and toward the end of the century the first automobiles powered by internal combustion engines were created in several industrially developed countries. By the first decade of the twentieth century automobile ownership was expanding at a rapid rate in the United States, and this pattern was followed in subsequent decades in many other parts of the world.

Cars gave people an unparalleled ability to go where they wanted, when they wanted, and with whom they wanted. In short, they promised freedom. Early motorists eagerly took advantage of this freedom, embarking on long journeys despite miserable road conditions and the uncertain reliability of their vehicles. By the 1920s automobile ownership had been *democratized* in the United States as manufacturing innovations dramatically lowered purchasing prices, giving rise to an era of mass motorization.

In the early-twenty-first century car ownership has expanded to such an extent that in many industrial nations the ratio of cars to people approaches or even exceeds one to two. Yet universal automobile ownership presents a paradox. Although the great virtue of the automobile lies in the freedom that it confers, the ownership and operation of a car has subjected its users to numerous restrictions. Traffic laws, registration and licensure requirements, vehicle inspections, insurance, and a significant financial burden put a serious crimp on feelings of unrestrained freedom. Individual freedom is also stifled by the sheer proliferation of automobiles; people acquire and use cars to enhance their mobility, but when they do so in large numbers the result is traf-

Safety demonstration on a Volkswagen car. Advances in automotive technology have contributed to the development of devices, such as seat belts and air bags, for the protection of the human occupants of automobiles. *(© Richard Olivier/Corbis.)*

fic-stopping congestion, and everybody's freedom of movement is diminished accordingly. In sum, the automobile is a prime example of how the aggregated pursuit of individual freedom can produce the opposite result—submission to numerous restraints, immobility, and frustration.

Clearing the Air

Many of the ethical quandaries posed by the automobile can be reduced to an overarching issue: achieving a balance between the individual freedom that comes with operating a car and the needs of society as a whole. The difficulty of doing so is exemplified by the forty-year-old campaign to reduce air pollution. A single car has a negligible effect on air quality, but 100,000 in a limited area can be the source of significant pollution. In recent years there have been substantial gains in air quality due to the application of technological fixes such as computerized engine management systems, reformulated fuels, and catalytic converters. But these never would have been developed and used if each motorist followed only his or her self-interest. Emission-control technologies add substantial costs to the purchase and operation of a vehicle, yet they do nothing to improve air quality if no other cars are similarly equipped. It is therefore necessary for an agency working on behalf of the collectivity, in most cases government at some level, to mandate that cars and the fuel they use produce fewer pollution-forming emissions. As long as everyone is required to meet similar regulations there is little cause for complaint. People may grumble about paying higher prices for cars and fuel, and they may resent the time absorbed by periodic smog checks, but few would want to return to the preregulation era when air pollution caused by uncontrolled vehicle emissions severely diminished the quality of life.

It has been relatively easy to mesh individual with collective interests in combating air pollution because the environmental consequences of operating an automobile are all too apparent to anyone who has to live in a gray-brown haze of smog. This in turn substantially increases public receptivity to the governmental actions

Emissions technicians evaluating automobile emissions in a garage. Technological advances in emissions systems, such as reformulated fuels and catalytic converters, have led to improvements in air quality. (*© Martha Tabor/Working Images Photographs.*)

taken to reduce emissions. It also helps that the available technological fixes do not require a massive overhaul of the transportation system; all that is needed are some modifications to automobile engines and the fuel they use. The same does not hold true when addressing another inescapable product of vehicle operation: the generation of carbon dioxide (CO_2). There are no easily applied technological fixes to reduce CO_2 emissions, which are the inevitable product of burning hydrocarbon fuels. The only likely solution entails the abandonment of the internal combustion engine in favor of battery-powered electrics, while hoping that battery performance can eventually be improved. Further in the future lies the possibility that fuel cells will become practical sources of power, but their adoption would necessitate major changes in the infrastructure that supports the automobile and the expenditure of billions of dollars. Moreover obtaining the hydrogen to power the fuels cells is problematic. The most feasible source is petroleum, and the energy costs of the conversion process would require the production and consumption of significantly larger quantities of this diminishing resource.

Even if alternatives to the internal combustion engine become available, the task of getting motorists to accept them remains, because CO_2 emissions do not have the immediate, all too apparent effects of ordinary smog. Because CO_2 is odorless and colorless, most drivers are unaware of the fact that, on average, they are pumping about a pound of it into the atmosphere for every mile they drive, and that vehicles account for about 30 percent of CO_2 emissions in the United States. Yet in the long run these emissions may be more harmful than the smog-forming by-products of internal combustion. If, as many atmospheric scientists believe, CO_2 accumulation in the atmosphere is a major cause of global warming, the long-term results of automobile operation could be disastrous. But global warming is still a controversial issue, and if it occurs will take a long time to manifest itself. Consequently it will be far more difficult to mandate the manufacture and operation of totally different kinds of vehicles, or to do away with the private automobile altogether.

Automotive Safety as an Ethical Issue

Setting aside the problem of controlling CO_2, the case for asserting the primacy of collective needs over individual freedom seems clear-cut in regard to automotive emissions controls. Somewhat more ambiguous is the issue of making safer cars. One may reasonably begin with the assertion that the most important determinants of the safe operation of vehicles are the actions and skills of their operators. When 30 percent of the more than 42,000 fatalities on U.S. roads in 2003 involved drivers whose blood-alcohol level was over the legal standard for driving under the influence, it is not reasonable to demand that cars should provide perfect protection from the consequences of individual irresponsibility. At the same time, however, some accidents may be unavoidable, and even when driver error is involved, death and injury cannot be considered appropriate penalties for momentary lapses.

For decades automobile manufacturers were convinced that *safety features were of scant interest to consumers* and they expended little or no effort to improve the ability of automobiles to protect their occupants in the event of an accident. This situation began to change dramatically in the 1960s, when Ralph Nader and other critics attacked the industry's indifference. Automotive safety became a salient cultural and political issue, and a combination of market demands and government regulations prodded manufacturers into making cars that did a much better job of protecting their occupants when accidents occurred.

Of all the safety improvements that ensued, the most important was the fitting of seat belts as standard equipment. Subsequent advances such as shoulder-and-lap belts made these restraints even more effective, but they were of no value when left unused. During the early 1970s only a small minority of U.S. drivers and passengers regularly used seat belts, so for the 1974 model year an effort was made to encourage their use by fitting cars with interlocks that prevented the vehicle from being started if all occupants had not buckled up. Vociferous protests caused Congress to repeal the requirement in short order.

Convinced that the majority of drivers and passengers could not be convinced to use seat belts, the federal government mandated the fitting of *passive restraints* to new cars. Some of these took the form of motorized harnesses that wound their way over an occupant's upper body, but far more popular was the airbag. By the mid-1990s driver and front-seat passenger airbags were virtually universal fittings on new cars. Airbag technology was predicated on the need to protect an unbelted male weighing 80 kilograms (175 pounds). Providing protection for a person of this size necessitated the design of airbags that inflated in milliseconds and reached speeds of up to 320 kilometers per hour (200 miles per hour), at which point they exerted 500 to 1,180 kilograms (1,100 to 2,600 pounds) of force on the upper body.

It soon became apparent that airbags deploying with this force could be lethal, especially for children and drivers under a certain height who had to sit close to the steering wheel. By mid-2003, 231 people (144 of them infants and children) had been killed by airbag deployments, some of them triggered by collisions occurring at very low speeds. In contrast to these airbag-related fatalities, there was an estimated 14 percent reduction of the risk of being killed in the event of an accident. But this is far less than the 45 percent reduction attributed to the use of seat belts. Used together, airbags and seat belts lower the risk of fatality by 50 percent. It is thus apparent that airbags are a useful supplement to lap-and-shoulder belts, but they are not a substitute. A majority of the driving public seems to have recognized this fact, and approximately 70 percent of drivers now use seat belts, far more than had been deemed possible when passive restraints were first decreed. This has allowed the installation of airbags that inflate with less force. Some cars are being designed with *smart* airbags that vary the force of deployment according to a number of variables, such as the weight of the driver or passenger. These improvements will make airbag deployment less hazardous, but the risk of some airbag-induced casualties still remains.

Assessing whether or not the lives saved by airbags have outweighed the deaths they cause is no easy task. It can be said with certainty, however, that no medicine with the airbag's ratio of deaths caused to lives saved would ever have been approved by government regulators.

Ethical Perspectives

Although the United States, with its long travel distances and individualist social values, has set a dominant pattern for automobile development and utilization, other countries have sometimes adopted public policies at variance with those of the United States. For example, in part because of smaller streets and roadways, cars in Europe are generally smaller in size than those in the United States. And because automobile ownership was for many decades largely restricted to upper-income individuals, European countries also have generally taxed gasoline at higher rates, with some of the revenues used to subsidize public transportation systems.

The issues engendered by automobile emissions and automotive safety hardly exhaust the ethical concerns posed by the automobile wherever it has taken hold. For example, important issues can be raised about the consequences of the automobile's ravenous consumption of energy. In addition to the environmental problems already mentioned, the massive demand for petroleum-based fuels has affected the distribution of wealth at both a national and international level. In many petroleum-producing countries the bulk of oil revenues has gone to a small segment of the population, contributing to a lopsided distribution of income and wealth, and exacerbating social tensions. For the world as a whole, high energy prices due in part to the ever-increasing demand for automotive fuels have made the efforts of poor countries to modernize their economies more difficult.

In the realm of international relations, important questions can be raised in regard to how foreign policies and military operations have been affected by the need to maintain access to, or even control of, oil supplies, especially in the Middle East. Finally the accelerating use of the world's petroleum supplies and their inevitable depletion should provoke questions regarding what, if anything, is owed to future generations by the present one. In sum, as befits an artifact that has shaped the modern world like few others, the automobile has generated a host of ethical issues that need to be addressed if reasonable and effective public policies are to be developed and implemented.

RUDI VOLTI

SEE ALSO *Environmental Ethics; Ford Pinto Case; Pollution; Safety Engineering: Practices; Roads and Highways*.

BIBLIOGRAPHY

Bryner, Gary C. (1995). *Blue Skies, Green Politics: The Clean Air Act of 1990 and Its Implementation*. Washington, DC: CQ Press. A narrative and analysis of the governmental processes that resulted in stiffer emissions regulations.

Flink, James J. (1988). *The Automobile Age*. Cambridge, MA: MIT Press. A comprehensive history of the automobile and its consequences.

Illich, Ivan. (1974). *Energy and Equity*. New York: Harper & Row. A provocative challenge to the dominance of automobile-based transportation and an energy-intensive society in general.

Nader, Ralph. (1972). *Unsafe at Any Speed: The Designed-In Dangers of the American Automobile*. New York: Grossman. The book that launched the crusade for improved automobile safety.

Volti, Rudi. (2003). "Reducing Automotive Emissions in Southern California: The Dance of Public Policies and Technological Fixes." In *Inventing for the Environment*, ed. Arthur Molella and Joyce Bedi. Cambridge, MA: MIT Press. A discussion of the interrelationship of policy initiatives and technological change.

AUTONOMOUS TECHNOLOGY

The term *autonomous technology* is associated with arguments that modern technology has grown out of control or develops independent of any particular human intention or plan. It is usually used to highlight undesirable aspects of technological society undermine human autonomy, thus signaling its ethical relevance. The clear ethical connotation of autonomous technology marks its difference from the notion of technological determinism, with which it is often associated.

Challenging the taken-for-granted notion of technology as simply an instrument or a tool, as well as the belief in human freedom, the concept of autonomous technology has been at the center of various controversies in the philosophy of technology, where it has functioned in three related contexts. First, it has served to articulate an uneasy feeling that has accompanied the mastery of nature and the fast pace of technological change since the Industrial Revolution. As early as the nineteenth century, stories were written about human beings being ruled by "their" mechanical creatures, which had gained autonomy. Mary Shelley's famous novel *Frankenstein* (1994 [1818]) is the best-known example. Second, the concept has been associated with those philosophers who stressed the alienating and dehumanizing aspects of modern technology. Examples include Martin Heidegger (1889–1976), Herbert Marcuse (1898–1979), and Lewis Mumford (1895–1990). Finally, third, are those who have popularized the term and made it a central theme in their analyses of technology. Here the natural reference is to Jacques Ellul and Langdon Winner.

Theories of Autonomous Technology

Ellul (1954) presents characteristics of modern technology such as automatism, self-augmentation, universalism, and autonomy—the last of which summarizes the rest. Ellul claims that modern technology, unlike traditional technology, is not bound by any heteronomous rules or principles, but develops according to its own rules. As its scale and pervasiveness increase, the development of technology (Ellul's term is *la technique*) is influenced neither by sociopolitical and economic changes, nor moral and spiritual values. Rather, technological change itself now defines the context of other aspects of culture such as capitalist competition for survival in the market. The pursuit of human well-being, presumably the purpose of technological development, is replaced by obsessive pursuit of efficiency, even though the exact meaning of efficiency is often unclear. Technological progress is assumed to be always beneficial, while dimensions of sacredness, mystery, and morality are minimized. Autonomous technology reaches fulfillment when people no longer feel uneasy about "mastering nature" that has come to contradict their own human autonomy.

Winner (1977) claims that autonomous technology is revealed most clearly in technological politics. Examples include the political imperative to promote technology, because problems from one technology require another to address it, and the phenomenon of reverse adaptation, in which an end is modified so that it fits the available means. Showing that technological artifacts have political implications (Winner 1980), Winner argues that modern technology should be perceived as legislation that shapes "the basic pattern and content of human activity in our time" (Winner 1977, p. 323) and as forms of life, which have become part of our humanity (Winner 1986). The dilemma of technological society is that decisions on technology are often necessitated by existing technologies (the technological imperative); examples include the nuclear power plant and nuclear waste storage. Furthermore, sometimes, the ends and means of technological enterprises are reversed (reverse adaptation), as one can see in the development of space projects. In this respect, Winner agrees with Ellul that "if one looks at the broader picture of how technique is welcome and incorporated into society, one can hardly be confident that the origins, activities, and results of social choice about technology are firmly in anyone's grasp at all" (Winner 1995, p. 67).

Nevertheless, while appreciating Ellul's analysis, Winner eventually criticizes Ellul for ignoring human agency in his conception of autonomous technology. For Winner, it was humans that have let modern technology grow out of control, by mistakenly ignoring its political dimensions. He argues that although technology is out of control or drifting without fixed direction, it is not fully self-determining, with a life of its own. Technology is only semiautonomous. Thus, the issue raised by autonomous technology is "what humanity as a whole will make of them" (Winner 1995, p. 71).

Criticism and Response

Concepts of autonomous technology have been subject to various criticisms and misunderstandings. First, autonomous technology is often accused of reflecting irrational technophobia. This view relies on the simple assumption that technology is a neutral instrument, and as such under full human control. Accordingly, autonomous technology is regarded as a self-contradicting term.

A second objection is that the history of technology shows that technological development is not autonomous. Social constructivists argue that technological developments are contingent, because they are shaped by various sociopolitical and economic influences. A famous example is how the bicycle came to have its current design (Pinch and Bijker 1987). In the nineteenth century, there was another competing design with a large front wheel. As time went by, the current design became the standard model, not because of any internal drive for efficiency but simply because people began to perceive the bicycle as a means of transportation rather than as something used for sport. Based on this thesis, some social constructivists have developed theories of public participation in technological decision making processes (Feenberg 1999, Bijker 1995).

A more serious challenge to autonomous technology is that the idea leads to technological determinism and pessimism. Technological determinism claims that technological development has a unilateral influence on all aspects of human life and follows a fixed path according to its inner dynamics. Consequently, there cannot be any meaningful effort to avert the situation. The concept of autonomous technology is often considered the most straightforward and pessimistic version of technological determinism that denies any hope for a better future in the technological society.

However, the idea of autonomous technology rests on an understanding of technology that is often overlooked by such criticisms. First, autonomous technology specifically refers to modern technology as opposed to traditional technology. Calling a hammer and a nuclear power plant "technology" in the same sense ignores technology as a *modern experience*. Second, the prime concern of autonomous technology is not individual technologies, such as the bicycle. For Ellul, technology (*la technique*) is the *ensemble* of individual technologies that compose a technological system. The particular development of the bicycle is thus irrelevant. Autonomous technology is not about the next step of individual technological development, but about the movement of the technological system at large, with its unintended socioeconomical, cultural, environmental, and political consequences. It is impossible for anyone to claim full control over technological change in this broad sense, which is always geared toward increased levels of technology or artifice in the human world.

When technology is viewed in this way, it is misleading to quickly identify autonomous technology and technological determinism. Autonomous technology does not claim that the evolution of individual technologies follows a fixed path, nor does it exclude possible sociopolitical interventions. On the contrary, Winner claims, "one can say that all technologies are socially produced and that technical devices reflect a broad range of social needs" (Winner 1995, p. 70). As aforementioned, the concept of autonomous technology should be seen in the broader context of technological society. Technological evolution would function like biological evolution, on its own terms but not in a wholly deterministic manner. Autonomous technology certainly allows superficial variances in technical processes, caused by sociocultural and economic factors, but the efficiency principle remains the driving force directing the all-embracing comprehensive technological enterprise, which human beings are not able to alter or stop. Carl Mitcham (1994) distinguishes Ellul's theory as a form of *qualified determinism*, contrasted with naive determinism.

Autonomous Technology and Human Freedom

Hence, the way in which autonomous technology undermines human autonomy is subtle and indirect. People can freely choose whether they will use this or that computer program, for example, but the decision is made based upon the belief in the inevitability of progress in computer technology, which no one can alter. The conviction that technological progress is inevitable and beneficial is the basis of virtually every political agenda and education system around the globe.

Is an escape possible? Does autonomous technology encourage pessimism by denying human freedom? It is undeniable that this concept is discouraging in the sense that it does not leave much room for a bright future or positive action toward change. Nevertheless, it is important to remember that this concept is proposed in the context of a social critique of the contemporary technological society, rather than being part of theoretical and neutral reflection on technology. Therefore, it is misleading to focus on whether technology is autonomous or not "by nature." The argument for autonomous technology remains strong, as long as people allow technology to increasingly dominate all aspects of their lives without any critical reflection.

Ellul (1988, 1989) sees little hope for reverting the movement of autonomous technology. He argues that the only chance—the only freedom—left for a human being in the face of autonomous technology is to acknowledge one's non-freedom and to practice an ethics of non-power, namely, deciding not to do everything one can do with technology. Because Winner (1977) views technnology as a political phenomenon, he denies the absoluteness of autonomous technology; he proposes new technological forms that can accommodate more public participation and flexibility, thus allowing the possibility of political intervention in the process of technological development. This suggestion was further developed in Richard E. Sclove's "design criteria for democratic technologies" (Sclove 1995). Winner (1977) says that autonomous technology is the question of human autonomy reiterated. This remark succinctly expresses the main concern of the concept, because, paradoxically enough, different theories of autonomous technology all emphasize the importance of human autonomy, whether they are encouraging or discouraging concerning the future of technological society.

WHA-CHUL SON

SEE ALSO *Artificial Intelligence; Automation; Autonomy; Critical Social Theory; Determinism; Ellul, Jacques; Frankenstein; Freedom.*

BIBLIOGRAPHY

Bijker, Wiebe E. (1995). *Of Bicycles, Bakelites and Bulbs: Toward a Theory of Sociotechnical Change*. Cambridge, MA: MIT Press.

Ellul, Jacques. (1954). *La Technique ou l'Enjue du Siècle* [The technological society]. Paris: Armand Colin. This work is the main reference to the notion of autonomous technology.

Ellul, Jacques. (1988). *Le bluff technologique* [The technological bluff]. Paris: Hachette. Ellul argues that discourses on technique since the 1980s make people feel comfortable with dehumanizing technique, completing the autonomy of technique. He also talks about the paradox of non-freedom as the only freedom left for people in technological society.

Ellul, Jacques. (1989). "The Search for Ethics in a Technicist Society." *Research in Philosophy and Technology* 9: 23–36.

Feenberg, Andrew. (1999). *Questioning Technology*. London: Routledge.

Micham, Carl. (1994). *Thinking through Technology: The Path between Engineering and Philosophy*. Chicago: University of Chicago Press.

Pinch, Trevor J., and Wiebe E. Bijker. (1987). "The Social Construction of Facts and Artifacts: Or How the Sociology of Science and the Sociology of Technology Might Benefit Each Other." In *The Social Construction of Technological Systems: New Directions in the Sociology and History of Technology*, ed. Wiebe E. Bijker, Thomas P. Hughes, and Trevor J. Pinch. Cambridge, MA: MIT Press.

Sclove, Richard E. (1995). *Democracy and Technology*. New York: Guilford Press.

Shelley, Mary. (1994 [1818]). *Frankenstein, or the Modern Prometheus, the 1818 Text*, ed. Marilyn Butler. Oxford: Oxford University Press.

Winner, Langdon. (1977). *Autonomous Technology: Technics-Out-of-Control as a Theme in Political Thought*. Cambridge, MA: MIT Press. Winner confronts with different versions of autonomous technology theory, while developing his own.

Winner, Langdon (1980). "Do Artifacts Have Politics?" *Daedalus* 109(1): 121–136. This is one of the most frequently cited work in philosophy of technology concerning the political aspect of technology.

Winner, Langdon. (1986). *The Whale and the Reactor: A Search for Limits in an Age of High Technology*. Chicago: University of Chicago Press.

Winner, Langdon. (1995). "The Enduring Dilemmas of Autonomous Technique." *Bulletin of Science, Technology & Society* 15(2): 67–72. The notion of autonomous technology is defended mainly against social constructivism.

AUTONOMY

Autonomy (from the Greek *autos* for self and *nomos* for rule, governance, or law) is defined as self-determination or self-rule. Its original use in ancient Greece referred to the sovereignty of states, but Immanuel Kant (1724–1804) and others in the modern period applied the term to individuals. For Kant, one is autonomous when one subjects oneself to moral rules recognized by the rational self. In contrast, one whose decisions and actions are shaped by others without critical reflection on the individual's part is heteronomous. Autonomy brings with it moral responsibility, and the autonomous person is open to charges of negligence or recklessness in the uses of science or technology if proper precautions against risk are not taken. Autonomy may also refer to the self-governing nature of professions or groups, such as the scientific community. Furthermore technology that operates without regular instruction from a person is sometimes called autonomous technology.

Conditions of Autonomy

Autonomy has many faces. Joel Feinberg (1989) points out at least four meanings: the capacity to rule oneself; the condition of ruling oneself; the virtuous ideal of ruling oneself; and the authority to rule oneself. Gerald Dworkin (1988) highlights eight common uses. One commonality is the idea that autonomy, like freedom, combines two aspects: the negative condition of freedom from external constraints and the positive condition of a self-determined will. Those barred from acting in accordance with their will, for instance, by physical constraints or coercive threats, are not able to act autonomously despite what they may internally will. Their will is either rendered impotent by force or limited to such an extent that a reasonable person could be said to have no choice. Someone who offers a wallet in response to a threat with a gun (Your money or your life) can be said to will such an action, but not autonomously, given the lack of reasonable alternatives. Yet a person may fail to act autonomously even without the existence of external constraints.

Harry Frankfurt (1989) famously argued that one cannot be said to choose freely unless one's first order desires (what one wants) are themselves chosen or affirmed by one's second order volition. That is, to be autonomous, one must want to want what one wants. Reluctant addicts who desire more heroin may wish that they did not want it, but nonetheless succumb to the strong first order desire for the drug. According to Frankfurt, they do not act autonomously, though they are free from external constraints. In contrast, rational agents who carefully reflect on their first order desires, identify with their preferred desires, and then act accordingly, *are* autonomous due to this vertical alignment of desires.

One problem with this view is that a person could have this vertical alignment of desires only as a result of undue interference from a third party (e.g., a hypnotist), making the identification inauthentic. This problem led Dworkin (1986) to add a procedural independence criterion to the concept of autonomy, meaning that to be autonomous one must identify with one's desires for reasons that are *one's own*. Yet some reasons that appear to be one's own may in fact be part of a larger system of values that has shaped the very person one becomes, and the desires one forms. For instance, a scientist's first order desire may be to receive a grant. The scientist may critically reflect on this desire, and approve of it, recognizing that grants are the way to succeed in science (they support work that leads to progress, publication, and future grants). So it appears that the scientist acts autonomously in applying for the grant. But the kinds of grants that a scientist may submit (or that have any likelihood of being funded), are in large part dependent on broader forces: governmental agendas, money-making prospects, and what counts as a hot issue. Has the scientist autonomously chosen the specific focus of the research? Traditional theories of autonomy do not allow much room for critique of background conditions that may unjustly or unduly shape an individual's desires and identification with those desires.

Relational Autonomy

The difficult question of determining when, if ever, *anyone* is truly free of external constraints that inhibit autonomy has led some feminist theorists to offer a theory of relational autonomy. Relational autonomy is built on the idea that our selves are relational and social, rather than essential and ontologically independent. Marilyn Friedman (1999) proposes an autonomy model that requires integration of first and second order desires, without putting priority on second order desires (sometimes a first order desire may be more authentic than what one has been socially shaped to believe one *ought* to want, especially under conditions of discrimination).

In a complementary manner, Diana Meyers (1999) argues that autonomy requires certain competency skills (e.g., self-definition, self-discovery, self-direction) that allow for sufficient critical reflection on one's desires and choices. If a social context impedes the development of competency skills for certain groups (e.g., women or racial minorities), then such people may never achieve full autonomy.

However Meyers (1999) allows for degrees and spheres of competency, which can result in partial autonomy. In the case of the scientist, investigation of the fairness of background conditions that determine the focus and availability of grants would be part of determining the individual's degree of autonomy (and resultant responsibility for actions). Contemporary life takes place against a large technological system (roads, electrical utilities, water systems, phone service, and others) that inevitably shapes the kinds of choices individuals can make. Relational autonomy theorists insist that the fairness of such background conditions be evaluated as part of our understanding of individual autonomy.

Significance of Autonomy for Moral Practice

Our theoretical understanding of individual autonomy will have significant effects on the use and meaning of autonomy in practical settings in medicine, law, scientific research, education,and more. In medical ethics, for instance, respect for autonomy is often considered the most important moral principle (Beauchamp and Childress 2001). It protects patients from paternalism, respects differences in individual values, and allows patients to refuse unwanted treatment. The principle of respect for autonomy includes rules regarding truth telling, promise keeping, and informed consent. Informed consent, in turn, consists of requirements of patient competency, disclosure of information, patient comprehension, voluntariness, and ongoing consent. Yet such conditions are often not guaranteed by simple informed consent documents, and even when fulfilled, they may "mask the normalizing powers of medicine" (Sherwin 1998, p. 28) that set the standards for competency, relevant information, and voluntariness.

Background conditions may also influence the degree to which one is autonomous in regard to new technologies. Available technologies can increase an individual's autonomy, for instance, when an insulin pump allows a diabetic person to avoid the constraints of dialysis, or a computer message board allows a patient with Lou Gehrig's disease to communicate preferences. Such technologies increase options, enhancing autonomy.

However some medical technologies, offered for the betterment of the individual, may in fact decrease autonomy, in that they override individuals' unpopular preferences. Some deaf individuals reject cochlear implant technology, some amputees refuse prosthetic replacements, and some intersexual people argue against sex-definition surgery. The available technologies, they warn, appear to increase options when in fact they eliminate other, less popular options, forcing individuals to fit the norm.

In the traditional models of autonomy, individual choice takes priority. But with relational autonomy, individual choices are only as valuable as their historical and relational precursors. Thus rather than taking a treatment request at face value, a relational autonomy model recommends the following:

1. lively dialogue, including critical questions regarding competency skills and the context of desire formation (our self-knowledge is in part social, and so engagement in dialogue should be seen as helpful rather than as a sign of disrespect) (McLeod 2002);
2. more respect for people who are only partially autonomous (e.g., children, individuals with mental retardation, mental illness, or senility);

3. recognition that patients may autonomously make decisions based on their familial situations (e.g., requesting assisted suicide because they do not want to be burdens on their families).

Indeed, on the relational autonomy model, making a choice *without* reference to our social context appears inauthentic rather than autonomous (Wolf 1996). Perhaps the most contentious issue of autonomy is determining when one's context undermines rather than engenders one's capacity for self-determination.

Autonomy in Science and Engineering

Professions or groups, as well as individuals, may be autonomous to the extent that they are self-governing. The autonomy of the scientific community has been defended as important for the preservation of free inquiry that results in knowledge production. Preserving that autonomy requires defining the boundaries and norms of the community. Free inquiry, for instance, may be stifled when academic scientists partner with private industry in order to gain grants that support the university as well as their own research. Such partnerships may decrease scientific autonomy by limiting the focus of investigation to what is marketable and/or profitable, and discouraging the sharing of results and methods in order to protect patents and preserve trade secrets. Scientific investigation will always be tied to funding, but must be protected from influences that threaten to corrupt the scientific process.

Yet with autonomy comes responsibility. Scientists who freely choose to develop nuclear weapons, or who experiment on genetically modified foods, retain some responsibility for the societal risks incurred in their work. The idea that science is value-free and that the responsibility for using or misusing scientific data rests with society at large rather than with the scientists who undertake the research is difficult to defend. Value-laden decisions are made throughout the scientific process. Scientists who retain autonomy in their profession must also accept the responsibility to avoid recklessness and negligence in respect to the risks created by their research (Douglas 2003).

Furthermore the value of free inquiry is limited when it threatens to undermine even more fundamental issues, such as access to free inquiry itself (Kitcher 2001). In defending this claim, Kitcher considers the work of sociobiologists and evolutionary psychologists who have attempted to support inegalitarian racial views that themselves threaten the ability of racial minorities to participate in scientific debates. Scientific autonomy, then, may also be limited by background conditions. A move to more democratic regulation of science (involving lay citizens) has been suggested as a possible remedy for these problems, highlighting again the relation between scientists and the broader community (Kleinman 2000).

Professional autonomy among engineers diverges from that of scientists in that engineers tend to have less individual autonomy on the job and more direct public impact in their work. Most engineers, at least in the United States, are employees rather than independent contractors, resulting in less opportunity for self-determination on the job, and setting up potential conflicts between their obligations as employees and their duties to exercise professional judgment. An employer that demands a sacrifice in safety precautions in the interest of profit or timeliness, for instance, may interfere with the autonomy of the engineer (Mitcham and Duvall 2000). Because engineering work often results in public technologies or structures (bridges, transportation, and others), failures of professional judgment can have widespread impact, as in the famous cases of the Challenger disaster, the American Airlines DC-10 crash of 1979, and the Hyatt Regency hotel walkway collapse (Whitbeck 1998). Whistleblowers may be required to sacrifice corporate loyalty (and job security) in the name of protecting the public good.

SARA GOERING

SEE ALSO *Autonomous Technology; Human Subjects Research; Informed Consent; Kant, Immanuel; Medical Ethics*

BIBLIOGRAPHY

Beauchamp, Tom, and James Childress. (2001). *Principles of Biomedical Ethics*, 5th edition. New York: Oxford University Press. Classic text in medical ethics.

Douglas, Heather. (2003). "The Moral Responsibilities of Scientists: Tensions Between Autonomy and Responsibility." *American Philosophical Quarterly* 40(1): 59–68.

Dworkin, Gerald. (1988). *The Theory and Practice of Autonomy*. New York: Cambridge University Press. Excellent discussion of the concept of autonomy and its significance.

Dworkin, Gerald. (1989). "The Concept of Autonomy." In *The Inner Citadel: Essays on Individual Autonomy*, ed. John Christman. New York: Oxford University Press.

Feinberg, Joel. (1989). "Autonomy." In *The Inner Citadel: Essays on Individual Autonomy*, ed. John Christman. New York: Oxford University Press.

Frankfurt, Harry. (1989). "Freedom of the Will and the Concept of a Person." In *The Inner Citadel: Essays on Individual Autonomy*, ed. John Christman. New York: Oxford University Press.

Friedman, Marilyn. (1999). "Autonomy, Social Disruption, and Women." In *Relational Autonomy: Feminist Perspectives on Autonomy, Agency, and the Social Self*, eds. Catriona MacKenzie and Natalie Stoljar. New York: Oxford University Press. The book contains a great collection of works on relational autonomy.

Kleinman, Daniel Lee. (2000). "Democratization of Science and Technology." In *Science, Technology and Democracy*, ed. Daniel Lee Kleinman. Albany: State University of New York Press.

McLeod, Carolyn. (2002). *Self Trust and Reproductive Autonomy*. Cambridge: MIT Press.

Meyers, Diana. (1999). "Intersectional Identity and the Authentic Self: Opposites Attract!" In *Relational Autonomy: Feminist Perspectives on Autonomy, Agency, and the Social Self*, eds. Catriona MacKenzie and Natalie Stoljar. New York: Oxford University Press.

Mitcham, Carl, and R. Shannon Duvall. (2000). *Engineering Ethics*. Upper Saddle River, NJ: Prentice-Hall.

Sherwin, Susan. (1998). "A Relational Approach to Autonomy in Health Care." In *The Politics of Women's Health*, ed. Susan Sherwin. Philadelphia: Temple University Press.

Whitbeck, Carolyn. (1998). *Ethics in Engineering Practices and Research*. Cambridge: Cambridge University Press. Excellent introduction to ethical issues in engineering.

Wolf, Susan. (1996). "Gender, Feminism, and Death: Physician-Assisted Suicide and Euthanasia." In *Feminism & Bioethics: Beyond Reproduction*, ed. Susan Wolf. New York: Oxford University Press.

AVIATION REGULATORY AGENCIES

Aviation regulatory agencies are charged with oversight of the aviation industry. Such agencies are primarily governmental or international organizations. The issue of safety is central to any such agency: Not only must the aviation industry be supervised, passenger aircraft must also be certified safe. How should this supervision and certification be accomplished? Is the most effective regulation done with a centralized system? What are the alternatives? What standards should be used? Aviation regulatory agencies, such as the Federal Aviation Administration (FAA) in the United States, have been established to address these issues.

The United States Federal Aviation Administration (FAA), in its draft Flight Plan 2004–2008, states that it regulates more than half of all air traffic. The FAA also certifies more than seventy percent of all large jet aircraft. Most countries around the world have their own civil aviation authorities to devise and implement regulations within their respective territories, but the FAA provides indirect or direct assistance to 129 countries to help improve their air traffic control systems. The International Civil Aviation Organization (ICAO) represents 188 independent civil aviation authorities, but the FAA is the largest intellectual and financial ICAO contributor.

History of the FAA

During World War I, the U.S. government expanded the aviation manufacturing industry, and Congress funded a postal program that would serve as the model for commercial air operations. In the early 1920s, many argued for federal regulation of the nascent commercial aviation sector to ensure public confidence, but others distrusted the government or wanted states to regulate aviation. Must the aviation industry be regulated? Prior to 1926, flyers of airplanes required no pilot's license, nor a license to carry passengers or materials, and took lessons from unlicensed schools or individuals. They generally took off and landed wherever they pleased (Komons 1978). But that year U. S. President Calvin Coolidge signed the Air Commerce Act (ACA) and federal oversight began under the direction of the Department of Commerce, which established safety standards and certification procedures for pilots and aircraft. The aviation industry was growing quickly and problems were being encountered with safety, aircraft route allocations, and airline formation. But was the public interest being protected or were the interests of the airline industry being served?

The response to this question was the 1938 Civil Aeronautics Act that transferred Federal civil aviation responsibilities to the Civil Aeronautics Authority (CAA), an independent agency. In 1940, the CAA was split into two agencies: One was the Civil Aeronautics Administration, responsible for air traffic control (ATC), airperson and aircraft certification, safety enforcement, and airway development. The other was the Civil Aeronautics Board (CAB), which became the specialized agency to regulate the airline industry. It was charged with assigning routes to air carriers and controlling the fares charged. In 1958 the Federal Aviation Act created the Federal Aviation Agency, which in 1996 was renamed the Federal Aviation Administration. The passage of this act was prompted by the development of jet airliners and a series of midair collisions, suggesting that greater centralization and standardization were necessary to ensure safety. The FAA absorbed the functions of both the CAA and the CAB (although CAB continued to exercise economic regulations of airlines until 1978) and acquired sole responsibility to operate a national ATC system and develop, initiate, and monitor standardized safety requirements for air travel. The FAA is charged with promoting safety and security and developing and maintaining an air traffic management system that is efficient, secure, and safe.

Deregulation

Is the FAA effective in exercising its responsibilities? Some have argued that the FAA is the cause of current problems in the United States because it resists change. Some of these problems are specific, such as runway incursions (critics maintain that the technology for surface navigation and communications has been inadequately developed), whereas others are more systematic, such as management issues that have led to cost overruns, schedule delays, and performance shortfalls. According to a Government Accounting Office Report it takes the FAA five to seven years to implement meaningful and lasting responses to challenges posed by increased capacity, safety, efficiency, and other demands.

In 1978 the Airline Deregulation Act ended the CAB, began the removal of government control, and opened the deregulated passenger air transport industry to market forces. Eventually, deregulation would benefit the consumer with lower ticket prices. By 1988 the number of people working in the industry had increased by thirty-two percent, with air traffic up by fifty-five percent and costs down seventeen percent. In 1998 ticket prices had been reduced by twenty percent and passenger numbers increased from 275 million to 600 million compared to ten years earlier.

Airline deregulation is nevertheless controversial (see Bailey et al. 1985). Deregulation and economic competition directly contributed to the bankruptcy of several major airlines. Prior to September 11, 2001, the projections were for the airline industry to grow 5 to 7 percent per year. Air travel declined over the next few years, putting airlines and aircraft manufacturers on the financial edge. With increasing economic pressures and rising fuel prices, the pressure exists to cut costs and take greater risks. It is the job of the FAA to insure that safety is not compromised.

Technology and Safety Regulations

People look to technology to solve many problems associated with the crowded ATC system. One concept being considered is the free flight system. Currently, major airlines use a hub and spoke model. Small commuter airlines feed into larger airports that allow major airlines to have a higher passenger density and reduce costs. In contrast, a free flight system allows people to fly direct from any nearby, small airport to an airport near their destination. This creates complexity and pushes the limits of technology, as aircraft will no longer use common routes. Only sophisticated navigation equipment and procedures would make this possible. The Global Positioning System (GPS) would allow ATC to track each airplane. Benefits of implementing such a free flight system would include time savings on trips of approximately 400 miles in length (Czajkowski 2002). However, one possible drawback is increased travel costs, which may further restrict air travel to the wealthy. This free flight concept might also include an airplane design such as the Advanced Flying Automobile that would make personal flights accessible to those who could afford the technology.

The FAA often partners with the National Aeronautics and Space Administration (NASA) on issues of technology development, including innovations to improve aging aircraft, prevent accidents caused by weather, and improve air traffic control operations. Collision avoidance is also being researched by the NASA Dryden Research Center. Early twenty-first century technology makes it possible for Unmanned Aerial Vehicles (UAVs) to fly in airspace with piloted aircraft (Degaspari 2003). Whether this will be allowed is up to the FAA. The FAA makes all final decisions concerning airspace, aircraft certification, aircrew certification, and airports. The main goal for these decisions is to ensure safety, but the FAA must also take technological and economic factors into consideration. Many technologies are not mature enough or the expense to the aviation industry is prohibitive. Decisions are made most rapidly when the public demands action due to safety concerns. However, swift decisions sometimes generate more controversy in the long term.

According to the National Transportation Safety Board (NTSB), human error has accounted for the greatest percentage of aviation accidents since the 1950s. Furthermore, increasing capacity and technological complexity at all levels of the aviation industry can exacerbate human error by introducing demands on limited cognitive capacities. Thus, the most important technological improvements to aircraft and ATC systems are those that can minimize human error. This also means that training and human resource management may be the best investment for regulatory agencies to fulfill their goal of improved safety. Regulatory agencies are also faced with a vast safety discrepancy between the top twenty-five airlines with the best safety records and the bottom twenty-five airlines with the worst. This suggests that the technologies and human resource management systems already exist to ensure greater safety. The challenge is in transferring these strategies and capabilities to other airlines and enforcing strict compliance with safety regulations by all airlines.

This raises the issue that not all segments of the aviation industry are regulated by the same set of standard rules. For example, general aviation (flights that are on-demand, that is, not routinely scheduled) accounts for seventy-seven percent of all flights in the United States, including the majority of pilot training flights. The bulk of fatalities occur in the general aviation sector, and the accident rate is many times greater than in the commercial sector. Both the Transportation Security Administration (TSA) and the FAA have different regulations for general aviation. New regulations were put in place after the terrorist attacks of September 11, 2001, because some of the terrorists utilized general aviation flight schools to learn how to steer aircraft. The seventeen general aviation associations comprising the General Aviation Coalition often work closely with regulatory agencies in crafting rules and best practice procedures.

Another sector of the aviation industry is ultralight aircraft, which are light weight (less than 150 lbs if not powered and less than 254 lbs if powered), single occupant, low-speed, recreational aircraft. The ultralight movement formed in the 1970s as operators began to attach small engines to foot launched hang gliders. In 1982, the FAA implemented ultralight regulations, and the Experimental Aircraft Association (EAA) develops and administers ultralight self-regulation programs.

Future Regulation

The FAA draft Flight Plan 2004-2008 outlines four goals. The first is increased safety, a top public-interest priority and economic necessity. People will fly only if they feel safe and are confident in the system. Increased capacity is the second goal: More passengers must be able to move quickly and efficiently through the system. The third goal is improved international partnerships to promote and enhance safety. The FAA works with other regulatory organizations such as the ICAO, the European Aviation Safety Agency, and the North American Aviation Trilateral. Lastly, the FAA seeks organizational excellence in all areas: strong leadership, fiscal responsibility, and performance-based management. The FAA also needs to simplify and clarify technical issues for the general public.

The challenges faced by the FAA and other aviation regulatory agencies may nevertheless inhibit achievement of such goals. Airline and aircraft manufacturing industries are having financial difficulties, and are thus reluctant to equip their aircraft with the latest technology to improve safety. If the technology for improved safety exists, should it be required on aircraft? This is a decision usually left to the FAA. An example illustrating this decision process is the post-2001 reinforcement of cockpit doors. An FAA regulation required the modifications, but the implementation time frame made the request quite reasonable. Most everyone could see the benefit of stronger cockpit doors and airlines agreed to spend the money. Some critics doubted it was enough to deter terrorists. The same is true of the decision to allow pilots to carry weapons in the cockpit. Critics argue that inexperience with handguns makes their use by pilots dangerous. But the FAA has allowed pilots to carry weapons under specific guidelines.

Although safety is a key element of the FAA Flight Plan 2004–2008 there are questions about whether it is doing everything possible. Former FAA Administrator Jane Garvey has stated that flying in a commercial aircraft is forty times safer than driving a car, but that does not clarify whether that level of safety is high enough to secure public trust. Prior to 2001, the FAA had the responsibility to deliver a safe system for passengers, not just a safe aircraft and competent pilot. The terrorist attacks highlighted several errors in airport security. Shortly after the attacks, the Transportation Security Agency (TSA) was formed and given the responsibility of protecting all transportation modes from terrorism and other criminal threats. Much money and effort has been expended to improve security at major airports. One result is increased passenger processing time before takeoff, which has resulted in many new federal security workers being added to the government payroll.

The FAA is also responsible for certification of new aircraft and engines. Airbus and Boeing are proposing the Airbus 380 and the 7E7, respectively, as large aircraft replacements for current civilian airliners. The Airbus 380, first shown to the public on January 18, 2005, will carry 550 people and have wingspans at the maximum allowable specification for current airport terminal requirements. Other designs are capable of achieving supersonic speeds that will require minimizing the shock wave generated by those aircraft. Should such aircraft be allowed to fly over populated areas? The original supersonic airliner, Concorde, was not allowed to fly supersonically over populated areas. Again, the FAA makes the final decision.

The FAA has the oversight of environmental issues concerning aircraft engines. One issue is noise pollution. Another is particulate emissions, which pollute the areas surrounding airports. While at altitude, emissions of NO_x and CO_2 are blamed for depleting the ozone layer and contributing to global climate change. The FAA has the power to regulate the concentrations of these substances found in engine exhaust emissions and is also able to modify the limits when target goals are not reached.

One major question is whether the FAA should be privatized. Canada, Great Britain, and New Zealand have already made this step. The advantages are clear in terms of possible cost savings to the government, but it is less clear if privatization is in the best interest of the customer. Many argue that the costs to the consumer will increase to pay for improvements to the system.

Aviation regulatory agencies are one response to the social and environmental dilemmas posed by aviation technologies. The public has come to rely on organizations such as the FAA to make decisions concerning equipment and cost which directly impact passenger safety. Is the FAA acting in the interests of the passenger and government or are they easily influenced by pressure groups from the aviation industry? Safety is the most important concern for air travel, but the public seems to have a blind trust in these agencies. The public should be more involved in these decisions, especially those concerning safety.

KENNETH W. VAN TREUREN

SEE ALSO *Airplanes; Global Positioning System; Regulation; Safety Engineering Practices; Security; Science, Technology, and Law; Terrorism.*

BIBLIOGRAPHY

Bailey, Elizabeth E., David R. Graham, and Daniel P. Kaplan. (1985). *Deregulating the Airlines (Regulation of Economic Activity)*. Cambridge: MIT Press.

Cleveland, Paul A., and Jared R. Price. (2003). "The Failure of Federal Aviation Administration Regulation." *Independent Review* 8(1): 53–64.

Czajkowski, Mike. (2002). *Benefits of Door-to-Door Travel Time Reduction to US Economy*. San Diego, CA: 2000 World Aviation Congress, October 10–12, 2000.

Degaspari, John. (2003). "Flying Blind." *Mechanical Engineering* 125 (7): 10–12.

Komons, Nick. (1978). *Bonfires to Beacons: Federal Civil Aviation Policy under the Air Commerce Act, 1926–1938*. Washington DC: DOT/FAA.

Oster, Clinton V., Jr.; C. Kurt Zorn; and John S. Strong. (1992). *Why Airplanes Crash: Aviation Safety in a Changing World*. Cambridge, UK: Oxford University Press.

Wells, Alexander T. (2004). *Commercial Aviation Safety*. New York: McGraw-Hill.

INTERNET RESOURCES

Boeing. (2003). "Jetliner Safety: Government's Role in Aviation Safety." Available from http://www.boeing.com/commercial/safety/pf/pf_government_role.html.

Federal Aviation Administration: *Draft Flight Plan 2004–2008*. Available from http://www.faa.gov/AboutFAA/FlightPlan.cfm.

Government Accountability Office (GAO). (2003). "Reform Effort Requires a More Strategic Result." February, GAO-03-156. Available from http://www.gao.gov/new.items/d03156.pdf.

Thierer, Adam D. (1998) "20th Anniversary of Airline Deregulation: Cause for Celebration, Not Re-Regulation." The Heritage Foundation Backgrounder Executive Summary, No. 1173, April 22. Available from http://www.heritage.org/Research/Regulation/BG1173.cfm.

AXIOLOGY

Axiology, according to its Greek etymology, means "theory of values." The term was introduced at the beginning of the twentieth century when it became a recognized part of philosophy. As a discipline distinct from science, axiology was sometimes even equated with the whole of philosophy, especially in Germany. The first books containing this expression are Paul Lapie's *Logique de la volonté* (1902); Eduard von Hartmann's *Grundriss der Axiologie* (1908); and Wilbur Marshal Urban's *Valuation* (1909).

The Concept of Value

This new branch of philosophy emerged as the concept of *value*, after having been treated almost exclusively in a technical sense in economics, began to be used in the plural (*values*) and to be an issue in philosophy. In response to the cultural imperialism of the sciences (including the so-called "human sciences"), philosophers defended their discipline and stressed that the "domain of values" was precisely a field that no science was able or entitled to treat, and was thus the exclusive responsibility of philosophy. Moreover, several philosophers argued that it was in the interest of science not to admit consideration of values into its own discourse. They advocated a neat separation of science and values, one that could be traced back to the famous clear-cut distinction between "being" and "ought to be" (*sein* and *sollen*) of Immanuel Kant: The realm of what is real is described by the sciences and has nothing to do with the realm of what ought to be, of what is worthy, which is determined by ethics. However, unlike Kant, these philosophers did not imply any rejection of a scientific—that is, rigorous and objective—treatment of the domain of values. Indeed, the neologism *axiology* indicated an intention to develop just such a treatment and to promote a more advanced and technically specific approach than the reflections on particular values that had been part of philosophy in the past.

In a very general sense, a *value* is whatever is positively appreciated; the concept usually indicates that positive characteristic for which something is appreciated, as well as the thing that carries this characteristic. Axiology considers only the first sense of value, conceived as an ideal object capable of exact study. The idea of positive appreciation can be made more precise by saying that a certain value attributed to something expresses the desirability of that thing by a certain subject: The value has the nature of a *relation* between an object and a desiring subject. This explains the early psychological trend in the theory of values, although this was soon superseded by those who maintained the *objectivity* of values (Franz Brentano, Max Scheler, Nicolai Hartmann, Wilhelm Windelband, Heinrich Rickert, and others). Therefore, not only does a value subsist independently of the fact of being or not being recognized, but it is possible to propose lists and classifications of values, on the basis of a specific access—typically an emotional intuition, according to Scheler.

However, axiology is nothing emotional; instead it aspires to be a strict *logic*. Edmund Husserl pointed out that it is possible to make a formal treatment of mental acts that are different from theoretical judgments, and "this has great significance, because it opens up the possibility of broadening the idea of formal logic to include a formal axiology and a formal theory of practice. Accordingly there arises what might be called a 'formal logic' of concrete values [der werte] and a formal logic of *practical* goods" (1969, p. 136). This approach allowed for a distinction between axiology and ethics that was not present in Kant. Indeed, as thinkers such as Hartmann and Scheler argued, although a value entails a duty in the moral sphere (i.e., the moral duty of the individual to satisfy the value), in a more general sense it implies *norms* that are not necessarily moral in character. Rickert, for example, argues that truth is also a value, because it imposes norms to be followed by those who are trying to attain it. The logic of values therefore includes only as a part the logic of truth, because there are not just epistemic and moral values, but also others such as aesthetic and religious values. Along this path it was natural to argue, with Scheler, that axiology is a logic and, as such, distinct from ethics, which is a theory of action. As a consequence, Scheler elaborated a formal theory of values, distinct from a formal theory of value-attitudes, and proposed an axiomatic treatment according to principles already outlined by Brentano. Axiology thus presented itself as a kind of rigorous discipline capable of meeting the requirements of exactness and even of formal rigor advanced by the sciences, though remaining within the realm of philosophy.

Axiology and the Social Sciences

Reference to values appeared as a specific characteristic of the epistemological structure of the historical and social sciences during the late-nineteenth- and early-twentieth-century debates that opposed them to the natural sciences. Values were seen as indispensable to *understanding* human actions in the social sciences, and as a necessary framework for historical and social scientific *explanations*. The most influential proponent of this view was Max Weber, who argued that although "reference to values" is indispensable in the social sciences, the social sciences must also be "value-free" (*wertfrei*), not only because values cannot be objectively affirmed, but also because there is a fundamental difference between ascertaining facts and evaluating how they "ought to be" according to a normative criterion:

> What is important from the methodological point of view is that the validity of a practical imperative as a norm, on the one hand, and the truth claims of a statement of empirical fact, on the other, create problems at totally different levels, and that the specific value of each of them will be diminished if this is not recognized and if the attempt is made to force them into the same category. (1978, p. 79)

This difference of levels entails

> the appreciation, quite simply, of the possibility that ultimate values might diverge, in principle and irreconcilably. For neither is it the case that 'to understand all' means 'to forgive all,' nor is there in general any path leading from mere understanding of someone else's point of view to approval of it. Rather it leads, at least as easily and often with much greater reliability, to an awareness of the impossibility of agreement, and of the reasons why and the respects in which this is so. (1978, p. 81)

Weber's argument may be clarified as follows. In order to understand and explain the conduct of human agents, the historian or social scientist must hypothesize that certain typical values inspired or guided their actions.

This hypothesis can be reinforced or modified by critical analyses of the objective evidence found in documents or other related empirical sources. Therefore, reference to values is not incompatible with objectivity. Nevertheless, historians and social scientists must refrain from expressing their own value judgments on the actions under consideration, that is, from making assessments of objectively recognized facts from the point of view of any value, because this would inevitably be a *subjective* assessment, which might even distort the objective representation of facts.

For example, a sociologist might objectively ascertain that vendetta is a value imposing certain norms of conduct within a given community, but the sociologist must refrain from expressing a judgment of approval or rejection regarding this value. This need becomes particularly clear when ideological or political values are involved in the understanding-explanation of historical or social events, because the personal value-options of the social scientist can easily induce an offer of a positive or negative portrayal of the objective situation by forcing its interpretation according to social scientist's sympathy with or hostility to the values actually followed by the people acting in this situation. This separation of objective, factual knowledge and value judgments is therefore an issue of intellectual integrity that also demands that scientists should not take advantage of objective results in their research to support their own (very legitimate) values, simply because these values are not a matter of objective knowledge. It is clear that this position is far from seeing axiology as a scientific assessment of values.

Challenges to Axiological Neutrality in Science

Weber's doctrine was widely accepted for decades: Science must be value-free, no mixture of science and values is legitimate, and the two spheres defend their legitimacies precisely by remaining clearly distinct. An initial challenge to this position occurred shortly after the middle of the twentieth century in disputes about the *neutrality of science*, or the extent to which science should and could properly remain independent from supposedly external powers and influences that might jeopardize its objectivity. Values, especially moral and political values, were included in this discussion, so that science was sometimes spoken of as "axiologically neutral." Advocates of neutrality admitted that it is often difficult to grant this requirement for science, but affirmed that it could and must be defended so as not to lose the most fundamental good of science—that is, objectivity. Others argued that the neutrality of science was impossible and not even desirable, and that so-called objectivity was only a fictitious mask placed on science for ideological and political purposes.

This debate may be adjudicated by noting that science is a complex phenomenon. Science as a system of *knowledge* must be distinguished from science as a system of human *activities*. Objectivity is the most fundamental feature of scientific knowledge, but several other motivations and values correctly concern the *doing* of science. Therefore, the real and challenging problem is that of not giving up scientific objectivity while at the same time recognizing that the scientific enterprise has to satisfy other values as well. For instance, society has much concern and expectation regarding the possibility of defeating AIDS, lending great support to biomedical and pharmaceutical research in this direction. Society's interest could not justify, however, inflating the objective purport of partial results obtained in AIDS research in order to respond to public expectation or to obtain more financial support. In another example, opposite parties in the ecological debate often force the interpretation of available scientific knowledge and information in order to make it subservient to their position, whereas a more appropriate attitude would be one of respect for the objectivity of scientific knowledge, using it as a basis for finding an equitable balance between the values of respect for the environment and technological progress.

A first admission of the presence of values in science occurred in a rather ambiguous form, in the discussion of the issue of theory comparison. Because neither empirical adequacy nor logical consistency are often decisive criteria for choosing between two rival scientific theories, a reasonable choice occurs by taking into account other criteria, such as simplicity, precision, generality, elegance, causal connection, fertility in predictions, and so on. These "virtues" (McMullin 1983) actually give rise to certain value judgments and in this sense it is said that one cannot dispense with values in science. It must be noted, however, that these values (and similar ones that have been discussed by Thomas Kuhn, Hilary Putnam, Larry Laudan, and others) are still related to the cognitive aspect of science. They are *epistemic values* and, as such, do not really respond to the question of whether non-cognitive values also have the right to be of concern in science.

The answer to this last question became irresistibly affirmative around the turn of the twenty-first century, owing to the increasing intensity and latitude of the debates regarding ethical and social problems posed by the development of technology and also of science, to the extent that these became inextricably nested and were called *technoscience*. The consideration of such non cognitive values is appropriate because it regards science and technology from the point of view of *action*. It has become clear that a broader range of values actually concerns the *doing* of technoscience, imposing a serious consideration of its axiological contexts that deserves to be included in the philosophy of science (formerly limited to a logico-methodological analysis of science), and even more significantly in a philosophy of technology. All this has implied a criticism of Weber's doctrine of value-free science that was developed especially by the Frankfurt School and also by several authors of different philosophical orientations (see, for example, Robert Proctor 1991).

In connection with its application to technoscience, axiology is finding again a rather broad circulation, not in the sense of a technically robust version of the philosophical theory of values, but in the more colloquial sense of a discourse concerned with values, a sense that is often better expressed in the forms of the adjective "axiological" or the adverb "axiologically" that do not strictly refer to a precise discipline. However, an in-depth discussion on values, their ontology, their logical relations, and their possible coordination is having an important revival, in particular in relation to science and technology, especially because one cannot escape the problem of making compatible the mutual respect of all such values. This discussion has given rise to certain technically-elaborated proposals, such as that of making use of the conceptual and formal tools of general systems theory (Agazzi 2004), or of a logical interpretation of values as non-saturated functions similar to the Fregean predicates (Echeverrìa 2002). This means that an axiology conceived as a rigorous theory of values, sensitive to applications to concrete issues, is among the intellectual needs of the twenty-first century, especially because this is deeply influenced by the presence of advanced science and technology.

EVANDRO AGAZZI

SEE ALSO *Fact/Value Dichotomy; Husserl, Edmund; Italian Perspectives; Values and Valuing; Weber, Max.*

BIBLIOGRAPHY

Agazzi, Evandro. (2004). *Right, Wrong, and Science*. Amsterdam and New York: Rodopi.

Echeverrìa, Javier. (2002). *Ciencia y valores*. Barcelona: Destino.

Husserl, Edmund. (1969). *Formal and Transcendental Logic*, trans. D. Caims. The Hague, Netherlands: Nijhoff.

Lapie, Paul. (1902). *Logique de la volonté*. Paris: Alcan.

Laudan, Larry. (1984). *Science and Values*. Los Angeles: California University Press.

McMullin, Ernan. (1983). "Values in Science." *Proceedings of the Philosophy of Science Association* 2: 3–26.

Proctor, Robert. (1991). *Value Free Science?* Cambridge, MA: Harvard University Press.

Putnam Hilary. (1981). *Reason, Truth, and History*. Cambridge, UK: Cambridge University Press.

Scheler, Max. (1973). *Formalism in Ethics and Non-Formal Ethics of Values*. Evanston, IL: Northwestern University Press.

Urban, Wilbur Marshall. (1909). *Valuation*. London: Sonnenschein; New York: Macmillan.

von Hartmann, Eduard. (1908). *Grundriss der Axiologie*. Bad Sachsa: Hermann Haacke.

Weber, Max. (1949). *The Methodology of the Social Sciences*. Glencoe, IL: Free Press.

Weber, Max. (1978). *Selections in Translation*. Cambridge, UK: Cambridge University Press.

B

BACON, FRANCIS

• • •

Francis Bacon (1561–1626) was born in London, England on January 22. His life combined politics and philosophy. As a politician, Bacon became a prominent lawyer, judge, member of Parliament, and adviser to the British monarch during the reigns of Queen Elizabeth I (1533–1603) and King James I (1566–1625). He reached the peak of his political power in 1618, when he was appointed Lord Chancellor, the highest judge in England. He fell from power in 1621 when he was impeached by Parliament for accepting bribes in his judicial cases, although he insisted there was no evidence that his judgments had been unfairly biased by the gifts he received. He died in London on April 9.

The idea that human beings should use science and technology to conquer nature for human benefit was first elaborated in the seventeenth century by Bacon. He supported that idea with five kinds of arguments—philosophical, theological, ethical, methodological, and political. Although the scientific and technological mastery of nature has become a fundamental idea in modern life, some people have challenged the wisdom of that idea by questioning Bacon's arguments.

Sir Francis Bacon, 1561–1626. A philosopher, statesman, and author, Bacon was the chief figure of the English Renaissance. His advocacy of "active science" influenced the culture of the English-speaking world. *(Source unknown.)*

Philosophy of Technological Science

As a philosopher, Bacon sought to move beyond traditional learning and establish a new intellectual world based on an observational and experimental science of nature that would give human beings power over nature for human benefit. In *The Advancement of Learning* (1605), he defended the pursuit of knowledge and surveyed the whole world of knowledge as it existed in his time. In *The Great Instauration* (1620), Bacon sketched a vast plan for his new scientific philosophy with technological powers, including the *The New Organon*, which proposed a new logic of inductive reasoning. Although he never completed this plan, he published many writings that worked out parts of it. In his *Essays*

(1625), his most popular work, he offered scattered but penetrating observations on human life. In *New Atlantis* (published posthumously in 1627), he wrote a utopian fable about a society ruled by what would today be called a technoscientific research institute.

Bacon's philosophical argument was that human beings needed to reconstruct all knowledge based on natural philosophy or physics, which required studying the laws of nature as physical regularities that can be established by observation and experimentation. Beginning with Socrates (470–399 B.C.E.), many philosophers have regarded natural philosophy as less important for understanding human life than moral philosophy and theology. But Bacon thought that natural philosophy should be regarded as "the great mother of the sciences" (Bacon 2000, pp. 64–65). In particular he praised the natural philosophy of Democritus (460–370 B.C.E.), who thought that everything in nature could be explained ultimately as caused by the physical motion of atoms (Bacon 2000, 2002). Such knowledge will give people both a theoretical understanding of nature and a practical or technological power over it, because understanding the causes will give them the power to produce effects. Human knowledge and human power will be combined. This power will be limited, however, by nature itself. "Nature is conquered only by obedience," Bacon declared. And "all that man can do to achieve results is to bring natural bodies together and take them apart; Nature does the rest internally" (Bacon 2000, p. 33).

Bacon's theological argument was that this new natural philosophy would be compatible with biblical theology, although the two needed to be separated. True science is the study of God's works as revealed in nature. True religion is the study of God's words as revealed in the Bible. The book of nature and the book of scripture are separated yet compatible. Through reason, people can discover the causal laws of nature. Through faith, they can ascend to God as the miraculous First Cause of nature's laws (Bacon 2002). Humans believe in miracles as a matter of faith. But this goes beyond natural science, because "miracles are either not true or not natural; and therefore impertinent for the story of nature" (Bacon 2002, p. 177). In using scientific knowledge of nature to exercise technological mastery over nature, people show a dominion over nature that manifests their dignity as the only creatures created in God's image (Bacon 2000, 2002).

Bacon's ethical argument was that this new science would be good both as an end in itself for the pleasure of understanding and as a means for its practical benefits. To know the truth about nature is satisfying in itself for those who choose a contemplative life, because such knowledge is "the sovereign good of human nature" (Bacon 2002, p. 342). Scientific knowledge also gives the power to control nature for human benefit through discoveries and inventions that make human life more secure. By thus securing "the empire of man over things," the new science will show a love for the good of humanity that expresses the Christian virtue of charity (Bacon 2000, p. 100).

Bacon's methodological argument was that the success of this new knowledge would depend on a rigorously inductive method of reasoning from observations and experiments. Humans will need a universal natural history that allows them to move from particular facts to general ideas that suggest experiments; and from these experiments they can move gradually to ever more general ideas, until they finally grasp the fundamental laws of nature (Bacon 2000). The theoretical understanding of these laws of nature as rooted in experimental science will then yield a practical mastery of nature through mechanical inventions and discoveries. Bacon pointed to printing, gunpowder, the compass, microscopes, telescopes, and other examples of technological discoveries of his time as illustrating the practical power of natural science (Bacon 2000).

Bacon's political argument was that the observational and experimental work required for the new science would necessitate the cooperative activity of many people over many years, which could be sustained only through public institutions devoted to scientific education and research. Bacon attempted to persuade Queen Elizabeth and King James to support his intellectual project (Bacon 2000, 2002). He suggested that political rulers should be guided by natural philosophers. For example, he thought that Aristotle's influence with Alexander the Great illustrated *the glory of learning in sovereignty*. In *New Atlantis*, he described an imaginary society organized to support a scientific research institute, which would produce discoveries and inventions that would benefit the whole society.

Influence and Critics

Bacon's proponents have included many of the leaders of modern science. In seventeenth-century England, scientists such as Robert Hooke (1635–1703) and Robert Boyle (1627–1691) undertook the cooperative experimental research advocated by Bacon. They set up the Royal Society of London in 1662 with a charter from King Charles II (1630–1685) to carry out Bacon's project. In the eighteenth century, Denis Diderot (1713–1784), Jean d'Alembert (1717–1783), and others in the French Enlightenment acknowledged the influence of Bacon in pointing them toward the promotion of the arts and sciences for human benefit. In America, Thomas Jefferson (1743–1826) praised Bacon as one of the three greatest human beings who ever lived (along with Isaac Newton and John Locke). In the nineteenth century, Charles Darwin (1809–1882) adopted Bacon's view of inductive science and his metaphor of the *two books* of God as showing how religion and science can be compatible. In the twentieth century, the increase in scientific discoveries and inventions from publicly supported research institutes seemed to vindicate Bacon's optimism. In *Consilience* (1998), Edward O. Wilson (b. 1929) sketched a program for the unification of all knowledge based on the physical laws of nature that would complete Bacon's project.

At the same time, since Joseph de Maistre (1753–1821) attacked him early in the nineteenth century, the number of Bacon's opponents has also grown. De Maistre was a French conservative who saw Bacon as a source for the morally corrupting atheistic materialism of the Enlightenment and the French Revolution. De Maistre argued that in basing all knowledge on physical causes, Bacon was denying the importance of moral and religious knowledge and undermining the dignity of the human soul as a spiritual power beyond the material world. Devout Christians such as Boyle had defended Bacon's science against the charge of atheistic materialism, and Bacon had written a "Confession of Faith" that conformed to the Protestant theology of John Calvin (1509–1564) (Bacon 2002). Yet de Maistre insisted that Bacon had hidden the atheistic implications of his scientific materialism through false professions of faith.

Since the twentieth century, Bacon's opponents have warned that his project for exploiting nature shows a disrespect for nature and nature's God, and a willful determination to replace the naturally given and divinely ordained with the artificially constructed and humanly manipulated. From C. S. Lewis (1898–1963) to Leon Kass, these critics worry that the abolition of nature through technology will remove the ethical limits on human will that come from nature or God. As biotechnology gives people the power to create new life forms and even redesign human nature, they might eventually find themselves in a totally artificial world empty of natural value.

Bacon's critics warn that to speak of humanity using science and technology to master nature for human benefit is vague in ways that hide inherent problems. To speak of *humanity* gaining such mastery suggests that all human beings will have equal power. But is it not inevitable that some human beings will have more of this power than others, and that they will use it to advance their selfish interests? Will the nations with the greatest access to scientific and technological power not use it to exploit those nations with less power? Can scientists and engineers be trusted to use their power for the good of all? If this power is publicly regulated, can the regulators be trusted to act for the common good?

To speak of the *human mastery of nature* suggests that human beings will have an unconstrained power that will set them apart from and above nature. But will that power not always be constrained by the potentialities of nature and by the limits of human knowledge? Will human beings not often change nature in ways that produce unanticipated consequences that are undesirable? And in changing nature, will human beings not change themselves as well? Does mastery of nature include mastery of human nature—meaning that some human beings will have mastery over the nature of other human beings, perhaps by genetically engineering the future generation of human beings? But would this not be the ultimate tyranny of some human beings over others? Even if individual human beings are free to use this power for changing their nature in whatever ways they desire, will this not create possibilities for foolishly choosing to use such power in dehumanizing ways? Might not the power of parents to manipulate the biological nature of their children deprive children of their dignity and freedom?

To speak of the mastery of nature for *human benefit* suggests that people have a clear grasp of the human goods about which they can all agree. But will people not often disagree about these human goods? And will these goods not often conflict with one another? Can one assume, as Bacon did, that biblical religion will guide understanding of the human goods to which human mastery of nature will be directed? Or do modern science and technology promote a materialistic and utilitarian view of the world that subverts religious belief while encouraging a hedonistic egoism? Can one still believe in the moral worth of human beings as spiritual creatures created in God's image? Or must science teach that human beings are only highly evolved animals? Even if Baconian science secures the technical means to master nature, can one trust that science to secure the moral ends of that mastery? Will human mastery of nature promote human nobility? Or will it produce a world of paltry pleasures and shallow souls? The future of science and technology as directed to the conquest of nature turns on how successful people are in thinking through such questions.

LARRY ARNHART

SEE ALSO *Atlantis, Old and New; Experimentation; Progress; Scientific Revolution; Utopia and Dystopia.*

BIBLIOGRAPHY

Bacon, Francis. (2000). *The New Organon*, trans. Lisa Jardine, and Michael Silverthorne. Cambridge, England: Cambridge University Press.

Bacon, Francis. (2002). *The Major Works*, ed. Brian Vickers. Oxford: Oxford University Press.

Jones, Richard Foster. (1982). *Ancients and Moderns: A Study of the Rise of the Scientific Movement in Seventeenth-Century England*. New York: Dover Publications. History of Bacon's influence on English science.

Kass, Leon. (1985). *Towards a More Natural Science*. New York: Free Press. Neoconservative attack on Baconian mastery of nature.

Leiss, William. (1994). *The Domination of Nature*. Montreal: McGill-Queen's University Press. Critical history of the idea of mastery of nature.

Lewis, C. S. (1947). *The Abolition of Man*. New York: Macmillan Publishing. Criticism of Baconian mastery by a Christian apologist.

Maistre, Joseph de. (1998). *An Examination of the Philosophy of Bacon*, trans. Richard A. Lebrun. Montreal: McGill-Queens University Press.

Merchant, Carolyn. (1980). *The Death of Nature: Women, Ecology, and the Scientific Revolution*. San Francisco: Harper & Row. Feminist history and criticism of the idea of mastery over nature.

Peltonen, Markku, ed. (1996). *The Cambridge Companion to Bacon*. Cambridge, England: Cambridge University Press. Surveys scholarly studies of Bacon.

Urbach, Peter. (1987). *Francis Bacon's Philosophy of Science*. LaSalle, IL: Open Court Publishing. Best study of Bacon's scientific method.

Wilson, Edward O. (1998). *Consilience: The Unity of Knowledge*. New York: Knopf.

Zagorin, Perez. (1998). *Francis Bacon*. Princeton, NJ: Princeton University Press. General survey of Bacon's thought.

BARUCH PLAN

The first atomic bombs were detonated in 1945. The Baruch Plan of 1946 served as the first proposal to control the spread and use of this awesome new power. President Harry Truman's original announcement about the bomb included a promise that it would not be used only for destructive purposes. In the words of the Baruch Plan, "Science, which gave us this dread power, shows that it can be made a giant help to humanity, but science does not show us how to prevent its baleful use. So we have been appointed to obviate that peril by finding a meeting of the minds and the hearts of our peoples. Only in the will of mankind lies the answer" (Baruch Plan, presented to the United Nations Atomic Energy Commission on June 14, 1946).

Background

At the end of World War II the United Nations passed a resolution to create a commission that would examine the use of nuclear energy and determine what institutional frameworks were needed to steer the technology toward peaceful uses. The creation of the United Nations Atomic Energy Commission (UNAEC) in January 1946 prompted the then U.S. secretary of state, James F. Byrnes, to convene a committee that would direct American policy on this issue. The committee was headed by Undersecretary of State Dean Acheson, who, in concert with a board of consultants that included leaders in business and science as well as members of the Manhattan Project, published the Report on the International Control of Atomic Energy (more commonly referred to as the Acheson-Lilienthal Report) on March 16, 1946.

The Acheson-Lilienthal Report proposed an American policy to create international frameworks to manage the use and dissemination of nuclear energy and technology. The main premise of the report was the creation of an international Atomic Development Authority that would control and monitor the use of atomic energy and its dangerous elements. The Acheson-Lilienthal Report did not propose to outlaw nuclear weapons but instead to globalize cooperation among states to encourage the use of the technology for productive and peaceful ends. This international body would promote research on and development of atomic energy innovation and be the sole owner of that technology. The Baruch Plan, the first proposal of the United States to the UNAEC, was drawn largely from the text of this report.

The Plan

Bernard M. Baruch, the U.S. representative to the UNAEC, submitted the report to the commission on June 14, 1946. The Baruch Plan, like the Acheson-Lilienthal Report, proposed the establishment of an Atomic Energy Development Authority that would control the development and use of atomic energy, beginning from the mining stage and including the development and implementation of atomic energy and its uses. The plan also demanded the termination of the development of the atomic bomb for use as weaponry and mandated an inspections team to investigate violations of that framework. The United States, at that time, was the sole possessor of nuclear weapons, although the Soviet Union was far along in the development process. The Baruch plan called for the immediate cessation of weapons development programs from all countries, and the close monitoring of peaceful nuclear programs in exchange for the United States giving the AEDA its nuclear devices. The purpose of the Baruch Plan was not to eradicate the use of nuclear energy from the world but to manage, monitor, and internationalize its peaceful benefits.

Immediately after the United States submitted its proposal to the UNAEC, the United States and the Soviet Union began deliberations on ways to implement the plan. The Soviet Union offered a counterproposal that differed from the U.S. version on several key points. The United States insisted on retaining control of its nuclear weapons while all fissile material was put under international control, while the Soviet Union demanded that the United States cede its weapons to international control before other countries gave up their fissile material. In addition, not only did the Soviet proposal mandate the cessation of the development, storage, and deployment of atomic bombs, it also directed that all preexisting weaponry be destroyed within six months of entrance into the convention.

The Soviet Union objected to several other points in the Baruch Plan. Another critical difference was the Soviet disagreement with the proposal that called for automatic sanctions for noncompliance with the proposed regulations. Discussions between the two countries lasted for several years, but it was evident early on that because of irreconcilable differences the Baruch Plan would never be implemented.

Legacies

While there is still debate on whether or not the United States ever seriously expected the Baruch plan to pass, it did leave the United States with a better understanding of its own moral responsibility in the cold war arms race. From 1946 on, Americans believed they had proven to the world their willingness and desire to eliminate nuclear weapons altogether, and blamed the Soviet Union for standing in the way of that goal. As long as there was a Soviet threat, the United States could feel that it was reluctantly but obligingly taking on the role of protector of the world.

Failure and Achievement

Although the Baruch Plan was never codified formally into international law, it put in place the basic tenets of the modern nonproliferation regime. The Acheson-Lilienthal Report that formed the contextual basis for the Baruch Plan never proposed a ban-the-bomb approach but instead was intended to create an international organization that would control every stage of nuclear energy development. Because the international agency would be the reigning authority and would have the authority to distribute the sites of nuclear energy processing around the world, it would create a global strategic balance. Many countries could profit from the peaceful benefits of nuclear energy. However, if one country tried to use its materials for malevolent purposes, other countries would be similarly equipped to defend themselves. These ideas led to many of the Cold War disarmament programs and treaties such as Atoms for Peace, the IAEA, and ultimately the nonproliferation treaty.

MARGARET COSENTINO
JESSICA COX

SEE ALSO *Atomic Bomb; International Relations; Military Ethics; Nuclear Ethics; Limited Nuclear Test Ban Treaty; Nuclear Non-Proliferation Treaty; Weapons of Mass Destruction.*

BIBLIOGRAPHY

Bailey, Emily; Richard Guthrie; Daryl Howlett; and John Simpson. (2000). *Programme for Promoting Nuclear Non-Proliferation Briefing Book.* Volume I: *The Evolution of the Nuclear Non-Proliferation Regime,* 6th ed. Southampton, UK: Mountbatten Centre for International Studies.

Sokolski, Henry. (2001). *Best of Intentions: America's Campaign against Weapons Proliferation.* Westport, CT: Praeger.

INTERNET RESOURCES

"The Acheson-Lilienthal Report." Available from http://www.atomicmuseum.com/tour/acheson.cfm.

"Early U.S. Arms Control and Disarmament Initiatives." Available from http://usinfo.state.gov/products/pubs/armsctrl/pt3.htm

"Glossary of Nonproliferation Terms." Available from http://www.cnsdl.miis.edu/npt/gloss/glossary.htm.

Schell, Jonathan. "The Unfinished Twentieth Century—Attempt to Find a Designation for the 20th Century." *Harper's Magazine,* January 2000. Available from http://www.findarticles.com/p/articles/mi_m1111/is_1796_300/ai_58509210.

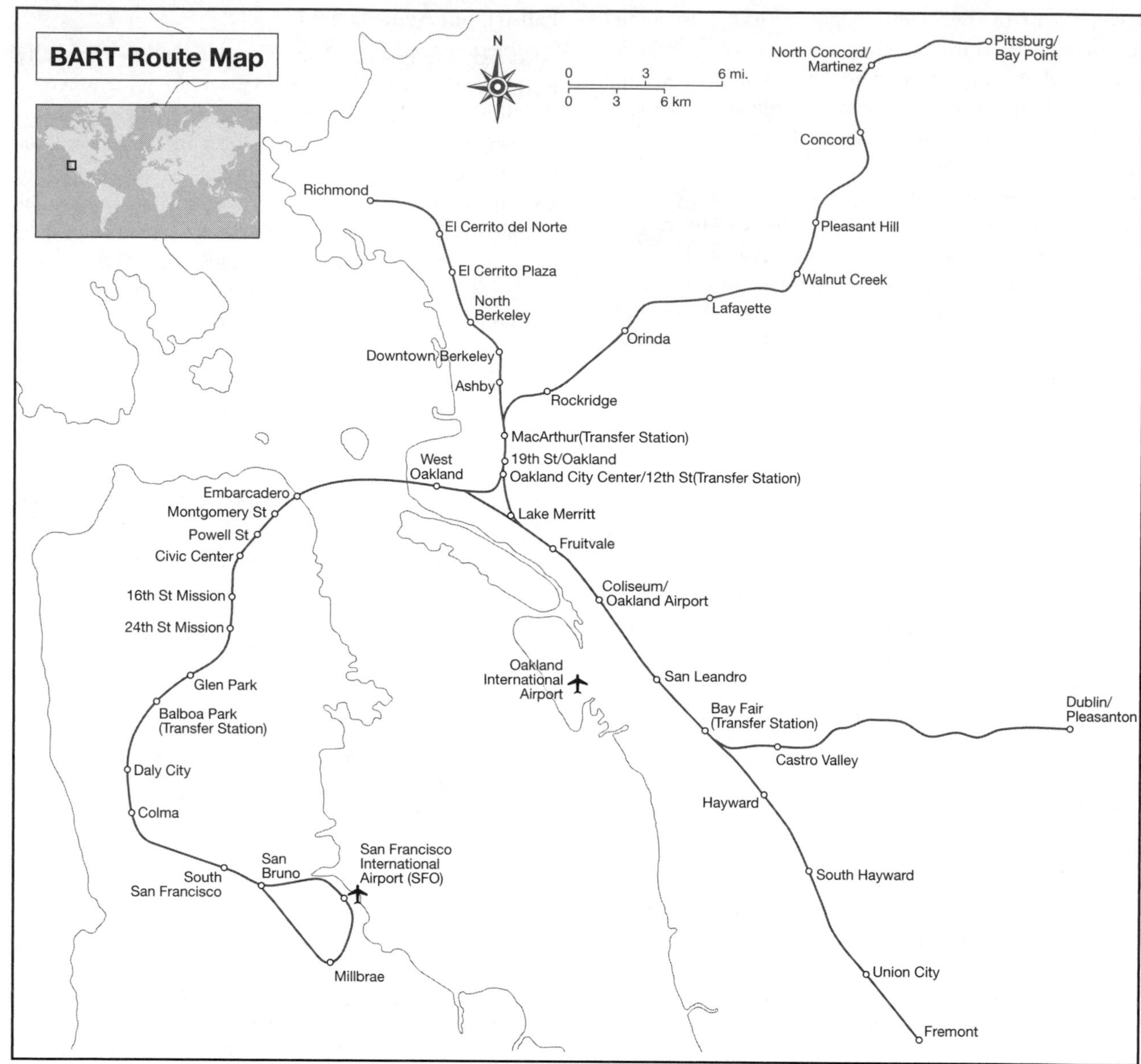

BAY AREA RAPID TRANSIT CASE

• • •

The Bay Area Rapid Transit (BART) system is a fast (eighty miles per hour top speed) commuter rail system serving three counties in the San Francisco Bay Area. It was authorized by public statute in 1957 and went into service in 1972. The prime contractor for the BART project was PBTB, a consortium of three large engineering firms, Parsons-Brinkerhoff, Tudor, and Bechtel. During the course of design and construction, three engineers undertook principled actions that played a significant role in advancing the development of engineering ethics in the United States.

The Engineers and Their Actions

Holger Hjortsvang, an experienced systems engineer, was involved with the Automated Train Control System (ATC). Max Blankenzee, a young programmer analyst, worked with Hjortsvang. They became concerned about the way the ATC subcontractor, Westinghouse Corporation, was doing its job. A principal issue with Hjortsvang was the absence of a systems engineering group to oversee the development of control and propulsion systems. Hjortsvang and Blankenzee reported their concerns to their managers, both orally and in writing. The response was "don't make trouble." Simultaneously electrical engineer Robert Bruder, monitoring the contractors installing and testing control and communications equipment, found that reports to his managers about sloppy work were ignored.

In November 1971 the three engineers brought their concerns in confidence to BART Board of Directors member Daniel Helix, providing him with written material. This led Helix to bring up the issues of ATC safety before a meeting of the Board. The Board, however, rejected the position of the anonymous engineers, as represented by Helix, by a large majority.

In short order BART management was able to identify the three engineers who had provided Helix with the information he brought to the meeting. Hjortsvang, Blankenzee, and Bruder were then fired without written cause or appeal. There are indications that their efforts to find new jobs were impeded by BART management. About a year later, they filed a wrongful discharge suit against BART.

Subsequent Events

Prior to the BART board meeting, Bruder, a licensed Professional Engineer, phoned William F. Jones, President of the California Society of Professional Engineers (CSPE), outlining the situation and requesting support. At Jones's request, CSPE Diablo Chapter members Roy W. Anderson and Gilbert A. Verdugo reviewed the situation and corroborated the essentials of the arguments made by Hjortsvang, Blankenzee, and Bruder.

Following the firings, Jones unsuccessfully tried to reach BART's general manager, B. R. Stokes. A meeting with Chief Engineer David Hammond was of no avail. BART management declined all requests to discuss the firings on the grounds of possible or pending legal action.

The CSPE then wrote a report about poor engineering at BART, which it sent to the California State Senate. This led to a staff study concluding that the BART project was not going well, but ignoring the plight of the three engineers whose action triggered the investigation.

The validity of the engineers' concerns was decisively confirmed on October 2, 1972, three weeks after BART began carrying passengers. A speed control command, corrupted by a short circuit in a transistor, caused a BART train to accelerate instead of slow down, resulting in a crash at the Fremont station. Fortunately there were no fatalities and only a few injuries.

The California State Senate commissioned a study by a three-member Blue Ribbon Committee of distinguished engineers that confirmed that the engineering of the ATC and some other aspects of the BART system were below par. Panel member Bernard Oliver, a past president of the Institute of Electrical and Electronics Engineers (IEEE), sent an incisive letter to a Westinghouse vice president specifying poor decisions that suggested to him that "the design [of the ATC] did not enjoy the attention of your top people" (Unger 1994, p. 252).

In November 1972, some CSPE officers, including, incredibly, Jones, charged the Diablo CSPE Chapter with unethical behavior in connection with their investigation of the BART project. They cited an ethics code provision against criticizing other engineers. This effort backfired when the CSPE Board of Directors, following the recommendation of the committee that adjudicated the case, not only rejected the charges, but commended the chapter for its efforts to protect the public safety, health, and welfare. However the CSPE faded out of the picture toward the end of 1972, apparently as a result of pressure from members employed by the consortium of large engineering firms running the BART project.

The IEEE Response

In September 1973 the IEEE Committee on Social Implications of Technology (CSIT) published an article in its newsletter describing the treatment meted out to the three BART engineers. The following March, the CSIT unanimously passed a two-part resolution addressed to the IEEE Board of Directors (BoD). Part (a) called for the establishment by the IEEE of mechanisms to support engineers whose acts in conformity to ethical principles may have placed them in jeopardy. Part (b) asked the IEEE to intervene on behalf of the BART engineers.

The BoD, advised by the IEEE U.S. Activities Committee (USAC), and an ad hoc committee that included Joel Snyder, Victor Zourides and Frank Cummings (USAC legal counsel), responded to part (b) by commissioning an amicus curiae brief to be presented to the court hearing the engineers' law suit. The brief was to enunciate general principles, rather than to side directly with the engineers. As ultimately drafted by Frank and Jill Cummings, the brief urged the court to determine that, if an engineer was discharged because of a bona fide effort to conform to an ethical obligation to protect the public safety, the termination should be considered a breach of an implied term of the employment contract. The brief was filed in January 1975. Shortly afterward, the engineers accepted an out-of-court settlement reported to be $75,000. The legal concepts argued have been used in subsequent cases, sometimes strengthened by a court's permitting the plaintiff to allege an *action in tort*, which opens the door to punitive damages.

One of the aluminum cars of the Bay Area Rapid Transit System. Problems with the system's development were revealed when one of the trains experienced a crash about a week after it began carrying passengers. *(John Dominis/Getty Images.)*

The response to part (a) of the resolution took longer. In 1978 procedures were implemented whereby IEEE members (later extended to include other professionals in fields covered by the IEEE) could appeal to the IEEE Member Conduct Committee for help if their careers were jeopardized in retaliation for acts in conformity to the principles underlying the IEEE Ethics Code.

The BART engineers underwent a painful ordeal that impacted their professional and personal lives. It took them between one and two years to get back on track professionally. Looking back, they felt that they could not have justified any other course of action. And the BART case became a major teaching tool for engineering ethics courses during the following decades.

STEPHEN H. UNGER

SEE ALSO *Engineering Ethics; Institute of Electronical and Electronics Engineers.*

BIBLIOGRAPHY

Anderson, Robert; Robert Perrucci; Dan Schendel; and Leon Trachtman. (1980). *Divided Loyalties—Whistle-Blowing at BART*. Ashland, OH: Purdue University Press. Mainly presents management point of view.

Friedlander, Gordon. (1972–1973). "The BART Chronicle" *IEEE Spectrum* 9, no 9 (September): 34–43; no. 10 (October): 60–72; and no. 11 (November): 41–54; 10, no. 3 (March): 32–37; and no. 4 (April): 40–44. A series of five articles on the BART system; includes detailed discussion of the Blue Ribbon Committee report.

Unger, Stephen. (1994). *Controlling Technology: Ethics and the Responsible Engineer*. 2nd edition. Hoboken, NJ: Wiley. Includes Oliver letter and amicus curiae brief.

BELL, DANIEL

Daniel Bell (b. 1919) was born in New York City on May 10, to an immigrant Jewish family; though religion would later play a central role in his sociological theorizing, he considered his Jewishness to be ethnic rather

Daniel Bell, b. 1919. A Harvard academic and prominent figure in the American Academy of Arts and Sciences, Bell is best known as one of the theorists of post-industrialism. *(The Library of Congress.)*

than religious. He graduated from City College of New York in 1938, and after a year of graduate study at Columbia University spent the next twenty years in journalism, writing and editing for the *New Leader*, *Fortune* (as labor editor), and *The Public Interest,* which he cofounded with Irving Kristol in 1965. In 1958 he became an associate professor at Columbia, where he received a Ph.D. in 1960 and was promoted to full professor in 1962. In 1969 he moved to Harvard University, where he received a Henry Ford II endowed chair in 1980, from which he retired in 1990.

Bell's importance is based primarily on three books: *The End of Ideology* (1960); *The Coming of Post-Industrial Society* (1973); and *The Cultural Contradictions of Capitalism* (1976). In these and related works, Bell defends a complex relation between science, technology, and ethics. On the one hand, he believes passionately in the science-based expertise of a technological elite; on the other, he clearly laments the loss of traditional cultural (including ethical) values in the anti-culture that accompanies technical elitism. As he explained in a new preface to the paperback edition of the third book just mentioned, he is "a socialist in economics, a liberal in politics, and a conservative in culture." In elaboration:

> (1) I am a socialist in economics. For me, socialism is not statism, or the collective ownership of the means of production. It is a judgment on the priorities of economic policy. I believe that *in this realm,* the community takes precedence over the individual. (2) I am a liberal in politics—defining both terms in the Kantian sense. I am a liberal in that, within the polity, I believe the individual should be the primary actor, not the group. And the polity has to maintain the distinction between the public and the private. (3) I am a conservative in culture because I respect tradition; I believe in reasoned judgments of good and bad about the qualities of a work of art. I use the term culture to mean less than the anthropological catchall and more than the aristocratic tradition which restricts culture to refinement and to the high arts. Culture, for me, is the effort to provide a coherent set of answers to the existential predicaments that confront all human beings. (Bell 1979, pp. xii, xiv, xv)

In a critical intellectual biography, Malcolm Waters (1996) questions all three self-characterizations by challenging the sociological distinctions in which they are grounded. Adapting the structural-functionalism of Talcott Parsons, Bell rejects any holistic understanding of contemporary society and instead distinguishes between three realms, each ruled by a different axial principle and displaying a different axial structure. In terms of their different central values, the techno-economic realm pursues material growth, the polity consent of the governed, and cultural novelty or originality. Each of these three realms may also be characterized by special relationships between the individual and the social order, basic processes, and structural problematics. Waters summarizes these distinctions in a grid supplied by Bell himself (see Figure 1).

Waters's criticisms—which are those of a friendly critic who is convinced that Bell is a major sociological theorist—are as follows. First, with regard to economic socialism, Bell's position is singularly weak. It entails no more than commitment to a minimum standard of living, for example in health care. A more robust socialist would question the capitalist ownership of the means of production. In fact in the economic sphere Bell is no more than a liberal.

Second, with regard to political liberalism, Bell is more convincing. "Bell makes explicit statements consistent with Jeffersonian democracy about individual rights, small government (notwithstanding a grudging

FIGURE 1

The General Schema of Society

Realm	**Techno-economic (social) Structure**	**Polity**	**Culture**
Axial principle	Functional rationality	Equality	Self-realization
Axial structure	Bureaucracy	Representation	Reproduction of meanings and artifacts
Central value-orientation	Material growth	Consent of the governed	Novelty and originality
Relationship of individual to social order	Role segmentation	Participatory	Sovereignty of the whole person
Basic processes	Specialization and substitution	Bargaining and legal reconciliation	Disruption of genres by syncretism
Structural problematics	Reification	Entitlements, meritocracy and centralization	Postmodernist anti-nomianism

SOURCE: Adapted from Waters (1996), p. 35.

approval of the New Deal) and the sanctity of the private sphere. [However] in [*Post-Industrial Society*] politics is not a source of last-resort interventions but rather an arena within which primary steering, namely planning, takes place" (Waters 1996, p. 168).

Third, with regard to cultural conservatism, Waters accepts this self-characterization but sees a problem with "his insistence that the three realms are [interdependent]. If he wants a return to authoritative standards in culture then there must be a source of such standards, and its only possibility is an illiberal state" (Waters 1996, p. 168–169).

Waters concludes that Bell is neither a neo-conservative, socialist, nor much of a liberal. "Despite all interest in the future possibilities of technology and post-industrialism Bell is an old-fashioned, traditionalistic, elitist conservative" (Waters 1996, p. 169). Bell might respond that Waters has simply misunderstood the nuances of his positions, while others, especially leftist critics, have good grounds for arguing that Bell is a neo-conservative despite his denials.

PAUL T. DURBIN

SEE ALSO *Critical Social Theory; Democracy; Industrial Revolution; Information Society; Socialism.*

BIBLIOGRAPHY

Bell, Daniel. (1960). *The End of Ideology: On the Exhaustion of Political Ideas in the Fifties*. Glencoe, IL: Free Press.

Bell, Daniel. (1973). *The Coming of Post-Industrial Society: A Venture in Social Forecasting*. Basic Books.

Bell, Daniel. (1976). *The Cultural Contradictions of Capitalism*. New York: Basic Books. Paperback edition with new preface was published in 1979; twentieth anniversary edition with a new 50-page afterward by the author was published in 1996.

Waters, Malcolm. (1996). *Daniel Bell*. London: Routledge. A critical but favorable intellectual biography.

BENJAMIN, WALTER

Walter Benjamin (1892–1940), a German-Jewish intellectual born in Berlin on July 15, was a cultural sociologist, literary critic, and translator of Charles Baudelaire and Marcel Proust. His works are informed by a mixture of Marxism and Jewish mysticism. Benjamin most often is associated with the Frankfurt School as well as with his friends and colleagues Teodor Adorno (1903–1969), Gerschom Scholem (1897–1982), and Bertolt Brecht (1898–1956), all of whom influenced his thought. Believing that the Gestapo was about to capture him, Benjamin committed suicide on September 27 at Port Bou on the French-Spanish border while fleeing from the Nazis. He left behind a large collection of notes and published and unpublished writings, most of which have been compiled, edited, and translated since his death.

Benjamin's books and essays deal with a multitude of subjects, with their most common themes being the degradation of contemporary experience and the need for a radical break with tradition and the past. Among his best-known works are *Einbahnstrasse* [One-way street] (1928), the essay "Das Kunstwerk im Zeitalter seiner technischen Reproduzierbarkeit" [The work of art in the age of mechanical reproduction] (1936), *Geschichtsphilosophische Thesen* [Theses on the philosophy of history] (1939 but published posthumously), and the monumental *Das Passagen-Werk* [*The Arcades Project*] (written between 1927 and 1940 and published posthumously). Among these works *The Arcades Project* is the most pertinent to science, technology, and ethics because it deals with the ways in which modern technology in the form of new architectural constructions altered human perception and experience.

Left unfinished at his death, *The Arcades Project* is an extended set of notes and quotations loosely arranged in thirty-six categories with titles such as "Dream City," "Baudelaire," "Fashion," and "Prostitution." For Benjamin the glass-enclosed streets of nineteenth-century Parisian arcades exemplified the commodification of experience and the distracted perception of reality. At home in these arcades is the *flâneur*, the "heroic pedestrian" or tourist who wanders aimlessly in the crowd, deriving pleasure from the exercise of what might be called a shopper's gaze. For the *flâneur* the city is a text to be read, but only from always changing vantage points and thus distractedly, with shifting glimpses of meaning in the kaleidoscope of signs. For Benjamin such distraction is the defining characteristic of contemporary perception, and some interpreters have argued that such perception has been extended in MTV-style editing, multitasking, channel and Web surfing, and the experience of cyberspace in general.

Benjamin also dealt with this issue in the essay "The Work of Art in the Age of Mechanical Reproduction," which considers how technology has altered not just aesthetic perception but the nature of art. For millennia even the most perfect artistic reproduction lacked the essential element of the original, "its presence in time and space, its unique existence at the place where it happens to be." That uniqueness bestowed authenticity. However, contemporary technologies of reproduction, especially sound recording, photography, and film, have undermined the traditional appreciation of originality and authenticity. Indeed, reproduction may favor the copies, which can be placed into situations impossible for the original: "The cathedral leaves its locale to be received in the studio of a lover of art; the choral production, performed in an auditorium or in the open air, resounds in the drawing room."

Among all technological media, Benjamin considered film especially significant for two reasons. First, like contemporary life, film is saturated by and dependent on technology, with the performance of a film actor mediated by a series of machines (camera, editor, projector). Second, it is film that best accommodates the distracted perception of the *flâneur*. At the cinema people simply sit back, relax, and watch the movie; they do not have to discipline themselves to pay attention: "The public is an examiner, but an absent-minded one." ("Work of Art in the Age of Mechanical Reproduction")

Benjamin's writings, including meditations on literature, history, philosophy, sociology, and art, are so broad that they have stimulated numerous fields of scholarship, and his meticulously crafted, indirect, and at times enigmatic style has influenced succeeding generations of reflections on technological culture. At the same time Benjamin has been criticized for a nostalgia that does not always appreciate the democratizing ethos at the core of the new forms of technological art he examined.

JAMES A. LYNCH

SEE ALSO *Consumerism; Science, Technology, and Literature; Tourism.*

BIBLIOGRAPHY

Benjamin, Walter. (2002). *The Arcades Project*, trans. Howard Eiland and Kevin McLaughlin. Cambridge, MA: Harvard University Press.

Benjamin, Walter. (2003). *Selected Writings*, Vol. 4: *1938–1940*. Cambridge, MA: Harvard University Press. Contains the third and final version of the essay here titled "The Work of Art in the Age of Its Technological Reproducibility."

Brodersen, Momme. (1996). *Walter Benjamin: A Biography*, trans. Malcolm R. Green and Ingrida Ligers. New York: Verso.

Rochlitz, Rainer. (1996). *The Disenchantment of Art: The Philosophy of Walter Benjamin*, trans. Jane Marie Todd. New York: Guilford Press. A critical study of Benjamin's thought and works.

Smith, Gary, ed. (1991). *On Walter Benjamin: Critical Essays and Recollections*. Cambridge, MA: MIT Press. Twelve critical essays by colleagues of Benjamin and contemporary scholars dealing with a wide range of Benjamin's interests.

BENTHAM, JEREMY

SEE *Consequentialism; Liberalism.*

BERDYAEV, NIKOLAI

Nikolai Alexandrovich Berdyaev (1874—1948) was born in Kiev, Russia, on March 6, and became a leading critic of positivism and scientism among the Russian intelligentsia. Forced into exile by the Communists in 1922, Berdyaev (also transliterated as Berdiaev, with the first name often anglicized as Nicholas) died in Clamart, France, on March 23.

Berdyaev's religious philosophy emphasizes human freedom and the person as distinct realities, not reducible to the empirical forms of choice behavior or individualism as described in the partial perspectives of the social sciences. On the basis of his personalism, Berdyaev argues against superficial pseudoreligious faith in the power of science and technology, a faith that he finds expressed in the ideology of materialistic determinism prominent among Russian intellectuals during the nineteenth and early-twentieth centuries. In *The Russian Idea* (1946), Berdyaev examines this century-long history of revolutionary intellectual culture that culminated in the Communist Revolution during his own generation, in an analysis that justifies his own odyssey from atheistic materialism to philosophical idealism and then back to a deepened religious faith in Orthodox Christianity.

In his earlier *The Meaning of the Creative Act* (1916), Berdyaev sees creativity as central to humanity and is stimulated by the biblical account of humans as created in the image of God to argue for a creative response to all aspects of life. The ground of meaning lies more with the free response to phenomena than with their objective descriptions. Indeed cognitive knowledge itself involves an intuitive symbolic realism akin to that of the orthodox experience of icons, which are understood as symbols that participate in the reality they symbolize, and in whose presence truth is revealed. Furthermore contrary to the philosophical traditions derived from Greek thought, Berdyaev sees being as part of a dynamic spiritual and revelatory process. From this perspective, world history is divided into three great epochs: one in which the existence of sin is revealed, another in which redemption from sin is made possible through divine adoption, and a third in which humans themselves become divinized cocreators of reality. What is important for Berdyaev is to recognize the ways in which creativity in science and technology can serve as false substitutes for spiritual cocreation in this third epoch.

In *The Destiny of Man* (1937) Berdyaev draws on the thought of the German mystic Jacob Boehme (1575–1624) concerning the *Urgrund* or nothingness from which God creates within eternity. The primordial uncreated freedom of human beings derives from the *Urgrund;* freedom is not created by God, although God freely participates with humans in the God-Human Christ and the tragic process of redeeming the world from evil, suffering, and death. Berdyaev likewise adapts Boehme's thought on *Sophia* to develop an arguably more orthodox theology than found, for example, in the erotic mysticism of Vladimir Solovyev (1853–1900). *Slavery and Freedom* (1939) contains Berdyaev's most extensive reflections on the person and the necessity of relation to others, while describing in detail human self-enslavement (Hegel's bad faith) to the various allures of nature and culture. Berdyaev's thought here parallels that found in *I and Thou* (1923) by the Jewish thinker Martin Buber (1878–1965).

Nicholas Berdyaev, 1874–1948. Berdyaev was a Russian philosopher and religious thinker. He was a leading exponent of Christian existentialism and bridged the gap between religious thought in Russia and the West. *(The Library of Congress.)*

As one of the earliest thinkers to recognize how science and technology can pose special problems for Christian culture, in an essay on "Man and Machine" (1934), Berdyaev argues that science and technology destroy the earth-centered, telluric or autochthonic forms of religious life, and threaten to ensnare human freedom in a depersonalized world. In such circumstances, the spiritual becomes more important than ever. Technical civilization calls for a spiritual renewal to challenge the limitations of science and technology just as science and technology challenged the limitations of nature.

Through his extensive writing Berdyaev gained an audience beyond the narrow Russian emigree circle in France. He became a *forbidden writer* widely read in the Soviet Union, and remains a vital source for critical reflection on science and technology. Perhaps because of this a Berdyaev revival has led to many of his writings being made available on the internet in both Russian and ongoing translations.

STEPHEN JANOS

SEE ALSO *Christian Perspectives.*

BIBLIOGRAPHY

Allen, Edgar Leonard. (1951). *Freedom in God: A Guide to the Thought of Nicholas Berdyaev.* New York: Philosophical Library.

Berdyaev, Nicholas. (1934). "Man and Machine." In *The Bourgeois Mind and Other Essays.* New York: Sheed and Ward. Collects four journal articles. Reprinted in *Philosophy and Technology*, eds. Carl Mitcham, and Robert Mackey (New York: Free Press, 1972).

Berdyaev, Nicholas. (1937). *The Destiny of Man,* trans. Natalie Duddington. London: G. Bles.

Berdyaev, Nicholas. (1944 [1939]). *Slavery and Freedom* trans. R. M. French. New York: C. Scribner's Sons.

Berdyaev, Nicholas. (1962 [1916]). *The Meaning of the Creative Act*, trans. Donald A. Lowrie. New York: Collier.

Berdyaev, Nicholas. (1992 [1946]). *The Russian Idea,* trans. R. M. French. Hudson, NY: Lindisfarne Books.

Lowrie, Donald A. (1974). *Rebellious Prophet: A Life of Nikolai Berdyaev.* Westport, CT: Greenwood Press.

Slatte, Howard Alexander. (1997). *Personality, Spirit, and Ethics: The Ethics of Nicholas Berdyaev.* New York: P. Lang.

BERLIN, ISAIAH

Isaiah Berlin (1909–1997), historian of ideas and political theorist, was born to Jewish parents in Riga, Latvia, on June 9, but spent most of his life after 1921 in Great Britain, studying and then holding various positions at Oxford University, where he served as professor of social and political theory (1957–1967) and founding president of Wolfson College (1966–1975). He also served as president of the British Academy (1974–1978), and was the recipient of numerous awards and honorary degrees, including a knighthood and the Order of Merit. After his biography of Karl Marx (1939), Berlin's published work consisted entirely of essays, one of which, "Two Concepts of Liberty" (1958) became one of the most influential expressions of liberal political theory of the latter half of the twentieth century.

Berlin saw scientific and technological advance as one of the dominant forces in the twentieth century. He followed developments in the philosophy of science, and was a close observer of the political domination of science in the Union of Soviet Socialist Republics (U.S.S.R.) during the later years of Joseph Stalin's reign. Although he did not write explicitly about the philosophy or morality of science and technology, Berlin's work provides significant insights into their ethical implications.

Isaiah Berlin, 1909–1997. British philosopher Berlin wrote widely on topics involving the history of ideas, political philosophy, and the relationship of the individual to society.

Berlin was opposed to the application of a single, dominant model to all subjects, arguing instead that different approaches are appropriate to different facets of experience. He recognized the validity of the scientific method in studying the natural world, but suggested that its application to the understanding of human beings (beyond the discoveries of the medical and biological sciences) was often mistaken, an example of pseudo-scientific ideology rather than genuine scientific knowledge. Berlin warned against the application of scientific models to the humanities and social sciences, which he believed should aim at capturing the unique qualities of particular human experiences, rather than the development of general laws and formulae (which he took to be the goal of science).

Berlin sought to explain, and seemed to endorse, the view that science is concerned with empirically discoverable facts, and with processes and relationships that can be explained in terms of identifiable rules or *laws*, while moral philosophy and politics are concerned not with facts about the way things are, but with values, or human beliefs about the way things should be. However, Berlin also argued that values are *objective*, deriving their validity from the realities of a common, universal human nature. This common nature encompasses great variety, is expressed differently in different cultures, and cannot be reduced to simple formulae. But it does allow people to understand one another, and places limits on the goals they can intelligibly and rightfully pursue.

Berlin insisted that science cannot tell people what to be or do; this they must decide for themselves, from among the possible, and often conflicting, values to which as human beings, they feel drawn. While he believed that the acquisition of scientific knowledge should be pursued as a goal in itself, Berlin believed that it would not point the way to any conclusions about ethics. The only way in which science might change thinking about ethics would be by transforming the understanding of human nature in such a way as to force human beings to change their ideas about morality. For instance, if science were to reveal that human beings lack free will, humanity would have to abandon its notions of individual moral responsibility. But Berlin warned against jumping to conclusions based on insufficient or inconclusive evidence, and the tendency to use science, or pseudo-science, as an excuse for evading moral responsibility.

On a practical level, Berlin was sharply critical of what he identified as a *managerial* approach to political problems. He reacted strongly against the vision of a final resolution of human conflicts through the application to human life of techniques of conditioning and management. Berlin did not deny the tremendous good produced by the advance of technology; but his writings reflect an anxiety that the very success of technology could be morally blinding, leading to a view of human beings as *material*, to be molded in such a way as to be conducive to social harmony. This opposition to blind devotion to technological advancement, which excluded moral considerations and ignored the dignity of individuals as free and unique beings, was an important influence in the development of Berlin's political thought.

Berlin's work is significant as a warning against the dangers of intellectual and practical misapplications of science, a critique of reductive understanding of human nature and experience, and a defense of individual liberty and dignity against technocratic control.

JOSHUA L. CHERNISS

SEE ALSO *Ethical Pluralism; Scientific Ethics.*

BIBLIOGRAPHY

Berlin, Isaiah. (1991). *The Crooked Timber of Humanity: Chapters in the History of Ideas*, ed. Henry Hardy. London: John Murray.

Berlin, Isaiah. (1996). *The Sense of Reality: Studies in Ideas and their History*, ed. Henry Hardy. London: Chatto and Windus.

Berlin, Isaiah. (1997). *The Proper Study of Mankind: An Anthology of Essays*, ed. Henry Hardy and Roger Hausheer. London: Chatto and Windus. This volume contains Berlin's major contributions to philosophy and political theory, as well as a sampling of his writings on the history of ideas.

INTERNET RESOURCE

The Isaiah Berlin Virtual Library. Available from http:://berlin.wolf.ox.ac.uk//. Contains a full listing of Berlin's published and unpublished works, as well as an extensive list of writings about Berlin and links to some of these, as well as photographs and reminiscences of Berlin.

BERNAL, J. D.

John Desmond Bernal (1907–1971), an eminent X-ray crystallographer and pioneer in the field of social studies of science and the movement for social responsibility in science, was born in Nenagh, County Tipperary, Ireland, on May 10, and died in London on September 15.

Life and Science

Following his education at Cambridge University, Bernal began his crystallography research at its Davy-Faraday Laboratory in London in 1923. After returning to Cambridge for a short period (1934–1937), he went to Birkbeck College, University of London, where he served as professor of physics (1937–1963), professor of crystallography (1963–1968), and professor emeritus (1968–1971). He initiated groundbreaking research on the crystals of sterols, proteins, and viruses and established the three-dimensional structures of nucleic acids, proteins, and viruses.

Bernal's work in molecular biology led to the conjecture that clays concentrated chemical compounds leading to the origins of life. He speculated in many directions and stimulated scientific research in many areas, arguing for the importance of space exploration and investigation of the possibilities of extraterrestrial life and was considered to be a founder of the field of astrobiology. In an early work, *The*

World the Flesh and the Devil (1929), he set out a futuristic sketch of further evolution, showing how scientific rationality could overcome obstacles in the physical, physiological, and psychological domains. A number of important women scientists worked in Bernal's lab, including Dorothy Hodgkin, with whom he made the first X-ray photograph of a protein (pepsin), and Rosalind Franklin, who did the empirical research that led to the discovery of the double helical structure of DNA.

During World War II, Bernal was a scientific adviser to combined Allied operations, serving in Lord Mountbatten's *department of wild talents*. After the war, he was active in the international peace movement. He was elected a fellow of the Royal Society in Britain in 1937 and, in the postwar period, became a member of the scientific academies of many eastern European countries. His awards included the Royal Medal of the Royal Society (1945), the Lenin Peace Prize (1953), and the Grotius Medal (1959).

Beyond laboratory results, it was Bernal's voluminous knowledge, breadth of vision, and conscientious activism that distinguished him. He led a complicated life, sitting on hundreds of committees and playing a leading role in many scientific and political organizations. He was a dazzling thinker and talker; indeed his contemporaries called him *Sage*. At the experiment level, however, he tended to generate seminal ideas while leaving the details to others. He was a mentor to several Nobel Prize winners.

J. D. Bernal, 1907–1971. Marxist in thinking and communistic in politics, Bernal is perhaps most well-known for his philosophical studies of the social aspects of science. He was also highly instrumental in the pioneering stages of x-ray crystallography and microbiology. *(Nat Farbman/Getty Images.)*

Science of Science

Although Bernal reached the heights of the academic establishment, he engaged in radical critique of its cherished assumptions and structures of power. Bernal was a Marxist in philosophy and a communist in politics. He participated in the Second International Congress of the History of Science and Technology in London in 1931, at which the unexpected arrival of a Soviet delegation created a great stir. Bernal was struck by the unity, philosophical integrality, and social purpose of the Soviet scientists, which contrasted with the undisciplined philosophies and remoteness from social considerations of their British colleagues.

In response Bernal became a leading force in a new movement for social responsibility in science that took a number of organizational forms, such as the Association of Scientific Workers and the Division for Social and International Relations of Science, a part of the British Association for the Advancement of Science. The movement had impact as well as opposition. John Baker's *Counterblast to Bernalism* (1939) led to formation of the Society for Freedom in Science (1940–1945), which devoted itself to the defense of pure science and rejected any form of social control of science.

Bernal argued for the necessity of a science of science. He saw science as a social activity, integrally tied to the whole spectrum of other social activities, economic, social, and political. His book *The Social Function of Science* (1939) quickly came to be regarded as a classic in this field. Based on a detailed analysis of science, under both capitalism and socialism, Bernal's dominant themes were that the frustration of science was an inescapable feature of the capitalist mode of production, and that science could achieve its full potential only under a new socialist order. According to Bernal, science was outgrowing capitalism, which had begun to generate a distrust of science that in its most extreme form turned into rebellion against scientific rationality itself. The cause of science was, for Bernal, inextricably intertwined with the cause of socialism. He saw science as the key to the future and the forces of socialism alone able to turn it.

For Bernal, the scientific method encompassed every aspect of life. There was no sharp distinction between the natural and social sciences. He regarded science as the starting point for philosophy. Science, philosophy, and politics were bound together in Bernal's highly integrated mind. He considered the Marxist philosophy of dialectical materialism to be the most suitable philosophy for science. Bernal saw it as a science of the sciences, a means of counteracting overspecialization and achieving the unity of science, which should reflect the unity of reality.

Bernal was unsympathetic to positivist philosophies of science, but also to criticisms of positivism that would undermine science itself; he thought of irrationalist and intuitionist currents as the backwaters and dead ends of human knowledge. He objected most to scientists, such as Arthur Eddington (1882–1944) and James Jeans (1877–1946), who brought irrationality into the structure of science by making what science did not know, rather than what it did know, the basis for affirmations about the nature of the universe. His enduring legacy is a defense of science that ties it inextricably to philosophy and politics.

HELENA SHEEHAN

SEE ALSO *Communism; Marxism; Science, Technology, and Society Studies.*

BIBLIOGRAPHY

Aprahamian, Francis, and Brenda Swann, eds. (1999). *J. D. Bernal: A Life in Science and Politics*. London: Verso.

Bernal, J. D. (1939). *The Social Function of Science*. London: Routledge.

Bernal, J. D. (1949). *The Freedom of Necessity*. London: Routledge and Kegan Paul.

Bernal, J. D. (1969). *Science in History*, 3rd edition. Cambridge, MA: MIT Press. 1st edition, 1954; 2nd edition, 1957.

Bernal, J. D. (1969 [1929]). *The World, The Flesh, and the Devil: An Enquiry into the Future of the Three Enemies of the Rational Soul*. Bloomington: Indiana University Press.

Goldsmith, Maurice. (1980). *Sage: A Life of J. D. Bernal*. London: Hutchinson.

Sheehan, Helena. (1985). *Marxism and the Philosophy of Science: A Critical History*. New Jersey: Humanities Press.

BETHE, HANS

Hans Albrecht Bethe (1906–2005) was a Nobel physicist and leader in efforts to promote social and ethical

Hans Bethe, (1906–2005). An Alsatian-born American theoretical physicist, Bethe is a prolific and creative contributor to several vital fields of nuclear physics. He also discovered the mechanism of energy production by stars, including the sun. *(AP/Wide World Photos.)*

responsibilities among scientists and engineers. Born in Strassburg, Germany (now Strasbourg, France), on July 2, Bethe received his doctorate from the University of Munich in 1928 and began teaching at Cornell University in 1935, where he continued throughout his career. Bethe died in Ithaca, New York, on March 6. In 1938 he published three papers on nuclear physics that became known as "Bethe's Bible," and he received the 1967 Nobel Prize for discoveries concerning energy production in stars.

During World War II, the U.S. government recruited Bethe to work on military technologies, and in 1943 he was made director of the theoretical physics division in the Manhattan Project at Los Alamos, New Mexico, where he helped develop the first atomic bomb. The use of nuclear weapons created a strong sense of social responsibility in Bethe, and during the Cold War he worked to reduce the danger posed by nuclear weapons.

In 1945 Bethe became one of the original supporters of the Federation of Atomic (later American) Scientists, which focused on educating others about the implications of nuclear weapons. Bethe also served as a member of the President's Science Advisory Committee from 1956 to 1964. Beginning in 1957, he headed a presidential study of nuclear disarmament, known as the "Bethe panel," and served the following year as scientific advisor to the U.S. delegation at the Geneva nuclear test-ban talks. Bethe was "one of the heroes" in the campaign that culminated in the limited nuclear test-ban treaty signed by the United States, Great Britain, and the Soviet Union on August 5, 1963 (Schweber 2000).

Complex Ethical Response to Nuclear Weapons

During this time, Bethe developed a complex response to the ethical dilemma created by his dual roles as an advisor to the Los Alamos National Laboratory and as a political and moral critic of the development of nuclear technologies—a tension that challenged many scientists and engineers. For fifty years, Bethe led the struggle to address such questions as: When should various nuclear technologies be developed? What is the proper role of scientists and engineers in a democracy? What moral and political responsibilities do they have for the use of the knowledge they create? Although Bethe believed that scientists should always feel responsible for the consequences of their work, he argued for no simple answers.

Bethe's response is founded on a distinction between pure and applied science and the criterion of political necessity. For Bethe, knowledge is a good in itself, and pure scientific research should proceed even when it might be used for immoral purposes. It is only at the point of application "that people should debate the question: Should we or should we not develop this? But the gathering of scientific knowledge preceding that debate, and certainly pure science itself should not be stopped" (Bethe 1983, p. 5).

Development in turn should be guided by necessity. For instance, during World War II, Bethe was convinced of the necessity of the atomic bomb because of the Nazi threat. The hydrogen bomb, however, was a weapon of such magnitude to be of little practical military value. "It was unnecessary. It should not have been done. And we would now be very much better off if [it] had never been invented" (Bethe 1983, p. 3).

Yet once Edward Teller (1908–2003) and Stanislaw Ulam (1909–1984) realized how to build the hydrogen bomb, Bethe believed that it needed to be developed before the Soviets. Caught in this dilemma, he wrote, "If I didn't work on the bomb somebody else would. . . . It seemed quite logical. But sometimes I wish I were more consistent an idealist" (Edson 1968, p. 125). He maintains that the only justification for the hydrogen bomb is to prevent its own use (Bethe et al. 1950).

Who Should Make Decisions about Controversial Projects?

Bethe was careful to distinguish between the duties of the individual scientist and those of the scientific community as a whole. He was aware that a single individual is powerless to change the trajectory of weapons development. When asked whether it is justified to participate in immoral research projects just because others will do the research anyway, he replied, "No, but that is just to save my own soul. My refusal does not save the world" (Bethe 1983, p. 7). A group of scientists, not a single individual, needs to make decisions about what research to pursue and which findings to publish. Especially within the cold war context, the scientific community should not refuse to work on weapons as a group, because that would set them up as a superpolitical body that is the sole judge of their actions.

According to Bethe, elected representatives should make decisions about weapons research and other controversial projects. But scientists ought to have a large influence in these decisions. "By working on these weapons one earns the right to be heard in suggesting what to do about them" (Schweber 2000, p. 170). This in turn creates a dilemma for scientists, because in order to earn the right to be heard they must be willing to work for the government in developing weapons systems. Decisions about the use of technology are both scientific and political in nature, and such decisions should not be driven solely by technical feasibility (Bethe 1983).

In the 1980s, Bethe argued against the Strategic Defense Initiative (SDI) (a system, dubbed "Star Wars" by opponents, proposed by president Ronald Reagan in 1983 that would use space-based technology to protect the United States from attacks by strategic nuclear missiles), claiming that it would be much easier to simply reduce nuclear arsenals rather than developing a massive missile defense shield. In 1995 Bethe published an open letter to all scientists claiming that a new political era had made the further development of nuclear weapons unnecessary. He called "on all scientists in all countries to cease and desist from work creating, developing, improving and manufacturing further nuclear weapons—

and, for that matter, other weapons of potential mass destruction such as chemical and biological weapons."

ADAM BRIGGLE

SEE ALSO *Atomic Bomb; Nuclear Ethics; Weapons of Mass Destruction.*

BIBLIOGRAPHY

Bethe, Hans. (1983). "The Ethical Responsibilities of Scientists." *Center Magazine* 16(5): 2–11. A collection of Bethe's thoughts in this area with comments and counterarguments by an interdisciplinary panel of scholars.

Bethe, Hans, et al. (1950). "Let Us Pledge Not to Use an H-Bomb First." *Bulletin of Atomic Scientists* 6(3): 75. Signed by Bethe and eleven other prominent physicists.

Edson, L. (1968). "Scientific Man for All Seasons." *New York Times Magazine* March 10, pp. 29, 122–127. A brief look at Bethe's accomplishments and concerns.

Schweber, Silvan S. (2000). *In the Shadow of the Bomb: Bethe, Oppenheimer, and the Moral Responsibility of the Scientist*. Princeton, NJ: Princeton University Press. The best source for Bethe's involvement in the social and political aspects of nuclear weapons, chronicling his and Oppenheimer's roles in postwar developments and restrictions of nuclear arms.

INTERNET RESOURCE

Bethe, Hans. "Letter to the Science Community." Pugwash Online. Available from http://www.pugwash.org/about/bethe.htm. A plea to end the development of all weapons of mass destruction, originally issued July 23, 1995.

BHOPAL CASE

In December 1984, a gas leak of approximately forty metric tons of methyl isocyanate (MIC) from a pesticide plant in Bhopal, India, resulted in as many as 3,000 deaths and injuries to thousands. MIC, an organic chemical used in the production of pesticides, is a volatile liquid that reacts violently with water. MIC is highly toxic to humans and short-term exposure can cause respiratory diseases, if not death, and can seriously affect reproduction. The circumstances and results of what was the industrial accident with the largest death toll in history has been widely used as a case study in engineering design and technology management.

Union Carbide of India, Limited (UCIL), a company controlled by U. S.-based Union Carbide Corporation (UCC), operated the Bhopal plant. UCC provided the basic plant design, supervised its engineering, and defined its operating procedures. Prior to the catastrophe, the plant had been losing money for several years due to weak demand in India for pesticides. This resulted in major personnel reductions, particularly in production and maintenance. At the time of the accident, the plant had been shut down for more than a month for a complete maintenance overhaul. Important safety devices were out of commission and personnel with no MIC training were in supervisory roles. Consequently, when a large amount of water entered an MIC tank due to a mistake during normal maintenance procedures (according to the Indian government version of events), the ensuing reaction caused a large gas leak; defects in the MIC unit and a lack of staff safety training prevented containment.

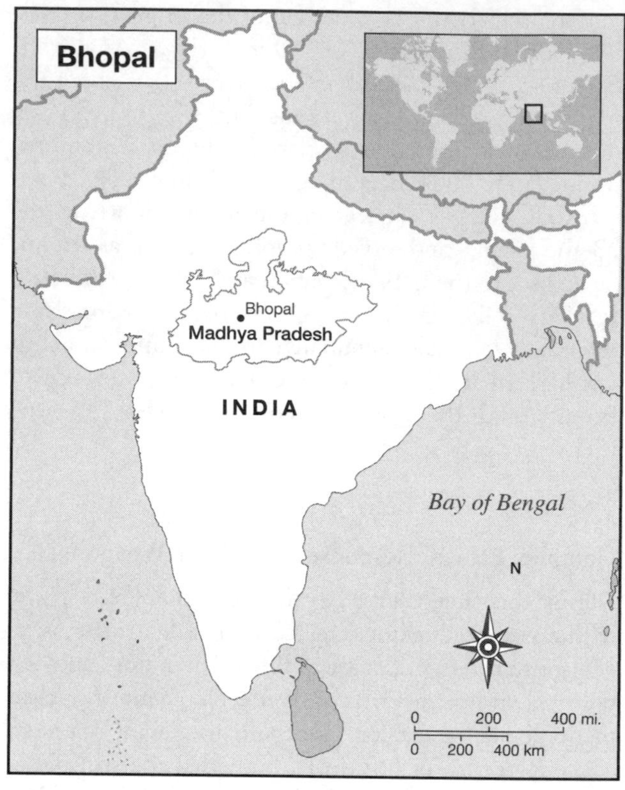

Developing countries often lack the infrastructure to safely support and maintain complex technologies. Companies based in countries such as India offer cheap labor and low operating costs for multinational corporations, but little incentive to promote environmental quality, safety procedures, and community investment (Bowonder, Kasperson, and Kasperson 1994). Increased risks posed by establishment of a MIC production unit close to slum colonies were never recognized by either UCIL or the Indian government.

UCC maintained safety standards at the Bhopal plant well below those at a sister plant in West Virginia; computerized data loggers, for example, were not employed at Bhopal. Furthermore, there was no attempt to follow up and implement safety recommendations of an Operational Safety survey conducted by a UCC team in 1982 (Shrivastava 1994). Specific safety problems that contributed to the disaster included: unreliable temperature and pressure gauges; the leaking MIC storage tank was filled beyond recommended capacity; a reserve storage tank for excess MIC already contained MIC; the community warning system had been shut down; a refrigeration unit that keeps MIC at low temperatures had been shut down; the gas scrubber designed to neutralize escaping gases had been shut down; the flare tower intended to burn off any MIC escaping from the scrubber had both a design defect and had been shut down; a water curtain intended to neutralize any remaining gas was too short to reach the top of the flare tower, where the gas exited (Patel 1997).

According to some observers, UCIL (and UCC) showed disregard for victims of the catastrophe, prolonging their suffering through failing to deal with their immediate needs. When MIC was released, the public alarm was not sounded until hours later. UCIL provided misleading information on treatment for toxic effects of MIC, resulting in inadequate treatment by local physicians. UCC blamed local workers for sabotage and conducted a media blitz to divert attention from the corporation (Morehouse and Subramaniam 1986).

The UCC strategy for negotiations focused on a fixed settlement. UCC fought hard to ensure the legal battle took place in India and the lawsuits filed in U.S. courts were rejected on the basis that the catastrophe occurred in India, the victims were Indian, and the plant was run by UCIL, an Indian subsidiary of UCC. In 1985, the Indian government passed the Bhopal Gas Leak Disaster Act, which made the government sole representative of all claimants. Later, using this act, the Bhopal Gas Leak Disaster Scheme emerged, further controlling registration, processing, and future compensation (Patel 1997).

UCC eventually settled out of court for $470 million, in the process denying any legal liability. To reciprocate, the Indian Supreme Court provided immunity from any future prosecution. A subsequent change in government prompted the court case to be reopened. Criminal proceedings against UCC and Warren Anderson (UCC Chairman at the time of the accident) have been pending in India since 1992. Under Indian law, the company has been deemed "fugitive" and India seized assets of UCIL to benefit victims of the catastrophe (Appleson 1999).

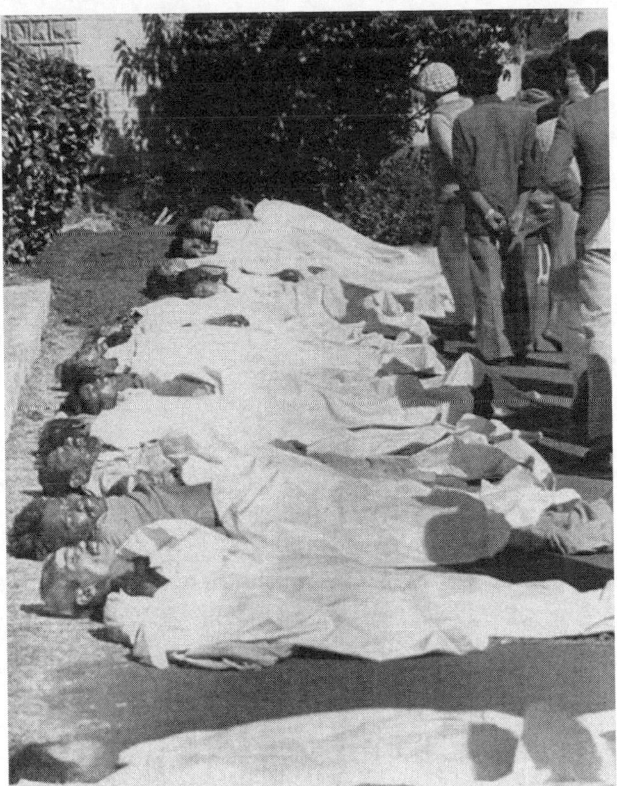

Bodies of victims of the Bhopal accident, waiting to be identified by relatives at the Hamidia Hospital. As many as 3000 people lost their lives in the tragedy. (*© Bettmann/Corbis.*)

The Bhopal disaster illuminates ethical issues throughout the chain of development of a technology, from the decision to build and operate a hazardous facility in a developing region that lacked the technical and institutional infrastructure to properly support it, to design decisions that compromised the plant's margin of safety, to failure to properly operate and maintain the plant. Perhaps the most troubling aspect from an ethical perspective is the failure of both industry and government to look beyond the legal issues and adequately confront the human suffering caused by the accident.

DEENA MURPHY-MEDLEY
JOSEPH R. HERKERT

SEE ALSO *Engineering Ethics.*

BIBLIOGRAPHY

Bowonder, B.; Jeanne X. Kasperson; and Roger E. Kasperson. (1994). "Industrial Risk Management in India Since Bhopal." In *Learning From Disaster*, ed. Sheila Jasanoff. Philadelphia: University of Pennsylvania Press. Review of basic issues in industrial hazards management and disaster prevention and the impact of the Bhopal catastrophe on industrial risk management practices in India.

Morehouse, Ward, and M. Arun Subramaniam. (1986). *The Bhopal Tragedy*. New York: Council on International and Public Affairs. Critical examination of the history and conditions that led to the Bhopal disaster and industry and government attitudes that put U.S. chemical workers at risk; strategies for preventing future disasters.

Shrivastava, Paul. (1994). "Societal Contradictions and Industrial Crises." In *Learning From Disaster*, edited by Sheila Jasanoff. Philadelphia: University of Pennsylvania Press. Uses Bhopal as case study to explore contradictory societal demands on technology and their role in causing and escalating social and economic crises arising from industrial accidents.

INTERNET RESOURCES

Appleson, Gail. (1999). "Bhopal Victims Sue Union Carbide Over '84 Disaster." Available from http://www.bhopal.org/newsstories/reuters.html. News story on victims' lawsuit; the site is maintained by a clinic in Bhophal sponsored by a charitable trust.

Patel, Trupti. (1997). "TED Case Studies: Bhopal Disaster." Available from http://www.american.edu/TED/bhopal.htm. Trade and Environment Database (TED) is an online compilation of environment-related case materials maintained by the Mandala Projects at American University.

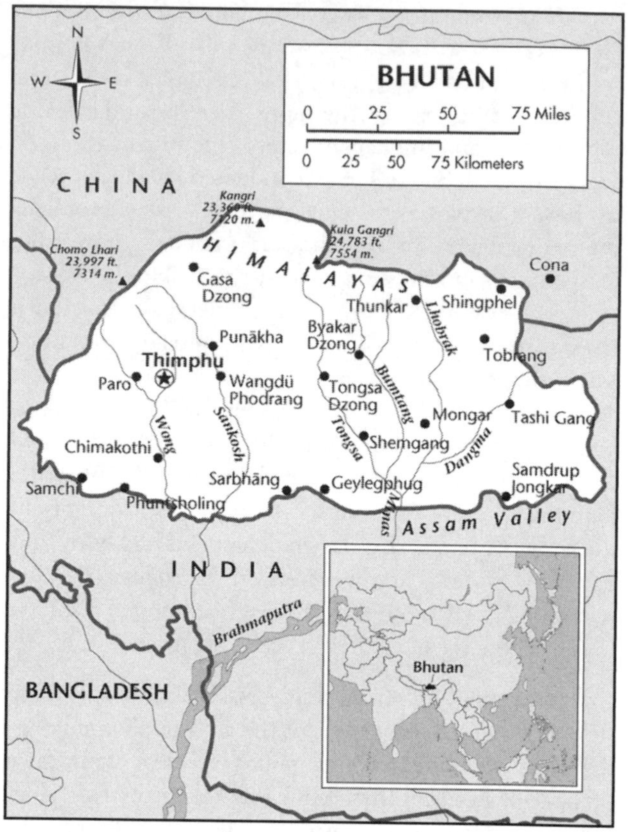

BHUTAN

Bhutan is a small landlocked country in the eastern Himalayas that is attempting to pursue an alternative to the common approaches to the relationships among science, technology, and ethics. Bordered on the north by Tibet and on the south by India, this Buddhist kingdom is approximately one-third the size of nearby Nepal, with a population estimated at around 1 million persons. In 1959, after the Chinese invasion of Tibet, Bhutan departed from a period of isolation that had lasted for centuries to accept assistance from India in building its first major road, thus initiating close diplomatic and economic ties with its southern neighbor. Despite its international ties, since 1960 Bhutan has pursued a cautious and circumspect approach to technology and development.

The vision guiding Bhutan's approach has emerged from the core values of Vajrayana Buddhism, specifically the Drukpa Kagyu and Nyingma lineages that dominate the country's spiritual landscape. The effect of those values on modern technological development is suggested in the frequently quoted maxim of Jigme Singye Wangchuck, the king of Bhutan: "Gross national happiness is more important than gross national product."

Ideas such as *ley jumdrey*, the law of *karma; tha damtshig*, the sacred commitment to interpersonal relationships; and the interdependence of all things are illustrated in the ubiquitous iconography of *thuenpa puenshi*, "the Four Friends," four animals that achieve a common good through thoughtful cooperation, an image that is painted on the walls of classrooms, government offices, hotels, shops, and homes throughout the country. Hagiographies of successful Buddhist practitioners convey the importance of self-discipline, the efficacy of ritual and contemplative practices, and the perfectibility of human beings, along with universal values such as honesty, compassion, harmony, and nonviolence. Divine madmen such as the antinomian folk hero Drukpa Kunley offer a corrective to pretentious, self-important authority and the soporific effects of habituation to mundane, consensus reality.

Guided by those core Buddhist values, Bhutan has approached the ideal of sustainable development, linking technological innovation, environmental conservation, cultural continuity, and good governance through development programs aimed at increasing human welfare rather than focusing only on industrialization and economic diversification. Conservation of the last remaining unspoiled forests in the Himalayan region is a national priority that is grounded in a preexisting indigenous conservation ethic. Protected conservation areas account for about 26 percent of the country's land area. Education in environmental science begins at the kindergarten level, and public banners reinforce that ethic with admonitions such as "Healthy Forest for a Healthy Environment, Let Us Maintain It." The Bhutan Trust Fund of Environmental Conservation, established in 1991, is widely acknowledged as the first national environmental trust in the world and has been a model for similar trusts in other countries.

Foreign exchange primarily involves tourism and hydroelectricity sold to neighboring India. Learning from the experiences of regional neighbors such as Nepal, Bhutan gradually opened its borders to foreign tourists but in 1974 adopted a policy of "high-value, low-volume" tourism to avoid the negative consequences of unrestrained tourism on the natural environment and the indigenous culture. A similar caution has been displayed in the development of hydroelectricity. According to 1996 estimates, only 2 percent of the hydroelectric potential of the nation has been tapped. In addition to the major dam at Chukha, many mini- and micro-hydroelectric projects are scattered throughout the country in order to avoid the watershed damage associated with larger projects while providing electricity directly to remote locales.

Perhaps the most dramatic and far-reaching technological change occurred in 1999 with the lifting of a government ban on broadcast television and the introduction of Internet access. The extent to which traditional Bhutanese values will be displaced by an ideology of consumerism and the values of an advertising culture remains to be seen.

JEFFREY R. TIMM

SEE ALSO *Buddhist Perspectives; Social Indicators.*

BIBLIOGRAPHY

National Environmental Commission, Royal Government of Bhutan. (1998). *The Middle Path: National Environment Strategy for Bhutan.* Bangkok, Thailand: Keene Publishing. This book was produced by the National Environment Commission and a policy statement of the Royal Government of Bhutan on conservation and sustainable development.

Priesner, Stefan. (1999). "Gross National Happiness—Bhutan's Vision of Development and Its Challenges." In *Gross National Happiness: Discussion Papers.* Thimphu, Bhutan: Centre for Bhutan Studies. This is the first book published by the Centre for Bhutan Studies and contains nine papers exploring various aspects of development *vis-á-vis* the "gross national happiness" concept.

Wangyal, Tashi. (2001). "Ensuring Social Sustainability: Can Bhutan's Education System Ensure Intergenerational Transmission of Values?" *Journal of Bhutan Studies* 3(1): 108–133. This article explores the role of public education in Bhutan in the transmission of Buddhist values and perspectives in face of increasing globalization of culture.

BIODIVERSITY

Life on earth began as bacterial cells at least 4,000 million years ago, and it has—with notable, but rare, catastrophic declines in diversity subsequently—expanded, evolved, and complexified across time. In the early-twenty-first century the earth teems with countless species arranged in many diverse patterns and relationships spread across varied landscapes. As human populations have expanded since the industrial revolution, with technologies becoming more powerful and increasingly capable of pervasive impacts, biodiversity is again in decline, this time as a result of human activities, especially the fragmentation of forests and other wild habitats. How to reverse the dangerous trend toward biological simplification has become one of the most urgent global environmental questions.

What Is Biodiversity?

Biodiversity, a contraction of *biological* and *diversity*, was introduced as a convenient abbreviation during preparations for a national symposium on the subject in the United States, which was sponsored by the Smithsonian Institution and the National Academy of Sciences (NAS) in 1986. This term, though technically difficult to define, has come to refer to the rich and textured web of life on earth. The term, and the concepts and ideas associated with it, gained world political prominence at the World Conference on Environment and Development in Rio de Janeiro in 1992, where the United Nations Convention on Biodiversity, a document that was subsequently ratified by a majority of nations, was passed. This convention called for the sustainable use and protection of the earth's biological resources, and the term biodiversity has become the rallying point for conservationists and others concerned about the rapid simplification of natural systems in the face of human development.

There are two approaches to defining biodiversity (Wood 1997, 2000). Perhaps the most popular approach is to define it as an *inventory* of diverse biological items.

One such inventory, which has been described as the *standard* definition of biodiversity, defines it as the sum total of species, genetic variation within species, and diversity of habitats and ecosystems in which species are found (Takacs 1996). Difference definitions, in contrast, define biodiversity as the total of differences among biological entities and processes (Takacs 1996; Wood 1997, 2000). These two approaches differ in that inventory definitions, which simply count elements of different types (species, genes, habitats), tend to count elements of the same type equally in their contribution to total diversity, whereas difference definitions highlight degrees of differentiation. Accordingly, under a difference definition, a species that is the lone member of its genus would be considered to contribute more diversity than a species that shares its genus with others (Solow et al. 1993, Weitzman 1998).

As noted, there are serious technical problems in defining biodiversity. It clearly would be desirable to have a definition that represents biodiversity as a measurable quantity—so that, for example, one could say that a given system is measurably more diverse than another, or that a given system is losing or gaining diversity at a specifiable rate. Unfortunately both inventory and difference definitions fail to provide a measurable index of biodiversity. Decisive arguments show that biodiversity cannot be represented as a list of countable and additive elements. This conclusion follows from the unavoidable fact that living nature can be organized into multiple, but incommensurable, hierarchies. For example, there is a phylogenetic hierarchy of species and genera, among others, as well as a spatial hierarchy of cells, organs, organisms, and ecosystems. Both hierarchies add significantly to the total diversity of life, but the elements of these hierarchies cannot be added together to create a meaningful sum. Similar arguments apply to difference definitions: Biological entities vary across many parameters and aspects, and these cannot be added together to represent a meaningful index of biological diversity (Wood 2000, Sarkar 2005).

This difficulty implies that biodiversity is too complex and multifaceted to be represented by a single measure or to be made a countable quantity, and has led to a search for *proxy* measures for biodiversity (Sarkar 2005). One popular proxy is simply to use species counts as conventional markers to represent total diversity, which has gained wide acceptance in practice because it is clear and allows relatively unambiguous measures. The consensus view of scientists, however, is that simplified measures such as this cannot capture the full richness and diversity of life. In the United States this simplification was nevertheless embodied in the Endangered Species Act of 1973 (ESA), which has become, despite its narrow focus on endangered species, one of the most important pieces of environmental legislation ever passed by a national legislature. The act concentrates heavily on avoidance of species extinction through a process that lists species as *threatened* or *endangered*, limiting damage to the listed species, subspecies, and special populations. Protection of habitat is mainly treated by the act as instrumental to the protection and recovery of endangered and threatened species.

Accordingly it has been suggested that the U.S. strategy—which can be referred to as a *rare species paradigm*—may be less effective than an alternative strategy developed by Australian practitioners, who develop algorithms that rank possible reserve designs according to their effectiveness, per area set aside, in saving biodiversity (defined in terms of a chosen proxy). The Australian approach, referred to as the *declining species paradigm* (Caughley 1994), has increasingly been applied in international settings. This approach is to develop and refine an algorithm that ranks various protection strategies according to their efficiency in using space to protect proxy variables chosen to represent managerial goals. This pragmatic approach—which emphasizes shared actions rather than abstract measures—can provide a rough operationalization of biodiversity: Biodiversity is what is saved by the actions of professionals who design reserves that are effective in responding to identifiable forces of simplification that are addressed in a real place (Sarkar 2005).

Speaking more generally, biodiversity can be thought of as the result of a magnificent and eternal process of change, as life has explored countless strategies for survival in countless climates and habitats. These explorations have led to an inexorable increase in diversity across time, because each increment in diversity opens up new possibilities and adaptations for other species, and to the hypothesis that diversity itself causes increases in diversity. This theory also has a negative side: Losses in diversity can increase the likelihood that further losses will occur as species are stressed by loss of mutualist species and populations (Whittaker 1970, Norton 1987). Thus whereas biodiversity has, in the big picture, increased over time, there have also been cataclysmic periods of species loss, and paleontologists speculate that there have been as many as six extinction events in which half or more species disappeared. At least some of these events are associated with meteor strikes on earth and, possibly, as a result of dust from enormous volcanic eruptions. Increasing rates of extinction and endangerment have led some scientists to speculate that the Earth is entering another such event, for the first time as a result of human activities. Whether the human species can survive such an event is not known, but the exponential effects of human activities are impacting the world at a scale previously produced only by global cataclysms.

Fear that the simplification of nature may cause an irreversible spiral of losses inspires scientists and conservationists to advocate strong measures to reverse simplification processes before it is too late. As noted, there exists a broad, practical consensus among experts about what actions are necessary to reverse, or at least slow, such processes such as establishing protected riparian corridors along rivers and developing core reserve areas while managing buffer zones around them. Whether the means, and the will, exists to rein in development that encroaches on wild habitats and drives species toward extinction remains uncertain. Conservationists agree that it is important to save as much biodiversity as possible, though there are seldom adequate resources to do even a fraction of the things that are widely recommended by experts for the protection of sensitive areas and diversity *hotspots*. Thus whereas success in protecting biodiversity is not assured, broad agreement in strategies to maximize biodiversity does inspire confidence that practitioners know what they are talking about—that the concepts used are *clear enough* to allow communication and cooperative action—even if no abstract definition of biodiversity can be considered to correspond precisely to any measurable quantity in nature (Norton 2005).

What Is the Value of Biodiversity?

Despite considerable agreement in conservation strategies and protective practices, there remain several cross-cutting disagreements regarding *why* biodiversity protection is important (Norton 1987, 1986). These are: (1) the *nature* of the value biodiversity has; (2) the *units* of diversity that should be valued; and (3) the appropriate *measures* of the value of biodiversity. These disagreements are important because they affect the prioritization given biodiversity protection in competition with other socially valued objectives, and also among various possible conservation objectives.

Disagreements regarding the *nature* of the value of biodiversity reflect differing theories of value. Monistic theories of value account for all value in nature according to a single measure. Utilitarians, economically oriented and otherwise, advocate decisions based on impacts on human well-being or satisfaction. Other monists have extended ethical concepts, usually applied only to humans, to other species and even to ecosystems, treating elements of nature as *ends-in-themselves*, as possessing *moral considerability*, and as having *goods-of-their-own* that compete with human welfare. The prominence of these two opposed, monistic theories, has resulted in a polarized discussion, often pitting economists against environmental ethicists, and no consensus regarding how to place measurable value on biological diversity has emerged.

The value of biodiversity is better captured by a pluralistic evaluative method, which treats the many social values derived from biodiversity as reinforcing each other. Actions that protect biodiversity protect complex natural systems, reduce soil erosion, promote aesthetic enjoyment and scientific interest, hold open options for economic uses, and support the values of the many individuals who value nature noninstrumentally.

Pluralism, though unpopular within academic disciplines, seems more consistent with the many ways that humans express their dependence upon, and love for, nature. Under a pluralist approach, multiple competing values must be balanced and prioritized against each other, but opportunities also arise to protect multiple social values simultaneously, opening up the possibility of win-win management policies through the protection of natural habitats as homes for biodiversity and many other values. The pluralist approach encourages a more political understanding of the value of biodiversity. Some authors conceptualize the problem of biodiversity protection as one of accepting responsibility for conveying a *trust*, or a gift from previous to subsequent generations as an obligatory legacy (Weiss 1989, Brown 1994). In a variant on the trust idea, other theorists argue that future generations have rights to a full complement of species and ecosystems, and that these rights should be protected by constitutional constraints that require governments to protect biodiversity (Schlickeisen 1994, Wood 2000). These trust doctrines and the constitutional amendment recommendation, built on a moral concern for the future, complement the idea of sustainable use and development of resources. The goal of protecting the evolved web of life, what scientists call biodiversity—whether for its possible uses in fulfilling human needs, the diverse aesthetic experiences it affords humans, or its noninstrumental value to the fulfillment of human needs—will be one of the great challenges of the future.

BRYAN G. NORTON

SEE ALSO *Biophilia; Deforestation and Desertification; Ecology; Environmental Ethics; Rain Forest; United Nations Environmental Program.*

BIBLIOGRAPHY

Brown, Peter G. (1994). *Restoring the Public Trust: A Fresh Vision for Progressive Government in America*. Boston: Beacon Press.

Caughley, Graeme. (1994). "Directions in Conservation Biology." *Journal of Animal Ecology* 63: 215–244. Outlines the differences between *rare species* and *declining species* paradigms.

Norton, Bryan G. (1987). *Why Preserve Natural Variety?* Princeton, NJ: Princeton University Press. A monograph surveying the science, values, and policies associated with biodiversity.

Norton, Bryan G. (2005). "What Are We Trying to Save? And Why? Toward a Policy-Relevant Definition of *Biodiversity*." In *The Endangered Species Act at 30*, ed. J. Michael Scott, Dale Goble, and F. Davis. Washington, DC: Island Press.

Norton, Bryan G., ed. (1986). *The Preservation of Species: The Value of Biological Diversity*. Princeton, NJ: Princeton University Press. An anthology providing a multidisciplinary review of nature protection.

Sarkar, Sahotra. (2005). *Biodiversity and Environmental Philosophy: An Introduction*. Cambridge, UK: Cambridge University Press. A highly recommended survey of issues and methods, written from the viewpoint of a philosopher of biology.

Schlickeisen, Rodger. (1994). "Protecting Biodiversity for Future Generations: An Argument for a Constitutional Amendment." *Tulane Environmental Law Journal* 8: 181–221.

Solow, Andrew; Stephen Polasky; and James Broadus. (1993). "On the Measurement of Biological Diversity." *Journal of Environmental Economics and Management* 24: 60–68.

Takacs, David. (1996). *The Idea of Biodiversity: Philosophies of Paradise*. Baltimore: Johns Hopkins University Press. A very useful examination of biologists' ideas on biodiversity.

Weiss, Edith Brown. (1989). *In Fairness to Future Generations: International Law, Common Patrimony, and Intergenerational Equity*. Tokyo: United Nations University; Dobbs Ferry, NY: Transnational Publishers. An analysis based in the theory of international law, of obligations to the future.

Weitzman, Martin L. (1998). "The Noahs's Ark Problem." *Econometrica* 66(6): 1279–1298.

Whittaker, Robert H. (1970). *Communities and Ecosystems*. New York: Macmillan. Whittaker introduced many of the concepts used to discuss biodiversity.

Wood, Paul M. (1997). "Biodiversity As the Source of Biological Resources: A New Look at Biodiversity Values." *Environmental Values* 6: 251–268. A focused argument for a difference definition, including also a discussion of the value of biodiversity as a *source* of biological resources.

Wood, Paul M. (2000). *Biodiversity and Democracy: Rethinking Society and Nature*. Vancouver, British Columbia: UBC Press. An examination of the moral and political obligations associated with protecting biodiversity.

BIODIVERSITY COMMERCIALIZATION

The Convention on Biological Diversity (CBD), which entered into force on December 29, 1993, established an international treaty for the conservation and sustainable use of biological diversity and set up a process for the further development of legal, policy, and scientific activities related to biodiversity. The treaty has been highly controversial, however, provoking strong differences in perspectives, especially between those claiming to speak for indigenous peoples and for commercializing enterprises.

Historical Background

Concerns about the global loss of biodiversity that emerged in the late 1970s took their initial legal form in the International Undertaking on Plant Genetic Resources voluntarily adopted by members of the Food and Agriculture Organization (FAO). This 1983 agreement, based on a proclaimed "universally accepted principle that plant genetic resources are a heritage of mankind and consequently should be available without restriction," aimed to "ensure that plant genetic resources of economic and/or social interest ... will be explored, preserved, evaluated and made available for plant breeding and scientific purposes."

Discussion of the costs and responsibilities for implementing such an agreement stimulated the United Nations Environment Programme (UNEP) in 1987 to establish an Ad Hoc Working Group of Experts on Biological Diversity to harmonize existing related conventions. Negotiations that produced the CBD began in 1990 among representatives from governments, corporations, and various interest groups including universities, research institutes, botanic parks and gardens, and community-based nongovernmental organizations (NGOs). The CBD was opened for signature at the United Nations Conference on Environment and Development in Rio de Janeiro, June 1992. According to the CBD itself, its objectives are "the conservation of biological diversity, the sustainable use of its components and the fair and equitable sharing of the benefits arising out of the utilization of genetic resources, including by appropriate access to genetic resources and by appropriate transfer of relevant technologies, taking into account all rights over those resources and to technologies, and by appropriate funding" (Article 1).

The forty-two articles of the CBD not only create substantive provisions for conservation, commercial development, scientific research, and education regarding biological diversity (articles 6–20), but also outline mechanisms for further development of these provisions through a Conference of Parties (article 23), Secretariat (article 24), and a Subsidiary Body on Scientific, Technical, and Technological Advice (article 25). One of the first actions of the Conference of Parties (COP) was to add a Protocol on Biosafety, negotiation on which began at a COP meeting in Cartagena, Colombia, in 1999, and continued in Montreal, Canada, in 2000, when agreement was reached. Although negotiations were concluded in Montreal, the results are still known as the Cartagena Protocol on Biosafety, which implements CBD article 19 with procedures for the "safe transfer, handling and use of any living modified organism resulting from biotechnology that may have adverse effect on the conservation and sustainable use of biological diversity."

Ethical Debates

As Kerry ten Kate and Sarah A. Laird (2001) have summarized them, there are two basic responses to the CBD and its issues. On the one side are those representing private commercializing enterprises (most prominently transnational agricultural and pharmaceutical corporations); on the other are those of indigenous or local interests from the source countries in the developing world.

BIOPROSPECTING. From the point of view of private corporations, they are involved in bioprospecting for what might be thought of as "green petroleum" in a process that will bring wealth to gene-rich but financially poor countries. Corporations argue that just as in the cases of other resources such as minerals, the commercialization of biological resources requires major capital investments in research and development over long periods of time with no guarantee of rewards. The only way a business enterprise can justify such investments is through an ability to patent those processes and products of its work. Moreover, the ultimate rewards will be in the long-term best interests not only of the corporations and their shareholders but of the source countries as well.

Demands by source countries for more up-front payments for raw biological resources access and for more explicit informed consent processes will ultimately destroy the bioprospecting market. Biological research and development work is in competition with genetic engineering of pharmaceuticals, bioinformatics, and new forms of synthetic and combinatorial chemistry including molecular biology and nanotechnology. Only if bioprospecting can remain competitive with such alternatives will it be pursued. Requiring that local populations be given extensive education about the biological resources to which they sell the rights, along with full disclosure of potential research and development trajectories, both negative and positive, only adds another level of costs that can easily drive corporations away from the kinds of investment that are ultimately beneficial to source countries.

BIOPIRACY. From the point of view of critics representing source countries, however, bioprospecting is better described as biopiracy. This term was coined in 1993 by the Rural Advancement Foundation International (RAFI), an NGO subsequently renamed the Action Group on Erosion, Technology and Concentration (ETC Group), and then widely disseminated when deployed as the title of Vandana Shiva's *Biopiracy: The Plunder of Nature and Knowledge* (1997). The word is part of the rhetorical critique of globalization or the anti-globalization movement, an equally controversial name for political and economic action that representatives themselves often prefer to describe as an alternate globalization (alter-globalization) or fair-trade (as opposed to free trade) movement.

According to the ETC Group, biopiracy involves the unjust appropriation of indigenous knowledge and genetic resources by individuals or institutions seeking control (usually patents or breeders' rights) over them, leading to the loss of control of their own resources by traditional peoples. In this sense, biodiversity commercialization is simply a new form of colonialization, in which developed countries through global corporations scour the world, extract genetic material, then patent these finds as their "discoveries." Colonization is now focused on life itself—plants, micro-organisms, animals, and even human organs, cells, and genes. From this perspective, the CBD may be used as a means to regulate access to biological resources in ways that lead to sharing with the communities the results of research and development and the benefits arising from the commercial utilization of genetic resources in a fair and equitable way. It may also function to protect diversity not only in biology but also in culture, not only facilitating advancements of knowledge in modern science but preserving the knowledge present in indigenous science.

IGNACIO AYESTARÁN
CARL MITCHAM

SEE ALSO *Agricultural Ethics; Biodiversity; Globalism and Globalization.*

BIBLIOGRAPHY

Kate, Kerry Ten. (1995). *Biopiracy or Green Petroleum? Expectations and Best Practices in Bioprospecting.* London: Overseas Development Administration.

Kate, Kerry Ten, and Sarah A. Laird. (1999). *The Commercial Use of Biodiversity: Access to Genetic Resources and Benefit-Sharing.* London: Earthscan.

Khor, Martin. (2002). *Intellectual Property, Biodiversity and Sustainable Development: Resolving the Difficult Issues.* London: Zed Books.

Secretariat of the Convention on Biological Diversity. (2001). *Handbook of the Convention on Biological Diversity.* London: Earthscan.

Shiva, Vandana. (1997). *Biopiracy: The Plunder of Nature and Knowledge.* Boston: South End Press.

BIOENGINEERING ETHICS

All of engineering can be viewed as a continuation of biology by other means—a metabiological activity. Bioengineering arose relatively recently with specific focus on living systems, for medical purposes in a close alliance with medicine, for scientific and industrial purposes and for other purposes.

A vast array of specializations and subfields have emerged, not always closely related and sometimes predating the overall recognition of bioengineering as a field. An ever expanding and at times confusing and overlapping taxonomy includes biomechanics (encompassing also biorheology and biofluid mechanics), instrumentation, biochemical engineering, bioastronautics, environmental engineering, biomaterials, tissue engineering, biological systems engineering, engineering of drug design and delivery, biotechnology instrumentation, bionanotechnology, and bioinformatics (Blanchard and Enderle 1999, Bronzino 1999, Fung 2001).

Bioengineering, as a field of research and applications, brings to bear not only engineering on medicine and biological organisms, but also a knowledge of biology on engineering designs. This helps assessing the meaning of engineering as the extender of biology and ultimately helps engineering develop a clearer sense of its own nature and address the ethical issues involved in its modification of nature and the creation of machines, that is, artifacts.

TABLE 1

The Interaction of Various Disciplines with Bioengineering Ethics

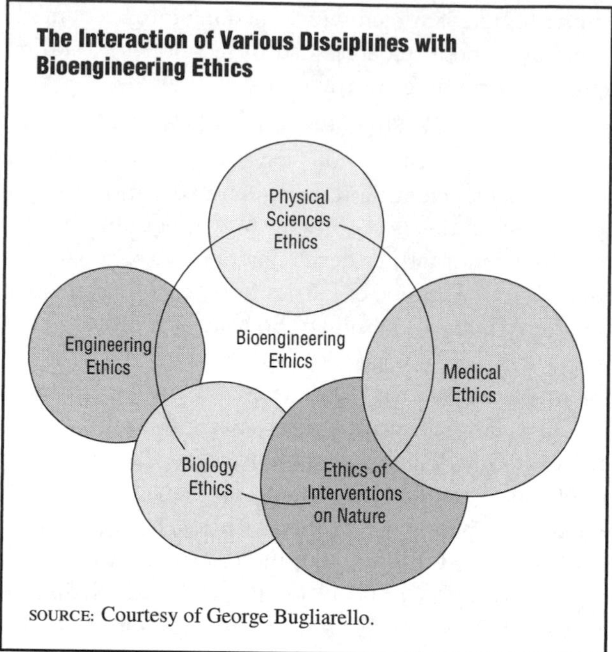

SOURCE: Courtesy of George Bugliarello.

The intersection of Bioengineering Ethics with Cognate Ethics.

Biomechanics

Biomechanics began to flourish in the 1960s, but interventions on the human body through artifacts have a long history that originated with prehistoric supports for fractured bones and skin decorations such as scarification, implanted rings, and tattoos. Daedalus with his mythical wings and the Tailor from Ulm with his arm-extending wings for gliding (Eyth 1885) were precursors of biomechanics, one in legend and the other in reality. After medieval times this process progressed to encompass eyeglasses, artificial teeth, and rudimentary artificial limbs. Eventually the interventions on the human body fulfilled other needs through diagnostic and curative tools and processes, from the application of bioengineering to bioastronautics starting in the 1960s (Konecci 1968) to X-ray visualization through computed tomography (CT), ultrasound scans, and magnetic resonance imaging (MRI), to hearing aids, surgical robots, autoanalyzers, DNA-sequencing machines, tissue engineering, and the application of engineering knowledge to the understanding of biological (Bejan 2000) and therapeutic processes. Most of these developments were strongly interdisciplinary, blending engineering, physiology, physics and mathematics. Interdisciplinarity continues to characterize the field.

Other bioengineering milestones include the first artificial organs. The artificial kidney was given practical form through the application of engineering principles by the Dutch physician Wilhelm Kolf in the 1940s, and the first heart pacemaker was implanted in 1958 in Sweden through the collaboration of the surgeon Åke Senning and the physician-inventor Rune Elmquist.

Pioneering studies of the brain were conducted by John Von Neumann and Walter Roseblith, and the study of neurons was initiated by Walter Pitt and Warren McCullogh. They opened a new domain for bioengineering, and also provided significant insight for the design of new kinds of computers.

An early example of the application of biology to engineering that has had an immense impact on human health is biological water and waste-water treatment processes. Biomimesis—the mimesis of biological designs, materials and processes—is another aspect of engineering applications that range from the creation of artifacts for medical and industrial purposes to genetic engineering and to ergonomics.

Other developments include the embryonic emergence of biomachines, as in the case of cardiac pacemakers and of bioelectrical sensors (biological sensors implanted on an electronic platform), and the biosoma concept of the integration of biological organisms and their two metabiological offshoots: society and machines (Bugliarello 2003).

Toward an Ethics of Bioengineering

Harmonization of the comprehensive ethical canons needed to address modifications of nature through the design and operation of artifacts and respond to conflicting views of the public good that engineering is committed to serve presents limitations and contradictions, as occurs when engineers develop products in which commercial motivation overshadows social goals. As a consequence, the flourishing of bioengineering as an offshoot of engineering has outpaced a focus on the ethical issues that confront it.

The complexity of formulating a bioengineering ethics arises from the need of bioengineering to be coherent not only with the ethics of engineering but also with those of biology, medicine, and the physical sciences, the fields with which bioengineering interacts most strongly (see Figure 1). Those specialized ethics, which are congruent with general ethics but distinct from it and complementary, must be rooted in the fundamental philosophical issues of each field: In the physical sciences, how do researchers obtain and verify knowledge? In biology, how can this be done in the context of living organisms and what is the nature of life, including the body-mind problem of consciousness? In medicine, what is the nature of disease? In engineering, what is the nature of the machine, why are there machines, and how far can humankind go with machines, for example, in making them self-reproducing?

The associated key ethical issues in physics and biology are concerned primarily with the purpose and conduct of research and the impacts and limits of research as exemplified by controversies in nuclear energy and cloning. In medicine, those issues relate to the limits of therapy, safety and risk, the Hippocratic imperatives, informed consent, and the role of the patient as well as the dilemma of individual versus societal benefit. In engineering, they have to do with the purposes and benefits of machines and interventions in nature, biosocial and environmental impacts, and risk and appropriate safety factors. The host of specific ethical issues associated with bioengineering arises from the need to incorporate the ethical questions of physics, medicine, and biology in addressing the domain, focus, and impact of bioengineering; its risks and safety factors; the views of nature that govern its activities; and the issues of activism and intellectual responsibility.

Domain, *focus*, and *impact* questions start with the positioning of the biomachine interface: Where should it be placed in the polarity between biological organisms and machines? To what extent should biomachines retain the essential characteristics of biological organisms *versus* those of machines? Also, should there be limits to biomimesis, the imitation of biology in creating devices? Are there potential dangers as well as benefits, and if so, what should guide the bioengineer? Should the ethical responsibility of bioengineering be exclusively humancentric, or should it extend to a broader biocentric domain with responsibility to other advanced life forms?

Relevant to urgent social needs are questions of prevention versus therapy. Historically, many medically oriented bioengineering activities have focused on therapy and very costly devices. This has improved medical capabilities, but to what extent should escalation of medical costs and principles of social equity make it an ethical imperative for bioengineering to focus more on prevention? Indeed, what should be the appropriate interface with medicine; what should be the specific role and responsibility of the bioengineer in a clinical environment? The dilemma of the individual versus society affects medicine and bioengineering alike and is at the core of the debate about health care: Should the focus be exclusively on the individual? To what extent should the cost to society also be taken into account?

The issues of medical versus industrial purposes, with their different motivations, also can be a source of contradictions and conflicts for bioengineers: Should they participate in a medical procedure or in the development of an industrial process merely for the technical challenge, without a clear understanding of the ultimate consequences? Should the imperative ethical requirement for bioengineers be to act as independent-minded professionals regardless of the pressure that may be put on them by a hospital, research laboratory, factory, or granting agency?

A closely related issue is the depersonalization of health care brought about by its increasing technicization. To what extent should bioengineers focus on the design of the clinical environment in which bioengineering machines are placed and processes are carried out and endeavor to reduce that depersonalization by taking into account the emotional component of human nature (a component that depends in turn on physiological factors, themselves amenable to medical and bioengineering research)?

What are acceptable *risks* and appropriate *safety factors* of bioengineering designs (a meeting point of the ethics of medicine and engineering with political, economic, and legal theories)? Do the efforts expended and the risks generated by a solution produce benefits that justify its development? A correlate ethical issue is the bioengineer's responsibility to follow up on the performance of a design or process, communicate the results whether they are positive or negative, and strongly advocate the adoption of satisfactory, safe, effective designs or processes and the elimination of dangerous and counterproductive ones.

Bioengineering interventions in natural processes must take into account the many basic and often conflicting values involved in *different views of nature*. These views range from utilitarian (an emphasis on the way in which humans derive benefits from nature) to the doministic (the drive to dominate nature for the sake of doing so) (Kellert 1996). Each view involves ethical dilemmas for bioengineering, starting with the basic issue of whether or to what extent to accept nature as is or to modify it teleologically; this can be thought of as an aspect of the conflict between biology (and at times religion) and engineering or medicine. The dilemma leads to different ethics—the ethics of discovery (science) versus that of design (engineering)—and to contemporary debates about genetic engineering (under what conditions should discovery lead to design?).

Activism and Intellectual Responsibility

In terms of *activism and intellectual responsibility*, to what extent should bioengineers intervene in the philosophical dialogue about the modification of nature, the future of humans and the human responsibility for other species? Should they participate actively in the political arena by pressing for new visions and their realization rather than seeing their role as a purely technical one? What is the ethical responsibility of bioengineers in projecting the potential modifications of nature that bioengineering can make possible and to inform society as to how beneficial modifications can be safely accomplished?

Provisional Answers

Even a cursory view such as the one presented here conveys the broad, complex, and fundamental nature of the ethical questions involved in bioengineering. Like all of ethics, bioengineering ethics deals with questions that are beyond the realm of the legal responsibility of bioengineers and may conflict with it. However, these are issues for which bioengineers should seek to define and enhance a professional conscience and behavioral guidance. So far only some of these questions have been addressed, and often only in a rudimentary way. Until a comprehensive bioengineering ethics has been formulated, a provisional set of tenets is needed. Those tenets might include the following:

- The *harm avoidance tenet* (essentially a restatement of the Hippocratic oath): to minimize the side effects of a design or intervention and devise something that bioengineers would use on themselves if necessary
- The *professional tenet:* to act as independent-minded professionals regardless of pressure from the environment in which bioengineers operate and intervene in professional and public discussions about engineering, medical, biological, and societal issues that bioengineering could illuminate
- The *approval tenet:* not to participate in medical procedures or in the development of industrial or military processes of which bioengineers do not personally approve no matter how technically challenging those procedures or processes are
- The *conflict of interest tenet:* not to advocate an unsafe, ineffective, or inferior design because one has a vested interest in it

- The *risk tenet:* to weigh the risks to human society and the environment of a bioengineering device or process
- The *effectiveness tenet:* to make the cost and risk of a design or intervention commensurate with the expected benefits
- The *responsibility tenet:* to assume the responsibility to follow up the performance of a design or process and communicate the results whether they are positive or negative
- The *finality tenet:* to attempt to expand the capabilities of humans, and, where appropriate, other biological organisms, being mindful of the metabiological nature of bioengineering as an activity that synthesizes two human drives: understanding nature and modifying it to preserve and enhance life

It is unrealistic to believe that a consistent and comprehensive bioengineering ethics will emerge rapidly from all the disparate elements and concerns that will contribute to its formation. A bioengineering ethics cannot be independent from the fundamental philosophical conceptions and ethics of the society in which bioengineering is embedded. These issues in turn are shaped and modified by advances in knowledge, social and political events, and the progress of bioengineering. It is, however, realistic and necessary to endeavor to establish some ethical principles that can guide the actions of bioengineers beyond their contingent legal obligations or at least to increase bioengineers' awareness of the ethical dilemmas that may confront them.

Ultimately, all forms of engineering are involved—directly or indirectly—in the modification of the biological world: For example, a highway, by bisecting a habitat, affects the ecology of that habitat and hence its biology. In the future, greater awareness and knowledge of biological processes resulting from advances in bioengineering will blur some of the boundaries between bioengineering and other fields of engineering, as in the creation of biomachines—intimate combinations of machines and biological organisms. This will add to the complexity of the ethical issues confronting the bioengineer and society.

GEORGE BUGLIARELLO

SEE ALSO *Bioethics; Biotech Ethics; Engineering Ethics; Medical Ethics.*

BIBLIOGRAPHY

Bejan, Adrian. (2000). *Shape and Structure, from Engineering to Nature.* New York: Cambridge University Press. Considers the application of engineering systems principles to patterns of biological forms and structures.

Berger, S. A.; W. Goldsmith; and E. R. Lewis eds. (2000). *Introduction to Bioengineering.* Oxford and New York: Oxford University Press. An introduction to mechanical, electrical, chemical, and biomehanial aspects of bioengineering.

Blanchard, Susan, and John Enderle. (1999). *Introduction to Biomedical Engineering.* San Diego: Academic Press. An overview of the field, including historical perspectives.

Bronzino, Joseph D. (2000). *Biomedical Engineering Handbook.* Vol. 2. Boca Raton, FL: CRC Press. A user-friendly, comprehensive handbook for bioengineering, including also ethical issues.

Bugliarello, George. (2003). *The Biosoma—Reflections on the Synthesis of Biology, Society and Machines.* New York: Polytechnic University. Considers the concept of machines (artifacts) as metabiological entities; covers the implications of their interactions with living systems and societal entities.

Eyth, Max. (1885). *Der Schneider von Ulm—Geschichte eines zweihundert Iahre zu früh Geborenen* [The tailor of Ulm: The history of someone born two hundred years too early]. Stuttgart, Germany: Deutsche Verlags-Anstalt. An account of one of the first non-mythical attempts at biomimesis.

Fung, Y.C. (2001). *Introduction to Bioengineering.* Singapore and River Edge, NJ: World Scientific, 2001. A biomechanics-focused treatise.

Kellert, Stephen R. (1996). *The Value of Life: Biological Diversity in Human Society.* Washington, DC: Island Press. A taxonomy of views of nature.

Konecci, Eugene B. (1968). "Impact of Bioastronautic Program Based Management on Planning Our Future Technological Base." In *Bioengineering: An Engineering View,* ed. George Bugliarello. San Francisco: San Francisco Press. An early view of the nature and significance of bioastronautics.

National Academy of Engineering. (2002). *The Bridge* 32(3). Washington, DC: Author. Contains articles detailing several views by engineers of the ethics of engineering.

BIOETHICS

• • •

Bioethics is a broad subject connecting advances in biological and medical science with moral concerns. Medical ethics is one large part of bioethics but by no means the only part. Bioethics has grown as a discipline precisely as science and technology have increasingly demonstrated that human beings are biological beings. Scientists have mapped the human genome and scanned the human brain. Researchers have evermore precisely shown the neural correlates of mental states, the genetic roots of behavior and illness. Through these developments, serious new ethical questions have been raised about studying and even modifying human biology. Bioengineering has also been used to replace parts of the human body that are no longer working or working well: dialysis kidney function, pacemakers stabilize irregular heartbeat, and respirators keep lungs pumping oxygen. Bioethics as a field is rooted in advances in technology just as is the case with the narrower field of medical ethics.

Broadly speaking, four sorts of issues in bioethics transcend the more restricted confines of medical ethics and the more global issues of environmental ethics. First are those that involve the tension between the needs of the few and possible risks to the many. The best example of this is biomedical research and the issues it poses of need, risk, consent, validity, and conflicts of interest. A second large set of issues relates the present to the future. How much and how quickly should humans change human biological nature with such technologies as cloning, stem cell research, genetic screening, selection, and therapy? A third set has to do with the response to pandemics such as AIDS and emerging "diseases of globalization" such as West Nile virus in the United States. The final set involves issues rooted in the way in which sophisticated technology enhances the disparity between rich and poor globally and provides opportunities for severe exploitation of the poorest of the poor.

Individuals and Society

Biomedical research, especially since the mid-twentieth century, has dramatically transformed medicine. The research itself, however, has been controversial and led to major national commissions reviewing the process of consent in the research setting as well as the establishment of federal oversight of clinical research funded by the government. The gold standard for research has always been the classic "double-blind" study in which matched sets of patients are given either the old standard treatment or the new treatment. Researchers do not know which patients are given which treatment so their conclusions about the efficacy of the new treatment supposedly cannot be biased by such knowledge.

But this can put health care professionals in a seriously compromising position. At some point before the end of the study a researcher may have reached a preliminary conclusion about which treatment is best. As a health care professional this person would appear obligated to provide the best available care for any patient. As a researcher, however, the individual should not destroy a study by stopping it too soon. Preliminary results are often superseded by longer-term studies. An example from the early 2000s is the case of hormone replacement therapy for postmenopausal women. The tension between individual and societal benefit in research is inevitable. What may benefit a few may raise risks for many.

For example, since the early 1990s many AIDS patients have demanded faster access to possible treatments, including vaccines that might give them short-term comfort. The AIDS community has argued that they have nothing to lose from looser access to unproven treatments or vaccines. They have a fatal disease and should, therefore, have access to any treatment that might, even hypothetically, offer some benefit even where no long term cure for AIDS is on the horizon. But mass access to unproven therapies can be dangerous for the many. A vaccine based on the use of live HIV could backfire and spread the disease. Treatments may work for a short time and encourage risky sexual behavior. If the virus then mutates around the treatment, then the result of not fully testing the treatment before widely using it may be increased suffering.

For the persons who are fatally ill, access to new therapeutic technology essentially adds no new risk to their situation. If the therapy does not work or even spreads the disease, they are no worse off. For society at large, however, the risk is much different.

The same tension between the individual and the group can also be seen in reverse. For example, if a mentally ill patient is doing well on a specific combination of medicine, it may be dangerous to this patient's health to change to a new experimental medication. Yet without studies that accurately compare older therapies with newer ones the larger community of patients that needs to be treated with psychotropic drugs will have to forgo any benefit from newer medication.

A third example of the tension between benefit to the individual and risks to the group comes from the emerging technology of xenotransplantation. This is the technology of animal-to-human transplants. Though tried sporadically since the 1960s, the use of animal organs to make up for the lack of human donor organs has never proven effective. The human body's rejection system rapidly recognizes that an animal organ, such as a pig liver, does not belong in a human body. The rejection of the foreign organ is immediate and complete.

New genetic technologies at least suggest a solution to these problems. Companies have created transgenic pigs that have two human genes. These are the precise genes that control the immediate rejection process. This means that morphologically compatible organs, such as those from pigs, might indeed be used as organs for human transplantation, at least until a human organ becomes available.

The problem is that organs from other species may carry new viruses or other diseases into the human population, diseases for which human beings have never developed any immunity. One such virus has been discovered in pigs, and others may be found. Here again is the tension between the individual and the group. No individual would accept a risky xenotransplant unless it was his or her last chance to prolong life. Persons high on the donor organ waiting would surely wait for a human organ. It is only the most desperate who would accept a transplant from transgenic pigs. For these persons the risk of a new infection is clearly outweighed by the certainty of death. For them the risk-to-reward ratio points in only one direction: Go for it. For society in general, however, the question is not nearly so easy. The general public is not terminally ill. For members of the public the risk of a new virus such as the notorious HIV is serious and likely outweighs the chance that they will need any transplanted organs in the future. For the sick individual a new technology carries one set of risk-to-reward ratios, whereas for a larger group the ratio reverses.

Can rules and policies be constructed that protect the group while providing opportunity for the desperate? At a minimum, perhaps, biological monitoring of patients and of those close to them should be required to uncover any new possible sources of disease. Should, therefore, the gravely ill be required to secure the consent of close family and friends to such monitoring before they can receive a transplant? What about possible rules for quarantine for those possibly exposed to an emergent new viral illness? These are some of the new bioethical questions that advances in genetics and transplant technology have raised.

A fourth area that displays the tension between benefits for the individual and risks to society concerns conflicts of interest within biomedical research. Research, to be valid, must remain rigorous and as far as possible objective. But care is not neutral or objective. Care focuses on one specific patient who needs help. Nowhere is the challenge of objectivity more serious than the evaluation of new drugs and other technologies that may enrich their inventors or discovers. Researchers studying the effect of a new drug may very well own stock in the company whose product they are evaluating. At the very least researchers hope to be funded again by their supporting companies. Can they really evaluate the results in a neutral way with such financial gains at stake? Can the heart surgeon who has perfected a new stent really be expected to ignore a potential windfall in evaluating this invention? But who else to go to for the best analysis of a new drug or technology? Not just any physician should be entrusted with such a serious evaluation. It seems obvious that the best specialists should perform the evaluation rather than just any individual with a medical degree. The specialists, however, are the very persons who will likely have the most to gain from positive evaluations. They are the ones whose knowledge of the field will allow them to invest wisely in just those companies whose cutting-edge technology they may very well be asked to evaluate. A positive evaluation may increase their wealth substantially. Even if they are not so invested, they certainly will want to continue doing substantial research for this company. If they offer too many negative evaluations, then they may not have their research funded in the future.

The problem of evaluating new biomedical technologies and their relationship both to individuals and to society is crucial as technology comes evermore to dominate the biological lives of humans. Specialists come to design, create, and evaluate the new technologies with less and less input from the public at large. This fact is not conspiratorial. It reflects the reality of increasingly specialized knowledge of technologies that influence human lives. The point made by Jacques Ellul in the 1950s, that humans live in a technological society from which they cannot easily abstract themselves, is nowhere better exemplified than in bioethics. Many people alive on the planet owe their lives to some biologically rooted technology, from vaccines to crops to gadgets. Without these technologies, many people would not be alive; with them humans exist as a result of technology, which transforms the biological face of the planet.

Bioethics and the Future

A second great area of bioethical inquiry concerns the use of technology to change the biological future of humanity. The first and most immediate question concerns human reproduction and the rising human population. The twentieth century witnessed a rapidly rising population not because the birthrate increased (in much of the world it actually declined) but because of increases in human longevity. The theoretical lifespan of a human being has not increased. Rather with technically improved sanitation, nutrition, and medical care, the average longevity of individuals has been dramatically increased. This increase has seriously outweighed any reduction in population from lower birthrates. The technology that has enabled this dramatic population increase has brought to the fore other questions about individual liberty to make reproductive choices and produced a rash of other technologies to control birthrates such as various forms of artificial contraception.

The development of technologies to control reproduction has produced rapidly declining birthrates in advanced countries such as Europe where declining births and aging populations has produced an "aging crisis" with too few workers to support the elderly and to supply workers for business. This sort of crisis would not be a problem if it were not for technology altering the rhythms of birth, life, disease, and death with which humans evolved for millennia.

A second form of technically driven effort to control and manipulate the human future comes in the form of attempts to screen out individuals with various forms of inborn, usually genetic, abnormalities. Of course, crude eugenics programs existed in the early part of the twentieth century whereby the "feebleminded" were permanently sterilized in an attempt to improve the biological future of humanity. Thirty thousand were sterilized in the United States. This process reached its horrible zenith in Nazi racial programs with their combination of ancient tribalism and modern technology.

More acceptable approaches to screening began in the early 1970s with the development of technologies enabling the screening of the unborn for abnormalities and of parents as carriers of genetic traits that when reproductively combined with a partner who had the same trait would produce a child with a genetic disease such as cystic fibrosis or sickle-cell anemia. These technologies provided parents ways to influence their own genetic offspring by selective abortion of any fetus that was abnormal. Over time, it could have substantial effects on the human future especially in technically advanced countries where pressures to have a "healthy child" are pronounced because it may be the only child that a specific couple has.

With the mapping of the human genome, scientists are increasingly able to pin down the specific genetic correlates of disease, from those cases in which a specific genetic abnormality causes a disease to cases in which genetics are only part of the cause of a human disease, or even to identify traits such as homosexuality. Because technology enables the identification of the genetic roots of many human traits, it increasingly empowers individuals to control their own fate and the fate of their progeny. If a woman knows, for example, that she has the BRCA1 breast cancer gene, and a first-degree relative actually has breast cancer, she then knows that she is very likely to get breast cancer. The data suggest that with these two factors, genetics and a case history, 85 percent of the time she will get breast cancer. With knowledge comes the opportunity for more rigorous screenings and the use of technology to avoid breast cancer. Knowledge of the gene changes her future and possibly that of her daughters.

Another example of how genetic knowledge changes the future is Huntington's Disease. This is a recessive genetic disorder that does not manifest itself until a person is in their late 30s After that the person progressively loses muscle control and eventually dies after a 5–7 year period. In the process they often need to be tied down to avoid hurting themselves with spastic movements. One can know even before birth whether one will have the disease or is a carrier. Knowledge of this fact surely will alter marriage, career, and family plans.

But the power of selection immediately raises the question of whether there is one "correct" sort of choice in various situations. Should some choices be encouraged, and others financially or otherwise discouraged? Should parents be encouraged to abort fetuses with some abnormality that will be costly to treat and denied insurance for future related treatments if they bear the child? In another actual example a young woman who had breast cancer in her family tree was considering the test for the breast cancer gene. Her insurer insisted that if they paid for the test they owned the results. If the results were positive, then it was highly likely that she would come down with breast cancer. The insurer made her an offer: They would pay for double radical mastectomy, or they would drop coverage for breast cancer from her policy. Knowledge changed a risk into a near certainty. It was no longer insurance against risk but a prepayment scheme for almost certainly needed services.

Genetic screening and testing thus raise direct and lively issues in the present. Issues that loom in the near future involve genetic engineering. Most authors reflect one of three possible responses: (1) passionate advocacy of human genetic engineering (Silver 1997, Stock 2002); (2) cautious acceptance (Buchanan et al. 2000); or (3) wary hostility (Kass 2002). Authors commonly begin arguing that the possibility of genetically designing human offspring is at hand. Actually, the capacity for genetic design is decades away if it is even possible. Many experts are increasingly doubtful that any rapid breakthroughs are likely. Despite several years of effort, cloning primates is turning out to be much more difficult than anticipated.

Supporters and critics of "redesigning humans" claim that whatever the current difficulties, it will eventually be possible to add or delete targeted genes. Combined with in vitro fertilization, this technology will allow people to choose the genetic destiny of their offspring. Because at some point in the future this scenario is likely to be possible, it should be the subject of discussion now. The technology would first be developed to treat genetic diseases such as Tay-Sachs or Huntington's chorea; no responsible parent could ever want such a disease to strike their descendants. But the technology that enables gene addition or deletion, that is, the "knockout" of something such as the specific gene for Huntington's or retinoblastoma, could just as easily be used to eliminate color blindness, male pattern baldness, or a tendency toward depression or addiction. But retinoblastoma, which leads to early blindness, seems clearly different than male pattern baldness, which is specifically genetic, or more loosely genetic dispositions to shyness or alcoholism. Both baldness and retinoblastoma are genetic but the argument for using knockout technology in the case of inherited blindness such as retinoblastoma seems much clearer than in baldness to which the term disorder or disease seems only loosely, if at all, to apply.

Some who have carefully studied these matters are moderate, voluntaristic optimists. They argue that, with care, patience, and thoughtfulness, humans can use technology wisely to eliminate Tay-Sachs or retinoblastoma from pedigree without committing to a complete redesign of human beings. Others such as Gregory Stock combine a sort of naive optimism with technological determinism. For them, the technology of redesign is fast approaching and will be used. So sign up to the inevitable future and go along for the ride.

The third group of writers, including conservatives such as Francis Fukuyama (2004) and Leon Kass (2002), or leftists such as Andrew Kimbrell (1998), seem like lonely fatalists who fear there is no realistic possibility of stopping the redesign of humanity. They seem fatalistic about the attempt and depressed at the prospect. Some have forsaken revealed religion as a means of guiding technology, so they cast their lot with human nature as a standard. Now, however, they seem to accept that Eden will be remade. Humankind's ability to do so, however, seems to undermine the very appeal to nature for guidance about the attempt. Hence, they are left rudderless in an ocean of uncertainty.

The same set of three views also appear in debates over human cloning. Passionate advocates such as Lee M. Silver (1997) and Stock (2002) see nothing wrong with the inevitable occasional practice of reproductive cloning. They think that in fact it will be used only occasionally, but are in principle not opposed to its widespread practice. Cautious acceptance is illustrated by Robert Wachbroit (1997), who argues that cloning can be used wisely and infrequently in cases of special need, for example, bone marrow for a child, without promoting widespread or general acceptance. Finally, Leon Kass (2002) and others, based on their conclusion that human cloning is an affront to human dignity, propose legal prohibitions on all such cloning, reproductive and therapeutic (arguing they cannot be separated and that the potential benefits of therapeutic cloning can be secured by other means).

Contemporary issues in bioethics thus pose a fundamental technological question with respect to the future: Does technology unleash the human passion for improvement in ways that reason cannot control? Is reason, as Thomas Hobbes (1588–1679) argued, a slave to the passions? If so, then technology, the supreme product of scientific reason, is only a tool to satisfy human desire for longevity, pleasure, and domination. Humans want a life of ease not disease. Technology thus aims to please by manipulating human biology to satisfy desires. Is this destiny or choice? If the former, then bioethical reflection is beside the point. If the latter, what choices should the collective bioethical wisdom of humankind encourage humans to make? Are humans now fated to a technological civilization from which they cannot escape as was argued by Martin Heidegger in his seminal essay "Question Concerning Technology" and by Ellul, Marcuse and others?

Bioethics and Globalization

A third area of bioethical inquiry especially related to science and technology concerns issues of globalization. Globalization is profoundly the result of technology. Technology has standardized production methods for low-skill workers in low-wage countries. It has increased information and travel networks to enhance information and capital flows across national and continental boundaries. Finally it has enhanced transportation of raw materials and finished goods from low-wage mines and factories to markets in the developed world. The first place where bioethics meets globalization is in the discussion of agricultural biotechnology and its impacts on peoples in developing countries.

But globalization and biology meet as well in the increasing flow of diseases around the world from those places where they have developed and coevolved with human and other species to new locations where they have created new problems for human life. The slave trade created an early instance of such problems. Sickle-cell anemia, which affects persons of African descent, carries no evolutionary advantage. Carrying the sickle-cell trait, which is recessive, does not produce the disease but nevertheless carries a resistance to virulent strains of malaria. In North America, which is malaria free, carrying the trait has no advantage, and a couple who both have the same recessive trait may conceive offspring with the disease. In sub-Saharan Africa, however, where as much as 40 percent of the population carries the trait, selective advantage is conferred. Thus moving the disease out of its evolutionary nest has raised issues for advanced countries, such as the need for screening programs for prospective parents of African descent, that would have not existed except for the global slave trade.

In another case, AIDS is a global pandemic that has grown rapidly with increased contact between human beings. HIV developed in Africa, but the effects have become global, and it has raised a number of serious new issues such as quarantine, the right to health care for those whose illness is the result of their own behavior, and a search for vaccines and specialized therapies that has consumed large amounts of research funds. Globalization has raised questions about competing needs to develop, for example, AIDS therapies versus an effective malaria vaccine—malaria being a disease that kills more persons who are much less responsible for their illness.

One final example is the appearance of West Nile virus in North America. As the name indicates, the historical location of this disease has been Africa and the Middle East. Borne by mosquitoes, it first appeared in the New York area in 1999. Over the next few years it spread virtually over the whole North American continent. It has become biologically fixed in this new location. It can be contained and treated, but will not be eradicated.

In a profound way, technology has become a part of the biological process of evolution. Technologies of globalization have spread disease from historic locations such as those of West Nile virus. Technology has become a sort of disease vector, a route by which new diseases travel to distant targets. If technology brings new populations into deep contact with what for them are new diseases, it also provides these same populations with means for evolutionary survival in the face of these and other diseases. Technology, for example, gives treatment for AIDS, means for tracking the spread of disease, and possibilities for other treatments. When technology is used to extend the power of humanity over a disease, the disease may become a serious one but one with which humanity can coevolve. Technology both causes the need for coevolution for North Americans with something such as West Nile virus and provides the means for such evolution, from spraying for mosquitoes to treatments, and if necessary to the development of vaccines. Technology thus becomes part of the Darwinian enterprise of evolutionary survival.

These problems have antecedents in the European colonization of the Americas where new diseases were brought by the settlers. But they now have more rapid global movement as a result of technology and technology can be aimed at providing cures or effective treatments of diseases of globalization.

Bioethical Justice

A final way in which global growth of technology both in medicine and transportation affects bioethics is by creating an emerging transnational trade in medical services. One example is the creation of a transnational market for so-called back-office operations. Billing has been outsourced to foreign low-wage countries for years. With information networks now available it is just as easy to bill insurers from Jamaica as it is from Kansas. The benefit is that Jamaicans or Indians will work for half or less of the U.S. minimum wage. But with increasingly sophisticated computer technology and education in less developed countries, even "back-office" physician or pharmacy services can be outsourced. Highly qualified radiologists in China could read standard X rays on their monitors for a third of the cost in the United States. Complicated readings might require a physician on the scene in the United States, but the yearly mammogram and similar procedures could be sent abroad. Billing is one thing, but how would patients personally discuss their test results with physicians halfway around the globe?

Pharmacy services will also increasingly be outsourced. With pharmaceutical prices in the United States still high and transportation increasingly efficient, it will become increasingly common for such drugs as Viagra to be made in China and shipped by anonymous clerks to U.S. addresses. The key issue here is the balance between price and safety. Can or should the government interfere to "protect" individuals from possibly unwise purchases of drugs from foreign sources that lack serious regulatory frameworks?

The transnational trade in medical services also includes highly technical services, which, being provided transnationally, are available only to those who can pay up-front. The best known and most troubling of these developments is the international trade in organ transplants, chiefly kidney or liver transplants. A person needs only one kidney to survive, and in some cases people have donated a kidney to save the life of a close relative. But enter a market made possible by technology: Highly qualified surgeons in India or China or elsewhere provide transplant services in fully staffed clinics primarily for other Asians with a desire for life and the wealth to pay. The surgeon and staff are well compensated. But in India, for example, the poorest of the poor are paid about $1,500 for a kidney. This amounts to a lifetime savings for the donor, but possibly no more than the cost of a plane ticket for the recipient. This raises enormous questions of justice and exploitation. Does money exploit the poorest of the poor who desperately need assistance? Does the whole practice raise questions of justice, where the rich can pay and the poor only suffer?

These questions also occur in increasing ways in the United States. At any one time hundreds of individuals in the United States are advertising a kidney for sale on the Internet. For the most part these are desperate lower or lower middle class people trying to avoid bankruptcy, home foreclosure, or property repossession. They see such a sale as one of the few ways to improve their fortune short of illegal activity or hitting the lotto. But does their very poverty make them subject to coercion and thus unable to give free and informed consent? In the United States researchers are forbidden from using prisoners for drug experiments because of the problems of coercion and lack of the ability to give informed consent. Would not the same argument apply to the desperate and the hopeless, who are ready to sell body parts via Internet technology? Technological power to commodify even the most personal of things, one's own body, creates bioethical issues that previous eras could avoid. Technological fatalism may overstate the case, but it does seem that the questions raised are inevitable.

Thus, technology may provide a means of evolutionary development in the face of changing biology. As such, technology develops around the fundamental biological and thus bioethical imperative of preserving human life. In the context of such a nexus between technology and Darwinism, bioethics provides both the comprehensive understanding of the problem and the subsidiary rules of honesty, disclosure, integrity, and justice that provide the moral ambit within which technology may be a morally acceptable vehicle for human well-being in a fundamentally Darwinian world.

What remains is the fundamental question of all technology. Can modern technology be contained within reason, or does the eternal passion for life and health overwhelm reason's capacity to moderate human desires within an ambit of moral principles and virtues?

RICHARD SHERLOCK

SEE ALSO *Agricultural Ethics; Bioengineering Ethics; Bioethics Centers; Bioethics Committees and Commissions; Biotech Ethics; Environmental Ethics; Genethics; Medical Ethics; Neuroethics; Posthumanism.*

BIBLIOGRAPHY

Brody, Baruch A. (1998). *The Ethics of Biomedical Research.* New York: Oxford University Press.

Buchanan, Allen; Dan W. Brock; Norman Daniels; and Daniel Wikler. (2000). *From Chance to Choice: Genetics and Justice.* New York: Cambridge University Press.

Ellul, Jacques. (1954). *La technique; ou, L'en jeu du siècle.* Paris: A. Colin. Translation by John Wilkinson as *The Technological Society* (New York: Knopf, 1964).

Fukuyama, Francis. (2004). "Transhumanism." *Foreign Policy* (September/October): 67–69.

Heidegger, Martin. (1977). *The Question Concerning Technology, and Other Essays.* New York: Harper and Row.

Kass, Leon R. (2002). *Life, Liberty, and the Defense of Dignity.* San Francisco: Encounter.

Kimbrell, Andrew. (1998). *The Human Body Shop.* Washington, DC: Regnery Books.

Land, Walter, and John B. Dosseter, eds. (1991). *Organ Replacement Therapy: Ethics, Justice, Commerce.* New York: Springer-Verlag.

Levine, Robert J. (1986). *Ethics and Regulation of Clinical Research,* 2nd edition. Baltimore: Urban and Schwarzenberg.

Loue, Sana. (1999). *Textbook of Research Ethics.* Dordrecht, Netherlands: Kluwer Academic.

Murphy, Timothy F. (2004). *Case Studies in Biomedical Research Ethics.* Cambridge, MA: MIT Press.

Nuffield Council on Bioethics. (1996). *Animal-to-Human Transplants: The Ethics of Xenotransplantation.* London: Author.

Robertson, John A. (1994). *Children of Choice.* Princeton, NJ: Princeton University Press.

Silver, Lee M. (1997). *Remaking Eden: Cloning and Beyond in a Brave New World.* New York: Avon.

Stock, Gregory. (2002). *Redesigning Humans: Our Inevitable Genetic Future.* Boston: Houghton Mifflin.

U.S. Institute of Medicine. (1996). *Xenotransplantation: Science, Ethics, and Public Policy*. Washington, DC: National Academy Press.

Youngner, Stuart J.; Renée C. Fox; and Laurence J. O'Connell, eds. (1996). *Organ Transplantation*. Madison: University of Wisconsin Press.

INTERNET RESOURCE

Wachbroit, Robert. (1997). "Genetic Encores: The Ethics of Human Cloning." Report from the University of Maryland Institute for Philosophy and Public Policy, Volume 17, Number 4. Available from http://www.ifgene.org/wachbroit.htm.

BIOETHICS CENTERS

Although there have been concerns involving medical ethics since ancient times, bioethics is an invention of the late twentieth century. The first freestanding center devoted to bioethics was created in 1969. By the beginning of the twenty-first century, most major American institutions of higher learning and most American medical schools had centers, institutes, or programs devoted to the study of biomedical ethics. The bioethics center is no longer a uniquely American institution but an international phenomenon, with new centers continuing to be established all over the world.

Hastings Center

The first bioethics center resulted from the work of a newly minted Harvard Ph.D., the philosopher Daniel Callahan. In the late 1960s, while writing a book on abortion, Callahan found himself engaging with a complex interdisciplinary literature that took him outside the boundaries of traditional philosophical inquiry. As a result of the sharp disciplinary boundaries of that time, Callahan was forced to investigate areas of law, social science, public policy, and medicine. Realizing that advances in science and medicine would continue to generate ethical dilemmas that would require interdisciplinary study and reflection, Callahan set out to create a place where those issues could receive serious, focused attention from multiple perspectives and academic disciplines.

Because that type of center would attempt to cross disciplinary boundaries, it had no natural academic home. To realize the vision of being truly interdisciplinary—bringing together individuals from the fields of theology, philosophy, law, medicine, and science—the new institute would have to be a freestanding institution that was not constrained by the boundaries of traditional academic disciplines. Callahan presented his proposal to a casual acquaintance and fellow resident of the town near New York City where he lived, Hastings-on-Hudson. The physician-psychoanalyst Willard Gaylin, a professor at the Columbia University College of Physicians and Surgeons, thought that the idea for a new institute was timely and appropriate. Together they sought financial support from individual donors and foundations to establish an institute that would examine ethics and the life sciences, and in 1969 the Hastings Center was founded. Originally called a Center for the Study of Value and the Sciences of Man, the Hastings Center opened in September 1970.

Kennedy Institute

In the same year a similar dialogue took place at Georgetown University in Washington, DC. Dr. Andre Hellegers, a faculty member in the department of obstetrics and gynecology in the School of Medicine, was concerned that discussions of the ethical issues in reproductive medicine were being relegated to conferences and professional meetings rather than being the subject of sustained and concentrated scholarship. He proposed the creation of a center to study reproductive ethics to the president of Georgetown, Reverend Robert Henle. In December 1970 they sought support from the Kennedy Foundation. In July 1971 the Kennedy Institute of Ethics opened at Georgetown University. Unlike the Hastings Center, which avoided academic ties for fear of losing its interdisciplinary orientation, the Kennedy Institute embraced its connection to Georgetown University. The institute established faculty chairs and a degree program run in conjunction with the university's philosophy department.

Although different in organizational structure, the Hastings Center and the Kennedy Institute quickly became crucial entities in the creation of the field of bioethics. Both institutions created libraries, issued publications, amassed grants, set out research agendas, and brought together scholars who became the early leaders in the field.

Expansions

Over the next thirty years dozens of bioethics centers and institutes were created. Almost all were housed within universities. By the 1980s many were established in academic medical centers.

Early bioethics centers were populated mostly by philosophers and theologians. In the 1970s those scholars were joined by lawyers and physicians as well as a few nurses, social scientists, and economists. The shift toward locating bioethics centers in academic medical centers reflected both the increasingly large role played by physicians in bioethics and the increasing legitimacy of bioethics as an area of inquiry important to the health sciences.

Beginning in the mid-1990s, a greater emphasis on what Arthur Caplan called empiricized bioethics emerged. Pressure to conform to the norms of academic medical centers meant that faculty members and students at bioethics centers had to be able to publish in leading medical and scientific journals. As a result, the empirical study of ethical issues and norms became a key aspect of the responsibilities assigned to bioethics centers. By the early 2000s social scientists and empirically trained clinicians held significant numbers of faculty positions in those centers, in some cases constituting the majority of their membership. Many bioethics centers continue to be shaped by the criteria for scholarship and promotion that prevail at medical schools in the United States and Europe. Whereas normative analysis once dominated bioethics discourse within and outside centers, many bioethicists have begun to speak in the language of descriptive facts, economic realities, and culturally based moral practices.

The location of bioethics centers in academic institutions has had another professionalizing influence on the field: the creation of professional degree programs. In 2003 there were over sixty master's programs in bioethics, and most of those degrees were granted through the centers in conjunction with the schools of which they were a part. Scholars who joined the field in its early days were all "immigrants," entering from disciplines as diverse as anthropology, sociology, philosophy, theology, medicine, law, public policy, and religion. Because of their institutional structure, centers provided appropriate homes for persons with very different disciplinary backgrounds. However, bioethics scholars in the future will be required to have specific bioethics credentials, either master's degrees or doctorates in the field. Increasingly, they may be employed in academic departments rather than in centers or institutes.

Assessment

The extent of the influence of bioethics centers on science, technology, and ethics is hard to gauge. Unlike traditional academic disciplines or centers whose goal is erudite scholarship, bioethics centers see as their mission not only the creation of new scholarly knowledge, but also engagement with professional groups, the public, and public officials who set policies. Bioethics centers commonly have elaborate outreach programs that include websites, newsletters, a strong media presence, public conferences, writings for the lay press, and distance learning programs. Many members of bioethics centers are public figures, scholars whose work extends beyond their academic base. They have shaped policy and public opinion on issues as far-ranging as informed consent, stem cell research, abortion, euthanasia, cloning, organ donation, research ethics, patenting, and genetically modified foods.

Bioethics centers first appeared as a response to emerging moral challenges, often technologically driven, in American health care. They became the locations where interdisciplinary work on complex moral problems could be done. Their future is uncertain. Bioethics has matured and become a discipline with journals, encyclopedias, awards, and book series. Although new ethical concerns continue to emerge in health care in the United States, in Europe, and internationally, the future of bioethics centers is not clear. With the emergence of a "professionalized" discipline that is both empirical and normative, it is likely that the work done in bioethics increasingly will be accomplished in academic departments. The success of the early bioethics centers and institutes may have created a field that has outgrown its older institutional structures.

ARTHUR CAPLAN
AUTUMN FIESTER

SEE ALSO *Bioethics; Bioethics Committees and Commissions.*

BIBLIOGRAPHY

Callahan, Daniel. (1973). "Bioethics as a Discipline." *Hastings Center Studies* 1: 66–73.

Fox, Daniel. (1985). "Who Are We?: The Political Origins of the Medical Humanities." *Theoretical Medicine* 6: 327–341.

Jonsen, Albert. (1998). *The Birth of Bioethics*. New York: Oxford University Press.

Post, Stephen G., ed. (2004). *The Encyclopedia of Bioethics*, 3rd ed. New York: Macmillan Reference USA.

Potter, Van Rensselaer. (1971). *Bioethics: Bridge to the Future*. Englewood Cliffs, NJ: Prentice-Hall.

Reich, Warren. (1994). "The Word 'Bioethics': Its Birth and the Legacies of Those Who Shaped Its Meaning." *Kennedy Institute of Ethics Journal* 4: 319–336.

BIOETHICS COMMITTEES AND COMMISSIONS

• • •

Since its inception in the 1970s, bioethics has been manifested not only in academic debate but also in committees or commissions directed toward the guidance of public discussion and policy making. In the research and clinical settings, Institutional Review Boards (IRBs), Institutional Biosafety Committees (IBCs), and Hospital Ethics Committees (HECs) serve the practical functions of bringing reflective expertise and modest public consensus to bear on ethical implementations of advancing scientific and technological forms of medicine and other biological research. At the state, national, and international levels, more general committees and commissions have sought to provide reflective consideration and policy guidance. These committees come in three types: term-limited, ad hoc, and permanent. The role these committees and commissions play in government and society depends on their structure and mission, the larger historical and social contexts, and trade-offs between broader, more fundamental inquiry and narrower, more policy relevant recommendations.

Bioethics Commissions in the United States

There are two broad classifications for federal bioethics commissions and committees in the United States: general and topic specific. General bioethics commissions have been appointed by Congress or the President to conduct inquiries into a diversity of issues and have both fostered wide-ranging public discussion and produced targeted policy recommendations. Topic-specific initiatives have in turn been created by different government agencies or the President to address specific technologies or aspects of scientific research. Other important elements in this context include the former Office of Technology Assessment (OTA) and other research and assessment agencies of government, state-level bioethics committees, and academic and nongovernmental bioethics centers and committees.

GENERAL FEDERAL BIOETHICS COMMISSIONS. Between 1974 and 2004, there were six general federal bioethics commissions (see Table 1 for a summary). The first public body on the national level to shape bioethics policy was the National Commission for the Protection of Human Subjects of Biomedical and Behavioral Research (National Commission). Created by the National Research Act of 1974 under Republican President Gerald Ford, the National Commission operated until 1978 and was administered by the Department of Health, Education and Welfare (DHEW). It contributed to the first federal regulations for the protection of human subjects of biomedical and behavioral research. The principles that served as the basis of these regulations were outlined in its 1978 Belmont Report, and the regulations became institutionalized in the form of Institutional Review Boards (IRBs). The National Commission also produced reports on research involving vulnerable subjects including prisoners, those institutionalized as mentally infirm, fetuses, and children.

One of the recommendations of the National Commission led to the creation of the Ethics Advisory Board (EAB) in 1978. During its approximately two-year existence, the EAB focused on issues involving fetuses, pregnant women, and human in vitro fertilization (IVF), but it had a broad charter that allowed it to investigate many bioethics issues. Originally intended as an ongoing standing board, the EAB was nonetheless disbanded by the Office of Science and Technology Policy in 1980 after producing four documents. Two major outcomes were the stipulation of criteria for federally-funded research in IVF and a pronouncement on human embryo research, which began a fifteen-year moratorium on such research.

One of the reasons the EAB was disbanded was because policy makers failed to distinguish its purposes from those of the President's Commission for the Study of Ethical Problems in Medicine and Biomedical and Behavioral Research (President's Commission) created by Congress in 1978 under Democratic President Jimmy Carter. The President's Commission had a broad mission and the authority to initiate its own reports on emerging issues judged important by its members. It was elevated to independent presidential status (by contrast, the National Commission had operated autonomously within the DHEW). Also unlike the National Commission, the President's Commission produced fewer specific recommendations targeted at federal agencies. Instead, it produced consensus reports that articulated mainstream views. These reports are highly regarded and "many have had sustained policy influence" (United States Office of Technology Assessment 1993, p. 12). Its report on foregoing life-sustaining treatments was most influential, and it led to the development of living wills. After a three-month extension, the President's Commission expired in March 1983 under Republican President Ronald Reagan.

The Biomedical Ethics Advisory Committee (BEAC) was the fourth government-sponsored general bioethics body. In 1986, Congress established the Biomedical Ethics Board (BEB), which was composed of six

Senators and six Representatives (this was modeled on the Technology Assessment Board, which oversaw the United States Office of Technology Assessment [OTA]). It took the BEB more than two years to appoint all the members of the BEAC, and in September 1988 (less than a week before it was originally scheduled to expire) the BEAC held its first meeting. Largely due to partisan politics around the abortion issue, BEAC's appropriations were frozen and it was unable to produce any reports before it officially expired in September 1989 under Republican President George H. W. Bush.

There followed an extended hiatus until Democratic President Bill Clinton signed an executive order to create the National Bioethics Advisory Commission (NBAC) in 1995. Chaired by Harold T. Shapiro, the NBAC held its first meeting in 1996, and its original mission was to investigate the two priority areas of human subjects research and genetic information. After the cloning of the sheep Dolly in 1996, however, President Clinton also requested a report on cloning. This became the NBAC's first report, which recommended that federal regulation be enacted to ban research using somatic cell nuclear transfer cloning to create children. It recommended that such legislation be crafted so as not to interfere with other uses of cloning that may not be as ethically problematic. The NBAC also produced reports on research involving biological materials, stem cells, and persons with mental disorders that may impair decision-making abilities. The NBAC recommended that federal funding be used only on stem cells derived from two sources: cadaveric fetal tissue and embryos remaining after infertility treatments. The NBAC expired in 2001.

The stem cell issue sparked the creation of the President's Council on Bioethics by George W. Bush (via executive order) in 2001. In his first national address, Bush created a new policy for the federal funding of stem cell research and announced the formation of the Council under the direction of Dr. Leon R. Kass.

TOPIC-SPECIFIC INITIATIVES. Other committees and commissions have been created by the U.S. government in order to provide topic-specific guidelines and recommendations (see Table 2 for a summary). The first noteworthy example is the Recombinant DNA Advisory Committee (RAC), which was created in 1976 in accordance with the National Institutes of Health (NIH) Guidelines for Recombinant DNA Research. The RAC is a permanent committee housed in the NIH that serves a threefold function: to provide a public forum for discussion about issues involving recombinant DNA, to make recommendations to the director of NIH, and to review certain individual research protocols. In this last role, the RAC often works in conjunction with IRBs and IBCs.

TABLE 1

General U.S. Bioethics Commissions

Name	Duration
National Commission for the Protection of Human Subjects of Biomedical and Behavioral Research (National Commission)	1974–1978
Ethics Advisory Board (EAB)	1978–1980
President's Commission for the Study of Ethical Problems in Medicine and Biomedical and Behavioral Research (President's Commission)	1978–1983
Biomedical Ethics Advisory Committee (BEAC)	1986–1989
National Bioethics Advisory Commission (NBAC)	1995–2001
President's Council on Bioethics (Council)	2001–

SOURCE: Courtesy of Adam Briggle and Carl Mitcham.

Most other topical committees have been temporary. In March 1988, the Assistant Secretary for Health directed the NIH to appoint an ad hoc panel that became known as the Human Fetal Tissue Transplantation Research Panel. The panel met three times and issued its final report in December 1988, which approved federal funding for research involving the transplantation of human fetal tissue from induced abortions. Although not a commission, the Ethical, Legal, and Social Implications (ELSI) research program marks a landmark investment in bioethics research by the federal government. ELSI was begun in 1989 by the NIH and the Department of Energy (DOE) as a joint project to fund research on the social implications of developments associated with the Human Genome Project (HGP).

The NIH formed the Human Embryo Research Panel in January 1994. This panel classified human embryo research into three categories and drafted guidelines for the review and conduct of acceptable research. Also in 1994, President Clinton created the Advisory Committee on Human Radiation Experiments, and charged it to investigate and report on the use of human beings as subjects of federally-funded research using ionizing radiation. The committee found the government blameworthy for not having procedures in place to protect the rights of human research subjects exposed to radiation without their consent. One final example of a topical commission is the Advisory Commission on Consumer Protection and Quality in the Health Care Industry. Created by executive order in 1996, this thirty-two-member commission focused on patient protections and consumer satisfaction in the health care industry. It developed the *Consumer Bill of*

TABLE 2

Topic Specific U.S. Bioethics Commissions

Name	Duration and Agency
Recombinant DNA Advisory Committee (RAC)	Permanent (created in 1976); NIH
Human Fetal Tissue Transplantation Research Panel	March–December, 1988; NIH
Ethical, Legal, and Social Implications (ELSI) program	Begun in 1989, the Human Genome Project expired in 2003 (but other ELSI programs continue); NIH and DOE
Human Embryo Research Panel	1994; NIH
Advisory Committee on Human Radiation Experiments	1994–1995; created by President Bill Clinton, reported to Cabinet-level group
Advisory Commission on Consumer Protection and Quality in the Health Care Industry	1996–1998; created through executive order by President Bill Clinton

SOURCE: Courtesy of Adam Briggle and Carl Mitcham.

Rights and Responsibilities in 1997, and issued its final report, *Quality First: Better Health Care for All Americans*, in 1998.

STATE LEVEL AND NONGOVERNMENTAL COMMISSIONS. Many state legislatures and executive branches must incorporate bioethics into their public policy making. Given this growing need, several states have created committees and commissions, most of which have been devoted to a single issue. Access to health care has been the single largest issue addressed by state-level committees. Some states, however, have created commissions designed to consider a broad range of issues. Two examples of state-level commissions are the New Jersey State Commission on Legal and Ethical Problems in the Delivery of Health Care, created in 1985 as a permanent legislative committee, and the New York State Task Force on Life and the Law, also created in 1985, with a broad mandate to make recommendations for policies involving medical technologies.

In addition to academic bioethics centers, several nongovernmental organizations in the United States have created bioethics centers or committees. For example, the American Medical Association, the nation's largest professional association of physicians, houses the Institute for Ethics, which studies ethical issues related to health care and biomedical research. Many churches and religious groups have also established bioethics committees. Two examples are the American Bioethics Advisory Commission, founded by the American Life League, and the Center for Bioethics and Human Dignity, founded by several Christian bioethicists.

International Bioethics Commissions

Before the term *bioethics* was used, the Nuremberg War Crimes Tribunal in 1945 made the treatment of human subjects in scientific research a major issue. Subsequent work by the World Medical Association led to the Declaration of Helsinki in 1964, which outlined ethical principles for medical research involving human subjects.

The first explicitly-named bioethics group on the international level was the Steering Committee for Bioethics (CDBI), which is a multidisciplinary ad hoc group created by the Council of Europe in 1983 (although it underwent name changes in 1985 and 1993). CDBI adopted the first international treaty on bioethics in 1996. The Commission of the European Union has also established bioethics committees, including the Working Group on Human Embryos Research; the Working Group on Ethical, Social, and Legal Aspects of Human Genome Analysis; and the Working Party on Ethical and Legal Issues Raised by New Reproductive Technology (also known as the Glover Commission), which produced the Glover Report in 1989.

On an even broader international level, the United Nations Educational, Scientific, and Cultural Organization (UNESCO) division of Ethics of Science and Technology created two bioethics advisory bodies in 1993 under the umbrella term of Bioethics Program: the International Bioethics Committee (IBC) and the Intergovernmental Bioethics Committee (IGBC). A major outcome of this program was the adoption of the Universal Declaration on the Human Genome and Human Rights by the General Conference, the only international instrument in the field of bioethics, endorsed by the United Nations General Assembly in 1998.

Bioethics Commissions Outside the United States

Susan Poland (1998) compiled a comprehensive list of bioethics committees and commissions around the world (see also Martinez 2003). Although dominated by the United States, Canada, and Europe, there have been commissions in the Philippines, Mexico, Japan, Turkey,

Russia, Israel, and elsewhere. What is most striking about this list is the diversity in structure, function, duration, context, and other variables. For example, although many commissions are temporary, there are some permanent and semi-permanent bodies. Canada and Australia have established permanent law reform commissions to make recommendations to parliament (Kasimba and Singer 1989, Williams 1989).

An example of a permanent committee more strictly focused on bioethics is the French National Consultative Committee on Ethics in the Biological and Medical Sciences (CCNE). Created in 1983, this agency is the first broad bioethics commission on a national level in France with the power not only to review research protocols but also to advise the government on appropriate legislative action (Isambert 1989). Another example of a permanent advisory body is the Human Genetics Commission in the United Kingdom, which is a non-statutory, independent advisory committee established in 1999. Its role is to advise Ministers on the appropriate response to developments in human genetics. Yet another example is the Standing Committee on Ethics in Experimentation established by the Medical Research Council of Canada (a grant-funding institution for health science research) in 1984. This committee aids in the development of federal policy as well. In 2004, Israel began formalizing plans for a National Council of Bioethics, which will serve as a governmental statutory authority, allowing it to monitor existing bioethics committees and giving it rather unusual legislative power for a bioethics panel.

Other bioethics commissions are special instantiations of a broader model of commission-based inquiry used by governments to investigate problems that face decision makers. Several European parliaments utilize the model of Enquete commissions, which are temporary bodies established to provide policy advice on vast range of issues. Many Enquetes have focused on bioethical issues; for example, the German commission studying "Law and Ethics of Modern Medicine" (2000; reinstated in 2003). Moreover, as in the United States, not all bioethics commissions are established by governments. For example, in Canada nongovernmental organizations such as the Canadian Medical Association and certain churches have formed bioethics committees.

Some bioethics commissions have exerted their influence on the future work of other commissions around the world. The Warnock Commission in the United Kingdom (chaired by philosopher Dame Mary Warnock) is one example. This fifteen-member committee met from 1982 to 1984 in order to examine the social, ethical, and legal implications of developments in assisted reproduction. Its report, *The Warnock Report on Human Fertilization and Embryology* (1984), is a landmark in the field because of its treatment of moral issues and its forthright explanation of the difficulties in seeking moral consensus. This distinguished it from previous reports (such as Peel [1972] and Black [1980]). Furthermore, the report was concise, readable, and showed respect for dissenting views (Campbell 1989). Both the process and product of this commission have influenced the work of other bioethics committees.

Historical and cultural contexts are crucial elements in determining the parameters for both the style and content of bioethics commissions. For example, in Japan there is a long tradition of paternalistic and authoritative relationships between medical professionals and patients and their families. Although there is a deep respect for elders in Japanese culture, there is also an ingrained research-oriented mentality that treats patients more as medical cases than persons (Kimura 1989). The culture is rapidly changing in Japan, but these traditions shape the challenges faced by bioethics commissions, because democratic deliberation and the "rights based" approach to medical ethics are both relatively new. In Germany, the Nazi legacy has left a "culture of remembrance" that vows to never again relive the horrors of state-sponsored eugenics and applied biology (Brown 2004). The protection of the sanctity of persons is written directly into its constitution, and Germany has a history of strict bioethics policies. Germany's unique history has impacted the way it structures inquiries into matters of bioethics. For the most part, German bioethics commissions have been conservative, control-oriented, paternalistic, and skeptical of scientific and technological developments (Sass 1989). The creation of the National Ethics Council by Chancellor Gerhard Schröder in 2001, however, signified a break in this dominant culture as once-taboo topics were made available for more serious discussion.

In contrast to the United States, many bioethics commissions in other nations have more limited public access policies. However, like the United States, most of these commissions include members who are not health care professionals or scientists.

Assessment

Bioethics commissions and committees have been created to serve a variety of purposes, including helping heterogeneous societies articulate common values and foster consensus about biomedical advances; serving as a crucial interface for science and politics; providing specific policy recommendations, technical advice, and even serving the judiciary; reviewing the implementation of existing laws; educating the general public about complex ethical issues arising from the rapid development of science and technology; serving as a forum for public participation in policy making; undertaking research; legitimizing action; and delaying action (see United States Office of Technology Assessment 1993; Walters 1989). Although they can be powerful due to their prestige and access to resources, no specific committee or commission can be all-encompassing. Trade-offs among the above functions are inevitable, perhaps the most important being between a wide-ranging, fundamental inquiry and a more topical, focused investigation geared toward the needs of decision makers. The wider commissions are more adept at educating the public and guiding long-term debates about basic ethical principles, whereas the narrower commissions tend to be more immediately policy relevant.

Maximizing the value of bioethics commissions requires utilizing relationships with bioethics centers, government, and society. A multitude of bioethics centers, professions, and organizations provides a widespread, pluralistic approach to bioethics debates, which promotes diversity of perspectives and propinquity to patients and researchers. Federal bioethics commissions can command the resources necessary to address nationwide issues, foster broad discussions, and articulate conflicting views, but can also be inflexible or captured by political interests. Understanding when to create permanent versus term-limited or ad hoc bodies is also an element influencing the utility of commissions and committees (see United States Office of Technology Assessment 1993).

Another important variable is membership composition, including the roles of different forms of expertise and public input. Membership is usually the most politically charged element of committees. Two examples are the U.S. President's Council on Bioethics and Israel's National Council of Bioethics, which have both been accused of being biased and captured by narrow political interests. In the former case, Chairman Leon Kass is seen as overly pessimistic about technology, while in the latter case Chairman Michel Ravel is seen as overly permissive of scientific research and its applications. Those who criticize these councils claim that common interest goals are not being served. This highlights the need to craft wise membership selection mechanisms in order to lend credibility to the commission.

An alternative path to institutionalizing bioethics is what Eric Juengst (1996) calls the "un-commission" model, best represented by the original design of the ELSI program, which adapted NIH mechanisms to create extramural grant support for research, education, and public participation projects on the social implications of genome research. The main critique of this program is that it could not affect policy, but Juengst argues that even national commissions are severely constrained in their ability to communicate policy recommendations effectively. He suggests that the "un-commission" model is better capable of providing adequate social-impact assessments to serve as a sound contextual base for policy making. This model of complementary research and public deliberation attached to scientific research funding provides another option for identifying and developing responses to emerging bioethics issues. The charge still stands, however, that such a model fails to immediately impact policy, and only adds "basic ethics research" to the basic science research, neither of which can truly aid decision makers or the public. Perhaps the best method is to provide distinct forums for both policy-relevant inquiry and basic ethical and social impacts research.

Commissions and committees gather interdisciplinary panels of experts to ponder questions that arise at the interface of science, technology, and society. However, most of these questions cannot be answered by specialists. In fact, delegating this decision-making responsibility to experts may undermine the public participation necessary to uphold strong democratic practices in the face of rapid changes. In this light, then, the proper role of bioethics commissions may be to clarify values and educate the public in order to ensure the "very possibility of a democratic future in the biotechnical age that is now upon us" (McClay 2004, p. 18). What bioethics commissions should provide are not final answers, but rather a clearer understanding of the questions and the consequences different answers may pose.

ADAM BRIGGLE
CARL MITCHAM

SEE ALSO *Enquete Commissions; President's Council on Bioethics; Royal Commissions.*

BIBLIOGRAPHY

Brody, Baruch A. (1989). "The President's Commission: The Need to be More Philosophical." *Journal of Medicine and Philosophy* 14(4): 369–384. Argues that greater philosophical clarity on terms such as "life" and "death" would have led to more coherent policies.

Brown, Eric. (2004). "The Dilemmas of German Bioethics." *The New Atlantis* 5(Spring): 37–53. Provides a historical overview of bioethics in Germany and assesses recent and potential future developments.

Campbell, Alastair B. (1989). "Committees and Commissions in the United Kingdom." *Journal of Medicine and Philosophy* 14(4): 385–402.

Isambert, Francois-Andre. (1989). "Ethics Committees in France." *Journal of Medicine and Philosophy* 14(4): 445–456.

Juengst, Eric T. (1996). "Self-critical Federal Science? The Ethics Experiment within the U.S. Human Genome Project." *Social Philosophy and Policy* 13(2): 63–95. Evaluates the ELSI experiment as a model for addressing policy issues raised by innovations in science. Illustrates the tensions inherent in the dual missions of providing policy advice and undertaking inquiry and the difficulty for any bioethics commission (or "un-commission") model in communicating policy recommendations.

Kasimba, Pascal, and Peter Singer. (1989). "Australian Commissions and Committees on Issues in Bioethics." *Journal of Medicine and Philosophy* 14(4): 403–424. Examines contributions of government commissions to ethical issues raised by medical and biological sciences.

Kimura Rihito. (1989). "Ethics Committees for 'High Tech' Innovations in Japan." *Journal of Medicine and Philosophy* 14(4): 457–464.

Martinez, Julio Luis, ed. (2003). *Comites De Bioetica* [Bioethics committees]. Madrid: Universidad Pontificia. A Spanish publication that provides an overview and evaluations of bioethics committees.

McClay, Wilfred M. (2004). "Science and Self-Government." *The New Atlantis* 4(Winter): 17–22. Gives a positive assessment of a report issued by the President's Council on Bioethics.

Poland, Susan Cartier. (1998). "Bioethics Commissions: Town Meetings with a 'Blue, Blue Ribbon.'" *Kennedy Institute of Ethics Journal* 8(1): 91–109. Available as a monograph from National Reference Center for Bioethics Literature, Kennedy Institute, Georgetown University, Washington, DC. A comprehensive listing of bioethics commissions around the world.

Sass, Hans-Martin. (1989). "Blue-Ribbon Commissions and Political Ethics in the Federal Republic of Germany." *Journal of Medicine and Philosophy* 14(4): 465–472.

United States Office of Technology Assessment (OTA). (1993). *Biomedical Ethics in U.S. Public Policy*. Reviews the history of four Federal bioethics initiatives: the National Commission for the Protection of Human Subjects of Biomedical and Behavioral Research, the Ethics Advisory Board, the President's Commission for the Study of Ethical Problems in Medicine and Biomedical and Behavioral Research, and the Biomedical Ethics Advisory Committee.

Walters, LeRoy. (1989). "Commissions and Bioethics." *Journal of Medicine and Philosophy* 14(4): 363–368. The lead paper in a special issue of the journal devoted to bioethics commissions.

Williams, John R. (1989). "Commissions and Biomedical Ethics: the Canadian Experience." *Journal of Medicine and Philosophy* 14(4): 425–444.

INTERNET RESOURCE

The President's Council on Bioethics. Available from http://www.bioethics.gov/. Provides a list of past U.S. commissions and committees with several links to reports.

BIOETHICS COUNCIL

SEE *Bioethics Commissions and Committees; President's Council on Bioethics*.

BIOLOGICAL WEAPONS

Biological weapons constitute an increasingly important ethical and political issue for science and technology. This entry examines that issue by defining biological weapons (BWs), reviewing the history of their use, considering efforts to deal with future threats, and analyzing the ethical and political aspects of BWs.

Definition

Biological warfare is the intentional use of disease-causing microorganisms or other entities that can replicate themselves—such as viruses, infectious nucleic acids, and prions—against humans, animals, or plants for hostile purposes. It also may involve the use of toxins, which are poisonous substances produced by living organisms, including microorganisms (such as botulinum toxin), plants (for example, ricin derived from castor beans), and animals (snake venom, for instance). The synthetically manufactured counterparts of those toxins are considered BWs when they are used for purposes of warfare.

Although biological agents have the potential to cause mass casualties, the numbers are often more a matter of scare mongering than real: when it is claimed, for example, that a pound of botulinum toxin can kill six billion people, which is not a real possibility. It nevertheless remains the case that one-quarter of all deaths worldwide and about 50 percent of all deaths in developing countries are attributed to infectious diseases. Although human beings have developed several physiological defenses against disease and in certain cases have acquired immunity through evolution, these natural defenses may be minimal in societies weakened by war or by famine, drought, stress, or other natural disasters.

Early Biological Warfare

Biological warfare may be as old as civilization. In the earliest forms it involved drawing enemy troops into disease-ridden areas on the basis of an etiological belief that epidemics were caused by inhaling air infected by particular telluric emissions. Animal and plant toxins also were used commonly in many societies to poison arrows and other kinetic weapons. In later times disease was spread by means of pollution of the environment (for example, dropping human or animal carcasses into wells or catapulting them into besieged cities), the use of kinetic weapons that were dipped into decaying corpses, and the distribution of objects contaminated by people with highly infectious illnesses such as smallpox.

However, it was not until the end of the nineteenth century that the propagation of disease and thus the effectiveness of such actions began to be understood. By 1914 microbiology had advanced considerably: Major bacterial disease-causing organisms had been isolated and cultivated; the existence of viral diseases had been discovered, although the pathogens were not yet well understood; and parasitic diseases were being studied. There was also an improved understanding of disease transmission, and that understanding contributed to better prophylaxis, prevention, and countermeasures. Not surprisingly, those insights and new techniques soon were applied for hostile purposes. World War I witnessed the first acts of sabotage (against animals) with cultivated disease-causing organisms.

Firefighters remove suspicious-looking packets from a post office distribution center. Harmful biological agents such as anthrax are sometimes distributed through mail. *(© Reuters NewMedia Inc./ Corbis.)*

During the 1920s and 1930s the fear of biological warfare increased significantly in parallel with scientific progress and as a consequence of experiences with the Spanish flu epidemic in 1918. In World War II only Japan actually used biological agents, employing them during military operations in China. Nazi Germany and the Allies did not produce an operational offensive BW before the end of the war apart from a limited British retaliatory capability to infect German cattle with anthrax.

The Cold War and Afterward

After World War II the Soviet Union and the United States, and to a lesser extent the United Kingdom, were the principal states continuing research, development, and production of offensive BWs. The United States formally halted its program in 1969 and then destroyed its existing BW stockpiles. An internal review had demonstrated the military utility of biological warfare, but the United States concluded that a BW capability would not contribute significantly to its existing security posture. The announcement of the termination of the offensive BW program was accompanied by the argument that BWs were of low military significance, which other countries were happy to adopt. To many diplomats a moral imperative became the driving force to achieve an international treaty, and the unilateral U.S. gesture thus helped pave the way for the 1972 Biological and Toxin Weapons Convention (BTWC). The Soviet Union, however, did not reciprocate and even accelerated its BW program despite being one of the three co-repositories of the BTWC, along with the United Kingdom and the United States. The program survived the 1991 breakup of the Soviet Union essentially intact, and despite assurances by the Russian leadership, there remain considerable doubts about whether Russia has terminated all prohibited BW activities.

BW proliferation became a major worry in the late 1980s in part as a consequence of the use of chemical weapons in the Iran–Iraq war. The concerns were heightened significantly in the 1990s when the United Nations Special Commission on Iraq (UNSCOM), which was set up after the liberation of Kuwait in 1991, revealed the advanced and extensive nature of Iraq's

BW programs. As the invasion of Iraq by American-led coalition forces in March 2003 illustrated, the mere assumption of the presence of BW can be highly destabilizing to international security. Countries such as China, Egypt, India, Iran, Iraq, Israel, North Korea, Pakistan, Russia, South Korea, and Taiwan are mentioned in connection with BW proliferation, but there is considerable uncertainty about whether those programs are offensive or defensive and about their level of sophistication.

Biological weapons involve dual-use technologies and processes that can be employed for both legitimate and prohibited activities. The ambiguities that result from the dual-use potential of those technologies are increased by the facts that (1) the active ingredient of the weapon (that is, the biological agent) is central to the making of the offensive weapon as well as to the development of some key means to protect against or manage the consequences of exposure to the biological agent (such as vaccines and medication) and (2) the final stage of the armament dynamic during which the applied technologies have no purpose other than weaponization may not become apparent until the biological agent is placed in a delivery system. As a consequence, the judgment of the true nature of certain activities comes down to a judgment of intent, and a country that has an antagonistic relationship with the state making the intelligence assessment is at greater risk of being called a proliferator than is one that has a friendly relationship. The perceived intent of a state is a major subjective component in the threat assessment.

Terrorism with pathogens became a primary concern in the 1990s after it was learned that the Japanese religious cult Aum Shinrikyo, which had conducted two deadly attacks with the nerve agent sarin in 1994 and 1995, also had unsuccessfully released BWs. Although another religious cult, the Rajneesh, had infected some 750 people with salmonella in an attempt to influence local elections in Oregon in the United States in 1984, the threat was not taken seriously until 2001, when an unknown perpetrator killed five people and infected seventeen more with anthrax spores delivered in letters. The fact that those attacks occurred in the wake of the terrorist strikes against the United States on September 11, 2001, heightened threat awareness around the world.

Future Threats and Ways to Deal with Them

The principal tool against biological warfare is the BTWC. The convention was the first disarmament treaty: It ordered the total destruction of all BW stockpiles, and it contains a comprehensive ban on the development, production, and possession of BWs. The core prohibition of the BTWC is based on the so-called General Purpose Criterion (GPC), which prohibits not specific objects as such (for instance, pathogens) but rather the objectives to which they may be applied (hostile purposes). The main advantage of the GPC is that its application is not limited to technologies that existed at the time of the conclusion of the treaty negotiation but to all innovations. This has proved critical in the light of the rapid advances in biology and biotechnology at the end of the twentieth century and the beginning of the twenty-first. As a result of the GPC the parties to the BTWC have been able to reaffirm the prohibition in the light of those technological developments at the periodic review conferences of the convention. However, the treaty lacks meaningful tools to verify and enforce compliance. Since its entry into force in 1975 there have been several allegations and some confirmed cases of material breaches, but the inability to deal with them under the treaty provisions has contributed to the perception of its weakness.

The BTWC also is being challenged by rapid developments in biotechnology and genetic engineering despite the availability of the GPC. Although these developments hold out the promise of improving the quality of life, much of the knowledge can be employed for hostile purposes by improving the stability and virulence of existing warfare agents or even by creating new agents based only on some components of an organism. The dual-use potential of many products, processes, and knowledge implies that any strengthened BTWC regime would require inspection rights in relevant scientific institutions and biotechnology companies. Many establishments are extremely reluctant to grant international inspectors access to their facilities for fear of losing propriety information.

As a consequence, efforts to strengthen the BTWC by means of a supplementary legally binding protocol have failed. The stalled multilateral negotiation process has shifted attention to a range of initiatives to be undertaken by individual states that are parties to the BTWC, including enhanced export controls, encouragement to establish ethical standards and professional codes of conduct, and the enactment of national legislation criminalizing activities contrary to the objectives and purpose of the BTWC by natural and legal persons and corporations.

Moral and Ethical Standards

The argument often is made that investments in technologies that contribute to the design and production of armaments are unethical because they ultimately contribute to the destruction of humans or consume resources that otherwise could have contributed to the improvement of humankind. Because of widespread moral aversion to biological warfare, involvement in BW development and production programs is condemned by many people.

The question of moral judgment is, however, complicated. First, work in the field of biology can be conducted without any link to the military establishment but still contribute to the development of biological weapons. Second, many activities are directed toward enhancing defence and protection against and the detection of biological warfare agents as well as toward the improvement of prophylaxis and the development of new pharmaceuticals. However, improvement in defence necessarily implies an understanding of the offensive characteristics of existing biological warfare agents as well as those of new pathogens, including genetically modified variants. The distinction between offensive and defensive research and development is difficult to make. In fact, the source of the complications with respect to moral judgment is the dual-use potential of most of the technologies involved.

Some scientists, researchers, and technicians, whether as individuals or as members of professional groups, have objected to participation in BW-relevant programs. However, international conventions do not always provide unambiguous moral guidance. International law governs behavior among states, not the conduct of individuals. In a narrow sense all state activities that fall outside the scope of an international prohibition are legal, contributing to a continuing tension between morality and legality.

This becomes clear in the justification of so-called biochemical nonlethal weapons despite the fact that both the BTWC and the Chemical Weapons Convention (CWC) prohibit any weapon that uses toxicity or infectivity whether or not its primary effect is incapacitating or lethal. Several states continue to pursue such weapon programs and justify them on humanitarian grounds. However, the use of a fentanyl derivative by Russian forces in the Moscow theater siege in October 2002 demonstrated that the margin between incapacitation and killing is very narrow. Fentanyl and its derivatives are obtained from opium-producing plants, and thus fentanyl is a biochemical toxicant that is covered by both disarmament treaties. Several U.S. agencies are actively pursuing several nonlethal technologies based on biochemical action. Since the 1920s the United States has systematically objected to the inclusion of harassing and incapacitating agents in the prohibitions against chemical and biological warfare.

Finally, the belief in the value neutrality of scientific activities and technology—the denial that the introduction of new insights or technologies has societal ramifications—held by many scientists constitutes a considerable obstacle to having discussions of ethical and moral issues. Especially if the potential negative societal effects are obvious and cannot be denied, the neutrality of science will be proclaimed (this does not happen if the societal benefits are clear). Indeed, many scientists feel actively discouraged to take part in ethical discussions and accept social responsibility for their work, convinced that research should be guided by its own thrust, independent from and indifferent to the outside political and social world. This view is sustained by early specialization and the lack of sufficient overlap and interaction between disciplines in teaching programs. Also, many scientists and professionals in the fields of biology and biotechnology are unaware of the existence of the BTWC.

The Future

In the early twenty-first century the BTWC, as well as the CWC with regard to toxins, is the main legal instrument to prevent biological warfare. However, an international treaty is subject to continuing pressures as a consequence of changes in the international security environment and technological developments that have a direct bearing on the objectives and purpose of the agreement.

Although the BTWC has a broad scope, the document governs only state behavior. Many developments relevant to the BTWC take place on substate (universities, research laboratories, and companies as well as terrorism) and transnational levels (transnational corporations and international organizations as well as terrorism). The responsibilities of these actors in supporting the goals of the BTWC is great but not well recognized. The impact of the convention on their economic activities is also great because certain transactions may be prohibited and certain goals are forbidden.

Both the research and industry sectors in the field of biology have a large stake in the successful implementation of the convention because otherwise their reputation could be tarnished. The introduction of ethical codes of conduct with respect to issues involving biological warfare in educational curricula and industry practices not only reinforces the treaty regime of the BTWC but also protects the economic interests of the research establishments and companies involved. To assess the moral or ethical aspects of their activities scientists and professionals must be aware not only of international rules and norms but also of how those rules and norms evolve.

JEAN PASCAL ZANDERS

SEE ALSO *Chemical Weapons; Just War; Military Ethics; Terrorism; Weapons of Mass Destruction.*

BIBLIOGRAPHY

Alibek, Kenneth, with Stephen Handelman. (1999). *Biohazard*. London: Hutchinson.

British Medical Association. (1999). *Biotechnology, Weapons and Humanity*. Amsterdam: Harwood Academic Publishers.

Dando, Malcolm R. (2001). *The New Biological Weapons: Threat, Proliferation, and Control*. Boulder, CO: Lynne Rienner.

Geissler, Erhard, and John Ellis van Courtland Moon, eds. (1999). *Biological and Toxin Weapons: Research, Development and Use from the Middle Ages to 1945*. SIPRI Chemical & Biological Warfare Studies no. 18. Oxford: Oxford University Press.

Lederberg, Joshua, ed. (1999). *Biological Weapons*. Cambridge, MA: MIT Press.

Roffey, Roger. (2004). "Biological Weapons and Potential Indicators of Offensive Biological Weapon Activities." In *SIPRI Yearbook 2004: Armaments, Disarmament and International Security*. Oxford: Oxford University Press.

Sims, Nicholas A. (2001). *The Evolution of Biological Disarmament*. SIPRI Chemical & Biological Warfare Studies, no. 19. Oxford: Oxford University Press.

Stockholm International Peace Research Institute. (1970–1975). *The Problem of Chemical and Biological Warfare*, 6 volumes. Stockholm: Almqvist & Wiksell.

Wright, Susan, ed. (1990). *Preventing a Biological Arms Race*. Cambridge, MA: MIT Press.

Zanders, Jean Pascal, ed. (2002). "Ethics and Reason in Chemical and Biological Weapons Research." *Minerva* 40(1): 1–91.

Zilinskas, Raymond A., ed. (2000). *Biological Warfare: Modern Offense and Defense*. Boulder, CO: Lynne Rienner.

INTERNET RESOURCES

American Society for Microbiology. "American Society for Microbiology Cautions That Scientific Publication Restraints May Have Negative Impact on Public Health and Safety." Available from http://www.asm.org/media/index.asp?bid=2813.

"Biological Weapons and Codes of Conduct." Available from http://www.ex.ac.uk/codesofconduct/Chronology/index.htm. Of particular interest are the sections "BW Codes" and "Conducting Research during the War on Terrorism: Balancing Openness and Security."

BIOMETRICS

Biometrics is the use of a person's physical or behavioral characteristics for the purpose of identification and verification. Leading biometric technologies based on direct imaging, measurement, and analysis of physical patterns are fingerprint recognition, eye and retinal scans, face (facial) recognition, and hand geometry. Biometric technologies that identify a person based on behavioral characteristics are voice (speech) recognition and signature recognition. DNA, body odor, and stride are all considered biometrics; however they are not deployed due to technical challenges in quantitative measurement and analysis.

Development and Uses

The term biometrics or *biometry* actually has an older meaning from the early twentieth century referring to the development of statistical and mathematical methods for data analysis in the biological sciences. In this sense the term has been largely replaces by *biostatistics*.

The development of biometrics in the sense relevant here if not the precise term can be traced back to the late nineteenth and early twentieth centuries when technologies of photographic portraiture, anthropometry, and dactyloscopy, as fingerprint reading and comparison was known, were used for purposes of bodily identification by law enforcement. In the 1930s, local laws required fingerprints and photo identification for birth certificates and driver's licenses. The next two decades witnessed widespread registration of personal identifiers through the expansion of passports and driver's licenses (Parenti 2003). Building on research conducted during the Cold War, the development and use of biometric technologies expanded significantly in the 1980s and 1990s.

Among biometric technologies, fingerprint recognition is the best known and most widely used. The development of fingerprinting goes back to the mid-nineteenth century when Sir William Herschel, a British colonial administrator in India, used inked handprints on contracts he made with the locals. In the 1870s, Henry Faulds, a British physician working in Japan, introduced a preliminary system of classification of human prints and proposed the use of fingerprinting for identification. In 1892 Francis Galton, father of eugenics, refined Faulds' system of classification and identified certain characteristics (minutia) for fingerprints.

Fingerprinting for law enforcement purposes was used for the first time in 1891 by an Argentine police officer, Juan Vucetich, who was able to arrest an offender based on a positive identification of the latter's fingerprints. Fingerprinting for criminal identification was introduced in the United States in 1903, and in a few years most major police departments started using the technique. In 1924 the fingerprint sections of the penitentiary at Leavenworth and the National Identification Bureau were consolidated to form the basis of the Federal Bureau of Investigation's (FBI) Bureau of Identification (Parenti 2003). With the introduction of the Automated Fingerprint Identification System (AFIS) in the early 1970s, criminal fingerprint records were computerized, enabling law enforcement to create and use searchable databases of prints.

Fingerprint recognition remains the most reliable biometric technology, but this has been challenged by some courts and researchers in recent years. For example, in 2002, a U.S. district court ruled that fingerprinting was not admissible as scientific evidence. Although the U.S. Court of Appeals modified this judgment, and researchers at the National Biometric Test Center (San Jose State University) did computer comparisons with exceedingly few errors attesting to the scientific validity of fingerprinting, there are still concerns. Security experts warn that wet, dirty, scarred, creased, or worn fingerprints might interfere with the scanning and recognition process. For example, in 2002 a Japanese researcher demonstrated that gelatin-based fake fingers could fool optical scanners.

Since the 1990s federal agencies, the intelligence community, and law enforcement have used hand geometry and fingerprint recognition to control access to facilities, identify criminals, check for false driver's license registrations, and maintain border security. Healthcare, financial, and transportation sectors use fingerprint and hand scans to eliminate badges, keys, and passwords and provide more secure and controlled access to facilities, computers, and databases.

In the early-twenty-first century lower costs and wider availability of biometric technologies together with a growing interest in convenience and security benefits have led to multiplication of biometric applications in varied contexts. For example, in the early twenty-first century schools increasingly use digitized fingerprints and/or hand scanners to enable students to pay for cafeteria meals, check out library books, and gain access to dormitories. The gaming industry deploys face recognition systems in casinos to identify card counters. In New York City low-risk probationers can report their whereabouts by scanning their hands at a kiosk instead of meeting with their probation officers. Plans to identify Medicaid patients at doctors' offices by fingerprint scans in order to eliminate healthcare fraud are underway in some states. Customers at some supermarkets and amusement parks will soon be able to make their payments with the touch of a fingerprint.

Spotlight Events

Although it is not entirely new, biometrics was thrust into the spotlight as a result of two early-twenty-first-century events: Super Bowl XXXV in January 2001, and the terrorist attacks of September 11, 2001. At the Super Bowl in Tampa, Florida, the police used video surveillance cameras equipped with face recognition technology to scan the faces of some 100,000 spectators in search of wanted criminals. Although it did not produce any significant results (only nineteen petty criminals were recognized), the surreptitious use of biometrics caused quite an outrage. The media dubbed Super Bowl XXXV the *Snooper Bowl*, a privacy rights group gave the City of Tampa the *2001 Big Brother Award for Worst Public Official*, and civil liberties advocates argued that the *digital police lineup* was a violation of the Fourth Amendment right to be free from unreasonable searches and seizures.

Months after the Super Bowl, the events of September 11 again focused attention on biometric technologies. In the face of growing security concerns, both governmental and nongovernmental entities (such as airports) turned to biometric technologies as part of their antiterrorism and homeland security efforts. For example, the U.S. Visitor and Immigrant Status Indicator Technology (U.S. VISIT) program and major airports, such as Logan International Airport in Massachusetts, Dallas/Forth Worth International Airport in Texas, and Palm Beach International Airport in Florida, use retina scan and/or fingerprint recognition systems to compare travelers against profiles of known or suspected terrorists in searchable databases.

Criticisms

Despite its touted benefits (security, convenience, protection of assets, and others), biometrics has been the subject of substantial criticism, which can be grouped into two categories. First, the use of biometric technologies presents certain technical challenges and limitations. Security experts note that fingerprint aging and changes in physical appearance such as hairstyle may undermine the reliability of fingerprint and face recognition systems, respectively. In terms of voice and signature recognition, experts warn that the discrepancies between the original identifier presented during enrollment may not correspond exactly to the one presented during verification and thus create difficulties in matching.

Second, biometric technologies present certain legal, ethical, and social implications as expressed by privacy and civil liberties advocates. Lawmakers and privacy experts direct attention to the inadequacy of legal protections regarding the collection, storage, and sharing of biometric data; and it is worth noting that the use of biometrics is not fully addressed in privacy legislation, and that there remain broad exemptions for law enforcement and national security purposes. In terms of ethical and social implications, some argue that biometric technologies turn the human body into nothing more than sets of data. Biometric systems, they contend, are dehumanizing because they are bureaucratic systems of identification and verification whereby people are subject to the control of others (Brey 2004). Biometric technologies can also limit freedom of movement and lead to social discrimination because they enable authorities to privilege or reject individuals based on biometric data (Lyon 2003). The most fundamental argument against biometrics relates to privacy invasion; this argument specifically targets face recognition technology.

Face recognition is the most contentious among biometric technologies because it is generally performed without one's knowledge. For fingerprinting or hand geometry to work, one must put the finger or hand under a scanner and thus is aware of being the subject of a biometric system of identification. However face recognition applications allow facial imagery to be captured without the consent or even the knowledge of the subject, and such technologies can be used for surveillance purposes. In this sense, one can argue that face recognition systems pose a plausible threat to privacy—the reasonable *control* an individual has over what information is made public, and what is not (Agre 2001).

Prior to implementing biometric technologies, policymakers, public authorities, and nongovernmental entities must consider the scientific basis, technical limitations, and possible negative consequences in order to analyze benefits and costs of biometric applications. If not, these implications might easily outweigh any security and convenience benefit, and challenge the free society in serious ways.

BILGE YESIL

SEE ALSO *Forensic Science; Police; Security; Terrorism.*

BIBLIOGRAPHY

Agre, Philip E. (2001). "Your Face Is Not a Barcode: Arguments against Automatic Face Recognition in Public Places." *Whole Earth* Winter: 74–77.

Brey, Philip. (2004). "Ethical Aspects of Facial Recognition Systems in Public Places." In *Readings in Cyberethics*, 2nd edition, ed. Richard A. Spinello and Herman L. Tavani. Sudbury, MA: Jones and Bartlett Publishers.

Lyon, David. (2003). *Surveillance after September 11*. Malden, MA: Polity.

Parenti, Christian. (2003). *The Soft Cage: Surveillance in America: From Slavery to the War on Terror*. New York: Basic Books.

BIOPHILIA

The term *biophilia* was coined by the Harvard entomologist Edward O. Wilson (born 1929) and used in the title of his book *Biophilia: The Human Bond with Other Species* (1984). It comes from the Greek *βιος*, "life," and *φιλια*, "love or affection," and means literally "love of life" or "life-loving."

Biophilia and Biodiversity

Wilson's thesis is that human beings have a deep, inbred psychological need for physical contact with a broad variety of other life forms. The concept of biophilia thus is closely linked with that of biodiversity (biological diversity). Although Wilson did not coin the term *biodiversity*—Walter G. Rosen did in the mid-1980s—he helped give it wide currency as editor of the 1988 book *Biodiversity*, the proceedings of the National Forum on Biodiversity held in Washington, DC, in 1986, sponsored by the National Academy of Sciences and the Smithsonian Institution. According to Wilson, biodiversity represents much more than a material resource for such things as medicines and genes; it represents a vital human aesthetic and psychological resource as well.

In *Biophilia* Wilson points out that *Homo sapiens* evolved in a biologically diverse matrix. That which most distinguishes humans from other species and that in which humanists take the most pride—intellect and cognitive skills—are, Wilson argues, an evolutionary adaptation to a natural environment replete with both opportunity and danger. Therefore, not only do people have as deep a psychological need for a biologically diverse environment as they do for such basic things as human companionship and conversation, the very identity of humans as a species was sculpted by interaction with other species.

The human bond with other species mentioned in Wilson's subtitle thus goes beyond a desire for aesthetic satisfaction and psychological well-being to the core characteristic of the human species, to the very essence of humanity. On the basis of this claim Wilson proposes a "deep conservation ethic" that remains nevertheless anthropocentric. If people bequeath an impoverished natural environment to future generations, they risk the intellectual degeneration—the devolution—of the human species. "Preparing for future generations," Wilson writes, "is an expression of the highest morality. It follows that the destruction of the natural world in which the brain was assembled over a million years is a risky step" (Wilson 1984, p. 121).

The Evolutionary Basis

Wilson's claim that the complexity of human intelligence reflects the complexity of the natural environment in which the human brain evolved was anticipated by the conservation biologist Paul Shepard (1925–1996). In *Thinking Animals: Animals and the Development of Human Intelligence* (1978) Shepard argued that as the progenitors of modern *Homo sapiens* were driven by climate change and competition from their ancestral arboreal habitat out onto the African savanna, they began first to scavenge and then to hunt animals as well as to forage for fruits, tubers, leaves, and seeds. They themselves were subject to predation by large carnivores. The ability to sort the animals encountered into general categories—prey of this kind or predator of that kind—Shepard suggests, was crucial to the survival and reproductive success of those "savanna waifs." Mentally classifying animals and plants into kinds was the origin of conceptualization, and the linking of those bioconcepts into webs of relationship was the origin of intellection.

Once early humans developed the ability to categorize—to conceptualize—that cognitive skill could be extended to other areas, such as meteorological and geological phenomena; kinship and other social relations; and gods, ghosts, and spirits. Shepard's title is a double entendre: Human beings became thinking animals (animals that think) by thinking animals (thinking about animals).

Wilson's claim that human beings require physical contact with a variety of other species for psychological health and well-being also was anticipated by Shepard. The early departure from the way of life (hunting and gathering) and the conditions of life (a diverse biological environment rich in other species) has produced, Shepard argues in *Nature and Madness* (1982), a kind of collective insanity that currently manifests itself in the form of a global environmental crisis. The shift first to an agricultural and then to an industrial relationship with nature has impoverished the range of human contact with nature. Moreover, the shift in social organization from small bands of peers making decisions by consensus to large hierarchical societies with leaders and followers inherent in the shift to an agricultural and then to an industrial mode of relationship with nature led, in Shepard's analysis, to an infantile demand for instant gratification of desire, ultimately at the expense of the natural environment and its other species.

Because the concept of biophilia is embedded in the theory of evolution—indeed, it is an element of evolutionary psychology—it could not have been anticipated before the advent of the Darwinian worldview. Before Shepard one finds notable intimations of biophilia in the marine works of Rachel Carson such as *Under the Sea Wind* (1941) and *The Sea Around Us* (1951) and in the montane works of John Muir such as *The Mountains of California* (1894) and *My First Summer in the Sierra* (1911).

The Biophilia Hypothesis

In the 1990s the concept of biophilia was expanded and transformed into the biophilia hypothesis, which states that "human dependence on nature extends far beyond the simple issues of material and physical sustenance to encompass as well the human craving for aesthetic, cognitive, and even spiritual meaning and satisfaction" (Kellert 1993, p. 20). Stated in the form of a hypothesis, biophilia becomes testable through standard scientific research procedures. As Wilson originally conceived it, biophilia was a largely positive "affiliation" with nature in all its biotic variety and splendor. Wilson also conceived biophilia as having in part a genetic basis. Obviously, the human need for things such as companionship and sexual intimacy is genetic: Companionship is necessary because the human species survives and reproduces most efficiently in cooperation with others, and only those who desire sexual intimacy pass their genes on to the next generation.

Wilson argues that the human need for contact with a diverse biota is also genetic, although less obviously, because that is the natural matrix in which the human species evolved. If this is true, the general biophilia hypothesis should have a qualifying aspect: biophobias of dangerous organisms. Research indicating universal biophobias—fears of certain life-forms that may be found in people irrespective of cultural differences—confirms the biophilia/phobia hypothesis. Narrowing that hypothesis down to specifics, for instance, the universal fear among humans of snakes and spiders, has been confirmed experimentally.

Biophilia is meaningful as a scientific hypothesis in the field of evolutionary psychology only if it is narrowed down to specifics. As Judith Heerwagen and Gordon Orians note, "There are fear and loathing as well as pleasure and joy in our experiences with the natural world. Thus the real issue is not whether biophilia exists, but rather the particular form it takes" (p. 139). Their research focuses on landscape aesthetics. Although the results of their testing of the biophilia hypothesis are nuanced, Heerwagen and Orians found through analysis of things as diverse as landscape painting, landscape architecture, and the selection of home sites by people who can afford to live wherever they choose that people prefer high, open ground with a wide vista overlooking water and not too far from trees. Such sites provided early humans with the ability to see from a safe distance predators and competitors approaching; the gravitational advantage of elevation for combat, if necessary; and the availability of animal and plant resources for eating and water for drinking and bathing.

Tendencies toward dichotomous thinking incline people to assume that if biophilia is inbred and genetic in origin, it is not a learned, culturally transmitted, socially constructed, and reconstructible response to nature. However, nature and nurture are more complementary than opposed. Most distinctively human traits that have a genetic basis—things that belong indisputably to human nature—are also strongly shaped by cultural context, idiosyncratic experience, education, and social conditioning. The uniquely human capacity to speak a language, for example, is genetically based, but which one of the world's thousands of languages a person learns to speak, how well, to whom to say what, and so forth, depends on history, cultural context, idiosyncratic experience, education, and social conditioning.

Consequences

Biophilia is not a human given but a human potential. Just as rhetoricians and poets maximally realize the human potential for language, natural historians such as Wilson and Carson maximally realize the human potential for biophilia. That potential can be generally fostered and nurtured or can be discouraged and stifled. The cost to a human being if the human potential to learn a language goes unfulfilled when an infant is raised in isolation from a linguistic environment is well known. What will be the cost to the human species as a whole if the biophilial potential of future generations is stanched by mass extinction and biological impoverishment? That is the millennial ethical question Wilson poses and ponders.

J. BAIRD CALLICOTT

SEE ALSO *Biodiversity; Environmental Ethics; Evolutionary Ethics.*

BIBLIOGRAPHY

Carson, Rachel. (1941). *Under the Sea Wind: A Naturalist's Picture of Ocean Life*. New York: Oxford University Press.

Carson, Rachel. (1951). *The Sea Around Us*. New York: Oxford University Press.

Kellert, Stephen R., and Edward O. Wilson, eds. (1993). *The Biophilia Hypothesis*. Washington, DC: Island Press.

Muir, John. (1894). *The Mountains of California*. New York: Century.

Muir, John. (1911). *My First Summer in the Sierra*. Boston: Houghton Mifflin.

Shepard, Paul. (1978). *Thinking Animals: Animals and the Development of Human Intelligence*. New York: Viking.

Shepard, Paul. (1982). *Nature and Madness*. San Francisco: Sierra Club Books.

Wilson, Edward O. (1984). *Biophilia: The Human Bond with Nature*. Cambridge, MA: Harvard University Press.

Wilson, Edward O., ed. (1988). *Biodiversity*. Washington, DC: National Academy Press, 1988.

BIOSAFETY COMMITTEES

SEE *Institutional Biosafety Committees*.

BIOSECURITY

Biosecurity involves preventing and minimizing intentional harm to people, crops, livestock, wildlife and ecosystems caused by biological agents that are either naturally occurring or human-made. Biosecurity technology research and development, policy formulation and operational practices principally pertain mostly to military weaponry, agriculture and medicine. The development and use of biological agents in these and related fields, such as aquaculture, are controversial primarily because they have intended and/or unintended positive or negative impacts on public health. For example, introducing naturally occurring biological agents into an ecosystem in order to control pests that are causing crop damage may have unintended negative impacts on unharmful organisms in addition to the positive impact of pest control. Consequently some leading experts distinguish biosecurity from "biosafety" which involves preventing and minimizing accidental harms caused by biological agents.

Biological Weapons and Warfare

Potential benefits and concern over threats caused by biological agents, and therefore the need for biosecurity, has existed through the ages and particularly with respect to their use as weapons in biological warfare. The first recorded instance of biological warfare occurred in 1346 when bodies of Tartar soldiers, who had died of plague, were catapulted over the walls of Kaffa (present-day Feodosiya, Ukraine) in order to infect the besieged residents. During the 1500s, Spanish conquest of South America and the Caribbean Islands spread infectious diseases to these unprotected regions. Similarly during the last of the French and Indian Wars, the English used blankets infected with smallpox to kill native populations in North America. In all such instances, naturally occurring biological agents were used to kill either enemies (during warfare), or native peoples, who were perceived as potential obstructionists to national expansionism.

Following discovery of the microbial basis of infectious diseases (i.e., germs) in the mid-to-late 1800s by European researchers Louis Pasteur (1822–1895) and Robert Koch (1843–1910), programs to research and develop chemical and biological weapons were conceived and implemented by several governments. Such weapons consist of a launching mechanism, artillery or missile delivery system, and an exploding canister or warhead capable of releasing chemicals or airborne pathogens. If inhaled, ingested or absorbed through the skin, such pathogens can cause diseases such as smallpox, anthrax, plague, or botulism, debilitating or killing people, livestock and/or wildlife.

Deaths of over 100,000 soldiers from Mustard Gas (a type of chemical weapon) during World War I heightened worldwide concern over the potential harm of biological weapons. The Geneva Protocol of 1925 banned the use of both chemical and biological weapons, although several countries including the U.S. and the former Union of Soviet Socialists Republic (USSR or Soviet Union) maintained "bioweapons" development programs and insisted on their right to use biological weapons in reprisal attacks if such devices were first used against them. During World War II, only Japan is known to have actually used biological weapons, namely during its battles against China: nevertheless, Britain, the USSR, and the United States all stockpiled biological weapons in the war's aftermath.

During the Cold War era (1945–1989) fear of biological (or germ) warfare was largely replaced with fear of radiological and nuclear weapons, although huge stockpiles of biological weapons were maintained by several nations. Offensive bioweapons programs in the U.S. were unilaterally halted by President Richard Nixon in 1972, just prior to an international convention to eliminate similar programs worldwide. Beginning in the 1990s the prospect of terrorist attacks involving chemical, biological, radiological, or even nuclear weapons of mass destruction became a new threat. Shortly after September 11, 2001, letters containing a refined preparation of dried anthrax spores were sent through the mail infecting more than twenty people and killing five individuals in cities across the Eastern U.S. Though the extent of this attack was limited, Jonathan B. Tucker notes that "it hinted at the mayhem that could result from the deliberate release of weaponized disease agents." Hence, according to a report published online by Michael Barletta in 2002, "bioterrorism — the deliberate use of microorganisms or toxins by non-state actors to sicken or kill people or destroy or poison food supplies upon which we depend — poses an uncertain but potentially devastating threat to the health and well-being of people around the world." In response there has been concerted interest in developing sensing technologies capable of detecting potentially harmful chemicals and pathogens in the environment.

Biological Threats to Livestock and Crops

Biological threats to plants and animals that are relied on by humans for food have existed since the beginnings of agriculture and domestication. Since that time there have been many instances in which biological agents have disrupted human food supplies. For example, the Irish Potato Famine (1845–1849) resulted in over 1 million deaths from starvation, a tragedy that came about because genetically invariant potato plants grown in Ireland at the time were susceptible to rapid infection by Phytophthora infestans fungi. Lack of genetic diversity limits natural defenses to disease and to biological agents that are intentionally introduced into an environment. In addition, new biological strains of livestock and plant pathogens can easily cause significant harm because they rapidly infect elements of ecosystems that have not developed immunities.

Controlling the spread of infectious diseases and associated harms may involve restrictions on growing plants or breeding animals, and controls on harvesting, shipping or processing of these for food or other purposes, as well as controlling the economic and ecological impacts of invasive alien species. Several nations including the U.S., as well as some states within the U.S. ban importation of certain types of fruits and vegetables. Most governments also require livestock owners to inoculate their animals against disease, such as foot-and-mouth disease (FMD). In 2002 a severe outbreak of foot-and-mouth disease in Britain required over 3 million animals in that country to be slaughtered. Controls to prevent the spread of infectious diseases may also need to involve quarantining livestock. The Paris-based World Organisation for Animal Health tracks infectious disease outbreaks in livestock, promotes animal health standards and makes recommendations for policy and legislation to governments throughout the world.

Government Oversight and Ethical Concerns

Introduction of naturally occurring or manmade genetically modified (e.g., recombinant DNA) viruses and experimental biotechnology into weaponry, livestock and plant and crops and medicine is controversial because, if not adequately controlled, these threaten the well-being of entire populations and ecosystems. For this reason government agencies in countries throughout the world impose health standards and carefully monitor and regulate experimental biotechnology research and development often as part of an overall biosecurity (and/or biosafety) policy. In the United States, the Department of Agriculture (USDA) has primary oversight of food production, processing, storage, and distribution; threats against the agriculture sector and rapid response to such threats; border surveillance and protection to prevent introduction of plant and animal pests and diseases; and food safety activities concerning meat, poultry, and egg inspection, laboratory support, research, education and outbreaks of food borne illness. Along with these responsibilities, the USDA also maintains a list of high consequence pathogens.

Also in the U.S. the Centers for Disease Control and Prevention (CDC) regulates several biosecurity matters and maintains a worldwide emergency biological threat response, assessment and control capability. Originally formed in 1946 to handle malaria outbreaks, the CDC now identifies and investigates outbreaks of disease and indicators of bioterrorism attacks through BioWatch. This program, which is co-sponsored by the U.S. Environmental Protection Agency and the U.S. Department of Homeland Security, includes over 4000 atmospheric monitoring stations located in cities throughout the United States, whose readings are constantly analyzed for evidence of harmful biological agents indicative of terrorist attacks.

The potential for dangerous microbes or their products being misused or mishandled and thereby causing harm to human beings and ecosystems on enormous scales also raises ethical concerns about their creation and management. Ethically, the potential for harm must be weighed against scientific, entrepreneurial or commercial freedoms to research and develop microbes for useful and even necessary reasons. Robert H. Sprinkle suggests that the classic "moral norm" shared among ethical scientists and physicians can be advanced by creating a "Biological Trust." Given ongoing invasions of ecological systems by alien species, as well as the potential for bioterrorism, other scientists including Laura A. Meyerson and Jaime K. Reaser concur that governments and scientists must work together to foster adequate and ethical policies and technological capabilities to prevent, detect, and respond to incidents involving microbes.

Today there is concern about whether or not professional ethics in science and engineering can adequately address biosecurity. Issues of particular concern pertain to international use of tax, trade and tariff policies to promote consistent biosecurity policies among nations; corporate investment in biosecurity research and development; and the fact that biosecurity in practice needs to be active and proactive for national deployments of sensing and monitoring technologies especially in unprotected metropolitan areas deemed most susceptible to potential harms caused by biological agents.

SAMUEL C. MCQUADE, III

SEE ALSO *Security; Weapons of Mass Destruction.*

BIBLIOGRAPHY

Barletta, M., (2002). *Biosecurity Measures for Preventing Bioterrorism.* Monterey, California: Monterey Institute of International Studies. Available from http://cns.miis.edu/research/cbw/biosec/pdfs/biosec.pdf

British Medical Association. (1999). *Biotechnology, Weapons and Humanity.* Report prepared under auspices of the Board of Science and Education of the British Medical Association. Amsterdam, Netherlands: Harwood Academic Publishers.

Guillemin, J. (2001). *Anthrax: The Investigation of a Deadly Outbreak.* Berkeley, California: University of California Press.

Meyerson, L. A., and J. K. Reaser. (2002). "The Biosecurity: Moving Toward a Comprehensive Approach." *Bioscience* 52: 593–600.

Meyerson, L. A., and J.K. Reaser. (2003). "Biosecurity, Bioinvasions and Bioterrorism." *Frontiers in Ecology and the Environment* 1(6): 307–314.

National Academies of Sciences. (2002). *Predicting Invasions of Non-indigenous Plants and Plant Pests*. Washington, DC: National Academy Press.

Society of Photo-optical Instrumentation Engineers. (2002). *Sensors, and Command, Control, Communications, and Intelligence (C3I) Technologies for Homeland Defense and Law Enforcement*. Conference Proceedings of the SPIE, April 1-5 (Orlando, Florida) Bellingham, Washington: SPIE.

Sprinkle, R. H. (2003). "The Biosecurity Trust." *Bioscience* 53: 270–278.

Tucker, J. B. (2003). *Biosecurity: Limiting Terrorists Access to Deadly Pathogens*. Peaceworks No 52. Washington, DC: US Institute for Peace.

BIOSTATISTICS

Biostatistics is the application of statistics to biology and medicine. It is concerned with the assessment of observed variation in living organisms, particularly human beings. It seeks better insight into the life process, with focus on the cause, treatment, and prevention of disease. It uses the theories and methodology of statistics, but has created specialized methods of its own.

The development of statistical inference in the late-nineteenth and early-twentieth centuries was motivated by problems in biology, and its growth stimulated by the subsequent explosion of research in science and technology and the advent of the electronic computer. Responding to challenges posed by large-scale biomedical research programs, biostatistics emerged as a vigorous distinct discipline. Its scope includes data collection and analysis pertaining to virtually all facets of the vast healthcare system. The study of health factors affecting populations, with emphasis on public health issues, is the realm of epidemiology, a closely related field using the theories and methods of biostatistics.

Experimentation on human subjects in clinical research involves both biostatistics and ethics, including ethical aspects of clinical trials. But the two fields also intersect on broader concerns related to medical uncertainty and complexity: poor understanding on the part of the public, conflicts of interest, manipulation by the market, and questions of responsibility. Greater awareness of these issues is needed to help address critical problems facing contemporary medicine.

Concepts and Methods of Biostatistics

In the field of descriptive statistics, biostatistics contributes to the preparation of official records characterizing the health of the nation. As participant in the biomedical research process, it provides study design based on theories of statistical inference, primarily the classical Neyman-Pearson theory of hypothesis testing. Applying a wide range of standard techniques, it considers the two types of error in testing, determines required sample size for desired power, and assesses the statistical significance of results. It estimates outcomes of interest with associated confidence intervals. Its best-known specialized technique is the randomized clinical trial (RCT) for controlled experiments. For observational research the chief methods are cohort and case-control studies.

HEALTH STATISTICS. An illustration of data provided by the National Center for Health Statistics is given in Figures 1 and 2, showing cancer death rates in the United States from 1930 to 2000 for the major sites, for males and females. Such records of health statistics are an important resource for public health policies and biomedical research, in this case for studies of the etiology, treatment, and prevention of cancer. For example, although lung cancer remains the leading cause of cancer death, the decreasing rate for males in the last decade reflects the decrease in the prevalence of smoking, with a plateau in the death rate seen thus far for women.

EXPERIMENTAL RESEARCH: THE RANDOMIZED CLINICAL TRIAL. A *clinical trial* is an *experiment* in which a selected group of patients is given a particular treatment (*intervention*), typically a drug, and followed over time to observe the *outcome*. In a *randomized clinical trial*, also called *randomized controlled trial* (both referred to as RCT), patients are assigned at random to one of two or more treatments to assess relative effectiveness. Individual differences among patients that may affect their response are assumed to be balanced out by the random assignment. Ethical mandates include *clinical equipoise* (lack of medical consensus on the superiority of any of the treatments) and *informed consent* (willing participation of fully informed patients). The *research protocol* describing the proposed trial must be approved by the local *Institutional Review Board* (IRB).

The study may conclude before an outcome is observed for each patient (for example, the patient is still alive when the outcome is death). Such patients are said to be still *at risk*, and have a *censored* observation. The graphic summary of results is the so-called *survival curve*, which shows the proportion of patients alive (or disease-free if the outcome is recurrence) at each point in time along the period of observation. It is based on the *life-table* or *actuarial method*, with time 0 representing the entry point of each patient into the trial. Showing two or more arms of a study on the same graph offers a visual comparison of treatment outcomes. Special techniques of *survival analysis* can compare groups with inclusion of censored observations. There are methods to test the hypothesis that there is no difference between treatments, including adjustment for observed patient characteristics that may affect outcome.

Figure 3 presents five-year results of a three-arm RCT comparing disease-free survival of breast cancer patients treated with total mastectomy, segmental mastectomy (lumpectomy), and segmental mastectomy with radiation therapy. All patients with positive axillary lymph nodes received adjuvant chemotherapy. The first graph shows lumpectomy to be just as effective as mastectomy; the other two indicate lumpectomy with radiation therapy to be significantly better than either surgical procedure alone.

OBSERVATIONAL RESEARCH: COHORT AND CASE-CONTROL STUDIES. The two main approaches to addressing questions for which experimentation is ethically not feasible or otherwise not practicable are the *observational designs* of *cohort* and *case-control studies*, the basic tools of epidemiology. They aim to discover or confirm an association between some *exposure* or *risk factor* and a disease, using specific criteria of statistical theory and methodology.

Cohort Study. This is usually a *prospective study* that identifies a large group (cohort) of individuals without the disease, but with information about the presence or absence of the risk factor under study. The cohort is then followed over time to observe for the occurrence of the disease. Smoking is a risk factor that cannot be studied in RCTs. In the hypothetical example shown in Table 1, a cohort of 2,000 adult males is followed to observe for a diagnosis of lung cancer; 500 of the men are smokers and 1,500 nonsmokers at the beginning of the study. As a possible outcome after twenty years, 24 percent of smokers and 2 percent of nonsmokers have contracted lung cancer. The measure of association used is the *relative risk* (RR) or *risk ratio*, 24/2 = 12.0.

Case-Control Study. This is a *retrospective* design, which identifies a group of people who have the disease (cases), selects a group as similar as possible to the cases except that they do not have the disease (controls), and then determines how many in each group were exposed to the risk factor. An actual example is shown in Table 2, in a study of the association between stroke in young adults and drug abuse, with 214 cases and 214 controls. It was found that seventy-three of the stroke victims had a history of drug abuse, compared with eighteen in the control group. The odds of drug abuse given the stroke are 73/214 to 141/214, and given no stroke, 18/214 to 196/214. The measure of association is the *odds ratio* (OR) as the estimate of relative risk, in this case .5177/.0918 = 5.64.

Comparison of Research Designs. The relationship between cohort and case-control studies is shown symbolically in Table 3. For both measures of association, RR and OR, a value of 1.0 indicates no association. There are statistical methods to test the hypothesis of no association and to provide a confidence interval for RR or OR. Confidence intervals that do not include 1.0 reflect a significant association; values less than 1.0 denote a protective effect of the factor being studied. The examples above (RR = 12, OR = 5.64) show strong associations. Media reports for a study claiming a 20 percent increase in relative risk, for example, would correspond to RR = 1.2, a weak association even if statistically significant.

Cohort studies permit careful selection of the study population and recording of the risk factor, and rates of disease can be calculated for both the exposed and unexposed group. But long observation of a large number of subjects is required, many may be lost to follow-up or their exposure may change, and the studies tend to be expensive. Case-control studies require fewer subjects, cost less, and can be completed in a relatively short time period. Instead of the risk of disease given the exposure, they estimate the odds of being exposed given the disease. But case-control studies rely on recall of past exposures that may be impossible to confirm and the selection of an appropriate control group is extremely difficult. A different group of controls, or just a change of a few, could completely alter the outcome. These are some reasons that so many conflicting results are reported in the medical literature. Others include small, improperly done clinical trials and those with short follow-up. But in any case, claims can only be valid for an association between exposure (or intervention) and disease. The assessment of causation is a lengthy, tentative process, with general guidelines to aid the research community (Hill 1967).

DIAGNOSIS AND SCREENING. Further uncertainties exist in the diagnosis of disease, and biostatistics provides methods to evaluate tests used in *diagnostic* and *screening* procedures. Most tests have an overlapping range of values for a healthy population and patients with the disease, so that in setting a *cutoff point* to distinguish *positive* from *negative* test results, two types of error may be made. The four possible outcomes are shown in Table 4, with the standard performance characteristics of diagnostic tests. *Sensitivity* is the ability of a

FIGURES 1–2

Figure 1: Age-Adjusted Cancer Death Rates, Males by Site, United States: 1930–2000

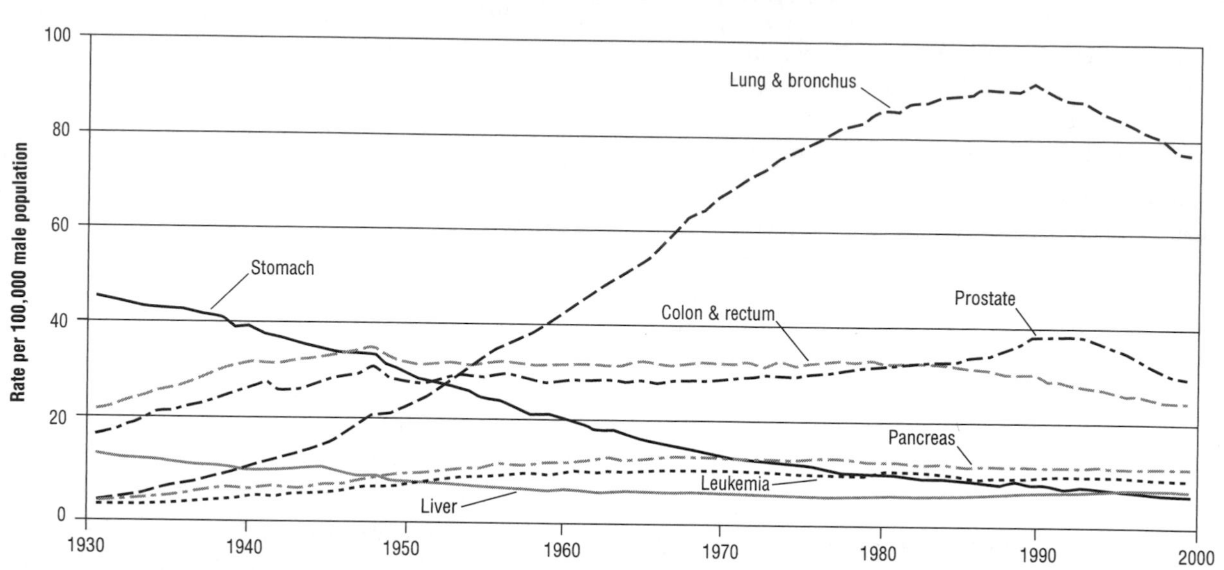

*Per 100,000, age-adjusted to the 2000 US standard population.
Note: Due to changes in ICD coding, numerator information has changed over time. Rates for cancers of the liver, lung & bronchus, and colon & rectum are affected by these coding changes.

SOURCE: American Cancer Society (2004), p. 2.

Figure 2: Age-Adjusted Cancer Death Rates, Females by Site, United States: 1930–2000

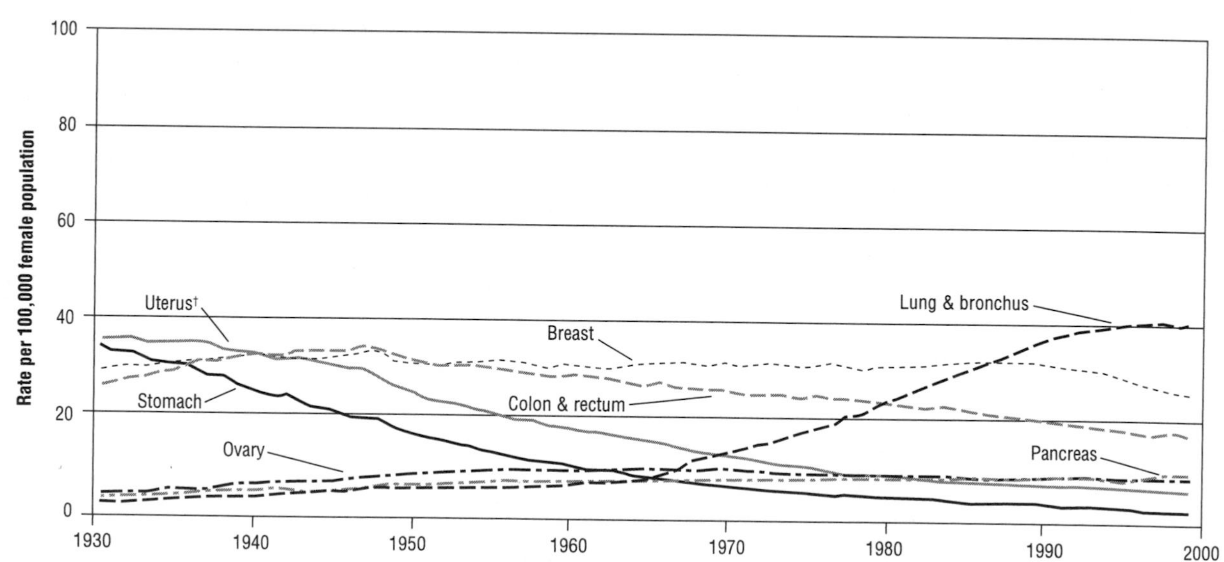

*Per 100,000, age-adjusted to the 2000 US standard population.
†Uterus cancer death rates are for uterine cervix and uterine corpus combined.
Note: Due to changes in ICD coding, numerator information has changed over time. Rates for cancers of the liver, lung & bronchus, colon & rectum, and ovary are affected by these coding changes.

SOURCE: American Cancer Society (2004), p. 3.

test to detect disease when present, and *specificity* its ability to indicate nondisease when none is present. The numeric example shows a test for fetal malformation with ultrasonography, which has reported 56 percent sensitivity and 99.5 percent specificity. The *prior probability* that a woman with poorly controlled diabetes has

FIGURE 3

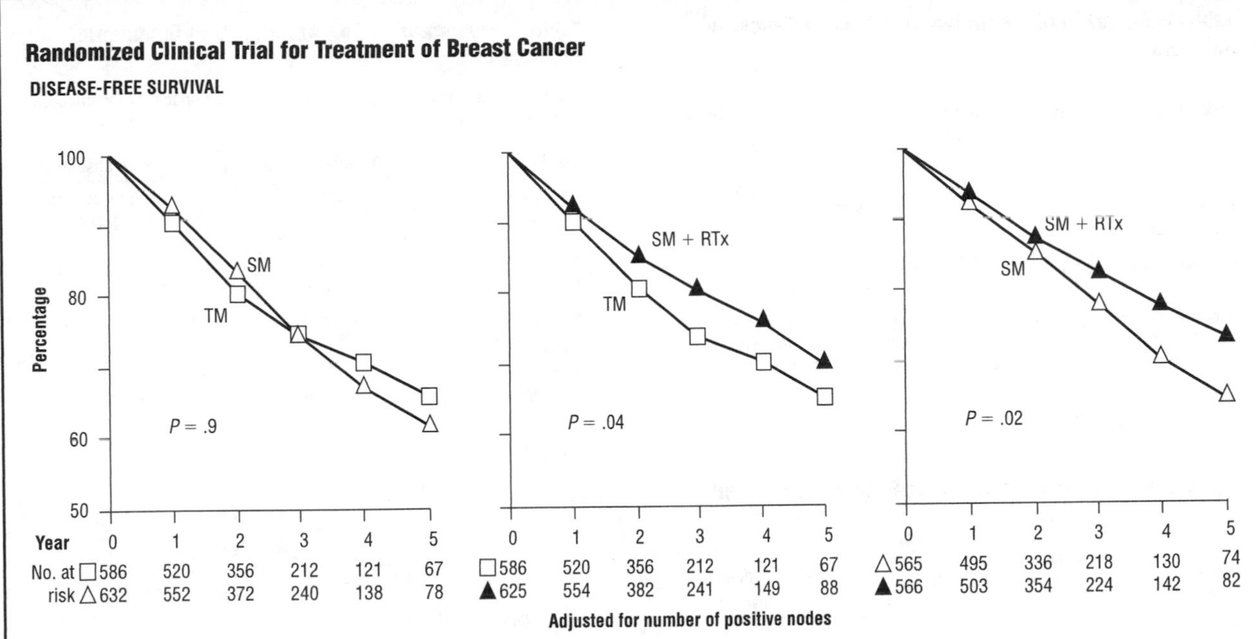

Comparison of survival for patients with total mastectomy (TM), segmental mastectomy (SM), and segmental mastectomy with radiation therapy (SM + RTx), using actuarial methods.

SOURCE: Dawson-Saunders and Trapp (1994), p. 204. Adapted from Fisher, Bernard, et al. (1985). "Five-Year Results of a Randomized Clinical Trial Comparing Total Mastectomy and Segmental Mastectomy With or Without Radiation in the Treatment of Breast Cancer." *New England Journal of Medicine* 312:665–673.

a malformed fetus is given as $P(D^+) = .20$. Using these numbers, one can apply a formula for conditional probabilities known as Bayes' Theorem to estimate the *predictive value* of the test, the *posterior probability* of malformation given a positive or negative test result. In this example a positive ultrasound yields a 96.6 percent probability that the fetus is malformed, and a negative result a 90 percent probability that it is normal.

In any one case a series of tests may be used to establish diagnosis, with the sensitivity and specificity of common tests established in previous studies. But there is inherent variation in the laboratory and imaging process itself, as well as the reliability of human raters. In addition, promising new markers for disease may present new uncertainties concerning cutoff points and criteria for treatment.

Decision to Treat: Prostate Cancer. For example, wide use of the test for prostate-specific antigen (PSA) has resulted in earlier diagnosis and decrease in the death rate from prostate cancer since the early 1990s (Figure 1). The test measures the blood level of PSA, a protein made by the prostate. It was defined as positive at 4.0 ng/ml, although higher levels also often indicate benign conditions. But a 2004 study reported prostate cancer on biopsy in 15 percent of 2,950 men with seven years of normal PSA levels and negative digital examinations. The prevalence of cancer was positively correlated with increasing PSA level from less than 0.5 to 4.0 ng/ml. Most of these cancers will not progress to life-threatening disease, and the question is whether to try to diagnose and treat these previously missed early cases.

Decision to Treat: Breast Cancer. About 50,000 cases of *ductal carcinoma in situ* (DCIS) are diagnosed in the United States each year, 20 percent of all breast cancers. These are cancers within the duct, not palpable, found on biopsy of suspicious regions identified by increasingly sensitive mammography. After lumpectomy an estimated 10 to 15 percent of DCIS will recur as invasive breast cancer. Prognosis is uncertain in individual cases, and variations of further treatment tend to be the recommended procedure. There has been a downward trend in breast cancer mortality (Figure 2), but breast cancer remains the second leading cause of cancer death for women, and a DCIS diagnosis creates vexing uncertainties for affected women.

NUMBER NEEDED TO TREAT (NNT): AN ESSENTIAL CONCEPT. Evaluating the results of a clinical trial, factors to consider include the type of patients studied, the length of follow-up, and the safety and effectiveness of treatment. The latter is especially important in *prevention trials*, when observed advances may involve

TABLES 1–5

Table 1: Cohort Study: Relative Risk of Lung Cancer in Smokers

Risk factor (Smoking)	Disease Yes	Disease No	Total	Risk of disease (Lung cancer)
Yes	120	380	500	120/500 = .24
No	30	1,470	1,500	30/1,500 = .02
Total	150	1,850	2,000	

Relative Risk or Risk Ratio: $RR = \frac{.24}{.02} = 12.0$

Hypothetical Example: A cohort of 2000 healthy men, of whom 500 are smokers and 1500 nonsmokers, is enrolled in the study and followed to observe for the development of lung cancer. Table shows the outcome after 20 years.

SOURCE: Courtesy of Valerie Miké.

Table 2: Case-Control Study: Odds Ratio for Stroke with History of Drug Abuse

Risk factor (Drug abuse)	Case (Stroke)	Control (No stroke)
Yes	73	18
No	141	196
Total	214	214

Odds of drug abuse in stroke patients $= \frac{73/214}{141/214} = .5177$

Odds of drug abuse in controls $= \frac{18/214}{196/214} = .0918$

Odds Ratio: $OR = \frac{.5177}{.0918} = 5.64$

Example of a case-control study to assess the relationship between drug abuse and stroke in young adults.

SOURCE: Adapted from Dawson-Saunders and Trapp (1994), p. 55.

Table 3: Symbolic Overview of Cohort and Case-Control Studies

Cohort → Risk factor	↓ Case Disease	Control No disease	Total	Risk of disease
Present	a	b	a + b	a/(a + b)
Absent	c	d	c + d	c/(c + d)
Total	a + c	b + d		
Odds of factor	$\frac{a/(a+c)}{c/(a+c)}$	$\frac{b/(b+d)}{d/(b+d)}$		

Relative Risk: $RR = \frac{a/(a+b)}{c/(c+d)}$

Odds Ratio: $OR = \frac{a/c}{b/d} = \frac{ad}{bc}$

Visual comparison of the two designs: The cohort study is prospective; it follows a group of subjects with known status of the risk factor (present/absent) and observes for the occurrence of disease. The case-control study is retrospective; it starts with cases who have the disease and controls who do not, and investigates the past exposure of each to the risk factor. The measure of association in cohort studies is the relative risk, which in case-control studies is estimated by the odds ratio.

SOURCE: Courtesy of Valerie Miké.

Table 4: Performance Characteristics of Diagnostic Procedures

Example: Ultrasonography to detect fetal malformation in cases with poorly controlled maternal diabetes.

Conclusion of test	Disease present (D^+)	Disease absent (D^-)
Positive (T^+)	True positive (.56)	False positive (.005)
Negative (T^-)	False negative (.44)	True negative (.995)

Sensitivity: Probability of true positive $= P(T^+|D^+) = .56$

Specificity: Probability of true negative $= P(T^-|D^-) = .995$

Prior probability of disease (*Prevalence*, best estimate before test) = $P(D^+) = .20$

Posterior probability of disease (*Predictive Value* of test) given by Bayes' Theorem.

• For positive test (PV^+):

$$P(D^+|T^+) = \frac{P(T^+|D^+)P(D^+)}{P(T^+|D^+)P(D^+) + P(T^+|D^-)P(D^-)}$$

$$= \frac{.56 \times .20}{(.56 \times .20) + (.005 \times .80)} = \frac{.112}{.116} = .966$$

• For negative test (PV^-):

$$P(D^-|T^-) = \frac{P(T^-|D^-)P(D^-)}{P(T^-|D^-)P(D^-) + P(T^-|D^+)P(D^+)}$$

$$= \frac{.995 \times .80}{(.995 \times .80) + (.44 \times .20)} = \frac{.796}{.884} = .90$$

The expressions above involve conditional probabilities. For example, sensitivity is the probability of a positive test (T^+) given the presence of disease (D^+). The formulas for positive and negative predictive value of a test require information on its sensititity and specificity, and the prior probability that the patient has the disease. Estimates of these may be generally known or be obtained form the literature, as in the present example.

SOURCE: Data from Dawson-Saunders and Trapp (1994), p. 232.

Table 5: Number Needed to Treat (NNT)

Example: Warfarin therapy to prevent stroke in patients with atrial fibrillation

Outcome	Control group	Experimental group
Annual risk of stroke	$p_c = .045$	$p_e = .014$

Absolute risk reduction: $p_c - p_e = .045 - .014 = .031$

Relative risk reduction: $= \frac{p_c - p_e}{p_c} = \frac{.031}{.045} = .69$

Number needed to treat: $NNT = \frac{1}{p_c - p_e} = \frac{1}{.031} = 32$

The number needed to treat (NNT) uses the same information as the other two expressions, but may be the most meaningful to consider. It indicates how many patients have to be treated for one to benefit from the treatment.

SOURCE: Data from Redmond and Colton (2001), p. 321.

long-term treatment of large populations. It is more informative to present NNT, the number of patients that have to be treated to prevent a single adverse event, than the usually reported relative percent reduction by the experimental treatment. For example, the anticoagulant warfarin was reported to achieve a 69 percent reduction in the annual relative risk of stroke in patients with atrial fibrillation. As shown in Table 5, the absolute reduction was 3.1 percent, from 4.5 to 1.4 percent, with its reciprocal as the NNT of thirty-two. This means that for every patient who benefits from the treatment, thirty-two on average have to be treated, with all thirty-two subject to side effects. For some low-risk patients the NNT is 145. People are far more critical in accepting treatment when results are expressed as NNT, rather than the large relative percent reductions heralded in promotions and the media.

Highlights of History

"One must attend in medical practice not primarily to plausible theories, but to experience combined with reason" (Hippocrates 1923, p. 313).This maxim appears in the *Hippocratic Corpus*, the writings collected under the name of the Greek physician Hippocrates (c. 460–c. 377 B.C.E.) that became the foundation of Western medicine. Nevertheless until a gradual change beginning around the mid-nineteenth century, medical practice was nearly always based on tradition and authority. Milestones in this transformation were discoveries made by two astute physicians who brought mathematics to medical investigation. One challenged the value of bloodletting, a common treatment dating back to antiquity. The other established the cause of childbed fever, a deadly disease of young mothers that was in fact an infection transmitted by physicians. The work of both met with hostility from the medical community.

PIERRE C. A. LOUIS AND THE NUMERICAL METHOD. By the early-nineteenth century there were large public hospitals in the major cities of Europe, and Paris was leading in the development of pathological anatomy, the use of autopsies to explore changes in the body caused by disease. The French physician Pierre Charles Alexandre Louis (1787–1872) spent years collecting and analyzing data on hospital patients, including the results of autopsies on fatal cases. He called his approach the Numerical Method, which involved tabulating data for groups of patients according to diagnosis and treatment received, and comparing their course of illness and survival patterns. In his major work on bleeding, published in 1835, he studied the effects of bloodletting in series of patients with different diagnoses and found essentially no difference in death rate or duration and severity of symptoms between patients bled and not bled and those bled at different stages of their disease. His findings completely contradicted the teachings of the day and met with sharp criticism, such as the argument that patients could not be compared in groups, because they differed in many respects. Louis reasoned that comparison was being made of essential features, abstracted from the general variability of other factors. The result was a systematic record of what was observed, not the anecdotal evidence of individual physicians who tended to remember the favorable cases. He developed guidelines for designing studies to evaluate different modes of treatment in his *Essay on Clinical Instruction* (1834).

Louis had great influence on the development of scientific medicine in the United States, because many young Americans were then studying medicine in Paris. One of these was Oliver Wendell Holmes (1809–1894), who in later recollections of Louis described the impact of the change he had observed: "The history of practical medicine had been like the story of the Danaides. 'Experience' had been, from time immemorial, pouring its flowing treasures into buckets full of holes. At the existing rate of supply and leakage they would never be filled; nothing would ever be settled in medicine. But cases thoroughly recorded and mathematically analyzed would always be available for future use, and when accumulated in sufficient number would lead to results which would be trustworthy, and belong to science" (Holmes 1883, p. 432).

IGNAZ SEMMELWEIS: A MEDICAL THEORY BASED ON MATHEMATICS. In July 1846 the young Hungarian physician Ignaz Semmelweis (1818–1865), trained at the medical school of Vienna, then the leading center of medicine in Europe, began work in the maternity clinic of its General Hospital. Confronted with the high death rates from childbed (puerperal) fever that would strike young women and often their babies shortly after childbirth, he undertook with passion to find the real cause of the disease. Occurring in hospitals throughout Europe and the United States, childbed fever was believed to have many different and vague causes, like cosmic-telluric-atmospheric influences and miasmas. In Vienna the first division of the maternity clinic, used for the training of medical students, had much higher mortality rates than the second division staffed by student midwives. Between January and June 1846 the death rate had ranged from 10 to 19 percent, compared with under 3 percent for the midwives, and it remained high as Semmelweis pursued his intense study of patient conditions and autopsies.

Two observations would fuse to spark the flash of insight in May 1847: (1) The staff of the first division, himself included, came to the maternity clinic directly from the dissection room where they had performed autopsies on the diseased patients (unlike the midwives); and (2) A colleague who had died of a wound sustained during a dissection revealed the same lesions on autopsy as the victims of childbed fever. Semmelweis's discovery entailed the recognition that the doctor and the women had died of the same cause, and the infectious material had been transmitted to the patients by the contaminated hands of the examining physicians. Semmelweis ordered all staff to wash their hands in chlorine of lime after autopsies, and immediately the death rate fell. When one woman with an ulcerating cancer of the uterus and another with an ulcerating knee injury gave birth, and in each case most of the patients nearby died of childbed fever, Semmelweis realized that the infectious material could also come from live tissue and be transmitted in the air, so that special precautions were needed for such cases. By 1848 the death rates were 1.27 percent in the first division and 1.33 percent in the second.

In his book *The Etiology, Concept, and Prophylaxis of Childbed Fever*, published in German in 1861, Semmelweis gave a detailed exposition of his theory, documented with extensive tables. For nearly forty years after the founding of Vienna's General Hospital, from 1784 to 1822, the death rate in the maternity clinic had averaged 1.27 percent. Between 1823 and 1840, after pathological anatomy studies were introduced, the rate rose to 5.9 percent. Then the clinic was split into two divisions, and between 1841 and 1846, the rate was 9.92 percent in the first division and 3.38 percent in the second, the pattern strongly implicating autopsies. Along these same lines, using careful observation, statistical evidence, and clear arguments, Semmelweis systematically eliminated the many other causes that had been proposed for childbed fever over the years.

Semmelweis held that invisible particles in decaying animal-organic matter were the universal necessary cause of childbed fever. Contrary to what had been claimed by others, childbed fever was a transmissible but not a contagious disease, like smallpox. Smallpox always caused smallpox, and every case of smallpox was caused by smallpox. Childbed fever was caused by resorption of decaying animal-organic matter of any source, and the latter could cause infection of any wound surface. Childbed fever was not a distinct disease, but a wound infection. The theory had complete explanatory power; it accounted for every case of the disease and its prevention. It established the etiologic approach to defining disease, the foundation of scientific medicine.

The Semmelweis theory was validated by the French chemist Louis Pasteur (1822–1895), founder of microbiology, who in 1879 identified streptococci as the chief microorganism causing childbed fever, and the English physician Joseph Lister (1827–1912), who introduced antiseptic methods in surgery. The germ theory of disease would follow. If *invisible particles in decaying animal-organic matter* is replaced by a current phrase containing *bacteria*, the Semmelweis theory remains valid and it has become a textbook case study in the philosophy of science (Hempel 1966).

Childbed fever is a tragic chapter in the history of medicine, not primarily because of the sad fate of Ignaz Semmelweis. (Suffering some sort of mental breakdown, he died abandoned, under suspicious circumstances, shortly after being committed against his will to a Viennese insane asylum.) Known from antiquity, childbed fever assumed serious proportions when childbirth became a hospital procedure, with doctors replacing midwives. Coupled with the rise of medical research in the autopsy room, progress cost the lives of hundreds of thousands of healthy young women who came to the charity hospitals to deliver. And the real tragedy was how long it took for the old theories to fade after the evidence was in, how long the debate went on about the causes of childbed fever as mothers went on dying. The problem of childbed fever was not definitively solved until the late 1930s, with the introduction of the sulfonamide drugs and then penicillin.

"Quels faits! Quelle logique!" was Pierre C. A. Louis's exasperated response as his critics proclaimed the merits of bloodletting. "Oh Logik!! Oh Logik!!" echoed Semmelweis in the closing paragraph of his great work, urging enrollment in a few semesters of logic before answering the noble call to argue the etiology of disease.

MODERN STATISTICAL INFERENCE. During the nineteenth century probability theory came to be used in the analysis of variation in astronomy, the social sciences, physics, and biology. The intense study of heredity, stimulated by the theory of evolution, spawned the birth of modern statistics around the turn of the twentieth century, associated with the names of Sir Francis Galton (1822–1911), Karl Pearson (1857–1936), and Sir Ronald Fisher (1890–1962). Formal statistical inference, with methods of hypothesis testing and estimation, was gradually introduced across a wide range of disciplines, including medicine.

In his work on the design of experiments in agriculture, Fisher proposed the idea of randomization, to make the experimental plots as similar as possible except for the treatment being tested. Applied to medicine, the approach led to the randomized clinical trial. The first strictly controlled clinical trial using random assignment of patients was set up by the British Medical Research Council in 1946 to evaluate streptomycin in the treatment of pulmonary tuberculosis. The trial was designed by the statistician Sir Austin Bradford Hill (1897–1991), who played a key role in bringing modern statistical concepts to medicine. In the United States randomized clinical trials were introduced in the mid-1950s when Congress authorized the National Cancer Institute to establish the Cancer Chemotherapy National Service Center to coordinate the testing of new compounds as possible anticancer agents. This launched the formation of national cooperative groups that became the mechanism for large-scale clinical trials, with funding provided for related research in statistical methodology.

Contemporary Biostatistics

Biostatistics is a strong academic discipline, with its professionals engaged in teaching and research, and working as consultants and collaborators throughout the healthcare field. The range of developments in theory and methodology—there is now a six-volume encyclopedia—as well as the increasing complexity of biomedical science and technology make the biostatistician an essential member of the research team.

In planning quality studies to assess risk factors of disease or the effectiveness of treatments, questions pertaining to research design, proposed controls, sample size, type of data to collect, length of study, and methods of analysis need to be guided by statistical considerations. *Historical*, rather than *concurrent controls*, may be appropriate for new treatment of a rare, usually fatal disease. In a randomized clinical trial, *stratified randomization* may be used, where patients are assigned at random within subgroups known to affect prognosis (for example, menopausal status in breast cancer). There are methods to assess the effect of multiple risk factors on outcome, such as *Cox regression*, *logistic regression*, and *loglinear analysis*. The essential means of modern analysis is provided by electronic database management and statistical software systems.

In approaches to statistical inference there is lively interest in *Bayesian methods* and *decision theory*. Within medicine there are the movements of *outcomes research*, to explore the effectiveness of medical interventions in the general population, and *evidence-based medicine*, to make more effective use of the medical literature in everyday practice. A related area is *meta-analysis*, which seeks to combine the results of published studies to obtain the best possible assessment of risk factors and treatments. Evaluating *alternative medicine* has become a pressing issue. The broader field of *health services research* also studies the *cost-effectiveness* of medical procedures.

Biostatistics and Ethics

The Hippocratic maxim, "Help or at least do no harm," has for 2500 years been the basis of medical ethics. How this can be done is explained by the Hippocratic precept cited earlier. To this end, medical practice should be based on experience combined with reason, namely, carefully collected observations (experience) analyzed with the tools of scientific methodology (reason). Biostatistics has assumed this function, and played a significant role in the great achievements of medical science and technology. Since the closing decades of the twentieth century, it has been faced with a crisis in U.S. (and Western) medicine, as the costs of health care spiral out of control.

Important advances include antibiotics and immunization, control of diabetes and hypertension, treatments for heart disease, cancer, and psychiatric disorders, diagnostic imaging, neonatal and trauma medicine, biomechanics, and organ transplants, with research continuing unabated on every front. But past successes have led many to unrealistic expectations of perpetual progress, putting them at risk for exploitation by a profit-driven healthcare industry. Medical technology tends to be oversold by the market, and an often poorly informed, vulnerable public is buying. Promotion in the media focuses on conditions that affect large segments of the population, such as chronic pain, which requires safe and effective individualized treatment for adequate control.

DEBATE OF MARKET VS. SCIENCE. In September 2004 the arthritis pain medication Vioxx, with sales of $2.5 billion in 2003, was withdrawn from the market by its manufacturer Merck because of findings of an increased risk of heart attacks and strokes. This triggered charges that the company had ignored earlier warnings, and the rival drugs Celebrex and Bextra, made by Pfizer, also came under scrutiny. Although helpful for many, these Cox-2 inhibitor agents did not claim greater effectiveness, only fewer gastrointestinal side effects than older alternatives like aspirin, ibuprofen, and naproxen. In the absence of adequate comprehensive studies, controversy continued concerning the relative risks and benefits of the various agents and the indications for their use. The larger debated issue is that of postmarketing surveillance (safety monitoring of drugs after release on the market), and the role of the Food and Drug Administration (FDA). The high cost of new drugs like Vioxx, challenged by medical critics, raises a further ethical concern. It is not only the physical harm done to so many, but the emotional and financial harm to all those struggling on limited means.

The individual must be more assertive in asking questions: Is this drug treatment necessary? What is the effectiveness (NNT) of the drug for a patient with the given characteristics? What are the side effects for this class of patient and how long is the follow-up of observation? Is there a less expensive, better-evaluated alternative? What are the interactions of the drugs the patient is taking? All drugs have side effects, and harmful effects of legally prescribed drugs are estimated to cause over 100,000 deaths in the United States each year. Ultimately it is up to the public to demand answers.

THE ETHICS OF EVIDENCE. An approach has been proposed for dealing with medical uncertainty, called the *Ethics of Evidence*. (Miké 1999, 2003). It can be expressed in two simple rules or imperatives: The first calls for the creation, dissemination, and use of the best possible scientific evidence as a basis for every phase of medical decision making. Complementing it, the second focuses on the need to increase awareness of, and come to terms with, the extent and ultimately irreducible nature of uncertainty.

There is a need for greater insight and closer involvement on the part of the public. Biostatistics can help to discern what is necessary, safe, and effective treatment, and should be fully utilized to produce the best available evidence. But even when it is properly used, uncertainties remain that are intrinsic to the techniques themselves and the limitations of medical knowledge. Most major diseases do not have a single cause, but result from the complex interplay of genetic and environmental factors. Systematic study of individual risk factors and their interactions must continue, in the search for better prevention and control. When Semmelweis made his great discovery, the numeric results were so dramatic that no formal statistical procedures were needed (and they did not yet exist). In the early twenty-first century it is a slow, incremental process to find and confirm small improvements. The real promise for medicine in the near future points to changes in lifestyle.

A study released in July 2004 estimates that 195,000 Americans die each year as a result of preventable medical error, and the data pertain only to hospitals. More open and direct participation of patients in their own treatment would help reduce error rates, keep in the forefront questions about the safety and effectiveness of proposed interventions, and curb the reflexive urge for malpractice litigation. An alert, educated public has a realistic view of medicine and does not expect it to solve all of life's problems. But it insists on well-funded biomedical research and its careful assessment, with effective government policies in place to ensure the best possible healthcare for all.

VALERIE MIKÉ

SEE ALSO *Meta-Analysis; Social Indicators; Statistics.*

BIBLIOGRAPHY

American Cancer Society. (2004). *Cancer Facts and Figures 2004*. Atlanta, GA: American Cancer Society. Annual report issued by the American Cancer Society, also available online at http://www.cancer.org/docroot/STT/content/STT_1x_Cancer_Facts__Figures_2004.asp.

Angell, Marcia. (2004). *The Truth about the Drug Companies: How They Deceive Us and What to Do About It*. New York: Random House. Documented critique of the pharmaceutical industry by a former editor of the *New England Journal of Medicine*.

Armitage, Peter, and Theodore Colton, eds. (1998). *Encyclopedia of Biostatistics*, 6 vols. New York: John Wiley & Sons.

Carter, H. Ballentine. (2004). "Prostate Cancer in Men with Low PSA Levels—Must We Find Them?" *New England Journal of Medicine* 350: 2292–2294.

Carter, K. Codell, and Barbara L. Carter. (1994). *Childbed Fever: A Scientific Biography of Ignaz Semmelweis*. Westport, CT: Greenwood Press.

Dawson-Saunders, Beth, and Robert G. Trapp. (1994). *Basic and Clinical Biostatistics*, 2nd edition. New York: McGraw-Hill/Appleton & Lange.

Evidence-Based Medicine Working Group. (1992). "Evidence-Based Medicine: A New Approach to Teaching the Practice of Medicine." *Journal of the American Medical Association* 268: 2420–2425.

Gehan, Edmund A., and Noreen A. Lemak. (1994). *Statistics in Medical Research: Developments in Clinical Trials*. New York: Plenum Medical Book. History of clinical trials, with emphasis on statistical aspects.

Hempel, Carl G. (1966). *Philosophy of Natural Science*. Englewood Cliffs, NJ: Prentice Hall. Includes case study of Semmelweis discovery.

Hill, Austin Bradford. (1967). *Principles of Medical Statistics*, 9th edition. New York: Oxford University Press. A classic text of biostatistics by one of its pioneers. Includes discussion of criteria for establishing causality in observational studies.

Hippocrates. (1923). "Precepts." In *Works*, Vol. 1, trans. W. H. S. Jones. London: Heinemann.

Holmes, Oliver Wendell. (1883). "Some of My Early Teachers: A Farewell Address to the Medical School at Harvard University, November 28, 1882." In *Medical Essays: 1842–1882*. Boston: Houghton, Mifflin.

Kohn, Linda T.; Janet M. Corrigan; and Molla S. Donaldson, eds. (2000). *To Err Is Human: Building a Safer Health System*. Washington, DC: National Academy Press. Includes discussion of large studies of medical errors in hospitals.

Louis, Pierre C. A. (1834). *An Essay on Clinical Instruction*, trans. Peter Martin. London: S. Highley. Original in French.

Miké, Valerie. (1988). "Modern Medicine: Perspectives on Harnessing Its Vast Emerging Power." *Technology in Society* 10: 327–338. Analysis of the need for better assessment of medical technology and related issues.

Miké, Valerie. (1999). "Outcomes Research and the Quality of Health Care: The Beacon of an Ethics of Evidence." *Evaluation & the Health Professions* 22: 3–32. Commentary by Edmund D. Pellegrino, "The Ethical Use of Evidence in Biomedicine," is included in this issue.

Miké, Valerie. (2003). "Evidence and the Future of Medicine." *Evaluation & the Health Professions* 26: 127–152. Further development of the Ethics of Evidence.

Miké, Valerie, and Robert A. Good. (1977). "Old Problems, New Challenges." *Science* 198: 677–678. Introduction to Birnbaum Memorial Symposium "Medical Research: Statistics and Ethics." International multidisciplinary conference with focus on ethical issues in biostatistics. Proceedings published in *Science*.

Miké, Valerie, and Ralph C. Marcove. (1978). "Osteogenic Sarcoma under the Age of 21: Experience at Memorial Sloan-Kettering Cancer Center." In *Immunotherapy of Cancer: Present Status of Trials in Man*, eds. William D. Terry, and Dorothy Windhorst. New York: Raven Press. Example of a clinical trial where use of a historical control was the appropriate procedure.

Miké, Valerie, and Kenneth E. Stanley, eds. (1982). *Statistics in Medical Research: Methods and Issues, with Applications in Cancer Research*. New York: John Wiley & Sons. Includes multidisciplinary discussion of clinical trials, including ethical aspects.

Redmond, Carol, and Theodore Colton, eds. (2001). *Biostatistics in Clinical Trials*. New York: John Wiley & Sons. Includes articles on Bayesian methods and NNT.

Shryock, Richard Harrison. (1969). *The Development of Modern Medicine: An Interpretation of the Social and Scientific Factors Involved*. New York: Hafner Publishing.

Semmelweis, Ignaz. (1983). *The Etiology, Concept, and Prophylaxis of Childbed Fever*, trans. K. Codell Carter. Madison: University of Wisconsin Press. Originally published in German as *Die Aetiologie, der Begriff, und die Prophylaxis des Kindbettfiebers* (1861).

Topol, Eric J. (2004). "Good Riddance to a Bad Drug." *New York Times*, October 2, p. A15. Op-Ed article by chairman of department of cardiovascular medicine at Cleveland Clinic, who in 1991 published results on cardiac risks of Vioxx.

INTERNET RESOURCE

Health Grades Inc. "Patient Safety in American Hospitals." Health Grades. Available from http://www.healthgrades.com/PressRoom/index.cfm?fuseaction=PressReleases. July 2004 study.

BIOTECH ETHICS

In the seventeenth century the philosophers Francis Bacon (1561–1626) and René Descartes (1596–1650) advocated a new way of doing science that would have the power to conquer nature for human benefit. (The old science had seemed to be more concerned with contemplating nature than controlling it.) In the contemporary world biotechnology is providing the technology for controlling and changing living nature, including human nature. However, because biotechnological power over the living world offers not only the promise for doing good but also an opportunity for doing evil, this has provoked an ethical debate over the modern scientific project for the mastery of nature through technology.

Biotechnology in History

Biotechnology can be defined as the technical manipulation of living organisms or parts of those organisms to provide products and services to satisfy human desires. If it is defined in this broad way, one can see that biotechnology has been employed throughout human history.

The history of biotechnology can be divided into three periods: ancient, modern, and contemporary. Ancient biotechnology began more than 10,000 years ago with the emergence of agriculture in ancient Mesopotamia. Modern biotechnology began in the nineteenth century with the development of industrial microbiology. Contemporary biotechnology began in the 1970s with new techniques for genetic engineering. In each period one can see the power humans have acquired to manipulate nature. But one also can see the natural limits of this power, which is constrained by the natural potentialities available in wild plants and animals and the natural complexities of behavioral traits in the living world.

Ancient biotechnology began when human beings started to domesticate plants and animals for human use. Throughout most of the history of the human species, spanning approximately six million years, human beings fed themselves by gathering wild plants and hunting wild animals. Then some people in a few parts of the world began to produce food by cultivating domesticated plants and herding domesticated animals. As a consequence those farmers and herders bred for and selected genetic modifications in domesticated organisms that were more suitable to human desires. Even in the early twenty-first century all of human civilization depends on this project in agricultural biotechnology.

The human power of domestication is limited, however, by the natural potentiality of wild plants and animals. Most plant and animal species in the wild are not suitable for domestication. For example, most wild plants are not good as a source of food because they are woody or do not produce fruit, leaves, or roots that are edible. Most wild animals are not susceptible to successful domestication because they cannot be bred and herded in a manner that makes them useful for human beings. Although advances in biological knowledge have increased human biotechnological power over living nature, that power will always be limited by the potentialities found in nature.

Modern biotechnology arose in the nineteenth century as growing knowledge in the biological sciences was applied to the technological manipulation of the living world for human purposes. For example, the chemist Louis Pasteur's (1822–1895) microbiological explanation of fermentation as resulting from the activity of microscopic organisms allowed improvements in the brewing of beer and other industries that depend on using fermentation by yeast to produce food and beverages. Pasteur also showed that infectious diseases are caused by disease-producing microorganisms and perfected techniques for vaccination that would create immunity to some of those diseases. Later, in the twentieth century, the discovery of the ways in which some fungi produce antibiotics such as penicillin revolutionized the medical treatment of bacterial infections. In the early 2000s there are hundreds of pharmaceutical agents derived from fungal fermentation.

However, even modern biotechnology shows the technical limits set by nature. Bacteria vulnerable to fungal toxins can evolve to become resistant to those toxins. Indeed, bacteria have been so successful in evolving tolerance to antibiotics that there is a growing fear in the medical profession that the age of antibiotic protection against infectious diseases is reaching its end. The power of this aspect of biotechnology for controlling living nature is great but limited.

The contemporary biotechnology that began in the last half of the twentieth century arose from a deeper knowledge of genetics and molecular biology and has provided humans with greater power over the living world. Even so, contemporary biotechnology is limited in its technical means by the physical and chemical limits of nature.

Contemporary biotechnology began in 1973 when Herbert Boyer and Stanley Cohen developed the technology for recombinant DNA, which allows scientists to alter DNA molecules and thus artificially create new forms of life. They did this by combining a number of discoveries. Bacteria protect themselves against certain viruses through the use of restriction enzymes that cut up viral DNA at specific sequences of nucleotide bases; this allows a scientist with the right restriction enzyme to cut out a specific genetic sequence. Bacteria contain plasmids, which are small loops of DNA that can pass from one bacterium to another. This allows bacteria to develop antibiotic resistance quickly if the genes for resistance are passed by plasmids. Boyer and Cohen showed how one could use a restriction enzyme to cut out a specific genetic sequence and then glue that sequence into a bacterial plasmid. That plasmid, with its new combination of genetic sequences, could be introduced into a bacterial cell. As the bacterial cell divided, it would produce copies of the recombinant plasmid, which then could be extracted from the bacteria.

An illustration of the value of this recombinant DNA technique is provided by the production of human insulin. People with diabetes do not have enough of the protein insulin to regulate blood-sugar levels. After the 1920s diabetic patients were treated with injections of insulin extracted from pigs and cattle. This is an example of modern biotechnology. Although pig and cow insulin is very similar to human insulin, there are enough differences that some people with diabetes have had allergic reactions. Contemporary biotechnology provided a solution to the problem by using recombinant DNA techniques. The human gene for insulin was identified and then could be inserted into a bacterial cell through a plasmid so that the bacterium would produce human insulin that could be harvested for use by human patients. In 1982 human insulin produced in genetically modified bacteria became the first drug of contemporary biotechnology to be approved by the U.S. Food and Drug Administration.

Contemporary biotechnology has developed hundreds of products with agricultural, environmental, and medical benefits. Agricultural biotechnology uses reliable techniques for genetic manipulation to produce new kinds of plants and animals to provide food that is cheaper and more nutritious. Environmental biotechnology is used to design genetically modified organisms that can clean up environmental pollution by consuming toxic materials. Medical biotechnology is used to devise new drugs and vaccines and therapeutic techniques that relieve or prevent suffering, cure disease, and enhance physical and mental well-being.

Ethical Issues

Despite its many benefits, biotechnology has provoked ethical controversy in six areas of moral concern: safety, liberty, justice, environmental nature, human nature, and religious beliefs.

SAFETY. Safety is a moral concern for opponents of biotechnology who worry that its power disrupts the complex balance in living nature in ways that are likely to be harmful. Individuals such as Jeremy Rifkin (1977) and groups such as Greenpeace have warned that genetically modified crops and foods could endanger human health as well as the health of the environment. Critics of medical biotechnology fear that biotechnology medicine alters the human body and mind in radical ways that could produce harmful consequences—perhaps far into the future—in ways that are hard to foresee.

Proponents of biotechnology such as James Watson (2003) and Michael Fumento (2003) argue that its techniques are so precise and controlled that it tends to be far safer than older forms of technology. Breeders of plants and animals have genetically modified organisms for thousands of years without understanding exactly what they were doing. But biotechnology in the early 2000s provides a better understanding of and greater power over genetic mechanisms so that it is possible to minimize the risks. In fact, there is no clear evidence that any human being among the hundreds of millions who have been exposed has become sick from eating genetically modified foods. Similarly, the risks to human health from medical biotechnology can be reduced by means of careful testing and new techniques for designing drugs and therapies that are designed specifically for individual patients with unique genetic traits. Nevertheless, the history of unforeseen harm from all technologies justifies a cautious approach.

LIBERTY. Liberty is a moral concern for those who fear that biotechnology will give some people tyrannical power over others. The history of eugenics, in which governments used coercion to eliminate those judged to be biologically "unfit," illustrates the danger of encroachments on liberty. Libertarian proponents of biotechnology such as Fumento and Virginia Postrel (1998) insist that there should be no threat to liberty as long as biotechnology is chosen freely by individuals in a free market economy. But conservatives such as Leon Kass (2002) worry that people could be coerced informally by social pressure, employers, and insurance companies so that they will feel compelled to adopt biotechnology products and procedures. Moreover, Kass and others suggest that biotech can give parents the power to control the nature and behavior of their children in ways that threaten the liberty of the children.

JUSTICE. Justice is a moral concern for people who anticipate that biotechnology will be so expensive that only the richest individuals will benefit from it so that the rich will have an unjust advantage over the poor. Even proponents of biotechnology such as Lee Silver (1998) worry that reproductive biotechnology eventually could divide humanity into two separate species based on the wealth or poverty of their ancestors: the "genrich" who would be genetically designed to be superior and the "genpoor" who would be left behind as biologically inferior beings. Of course in some ways this problem is not unique to biotechnology because rich people always have unfair advantages over the poor, but the libertarian defenders of biotechnology foresee that in a free-market society prices for biotechnology products and services eventually will decline as a result of competition, and this will lessen the advantages of the rich over the poor. Similarly, critics of biotechnology argue that the rich nations of the world will benefit more from this new technology than will the poor nations, yet libertarians predict that international free trade will spread the advantages of biotechnology around the world.

ENVIRONMENTAL NATURE. Environmental nature is a moral concern for environmentalists such as Rifkin and Bill McKibben (2003). Those environmentalists predict that biotechnology will promote the replacement of the natural environment with a purely artificial world and that this will deprive human beings of healthy contact with wild nature. They also fear that introducing genetically modified organisms into the environment will produce monstrous forms of life that will threaten human beings and the natural world.

Proponents of biotechnology respond by noting that beginning with agriculture, human beings have been creating genetically modified organisms that transform the environment for thousands of years. All organisms modify their environments, sometimes with global effects. For example, the oxygen in the earth's atmosphere has been produced over billions of years by photosynthetic organisms. Biologists such as F. John Odling-Smee (2003) have called this "niche construction." So human beings are not unique in their capacity for changing their environments. Although this sometimes has produced disasters such as the extinction of plants and animals and the emergence of new disease-causing agents, people have learned to adjust to these dangers, and contemporary biotechnology provides more precise knowledge and techniques to recognize and avoid such dangers. Moreover, environmental biotechnology is developing new organisms, such as bacteria genetically engineered to metabolize toxic wastes, to restore dangerous natural environments to a condition that is safe for human beings.

HUMAN NATURE. Human nature is a moral concern for anyone who fears that biotechnology could change or even abolish human nature. Both environmentalists such as Rifkin and McKibben and conservatives such as Kass and Francis Fukuyama (2002) worry that the biotechnological transformation of human nature will produce a "posthuman" world with no place for human dignity rooted in human nature. On the other side of this debate Nick Bolstrom (2003) and others in the World Transhumanist Association welcome the prospect of using biotechnology to move toward a "transhuman" condition. More moderate proponents of biotechnology dismiss both positions for being based on exaggerated views of the power of biotechnology.

In a report by the President's Council on Bioethics (2003) Kass and other members of the council contend that biotechnology expresses a willful lack of humility in pursuing a scientific mastery of nature that carries out the modern scientific project first described by Francis Bacon. When a physician uses medical therapy to restore the health of a patient, the physician cultivates the body's natural capacity for healing to serve the natural goal of health. Such medical treatment is guided in both its means and its ends by nature. But when biotechnologists use genetic engineering or psychotropic drugs to extend human bodily or mental powers beyond their normal range, they act not as nature's servant but as nature's master because they are forcing nature to serve their own willful desires.

As an example Kass and other members of the council point to the use of psychotropic drugs such as Prozac that alter the biochemistry of the brain to elevate mood. Using such drugs to cure severely depressed patients can be justified as therapy directed toward restoring normal mental health, but their use to change human personality radically—perhaps by inducing feelings of contentment that never yield to sadness—would violate the normal range of human mental experience set by nature. The ultimate aim of such a psychopharmacological science would be a drug-dependent fantasy of happiness that would be dehumanizing. Furthermore, scientists such as David Healy (2004) have warned that any drug powerful enough to change human personality is likely to have severely harmful side effects.

The President's Council (2003) warns against the excessive pride inherent in Bacon's project for mastering nature, which assumes that nature is mere material for humans to shape to their desires. Rather, it urges people to adopt an attitude of humility and respect and treat the natural world as a "gift." To respect the "giftedness" of the natural world is to recognize that the world is given to humans as something not fully under their control and that even human powers for changing the world belong to human nature as the unchanging ground of all change (Kass 2003).

Proponents of biotechnology could respond by defending Bacon's project as combining respect for nature with power over nature. At the beginning of the *Novum Organon* Bacon observed that "nature to be commanded must be obeyed" because "all that man can do is to put together or put asunder natural bodies," and then "the rest is done by nature working within" (Bacon 1955, p. 462). Kass has used the same words in explaining how the power of biotechnology is limited by the potentialities inherent in nature (Kass 1985).

Throughout the history of biotechnology—from the ancient Mesopotamian breeders of plants and animals, to Pasteur's use of microorganisms for fermentation and vaccination, to Boyer and Cohen's techniques for gene splicing—people have employed nature's properties for the satisfaction of human desires. Boyer and Cohen did not create restriction enzymes and bacterial plasmids but discovered them as parts of living nature. They then used those natural processes to bring about outcomes, such as the production of human insulin for persons with diabetes, that would benefit human beings. Biotechnology has the ability to change nature only insofar as it conforms to the laws of nature. To command nature people must obey it.

Baconian biotechnology is thus naturally limited in its technical means because it is constrained by the potentialities of nature. It is also naturally limited in its moral ends because it is directed toward the goals set by natural human desires. Kass and the President's Council

(2003) acknowledge this by showing how biotechnology is employed to satisfy natural desires such as the desire of parents for happy children and the desire of all human beings for life and health. As they indicate, it is not enough to respect the "giftedness" of nature because some of the "gifts" of nature, such as diabetes and cancer, are undesirable. People accept some of nature's gifts and reject others on the basis of the desires inherent in human nature.

RELIGIOUS BELIEFS. To appreciate life as a gift that should elicit a feeling of humility rather than mastery is a religious emotion. Some of the moral concerns about biotechnology express the religious attitude that life is sacred and therefore the biotechnological manipulation of life shows a lack of reverence for the divinely ordained cosmic order. The biblical story of the Tower of Babel (*Genesis* 11:1–9) suggests that the human lust for technical power over the world provokes divine punishment.

In 1977 the environmentalist Jeremy Rifkin wrote a book attacking biotechnology with the title *Who Should Play God?: The Artificial Creation of Life and What It Means for the Future of the Human Race*. The title conveys the direction of his argument. The "creation of life" is proper only for God. For human beings to create life "artificially" is a blasphemous transgression of God's law that will bring punishment upon the human race. Rifkin often uses the imagery of the Frankenstein story. Like Doctor Frankenstein, biotech scientists are trying to take God's place in creating life, and the result can only be the creation of monsters. When people such as Rifkin use the phrase "playing God," they evoke a religious sense that nature is a sacred expression of God's will and therefore should not be changed by human intervention. Rifkin has said that "the resacralization of nature stands before us as the great mission of the coming age" (Rifkin 1983, p. 252).

In contrast to Rifkin, Bacon thought that regarding nature as sacred was a pagan idea contrary to biblical religion. In pagan antiquity the natural world was the sacred image of God, but the Bible teaches that God is the transcendent Creator of nature; therefore, God's mysterious will is beyond nature. Although nature declares God's power and wisdom, it does not declare the will and true worship of God. Bacon believed that true religion as based on faith in biblical revelation must be separated from true philosophy based on the rational study of nature's laws (Bacon 1955).

Some biblical theologians, such as Philip Hefner (2003) and Ted Peters (2003), have restated this Baconian claim that the biblical conception of God as the supernatural creator of nature separates the sacred and the natural and thus denies pagan pantheism. They argue that because human beings have been created in God's image and God is the Creator, human beings must share somehow in God's creativity. The Bible declares that when God made humanity in his image, this was to include "dominion" or "mastery" over all the earth, including all the animals (Genesis 1:26–28). Hefner reads the Bible as teaching that human beings are "created cocreators." As "created," humans are creatures and cannot create in the same way as God, who can create *ex nihilo*, "from nothing." However, as "cocreators" people can contribute to changes in creation. Of course, Hefner warns, people must do this as cautious and respectful stewards of God's creation, but it is not appropriate to worship nature as sacred and thus inviolable.

The theological idea of human beings as cocreators was affirmed by Pope John Paul II in his 1981 encyclical *Laborem Exercens* and criticized as a "remarkably bad idea" by the Protestant theologian Stanley Hauerwas (Houck and Williams 1983). In his 1991 encyclical *Centesimus Annus* the Pope stressed the importance of human technological knowledge in improving the conditions of life (Novak 1993).

That God transcends nature, that nature is thus not sacred, that human beings as created in God's image share in God's creative activity, that human beings have the power and the duty to master nature by artful manipulation, and that they have the moral duty to do this as an activity of charity for the improvement of human life—all the precepts Bacon drew from the Bible to support his view of the new science—have been accepted by some biblical believers. But many of those believers worry that modern science promotes an atheistic materialism that denies the dignity of human beings and of the natural world generally as God's Creation. In particular they worry about whether biotechnology expresses an unduly willful attitude toward the world as merely raw material for human manipulation and survival.

LARRY ARNHART

SEE ALSO *Bioengineering Ethics; Bioethics; Biological Weapons; Food Science and Technology; Genethics; Genetically Modified Foods; Nanoethics.*

BIBLIOGRAPHY

Arnhart, Larry. (2003). "Human Nature Is Here to Stay." *The New Atlantis* 2: 65–78. Argues that biotechnology will not abolish human nature.

Bacon, Francis. (1955). *Selected Writings of Francis Bacon*. New York: Random House.

Bud, Robert. (1993). *The Uses of Life: A History of Biotechnology*. Cambridge, UK: Cambridge University Press.

Diamond, Jared. (1997). *Guns, Germs, and Steel: The Fates of Human Societies*. New York: Norton. A history of how agricultural biotechnology has shaped human civilization.

Fukuyama, Francis. (2002). *Our Posthuman Future: Consequences of the Biotechnology Revolution*. New York: Farrar, Straus, and Giroux. A neoconservative argument for legally restricting biotechnology.

Fumento, Michael. (2003). *Bioevolution: How Biotechnology Is Changing Our World*. San Francisco: Encounter. A detailed survey of all forms of biotechnology by a libertarian proponent.

Healy, David. (2004). *Let Them Eat Prozac: The Unhealthy Relationship between the Pharmaceutical Industry and Depression*. New York: New York University Press.

Hefner, Philip. (2003). *Technology and Human Becoming*. Minneapolis, MN: Fortress Press.

Houck, John, and Oliver Williams, eds. (1983). *Co-Creation and Capitalism: John Paul II's Laborem Exercens*. Washington, DC: University Press of America.

Kass, Leon. (1985). *Toward a More Natural Science*. New York: Free Press.

Kass, Leon. (2002). *Life, Liberty, and the Defense of Dignity*. San Francisco: Encounter.

Kass, Leon, ed. (2003). *Beyond Therapy: Biotechnology and the Pursuit of Happiness: A Report of the President's Council on Bioethics*. Washington, DC: Dana Press.

McKibben, Bill. (2003). *Enough: Staying Human in an Engineered Age*. New York: Times Books. An environmentalist attack on biotechnology.

Novak, Michael. (1993). *The Catholic Ethic and the Spirit of Capitalism*. New York: Free Press.

Odling-Smee, F. John; Kevin N. Laland; and Marcus W. Feldman. (2003). *Niche Construction: The Neglected Process in Evolution*. Princeton, NJ: Princeton University Press.

Peters, Ted. (2003). *Playing God?: Genetic Determinism and Human Freedom*, 2nd ed. New York: Routledge.

Postrel, Virginia. (1998). *The Future and Its Enemies*. New York: Free Press. A libertarian defense of technological innovation as the basis for human freedom.

Rifkin, Jeremy. (1977). *Who Should Play God?: The Artificial Creation of Life and What It Means for the Future of the Human Race*. New York: Delacorte Press.

Rifkin, Jeremy. (1983). *Algeny*. New York: Viking.

Rifkin, Jeremy. (1998). *The Biotech Century: Harnessing the Gene and Remaking the World*. New York: Penguin Putnam.

Ruse, Michael, and David Castle, eds. (2002). *Genetically Modified Foods: Debating Biotechnology*. Amherst, NY: Prometheus Books. An excellent collection of statements on all sides of the ethical debate over agricultural biotechnology.

Sherlock, Richard, and John D. Morrey, eds. (2002). *Ethical Issues in Biotechnology*. Lanham, MD: Rowman and Littlefield. An excellent anthology of material on the ethics of biotechnology.

Silver, Lee. (1998). *Remaking Eden: Cloning and Beyond in a Brave New World*. New York: Avon. A libertarian celebration of the power of biotechnology for changing the world.

Watson, James D. (2003). *DNA: The Secret of Life*. New York: Knopf. A broad history of genetics and biotechnology by the codiscoverer of DNA.

INTERNET RESOURCE

Bolstrom, Nick. (2003). *The Transhumanist FAQ*. Available from http://www.transhumanism.org/resources/faq.html.

BIOTECHNOLOGY ETHICS

SEE *Biotech Ethics*.

BIRTH CONTROL

Birth control, or contraception, is the practice of preventing or reducing the probability of pregnancy without abstaining from sexual intercourse. In premodern texts references to the enhancement of fertility and birth outweigh references to their restriction, and the development of contemporary contraceptive technologies emerged from work on fertility enhancement. Today, however, one of the most common ways in which scientific and technological advances are experienced is through people's control of fertility and birth.

History of Birth Control

The desire to control fertility has always existed in tension with the desire to procreate and with social motives to preserve population sizes. Infanticide and abortifacients were used frequently in premodern and early modern societies to control the number of offspring. However, diverse contraceptive techniques also existed, including the natural rhythm method (avoiding intercourse during ovulation), coitus interruptus (withdrawal before ejaculation), coitus obstructus (using pressure to block the male urethra), and coitus reservatus (avoiding ejaculation). Other methods included suppositories such as crocodile dung in ancient Egypt, cervical barriers, and intrauterine devices (IUDs).

Neither the ancient Greeks nor the Romans consid-ereed contraception immoral. That also was the case among the Germanic, Celtic, and other non-Mediterranean peoples in much of the medieval period. It is not surprising that the Christian Church had difficulty enforcing rules and moral norms against contraception. Early Church fathers made the moral standing of sexual intercourse an important feature of their teachings. Most important, Augustine (354–430) saw the procreation of children as one of the three justifications for Christian marriage. If sexual intercourse was performed with the specific intent of engendering offspring, it was done without sin. Augustine's views influenced subsequent treatments of contraception in the Catholic Church (Dupré 1964), and certain medieval canons criminalized contraception.

Life in modern industrial societies removed the agrarian incentive to produce numerous children. Emerging individual perspectives on procreation clashed with received social norms and many religious teachings. Technological improvements in contraceptive techniques decreased their cost and increased their availability. For example, the vulcanization of rubber in the mid-nineteenth century by Charles Goodyear (1800–1860) led to the mass production of condoms, which were made from animal intestines in seventeenth-century Europe, and other birth control devices.

Although most Catholic authorities reacted with renewed criticism of contraception, several groups that were promoting birth control challenged them. For example, neo-Malthusians in England in the early nineteenth century wanted to increase the standard of living of the poor by reducing birth rates. Others argued that birth control techniques promoted greater sexual freedom or aided eugenic attempts to improve the hereditary "stock." Many women went to extreme lengths to avoid pregnancy because of the disproportionate burden it placed on them. Those efforts were made more difficult by the declining authority of midwives in the nineteenth century in favor of male doctors, many of whom did not recognize the right of women to terminate or prevent pregnancy.

By the end of the nineteenth century many people were interpreting the increasing prevalence of birth control as a sign of social decadence and moral degradation. Some people in the United States argued that women, especially upper-class women, were shirking their "patriotic duty" to have children, sinning against nature, and committing "race suicide" (Reed 1978). Anthony Comstock (1844–1915) became the most eminent crusader against the dissemination of contraception literature. In 1873 Congress passed the Comstock Act, which defined information about contraception as obscene and prohibited the dissemination of contraceptives through the mail or across state lines. Several states also banned or restricted the dissemination of contraceptives. The strictest laws were passed by Connecticut, where married couples could be arrested for using birth control.

The most common arguments against birth control were that it promoted lewd or sinful behavior, weakened the stability provided by large families, signified a rebellion by women against their primary social role of motherhood, and undermined certain racial ideals. By contrast, those in favor of birth control argued that it promoted autonomy for women, stronger families and marriages, economic equality, and environmental health.

In the early twentieth century Margaret Sanger (1879–1966), an advocate for contraceptives who coined the phrase *birth control*, attempted to increase access to birth control by using arguments based primarily on socioeconomic justifications (Reed 1978). She crusaded against the Comstock Act, beginning with the creation of a birth control clinic in 1916. Sanger popularized the image of birth control as a means of individual freedom, self-determination, and gender equality. Legislative changes slowly followed, along with the growing legitimization of birth control methods by much of society, especially the medical community. Sanger's American Birth Control League and other organizations became known as Planned Parenthood in 1942.

In the 1960s population control became a popular movement to reduce poverty and conserve natural resources. Some anthropologists argued that irresistible reproductive pressures arising from the lack of safe, effective contraception had led all past cultures into a self-destructive pattern of production intensification and environmental degradation. Modern contraceptive technologies, however, offered an opportunity to alter that perennial pattern by lowering fertility rates (Segal 2003). The new emphasis on birth control in response to concerns about the disparity between lowered death rates and continued high birth rates in the developing world was made clear in the "Proclamation of Teheran" (paragraph 16) by the 1968 International Conference on Human Rights.

In the United States anticontraceptive laws remained in effect until the U.S. Supreme Court struck down the Comstock Act as unconstitutional in 1965. Until that time most pharmaceutical companies had refrained from investing in birth control technologies because of those laws and fear of religious objections, especially from the Catholic Church. The independent development of synthetic progesterone in the early 1950s by Frank Colton, a chemist at J. D. Searle Pharmaceutical, and Carl Djerassi, working for Syntex, a pharmaceutical company based in Mexico, allowed Gregory Pincus to create what would become known as the birth control pill. That development sparked a revolution in contraception.

The pill received approval from the U.S. Food and Drug Administration (FDA) as a contraceptive in 1960 after controversial research was done on women in third world nations. Five years later more than 6.5 million U.S. women were taking oral contraceptives. In the 1970s and 1980s contraceptive technologies continued to develop, including lower-dose birth control pills (the initial doses were found to be ten times higher than the necessary amount, causing many dangerous side effects) and a T-shaped IUD. The IUD fell out of favor because it was linked to pelvic inflammatory disease. In the 1990s the FDA approved the first hormone injections and emergency contraceptives.

The twenty-first century continues to bring new contraceptive technologies, including the birth control patch, continuous birth control pills that schedule fewer menstrual cycles per year, and male birth control pills. Despite the increased use of these technologies contraception still stimulates a wide range of ethical judgments that range from mortal sin to moral imperative. It also spans the legal and policy spectrum from laws that ban birth control to those, such as the 1979 "one child per couple" policy in China, that practically mandate it.

Issues of birth control and reproductive rights remain highly controversial elements of modern politics. Hence, whereas rising rates of teenage pregnancy lead many people to applaud the greater use of birth control, others have promoted abstinence. However, there were increasing debates about the abstinence-only education programs encouraged by the administration of U.S. President George W. Bush. Many critics argued that the administration was misusing science to promote an anticontraception moral agenda (Union of Concerned Scientists 2004).

Technological Methods

Contraceptive techniques can be divided into three categories: blockage of sperm transport to the ovum, prevention of ovulation, and blockage of implantation. Both men and women can use methods in the first category, whereas those in the latter two categories are available to women only. Each technique presents different tradeoffs among variables such as comfort, price, availability, safety, and effectiveness.

BLOCKAGE OF SPERM TRANSPORT TO THE OVUM. Natural contraception, also known as the rhythm method of birth control, relies on abstinence from intercourse during a woman's fertile period. Carefully tracking menstrual cycles and/or monitoring fluctuations in body temperature can predict ovulation. Neither method is very effective (average failure rates range from twenty to thirty annual pregnancies per hundred women) because of variability in ovarian cycles. Coitus interruptus has a similar failure rate.

Other techniques in this category involve chemical contraceptives such as spermicidal foams, sponges, creams, jellies, and suppositories. When inserted into the vagina, those contraceptives can remain toxic to sperm for roughly an hour. These techniques are usually not very effective and are used mostly in conjunction with barrier methods that mechanically prevent sperm transport to the oviduct. Those methods include condoms (thin, strong rubber or latex sheaths), which are available for both male and female use. Females also can use the diaphragm, which is a flexible rubber dome positioned over the cervix. An alternative to the diaphragm is the cervical cap, which is smaller and is held in place by suction. Sterilization is a more permanent and highly effective method of birth control. It involves the surgical disruption of the ductus deferens (vasectomy) in men and the oviduct (tubal ligation) in women.

PREVENTION OF OVULATION. Oral contraceptives, or birth control pills, function by manipulating the complex hormonal interactions in the ovarian cycle. They contain synthetic estrogen-like and progesterone-like steroids and are taken for three weeks and then discontinued for one week. The steroids inhibit the secretion of certain hormones, preventing follicle maturation and ovulation. The one-week period of discontinuation allows menstruation to occur, although without the presence of an ovum. Recent developments prolong the length of the menstrual cycle and thus can reduce the annual number of menstruations. Oral contraceptives also prevent pregnancy by increasing the viscosity of cervical mucus, making the uterus less likely to accept implantation, and decreasing muscular contractions in the female reproductive tract.

Birth control patches also have been developed. They are applied directly to the skin and secrete synthetic steroids that work in the same way as do those in the contraceptive pill. Also available are long-acting subcutaneous contraceptives such as Norplant®. Norplant® consists of six matchstick-size capsules that gradually release progestin. The patches are inserted under the skin in the inner arm above the elbow. Once implanted, these contraceptives are effective for roughly five years. Additionally, injectable time-release synthetic hormones, which provide contraceptive effects for one to three months depending on the product, can be obtained. In the United States all these methods are available only with a prescription and are quite effective, with average failure rates of less than one annual pregnancy per hundred women.

BLOCKAGE OF IMPLANTATION. These are the most controversial techniques because they act after fertilization has taken place by preventing the implantation of a fertilized ovum in the uterus. The most common technique in this category is the IUD, which is inserted into the uterus by a physician. The mechanism of action of the IUD is not completely understood, but evidence suggests that the presence of this foreign object in the uterus produces a local inflammatory response that prevents implantation of the fertilized ovum. Early IUD techniques were associated with serious complications. More recent methods are much safer, but the popularity of IUDs has waned.

Implantation also can be blocked by emergency contraception, or "morning-after" pills. These pills can prevent pregnancy when they are taken within seventy-two hours after intercourse. Often used in the case of rape, emergency contraceptive kits usually involve high doses of hormones that either suppress ovulation or cause premature degeneration of the corpus luteum. The latter effect removes the hormonal and nutritive support required by a fertilized ovum. The controversial "abortion pill" RU 486 (Mifepristone®) blocks the female hormone progesterone, making it impossible for the body to sustain a pregnancy. The association of this pill with abortion explains why it took twenty years after its invention in 1980 by a French pharmaceutical company for the FDA to approve it in 2000.

CURRENT RESEARCH. Research continues in all these categories, partly because unplanned pregnancies continue to present personal and public health problems (Institute of Medicine 2004). Advances in genome sequencing, materials science (a multidisciplinary field focused on the properties of functional solids), and drug delivery are important factors in new techniques. Longer-lasting hormone-releasing IUDs are being developed along with improved methods for inserting and removing them. Other techniques target chemical reactions between ova and sperm or manipulate the pituitary secretion of certain reproductive hormones in both males and females.

In 2005 researchers in the United States partnered with a European biotechnology company to develop a male contraceptive pill. Such contraceptives could be based on a variety of techniques, ranging from inhibiting spermatogenesis to disabling the motility of sperm. Research involving reversible chemical sterilization also is being carried out.

Additionally, efforts are under way to develop immunocontraception that would allow the use of vaccines that prod the immune system to produce antibodies targeted against a protein that is critical to the reproductive process (Ada and Griffin 1991). Such vaccines would work for both males and females. In males vaccines would create antibodies against the production of gonadotropin-releasing hormone (GnRH), which is essential for sperm production. In this case supplemental testosterone injections would be needed because of the loss of GnRH. In females some vaccines that are being tested induce the formation of antibodies against the creation of human chorionic gonadotropin (hCG), which is essential for supporting the corpus luteum during pregnancy. These techniques present concerns about endocrine disruption and autoimmune pathologies. Immunocontraception is fairly commonly used as a strategy for the control of wildlife populations. Although research on human applications has proceeded since a special working group was formed by the World Health Organization (WHO) in 1973, no safe and effective methods had been developed by 2004. Clinical trials continue.

Ethical and Political Issues

The association of contraceptive practices with prostitution, extramarital affairs, and the perceived breakdown of sexual mores is related directly to the discomfort with which most religious traditions have responded to these methods. Today, however, most laypeople, along with most scholars in different traditions, accept the morality of contraception within marriage. However, that acceptance has not extended to all religious traditions.

The clearest example of continuous opposition to the use of artificial birth control methods comes from official Roman Catholic teachings. Catholic teachings on contraception remain important for contemporary debates, especially the 1930 encyclical issued by Pope Pius XI titled *Casti Cannubii* [On Christian Marriage], which called birth control a sin and opposed birth control by artificial means. In 1968 Pope Paul VI condemned contraception but permitted the use of natural rhythm methods. Today, although Catholic doctrine still advocates the use of natural methods such as abstinence during fertile periods, it completely condemns the use of artificial contraception or voluntary sterilization. The grounds for this rejection are related to what is claimed to be an inseparable connection between the sexual and procreative acts. Because many developing countries have large Catholic communities, many have criticized the official position of the Catholic Church as insensitive to overpopulation problems and to the effects of continuous childbearing on the well-being of women and children. The spread of HIV and AIDS in many developing countries has provided an important reason for criticizing Catholic opposition to methods that can be effective in preventing the spread of a deadly disease.

In spite of Catholic opposition to artificial contraception many other Christian churches have become more accepting of the role of birth control within marriage. In most cases the reasons for that openness are related to the consequences unlimited procreation can have on a marriage, other children, or the community in general. For many Christian denominations the use of both natural and artificial contraceptives methods is a way to express responsible parenthood. Other religion traditions, such as Islam, Orthodox Judaism, and Hinduism, also accept the morality of contraception as long as it is not harmful to the persons involved. Islamic teachings, for example, historically have been fairly tolerant of contraception. That allowed discussion and development of birth-control techniques by medieval Arabic writers, including the Muslim physician Ibn Sina (980–1037). The Jewish tradition also tends to support birth control, although with many qualifications, and makes it primarily the responsibility of women (Feldman 1968).

Feminists' attitudes toward artificial birth control methods are, as with many other reproductive technologies, ambivalent. On the one hand, contraception has freed women from unlimited reproduction, facilitated their incorporation into the labor force, and allowed them to make autonomous choices about whether and when to have children and about how many of them to bring into the world. On the other hand, birth control methods are developed, implemented, and used in the context of patriarchal societies that still are involved in controlling women's lives and in many cases continue to show little interest for women's well-being.

In this context the fact that most contraceptive methods have been developed for women is a matter of concern, especially because women rarely have been involved in making decisions about what technologies to develop. Also a matter of feminist concern is the fact that many contraceptive methods, such as those involving hormones, appear to have been developed with more interest in their efficacy than in their safety. Similarly, although male reproductive biology seems to be more difficult to interrupt, it appears that part of the scarcity of research in that area can be attributed to fear of affecting the male libido, a concern that has not affected research on female contraception.

Many feminists have objected to the testing of new contraceptives on women in developing countries and have expressed worries about possible social abuses in both industrialized and nonindustrialized nations arising from the use of long-acting implantable contraceptives such as Norplant®. Once implanted, Norplant® can be removed only surgically. That makes this contraceptive far more effective than many others in which compliance can be a problem. These worries are not easily dismissible in light of the fact that in the United States, for example, several state legislatures have considered regulations that would pay women on welfare to use Norplant®. Some judges have imposed the use of this drug as an alternative to a lengthy prison sentence for women convicted of child abuse.

In developing countries the likelihood of abuses resulting from the use of this type of contraceptive is even more obvious. Powerful population control interests can result in subtly or clearly coercive methods to assure women's use of birth control. The fact that Norplant® requires surgery, together with the scarcity of health care resources, makes concerns about the possibility of coercion even more pressing.

Also feeding feminists' worries about possible abuses of birth control methods were attempts by members of eugenics movements in the early twentieth century to control the reproductive activities of those considered undesirable. In most cases involuntary sterilization was the method of choice to prevent those with mental problems, criminals, immigrants, and poor and minority women from reproducing under the idea that if they were not stopped, lower-class offspring would outnumber the upper classes' progeny.

New demographic trends such as below-replacement birth rates in some European nations, together with what appears to be an environmentally caused decline in fertility among both men and women in industrialized countries, may put discussions of birth control in a different framework in the future, especially in nations with strong welfare systems. In those nations the aging population has been putting a serious strain on public resources. In this context some might argue for the need to encourage births rather than control them.

INMACULADA DE MELO-MARTÍN
ADAM BRIGGLE

SEE ALSO *Bioethcs; Eugenics; Population.*

BIBLIOGRAPHY

Ada, G. L., and P. D. Griffin, eds. (1991). *Symposium on Assessing the Safety and Efficacy of Vaccines to Regulate Fertility.* Cambridge, UK, and New York: Cambridge University Press. Sixteen essays collected by the Special Program of Research Development and Research Training in Human Reproduction at the World Health Organization.

Bullough, Vern. L., ed. (2001). *Encyclopedia of Birth Control.* Santa Barbara, CA: ABC-CLIO.

Dupré, Louis. (1964). *Contraception and Catholics: A New Appraisal.* Baltimore: Helicon. Examines the arguments that the Catholic Church has proposed against the use of artificial birth control.

Feldman, David M. (1968). *Marital Relations, Birth Control, and Abortion in Jewish Law.* New York: Schocken.

Holmes, Helen; Betty Hoskins; and Michael Gross, eds. (1980). *Birth Control and Controlling Birth: Women-Centered Perspectives.* Clifton, NJ: Humana Press.

Institute of Medicine. (2004). *New Frontiers in Contraceptive Research: A Blueprint for Action.* Washington, DC: National Academies Press. Provides recommendations on how science can best address concerns of fertility regulation and reproductive health.

Moskowitz, Ellen, and Bruce Jennings, eds. (1996). *Coerced Contraception? Moral and Policy Challenges of Long-Acting Birth Control.* Washington, DC: Georgetown University Press.

Reed, James. (1978). *From Private Vice to Public Virtue: The Birth Control Movement and American Society since 1830.* New York: Basic Books. Focuses on the strategies and efforts of a small group of U.S. citizens in the birth control movement.

Segal, Sheldon J. (2003). *Under the Banyan Tree: A Population Scientist's Odyssey.* New York: Oxford University Press. Stresses the importance of education in achieving population control and poverty reduction and chronicles personal experiences with international family-planning efforts to increase understanding about population issues.

INTERNET RESOURCE

Union of Concerned Scientists. (2004). *Scientific Integrity in Policy Making: An Investigation of the Bush Administration's Misuse of Science.* Cambridge, MA: Author. Available at http://www.ucsusa.org/documents/RSI_final_fullreport.pdf.

BLACKETT, PATRICK

Patrick Maynard Stuart Blackett (1897–1974) was born in Kensington, London, on November 18, and became a Nobel Prize–winning physicist who at once promoted scientific research to defeat Nazism and criticized the World War II Allied bombing of cities. After serving in the Royal Navy during World War I and establishing a successful career in physics, he became a science advisor on military matters during World War II and later to both the Indian and British governments on science and technology policy. He died in London on July 13, as a leading figure in the British scientific community and a defender of science in the service of socialist political ideals and of "small science" practiced independent of large government grants.

Physics

After earning a Ph.D. in physics in 1921 from Cambridge University, Blackett did postdoctoral work in the Cavendish Laboratory and was appointed professor at the University of Manchester in 1937. He developed an international reputation for masterful experimental work in cosmic-ray and particle physics using cloud chambers, Geiger counters, and magnetic fields. He also made important contributions to the study of nuclear transformations as the first to photograph the mutation of one element into another (nitrogen into oxygen after bombardment by an alpha particle) and matter arising out of energy (electrons and positrons from gamma rays). In 1933 Blackett and the Italian physicist Giuseppe Occhialini confirmed the existence of the positively charged electron or positron, but were cautious in publishing the results.

When the 1936 Nobel Prize in physics was awarded to the American scientist Carl Anderson for the discovery of the positron, many argued that Blackett deserved equal credit. But Blackett himself never engaged in disputes on this issue and emphasized instead the importance of Anderson's work. Such conduct highlighted his integrity and collegiality in the scientific community as well as his cautious and disciplined style of research. He subsequently received the 1948 Nobel Prize in Physics "for his development of the Wilson cloud chamber method, and his discoveries made therewith in the fields of nuclear physics and cosmic radiation."

Blackett began defense related research even before the outbreak of World War II by helping build an air defense network through the establishment of radar stations and antisubmarine research for the Royal Navy.

He was central to the development of operations research, which for him meant the analysis of data in such a way as to provide advice to military and political decision makers.

After the war, Blackett returned to Manchester where he took up research on the origins of interstellar magnetic fields and those of Earth. When his hypothesis that the magnetic fields of large bodies were a fundamental property of their rotating mass failed to be supported by the evidence, he readily acknowledged his error. Blackett later researched the magnetism of rocks and continental drift. In 1953 he was appointed head of the Physics Department at the Imperial College of Science and Technology in London. In addition to his focus on integrity and patience in research, he crafted his laboratories according to the ideal of small science performed with modest-sized instruments, which ran contrary to the postwar practice of "big physics" with massive instruments. He ended his career serving as the official representative of the British scientific community as president of the Royal Society from 1965 to 1970.

Ethics and Politics

Although there is unanimous agreement on Blackett's contributions to physics, his engagement in public affairs caused controversy concerning the proper role of scientists in politics and tensions between the ideals of science as objectively removed from society and science as a means to serve or even shape societal goals. Blackett's life is a study in "how (and at what price) one can reconcile a scientific career with political activism" (McCray 2005, p. 186). Most mainstream scientists emphasized the freedoms that allowed for scientific autonomy. But fueled by his belief that science can provide societal benefits by being more thoroughly integrated with politics, Blackett spoke out for more government investment in science, greater science education, and tighter links between science and industry. For his biographer, Mary Jo Nye, "Achieving these aims required cultivating popular interest in science and taking on the role of public scientist, no matter how uncomfortable or inconvenient this role might become" (2004, p. 6). As he grew older, Blackett devoted more and more time to political matters.

He maintained that the best relationship between knowledge and governance would unfold under socialism, and he allied himself with the scientists for social responsibility movement (known as "Bernalism" in Great Britain) that held that a scientifically oriented socialism could solve economic and political troubles. Blackett's career showed how the external ethics of science relates not only to questions of scientists' responsibilities for applications of their work, but also to larger questions about scientists' roles in shaping public policies more generally.

Blackett was not a pacifist and argued that it was the duty of scientists to engage early in the war efforts to defeat Germany. He was one of the pioneers in the newly emerging role of scientists as advisors to political and military decision makers, choosing both to perform scientific work in support of the war and to join the forum of political debates about the war. He criticized the Allied wartime civilian bombing strategies as both immoral and ineffective. It dehumanized victims and perpetrators, and led to postwar atomic policies, which seemed to countenance further brutalization as a normal course of political and military policy.

An early proponent of international control of atomic energy, Blackett opposed British development of atomic weapons, favored a neutralist foreign policy and greater cooperation with the Soviet Union, and proposed bilateral disarmament strategies for both atomic and conventional weapons. He also found the application of game theory to nuclear war scenarios morally repugnant and another sign of the dehumanizing consequences of weapons of mass destruction. His views ran contrary to mainstream attitudes and were often dismissed as dangerous because of his sympathy toward the Soviet Union and participation in socialist organizations such as the World Federation of Scientific Workers.

Blackett published his unpopular and contentious criticisms of U.S. and British policies in *Military and Political Consequences of Atomic Energy* (1948), which appeared in the United States under the title *Fear, War, and the Bomb* (1949). Most controversial was his notion that the bombing of Hiroshima and Nagasaki were the first acts of the Cold War, carried out to intimidate the Soviet Union. Many critics attacked Blackett's expertise and legitimacy to discuss matters of politics, arguing that he misused his prestige as a scientist to bolster a political agenda. But attitudes changed over the following decade, and Blackett's *Studies of War* (1962), which presented the same basic argument as his earlier publications, received praise from scientists as well as politicians.

Blackett was later instrumental in the development of the Ministry of Technology (serving as its advisor from 1964 to 1969) and more general science and technology policies for the British government. He also advised the Indian government on research and development strategies, especially for the military. Jawaharlal Nehru, the first prime minister of independent India, and Blackett agreed that modern science and technology were crucial for the future of India, and that atomic weapons should be banned but atomic energy should be used for electricity generation in developing countries. Blackett favored applied research in developing countries (based on technology transfers from the West) rather than the development of basic research institutions. This recommendation was widely attacked as a form of outdated colonial prejudice (Nye 2004).

Along with other prominent scientists in the post–World War II era, he helped forge a new identity of the twentieth-century scientist as public citizen (Nye 2004). This identity remains controversial as modern science and technology continue to influence so many facets of life. Blackett's career serves as a sounding board to explore important questions about the role of scientists in politics and the nature of their social responsibilities.

ADAM BRIGGLE

SEE ALSO *Bernal, J. D.; Military Ethics; Operations Research.*

BIBLIOGRAPHY

Blackett, Patrick. (1949). *Fear, War, and the Bomb: Military and Political Consequences of Atomic Energy.* New York: McGraw-Hill. First published in 1948 in London by Turnstile Press. Critiques U.S. and British atomic energy policies after World War II as militarily unrealistic and politically dangerous and concludes with his recommendations.

Blackett, Patrick. (1962). *Studies of War: Nuclear and Conventional.* New York: Hill and Wang.

McCray, Patrick. (2005). "Political Science." *American Scientist* 93(2): 186–187. A review of the Nye biography.

Nye, Mary Jo. (2004). *Blackett: Physics, War, and Politics in the Twentieth Century.* Cambridge, MA: Harvard University Press. The most comprehensive source on Blackett's life, physics, and politics.

BODY

• • •

The role of the body in relation to science and technology is complex. The first book on the philosophy of technology, Ernst Kapp's *Grundlinien Einer Philosophie der Technik* (1877), analyzes technologies in terms of body parts and organs. Stoves are technological "stomachs," machines are extensions of "arms and legs," and so forth. In contemporary times bodies and embodiment have become increasingly important. There is a great deal of discussion about "posthuman" and disembodied development with respect to "cyberspace" and electronic systems of communication such as the Internet and other virtual processes. Ironically, this discussion has brought the role of human bodies back into consideration.

From Ancients to Moderns

Twenty-first century discussions echo much older traditions with respect to the human body. Ancient Greek philosophers often distinguished between body, soul, and spirit (Plato), with the strongest distinction being made between the materiality of the body and the immateriality of soul and spirit. In early modernity those distinctions were simplified into variations on a body-mind dualism (René Descartes) that continue to motivate much philosophical debate.

In antiquity religio-ethical ideas also were associated with the distinction between bodily materiality and soul-spirit immateriality. Generally speaking, materiality was conceived of as being of lesser worth, clearly finite and mortal and perhaps evil. Whether merely restrictive of the higher tendencies of the immaterial soul-spirit or deceptive and actively negative, the materiality of body carried negative associations. The Platonist trajectory emphasized a learning process that involved movement from a kind of captivity in the body, its deceptive senses, and the "body as prison" to an ascent toward the ideal realms of the good, the true, and the beautiful.

The early modern simplification of that trajectory weakened the ancient religio-ethical associations and replaced tainted materiality with the "mechanical" as the interpretation of the body. Body becomes the mechanical means by which motion is possible, but mind is enclosed "inside" a body object as a subject aware only of its impressions, sensations, or ideas caused by things that are external to itself. The model for this notion of body, used by the philosophers Descartes (1596–1650) and John Locke (1632–1704), was the *camera obscura*, in which the body was a dark room inside of which was the subject or mind that could view the images or representations cast on the *tabula rasa* inside. In this formulation the mind was situated inside a mechanical contrivance and could know or experience only its own sensations or representations.

Later modernity began to develop two less dualistic concepts of the body. One direction was physicalist and attempted to reduce all mental phenomena to physical ones (Ryle 1949) and the other was existentialist, using phenomenology to analyze a "lived body" or experiential body (Husserl 1970, Merleau-Ponty 1962). Both schools of thought lessen or deny a body-mind distinction and drive the analysis toward oneself *as* body. However, physicalism retains a basically mechanistic view of body, whereas phenomenology elevates bodily experience to include materiality. In the phenomenological sense all intelligent behavior presupposes bodily activity.

Body in Science

With the rise of early modern science the role of body began to take on a different significance. After the seventeenth century science was both technological and observational; those dimensions usually were termed experimental: Science practice included devices both for measurements and for making new discoveries achieved through (perceptual) observations displayable.

Galileo Galilei's (1564–1652) optics—telescopes but also microscopes—were the means by which new celestial phenomena were sighted, inclined planes were used to measure acceleration, and experiments were developed as proofs of specific scientific insights. Through the use of the telescope sun spots, Jupiter's satellites, the phases of Venus, and the mountains of the moon became new phenomena for emergent science. However, the instrumental means were also those which mediated perceptions, in this case vision. Although as a scientist Galileo paid little attention to the body itself, he did proclaim the new vision made possible by the telescope to be superior to that of the body by itself. Scientific vision was enhanced vision, but it also was mediated by means of instruments.

The body in this sense remained a background phenomenon but one that nevertheless had to be taken account of. In contemporary science this is even more important. For example, in contemporary technologized observations, only since the twentieth century have imaging technologies been able to present phenomena that lay far beyond the limits of unmediated human perception. In astronomy wave frequencies ranging from gamma waves to radio waves can be imaged, whereas until the twentieth century only optical light imaging was available.

In the early twenty-first century, however, all such imaging must implicitly take account of human perception insofar as "false color" imaging, the transformation of data into images, and simulations and modeling with computerized tomography all produce visualizations that translate data into visual gestalts that are available for human perception. The body is thus the background referential focus for science imaging. Increasingly, philosophers of science have begun to take visualizations into account (Galison 1997, Ihde 1998).

Bodies in Technology

The role of body with respect to technologies is even more ancient. When Kapp analogized technologies by using organ and body-part metaphors, he was drawing on a much older convergence of body roles. The Medieval thinker Roger Bacon (1220–1292) began to imagine machines that could fly, go under water, and be protected with armor from arrows and missiles; those fantasy machines were visualized much later in Leonardo da Vinci's technical drawings. Many of those imaginary machines utilized amplified human bodily powers (and thus could not actually work) because engines and motors had not yet been invented. However, those fantasy machines also reflected a new attitude toward bodily work. Those which could work on the basis of ancient physics—the simple machines of screw, wedge, levers, and pulleys—did magnify bodily powers, and with that magnification one could do more than unaided bodies could.

As the historian Lynn White, Jr., pointed out, by medieval times technologies such as cranes, lifting devices, gears, and above all mechanical clocks had begun to transform what was possible through machine-aided work. Windmills pumped out the lowlands of Holland and cathedrals of astonishing heights were built with weight-lifting machinery that magnified human bodily power, but more powerful animal bodies also were enlisted. One can still see the large drum-powered lifting device in Mount Saint Michel, which used donkeys to make it rotate. Later still came the artificial engine that launched yet another revolution: the steam engine.

Here, as in science, the measure of the human body, extended technologically, lay in the background. Machines now produced work, leaving the felt sense of effort and power on the sidelines. The previous multiplication of powers through the use of slaves could take a different direction through the use of technologies. In this case the ethics related to bodies is a social-political ethics. From slavery to the working class, bodies are embedded in work practices that are mediated by technologies. Clocks were used to regulate social time, and the panopticon was used to regulate prison behavior (Foucault 1977).

Body in Medicine

In yet another dimension bodies play other roles, particularly in medical practices. Here the interplay between bodies as objects and subject bodies often becomes focal. Historically, as with early modern science, medical practice underwent significant changes precisely by displaying the body as object, particularly as *visualizable* object. Leonardo da Vinci (1452–1519), later followed by Andreas Vesalius (1514–1564), depicted bodies as visualized objects. Dissections and autopsies became favorite matters for those depictions. Corpses showed bodily biological structures. That knowledge could be used indirectly to treat living bodies. However, the delicate problem that led to technological trajectories involved finding a way to observe what was going on physiologically without destroying or making into a dead object a living body that was under investigation.

One can trace the history of changes in diagnostic techniques, beginning with direct hands-on examinations, which were late to arrive in modernity (eighteenth century), proceeding to perceptual mediating instruments such as the stethoscope, which produced ausculatory imaging through sound (nineteenth century), and ending with contemporary largely visual imaging (from X-rays to magnetic resonance imaging and positron emission tomography scans).

This trajectory culminates in techniques that are used to display the internality of the body without using a physically invasive process. The preservation of health within this trajectory is one that recognizes that only a subject, or lived body, is the ethical object of therapeutic medical practice. The ethical considerations in this case involve the need to evaluate and preserve levels of healthfulness through the application of knowledge. However, respect and care for living bodies remains the implicit central focus.

In addition to the changing notions of the human body noted above, contemporary studies related to feminism are of importance. In early modern science visualism was prominent. Feminists have joined phenomenologists in taking account of perspectivalism and situatedness. Some authors, however, also have pointed out that observation not only is objectivistic but may include aspects from the human biological heritage; even scientific curiosity may harbor a *predatory* dimension (Haraway 1991). Moreover, vision may entail gendered differentiations, with the "male gaze" being a form of perception that is constructed differently from those found in other human gendered practices (Bordo 2004, Butler 1999). Here the questions of gender relations with associated questions of mutual respect and interpersonal relations move to the forefront of ethical concerns.

Response

Returning to the topic of technologies and human bodies, with the massive impact of transportation, information, and imaging technologies it becomes obvious that what is often a background role for bodies takes on more explicit form in the uses of those technologies.

The bodily-perceptual experiences of space-time transformations are perhaps the most dramatic. In science imaging the near distance of observation, made ordinary with the close-up imaging of Mars and Saturn, has changed the sense of "apparent distance," providing a near distance to those planetary bodies. In medicine the development of distance surgery that calls for eye-hand coordination using robotics and visualizations has changed the way in which bodily skills are utilized and thus has implicated body-technology relations. Even in debates about artificial intelligence and related neurological studies the role of bodily motility has become a prominent issue, one that also is related to contemporary robotics studies (Dreyfus 1992). With electronic and virtual communications the role of the human body has taken on yet different experiential qualities. Experiments with virtual reality equipment and later with augmented reality equipment have made the role of whole body movement, balance, and kinesthesia newly important so that cognitive science has become aware of how action is experienced at a distance through prostheses and other material extensions of technologies.

The overall result has been a renewed emphasis on studies of the body. Many disciplines show this, including philosophy, women's studies, cognitive sciences, and robotics, as well as new forms of sociology, anthropology, and cultural studies.

DON IHDE

SEE ALSO *Bioethics; Cosmetics; Phenomenology; Virtual Reality.*

BIBLIOGRAPHY

Bordo, Susan. (2004). *Unbearable Weight: Feminism, Western Culture, and the Body*. Berkeley: University of California Press. Bordo's work draws from phenomenology and is sensitive to both sensory and social aspects of bodily experience.

Butler, Judith. (1999). *Gender Trouble: Feminism and the Subversion of Identity*. New York: Routledge. Butler is a radical thinker who has been able to introduce gender construction into many philosophical debates.

Dreyfus, Hubert. (1992). *What Computers Still Can't Do: A Critique of Artificial Reason*. Cambridge, MA: MIT Press.

Dreyfus's work has been particularly important for a current generation of computer designers interested in bodily motion and robotics.

Foucault, Michel. (1977). *Discipine and Punish: The Birth of the Prison*, trans. Alan Sheridan. New York: Pantheon. Foucault has been a primary author influencing those who would understand the institutional and social dimensions of discourses.

Galison, Peter. (1997). *Image and Logic: A Material Culture of Microphysics*. Chicago: University of Chicago Press. Galison's work in the history and philosophy of science is sensitive to the instrumentation and technologies of science, as well as its contemporary interdisciplinary social structures.

Haraway, Donna. (1991). *Simians, Cyborgs and Women: The Reinvention of Nature*. New York: Routledge. Haraway, whose background was originally in biology, has become a foremost "technoscience" thinker. Her figure of the "cyborg" has inspired a very large number of thinkers.

Husserl, Edmund. (1970). *The Crisis in European Science and Transcendental Phenomenology: An Introduction to Phenomenonological Philosophy*, trans. David Carr. Evanston, IL: Northwestern University Press. Husserl's last major work became the inspiration for a number of contemporary styles of thought, including what became known as the "social constructionist" movements.

Ihde, Don. (1998). *Expanding Hermeneutics: Visualism in Science*. Evanston, IL: Northwestern University Press. This book begins a reframing of philosophy of science in hermeneutic terms by analyzing the "visual hermeneutics" employed in science imaging.

Kapp, Friedrich. (1877). *Grundlinien Einer Philosophie der Technik: Zur Ensehungsgeschichte der Cultur au neuen Gesichtspunkten*. [Foundations of a philosophy of technology]. Braunschweig, Germany: G. Westermann. The first major philosophy book to use the term, philosophy of technology.

Merleau-Ponty, Maurice. (1962). *Phenomenology of Perception*, trans. Colin Smith. New York: Humanities Press. Merleau-Ponty's best known book which emphasizes the role of perception within bodily activity, taken as the basis for all intelligent behavior.

Mumford, Lewis. (1962). *Technics and Civilization*. New York: Harcourt, Brace. Mumford was an early twentieth-century intellectual historian, one of the few who focused upon the role and development of technologies in relation to human history.

Ryle, Gilbert. (1949). *The Concept of Mind*. London and New York: Hutchinson's University Library. This is one of the first books to attack the doctrine of dualism as proposed by Descartes; it was influential on later analytic philosophy and the physicalist directions taken.

White, Lynn, Jr. (1962). *Medieval Technology and Social Change*. Oxford: Clarendon Press. This book contended that there was a massive "technological revolution" in Europe that preceded the Renaissance and early modernity.

BOMBS

SEE *Atomic Bomb; Weapons of Mass Destruction.*

BOVINE SOMATOTROPIN

SEE *Agricultural Ethics.*

BOYLE, ROBERT

• • •

Born at Lismore Castle in Munster, Ireland, on January 25, 1627, Robert Boyle (1627–1691) was an experimentalist who made fundamental contributions to chemistry, hydrostatics, philosophy of science, and the relationship between science and religion, including morality and natural theology. Before he penned his first work on natural philosophy, the deeply pious Boyle wrote several essays and treatises on religious themes, and his early interests in morality, theology, and casuistry remained undiminished throughout his life. In some of his most important mature works he linked his religious interests explicitly with his scientific pursuits, but implicit connections are often just beneath the surface in many of his writings.

The intensity of Boyle's interest in moral philosophy is readily seen in his earliest treatise, the *Aretology*, an unpublished work on ethics, vocation, and self-knowledge. This work reflects influences from Aristotle and the Christian humanist tradition, especially the German theologian Johann Alsted (1588–1638), whose enormous *Encyclopedia* (1630) served Boyle as a quarry to mine. Boyle's first published essay was dedicated to Samuel Hartlib (1600–1662), a Prussian-born disciple of the Czech educational reformer Johann Comenius (1592–1670). Its theme—that physicians should disavow secrecy and openly disseminate recipes for effective medicines, as an act of Christian charity—would be repeated numerous times in other works. An ethical impulse to improve the human condition through the application of chemistry to medicine motivated Boyle, as much as anything else, to become a scientist. A further motivation came from his conviction that nature was the third divine book in the human library—scripture and conscience were the others. The study of nature was divinely mandated, and the knowledge it produced would point unambiguously to the creator.

The Bible and the Christian Experiment

No influence was more important, however, than the Bible. Although he was not a Puritan himself, Boyle

Robert Boyle, 1627–1691. A chemist, physicist, and natural philosopher, Robert Boyle was a leading advocate of "corpuscular philosophy." He made important contributions to chemistry, pneumatics, and the theory of matter. *(The Library of Congress.)*

sought before all else to be biblical in everything he did, like the Puritan divines he counted among his friends. His devotion to the Bible, which he read daily in Hebrew and Greek, was nothing short of profound. At the urging of biblical scholar James Ussher (1581–1656), Boyle wrote *Some Considerations Touching the Style of the Holy Scriptures* (1661), in which he rejected the claims of courtly "wits" that biblical language was too poorly chosen for a divinely authored book. He also rejected courtly mores, which promoted the sinful vices of vanity, promiscuity, and greed, rather than the biblical virtues of humility, chastity, and charity.

Boyle sought to bring such virtues not only to his private life as an anonymous giver of alms, but also to his public life as the leading English natural philosopher of his generation. His stated policy was "to speak of Persons with Civility, though of Things with Freedom," instead of "railing at a man's Person, or wrangling about his Words," for "such a quarrelsome and injurious way of writing does very much mis-become both a Philosopher and a Christian" (Hunter and Davis 1999–2000, Vol. 2, p. 26). In an age known for the strongly negative tone of its scientific controversies, Boyle was remarkable for his consistent avoidance of derision. In his last major theological work, *The Christian Virtuoso* (1690–1691 and 1744), he reflected on other ways in which Christianity mirrored the moral attitude and experience of the scientist (the *virtuoso*). Living the Christian life, he argued, is like "trying an experiment" that leads to personal peace and happiness, in this world as well as in the world to come. Just as "personal experience" could show the evil consequences of "a vicious course of life," so the same experience could "assure him of the practical possibility of performing the duties and functions of a Christian." Likewise, "heedful observations" would "satisfy a man of the vanity of the world, and the transitoriness of… sinful engagements, and of the emptiness of those things, for which men refuse the ways of piety and virtue" (Hunter and Davis 1999–2000, Vol. 12, pp. 431–432). The Christian virtuoso, Boyle claimed, would put truth over personal gain; cultivate humility, generosity, and trustworthiness; promote open communication over secrecy (as far as possible, given his vital interest in alchemy); and show devotion to scientific work as a kind of religious vocation. In short, it is no accident that Boyle considered himself a "priest" in the "temple" of nature.

The Mechanical Philosophy and Natural Theology

Although Boyle often spoke of nature as a temple, his favorite metaphor was much more impersonal. The world was "a great piece of Clock-work" (Hunter and Davis 1999–2000, Vol. 8, p. 75), containing numerous smaller engines—the bodies of animals, sometimes likened to "watches," and of humans—with God the clockmaker. By the mid-seventeenth century, artisans could build and repair a great variety of clockwork mechanisms that were capable of following the motion of the heavens and imitating the motions of animals and humans. This encouraged natural philosophers to think that the universe and its parts could best be explained in terms of matter and motion, giving rise to what Boyle himself first called the *mechanical philosophy*. Although he saw the possibility that some would have the great clockwork run on its own, without divine involvement or supervision, Boyle nevertheless found the new mechanical science theologically superior to the prevailing Aristotelian concept of nature. His subtle book on the doctrine of creation, *A Free Enquiry Into the Vulgarly Received Notion of Nature* (1686), argued that the *vulgar* (i.e., commonplace) view was idolatrous for the way in which it personified nature—for example, *nature abhors a vacuum*, or *nature does nothing in vain*—effectively placing an intelligent, purposive agent, "much like a kind of 'Goddess'" (Hunter and Davis

1999–2000, Vol. 10, p. 456) between the creator and the creation. It was far more appropriate, Boyle believed, to explain phenomena in terms of impersonal, "mechanical" properties and powers created by a personal God. In this way the sovereignty of God would be underscored—and people would be more likely to worship their creator, the real source of intelligence and purpose in nature.

For Boyle, as for many of his contemporaries, science had a central religious function: to make plain the signature of God in creation. Echoing his own lifelong struggle with religious doubt, Boyle saw the design argument, especially but not exclusively in its biological form, as a powerful foil against unbelief. He did not seek merely to confute philosophical atheism, which he realized was rare in his day, but fully to persuade people of the existence of the divine creator and legislator, that they might thereby live piously in the full sight of God. Changed lives and hearts, not just changed minds, were his goal. Here again, the Christian virtuoso had much to contribute. It is "very probable," Boyle noted, "that the world was *made*, to manifest the existence, and display the attributes of God; who, on this supposition, may be said to have made the world for the same purpose, for which the pious philosopher studies it" (Davis and Hunter 1999–2000, Vol. 12, p. 483). In keeping with this attitude, Boyle left funds in his will to establish a lectureship for "proveing the Christian Religion against notorious Infidels [and] Atheists," including even Jews and Muslims, although lecturers were expressly forbidden from discussing "any Controversies that are among Christians themselves" (Madison 1969, p. 274). Ultimately, however, Boyle believed that the best evidence for the truth of Christianity came not from the testimony of nature, but from the testimony of those who had witnessed the miracles of Jesus and his disciples. Through the eyes of the biblical authors one could have a trustworthy vicarious experience, sufficient to establish the authenticity of the gospels as a divine revelation. Although a systematic treatment of this topic remained unfinished at his death, Boyle's published works contain much information about his views on miracles, including their consistency with the mechanical philosophy.

However, the mechanical philosophy, especially as it was articulated by the French philosopher René Descartes (1596–1650), also had a darker side. Animals were typically seen as little or nothing more than complex machines, with full rationality and sensitivity reserved only for humans, angels, and God. When coupled with a nearly universal desire to improve the human condition by advancing the knowledge of anatomy and physiology, the temptation to engage in animal experimentation was often too great to resist. Boyle, who sought as much as anyone to enhance what he called "the Empire of Man over Other Creatures" (Hunter and Davis 1999–2000, Vol. 3, p. 193), carried out numerous diverse experiments involving both vertebrate and invertebrate live animals—dogs, cats, birds, butterflies, worms, and many others. Yet he did so with considerable sympathy and even regret; on several occasions, he even released animals that had survived one experiment precisely in order to spare them further suffering. Unlike Descartes Boyle was not convinced that animals lack sensation, and he considered gratuitous cruelty to animals blasphemous, since all creatures belonged to God. At the same time he believed that God intended the creatures to serve humankind, thus sanctioning a certain amount of animal experimentation.

Boyle's Legacy

Boyle's influence on subsequent thinking about science, religion, and morality has been larger than many writers realize, much larger (for example) than that of Isaac Newton (1642–1727)—who actually published very little of importance about religion, although he devoted many years to the study of theology and church history. The Anglo-American tradition of natural theology derives substantially from Boyle's extensive treatment of the subject, and his outstanding example of a pious scientist writing about the Bible and morality has been much imitated.

EDWARD B. DAVIS

SEE ALSO *Christian Perspecitves; Scientific Revolution.*

BIBLIOGRAPHY

Davis, Edward B., and Michael Hunter, eds. (1996). *Robert Boyle: A Free Enquiry into the Vulgarly Received Notion of Nature.* Cambridge: Cambridge University Press.

Harwood, John T., ed. (1991). *The Early Essays and Ethics of Robert Boyle.* Carbondale: Southern Illinois University Press.

Hunter, Michael. (1993). "Casuistry in Action: Robert Boyle's Confessional Interviews with Gilbert Burnet and Edward Stillingfleet, 1691." *The Journal of Ecclesiastical History* 44(1): 80–98.

Hunter, Michael, and Edward B. Davis, eds. (1999–2000). *The Works of Robert Boyle,* 14 vols. London: Pickering & Chatto.

Madison, R. E. W. (1969).*The Life of the Honourable Robert Boyle* London: Taylor & Francis.

Oster, Malcolm. (1989). "The 'Beame of Diuinity': Animal Suffering in the Early Thought of Robert Boyle." *British Journal for the History of Science* 22: 151–180.

Shapin, Steven. (1994).*A Social History of Truth: Civility and Science in Seventeenth-Century England*. Chicago: University of Chicago Press.

BRAIN DEATH

Physicians could not reliably diagnose death in all cases until the early nineteenth century when a new technology, the stethoscope, was invented and medical scientists began to understand cardiorespiratory anatomy and physiology. Ironically, it was the introduction in the late twentieth century of more new technologies, such as the mechanical ventilator, that once again caused uncertainty about the definition and determination of death.

Before life-sustaining technology was introduced, critical vital functions such as heartbeat, breathing, and brain activity were so interdependent that when one function ceased, they all did. For example, when a person suffered a massive heart attack and cardiac arrest, breathing and consciousness were lost almost simultaneously because the heart pumps nutrient rich, oxygenated blood to the brain and the rest of the body. If a person stopped breathing, say from drowning, heartbeat and consciousness were almost immediately lost for the same reason—no oxygen reached the brain and heart. Similarly, when a massive brain injury occurred, consciousness and spontaneous breathing stopped because of destruction of the respiratory center in the brain stem. There was thus no need to choose between cardiac, respiratory, and brain function as the unique function whose loss signaled the transition from human being to corpse.

With the introduction of the mechanical ventilator and the modern intensive care unit (ICU), patients with severe head injuries, who previously would have died, were sustained with beating hearts and healthy functioning of all other organs such as kidney, liver, and pancreas. These patients, when they have lost all brain function, are termed "brain dead." In 1968 an ad hoc committee at Harvard Medical School proposed that such patients, who were legally and medically considered to be alive, be classified as dead.

Two Types of Death

Although brain death as death was quickly accepted in the United States by legal and medical communities and, seemingly, by the public at large, new debates began, at least in academic circles, about just what made these warm, pink, heart-beating patients dead. Interestingly, the Harvard Ad Hoc Committee did not address this issue. Rather, they gave two utilitarian reasons to reclassify brain-dead patients as dead. Brain death is relatively easy to diagnose and the prognosis is dismal: No person accurately diagnosed has ever recovered consciousness and, at least in the first decade, brain-dead patients were very unstable and would suffer cardiovascular collapse and "traditional" death within hours or days. Therefore, many people saw no point in keeping brain-dead patients "going" by mechanical ventilation. But in 1968, U.S. society had no experience with the removal of life-sustaining treatment (something that in the twenty-first century happens daily in leading hospitals), and physicians feared they would be charged with homicide if they turned off the ventilator. The Ad Hoc Committee suggested that by declaring such patients dead, this fear would be removed.

The second reason given by the Ad Hoc Committee had to do with organ transplantation, which was becoming an increasingly effective treatment for end-stage organ failure. Because of all the *life* remaining in brain-dead patients, they were potentially an excellent source of organs. But taking their vital organs would violate the so-called dead donor rule that forbids killing patients by removing their organs. Classifying them as dead would avoid this problem and quell any controversy.

It was not until 1981 that a coherent philosophical or conceptual argument was put forth to explain why brain-dead patients were actually dead. In that year, in a landmark article, James L. Bernat and his colleagues at Dartmouth College proposed that the integrating function of the brain stem was the critical one whose loss marked the transition from life to death. Bernat went on to explain that loss of integration meant the permanent cessation of functioning of the "organism as a whole"—that is, the loss of "spontaneous and innate activities carried out by all or most subsystems" and "the body's ability to organize and regulate itself" (Bernat, Culver, and Gert 1981, p. 390). He gave as examples neuroendocrine control, temperature regulation, and the ability to maintain blood pressure and fluid and electrolyte balance. Bernat gave no significance to another important brain function, consciousness and cognition.

In simple terms, the brain has two major functions. The integrative function, which Bernat found critical, resides primarily in the brain stem, the primitive part of the brain that lies buried under the much larger cerebral hemispheres, which are most developed in higher animals, especially primates. Consciousness and cognition reside primarily in the cerebral hemispheres.

Although brain death was quickly accepted legally and clinically throughout the United States, many philosophers (Veatch 1976, Bartlett and Youngner 1988, Gervais 1986) argued that consciousness and cognition were the critical functions that distinguished a living from a dead person. Their criticism was twofold. First, the integrative function was not actually lost; it was merely taken over from the brain stem by machines and ICU personnel who kept patients *alive* by breathing for them and maintaining blood pressure and other vital activities. Second, consciousness and cognition more accurately reflect what is unique about human beings—the function without which they are dead. In contrast to Bernat, who argued that loss of integrative function is what humans have always meant by death, his critics argued what people really care about is whether or not *there is anybody home*.

Practical Problems

In fact, studies of health professionals have indicated that while some accept brain death as death because of loss of integrative function, an equal number do so because the patient has permanently lost consciousness and cognition (Youngner 1989). Interestingly, these studies also demonstrate that many health professionals do not really consider brain-dead patients to be dead, but rather *good as dead* because they will die soon despite intervention and have an unacceptable quality of life. A later study demonstrated a similar diversity of opinion and belief among the general public (Siminoff, Burant, and Youngner 2004).

Other problems with brain death have emerged. First, although the clinical and legal criteria inevitably call for loss of *all* brain functions, it turns out that clinical tests commonly used to assure the criterion has been fulfilled simply do not test for some functions that often remain (Halevy and Brody 1993). For example, the production of vasopressin, a hormone essential for maintaining fluid and electrolyte balance, continues in many patients declared brain dead. Bernat responded to the dilemma by saying that it is only *critical* functions that count, but gave little guidance about how to distinguish critical from noncritical ones (Bernat 1998).

A second problem with brain death is that the clinical course of patients who have been declared brain dead is not as certain as when the syndrome was first encountered in the 1960s. Then, patients who were brain dead were notoriously unstable and suffered cardiovascular collapse and cardiac arrest within hours or days. Now, with more clinical experience and more sophisticated interventions, brain-dead patients can *survive* the period of instability to enter a chronic state in which they can be maintained at home with little more than ventilatory support. Some have continued in this state for months and years (Shewmon 1998). An editorial in a prominent neurology journal proclaimed "even the dead are not terminally ill anymore" (Cranford 1998, p. 1530), an ironic statement that captures much of the ambiguity surrounding clinical states in which some, but not all, vital functions remain.

Practical Acceptance

Despite the ambiguities about brain death and how poorly it is understood by the public, acceptance of brain death at the public policy level seems fairly solid. The prognosis for brain-dead patients is uniformly bleak, even for those retaining residual brain functions such as the production of vasopressin. None ever recover consciousness, and most die traditional deaths within days. Moreover, unlike abortion, brain death remains off the radar screen of the religious right, which is very concerned about a *culture of death* in the United States that reduces human dignity and value. Perhaps brain death was "grandfathered in" before the religious right was politically galvanized by *Roe v. Wade* in 1973.

Interestingly, while brain death was quickly accepted and remains relatively uncontroversial in the United States, the situation is quite different in some other countries, most notably Japan, where brain death was not recognized by law until 1997. Patients who have lost brain function may be declared dead only for the purpose of organ transplantation, and then only if both the patient, when living, and the family, after death, have signed written documents. Unlike in the United States, brain death has been the subject of much public discussion and controversy for more than four decades, including the publication of more than 100 books on the subject for the general public and its inclusion as subject matter in popular comic books for children (Lock 2002). While not as contentious as it is in Japan, the debate over brain death in Denmark and Germany has been much stormier than that in the United States.

Because of the growing gap between the demand and supply for transplantable organs, it is unlikely that brain death will become a subject of controversy in the United States. Controversy is more likely to come if desperate patients and transplanters try to expand the current definition of death to include patients with brain injuries less severe than brain death.

STUART J. YOUNGNER

SEE ALSO *Bioethics; Death and Dying; Persistent Vegetative State; Science, Technology, and Law.*

BIBLIOGRAPHY

Ad Hoc Committee of the Harvard Medical School to Examine the Definition of Brain Death. (1968). "A Definition of Irreversible Coma." *Journal of the American Medical Association* 205(6): 337–340. This landmark article introduced the concept of brain death to the medical community and lay public nearly 40 years ago.

Bartlett, Edward T., and Stuart J. Youngner. (1988). "Human Death and the Destruction of the Human Cortex." In *Death: Beyond Whole-Brain Criteria*, ed. Richard M. Zaner. Dordrecht, Netherlands: Kluwer Academic.

Bernat, James L. (1998). "A Defense of the Whole-Brain Concept of Death." *Hastings Center Report* 28(2): 14–23.

Bernat, James L.; Charles M. Culver; and Bernard Gert. (1981). "On the Definition and Criterion of Death." *Annals of Internal Medicine* 94: 389–394.

Cranford, Ronald. (1998). "Even the Dead Are Not Terminally Ill Anymore." *Neurology* 51(6): 1530–1531.

Gervais, Karen Grandstrand. (1986). *Redefining Death*. New Haven, CT: Yale University Press.

Halevy, Amir, and Baruch Brody. (1993). "Brain Death: Reconciling Definitions, Criteria, and Tests." *Annals of Internal Medicine* 119: 519–525.

Lock, Margaret. (2002). *Twice Dead: Organ Transplants and the Reinvention of Death*. Berkeley and Los Angeles: University of California Press.

Shewmon, D. Alan. (1998). "Chronic 'Brain Death': Meta-Analysis and Conceptual Consequences." *Neurology* 51(6): 1538–1545.

Siminoff, Laura; Burant, Christopher; and Youngner, Stuart J. (2004). "Death and Organ Procurement: Public Beliefs and Attitudes." *Social Science & Medicine* 59(11): 2325–2234.

Veatch, Robert M. (1976). *Death, Dying, and the Biological Revolution: Our Last Quest for Responsibility*. New Haven, CT: Yale University Press. One of the first discussions about brain death.

Youngner, Stuart J.; C. Seth Landefeld; Claudia J. Coulton; et al. (1989). "Brain Death and Organ Retrieval: A Cross-Sectional Survey of Knowledge and Concepts among Health Professionals." *Journal of the American Medical Association* 261(15): 2205–2210.

Youngner, Stuart J.; Robert M. Arnold; and Renie Schapiro, eds. (1999). *The Definition of Death: Contemporary Controversies*. Baltimore: Johns Hopkins University Press. A collection of essays about problems in the diagnosis and determination of death from philosophical and clinical perspectives.

BRAVE NEW WORLD

One common way to evoke unease about modern science and technology is to say that humanity is headed toward a "brave new world." Aldous Huxley's novel *Brave New World*, first published in 1932, depicts a World State in which biological technology and psychological conditioning were used to make everyone feel happy all the time, but this was achieved by creating a mechanized world in which people were reduced to soulless animals. Much of the debate over science and technology has centered on the question of how to avoid such a "brave new world."

Huxley (1894–1963) was a prominent English novelist and essayist. Of his many novels, *Brave New World* is the one that is best known in the early twenty-first century. It reflects his interest in biological science, which he shared with his grandfather Thomas Henry Huxley (1825–1895), his brother Julian Huxley (1887–1975), and his friend J. B. S. Haldane (1892–1964), all of whom were prominent biologists.

The New World State

Brave New World is about an imaginary World State in the future where a combination of genetic manipulation and social conditioning has produced a stable industrialized society governed by the political slogan that "everyone belongs to everyone else." Human eggs are fertilized in laboratories and then incubated under varying conditions for the mass production of people, who are shaped to fill their social caste roles as Alphas, Betas, Gammas, Deltas, or Epsilons. Some people have been cloned from the same fertilized egg, so that they are genetically identical. The higher castes fill managerial roles, and a few of these become Controllers ruling over the World State. The lower castes fill menial roles. There are no parental or familial attachments. The idea of being born to a mother after developing in her womb is considered obscene and primitive. People are thus freed from the emotional conflicts of family life. Because everyone is conditioned to fill an assigned role, they all feel happy doing what they do, and there is no class conflict. There are many amusements to keep people happy, including the "feelies," movies that arouse audiences not only visually and audibly but also tactually. Sexual promiscuity is a social duty, and people derive recreational pleasure from having hundreds of sexual partners. Anyone who might feel a little anxious or sad takes the drug *soma*, which induces blissful euphoria and allows people to "escape from reality" for long periods without any painful aftereffects. Medical science preserves the

Scene from the 1980 TV movie version of *Brave New World*. A subsequent film depiction of Huxley's scientific utopia was made in 1998, featuring Leonard Nimoy as Mustapha Mond. *(The Kobal Collection. Reproduced by permission.)*

youthful vigor of everyone until death. There is no interest in traditional art or religion, because people have never felt the intense suffering or conflicts that are presupposed by art and religion.

A few individuals rebel against this social conformity and emotional shallowness. They desire the intense emotions of romantic love, art, religion, or pure science. If they become too disruptive, they can be exiled to distant islands. One of the rebels is John the Savage, who originally was born to a woman and raised on an Indian reservation in New Mexico before being brought to London. The Savage has educated himself by reading William Shakespeare's plays, which give him poetic language to express his deep longings. The Savage meets Mustapha Mond, the World Controller for Western Europe, who shares his interest in art and religion. Mond has also been moved by a love of pure science for its own sake that cannot be satisfied by the applied science and technology promoted in the World State. As a young man, Mond could have been exiled to an island for rebels, but he decided to sacrifice his personal happiness to become a Controller who would rule for the greater happiness of the World State.

Antecedents and Consequents

Huxley's novel thus depicts the sort of scientific utopias that were predicted by people such as Haldane. The artificial production of children, the genetic engineering of character traits, the abolition of family life, recreational sex separated from reproduction, the use of new psychotropic drugs to induce euphoric moods, the prolongation of youthful health into old age—these and other innovations in Huxley's novel had already been predicted by Haldane in his book *Daedalus; or, Science and the Future*, first published in 1923. Haldane foresaw that these changes in scientific technology would bring changes in morality. So that what was traditionally thought to be bad would be regarded as good. Welcoming this prospect as moral progress, Haldane suggested: "We must learn not to take traditional morals too seriously" (1995, p. 49). In contrast to Haldane's optimistic attitude, Huxley's novel elicits the fear that Haldane's utopia would be dehumanizing.

Huxley takes his title from Shakespeare's assessment of utopian aspirations in *The Tempest* (1610), near the end of which the young woman Miranda marvels concerning her island home, "O brave new world, that has such people in it" (Act 5, Scene 1). The original allusion was to the New World of the Americas that was in the process of being colonized. Jamestown, the first permanent English settlement in the New World was founded in 1607, although from the perspective of the indigenous inhabitants the new world was precisely that which was created by the transplantation of European culture. The phrases "new world" and "brave new world" have thus become synonymous with major cultural transformations, especially those dependent on modern science and technology. Popular adaptations include one for radio (1956, with Huxley himself narrating), two television movies (1980 and 1998), a feature-length film *Demolition Man* (1993) with numerous allusions, and a heavy-metal music album (by Iron Maiden, 2000). "Brave New World" was also the title of a four-day New York theater event in 2002 responding to the terrorist events of September 11, 2001.

In his 1958 collection of essays *Brave New World Revisited,* Huxley said that the world described in his novel was contrary to "man's biological nature," because it treated human beings as if they were social insects rather than mammals. Social insects such as bees, ants, and termites naturally cooperate because the good of the social whole is greater than its individual members. But mammals are only "moderately gregarious," Huxley observed, in that they can cooperate with one another, but they will never subordinate their individual interests totally to the community. In social insect colonies, reproduction is communal (through the queen), so that most of the insects do not reproduce and thus do not feel any individual attachment to offspring. Among mammals, however, individuals produce offspring directly and feel a parental attachment to them. As large-brained mammals, human beings must devise arrangements for balancing social order and individual freedom. *Brave New World* shows how dehumanizing it would be for human beings to be so designed that they gave up individual freedom for the stable order of something like a social insect colony.

The very fact that people in *Brave New World* need *soma* as an "escape from reality" indicates that the World State has not succeeded in abolishing their mammalian nature and turning them into social insects. Any careful reader of Huxley's novel can see intimations of all those natural desires that distinguish the human species. These desires are expressed in the many individuals who have to be sent into exile on remote islands. Even a World Controller such as Mond feels those desires, which leaves the reader wondering why he would take a ruling office that makes him unhappy.

Critics and Criticism

Critics of modern technology—such as C. S. Lewis (1898–1963), Lewis Mumford (1895–1990), and Leon R. Kass (2002)—see the world depicted in Huxley's novel as the final stage in the modern project for the technological conquest of nature, in which human nature itself will be conquered by being abolished. Once human beings become merely raw material for technological manipulation—particularly through human biotechnology—then human beings will be replaced by "posthuman" artifacts. This will be the ultimate tyranny because humans will have absolute power over those whose nature is to be remade. These critics worry that if human nature is abolished as a given, and thus there is no natural ground for moral judgment, there remains no clear standard for judging the moral uses of technology beyond the arbitrary impulses of those who control the technology. After being advised by Kass about the moral dangers in harvesting stem cells from human embryos, president George W. Bush delivered a nationally televised speech on August 9, 2001, in which he warned that "we have arrived at that brave new world" described by Huxley (Bush 2002, p. 308).

Libertarian proponents of modern technology—such as Lee M. Silver (1997) and Virginia Postrel (1998)—reject this dark view by arguing that what is wrong with the society in Huxley's novel is its rule by a coercive World State that has eliminated individual liberty. From a libertarian position, the biotechnological conquest of nature is not harmful as long as it occurs through individual free choice. So, for instance, if parents want to use the latest reproductive technology to promote the health and happiness of their children, they should be free to do so, with the hope that parental love will move them to act for the best interests of their children. People will make mistakes, but in a free society they will learn from their mistakes.

In response, conservatives such as Kass warn that leaving biotechnology to individual choice could still lead to a "brave new world," because parents and others might be seduced into using biotechnology in ways that would bring about a degrading, dehumanized world. For example, parents with the best of intentions might choose to genetically design their children to have desirable traits without realizing how this would turn children into artificial products of human will and thus deprive them of human dignity. Or the pursuit of happiness might induce people to become dependent on mood-brightening drugs without considering the degradation in such illusory contentment.

Assessing the prospect of a "brave new world" requires judging both the technical possibility and the moral wisdom of the technological mastery of nature as extended to the mastery of human nature.

LARRY ARNHART

SEE ALSO *Huxley, Aldous; Science Fiction; Science, Technology, and Literature; Utopia and Dystopia.*

BIBLIOGRAPHY

Bush, George W. (2002). "Address To The Nation." In *The Future is Now: America Confronts the New Genetics*, ed. William Kristol and Eric Cohen. Lanham, MD: Rowman and Littlefield.

Firchow, Peter Edgerly. (1984). *The End of Utopia: A Study of Aldous Huxley's "Brave New World."* Lewisburg, PA: Bucknell University Press. Best study of Huxley's novel.

Fukuyama, Francis. (2002). *Our Posthuman Future: Consequences of the Biotechnology Revolution*. New York: Farrar, Straus and Giroux. Argues for restricting biotechnology to avoid a "brave new world."

Haldane, J. B. S. (1995). "Daedalus; or, Science and the Future." In *Haldane's "Daedalus" Revisited*, ed. Krishna R. Dronamraju. Oxford: Oxford University Press.

Huxley, Aldous. (1965). *"Brave New World" and "Brave New World Revisited."* New York: Harper and Row.

Kass, Leon R. (2002). "Preventing a Brave New World." In *The Future Is Now: America Confronts the New Genetics*, ed. William Kristol and Eric Cohen. Lanham, MD: Rowman and Littlefield.

Lewis, C. S. (1947). *The Abolition of Man*. New York: Macmillan.

Mumford, Lewis. (1970). *The Myth of the Machine*, Vol. 2: *The Pentagon of Power*. New York: Harcourt Brace Jovanovich. Argues that modern technology has already created a "brave new world."

Postrel, Virginia. (1998). *The Future and Its Enemies*. New York: Free Press.

Silver, Lee M. (1997). *Remaking Eden: Cloning and Beyond in a Brave New World*. New York: Avon. An optimistic view of the liberating effects of a "brave new world" of biotechnology.

BRECHT, BERTOLT

German playwright, poet, and theatrical reformer Eugen Berthold Friedrich Brecht (1898–1956) developed theatre as a forum for critical reflection on society in order to advance his Marxist beliefs. Born in Augsburg, Bavaria, on February 10, Brecht studied medicine in Munich and briefly served at an army hospital in World War I. During the early 1920s, he developed an antibourgeois attitude and studied Marxism. Brecht lived in Berlin from 1924 to 1933, where he collaborated with composer Kurt Weill (1900–1950) and developed his theory of "epic theater" and his austere, irregular verse. In 1933, Brecht went into exile, spending six years in the United States (1941–1947), where he did some film work in Hollywood. During exile, Brecht wrote most of his great plays, essays, and poems, while his work was being burned in Nazi Germany. In 1949, he moved back to Berlin and despite the controversial communist ideals of his work, he enjoyed great success. Brecht died of a heart attack in East Berlin on August 14.

Bertolt Brecht, 1898–1956. Brecht has been called one of the greatest German playwrights of the 20th century. His works reflect his thoughts on the technologies of film and radio, which were newly emerging during his time. *(The Granger Collection Ltd.)*

Technology and Communication

Brecht realized that the emerging technologies of film and radio provided important opportunities for rethinking the formal properties of communication. He was aware of the ways in which new technologies construct their audiences in modes of reception ranging from passive, which he disliked, to active and participatory, which he favored and encouraged. Reception and representation were key to Brecht's idea of what he termed "communication with consequences." He believed that audiences perceive the real causality of the story being told only if the devices of the media solicit active inquiry.

Although he felt the new media had great potential to liberate people, Brecht also maintained that radio ignored the possibilities of organizing its listeners as suppliers of ideas. If radio were to change its focus from distribution to communication, turning listeners also into speakers, then it might generate positive social change. He did not foresee the use of radio for propaganda by right-wing (as well as leftist) ideologues. Brecht, like director Erwin Piscator (1893–1966), felt that film could be used positively within theater, and he was interested in the way new technologies of communication reconfigured content. Developments within filmmaking, for example, inspired his notion of *Gestus*, actions that are both simply themselves and emblematic of larger social practices.

In some of his productions, Brecht projected subtitles in advance of scenes to announce the plot to the audience. By abandoning the tension and surprise, this "communication with consequences" focused the audience on the more important task of thinking critically, socially, and politically. Distancing the audience from his plays was also crucial to his Marxist drama. Unlike the Aristotelian premise that the audience should be made to believe that what they are witnessing is happening here and now, the Marxist premise that human nature is historically conditioned required an "epic theater," which gave the audience critical detachment. This was Brecht's *Verfremdungseffekt* (alienation effect) that portrayed action in a "scientific spirit" and reminded the viewer that theater is not reality.

Critical inquiry that exposed the oppression and inequalities of capitalist production was central to Brecht's view of the potential of new technology. Spectators were able to regard the situations of the characters and the actions of the dramas as indicative of class warfare, thus underscoring the social, rather than psychological, genesis of the human condition.

Changing Views About Science and Technology

In a radio speech on March 27, 1927, Brecht stated, "It is my belief that [man] will not let himself be changed by machines but that he will himself change the machine; and whatever he looks like he will above all look human." In the same talk, he argued that this new human would be acutely aware that guns can be used for him or against him, houses can shelter or oppress him, and that live works can discourage or encourage him. To this neutralist position, Brecht added a general element of optimism. He argued that science could change nature and make the "world seem almost habitable," by overthrowing the oppressive religious mystification of experience that taught people to tolerate their fate.

Brecht realized that developments in science and technology were driving and shaping society, and he believed that these changes had to be reflected in the theatrical presentation of human transactions. His epic and dialectical theater with its emphasis on critical inquiry highlighted the increased responsibility created by new technological powers. Brecht's characters were never products of metaphysical forces, and their actions were not fated. Rather, they grappled with personal responsibilities shaped and conditioned by the larger world.

Brecht's *Leben des Galilei* (Life of Galileo) shows not only this fallible, striving quality of his characters, but also captures his growing unease about the human and social consequences of modern science and technology. The original 1938 version of the play portrays Galileo as a cunning, noble, and brave seeker of truth who brings light to an age of darkness. The bombing of Hiroshima in 1945, however, caused Brecht to revise the play. In this later version, Galileo is portrayed as a coward who quickly recants the truth at the sight of torture devices. He practices science only for his own gain, without regarding the possible harms or benefits to humanity. Brecht, despite his deep distrust of religion, even allows the Church to eloquently and persuasively defend its position. Ultimately, Galileo is portrayed as the initial instigator of a tradition that leads to the horrors of atomic weapons. In the play's final scene, Galileo denounces himself, because he sought knowledge for self-aggrandizement and not for the good of humanity. Brecht shows that the pursuit of truth absent considerations of the good led to the split between science and society that culminated in the use of atomic weapons on civilians. Science brings darkness rather than enlightenment.

Brecht saw the unbridled quest for knowledge and its potentially destructive consequences as a pressing concern of his age. Just as he satirized the "resistible" rise of Hitler, Brecht wanted to show how the exercise of critical thinking and personal responsibility could resist the rise of destructive technologies. Using irony, humor, and skepticism, he cautioned that human society must morally progress in order to understand and wisely direct the rapid advances in science and technology. As Brecht wrote in *Leben des Galilei*:

May you now guard Science's light
Kindle it and use it right
Lest it be a flame to fall
Downward to consume us all

CAROL MARTIN

SEE ALSO *Science, Technology, and Literature*.

BIBLIOGRAPHY

Brecht, Bertolt. (1981 [1938]). *Leben des Galilei* [Life of Galileo], eds. H. F. Brookes and C. E. Frankel. London: Heinemann. An examination of the problems that face scientists and the spirit of free inquiry when challenged by the requirements of government and official ideology.

Brecht, Bertolt. (1970–1987). *Collected Plays* 7 vols., eds. John Willett and Ralph Manheim. London: Methuen.

Esslin, Martin. (1971). *Brecht: The Man and His Work*, rev. ed. Garden City, NY: Anchor Books.

Willett, John. (1959). *The Theatre of Bertolt Brecht*. London: Methuen.

BRENT SPAR

The Brent Spar was an oil storage buoy built and owned by Royal Dutch Shell (Shell Oil) in 1976. The spar (or large cylindrical storage buoy), 147 meters tall, was used in the North Sea to temporarily store crude oil. A new pipeline made the spar unnecessary and over time Shell Oil chose to dispose of the spar by sinking it in deep water off the west coast of Great Britain. During the mid-1990s this proposal became a major environmental issue in Europe.

Disposal Options

Sinking was the cheapest (approximately $18 million) and safest option for the workers who would be performing the task. Other options, however, existed. At a greater expense, the spar could have been refurbished to perform other functions. At two to four times the cost of sinking it, the spar could have been cleaned and dismantled, with the steel then recycled. Dismantling operations, however, posed up to six times more risks to workers and the immediate coastal environment where the dismantling would be performed.

Shell Oil chose to dispose of the spar in more than 2 kilometers of water and received permission to do so from the British government in 1994. Both Shell Oil and the British government agreed that the potential damage to the local environment from oils, waxes, and other materials still inside the spar would be limited to the immediate area and that the impact would be short lived.

In April of 1995, Shell Oil began towing the Brent Spar to its deep-water burial at which time protesters associated with Greenpeace boarded the platform. The protesters demanded that Shell Oil cease its dumping plan in favor of what they contended were more environmentally benign choices and argued that disposal at sea was wrong on principle. Greenpeace and other environmental groups called for the boycott of Shell Oil gas stations across Europe and in some places sales at those stations fell by half. Two such stations in Germany were attacked with fire bombs.

On June 20, 1995, due to intense public pressure and negative publicity, Shell Oil temporarily halted its deep-sea disposal operations. Over the following years, the company evaluated a number of different disposal options, finally dismantling the Brent Spar in a deep bay in Norway, beginning in January 1998. Sections of the spar were recycled in the construction of a new ferry terminal in Norway. Total disposal cost was approximately $96 million.

During the protests, Greenpeace claimed the spar contained large amounts of dangerous chemicals that would cause serious harm to the environment. Shell Oil and the majority of independent scientists argued that deep-sea disposal was in fact the safest option. After the decision to cancel the disposal in 1995, Shell Oil hired an autonomous firm, Det Norsk Veritas, to assess the alternatives. The firm determined that the actual amount of residual oil and some heavy metals still inside the spar was slightly higher than originally claimed by Shell Oil, but significantly lower than the amount claimed by Greenpeace. Media reports discovered other inconsistencies in the organization's arguments. Greenpeace was successful in stopping the disposal operation, but lost legitimacy after its story began to unravel. The debate also left Shell Oil's reputation with the public significantly damaged.

Ethical and Policy Lessons

The Brent Spar incident has a number of ethical and policy implications. Disposal of the spar could have set a precedent for disposal of other oil facilities, and potentially caused environmental damage. Some argued that Shell Oil's risk-benefit analysis could not adequately gauge the effects of disposal. At issue was the company's ability to determine environmental harm versus its bias toward monetary benefits to it and its shareholders.

Furthermore some saw trade-offs between harm to the environment and benefit to the company as completely illegitimate and nonfungible. The feasibility of the business ethic of the triple bottom line of business, society, and environment, in which corporations consider all three outcomes in their decision making, was also at stake.

Finally a number of ethical issues arise concerning the dialogue itself. Did Greenpeace have standing to protest a legal action by Shell Oil? Was Greenpeace a legitimate speaker for the environment? Was Shell Oil obliged to speak with different stakeholders or groups, and what process should the company have pursued? These questions highlight the difficulty of convening legitimate, representative groups, and carrying out group decisions when all parties are free to opt out or otherwise dissent.

The saga of the disposal of the Brent Spar combined debate over scientific information with a political dispute over environmental values. Greenpeace was able to use inaccurate scientific information to buttress an ethics argument against dumping waste in the sea. It also argued that dumping the spar would allow Shell Oil to avoid the full cost of the spar's use and disposal. Shell Oil disputed the scientific information Greenpeace presented, but failed to adequately counter the ethics argument. The public and media largely failed to grasp the scientific dispute, and sided with Greenpeace on ethical grounds. The Brent Spar incident illustrates the difficulty of introducing scientific evidence into essentially political arguments.

ELIZABETH C. MCNIE
TIND SHEPPER RYEN

SEE ALSO *Engineering Ethics; Environmental Ethics; Nongovernmental Organizations; Oil.*

BIBLIOGRAPHY

Entine, John. (2002). "Shell, Greenpeace and Brent Spar." In *Case Histories in Business Ethics*, ed. Chris Megone and Simon J. Robinson. New York: Routledge.

INTERNET RESOURCES

Brent Spar Dossier. Shell.com. Available from http://www.shell.com/brentspar/. Shell Oil's history of the incident.

Greenpeace Brent Spar Protest in the North Sea. Greenpeace. Available from http://archive.greenpeace.org/comms/brent/brent.html. An account of the incident from the organization's perspective.

BRIDGES

Bridge building as a human activity predates recorded history, and bridges are among the earliest structures described in the historical record. In the fifth century B.C.E. Herodotus reports on a bridge over the Euphrates River made of timber resting on a stone foundation. Roman stone bridges at Segovia (Spain) and Nîmes (France) are still standing 2,000 years after their construction. In the Middle Ages, bridge building became the province of specialist monastic orders. Medieval bridges were conceived as places to live, not just as a means of passage from one side of a river to another. London Bridge in 1594 supported 100 houses and shops.

Bridge Engineering

In the nineteenth century, bridge building became a scientific discipline, after a backlash brought about by notorious disasters in which bridges failed to endure mathematically predictable loads. A fascinating 1887 monograph by George L. Vose (1831–1910) reflects the period in which bridge building crystallized into a scientific and mathematical discipline. Vose complained that any charlatan could proclaim himself a bridge builder and find customers, while ignoring the mathematics that made the calculation of safety margins simple. "There is at present in this country absolutely no law, no control, no inspection, which can prevent the building and the use of unsafe bridges" (p. 12). He pointed out that the science of bridge loads was well understood: A dense crowd of people creates a load of up to 140 pounds per square foot, while soldiers walking in step double the strain; snow and ice can create a load of 10 to 20 pounds per square foot, while heavily loaded freight trains can create a strain of 7,000 pounds per square foot.

Vose was a pioneering proponent of safety margins. He argued that bridges should be designed to carry a load four to six times greater than the actual loads they are likely to carry under any foreseeable circumstances. Many existing bridges did not meet these standards; some, in fact, were capable of carrying only the predictable load. Of these, Vose acerbically noted that such a bridge is warranted "to safely bear the load that will break it down" (p. 55). The country, in his estimation, was full of highway bridges "sold by dishonest builders to ignorant officials" and awaiting only "an extra large crowd of people, [or] a company of soldiers" to collapse (p. 16).

According to the structural engineer David P. Billington, however, a second transformation occurred when bridges (along with tall buildings) became uniquely modern works of art by exploiting the properties of new structural materials such as steel and reinforced concrete. In the period after 1880 engineers began "to explore new forms with these materials," the first maturity of which occurred in the period between the two world wars (1983, p. 7). The bridge designs of the Swiss engineer Robert Maillart (1872–1940) are archetypical achievements of this new era.

In the contemporary world the Clifton (Bristol, 1864), Brooklyn (New York, 1883), Golden Gate (San Francisco, 1937), and Tsing Ma (Hong Kong, 1997) Bridges are indeed considered works of art, objects whose function is intertwined with their beauty. For the engineer Henry Petroski "there is no purer form of engineering than bridge building" (1995, p. 14). Whereas houses and buildings are designed for appearance, and then engineered, the process followed in bridge construction is the opposite. A bridge must be designed to perform its function successfully; its beauty emerges from the engineering.

Ethics and Bridges

The ethical issues pertaining to bridges span a range of questions. Is a particular bridge really needed? What impacts do bridges have on the social and natural environments where they are constructed? What levels of safety are appropriate in bridge design?

NEEDS. Insofar as they are major public works projects, the need for bridges has to be obvious and they often have to pass a hurdle of criticism before being constructed. At the same time bridges are sometimes built so that powerful politicians can create jobs and funnel money to their districts, or reward political contributors. According to environmental groups in Alaska, the proposed Gravina Island Bridge is an example. Designed to be 1.6 kilometers long and 24 meters higher than the Brooklyn Bridge, the $200 million structure would link the depressed town of Ketchikan (and its 7,500 residents) to an island that has fifty residents and an airport with six flights a day in the busy season. The island is already well served by ferry, and the bridge would bisect a channel used by shipping and floatplanes.

SOCIAL AND ENVIRONMENTAL IMPACTS. Most people do not want a bridge in their own backyards, with the concomitant loss of views and increases in local traffic, leading to a decrease in property values. Illustrating the NIMBY (not in my backyard) syndrome, even citizens who will benefit prefer that a bridge be sited in someone else's neighborhood. The site originally studied for the George Washington Bridge in New York City was at West 110th Street in Manhattan. Two powerful local institutions, St. Luke's Hospital and Columbia University, strenuously opposed this location. Columbia's president, Nicholas Murray Butler, said that the proposed site was "little short of vandalism" (Petroski 1995, p. 242). The bridge was eventually built (1927–1931) on unused land much further north at West 179th Street.

Robert Moses (1888–1981), the motivating force behind many of New York's best-known bridges and parks, is famous for his ruthless treatment of opponents and of local communities that stood in the way. His beautiful Verrazano-Narrows Bridge, built 1959–1964 with either end in a highly populated neighborhood, caused the seizing and demolition of 800 buildings in Bay Ridge, Brooklyn, displacing 7,000 people. On the Staten Island side, 400 buildings were taken by eminent domain, displacing 3,500 residents.

Moses's determination, and his willingness to counter his opponents in the same visceral language they used to attack him, is evident in a series of monographs issued at his direction. In 1939, when the *New York Herald-Tribune* opposed his proposed Brooklyn Battery Bridge, Moses had the Triborough Bridge Authority publish a brochure entitled "Is There Any Reason to Suppose They Are Right Now?" It ridiculed the *Herald-Tribune*, excerpting two decades of editorials opposing previous Moses park and highway projects. Moses painted the newspaper as the voice of millionaires who did not want their neighborhoods tainted by projects that would benefit the common folk.

Another organization opposing the Brooklyn Battery Bridge was the Regional Plan Association, which argued that it was not a natural site for a bridge and would deface the land- and cityscape. In his counterattack, Moses noted that the association had backed a proposal for the construction of a 200-meter obelisk in the Battery, which Moses claimed would obscure the view much more than his proposed bridge. In the end, however, Moses lost the battle, and a tunnel was built in lieu of the bridge. Tunnels are frequently proposed as alternatives to bridge projects; underground, they have the virtue of not being seen, but tend to be more expensive to build and are of necessity narrower, carrying less traffic and freight.

BRIDGE SAFETY. Bridges collapse for one of two reasons. Either their design and construction fail to meet contemporary industry standards, or those standards are inadequate to ensure safety in the face of unexpected circumstances. An example of negligent construction was West Gate Bridge, in Melbourne, Australia, which fell while being erected on October 15, 1970. Thirty-five workers were killed in the collapse. The bridge was being assembled in sections, which were elevated and then bolted to one another. It was discovered that two adjoining sections were not flush with one another as designed; the difference in "camber" was about 3 inches, while the specifications called for a difference of no more than 1 inch. In order to fix the problem, the builders should have lowered the two pieces to the ground again, but this would have caused a delay and a cost overrun, so instead they decided to fix them in place.

They applied a very primitive solution, one of placing 8-ton concrete blocks on the higher span, to push it back into line with the other one. This then caused the steel plates to buckle out of shape by as much as 15 inches. In an ill-fated and foolhardy attempt to eliminate the buckling, the builders decided to remove the bolts holding the steel plates in place. After the first sixteen bolts had been removed, the plates had slipped so much that the remaining bolts were jammed and could not be unscrewed. The workers then tightened each of these until they broke, removing the pieces. Like a man sawing off a tree limb upon which he is sitting, they continued removing bolts, until the entire structure collapsed, killing many of them. A Royal Commission appointed to investigate the disaster concluded that what had happened was "inexcusable" and that the builder's performance "fell far short of ordinary competence" (Royal Commission 1971, p. 97).

An example of a structure that arguably was designed acceptably by contemporary standards, but that fell anyway, was the Tacoma Narrows Bridge (built 1938–1940), popularly known as "Galloping Gertie" because of the alarming way it flailed around under high winds before eventually tearing apart. While most bridge disasters occur when a load crosses the bridge that exceeds its carrying capacity, the Tacoma Narrows Bridge had more than an adequate margin of safety for any traffic load. What the architect had failed to anticipate was that the long and thin bridge had "aerodynamic qualities somewhat like the wing of an aeroplane" (Rastorfer 2000, p. 33). Buffeted by heavy winds on November 7, 1940, the whole span began to twist. Finally, hours after Gertie began its last gallop, the bridge tore itself apart and fell.

Petroski notes that bridge failures follow an approximately thirty-year cycle. A notorious failure leads to the use of a new model, which at first is designed conservatively, but then extended and overextended, until a new failure results, and then a new model emerges. The "high girder" design led to the collapse of the Tay Bridge (Dundee, Scotland, 1879), which resulted in the new cantilevered design, which was responsible for the double collapse (in 1907 and 1916) of the Quebec Bridge, which brought about the suspension model, of which Galloping Gertie was an example. In this sense bridges may illustrate a general dynamic, one society should always take into consideration when attempting to make informed ethical use of science and technology.

JONATHAN WALLACE

SEE ALSO *Dams; Water.*

BIBLIOGRAPHY

Billington, David P. (1979). *Robert Maillart's Bridges*. Princeton, NJ: Princeton University Press. Billington has written two other books exploring the work of this famous bridge engineer.

Billington, David P. (1983). *The Tower and the Bridge: The New Art of Structural Engineering*. New York: Basic.

Kharbanda, O. P., and Jeffrey K. Pinto. (1996). *What Made Gertie Gallop? Lessons from Project Failures*. New York: Van Nostrand Reinhold.

Mock, Elizabeth B. 1972 (1949). *The Architecture of Bridges*. New York: Arno Press, published for the Museum of Modern Art.

Petroski, Henry. (1995). *Engineers of Dreams: Great Bridge Builders and the Spanning of America*. New York: Knopf.

Rastorfer, Darl. (2000). *Six Bridges: The Legacy of Othmar H. Ammann*. New Haven, CT: Yale University Press.

Royal Commission (E. H. E. Barber, F. B. Bull, and H. Shirley-Smith). (1971). *Report of Royal Commission into the Failure of West Gate Bridge*. Melbourne, Australia: C. H. Rixon, Government Printer.

Triborough Bridge Authority. (1939). "Is There Any Reason to Suppose They Are Right Now?" New York: Author.

Vose, George L. (1887). *Bridge Disasters in America: The Cause and the Remedy*. Boston: Lee and Shepard.

BROOKINGS INSTITUTION

SEE *Public Policy Centers.*

BUDDHIST PERSPECTIVES

• • •

Buddhism arose around 500 B.C.E. as a practical response to the trouble and suffering that characterize the human condition. Uniquely among traditions concerned with those issues, Buddhism has never offered a final description of ultimate reality; it also has not proposed a universal fixed solution to the persistent and concrete problems of solely human trouble and suffering. Instead, Buddhism has developed a general yet systematic strategy for generating truly sustainable resolutions of the trouble and suffering that afflict all sentient beings in their specific contexts.

Significant common ground with the traditions of science and technology, particularly as they have developed in the West, is suggested by Buddhism's commitments to developing insight into patterns of causal relationship; challenging both common sense and other, more sophisticated forms of presupposition and authority; construing knowledge as a cumulative and consensual process; and devising concrete interventions to redirect patterns of human activity. However, Buddhism traditionally also has avoided any form of reductionism (materialist or otherwise), countering claims of both privileged subjectivity and absolute objectivity, inverting the presumed priority of facts over values, identifying the limits of (especially instrumental) rationality, and cultivating limitless capacities for emotionally inflected relational transformation. These commonalities and differences suggest that Buddhism is well positioned to complement but also critically evaluate science and technology as epistemic (knowledge-centered) and practical enterprises.

Historical Background

Originally promulgated in what is now northern India by Siddhartha Gautama (likely 563–483 B.C.E.), who became known as the Buddha, or "Enlightened One," the teachings of Buddhism quickly spread across the subcontinent and, over the next half millennium, throughout central, eastern, and southeastern Asia. Its emphasis on the need for context-specific responses and resolutions tailored to each new linguistic and cultural environment resulted in a distinctive pattern of accommodation and advocacy through which Buddhism steadily diversified, resulting over time in a complex "ecology of enlightenment."

Traditionally, Buddhist teachings and practices have been classified into three broad evolutionary streams: the *Hinayana* ("Small Vehicle") stream, which is prevalent today in southeastern Asia and more commonly is called the *Theravada*, or "way of the elders"; the *Mahayana* ("Great Vehicle") stream, which is most prevalent in eastern Asia; and the *Vajrayana* ("Diamond Vehicle") stream, which is associated primarily with Tibet and the societies and cultures of north-central Asia. None of these streams has a universally central text such as the Confucian Analects, the Christian Bible, or the Muslim Qur'an. There also are no globally fixed Buddhist institutions or centralized authorities. Although the analogy is not precise—especially because Buddhism is not a theistic tradition and does not advocate a pattern of belief in a supreme deity or deities—one can compare the breadth of Buddhist teachings and practices with that of the "Abrahamic" religions of Judaism, Christianity, and Islam.

A coherent axis of critical insights and practical strategies has remained constant in the course of the historical development of Buddhism. This axis is expressed most succinctly in the so-called *Four Noble Truths*, the fourth of which has come to be known as the *Eightfold Path:* All *this* is suffering, troubled or troubling (Sanskrit: *duhkha*); suffering or trouble arises with particular patterns of conditions; suffering or trouble ceases with the dissolution or absence of those patterns; and those patterns of conditions can be dissolved through the cultivation of complete and appropriate understanding, intentions, speech, action, livelihood, effort, mindfulness, and attentive virtuosity. The insights and practices summarized in the Four Noble Truths traditionally have been referred to as the *Middle Way*, a brief examination of which can introduce Buddhism's distinctive stance with respect to science and technology.

THE MIDDLE WAY: THE ONTOLOGICAL PRIORITY OF AMBIGUITY. Buddhism originated at roughly the time when early Greek thinkers were developing the precursors to natural science and philosophy. As in Greece, the intellectual terrain in India in the first millennium B.C.E. was extremely fertile. If anything, the range of Indian beliefs and debate regarding the nature of ultimate reality, its relationship to the world of experience, and the meaning and purpose of the good life exceeded that which developed on the Peloponnesian peninsula and in Asia Minor.

Recognizing the interdependent origins of all things, the Buddha saw that each individual view in the spectrum of beliefs failed to resolve the trouble and suffering afflicting all sentient beings. Moreover, he realized that the entire spectrum—encompassing a range of metaphysical and ethical positions running from hard materialist reductionism and hedonism at one end to

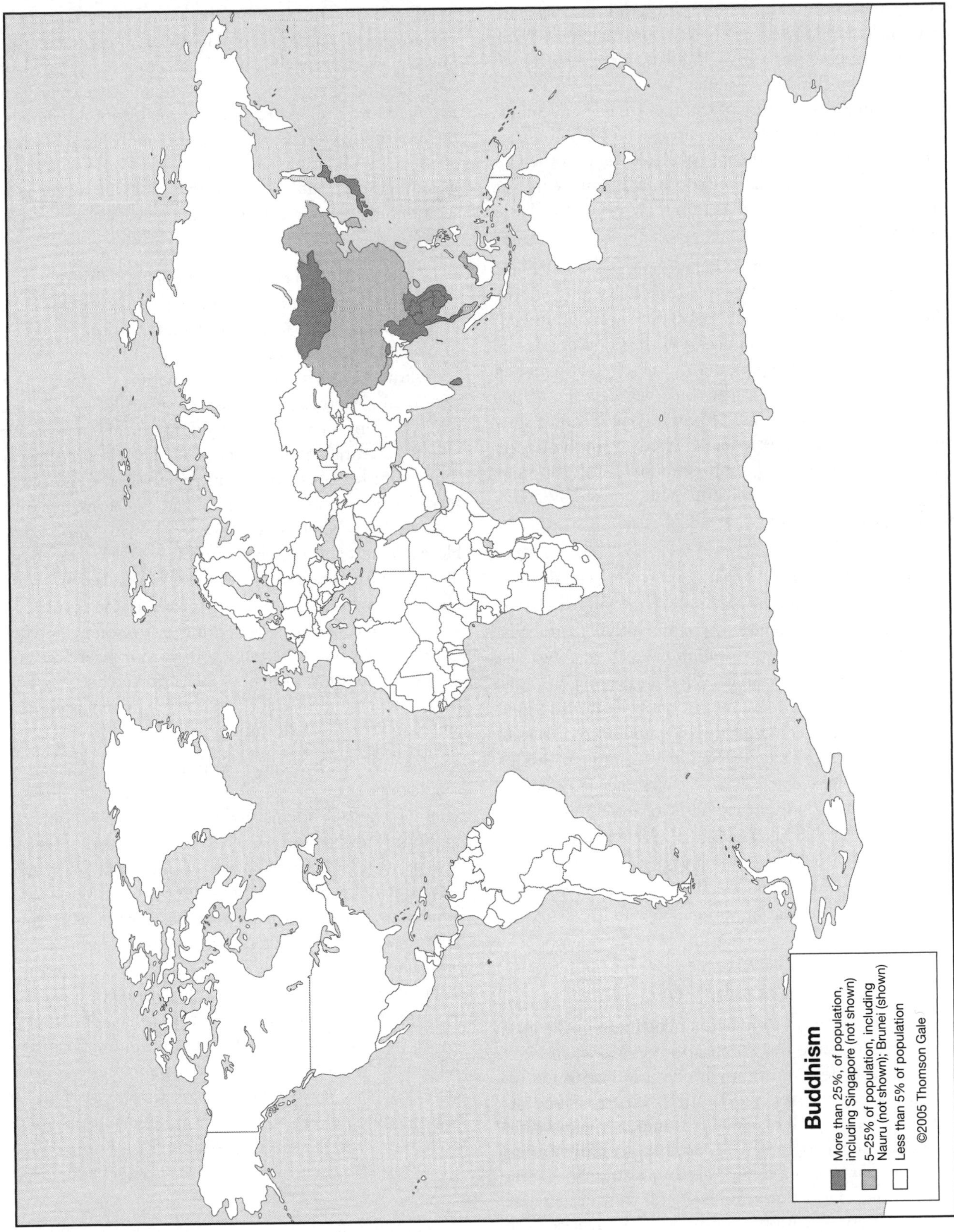
Buddhism
More than 25%, of population, including Singapore (not shown)
5–25% of population, including Nauru (not shown); Brunei (shown)
Less than 5% of population
©2005 Thomson Gale

theistic monism and asceticism at the other—was similarly inadequate. The very conviction that some independent ground (matter or spirit, for example) or grounds (as in the case of metaphysical dualism) underlies all things was a primary cause of trouble and suffering. Equally conducive to suffering was the belief that individual things exist independently of one another. In actuality, the Buddha realized, nothing literally exists or "stands apart" from all other things. What is most basic is relationality.

Rather than being a compromise position or a synthesis of a variety of contrasting views, the Middle Way consisted of the process of critically countering *all* epistemic and practical stances and the "horizons" associated with them. It represents a return to that which is prior to the exclusion of the "middle" between "this" and "that," between what "is" and what "is-not." This process is modeled most concisely perhaps in the teaching of the three marks, an injunction to see all things as troubled or troubling, as impermanent, and as having no self or fixed essence and identity.

THE TEACHING OF THE THREE MARKS. The distinction between is and as—that is, between existential claims and strategic claims—is particularly important in the imperative to see all things as characterized by *duhkha,* or suffering and trouble. Whereas claiming that all things are troubled or suffering can be shown to be empirically false, seeing all things as troubled or suffering causes one to perceive how even the moments of greatest happiness come at a cost to someone or something. Far from being an exercise in pessimism, seeing all things as troubled or troubling helps a person understand his or her situation from another person's perspective. In effect, this entails opening up connections that allow people to realize an ethically shared presence. It means becoming aware that in some way all people make a difference to one another and have a responsibility for asking, "What kind of difference?"

Seeing all things as impermanent (Sanskrit: *anitya*) makes it impossible for people to assume or even hope that they can hold on to anything forever. This undercuts the kinds of expectation that lead to disappointment and suffering. It also makes it impossible to sustain the belief that people can do nothing to change their current circumstances. Seeing all things as ceaseless processes means seeing that no situation is truly intractable. Because every situation continuously evidences both energy and movement, debate cannot center on whether change is possible but only on what direction it should take and with what intensity.

Finally, seeing all things, including humans, as lacking any essential nature or identity renders impossible any claims that specific people are inherently good or bad. This dissolves the primary, prejudicial grounds for racial, ethnic, religious, and political conflict; it also undercuts any pretense that people simply are who they are. Seeing all things as *anātman* (Sanskrit)—literally, as having "no-self"—forfeits the basic conditions of maintaining chronic conflicts and opposition.

It also entails abandoning any justification for separating spirit and nature, the human and the animal, the individual and its environment, and consciousness and matter. The teaching of no-self thus came to be associated with the practice of seeing all things as empty (Sanskrit: *śūnya*), that is, as a function of horizonless relational patterning. For this reason, in later Buddhist usage emptiness (Sanskrit: *śūnyatā*)—the absence of any abiding essential nature—often has been equated with fullness. Instead of signifying its privation, the emptiness of a thing consists in its unique way of bringing into focus and contributing to all other things. An observable example of this is the way species contribute both directly and indirectly to one another's welfare in a sustainable ecosystem, with each species uniquely processing, circulating, and augmenting the resources of the system as a whole. As put by the second-century C.E. Indian Buddhist philosopher Nāgārjuna (ca. 150–250 C.E.), understanding emptiness means appreciating the mutual relevance of all things.

Doing this, however, also entails realizing that what people refer to as separate, individual "things"—whether plants, animals, human beings, or histories—are nothing more than people's own editions of the total pattern of relationships that they focus and to which they contribute. For example, what people take a dog *to be* reflects their own values—the horizons of what they believe (or will allow) to be relevant—and this varies with whether a person is a laboratory worker, an only child living on a farm, or an elderly person confined to a small apartment. Because the particulars of people's experiences are conditioned by their values and intentions, people's day-to-day experiences cannot provide complete or objective pictures of their situation. In actuality, what people customarily assume to be independently existing objects are compounded or put together (Sanskrit: sa l sk l ta) out of habitual patterns of relationship.

Although many of these habits—and thus the nature of people's experience—reflect relatively individual values, intentions, likes, and dislikes, they also are conditioned by the values, goals, and desires embodied in families, communities, social and political institutions, and cultures. In Buddhist terms, the human world arises as an expression of people's karma and any practice directed at resolving the suffering or trouble that occurs in it must be karmically apt.

THE TEACHING OF KARMA. According to the Buddhist (as opposed to Hindu) teaching of karma, people should not see the topography of their life experiences as a simple and objective outcome of the intersection of their actions and the operation of universal moral law and/or divine will. It also should not be seen as a simple function of "natural law" and/or "chance." Instead, individual and communal experiences should be seen as reflecting ongoing and always *situated* patterns of consonance and dissonance among people's values and intentions. In light of the emptiness and impermanence of all things, karma can be understood as a function of sustained acts of disambiguation, a pattern of values-intentions-actions that constitutively orders the world and the individual's experienced place in it. Thus, not only do people have and share responsibility for the direction in which things are headed, the *meaning* of the human situation as a whole is continuously open to revision. The Buddhist cosmos may be described as irreducibly dramatic, a place in which all things are at once factually and meaningfully interdependent.

The Buddha most commonly discussed karma in terms of basic relational orientation: an orientation toward chronic and intense trouble and suffering (Sanskrit: *samsara*) and another toward liberation from those states (Sanskrit: *nirvana*). Orienting the individual and communal situation away from *samsara* and toward *nirvana* cannot be done through independent exertions of will aimed at bringing about the world people want. Understood karmically, controlling one's people's circumstances so that one experiences what one wants causes one to live increasingly in want, in circumstances increasingly in need of further control. Skillfully and sustainably directing one's situation away from trouble and suffering depends on seeing all things as thoroughly interdependent in a world in which differences truly make a difference and freedom is not a state of limitless choices or autonomy but a horizonless capacity for relating freely. Buddhist freedom does not pivot on matters of fact but on meaning; it is a matter not of controlling consequences—the victory of "free will" over "chance" and "determinism"—but of demonstrating appreciative and contributory virtuosity.

PRAJÑĀ, SAMĀDHI, AND ŚĪLA: WISDOM, ATTENTIVE MASTERY, AND MORAL CLARITY. All Buddhist practice thus can be seen as directed toward healing the "wound of existence." Traditionally, this was understood as requiring three dimensions of sustained capacity building: *prajñā*, *samādhi*, and *śīla*, that is, insight into the irreducible relationality of all things; attentive mastery, a function of meditative training that implies both perceptual poise and responsive flexibility; and moral clarity arising from attunement to the currents of value and meaning constitutive of any karmically inflected situation and a capacity for discerning how to orient them away from *samsara* and toward *nirvana*.

Thus, Buddhist practice is always both a *critique of self* and a *critique of culture*. Neither of these as aspects entails a general rejection of personal or social norms and institutions. However, both necessitate continuous and context-sensitive evaluation of those norms and institutions and the material processes through which they are realized. The relative balance of these dimensions of Buddhist practice of course have varied historically. In light of the nature of contemporary societies, they entail a readiness to engage science and technology critically.

Buddhism in Relation to Science and Technology

There have been robust traditions of science and technology in many Buddhist cultural spheres, particularly in India and China. In general, those traditions were not subject to direct critical attention and did not play significant roles in shaping the patterns of accommodation and advocacy that characterized Buddhism's adaptation to its changing cultural, social, and historical circumstances. Although there are passages in early canonical teachings that indirectly address the place of technology in governance and the furthering of social good (e.g., the *Cakkavatti Sihanda Sutta*), Buddhist critiques of scientific knowledge and considerations of the ethics of technology are only implied in broader critiques of religious, philosophical, and commonsense views. This was true throughout the first two millennia of Buddhist history even when Buddhist universities were the largest and most comprehensive in the world (roughly 600–900 C.E.), with faculties of as many as 2,000 teaching international student bodies in excess of 10,000.

A major shift occurred with the rapid expansion of European colonialism from the sixteenth through the late nineteenth centuries. Resting on interwoven scientific and technological advances, the colonial era brought Buddhism to the attention of the West and also brought modern Western traditions of science and technology to the attention of the Buddhist world.

Two primary currents of interaction emerged at the beginning of the twentieth century and have remained strong since that time. The first involves Buddhist accommodations of scientific and technical knowledge, initially in the colonial states of southern and southeastern Asia. Reflecting on the course of events on the Indian subcontinent, Buddhist leaders concluded that to the extent to which Buddhism was positioned as a religion based on revelatory insights and "unscientific" practices, it would undergo rapid and probably fatal erosion. Those leaders thus began to find textual evidence that would support the claim that Buddhism was in fact a rational and empirically grounded tradition that in many ways prefigured the role of science in the modern West. This "Protestant Buddhism" positioned itself as scientifically rational, logical, and devoid of the sorts of superstitions, myths, and mysticism that were a severe liability in Western eyes. The legacy of those "reform" movements can be seen today in the "globalization" of Tibetan Buddhism.

The second current of interaction developed largely as a result of the rise of science as the West's intellectual sovereign, the associated corrosive effects on European and American religious faith, and the breakdown of classical Newtonian physics. Asian traditions, Buddhism in particular, appeared as complementary systems that could provide scientific reality with a cogent ethical dimension, with scientists and philosophers such as Albert Einstein (1879–1955), Alfred North Whitehead (1861–1947), Betrand Russell (1872–1970), and Robert Oppenheimer (1904–1967) hailing Buddhism as the religion of the future and the appropriate partner of science.

In the final third of the twentieth century, as Western knowledge about Buddhism increased, there came to light—especially in cosmology, physics, biology, ecology, and the computational sciences and neuroscience—patterns of uncanny resonance with Buddhist teachings that caused many people to conclude that they demonstrated the prescient, "postmodern" nature of Buddhism and its "anticipation" of, as well as potential for contributing to, contemporary science. More cautious commentators have seen the encounter between Buddhism and contemporary science—particularly in psychology, medicine, the biology of communication and perception, and behavioral science—as extremely fertile and mutually beneficial, with each tradition being assisted in its pursuit of truth.

Some Buddhists question the logic and wisdom of the marriage of Buddhist and scientific approaches to truth. It has been pointed out, for example, that legitimizing Buddhist teachings on the basis of their anticipation of current scientific truths is counterproductive. In light of the fact that the history of scientific change can be described as a "punctuated" evolution of essentially broad and incompatible research paradigms, many contemporary scientific truths will have no place in the science of the next decade, much less in that of the next century. Identifying Buddhism with current scientific paradigms runs the risk of discrediting Buddhism as they are replaced.

Moreover, it has been argued that although science often has been characterized as explicitly eschewing questions of meaning and claims neutrality with respect to the uses of scientific knowledge, Buddhism is centrally concerned with fostering directed revisions of the interdependence of all beings and stresses the union of knowledge and compassionate engagement.

Prospects for Critical Interaction

This suggests an opportunity for a "third stream" that would restore and enhance Buddhism's traditional role of examining patterns of belief and conduct and disclosing how they are limited and/or counterproductive in terms of understanding and resolving trouble and suffering.

Until recently most Buddhist work along these lines focused on the roles of science and technology in industrial and postindustrial patterns of economic development that have induced a drift toward materialism, consumerism, and fractious individualism. It has been noted that science and technology have played into global historical processes through which diverse patterns of sustainable interdependence have been replaced with patterns of simple coexistence. This systematic translation of diversity into mere variety has been criticized as resulting in a decrease of responsive and contributory capacity that is particularly apparent at the community level, with entire villages having been rendered unsustainable through incorporation into the global market economy. Here primary ethical attention has been given to the uses of science and technology to further elite, corporate, and national interests over and often against those of particular populations and the natural environment.

CHALLENGING THE VALUE-NEUTRAL STATUS OF SCIENCE AND TECHNOLOGY. Some Buddhist critics have begun to question whether the moral valence of science and technology can be restricted to the way in which they are used. When considered in the context of interdependence and karma, it is apparent that

Western-style development both drives and is driven by scientific and technological activity and that this symbiotic relationship is not accidental. In actuality it reveals deeply and continuously shared values. Because Buddhist ethics is concerned foremost with how both intentions and values shape human circumstances and experience, this recognition entails admitting that science and technology have a moral influence apart from any particular uses to which they are put.

At least since the time of Galileo (1564–1642), Western (and now global) science and technology have coevolved, embodying a constellation of values that include precision, predictability, objectivity, universality, power, and independence, all of which can be said to depend on the values of control and autonomy. These core values have proved to be highly compatible with short-term positive consequences in responding to trouble or suffering. Promoting these values means promoting the freedom to experience what people want in circumstances they prefer. From within a linear causal framework there is little reason to expect that the same situation will not hold in the long term.

However, in terms of the recursive processes of karmically ordered causation and change, control and autonomy—when expressed with sufficient commitment and/or on a sufficient scale—generate ironic effects and intensifying cycles of perceived trouble or suffering. For instance, a sustained commitment to control leads to increasing capacities for control but also creates circumstances that are both open to *and* in need of control. Because control always is exerted over and against another person or situation and cannot truly be shared, its widening instantiation engenders increasingly steep slopes of advantage/disadvantage, with a prime example being the income and wealth disparities endemic to technology-permeated global markets.

DISPLACING THE INDIVIDUAL AS THE UNIT OF ANALYSIS IN EVALUATING SCIENCE AND TECHNOLOGY. Although autonomy or the freedom to choose or control the nature of one's experienced circumstances may appear to be a simple ethical good, this is true only insofar as *individual* needs, desires, and values are taken as an evaluative basis or unit of analysis. In the absence of universal agreement about the desired nature of shared circumstances and the meaning of the good or the effective isolation of disagreeing parties, multiple exercises of autonomy within a population necessarily result in conflict.

The dominant Western ethical responses to this dilemma—utilitarianism and communitarianism—have not challenged the assumption that individually existing beings are the basic unit of both ethical analysis and communities. Those schools of thought thus have remained compatible with unabated commitments by both individuals and communities to technological development biased by an orientation toward control and autonomy. By contrast, the ethics associated with the Buddhist teachings of emptiness, interdependence, and karma require that qualities of relationship be taken as the basic unit of consideration. Generally stated, granted that the individual, independently existing, and rightfully autonomous self is a pernicious fiction, using the individual as the unit of analysis in evaluating science and technology can only lead to ironic consequences.

From this perspective it has been argued that control- and autonomy-biased technological development leads to mediating institutions, such as global commodity markets and mass media, that allow meaningful differences to be nullified while distracting attention from immediate personal, communal, and environmental relationships. This brings about a systematic erosion of diversity and situational capacities for mutual contributions to shared welfare. Thus, whereas control- and autonomy-biased technologies are conducive to ever-widening *freedom of choice*, they are correlated with an increasingly compromised capacity for *relating freely* and thus with ever more intense and chronic patterns of ignorance, trouble, and suffering.

In more general terms Buddhist ethics cautions against blurring the distinction between tools and technologies. Tools should be evaluated in terms of their task-specific utility for individual users (persons, corporations, or nation-states) and should permit the exercise of "exit rights," that is, choosing *not* to use them. Technologies, however, never are used in a literal sense. Instead, they consist of broad patterns of conduct that embody systems of strategic values and encompass activities that range from resource mining and tool manufacturing to marketing and the innovation of new cultural practices. Although one may choose not to use the tools associated with a particular technology, the world in which one lives continues to be shaped by that technology. With respect to technologies, there are no real exit rights.

From a Buddhist perspective technologies and the sciences with which they symbiotically develop systematize the way people conceive and promote their ends, conditioning the meaning of things, and thus can be evaluated only in terms of the ways in which their core values affect the quality of people's conduct and relationships. In Buddhist terms this entails critically assessing how and to what extent these values are consonant with the core Buddhist practices of cultivating wisdom, attentive virtuosity, and moral clarity for the purpose of realizing liberating patterns of interdependence.

It generally is agreed among Buddhists that scientific advances in people's understanding of factual processes—for example, the dynamics of climate change—should inform efforts to resolve current and future trouble and suffering sustainably. It also is agreed that scientific and technological research should be undertaken in ways that contribute not only to human welfare but to the welfare of all sentient beings. In combination, these commitments make imperative a deepening of the historically arranged "marriage" of Buddhism, science, and technology and promise an increasingly skillful furthering of the Middle Way.

PETER D. HERSHOCK

SEE ALSO *Bhutan; Chinese Perspectives; Hindu Perspectives; Indian Perspectives; Japanese Perspectives; Virtue Ethics.*

BIBLIOGRAPHY

Anonymous. (1995). *Cakkavatti Sihanda Sutta*, trans. Maurice Walsh. In *The Long Discourses of the Buddha*. Boston: Wisdom Publications. Authors are ancient and unnamed. This short text is a provocative commentary on the forces leading to social collapse and the importance of values in framing adequate responses to those forces.

Goleman, Daniel. (2003). *Destructive Emotions: A Scientific Dialogue with the Dalai Lama*. New York: Bantam Doubleday Dell. This is a fine introduction to some of the work being done at the interface between Western psychology and psychotherapy and traditional Buddhist psychology and self-cultivation.

Gombrich, Richard. (1995). *Theravada Buddhism: A Social History from Ancient Benares to Modern Colombo*. London: Routledge. This short volume includes a succinct and careful introduction to the teachings and history of early Buddhism and its historical development into the Theravada tradition, one of the three major divisions of contemporary Buddhist thought and practice.

Hershock, Peter D. (1999). *Reinventing the Wheel: A Buddhist Response to the Information Age*. Albany: State University of New York Press. This book examines the karmic implications of the core values embodied by the predominant technological lineage, considering its historical, social, political, and economic precedents, with particular emphasis on its colonial legacy and effects on consciousness and community.

Jones, Ken. (1993). *Beyond Optimism: A Buddhist Political Ecology*. Oxford: John Carpenter. This monograph provides a sustained account of the implications of using Buddhist resources to understand and reframe contemporary political, economic, and environmental institutions.

Lopez, Donald, Jr. (2002). *A Modern Buddhist Bible: Essential Readings from East and West*. Boston: Beacon Press. This is a collection of historical essays and articles by a wide range of seminal figures in the meeting of Buddhism and the contemporary West, drawn from throughout the twentieth century.

Loy, David. (2002). *A Buddhist History of the West: Studies in Lack*. Albany: State University of New York Press. A methodical reading of historical, scientific, political, and cultural changes in the history of the European and American West through the conceptual lens of early Buddhist thought.

Payutto, Bhikkhu P. A. (1993).*Towards Sustainable Science: A Buddhist Look at Trends in Scientific Development*, trans. B. G. Evans. Bangkok, Thailand: Buddhadhamma Foundation. This is an excellent introduction to contemporary Thai Buddhist critical encounters with Western science, written by one of the leading exponents of Buddhist engagement with contemporary issues.

Varela, Francisco J.; Evan Thompson; and Eleanor Rosch. (1991). *The Embodied Mind: Cognitive Science and Human Experience*. Cambridge, MA: MIT Press. Deeply informed by Buddhist teachings and practices, this is a masterly scientific assessment of the biology of consciousness, thought, and behavior by leading scientific researchers in cognitive science.

Waldron, William S. (2000). "Beyond Nature/Nurture: Buddhism and Biology on Interdependence." *Contemporary Buddhism* 1(2): 199–226. A concise look at issues surrounding nature and nurture in the development of living systems through a close examination of the implications of Buddhist readings of interdependence.

Wallace, B. Alan. (2003). *Buddhism and Science*. New York: Columbia University Press. A fine collection of papers and essays that range across the sciences, providing an excellent introduction to "second stream" encounters of Buddhism (especially Tibetan Vajrayana Buddhism) and contemporary science.

Watts, Jonathan, and David Loy, eds. (2002). *Spiritual Responses to Technology*. Special issue of *ReVision Journal* 24(4). This is a good introduction to contemporary Buddhist thought on technology and its personal, social, and cultural implications.

Williams, Paul. (1989). *Mahayana Buddhism: The Doctrinal Foundations*. New York: Routledge. This is a very good introduction to the history and thought of the Mahayana and Vajrayana traditions of Buddhism.

BUILDING CODES

Building codes are extratechnological laws that govern the design and construction of structures. They can be placed within a hierarchy that begins with metaethics, and includes ethics, laws, codes, ordinances, standards, and operating practices. A typical code provision is, for example, the government enforced specification that the exterior doors of public buildings must open outward (International Conference of Building Officials [ICBO]), or that the vertical rise of steps and stairs shall not be less than four inches nor more than seven inches (ICBO). These requirements are, however, social rather than technological in origin because they are intended to mediate human behavior in the case of emergencies such as fires in buildings.

In general one can say that building codes both reflect and enforce social values. They are, then, an historical index of how social values regarding the safety, health, and welfare of individuals are materialized as the built world. Because the ethical significance of building codes must be understood within the context of their evolution and development, a historical view of this topic is helpful.

Historical Development

The first building code is generally credited to be Article 229 of the Code of Hammurabi (Mesopotamia, 2250–1780 B.C.E.), which requires that "If a builder build a house for someone, and does not construct it properly, and the house which he built falls in and kill its owner, then that builder shall be put to death." (Harper 1904, p. 81) The ethical principle behind this code is an *eye for an eye*—the deontic idea that justice is absolute and unchanging, never moderated by local conditions or human situations.

In contrast to such moral absolutism the Greeks, Romans, and early Islamic societies developed more complex or nuanced building codes. These may be said to be of three types: *tacit codes* that regulate cultural production, *legislative codes* that regulate public resources, and *industrial codes* that regulate modern material and labor standards.

Tacit or unspoken codes are those that bind citizens to the customary practices of their community. Anthropologists argue that the way cultures build—what Kenneth Frampton (1995) calls *tectonic culture*—is as important and distinct as the way they speak. Tacit building codes are systems of ordering and inhabiting the world in a manner that is consistent with cosmological order as the community interprets it. To build well means to construct one's house and dwell righteously—in a manner consistent with divine order (Norberg-Schulz 1979). To depart too far from the tectonic order of one's culture would be to offend the god(s), or those forces responsible for ordering the universe. Tacit codes are a powerful part of vernacular societies but diminish in their influence with the self-conscious invention of modern design and construction practices. The ethical principle behind tacit or vernacular building codes is *sin* against divine authority.

Legislative codes are explicit civil laws concerned with maintaining equity and justice between private parties and that guard public resources such as streets against private exploitation or carelessness. Early examples of this type are the Byzantine Roman Treatise of Julian of Ascalon (533 C.E.) and the codes of the Prophet Mohammed during his reign in Medina (622–632 C.E.) (Hakim 1986). These codes make explicit both the rights and obligations of citizens building within previously tacit conventions. A typical example was a law regulating the construction of *party walls*, a single wall that separates and supports two houses. According to architect and planner/historian Besim Hakim, Mohammed said that "a neighbor should not forbid his neighbor to insert wooden beams in his wall" (Hakim 1986, 2003). In the context of desert dwelling, party walls are private resources that enable a public way of life by aggregating individual dwellings into an urban form that shields the community as a whole from inhospitable natural conditions created by too much sun and wind.

The ethical principle that informs these early codes is not, however, conceptually different from those that developed in England on the basis of legislative action, first in 1189 and most significantly in 1676 in response to the great London fire of 1666 (American Institute of Architects [AIA]). These ordinances were principally fire protection measures that ultimately rely upon what nineteenth-century utilitarian philosophers referred to as the *greatest happiness principle*—the notion that right actions are those that cause the greatest amount of happiness and the least amount of pain (Bentham 1962). The conditions of rapid industrialization and urban population growth in mid-nineteenth-century Britain certainly lent urgency to the development of explicit codes that suppressed some individual rights, such as the freedom to construct one's roof of highly flammable thatch, in the name of the public good. In the view of utilitarian philosophers, principally Jeremy Bentham (1748–1832), such suppression of individual rights was justified for the overall health of the *civic economy*, the ability of the society to provide for the general well being through preventative measures (Chadwick 1965). The greatest happiness principle was quickly expanded in Europe and North America to regulate not only fire, but those unsanitary conditions associated with rapid urbanization and industrialization that threatened general public health (Melosi 2000).

Industrial codes were developed by government and industry to standardize modern building materials and processes. As new building components such as glass and iron became increasingly available in the late-nineteenth century, it became progressively inconvenient and uneconomical for builders in different locations to employ materials of differing thicknesses, lengths, and strengths. In 1901 the National Bureau of Standards was created by an act of Congress to conduct research and aid small business by creating universal standards of production. In the early-twentieth century, manufacturing organizations, comprised and funded by competing producers such as the American Institute of Steel Construction (AISC, founded in 1921), recognized that it was in their common interests to self-regulate standard measures of size and quality before government did so. Without such standard codes of production, it would be very difficult, for example, to use steel produced in Pittsburgh in a building designed in Chicago to be constructed in San Francisco. Economic and political interests inspire these codes and standards. They are designed to optimize exchange value across political jurisdictions, and are linked to the general process of modernization in which the tacit knowledge of the artisan is supplanted by the formal knowledge of the engineer.

Authorization and Conflict

In the European Union and much of the world, building codes are national and international in scope. This situation has developed from the familiar historical process of modernization. In the United States, however, the legislation of building codes is a state or municipal responsibility resulting in the existence of no fewer than five major building codes and a multiplicity of municipal codes in large cities such as New York.

In 1994 the International Code Council (ICC) was established by the three dominant not-for-profit organizations responsible for the writing of model codes in an attempt to further standardize building codes throughout the Americas. Based near Washington, D. C., the ICC provides a wide range of services to its members through its sixteen regional offices in the United States. Although the ICC's International Building Code (IBC) has been approved for use by forty-four states, individual local jurisdictions are only slowly adopting and enforcing it. This effort may eventually lead to the adoption of a comprehensive building code for the hemisphere, but success will depend upon the speculative possibility of resolving the long-entrenched interests of local industries, labor unions, architects, and building engineers. Toward this end the ICC has established a quasi-democratic process for code development in which each of the dominant model code groups are equally represented.

Building codes exist within a now complex matrix of legislation from all levels of government. Strictly speaking, building codes regulate only the safety of a building structure, its materials, and the environmental systems that render architecture habitable. They are, however, closely related to other types of codes, such as federal, state, and municipal environmental laws (which regulate emissions and impacts on air, water, and land); zoning ordinances (which regulate such urban concerns as land use, drainage, density, and signage); historic preservation ordinances (which stipulate criteria and processes for mandating the preservation of private property); and design review ordinances (which stipulate criteria and processes for regulating the aesthetic compatibility of new structures in existing districts). These vary significantly from nation to nation, state to state, and city to city.

The social production of codes tends to reinforce the interests of codemakers. Historically the manufacturers of building products and systems such as Willis Carrier (1876–1950), the entrepreneur-developer of modern air conditioning, have competed for control of code making with the publicly employed professionals who now dominate the field. For this reason the authorship of building codes is the principal conflict associated with them. This lingering question fuels conflict between governmental regulators, property owners, and the construction industry. In the social democracies of the European Union or the centrally planned economies of Asia or South America, the property rights of individuals and the technological practices of industry are significantly restricted by a broad definition of the public good. In the United States, however, the public good tends to be narrowly defined through scientific criteria generally limited to human safety and health. Behind these differing approaches to the social construction of building codes is a fundamental question of political trust. In the Netherlands, for example, planners and government technocrats are generally respected and trusted to make decisions that reflect the interests of citizens. In the United States, however, citizens tend to trust the market and their own judgment over that of government. Judged on the criterion of the *sustainable development* of cities (Campbell 1996), Dutch codemakers tend to be more effective than those in the United States because citizens tend to understand building codes as a moral obligation to fellow citizens rather than as an imposed restriction on individual property rights.

Assessment

The development of tacit, legislative, and industrial building codes was never a simply a matter of economics, science, or ethics. Rather their formulation is a highly social and contentious process through which some interests are suppressed and others reinforced. In theory one may distinguish how a priori economic, scientific, or moral logic might define a building code. In practice, however, these logics are conflated by the social situation—usually a catastrophe—that mandates changed building practices.

Langdon Winner argued that "... we do not use technologies so much as live them" (Winner 1997, p. 202). His logic suggests that free democratic societies should promote citizen participation in articulating the technical codes that strongly influence the landscapes of daily life. According to Francis Ventre, "...it is the state of knowledge ... [moral, political, and practical] that drives regulation's juggernaut. But whose knowledge? The regulatory expansion after the 1920s seems to owe more to a public will rallied and given form by the cultural preferences and superior technical knowledge of articulate minorities who could link that preference and knowledge to wide social concerns" (Ventre 1990, p. 56) Employing similar logic, Andrew Feenberg proposes that the development of technical codes is the discursive process through which societies modify their fundamental values. It is important to recognize that such *civilizational change* is not what economists would call a trade-off in which an economic good is sacrificed for an environmental or public safety good. Rather such revision of technical codes redefines the cultural values within which economic activity takes place (Feenberg). From both an ethical and historical perspective Americans are no more likely to retreat from emerging environmental standards, for example, than from the Americans with Disability Act (1990), the New York City legislation requiring buildings to have fire exits (1860), or the abolition of slavery (1862).

The historical process of regulating how structures are built is indistinguishable from the social process of deciding how human beings will live together—there will be as many building codes as there are distinct societies. This is one reason why the internationalization of building codes, as proposed by the ICC, raises ethical and environmental questions related to technological colonization. The citizens of Mexico, for example, increasingly resist attempts by global institutions to standardize local building practices that sustain unique cultural practices and ecological conditions. The process of modernization does tend toward the standardization of building codes across countries and continents, but distinct tectonic cultures are not likely to disappear anytime soon. A more important question may be the degree to which citizens of any given society participate in the articulation of building codes, because it is through citizen involvement that government technocrats become accountable for how the community lives, citizens come to trust codemakers, and codes are lived as moral obligations.

STEVEN A. MOORE

SEE ALSO *Architectural Ethics; Building Destruction and Collapse; Engineering Ethics; Modernization; Science, Technology, and Law.*

BIBLIOGRAPHY

American Institute of Architects. (1990). *An Architect's Guide to Building Codes & Standards*. Washington, DC: Author. This source is both a historical and practical guide.

Bentham, Jeremy. (1962 [1838]). *The Works of Jeremy Bentham*, Vol. 1 *Introduction: An Introduction to the Principles of Morals and Legislation*, ed. John Bowring. New York: Russell & Russell. Bentham is generally considered the projenitor of the utilitarian ethics articulated in this text.

Campbell, Scott. (1996). "Green Cities, Growing Cities, Just Cities?: Urban Planning and the Contradictions of Sustainable Development." *Journal of the American Planning Association* 62(3): 296–311. Campbell's diagram of the "planner's triangle" is, perhaps, the most elegant description of sustainable development as envisioned by the authors of "Our Common Future" (1987), World Commission on Environment and Development.

Carrier, Willis Haviland. (1914). *Engineers Hand-book of Tables, Charts ....* Buffalo, NY: Buffalo Forge Co. Carrier's "Hand-book," published as a practical guide, is based upon assumptions that are considered highly ideological by twenty-first century standards.

Chadwick, Edwin. 1965 (1842). *Sanitary Conditions of the Labouring Population of Great Britain*. Edinburgh: Edinburgh University Press. This report to Parliament was Chadwick's major accomplishment. It is historically significant because the author employed empirical and statistical methods rather than the then dominant deductive method of scientific inquiry.

Duany, Andres, and Elizabeth Plater-Zyberk. (1991). *Towns and Town-making Principles*, ed. Alex Krieger with William Lennertz. Cambridge, MA: Harvard University Graduate School of Design; New York: Rizzoli. Although not the most well known treatise on "new urbanism," this volume makes a good case for the development of morphological, or formal, architectural codes.

Feenberg, Andrew. (2002). *Transforming Technology: A Critical Theory Revised*. New York: Oxford University Press. This is a revised and updated version of *Critical Theory of Technology* (1991), New York and Oxford: Oxford University Press.

Frampton, Kenneth. (1995). *Studies in Tectonic Culture*. Cambridge, MA: MIT Press. Frampton's work, including his critical regionalism hypothesis of 1983-1991, should be credited with renewing the academic study of technology within architecture and art history.

Ferguson, R. S. (1974). *The Development of a Knowledge-Based Building Code*. Technical Paper NRCC 014355. Ottawa: National Research Council Canada, Division of Building Research.

Hakim, Besim S. (1986). *Arabic-Islamic Cities: Building and Planning Principles*, 2nd edition. London and New York: KPI.

Harper, Robert Francis, ed. (1904) *The Code of Hammurabi: King of Babylon*, 2nd edition. Chicago: University of Chicago Press.

International Code Council. (2003). *International Building Code*. Falls Church, VA: Author.

International Conference of Building Officials. (1997) *Uniform Building Code*. Vol. 1. *Administration, Fire—and Life-safety and Field Inspection Provisions*. Whittier, CA: Author.

Melosi, Martin. (2000). *The Sanitary City: Urban Infrastructure in America from Colonial Times to the Present*. Baltimore, MD: Johns Hopkins University Press.

Norberg-Schulz, Christian. (1979). *Genius Loci: Towards a Phenomenology of Architecture*. New York: Rizzoli.

Ventre, Francis T. (1990). "Regulation: A Realization of Social Ethics." *VIA* 10: 51–62.

Winner, Langdon. (1977). *Autonomous Technology: Technics-out-of-Control as a Theme in Political Thought*. Cambridge, MA: MIT Press. This is an early and still significant study by the author that examines how technological choices are really social choices disguised in the name of efficiency.

INTERNET RESOURCE

Hakim, Besim S. (2003). "Style and Urbanism: New Urban Codes and Design Guidelines. *Council Report* III/IV (April 2003). Available at http://www.tndtownpaper.com/council/Hakim.htm.

BUILDING DESTRUCTION AND COLLAPSE

Engineers and architects design buildings to stand, and the vast majority of them do so without major incident. Yet occasionally a building does collapse, bringing with it questions about the science, technology, and ethics of structures. Though they happen for a variety of reasons, collapses can be clustered into three groups: those resulting from natural disasters (earthquakes, mudslides, tornadoes, and the like); inadvertent collapses (because of flaws in design, use, and/or maintenance); and intentional destruction (including both planned demolition and malevolent attacks). Each type raises different, if related, ethical questions.

Two types of explanation exist for collapses. The first is focused on the mechanics or physics of the destruction; it asks what forces were acting on (and being produced by) what parts of the structure and in what fashion. The lessons drawn from such analyses will be, necessarily, structural or mechanical in nature. Matthys Levy and Mario Salvadori (2002), for instance, declare that collapses are always due to structural failure, though this failure may come about in a variety of ways (and, though they do not explicitly say so, may or may not be accidental).

A second type of explanation focuses on what might be termed social—rather than physical—dynamics. Here, the forces are those of the designers and others involved in determining whether and how to erect (or destroy) a structure. Such forces are more difficult to analyze and impossible to quantify, but they are as much a part of building success and failure as are the physical laws that allow them to stand or fall. These two kinds of explanations often have different relative weights in examinations of natural, inadvertent, and intentional destructions.

Natural Disasters

Building destructions caused by natural disasters are the most deadly and devastating kind. The 1923 earthquake near Tokyo, Japan, measured 8.3 on the Richter scale and left 100,000 dead; the 1995 Kobe, Japan, earthquake, rated 7.2, was the costliest ever, causing an estimated US$150 billion in damage and destroying nearly 100,000 structures. Tornadoes (including the 148 that formed the Super Outbreak of 1974, killing 315) and hurricanes (such as Camille of 1969, which killed 200 and caused billions of dollars in damage) can cause massive devastation as well.

Although the basic cause of the building collapses in these disasters is structural failure (as is true in any collapse), such widespread collapses pose the immediate challenge of disaster response in the face of damaged (or even nonexistent) infrastructure. Is the community able to cope (on its own or with outside assistance) when communication, rescue, and medical systems have been damaged or destroyed?

Secondary challenges emerge as investigators study which structures failed and which survived, in an effort to learn lessons for future construction. These studies may confirm existing knowledge (e.g., the Kobe Report's confirmation that newer structures survived because of their more sophisticated designs), may point to a need for new knowledge or regulation (as in the 1923 Tokyo quake, which led to Japan's first building code), or may uncover flaws in applying existing knowledge, either because that knowledge is not sufficiently detailed or because it has been inexpertly applied (as turned out to be the case with earthquakes in Mexico City in 1985 and Turkey in 1999). The causes of devastation here are clearly beyond the scientific; cultural and economic factors play significant roles, as do settlement and development patterns. Resulting questions have to do with building standards and where (and how well) they are applied, and economics (decisions about how much safety is worth).

Once an immediate crisis has passed and investigations have been completed, then comes the most challenging phase: deciding what to do next. When the lessons are scientific, they can be codified and shared. When the lessons are cultural or economic, they are harder to learn or apply. Often the issue becomes one of conflict between governmental control and citizen freedom. How much control should local or national governments have over private construction, and how many federal dollars should go toward relief if, say, people build in known flood plains or tornado alleys, while failing to take precautions (or neglecting to purchase appropriate insurance)?

The Murrah Federal Building in Oklahoma City, Oklahoma, after the 1995 bombing. The incident prompted new levels of concern for building standards. *(© James H. Robinson/Photo Researchers.)*

Inadvertent Collapses

The effects of the power of nature may be more deadly, but the effects of the fallibility of human nature provoke a stronger urge to assign responsibility. In 1922 the Knickerbocker Theatre in Washington, DC, suffered a partial collapse, killing ninety-five people. A severe snowstorm that evening both precipitated the collapse and prevented a larger death toll, but was not the underlying cause of the collapse. Subsequent investigations uncovered shoddy design and materials, but charges against the designers and builders were dismissed, and the resulting call to institute district-wide licensing requirements for architects and engineers went unheeded until 1950 (after every other state in the union had adopted licensing laws for engineers). Twenty other states had already passed such laws at the time of the Knickerbocker collapse, seventeen of them in the four years prior to that disaster. New York—home of the American Society of Civil Engineers (ASCE)—was one of those states, passing its law in 1920, after a decade of heated debate and resistance by the ASCE.

When two walkways in the lobby of the Hyatt Regency Hotel in Kansas City, Missouri, collapsed in 1981 during a crowded dance contest, 114 people died. The Hyatt disaster challenged the resolve of a profession that, in its codes of ethics, had recently declared public safety to be the paramount goal. Licensing laws had been in place for over thirty years, but the Hyatt case posed the first test of such regulation in the face of a collapse. Disasters such as the Knickerbocker had encouraged the call for licensing, but once passed, such laws were used primarily to deal with unethical business practices. After five years of investigations and negotiations, two engineers who had supervised the design of the hotel lost their licenses, a decision decried by many of their colleagues as inappropriately harsh given the complex chain of events and professionals involved in the design and collapse of the structure. That criminal charges had been dismissed for lack of evidence strengthened such opposition.

If news of the Hyatt collapse challenged the engineering profession, the story of the Citicorp building in Manhattan renewed its faith and confidence. A 1995 *New Yorker* magazine article revealed that in 1978—a year after Citicorp Center opened—the structural engineer discovered a fatal flaw in the fifty-nine-story building. William LeMessurier blew the whistle on himself and in collaboration with the building owners, insurance agencies, and city officials devised a plan for retrofitting the building to ensure its safety. To avoid a public panic, the building tenants were not informed of the repairs being made to the structure. The case is frequently cited as an exemplar of ethical behavior on the part of those involved, most notably LeMessurier himself, yet the secrecy of the case raises questions about the public's right to know the risks they face and to decide what risks they are willing to assume.

When mercifully vacant buildings collapse, as in the cases of the Hartford Coliseum (1978, Connecticut) and Kemper Arena (1979, Kansas City, Missouri), the effects are dramatic, but far less wrenching for the public as well as for the building profession. In these two collapses, multiple factors combined in unexpected and unfortunate ways. Heavy rains and high winds exploited previously unnoticed weaknesses in the Kemper Arena roof design. In the Hartford collapse, early deformations in the structure were dismissed as insignificant for years, only to compound into the collapse of the roof just hours after an event that had drawn some 5,000 spectators. Hundreds of roof and structure collapses occurred during that winter of record snowfalls, but none so memorable as the one in Hartford. These cases (and the snow-induced Knickerbocker collapse) point to the interplay of natural and human causes in some major collapses, which complicates the matter of assigning responsibility.

As with natural disasters, accidental collapses lead to investigations. Designers strive to derive lessons about design in an attempt to extract some good from the rubble. The easier lessons to learn or reinforce about design and building practice are the scientific ones. Updating building codes and reminding designers of the need for structural redundancies are straightforward actions. The harder lessons are those related to responsibility. How far should the responsibility of a designer extend and to whom? Changes in liability and licensing in the United States over the past century have at once increased designers' authority and their obligations. That tradeoff is the underlying principle of modern professional ethics—professionals possess highly specialized knowledge, which can be used for good or ill, and the public invests professionals with the authority to make decisions and to self-regulate in exchange for a promise to serve the public granting that authority.

Intentional Destruction

In contrast to natural and human disasters are building destructions brought about intentionally, whether through intent to protect or to harm. As buildings age and congestion increases, some owners opt for planned demolition, often to clear the way for newer, safer, or larger structures. Controlled Demolition, Inc., operated by The Loizeaux family of Maryland has become famous for its skill at bringing a structure the size of Three Rivers Stadium (2001, Pittsburgh) down to the ground without harming people or the new stadium rising next door. Robert Moses was perhaps the most prolific developer of the twentieth century, yet he was, reflexively, the most prolific demolisher as well, and has as a result been both praised and vilified for his role in altering the New York cityscape. Whether controlled demolition is large or small, the collapse of each structure marks the end of potentially heated negotiations over preservation and land use.

Whether or not general agreement exists on such demolitions, they are at least planned publicly. Covert acts of intentional destruction exist as well—in the forms of arson, war, and terrorism. Ironically, the World Trade Center (WTC, 1993 and 2001, New York City) and the Murrah Federal Building (1995, Oklahoma City, Oklahoma) act as links between the public and the secret types of building destruction. The WTC began with the planned demolition of the commercial district known as Radio Row and was itself demolished by terrorists. The birth and the death of the WTC both produced victims—those in the former were fortunate to escape with their lives, if not their livelihood. The Murrah building, damaged beyond repair by U.S. terrorists, was eventually brought down by the Loizeaux family firm.

Intentional destruction, though it may be less deadly than other types, is most unsettling because it pits one group of people against another. Although the collapse of the WTC towers was probably an unplanned result of the terrorist airplane attacks, the military does study how to destroy buildings and is even designing "bunker-busting" bombs to attack special fortifications. Yet even in the civilian arena, it is common to debate who properly controls or decides acceptable tradeoffs. In both publicly and privately planned demolition, those making the decisions are rarely those who will be affected by them.

The Oklahoma City bombing ushered in a new era of concern for building standards, though it was not the first terrorist attack on U.S. soil (which dates at least to the deadly 1920 bombing of the Morgan Bank in New York City). If the Murrah bombing was a chink in the armor of U.S. confidence, that crack became a gaping hole with the destruction of the WTC. The investigations into the Oklahoma and New York cases were unusual in that they began by exploring nonmechanical causes, focusing appropriately on the role of the terrorists. But in the WTC case, behind the calls for vengeance and war were whispers asking whether the towers should have stood longer once they had been attacked. The comparatively minor damage suffered by the Pentagon during the same attack vividly demonstrated how important a role building design plays in building performance. How far does a designer's obligation to build a "safe" building extend? The two investigations converged around questions about how best to design future structures to preserve freedom and access while protecting building integrity and security.

Several stages of response are common across these three types of building destruction: the search for lessons, the discovery of complexity in the causes, the proposal to change current practice, and the reluctant acceptance that actual changes will be less sweeping than those proposed. Among the challenges faced by those responding to building collapses, two are continual. The first is that, hard as it may be to identify the causes of a particular collapse, it is inestimably harder to identify solutions that will prevent a whole category of future collapses. The second challenge is to achieve a balance between studying past failures and designing for future successes.

SARAH K. A. PFATTEICHER

SEE ALSO *Architectural Ethics; Building Codes; Design Ethics; Engineering Ethics; Fire; Hazards; Terrorism.*

BIBLIOGRAPHY

Caro, Robert A. (1975). *The Power Broker: Robert Moses and the Fall of New York.* New York: Vintage. Pulitzer-prize winning biography of developer Robert Moses detailing his far-reaching effects on the infrastructure of the city and state of New York.

Herring, Susan Davis. (1989). *From the Titanic to the Challenger: An Annotated Bibliography on Technological Failures of the Twentieth Century.* New York: Garland. Detailed, reliable guide to key primary and secondary sources on wide range of failures.

Levy, Matthys, and Mario Salvadori. (2002). *Why Buildings Fall Down: How Structures Fail,* rev. edition. New York: Norton. Accessible introduction to the physics of building destruction, written for a lay audience; sequel to *Why Buildings Stand Up*.

Liss, Helene. (2000). *Demolition: The Art of Demolishing, Dismantling, Imploding, Toppling, and Razing.* New York: Black Dog and Leventhal. The story of the Loizeaux family and their company, Controlled Demolition, with extensive photographs.

Morgenstern, Joe. (1995). "The Fifty-Nine-Story Crisis." *New Yorker* 71(14): 45–53. The article that broke the story of the averted crisis at Citicorp, after nearly twenty years of secrecy.

Petroski, Henry. (1994). *Design Paradigms: Case Histories of Error and Judgment in Engineering.* One of several books by Petroski using historical case studies to demonstrate the technical lessons engineers learn from failure. New York: Cambridge University Press.

Schlager, Neil, ed. (1994). *When Technology Fails: Significant Technological Disasters, Accidents, and Failures of the Twentieth Century.* Detroit: Gale Research. This encyclopedia of more than one hundred technological disasters includes sixteen building and structural collapses, each with a brief bibliography.

Wearne, Phillip. (1990). *Collapse: When Buildings Fall Down.* New York: TV Books. An introduction to the findings of forensic engineering investigations into 11 major structural collapses of the last half-century.

INTERNET RESOURCE

"Why the Towers Fell." NOVA Online. Available from http://www.pbs.org/wgbh/nova/wtc/. This Internet site supplements the PBS program of the same name, exploring the reasons for the collapse of the World Trade Center.

BUSH, VANNEVAR

Inventor and adviser to U.S. presidents during World War II, Vannevar Bush (1890–1974), was born in Everett, Massachusetts, on March 11, and became a major architect of postwar science policy. He earned doctorates from both Harvard University and the Massachusetts Institute of Technology (MIT), where after a few years in industry he became professor and then dean of engineering. At MIT he also contributed to development of the "differential analyzer," a precursor of the computer. In 1938 he was elected president of the Carnegie Institute of Washington, DC, and then served as director of the U.S. Office of Scientific Research and Development (OSRD), which provided oversight for federal science support from 1941 to 1947. Bush later

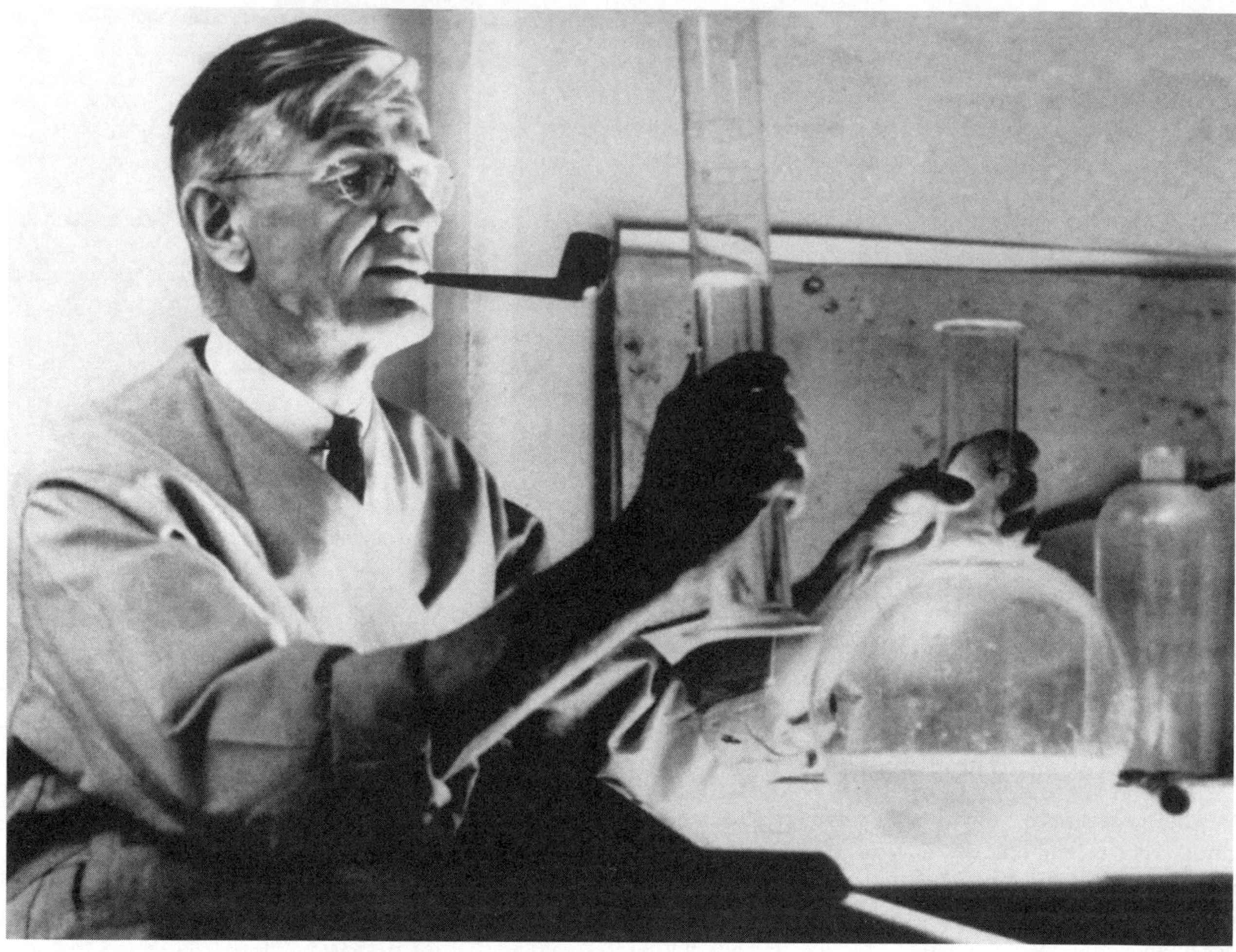

Vannevar Bush, 1890–1974. Bush was a leader of American science and engineering during and after World War II. He was instrumental in the development of the atomic bomb and the analogue computer, as well as an administrator of government scientific activities. *(The Library of Congress.)*

became involved in the private sector, serving as honorary chairman of the MIT Corporation from 1959 to 1971. He died in Belmont, Massachusetts, on June 30.

Policy Achievements

In 1940 Bush persuaded President Franklin D. Roosevelt to create the National Defense Research Committee, which was later subsumed under the OSRD. Arguing that success in World War II would depend largely on innovations in military technologies, Bush led the OSRD in coordinating the relationship between science, the military, and industry. Under his leadership, scientific research yielded vast improvements in military technologies such as the submarine and radar. Bush was also the top policy advisor to President Roosevelt for the Manhattan Project to create the atomic bomb. Although much OSRD work was top secret during the war, Bush obtained near celebrity status, with an article in *Colliers* magazine heralding him as the "man who may win or lose the war" (Ratcliff 1942).

In 1945 Bush wrote two works that pointed toward the future of science and technology. The first was a report titled *Science, the Endless Frontier*, addressed to President Harry S Truman. The impetus had come from President Roosevelt, whose letter of request saw in the wartime collaboration "new frontiers of the mind" to be pioneered for creating "a fuller and more fruitful America" (Bush 1945b, p. viii). In response, Bush argued that scientific progress is essential to the well-being of the nation, specifically addressing the potential of research to promote the public good by preventing and curing disease, supporting economic progress, and improving national security. Bush recommended creation of a "National Research Foundation," arguing that the government "should accept new responsibilities for promoting the creation of new scientific knowledge and the development of scientific talent in our youth" (p. 4). This idea was realized in 1950, after modification by the Steelman Commission, as the National Science Foundation (Steelman 1980 [1947]). But Bush also recognized that "progress in other fields such as the social sciences and the humanities is likewise important" (Bush 1945b, p. v).

Bush's second 1945 publication was a prescient essay, "As We May Think," that established him as a pioneer of the information age. He had been working on his differential analyzer (an analog computer) since the 1920s. This article reflected on the profound implications of such work. The specialization of the sciences had produced a glut of information that was difficult to organize, access, and share. In order to continue the expansion of the knowledge base, Bush outlined a system for storing, retrieving, and linking information. Toward this end, he imagined the *memex*, a mechanical device for storing information that could be consulted rapidly and flexibly.

A precursor to the personal computer, the memex desk was envisioned as using microfilm as an information storage device and having the ability to navigate and form associative linkages or "trails" within vast stores of information. This foreshadowed the notion of the "link" nearly fifty years before its popular usage, thus enabling Bush to be thought of as a conceptual creator of the Web and hypertext systems.

One other key contribution to the industrial development of science in the United States is that Bush instilled in one of his graduate students, Frederick Terman, a belief that regional economies would come to depend on strong relationships between business entrepreneurs and scientific researchers. Terman was later instrumental in forming Silicon Valley, one of the greatest concentrations of high-tech power in the world (Zachary 1997).

Policy Fallout

Bush is credited as an original defender of what has come to be called the "linear model" of science–society relations: give scientists money, and they will just naturally produce socially beneficial results; pure science leads to technology and innovation. Beginning in the decade of his death, however, such a theory was subject to increasing criticism. The economic decline of the late 1970s and 1980s, the end of the cold war in the early 1990s, and the ballooning federal budget deficits of the same period combined to stimulate a rethinking of post–World War II governmental policies toward the funding of science. Although the United States claimed the largest number of Nobel Prizes in science, its economy was in many sectors being bested by Japan, Germany, and other nations. The end of the cold war and the absence of an opposing superpower removed a major justification for continued U.S. investment in more and better high-tech weapons systems. Economic stagnation and budget deficits further called into question the effectiveness of federal investments in science.

Parallel to such political and economic questions, social studies of science challenged the idea of the purely nonpolitical character of science. For example, feminist criticisms of investments in cancer research (more money for prostate cancer than for breast cancer, despite more people dying of breast cancer) clearly illustrated how the interests of scientific researchers (mostly males) could influence the directions of science. Taken together these three types of questioning conspired to sponsor a broad reassessment of U.S. science policy—a reassessment whose most prominent feature has been increasing engagement with the social sciences.

Public science funding continues to be criticized for propagating the linear model that separates the production of scientific knowledge from society. Policy theorists are calling for a new "social contract for science" that would make science more directly accountable to benefits in health care, economic productivity, and national security.

Yet Bush himself was deeply aware of the societal context of science and technology. For example, in 1944 he proposed creation of an advisory committee on postwar U.S. nuclear legislation in order to deal with the threat that this new technology posed to international peace. In *Science, the Endless Frontier,* he argued for interdisciplinary science: "Science can be effective in the national welfare only as a member of a team" (1945b, p. 1). He furthermore stated that "It would be folly to set up a program under which research in the natural sciences and medicine was expanded at the cost of the social sciences, humanities, and other studies so essential to national well-being" (p. 18). In *Modern Arms and Free Men* (1949), Bush tackled important questions about the role of science in a democracy. The culmination of his understanding of science as an agent of social betterment comes in the form of his aptly titled collection of essays, *Science Is Not Enough* (1967). Insofar as American science policy has become isolated from its social context, it has done so against Bush's own vision for the proper relationship between science and the state.

ADAM BRIGGLE
CARL MITCHAM

SEE ALSO *Science Policy.*

BIBLIOGRAPHY

Bush, Vannevar. (1945a). "As We May Think." *Atlantic Monthly* 176(1): 101–108. Argues that postwar science should focus on making the inherited store of knowledge more accessible and imagines the memex as a device for achieving this goal.

Bush, Vannevar. (1945b). *Science, the Endless Frontier.* Washington, DC: U.S. Government Printing Office. A foundational document in U.S. science policy that claims science is crucial for national well-being specifically in the areas of medical research, economic progress, and national security.

Bush, Vannevar. (1949). *Modern Arms and Free Men: A Discussion of the Role of Science in Preserving Democracy.* New York: Simon and Schuster. Presents an optimistic account of the future of freedom and individuality despite the significant threats posed to democratic values by emerging weapons technologies.

Bush, Vannevar. (1967). *Science Is Not Enough.* New York: Morrow. A collection of essays that discusses the limits of science both in the quest for understanding and in the fulfillment of human values.

Ratcliff, J. D. (1942). "War Brains." *Colliers* 109(3): 28, 40. Describes Bush's role in administering the military scientific research during World War II.

Steelman, John R. 1980 (1947). *Science and Public Policy.* 5 vols. New York: Arno Press. Presents a detailed analysis of the us research system in order to permit the government to manage its growing research and development operations and to coordinate the activities of government, industry, and academia. It was instrumental in shaping U.S. science policy by supporting some aspects of *Science—the Endless Frontier* and challenging others.

Zachary, G. Pascal. (1997). *Endless Frontier: Vannevar Bush, Engineer of the American Century.* New York: Free Press. A biography that details Bush's life and evaluates his legacy.

INTERNET RESOURCES

Bush, Vannevar. (1945a). "As We May Think." *Atlantic Monthly* 176, no. 1. Available from http://www.theatlantic.com/unbound/flashbks/computer/bushf.htm.

Bush, Vannevar. (1945b). *Science, the Endless Frontier.* Available from http://www.nsf.gov/od/lpa/nsf50/vbush1945.htm.

BUSINESS ETHICS

Business ethics names both a *phenomenon* (the ethics espoused and practiced in business) and the *field of study* of that phenomenon (the serious study of business ethics). As a branch of *ethics* (or moral philosophy), the field of business ethics is interested in how judgments of right and wrong, good and bad, moral obligation and responsibility, rights and duties, and the like, are made and justified. As a branch of *applied ethics* it explores how these judgments are carried out in the specific domain of work, commerce, and economic activity.

As a *descriptive* enterprise, business ethics is an analytical exercise in understanding and explaining how people and organizations make their ethical judgments and decisions. As a *prescriptive* enterprise, business ethics seeks to arrive at defensible, normative, moral judgments of business matters in ways that are helpful to the actual practice of business. Business ethics overlaps significantly with what is often called *corporate social responsibility*—a movement calling on corporations to be responsible not just to shareholders but to the society (and the ecosystem) in which it operates. The field of business ethics is interested in more than just social and environmental responsibilities but those are certainly critical component areas.

Science and technology share a long, close, and mutually-influential relationship with business. Business needs and opportunities drive much scientific research and technological development, on the one hand, while discoveries and technological innovations transform business, on the other (Burrus 1993, Martin 1996, Tapscott and Caston 1993). Technology is widely accepted as the primary, dominating force that has transformed business around the world with rising intensity since the 1950s. Business ethics, as a reflective and sometimes reactive discipline, has typically lagged behind business changes and began to address this technological transformation only in the late-twentieth century (Gill 1999).

Historical Development of the Field

The basic questions of business ethics (for instance, fairness in wages and prices, responsibility for defective or dangerous products, fulfillment of contractual agreements, and morality of interest rates) have been of interest throughout human history and throughout the world. For example, the Jewish and Christian scriptures and the ancient Greek philosophers pay considerable attention to issues of wealth and poverty, honesty in transactions, liability for injury, justice in compensation, and other matters generally considered to be in the business ethics domain. So too, Buddhist tradition provides guidance about *right livelihood.* Medieval Catholicism considered the morality of usury and interest on loans. Karl Marx put capitalist economics on trial and called for justice and freedom for workers. Sociologist Max Weber famously studied the *Protestant ethic and the spirit of capitalism.* Thus while the constraints of nature and of social tradition have determined the work and economic experiences of most people throughout history, there have been recurring discussions of whether various aspects of this experience are right or wrong.

The rise of modern industry and the factory system, along with the great migrations of peoples across oceans and continents, especially during the nineteenth and early-twentieth centuries, brought major changes and disruptions to the ways people worked and the ways business was carried out. Business moved from a rural, agricultural, and familial base to an urban, industrial, and organizational one. The impact of these changes on individual workers, on families and communities, and on the environment, and the rise of a new class of wealthy business leaders—and of new forms of poverty—provoked intensified ethical debate not just among academic professionals but writers, politicians, preachers, poets, and populists.

Nevertheless as a discrete, self-conscious, academic field, business ethics emerged only during the 1960s and 1970s and grew steadily through the 1980s and 1990s and on into the twenty-first century. The rapid emergence of this field during the last quarter of the twentieth century was truly remarkable. Business schools created courses in business ethics; students began pursuing Ph.D degrees in the field; and centers for business ethics sprang up at many campuses. Associations, such as the Society for Business Ethics, Business for Social Responsibility, and the Ethics Officers Association, were formed to bring together scholars and practitioners in the new field. Journals were launched, such as *Business and Professional Ethics Journal* in 1981, the *Journal of Business Ethics* in 1982, and *Business Ethics Quarterly* in 1991. The quantity and quality of textbooks, monographs, and other literature on business ethics was first impressive, then daunting to those wishing to keep up with it. In the corporate arena itself, companies increasingly created ethics codes, statements, and training programs. By the turn of the twenty-first century, business ethics had won a respected and significant place in virtually all business education programs and in the consciousness of business managers (Freeman 1991, Werhane 2000).

The impetus for the development of business ethics as a field of study and of professional practice has come from several factors: First the rapid development of technology and its multifaceted deployment in business has modified and intensified the traditional list of business ethics challenges. Technology amplified old problems, created new ones, and complicated and speeded everything up.

Second social and cultural developments, in the 1960s and since, gave rise to a widespread questioning of traditional ethical authorities. Demands for recognition and equal treatment by students, women, and ethnic minorities, a new sense of urgency to care for the environment, and a growing ethnic, religious, and cultural diversity in the workplace all helped to put in question traditional ways of running businesses and of thinking about ethical right and wrong. Thus just as the technology-enhanced business ethics challenge was increasing, the assumption of a widely-shared consensus on values and ethics was becoming untenable.

Third across the intellectual and academic horizon, academic specialization grew, fueled partly by the scope and complexity of various old and emerging fields of research and partly by an explosion in the quantity of data available for consideration. The development of a specific field of business ethics (just like that of medical/bioethics) became logical, possible, and necessary. The growth of the business ethics challenge combined with the loss of a common set of values and ethics to create a fertile field of inquiry and service for a new academic specialization.

Fourth a growing number of high profile business ethics crises and scandals provoked calls for both better government regulation and oversight of business, on the one hand, and for better business ethics education and practice, on the other. Among the high profile ethics cases were trading, accounting, and financial scandals; the manufacture and sale of dangerous products (automobiles, tires, drugs); the use of child labor and sweat shops; ecological disasters (the Exxon Valdez, Bhopal); industrial pollution and depletion of natural resources; and vastly growing inequalities in wages and compensation for executives and workers. The 1991 U.S. Federal Sentencing Guidelines for white-collar criminals specified that law-breaking companies could reduce their penalties by up to 40 percent if they instituted compliance and ethics training programs.

Business Ethics: The Central Issues

The organizing question in business ethics is how to do the right thing (not just the profitable or possible or popular or even legal thing). Various philosophies, religions, and individuals answer the *what is right and how does one know it?* question in different ways, but there is widespread (if not universal) agreement that at its core, something becomes *wrong* when it *harms* (or seriously risks harm to) people. The Hippocratic Oath argued that the first duty of medical ethics was to *do no harm*. The same is true with respect to business ethics: An ethical business is one that seeks to avoid harm. What is ethically right and good is what can help people toward a free, healthy, and fulfilled human life. Obviously harm and help are elastic and debatable concepts but thinking about ethical right and wrong in these simple, historic, classic terms helps focus the ethical enterprise around a common language and concern in an important way.

In raising its questions of right and wrong, the scope of business ethics is as broad as business itself. Business ethics, perhaps because it is such a young field, has no single dominating method or paradigm. To arrive at a relatively inclusive understanding of the field, business ethics can be approached from five different perspectives. The first is a review of the range of typical *ethical dilemmas and problem cases* that arise across the business spectrum. The second briefly examines the *ethical values and methods of analysis* typically used to address the range of business ethics dilemmas. The third perspective is an analysis of the major *stakeholders* in business ethics so as to understand *who* is involved and what their ethical interests might be. A fourth perspective examines the basic components in a comprehensive organizational ethics. And finally, while the interaction of science, technology, and business ethics will be discussed as appropriate throughout this entry, a summary of business ethics will be drawn from the science/technology viewpoint.

Ethical Dilemmas and Critical Cases

One way to approach business ethics is by an analysis of specific problem cases or dilemmas (quandaries). An ethical dilemma arises when there is a question of determining the right thing to do. It often occurs because of a conflict of moral values or principles either within an individual or between two or more agents. Focus on the case method is called *casuistry* (Jonsen and Toulmin 1988; Brown 2003; Goodpaster and Nash 1998; Jennings 1999; Ferrell, Fraedrich, and Ferrell 2000). Casuistry analyzes ethical dilemmas and quandaries to aid in wise decision making and right action.

CLASSIFYING ETHICAL ISSUES. Ethical dilemmas and problem cases can be classified in several different ways. A threefold distinction can be made among (a) personal, *micro-ethical* issues; (b) organizational, *organizational* issues; and (c) systemic, *macro-ethical* issues. Another categorization can follow the functional areas of business, such as management, finance, accounting, human resources, marketing and advertising, supply chain management, sales, manufacturing, and more. Still another approach could focus on cross-cutting, thematic areas such as technology, communications, meeting, relationships, and the like.

Conflict of interest cases are often at the root of ethical dilemmas in these categories. For example, one's personal interest (for instance, a bonus for meeting a sales target or a personal gift) may conflict with one's professional responsibility (such as serving client needs and employer standards). A business interest in a foreign country may conflict with the social or environmental interest there. Bribes, kickbacks, insider trading, inappropriate use of company information, resources, or contacts to advance personal/noncompany interests, or hiring a talented friend are all examples of possible conflict of interest.

Dilemmas about truthfulness and accuracy in communication are also to be found throughout the business arena. Internal communications up and down the line, press releases and public relations, advertising and product labeling, financial reporting, and handling proprietary information and intellectual property, among other business activities raise difficult questions of ethical communication. How much information is owed and to whom? While it is clearly not right to publish immediately and fully all information one has to all people who ask for it, falsehood, deception, and evasion undermine trust and are often harmful.

Justice and fairness in policies and relationships are also a recurring ethical challenge throughout organizations. Relationships among employees at various levels and in different areas of the company may be disrespectful, inequitable, unfair, and harmful. Hiring practices, compensation, promotion, and workload differences might be unfair. Suppliers and business partners may not be treated fairly and honestly. The community may be unjustly burdened with the costs of an environmental cleanup due to a company's decision not to manage its wastes responsibly.

Technology has had a major impact on the ethical dilemmas faced in business. As the technological tools become more powerful, ever more vigilance is required to make sure they are used for good and not evil. Technologies also produce unanticipated consequences, *bite back* effects, that ethics must review (Tenner 1996). Old practices present new challenges when technology is introduced. Marketing and advertising ethics must now evaluate e-marketing practices. Customer data issues have become important as computerization makes possible tracking, profiling, and commoditization of what customers may assume is their private information.

Relationship issues are given a radical new spin when distant, extended enterprises, enabled by technology, become the order of the day. E-mail as the primary form of communication, the expectation of anytime/anywhere connectedness, and the management of employees in multiple, extremely diverse political-social settings around the world are technology-driven challenges that beg for ethical perspective.

RECOGNIZING, ANALYZING, AND RESOLVING ETHICAL DILEMMAS IN BUSINESS. A focus on ethical problem cases requires, first of all, determining whether a truly serious ethical dilemma that requires attention exists. Two compliance-oriented questions will often (though not always) identify a serious dilemma: (a) Is there a serious question of illegality? and (b) Is there a possible violation of the ethics and standards spelled out by the business's organizational code or by a related professional association? If the answer to either of these is positive, the issue is probably of serious ethical concern.

Some ethically important situations may slip under the radar of the two compliance test questions so four others must also be considered: (c) Is someone liable to be harmed by this? (d) Would individuals want this done to them or their loved ones? (e) Does this really bother human conscience and values? and (f) Would this continue if it were publicized in the evening news or on the front page of a newspaper?

If the answers to some or all of these questions are positive, the next stage is to analyze the case carefully. The facts of the situation must be clarified. Who is involved? What has happened? What are the ethical values and principles at stake? (The ultimate decision will need to be justified by appealing to such values). What are the options for response and the likely consequences of each response, short- and long-term? What help can others provide (colleagues, experts, veterans of similar cases) in analyzing and understanding this dilemma?

The third stage (after recognize and analyze) is to *resolve* the dilemma by choosing the best possible option available, acting on it with courage, and then following through, fully and responsibly. Not only the immediate decision and action but longer-term reforms might be appropriate to minimize recurrence of such dilemmas.

Casuistry is certainly an important part of business ethics. If ethics remains only a set of ideals or an abstract theory, unapplied (or inapplicable) to particular cases, it has failed. One of the virtues of casuistry is that it can quickly focus the participants' attention on something concrete, specific, and shared: the problem. Trying to begin with an agreement on abstract, general principles and values is often much more elusive. On the other hand a focus on cases alone can reduce business ethics to a reactive *damage control*. Decision making and action in response to extreme cases must not be allowed to become the whole enterprise. Even if one starts with concrete cases, part of the follow-through after responding to the case at hand is to move *upstream* in the organization and its practices to locate the sources and contributing factors to those downstream dilemmas.

Ethical Values, Principles, and Methods of Analysis

A second way into business ethics is to equip oneself with theories and insights from moral philosophy and carry these tools into the business domain (Beauchamp and Bowie 2001, DeGeorge 1999). Business ethics courses and textbooks, which frequently are designed and taught by people trained in philosophy, typically present two or more options in moral philosophy as potential tools for determining the right thing to do in business.

The two most common theories are the consequentialist utilitarianism of Jeremy Bentham and John Stuart Mill, and the non-consequentialist deontologism of Immanuel Kant. In addition to these two prominent options in Enlightenment modernity, business ethicists sometimes add brief discussions of ethical relativism, egoism, a feminist *ethics of care*, and some account of virtue (character) ethics. It is also common to include discussion of theories of justice (economic or distributive justice), often including the work of John Rawls and Robert Nozick.

After sketching such options in basic moral philosophy, business ethics textbooks of this type then counsel readers to choose one of these ethical theories to help moral philosophy to help decide ethical questions." Of course, virtually every moral philosophy (and moral theology) has some valuable insight to contribute to business ethics. Just as it can be useful to ask questions to identify an ethical dilemma, it can be helpful rather than confusing to examine one's ethical options from the perspective of several of these theories. With the utilitarians one could ask which possible response to the ethical problem would produce the best consequences for as many people as possible. With the Kantians one would ask how individuals would respond if they thought all people in comparable circumstances would copy the response. One could ask the *egoist* question—What is truly in the individual's best interest?—and, so too, questions about genuine caring, about the guidance of conscience and feeling, and about what surrounding culture thinks is right. Every insight and every theory is not equally insightful in every case, of course, so wisdom and discernment are always called for.

By focusing on moral philosophy in this way business ethics is actually showing its historic debt to Enlightenment thought. Kant and Mill and their contemporary philosophers were products of the modern scientific revolution of Isaac Newton and his colleagues, in which the physical universe was redescribed in terms of rational, universal, objective *laws*. In the footsteps of the scientists, the philosophers wished to discover moral laws of a universal, rational, objective character, independent of any notion of purpose or particularity of community. While this way of thinking about rational, universal, disinterested, objective laws contributes some helpful insights to the moral life, it has proven to be insufficient by itself (MacIntyre 1984, 1990). The young business ethics guild has slowly been waking up to the failure of Modern ethics. Viewed negatively, the Postmodern rejection of Enlightenment styles of moral philosophy points away from certainty and toward relativism or even nihilism.

Viewed more positively, the path has been opened up to explore new ways of thinking about business ethics that draw together the ethical insights of many voices and that more closely fit the actual ethical experiences of people in business. The success of some efforts to bring people together to formulate and implement business ethics principles, such as the Caux Round Table Principles, has been promising.

Business Ethics Stakeholders: Who Matters?

Business ethics can be approached by a problem focus, a theory focus, or, thirdly, a *people* focus, often called *stakeholder analysis*. To the traditional term shareholder (stockholder or investor/owner) has been added the term stakeholder (Freeman 1984; Weiss 1998; Post, Lawrence, and Weber 1999). A stakeholder is anyone affected by, or having a significant interest in, a business. They may not own financial shares of stock but they still have a significant stake, an interest, in what the business does. The assumption is that people have a moral right to some say in decisions that significantly affect their lives. In stakeholder relationships, the ethical questions concern the rights and responsibilities appropriate to each party to the relationship. Stakeholder analysis emerged from a realization that some parties were bearing costs (or reaping benefits) from business operations without being recognized. The following is a brief discussion of six major stakeholder groups.

OWNERS. One well-known view has it that the only responsibility of business is to maximize profits for its owners, provided this is done without fraud or other illegality (Friedman 1970). Certainly the owners (investors, shareholders, and financiers) of a business have a right to have their investment managed in their financial interest. It is not true, though, that profits are the only concern, even for the owners. Owner/investors also have a legitimate claim to adequate, accurate information about the business and its financial affairs.

What are the ethical rights and responsibilities of business owners in various circumstances? How does this differ under different ownership structures? What responsibility and accountability do business owners have toward other stakeholders? Are there ways of evaluating the legitimacy, fairness, and appropriateness of the owners' return on investment relative to what employees, customers, executives, and other employees receive? A stakeholder analysis approaches the business ethics arena with this sort of wider and deeper interest.

Technology has affected the ownership of business by facilitating complex, vast, high-speed new ownership patterns in the marketplace. Mutual funds own large percentages of many businesses. Under these fluid and impersonal circumstances, who are the owners to be held responsible for a business's behavior? How do small investors assume any of that responsibility even if they would like to? Perhaps the answer will become clear as information and communication technology renders the operations of both corporate management and fund management more fully transparent and as Internet-based movements organize small investors into effective lobbyists for reform (Tapscott and Ticoll 2003).

EMPLOYEES. If anyone has a clear stake in a company, it is the employees whose livelihood and vocation lies there. Business ethics pays attention to employees (including management) in several ways. First most of the ethical cases and crises that come along involve employee participation. The ethical analysis of employee choices, communications, and behavior occupies a good deal of the attention of business ethics. How managers and owners treat employees is another ethical concern. Job security, compensation, safety, harassment, prejudice, and even the quality of employee work experience, are ethically important. How should the personal ethical convictions of an employee be expressed (or not) in the workplace? How are employees trained in the company's ethics? How are ethical responsibilities related to various business roles?

Technology has modified the spectrum of ethical problems faced by employees. Perhaps the most striking impact of technology is when it eliminates employee jobs, either by replacing workers with robots and machines or by enabling jobs to be moved to locations where employees cannot follow. Is there a moral responsibility to help displaced employees to find other work?

Technology can be used or abused in monitoring employee communication and activity. Privacy must not be violated. Confidentiality must be protected. New stress-related injuries have emerged among computer users. Computers and the Internet have enabled some employee abuses such as game playing, pornography downloading, excessive personal use, and distribution of vulgar, hateful, or time-wasting messages to other employees. The same technology, however, allows for telecommuting from a home workstation, assisting a parent tending to a sick child. New issues of health and ethical management also arise concerning possible employer expectations of employees to be connected to their work anytime, anywhere.

CUSTOMERS. The most cynical non-ethical stance toward customers in the past was characterized by the Latin phrase caveat emptor—*let the buyer beware*. Viewed by stakeholder analysis, however, business ethics explores customer-related issues in marketing, advertising, product pricing, safety, quality, service, and support. What are the rights and responsibilities of customers vis-à-vis a company? Technology has made a huge impact on the development of products and services available to customers in the early twenty-first century. It also has modified marketing and advertising, as well as sales and service, by utilizing electronic media for all of these activities. Customer service and support and the privacy of customer data are among the ethical issues raised in new ways by technology. The Internet has also enabled some customers to help support each other in various user groups.

BUSINESS SUPPLIERS AND PARTNERS. Business-to-business relationships have become even more important and challenging in an era of outsourcing, complex supply chains, and virtual corporations. Government regulations and legal contracts simply cannot guarantee integrity in these relationships. The essential ingredient is trust, which depends on voluntary adherence to shared values and ethics (Fukuyama, 1995). What are the ethical responsibilities of business partners to each other? As technology enables businesses to create working relationships in distant and culturally-diverse settings where laws and local ethical values may permit child or slave labor, discrimination based on gender or religion, bribery, and environmental pollution—or where Euro-American business practices may be viewed as hopelessly corrupt, vulgar, and unjust, the challenge to business ethics is to figure out the ethically right thing to do in relation to the business partner stakeholders.

GOVERNMENT. As the presumptive guardians of the law, justice, order, and the well-being of nations, governments are also important stakeholders in business. This is true of all business-to-government interaction but in the economy of the twenty-first century, business's capacity to have both positive and negative impacts on states and their populations is extraordinary. Several multinational corporations have larger annual budgets than most nations in the world. The kind and extent of governmental regulation and oversight of business results in part from ethical values and choices. The influence of business on government (lobbying, campaign contributions) also is, and needs to be, subject to moral debate. In an era of globalization of business, earlier understandings of the proper relationship of governments to businesses must be rethought.

COMMUNITY. Communities often benefit both directly and indirectly from business. A strong business climate can bring jobs, income, and skills to communities. Even those who are not investors, employees, or customers of a business can benefit from its presence. But costs of the business are often *externalized* into the host community. Traffic congestion and environmental cleanup are two examples of costs to communities. A community may grow up around a business, creating schools, roads, and other cultural and social infrastructure that make it possible for that business to recruit good workers and thrive economically. If the business then relocates to China, based on investor demands for higher profit margins, an ethical issue arises. Communities have a stake in business.

Clearly there are other potential stakeholders in a business, such as professional associations, non-profit organizations, and schools. The strategy is to identify the relevant stakeholders and put the ethical focus on their respective rights and responsibilities.

The Basic Components of an Organizational Ethics

A fourth approach to business ethics is to work from a practical analysis of the way values actually work in organizations and communities (Solomon 1992; Batstone 2003; Trevino and Nelson 1999). This approach draws from historical and social scientific studies of business and other organizations, as well as from classical philosophical and theological approaches to ethics and values. The goal is to understand business ethics in a way that is simultaneously holistic, integrative, deep, and practical. In this approach six components in a holistic organizational ethics can be identified.

MOTIVATION. WHY BE ETHICAL IN BUSINESS? It is not at all self-evident why businesses should be run in ethically. The argument for doing so must be made in a way that will motivate business leaders and employees to make ethics a priority. A complete argument for operating a business in an ethical manner includes the following: (a) avoidance of litigation and the penal system (ethical companies generally steer clear of breaking the law; legal compliance is a sort of minimum standard of ethics); (b) regulatory freedom (increased laws and regulations result from patterns of unethical behavior); and (c) public acceptance (unethical businesses are often punished by journalistic exposes, citizen watchdog groups, and bad reputations).

In addition to the preceding three *external* reasons, having to do with the political and cultural environment in which business operates, there are four *internal* reasons to be ethical, connected to the four basic parts of any business in the early 2000s: (d) investor confidence (financial resources will be withheld from untrustworthy businesses); (e) partner/supplier trust (more than ever in the era of extended enterprise, business partnerships depend on trust, ethics, and integrity); (f) customer loyalty (customers avoid businesses that treat them in an unethical manner and also avoid brands that are associated with the unethical treatment of workers); (g) employee recruitment and performance (good employees are attracted by ethical employers; especially in the *knowledge economy*, employee sharing and teamwork flourish best in an atmosphere of trust and ethics).

Finally there are three *deep* reasons for running an ethical business: (h) personal and team pride and satisfaction (business success that comes by virtue of ethical behavior is rewarding to the individual; being ethical aligns with human nature and conscience in important ways); (i) intrinsic rightness (individuals and organizations should be ethical simply to be in alignment with a moral universe—God, reason, and human tradition argue for doing the right thing even when there is no immediate or direct payoff); and (j) missional excellence (being ethical is fundamentally about the essential values woven into the fabric of an excellent organization; ethics is less an external measuring stick than an internal set of traits).

CORPORATE MISSION AND PURPOSE. Assuming a business organization is adequately motivated to operate in an ethical manner, the next priority is to clarify the core mission and purpose of the organization. This is an Aristotelian, biblical, and traditional starting point for ethics. "The values that govern the conduct of business must be conditioned by *the why* of the business institution. They must flow from the purpose of business, carry out that purpose, and be constrained by it" (Sherwin 1983, p. 186). The first focal point in the positive construction of a sound business ethics is to clarify the *telos* of the business. An inspiring, unifying business mission that taps into basic human drives (e.g., to be creative or to be helpful to others) can leverage and guide sound ethics in an organization. For Aristotle, things, people, and organizations are embedded with *final causes*, purposes, and destinies to fulfill, and ethics is about how to achieve these. For biblical ethics, the determination of *who is God* (the First Command) is decisive for the ethical standards related to that choice (Commands Two through Ten). For great and enduring businesses, preserving the core mission and values is of primary importance (Collins and Porras 1994).

CORPORATE CULTURE AND VALUES. Given a clear and compelling mission, the next focal concern of a sound business ethics is the formal and informal corporate culture. Does the culture empower or impede the achievement of this mission? Corporate culture is not a neutral or arbitrary construction as far as ethics is concerned. No matter how excellent the mission and no matter how impressive the ethics code of a company, a defective or misaligned culture will present an insurmountable obstacle to sound ethics and business excellence. The formal systems of review, promotion, recognition, and discipline—and the informal culture of communication styles, office set-up, and so on—are what enable or disable the mission. The positive traits that assist the mission are the virtues, the values that must be embedded in what the organization *is*, not just what it *does*.

BUSINESS PRACTICES AND GUIDING PRINCIPLES. But businesses not only *are*, they *do*. After the culture, business ethics focuses on the *practices* of the company, the basic things the company needs to do, how its people spend their time and energy. The business must identify its basic practices (specific areas such as marketing, accounting, and manufacturing as well as cross-cutting activities like communicating and meeting). For each area of business practice, the company must decide which ethical principles should guide. Ethical principles and rules establish negative boundary conditions that must not be transgressed and positive mandates and ideals to pursue. Leaving important areas of practice with inadequate guidelines undermines the capacity of the business to achieve ethical excellence, the importance of the company ethics code.

ETHICS TROUBLESHOOTING AND CRISIS MANAGEMENT. Even in the best of circumstances, ethical dilemmas and crisis cases will emerge from time to time. It is therefore essential to create a method and framework for managing crises effectively. Making damage control and ethical crisis management the focal point of business ethics can unwittingly serve as an invitation to an unremitting succession of such crises. But as a component subordinated to a broader, more holistic business ethics, the crisis management, dilemma resolution part of the ensemble is essential. Corporations are increasingly creating ombudsmen, ethics and compliance offices, ethics hotlines, confidential means of raising questions or reporting questionable activities, whistle-blowing protocols, and the like. It is essential that businesses make clear what their employees and other stakeholders should do when apparent ethics questions and problems arise.

ETHICAL LEADERSHIP. Finally business ethics requires that attention be focused on leadership and management. Exemplary ethics does not exist without leadership. Ethics and values leadership must come from the executive and board levels of a company in the form of communication as well as action. Leaders must be heralds of the values and ethics that matter. They must exemplify the highest ethics in their own behavior and they must create systems, structures, and policies that support and reward ethical excellence and sanction unethical actions. Business leaders must create and maintain ethics training and evaluation programs throughout the organization. Without good leadership, good business ethics cannot be created and sustained.

The Impact of Science and Technology on Business and its Ethics

While business has often been conducted in a non-scientific and non-technological, traditional manner, ambition, competition, and the pressing need to solve business challenges of all kinds have encouraged businesses to learn from, and even sponsor, scientific and technological work. Since the eighteenth century, particularly, business, science, and technology have worked closely together. Manufacturing, construction, and transportation technologies decisively reshaped modern business beginning with the Industrial Revolution. Communication and information technologies have been the center of the most influential developments since the mid-twentieth century. Biotechnologies may be the most significant arena for business/science/technology interaction in the twenty-first century.

Science and technology have affected business and its ethics in several important ways. First they introduced radical change in the products of business. Technological products dominate virtually every area of people's lives, virtually every hour of the day. A host of specific ethical questions may be raised about these technological products, regarding their safety, reliability, cost and value, appropriateness, and side effects. Is their manufacture, usage, and disposal conducted in an environmentally responsible way? Are the trade-offs, the winners and losers, and the side effects, ethically appropriate and justifiable?

Science and technology have also transformed the workplace in important ways. The mechanization and automation of the workplace has continued unabated since the beginning of the nineteenth century. Information technology has enabled businesses to extend their operations all over the world and around the clock. How should people evaluate the outsourcing and exporting of jobs and the disruption of local economies by technologically-enabled global business? How do traditional safeguards against unethical acts by the powerful, such as national borders, local customs, and face-to-face, human-scale accountability relationships, get replaced in the early 2000s? What are the ethics of allowing, or even encouraging, workers to stay connected and available to their work twenty-four hours per day, seven days per week?

Technology acts as an amplifier of both problems and possibilities (for instance, the greater accessibility of medical records has both positive and negative sides). It also creates greater speed, reducing the time that individuals can devote to careful ethical reflection, which is required by the growing scale of the problems. Technology is much better at increasing the quantity of information and communication than the quality of knowledge and the wisdom of relationships. Technology creates many new opportunities for diversity, but also fosters standardization and repetition. Technology produces significant democratization of knowledge even as a new *digital divide* is emerging around the world.

In 1911 Frederick W. Taylor's *Principles of Scientific Management* promoted a new way of thinking about business management that privileged expert, technical judgments over those of ordinary workers and citizens. Taylor argued that efficiency was the primary goal of human thought and labor and that what could not be measured did not count. Henry Ford's automobile assembly line famously applied this kind of thinking. Workers became virtual appendages of machines. While there were certain gains in production from this approach, by the 1970s it became clear that even greater productivity was possible through the humane and respectful treatment of workers.

What is sometimes overlooked in discussions of business and technology is the way that technology itself is embedded with certain basic values, such as efficiency, quantifiability, power, speed, repetition, predictability, rationality, and so forth. As long as technology is viewed as a set of tools and methods to help a business achieve its mission, those technological values can be located in a richer cultural context that also preserves values such as openness, innovation, risk, human caring, beauty, and quality. If technology is put in the driver's seat rather than the toolbox of business, it will eventually come into conflict with human values, at a considerable (if not total) cost to workers, businesses, and the larger economy. In short business ethics in the coming years will need to pay serious attention not just to the complexities of particular technological innovations but to their collective impact on the mission and culture of businesses and their surrounding communities (French 1995).

DAVID W. GILL

SEE ALSO *Economics and Ethics; Entrepreneurism; Work.*

BIBLIOGRAPHY

Batstone, David. (2003). *Saving the Corporate Soul.* San Francisco: Jossey-Bass. Excellent review of ethical values and perspectives related to key aspects of contemporary business.

Beauchamp, Tom L., and Norman E. Bowie. (2001). *Ethical Theory and Business*, 6th edition. Upper Saddle River, NJ: Prentice-Hall.

Bowie, Norman E. (1999). *Business Ethics: A Kantian Perspective.* Oxford: Blackwell. Norman Bowie is the leading Kantian business ethicist.

Brown, Marvin T. (2003). *The Ethical Process: An Approach to Disagreements and Controversial Issues*, 3rd edition. Upper Saddle River, NJ: Prentice-Hall.

Burrus, Daniel. (1993). *Technotrends: How to Use Technology to Go Beyond Your Competition.* New York: HarperBusiness.

Collins, James C., and Jerry I. Porras. (1994). *Built to Last: Successful Habits of Visionary Companies.* New York: HarperBusiness. Influential study of successful companies shows that corporate mission and values ("more than profits") play key role in long-term success.

DeGeorge, Richard T. (1999). *Business Ethics*, 5th edition. Upper Saddle River, NJ: Prentice-Hall.

Ferrell, Odies Collins, John Fraedrich, and Linda Ferrell. (2002). *Business Ethics: Ethical Decision-Making and Cases*, 5th edition. Boston: Houghton Mifflin.

Freeman, R. Edward. (1984). *Strategic Management: A Stakeholder Approach.* Boston: Pitman. Freeman is the pioneer of business ethics stakeholder approaches.

Freeman, R. Edward, ed. (1991). *Business Ethics: The State of the Art.* New York: Oxford

French, Peter. (1995). *Corporate Ethics.* Orlando, FL: Harcourt Brace.

Friedman, Milton. "The Social Responsibility of Business is to Increase its Profits." *The New York Times*, September 13, 1970, Magazine section. A now-classic argument against corporate moral and social responsibility.

Fukuyama, Francis. (1995). *Trust: The Social Virtues and the Creation of Prosperity.* New York: Free Press. A thorough comparative historical study of high-trust and low-trust societies and their business and economic consequences.

Gill, David W. (1999). "The Technological Blind Spot in Business Ethics." *Bulletin of Science, Technology, & Society* 19(3): 190–198.

Goodpaster, Kenneth E., and Laura L. Nash. (1998). *Policies and Persons: A Casebook in Business Ethics*, 3rd edition. New York: McGraw-Hill.

Jennings, Marianne M. (1999). *Business Ethics: Case Studies and Selected Readings*, 3rd edition. Cincinnati, OH: West.

Jonsen, Albert R., and Stephen Toulmin. (1988). *The Abuse of Casuistry: A History of Moral Reasoning.* Berkeley, CA: California. The leading exponents of casuistry in business ethics.

MacIntyre, Alasdair. (1984). *After Virtue: A Study in Moral Theory*, 2nd edition. Notre Dame, IN: University of Notre Dame. Macintyre's works have influenced many to abandon Kant, Mill, and modern ethical theory in favor of Aristotle and a more holistic approach.

MacIntyre, Alasdair. (1990). *Three Rival Versions of Moral Enquiry.* Notre Dame, IN: University of Notre Dame.

Martin, James. (1996). *Cybercorp: The New Business Revolution.* New York: Amacom. How technology changes businesses.

Post, James E.; Anne T. Lawrence; and James Weber. (1999). *Business and Society: Corporate Strategy, Public Policy, Ethics*, 9th edition. Boston: McGraw-Hill.

Sherwin, Douglas S. (1983). "The Ethical Roots of the Business System" *Harvard Business Review* 61(6): 183–192.

Solomon, Robert C. (1992). *Ethics and Excellence.* New York: Oxford. Solomon is the leading Aristotelian, virtue business ethicist.

Tapscott, Don, and Art Caston. (1993). *Paradigm Shift: The New Promise of Information Technology.* New York:

McGraw-Hill. Tapscott and his various co-authors are superb analysts of the impact of technology on business.

Tapscott, Don, and David Ticoll. (2003). *The Naked Corporation: How the Age of Transparency Will Revolutionize Business*. New York: Free Press.

Taylor, Frederick W. (1911). *Principles of Scientific Management*. New York: Harper and Brothers. The classic argument for subordinating business to science and technology.

Tenner, Edward (1996). *Why Things Bite Back: Technology and the Revenge of Unintended Consequences*. New York: Alfred A. Knopf.

Trevino, Linda Klebe, and Katherine Nelson. (1999). *Managing Business Ethics*, 2nd edition. New York: John Wiley. Excellent holistic, practical approach to business ethics.

Weiss, Joseph W. (1998). *Business Ethics: A Stakeholder and Issues Management Approach*, 2nd edition. New York: Dryden.

Werhane, Patricia H., ed. (2000). *Business Ethics Quarterly* 10(1). Special tenth anniversary edition. Leading figures in the field of business ethics take a retrospective look at the field.

BUTLER, SAMUEL

Samuel Butler (1835–1902) was born in Nottinghamshire, England, on December 4. He was an early critic of evolutionary theory and was among the first to raise philosophical questions about human-machine relations. After being educated at Cambridge University Butler decided to forgo an anticipated ordination and moved to New Zealand to become a sheep rancher (1859–1864). There he read the biologist Charles Darwin's (1809–1882) *On the Origin of Species* (1859), whose theory of evolution became an obsession. Butler died in London on June 18.

At first Butler was convinced by the theories of Darwin; the two corresponded, and Butler became close friends with Darwin's son, Frances. Upon returning to England, Butler was initially a staunch defender of evolution. As a contribution to that defense he began a book to supplement Darwin's theory by elucidating the role of habit in relation to inheritance. However, while doing research Butler discovered the theory of the inheritance of acquired characteristics of Jean Baptiste de Lamarck (1744–1829) as well as the biologist St. George Jackson Mivart's (1827–1900) critique of natural selection, *Genesis of Species* (1871). Now that he was convinced that Darwin was wrong, Butler's book, *Life and Habit* (1878), became an attack. It was followed by a series of other critiques that did not have wide influence: *Evolution, Old and New* (1879), *Unconscious Memory* (1880), and *Luck, or Cunning?* (1887).

Best known in the early 2000s for his novel *The Way of All Flesh* (published posthumously in 1903), Butler achieved literary and financial success during his life from two satirical novels that often are described as Swiftian: *Erewhon* (1872) and its sequel *Erewhon Revisited* (1900). Those works, which originated in an essay titled "Darwin among the Machines" (1863) and continued his lifelong preoccupation with evolution, are of particular interest in regard to the ethics of technology.

The books whose titles are the word *nowhere* spelled backward envision a dystopian society in which machine development has been limited consciously and severely. In the first novel an unnamed narrator accidentally visits Erewhon, a land ruled by philosophers and prophets who equate morality with beauty and health and illness with crime. In Chapters 23 to 25, collectively called "The Book of the Machines," the narrator (whose name, Higgs, is revealed in the continuation) reads a treatise that considers the possible evolution of machine consciousness and details the Erewhonian revolution that led to the prohibition of machines to prevent their domination of the human race. The author argues that the rapid evolution of "higher machines" will lead to their consciousness if steps are not taken "to nip the mischief in the bud and to forbid them further progress."

The narrative further speculates on the nature of consciousness and offers a prescient description of modern DNA testing, anticipating a time when it may "be possible, by examining a single hair with a powerful microscope, to know whether its owner could be insulted with impunity." There is also a linking of machine consciousness with miniaturization and a consideration of whether human life may be merely a step in machine evolution. Chapter 23 concludes:

> We cannot calculate on any corresponding advance in man's intellectual or physical powers which shall be a set-off against the far greater development which seems in store for the machines. Some people may say that man's moral influence will suffice to rule them; but I cannot think it will ever be safe to repose much trust in the moral sense of any machine.

Selections from "The Book of Machines" have been reprinted frequently and often are used to initiate discussions of issues that remain fundamental to the ethics of technology.

JAMES A. LYNCH

SEE ALSO *Utopia and Dystopia.*

BIBLIOGRAPHY

Breuer, Hans-Peter. (1975). "Samuel Butler's 'Book of the Machines' and the Argument from Design." *Modern Philosophy* 72: 365–383. Discusses Butler's reading of Darwin's theories.

Butler, Samuel. (1970). *Erewhon*, ed. Peter Mudford. New York: Viking. A standard reprint.

Pauly, Philip. (1982). "Samuel Butler and his Darwinian Critics." *Victorian Studies* 25: 161–180. A history of the rift between Butler and Darwin.

C

CANADIAN PERSPECTIVES

Canada, by landmass the largest country in North America, is the smallest in population, at just over 32 million inhabitants. Despite this relatively small population, Canada has made a number of distinctive contributions to discussions of science, technology, and ethics. Among these is, notably, the Genomics, Ethics, Economics, Environment, Law, and Society (GE^{3}LS, pronounced *gels*) program, part of Genome Canada, that has supported more than seventy investigators and as many graduate students to investigate issues related to genomics research. As the name indicates, the goal is to promote social context research and education related to new developments in genetics.

General Background

While many would argue that ethical, economic, and social aspects have always been embedded in the management of science in Canada, these have not always been present in a formal sense. In earlier years companies, governments, and scientific researchers often consulted, in an ad hoc fashion, with social scientists and humanists about the impacts of their plans. In the 1970s, however, the demand for formal review arose (e.g., in environmental assessments, which frequently considered socioeconomic impact statements) at the same time as the supply of social scientists expanded and new university research institutes and degree programs related to applied ethics, human rights, environmental economics, risk studies, and science, technology and society (STS) studies were introduced.

The professional and academic efforts to investigate, consider, and implement ethical, economic, environmental, legal, and social studies related to genomics and other life-science research in Canada evolved in tandem with related efforts in the United States. Many of the researchers now leading GE^{3}LS teams previously participated in the Ethical, Legal, and Social Implications (ELSI) program initiated in 1990 by the Human Genome Project (HGP), which was based in the U.S. Department of Energy and the U.S. National Institutes of Health (NIH). The combined ELSI efforts constituted the largest bioethics program in the world and as such has been internationally influential, especially in Canada.

The Canadian efforts became more organized once the demand began to become more formal. Perhaps the first significant requirement for comprehensive social science analysis arose in the context of the evolving environmental legislation in the provinces and at the federal level. This culminated with the passage of the new Canadian Environmental Protection Act (CEPA) in 1992, which required assessment of any environmental effect on health, socioeconomic conditions, physical and cultural heritage, and aboriginal, historical, archaeological, paleontological, and architectural interests.

Shortly thereafter research into the human and various plant, animal, and microbial genomes accelerated. The scientific research efforts surrounding genomics is aimed at decoding all of the genetic information of an organism. This revolutionary research has given rise to a number of social, ethical, legal, and environmental issues.

Specific Initiatives

At about the same time, two independent processes arose to address the need for more social, ethical, legal, economic, and environmental review of Canadian science. One emerged from political discussions, the other from the research community.

In 1983 the federal government adopted its first Canadian Biotechnology Strategy, with an informal group of representatives from industry, consumer groups, and academia providing recommendations to the Canadian Minister of Industry. In 1998 the government concluded that if Canada were to become a leader in biotechnology research, it would need an advisory body with a wider membership base in order to examine and reflect on the changing role of science in society. This led, in 1999, to the establishment of the Canadian Biotechnology Advisory Committee (CBAC) as a part of a renewed Canadian Biotechnology Strategy. The CBAC consists of up to twenty members appointed for three-year terms, and is supported by an executive director with a small staff. Its mandate is to provide comprehensive advice on current policy issues associated with the ethical, legal, social, regulatory, economic, scientific, environmental and health aspects of biotechnology and to provide Canadians with easy-to-understand information and opportunities to voice their views. It is the CBAC that provides both a market for GE^3LS studies and a conduit for promoting the results to a broader audience.

In 1998 the nation's three peer-reviewed granting councils—the Social Sciences and Humanities Research Council, the Natural Sciences and Engineering Research Council, and the Canadian Institute of Health Research—released a tri-council policy statement titled "Ethical Conduct for Research Involving Humans." This statement laid out a series of policies related to confidentiality, consent, balance between benefits and harms, and respect for human dignity and the vulnerable. Universities, public labs, and industry responded by developing internal processes to conform to these ethical standards for both new and ongoing research.

In 2001 the three councils created an Interagency Advisory Panel on Research Ethics to support the development and evolution of collaborative ethics research following the 1998 statement. The advisory panel, composed of twelve volunteer members whose backgrounds span several disciplines including the social sciences, natural sciences, law, and commerce, meets regularly to examine and recommend policies related to council practices for life-science research. Once the direction was set, most national research efforts conformed and adopted ethical and socioeconomic reviews as a formal part of their structure.

The Canadian Networks of Centers of Excellence (NCE) program, started in 1990 by Industry Canada and the three granting councils to fund long-term discovery research networks involving industry, academia, and government, initially had little or no role for socio-ethical review. Once the tri-council guidelines related to research ethics involving humans were developed, however, the NCE program incorporated them into their projects and, in the competition round completed in 2002, formally included a GE^3LS research component and incorporated dedicated funding for GE^3LS programs. For example, the Advanced Food and Materials Network that began in 2002 spends C$22.2 million and involves eighty-eight investigators; C$3.5 million of the budget goes to GE^3LS studies, which fund eighteen investigators.

The single largest public investment in GE^3LS has been through Genome Canada. In the first two rounds of competition (in 2001 and 2003), Genome Canada funded five GE^3LS projects (one each in British Columbia, the Prairies, and Quebec and two in Ontario) and supported one GE^3LS investigation in a science project (related to potatoes). Those six projects had a total budget of more than C$16 million (equal to about 8% of the total investment of more than C$600 million by Genome Canada) and involved more than seventy investigators and at least as many graduate students. In 2005 the third competition for projects was underway and Genome Canada solicited dedicated GE^3LS projects and instructed all science projects to incorporate GE^3LS components. A brief review of a number of science projects suggests project proponents intend to invest on average 1 to 3 percent of their total requested funds in GE^3LS activities.

PETER W. B. PHILLIPS

BIBLIOGRAPHY

INTERNET RESOURCES

Canadian Biotechnology Advisory Committee (CBAC). Available from http://cbac-cccb.ca/epic/internet/incbac-cccb.nsf/en/Home. The home page of the CBAC.

Interagency Advisory Panel on Research Ethics. Available from http://www.pre.ethics.gc.ca/english/policystatement/policystatement.cfm. This site contains the tri-council policy statement, "Ethical Conduct for Research Involving Humans," first issued in 1998 and revised in 2000 and 2002.

CANCER

During the twentieth century wealthy countries underwent a transition in mortality from acute, infectious diseases such as pneumonia to chronic diseases such as cancer. By the late twentieth century the lifetime risk of a person receiving a cancer diagnosis in the United States had climbed above one-in-three. The quest for an elusive "cure" for cancer became a policy imperative, and by the first decade of the twenty-first century U.S. government expenditures on cancer research had reached three billion dollars per year. Notwithstanding decades of heavy research funding, advances in long-term survival for many of the common types of cancer have remained insignificant, and critics have charged that research funding has been too narrowly focused.

Etiologies

The ancient Greeks and Romans understood cancer and other diseases in terms of the bodily humors of phlegm, blood, black bile, and yellow bile (Rather 1978). When the humors were out of balance, such as an excess of black bile in the case of cancer, a disease could erupt. Similar humoral approaches characterized other Old World medical systems, such as the traditional medicines of east and south Asia. Although the rise of scientific biology displaced humoral thinking from the medical sciences, humoral approaches to disease can still be found in some *complementary and alternative medicine* (CAM) approaches to cancer, such as macrobiotic, Ayurvedic, and other traditional Asian medical systems, as well as in general notions of *rebalancing* the body. Ancient physicians also identified diet and trauma as two possible environmental sources of cancer, and those ideas continue to be relevant to thinking on the etiology (causes) of cancer in the early-twenty-first century.

By the beginning of the twentieth century medical researchers were pursuing diverse approaches to cancer etiology. In the wake of the bacteriological revolution, many researchers thought that cancer was an infectious, bacterial disease. Although bacterial theories and therapies were on the wane by the 1920s, throughout the twentieth century a marginal network of researchers kept the approach alive, and they developed dietary and vaccine-based therapies (Hess 1997). At the end of the twentieth century, bacteria were gaining some general recognition as a risk factor for digestive tract cancers. Viral oncology, which had a peak of popularity during the 1960s, had also won general acceptance for viruses as the cause of some human and animal cancers.

At a popular level, laypeople in early-twentieth-century Western countries frequently believed that trauma was a significant cause of cancer (Clow 2001), and the belief is still widespread in some countries. The medical profession recognized a related risk factor of tissue irritation from sources such as tobacco or childbirth. The interest in tissue irritation gradually developed into research programs on chemical carcinogenesis. In the eighteenth century the relationship between creosote tar and scrotal cancer in chimney sweeps had been identified. By the end of the twentieth century, a wide range of chemicals, as well as some forms of electromagnetic radiation, were acknowledged as risk factors, including especially the carcinogens in cigarette smoke.

At the beginning of the twentieth century, some medical researchers also drew attention to the role of internal biological processes in cancer etiology. One theory assumed that embryonic cells remained embedded in differentiated tissues and that they could develop into cancer under some conditions. The theory did not win widespread acceptance, but therapies based on enzymes and other dietary modifications continued as part of the field of CAM cancer care. Furthermore the theory drew attention to the role of growth hormones in cancer, which became part of mainstream cancer research. By the middle decades of the twentieth century, research programs were also emerging on the role of sex hormones in some cancers.

Another development during the twentieth century was research on inherited susceptibility to cancer, which developed from longstanding beliefs about heredity and cancer. Animal experiments in the early decades of the twentieth century confirmed the role of heredity, and by the late-twentieth century it became clear that some types of inherited gene variations (alleles) carried very high risk for some types of cancer, such as the BRCA1 and BRCA2 genes for breast cancer. However, epidemiologists at the end of the twentieth century generally believed that heredity explained only a minor percentage of the variation in the aggregate incidence of cancer and in its growth in incidence.

The development of molecular biology in the second half of the twentieth century allowed a synthesis of various risk factors (for example viruses, chemical carcinogens, radiation) at the molecular level of genetic damage and the expression of genes related to cancer (oncogenes and tumor suppressor genes). However, heredity as a risk factor needs to be distinguished from the understanding of carcinogenesis at a molecular level. Epidemiologists have increasingly given priority to environmental and lifestyle factors, of which diet and exposure to carcinogenic chemicals are generally seen as the central risk factors. In the early-twenty-first century, other recognized risk factors include reproductive behavior, obesity, viral infection, and excessive exposure to sunlight.

Ethics in the Clinical Setting

Most discussions of ethics and cancer focus on the doctor-patient relationship and the various types of ethical problems that emerge in cancer diagnosis and treatment (Angelos 1999). A key issue involves the communication of information to the patient. In some countries physicians have historically informed family members of the diagnosis but have concealed the diagnosis from the patient, even if the patient asks for the information. The practice appears to be changing, but other questions remain. For example, should a physician inform the patient of the diagnosis and/or prognosis, even if the patient asks not to be informed? Likewise should a clinician volunteer statistical information about prognosis even when only more general information is requested?

A related but in a sense inverted problem involves the disclosure to kin of a known genetic mutation that is related to cancer, such as the BRCA1/2 mutation. Patients who undergo such testing often do not expect to benefit personally from it, but they hope that the information will be helpful to kin. As a result, questions have been raised about informed consent regarding the autonomy of the patients who undergo testing, who may feel compelled by responsibility toward kin as a reason for undertaking the testing, as well as the autonomy of kin, who may not want to know such information or may fear genetic discrimination (Hallowell et al. 2003).

A second issue in the doctor-patient relationship involves the ethics of physician reactions to decisions by patients to withdraw from treatment. Sometimes patients decide that the side effects of conventional treatments, such as chemotherapy, are too severe in comparison with the potential benefits (long-term remission) for their particular type of cancer. Patients may combine the decision to withdraw from treatment with a decision to opt for a CAM treatment, but sometimes they simply forego chemotherapy for reasons other than pursuing a successful treatment. For example, patients may decide that there is no hope for recovery and that they are ready to die, or they may feel healthy and may want to work until they no longer can. However oncologists may not recognize nonmedical reasons as *good reasons* for refusing treatment, or they may reject the patient's assessment of the relative risks and benefits of various options, and consequently a communication gap may emerge when oncologists refuse to continue to monitor patients who refuse treatment (Huijer and Leeuwen 2000).

When parents make similar decisions for children, the cases can end in bitter conflicts. In some cases doctors have called in state agencies to take children away from their parents and forcibly deliver conventional therapies. Presumably some calculation of the benefits and risks of both the proposed conventional therapy (including no treatment) and the alternative treatment option (including no treatment) pursued by the parents inform decisions about whether to support the parents or take their child away. As a result, in some cases doctors may support the parents' decision. For example, a child was diagnosed with a type of brain tumor for which conventional therapies offered no possibility of cure. The parents decided to try antineoplastons, an experimental therapy that had only limited supporting evidence at the time but held some risk associated with the insertion of an intravenous catheter. In this case the doctors and hospital opted to insert the catheter and follow the patient, but they also informed the parents of their skepticism that the therapy would be beneficial (Jackson 1994).

Ethics and Research Funding

Ethical issues have also emerged around the politics of funding. One key area has been research funding on chemical carcinogenesis. For years, evidence that smoking is a substantial risk factor for lung cancer (as well as some other types of cancer) was suppressed, and epidemiologists who sought funds for and produced evidence on the role of smoking faced a long battle for recognition. In the early-twenty-first century a younger generation of epidemiologists faces a similar battle to gain acceptance for claims that military and industrial pollution is a major risk factor (Davis and Webster 2002). Historically researchers who have attempted to document risks from industrial pollutants such as ionizing radiation have faced suppression, and industry support groups also have produced scientific dissensus by funding studies that questioned the risks associated with industrial pollutants (Proctor 1995).

In addition to the politics of funding for research on etiology, ethical issues also have emerged around funding choices for research that evaluates or develops therapies. In the early twentieth century surgery was the only mainstream therapy for cancer, but radium-based therapies gained currency by the 1920s, and chemotherapy emerged after World War II. Surgeons and physicians who owned radium or advocated chemotherapy actively opposed the vaccine-oriented therapies developed by researchers who adopted immunological or biological approaches (Hess 1997). Similar suppression has been documented for nutritional therapies and a range of other CAM approaches to cancer (Moss 1995).

As cancer treatment developed during the twentieth century, medical subspecialties and cancer-related treatment industries opposed radical changes in treatment that threatened to undercut the profits of surgery, radiation therapy, and patented drugs. Although biological/immunological therapies for cancer (such as the use of interleukins and drugs that block the formation of blood vessels) are gaining ground in the early-twenty-first century, those developments take place through the mechanism of patented drug development. Researchers who investigate therapies that rely on unpatented products derived from plant or animal substances have been unable to obtain the level of private sector investment that is necessary to become competitors in the field of cancer therapy, which after the early 1960s involved a very costly drug approval process. As a result, a wide range of potentially lifesaving therapies has remained underinvestigated. Public funding agencies in the United States and other countries that could have stepped in to provide research funding for orphaned, unpatented therapies did not do so until the late-twentieth century, and even then the funding remained very minimal. (The term "orphaned" refers to therapies that lack sufficient research funding to be brought to market, because private firms cannot recuperate research costs in future sales due to lack of patentability or size of market.)

Another way in which research on unpatented products can hit a dead end is due to the way that the ethics of clinical trials has developed. Ethicists have argued in favor of *equipoise*, that is, the condition that study and treatment arms in a clinical trial have equal risk/benefit profiles. As a result, in cancer research placebo controls are rarely used; instead an experimental treatment is compared to the treatment standard. Frequently the experimental treatment is the standard treatment plus an additional drug. The standard of equipoise protects patients with life-threatening diseases from research that would put them at risk of receiving completely inefficacious treatment. However because funding is absent to generate preliminary human data, unpatented therapies can be locked in a limbo that prevents head-to-head testing against standard therapies. In this way ethical considerations at one level (patient rights) can negatively impact ethical considerations at another level (investigation of orphaned or unpatented therapies).

In short, significant ethical issues remain unaddressed regarding research funding for both etiology and treatment. Industrial interests external to cancer research and treatment, such as industries that generate significant pollution with suspected carcinogens, have opposed research that might lead to costly changes in materials or production processes. Likewise industrial interests internal to cancer research and treatment, such as medical subspecialties and the pharmaceutical/biotechnology industries, have opposed research that might open the door to competition from unpatented products. After decades of publicly supported research that have followed President Nixon's declaration of the war on cancer in 1971, for many patients therapeutic options remain limited and long-term prognosis remains dismal.

DAVID HESS

SEE ALSO *Death and Dying; Health and Disease; Medical Ethics.*

BIBLIOGRAPHY

Angelos, Peter. (1999). *Ethical Issues in Cancer Patient Care.* Dordrecht, The Netherlands: Kluwer. Good survey of the bedside ethical dimensions of conventional cancer care.

Clow, Barbara. (2001). *Negotiating Disease.* Montreal McGill-Queen's University Press. Survey of popular and academic understandings of cancer etiology in the twentieth-century.

Davis, Devra Lee, and Pamela Webster. (2002). "Epidemiology and Science: The Social Context of Science." *Annals of the American Academy of Political and Social Science* 584(13): 13–34. Example of the younger generation of epidemiologists that is developing the case for environmental etiologies.

Hallowell, Nina; Foster, C.; Eeles, R., et al. (2003). "Balancing Autonomy and Responsibility: The Ethics of Generating and Disclosing Genetic Information. *Journal of Medical Ethics* 29(2): 74–79. A leading journal for those interested in medical ethics.

Hess, David. (1997). *Can Bacteria Cause Cancer?* New York: New York University Press. A multidisciplinary analysis of the history of one type of alternative theory and therapy for cancer, analysis of its suppression, evaluation of the evidence, and discussion of policy remedies.

Huijer, Marli, and Evert van Leeuwen. (2000). "Personal Values and Cancer Treatment Refusal." *Journal of Medical Ethics* 26(5): 358–362.

Jackson, Jennifer. (1994). "Unproven Treatment in Childhood Oncology: How Far Should Pediatricians Co-Operate? Commentary." *Journal of Medical Ethics* 20: 77–79.

Moss, Ralph. (1995). *The Cancer Industry.* Brooklyn, NY: Equinox Press. Classic study of the suppression of alternative cancer therapies and researchers.

Proctor, Robert. (1995). *Cancer Wars*. New York: Basic Books. Review of research controversies surrounding environmental etiologies of cancer.

Rather, L. J. (1978). *The Genesis of Cancer: A Study of the History of Ideas*. Baltimore, MD: Johns Hopkins University Press. Examples the early theories of cancer.

CAPITALISM

Capitalism is both a special kind of self-organizing system for structuring economic activity and a historical movement in support of such a system. Its first full development is generally taken to have occurred in the late-eighteenth and early-nineteenth centuries in England, but its ideals of private property and open markets have been variously manifested and defended since. Capitalism is also coupled to a distinctive ethical view of the world, linked closely with developments in modern science and technology, and a source of challenges to other alternative ethical and political perspectives.

Historical Origins

The root of the abstract noun capitalism is the Latin *capitalis*, from *caput*, meaning head, from the hypothetical Indo-European *qap-ut*, by which cattle (another related term) are counted and thus in many preindustrial societies wealth measured. A popular but mistaken belief views capitalism as a transcultural phenomenon that "only needs to be released from its chains—for instance, from the fetters of feudalism—to be allowed to grow and mature" (Wood 2002). In reality, however, capitalism depends on special cultural conditions, including ethical commitments to the primacy of the individual and the importance of material welfare.

The political economist Adam Smith (1723–1790), who is often taken to be the father of modern capitalism, analyzes the accumulation of *capital* promoted by free markets and the productive efficiencies of increased divisions of labor. But the resulting economic order is what he calls a *system of natural liberty*. Even the extended critique of political economy found in the work of Karl Marx (1818–1883) prefers the more concrete *das Kapital* and *der Kapitalist* over *der Kapitalismus;* Marx's opposition is to the *capitalist production system* not capitalism. The English word capitalism first appears in print in British novelist William Makepeace Thackeray's *The Newcomes* (serial publication, 1854). It was left to later economists such as Werner Sombart (*Modern Capitalism* [1902]) and Max Weber (*The Protestant Ethic and Spirit of Capitalism* [1904]) to make capitalism as economic system and political ideology the center of debate.

As Weber, Sombart, and others make clear, capitalism as an economic system is closely associated with but not precisely the same as a market system. Free markets are possible on small scales, but capitalism presumes larger-scale industrial enterprises resting on both modern technology and a legal system that gives corporations the status of a person, thus creating a buffer between corporate and personal wealth and responsibility. Capitalism describes an economic system in which property resources are privately owned, but in a form not identical with individual wealth, with interactions between the supply and the demand for goods and services used to direct and coordinate economic activities. Once provided with a legal structure of private ownership enforced by the state and open markets, capitalism is self-organizing as if by means of what Smith once called an *invisible hand*. The result, it is claimed, is efficiency in two forms: technological (producing a given amount of goods with the minimum amount of resources) and allocative (distributing resources in the best way possible).

Science, Technology, and Capitalism

The role of science and technology in well-functioning capitalist economies is essential to their success. Continuing economic growth helps promote acceptance of inequalities, with such growth depending on increases in worker productivity, which is in turn supported by improvements in technology. Unfettered movement of capital, or access to capital, helps spur investment in research and development. This investment leads to scientific discovery and technological innovation, albeit sporadically.

It is also important to note that the free movement that is associated with capital and labor under capitalism has been more or less closely coupled with democratic politics. Indeed defenders of capitalism such as Michael Novak (1982) have argued that *democratic capitalism* must be distinguished from all attempts at centralized state control of science, technology, and capital.

However the relation between science, technology, and capitalism is two-sided. Not only does capitalism tend to promote science and technology, but science and technology have been argued to promote capitalism. Joel Mokyr's broad overview of economic and technological progress (1990) and of science and wealth (2002) place as much stress on how inventors and scientists have contributed to capitalism as how capitalists have funded technology and science. According to Mokyr, the development of systematic means for knowledge production and invention, and their institutionalization, went hand in hand with the development of systematic or industrial means of material production.

Scientists such as Michael Polanyi (1951) take this argument one step further and argue that the organization of science provides a model for democratic capitalism. It is in the scientific community that equality finds its strongest exemplar, and that the free flow of knowledge together with division of labor leads to an expansive production of knowledge that can serve as positive influence on and support for economics and politics.

At the same time the costs of some scientific and technological projects have on occasion been beyond the means of private capital formations. Historically states, not private corporations, have been required to pioneer public water systems, nuclear energy, major advances in airplane propulsion and design, cancer research, space exploration, the Internet, and decoding of the human genome. But it is mostly democratic capitalist economies that have provided the tax base and public support for such large-scale, big science efforts without measurable costs in consumer welfare—in the hopes that such expensive research and development projects would eventually contribute to greater public benefit.

For Mokyr, the roots of twentieth-century prosperity were the capitalist industrial revolutions of the nineteenth century, which were precipitated in part by the scientific revolution and Enlightenment of the seventeenth and eighteenth centuries. "To create a world in which 'useful' knowledge was indeed *used* with an aggressiveness and single-mindedness that no other society had experienced before was the unique Western way that created the modern material world" (Mokyr 2002, p. 297). Moreover in an increasingly knowledge-driven economy, scientists and engineers are themselves more often becoming entrepreneurial capitalists. The opportunities not just for profit but for conflicts of interest and other failures in professional ethics are nevertheless not to be minimized.

An Ethical Kaleidoscope

One key question imposed on policymakers in capitalist systems concerns the justice of those inequalities that capitalism promotes, and whether there might be appropriate remedies for such inequalities or alternative, more equitable systems of production. Capitalists typically argue both that property rights are grounded in human rights and that some level of inequality is beneficial to all because it stimulates productivity. Beyond social justice are other issues of professional ethics and cultural conflict that also deserve acknowledgement.

SOCIAL JUSTICE. The historical development of social justice issues can be traced to the classic period of the Industrial Revolution in England (c. 1750–1850). Social critics of the associated economic individualism among capitalists argued for an alternative of social solidarity among the workers and for some degree of common ownership of the means of production. Although specific mechanisms varied, a general term for this alternative is *socialism*.

The ideal of socialism, like capitalism, is a theoretical construct—a fiction. The spectrum of economic systems is bracketed by capitalism and socialism, but all economies in the early twenty-first century are in fact mixed, that is, lie somewhere in between these two extremes. On the capitalist end, economic individualism and the right to property are paramount, leading sometimes to major social inequalities. On the socialist end, communal ownership and collectivist values play a significant role, often leading to bureaucratic inertia. The historical response to capitalist failures has been state intervention, and to socialist shortcomings privatization.

THREE CONTINUING CRITICISMS. The issue of social justice has led to three general criticisms of basic assumptions of capitalist systems. The first basic assumption is that profits serve as the driving force for social as well as economic actions. If profits are not present, individuals will not have any incentive to act. But this view of humans as calculating, optimizing individuals may promote morally objectionable behavior. Defenders of capitalism respond that the profit motive is simply a reality of human nature (although the scientific evidence for this is at best ambiguous), appeal to the virtues of freedom of choice, and express faith in the ability of nongovernmental institutions to develop ethical protocols for behavior among individuals.

A second basic assumption is the sanctity of individual property rights. The criticism is that property itself is a kind of social fiction that in too strong a form may easily undermine equity or the collective good. In response, property rights are defended as basic human rights. Strong property rights are further argued ultimately to promote a productivity that benefits all, even though it may selectively benefit some more than others. An expanding pie gives even those with small pieces more to eat.

A third basic assumption is the value of free markets in both goods and services (the output market) and in the factors of production (the input market). Insofar as the input market is focused on material resources, liquid capital (money), and fixed capital (plant and machinery), this assumption is challenged only by environmentalists who argue that some natural resources may be undervalued. But the free market in labor input has a tendency, others argue, to treat humans as commodities. An unregulated labor market may lead to the violation of basic human rights—rights that, in other contexts, capitalism purports best to serve. The degree to which a society protects workers from the vagaries of the labor market is one strong measure of the influence of the socialist end of the capitalism-socialism spectrum.

The second and third assumptions—that capitalist systems have well-defined private property rights and input-output distributions guided by free markets—depend on ideal conditions that are unlikely to obtain fully in the real world. When property rights are weak or prices provide unreliable signals to market participants, a capitalist system may fail to realize its potential for good. In this regard, economists have identified four types of market failures under capitalism: (a) excessive market power where individual buyers or sellers have significant control over output, price, or both; (b) externalities where one economic agent imposes costs or benefits on another without the latter's knowledge or consent; (c) public goods where markets are either nonexistent or the good will be underproduced because there is little or no incentive for private property owners to provide a good that others can use without paying for it; and (d) asymmetric or incomplete information where either the buyer or the seller lacks sufficient information to make a free and rational decision.

Failures (b), (c), and (d) offer special challenges for science and technology in capitalist economic systems. Scientists and engineers almost always know more than others about what they may be providing by way of productive inputs or outputs. In many instances scientific research and technological development take place at the leading edge of economic activity where there is not yet and may never be any market sufficient to support it. And certainly the requirements of free and informed consent in human subject research can dramatically illustrate asymmetries in information between scientists and nonscientist participants.

CULTURAL CONTRADICTIONS. Finally there are ethical issues associated with what sociologist Daniel Bell has termed *The Cultural Contradictions of Capitalism* (1976). According to Bell, contemporary society can be organized into three distinct realms: the technoeconomic structure, the polity, and culture. The technoeconomic order is concerned with the production of material goods and services; the polity with social justice, the proper use of force, and the regulation of conflict; and culture with the meaning of human existence as expressed in various imaginative forms. At any one historical period each further exhibits distinctive norms and follows its own rhythm of change, with complex interactions that may be mutually reinforcing or subtly undermining. From the perspective of this framework, one of the general challenges of capitalist modernity is the way in which drives for change in the technoeconomic structure threaten to undermine traditions of cultural meaning on which all social orders ultimately rest.

In the contemporary capitalist world the three realms are ruled by antagonistic principles: competitive efficiency for the capitalist economy, liberty and equality in the polity, and self-realization or self-expression in culture. Bell's particular argument is that not only are there tensions between the contemporary norms (which he interprets somewhat differently) operative in each of these three realms, but also within the modernist, self-expressive culture itself. Together such antagonisms may destabilize the whole social order or particular regions within it. Certainly between the special cultures of science and technology and the general culture of self-expression yoked to capitalist productivity there are tensions that threaten the stability of science, for example, when scientists hype their results or shape them to fit economic interests. The globalization of capitalism, as a carrier of science, technology, and particular cultural values, no doubt provides further opportunities for cultural conflicts.

WILLARD DELAVAN
CARL MITCHAM

SEE ALSO *Conservatism; Critical Social Theory; Economics and Ethics; Market Theory; Marx, Karl; Smith, Adam; Work.*

BIBLIOGRAPHY

Bell, Daniel. (1996). *The Cultural Contradictions of Capitalism*, 20th anniversary edition. New York: Basic Books. First published in 1976.

Berger, Peter L. (1986). *The Capitalist Revolution*. New York: Basic Books. Sociologist Peter Berger maintains that capitalism transforms all aspects of social being and that capitalism provides the most successful mechanism yet devised for improving the material wealth of humankind. He provides a provocative set of testable hypotheses about the relationship between the new capitalism and human values.

Heilbronner, Robert. (1993). *Twenty-First Century Capitalism*. New York: W. W. Norton. A fascinating extension of lectures examining the relationship between capitalism and the notion of progress encompassing capitalism as both political and social system and its future potential for fostering economic growth.

Hutton, William, and Anthony Giddens, eds. (2000). *Global Capitalism*. New York: Free Press. Explores the impact of globalization on capitalism. It is a rich source of new perspectives on the new capitalism, its connection to social cohesion and justice, and comparative economic and political systems.

Mokyr, Joel. (1990). *The Lever of Riches: Technological Creativity and Economic Progress*. New York: Oxford University Press.

Mokyr, Joel. (2002). *The Gifts of Athena: Historical Origins of the Knowledge Economy*. Princeton, NJ: Princeton University Press.

Novak, Michael. (1982). *The Spirit of Democratic Capitalism*. New York: Simon and Schuster.

Polanyi, Michael. (1951). *The Logic of Liberty: Reflections and Rejoinders*. Chicago: University of Chicago Press.

Schumpeter, Joseph A. (1950). *Capitalism, Socialism, and Democracy*. New York: Harper & Row.

Soto, Hernando de. (2000). *The Mystery of Capital*. New York: Basic Books.

Wood, Ellen Meiksins. (2002). *The Origin of Capitalism: A Longer View*. New York: Verso.

CARE

SEE *Ethics of Care*.

CARSON, RACHEL

For post-World War II America, scientist and writer Rachel Louise Carson (1907–1964), born in Springdale, Pennsylvania on May 27, popularized the idea that ethical discussions of science and technology should consider environmental concerns. Using the insights of ecology, Carson pointed out that humans and nature were inextricably, even physically connected; for example, they were subject to similar dangers from industrial chemicals in the environment. Therefore, Carson

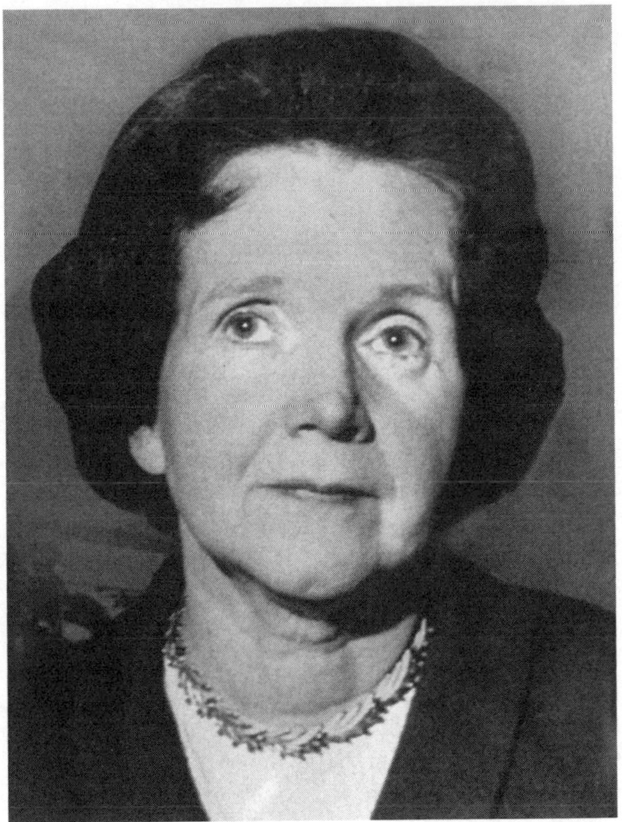

Rachel Carson, 1907–1964. Carson was an American biologist and writer whose book *Silent Spring* aroused an apathetic public to the dangers of chemical pesticides. *(The Library of Congress.)*

argued, humans should try to respect rather than dominate nature. This argument culminated in her international bestseller, *Silent Spring* (1962), published shortly before her death from breast cancer on April 14.

Early Work and Writings

Raised in a rural but rapidly industrializing area of Pennsylvania, Carson attended Pennsylvania Women's College (now Chatham College) from 1925 to 1929, where she majored in biology. From 1929 to 1934 she attended Johns Hopkins, graduating with a master of science in zoology. Due to the Depression, Carson could not afford to stay in school and earn her Ph.D. Instead she found a job as an editor and science writer with the U.S. Fish and Wildlife Service. She worked there until 1952, when the international success of her second book, *The Sea Around Us* (1950), finally made it possible for her to quit and write full-time.

Carson's professional background gave her a strong grounding in the latest research from several different scientific disciplines. As well as editing the work of other scientists, her job was to synthesize and publicize scientific information for the public. In addition, before ecology became a well-known approach, Carson had embraced an ecological perspective. (Ecology is the science that studies the interactions of organisms in the natural world.) Her first book, *Under the Sea-Wind* (1941), traced the many complex layers of marine ecosystems. During her employment, Carson also became concerned with the impact of various new postwar technologies on the wildlife and environment—among them, the pesticide dichlorodiphenylthrichloroethane (DDT), a wartime technology released into the consumer market in 1945.

As Carson's career as a writer began to gather momentum, so did her ideas about science, technology, and the environment. Repeatedly she emphasized the need to educate the public about science. She also challenged the idea that "*science* is something that belongs in a separate compartment of its own, apart from everyday life" (Brooks 1972, p. 128). Carson's developing critique of science targeted restricted circles of experts who isolated their knowledge of the natural world from the public. Her next book, *The Edge of the Sea* (1955), strove to make scientific information about the seashore accessible to the general reader. She also encouraged her readers to engage in firsthand experience with the environment to give them a reference point for evaluating scientific knowledge and discoveries.

Silent Spring

The United States's development of the atomic bomb proved to be a crucial turning point in Carson's thinking about the interactions of humans and their environment, and the consequences of science and technology. As she remembered, the possibility of humans being able to destroy all life was so horrible that "I shut my mind—refused to acknowledge what I couldn't help seeing. But that does no good, and I have now opened my eyes and my mind. I may not like what I see, but it does no good to ignore it..." (Lear 1997, p. 310). Instead Carson faced man's destruction of his environment. In particular she focused on synthetic chemical pesticides.

In *Silent Spring* Carson argued that science and technology had largely ignored the environmental consequences of pesticides in disturbing the *balance of nature*. This metaphor referred to the ecological interactions of species in the natural world, and Carson showed how pesticides interrupted these complicated relations. The widespread use of persistent synthetic chemical pesticides endangered birds, wildlife, domestic livestock, and even humans. Residues from DDT, aldrin, dieldrin, heptachlor, and other chemicals contaminated most water, soil, and vegetation. The federal government had not only failed to protect citizens from these dangers, but by carrying out aerial spraying attacks on the fire ant and the gypsy moth, it had committed some of the worst offenses. Chemical dangers even penetrated suburbia, where people intensively sprayed their homes and gardens. Carson discussed both the immediate consequences for human health and the possible long-term hazards, including genetic damage and cancer. In particular she blamed scientific experts (economic entomologists and agronomists, among others) who supported the chemical-based technologies of industrialized agriculture. For Carson agribusiness epitomized the industrial mindset of man dominating nature for the interests of private economic gain.

Silent Spring resulted in an enormous public uproar. The book raised issues that extended far past the debate on pesticides. Ultimately it questioned how modern, industrialized society related to the natural world. Pesticides were but symptoms of the underlying problem: the idea that humans should dominate and control nature. Carson wrote that the "*control of nature* is a phrase conceived in arrogance, born of the Neanderthal age of biology and philosophy, when it was supposed that nature exists for the convenience of man" (Carson 1962, p. 297). However, many readers disagreed. Criticizing Carson's idea of the balance of nature as too static, they argued instead that nature was inherently unbalanced. Man had to use pesticides and dominate nature in order to ensure his own survival. In fact Carson's understanding of the balance of nature was complicated: The phrase implied stasis, but she also portrayed nature as an active entity capable of great change.

Altogether Carson put forth an environmental ethic based on the physical, ecological connections that existed between humans and their environment. She insisted that science and technology be evaluated according to this ecological standard, where humans and nature merged as one. Moreover as part of the fabric of life, humans had no right to put the entire biotic community at risk. By popularizing ecological ideas, Carson treated her readers as capable of understanding and participating in scientific debates. She also redefined calculations of risk: Decisions on environmentally hazardous technologies should take into account public environmental values as much as scientific findings of harm. Moreover scientists and industries should bear the burden to prove their products safe, rather than the public having to prove them dangerous.

In *Silent Spring*, Carson set the foundation of the environmental movement that began in the late-twentieth century. The insight that humans and nature were ecologically linked gave people new ways to conceive of environmental issues. The environment existed not only in the wilderness and the national parks, but in the immediate, intimate surroundings of home, garden, workplace, and even the health of the physical body. Carson also sparked the ongoing public debate about how to best consider environmental issues in making ethical decisions about science and technology. She was especially significant for her grassroots appeal—making everyday people aware of their role in preserving their environment.

MARIL HAZLETT

SEE ALSO *Agricultural Ethics; DDT; Ecology; Environmental Ethics; Environmentalism; Waste.*

BIBLIOGRAPHY

Brooks, Paul. (1972). *The House of Life: Rachel Carson at Work.* Boston: Houghton Mifflin. Brooks, Carson's editor, wrote this book after her death in order to shed some light on Carson's private life, her writing process, and the history of *Silent Spring.*

Carson, Rachel. (1962). *Silent Spring.* Boston: Houghton Mifflin. Details the dangers of the first generation of synthetic chemical pesticides produced in the postwar era.

Lear, Linda. (1997). *Rachel Carson: Witness for Nature.* New York: Henry Holt and Co. The definitive biography on the life of Rachel Carson.

CARTESIANISM

SEE *Descartes, René.*

CENTRAL EUROPEAN PERSPECTIVES

Although the countries of Central Europe (CE) have a long tradition of critical reflection on science and technology, this tradition was severely curtailed from World War II to the end of the Cold War. Only since the early 1990s have discussions emerged that might be described as contributing to bioethics, environmental ethics, computer ethics, and related fields of science, technology, and ethics. Other traditions of scholarship nevertheless have developed in ways that may be related to these fields, and deserve consideration, especially when placed within a larger historical and philosophical context.

Boundary Issues

CE has been defined according to different criteria. A variety of factors—geographic, religious, linguistic, strategic, ethnic, historical, sociopsychological, and developmental—have shaped the dividing lines of the lands located between Russia and the German-speaking countries. In some conceptions, even Russia and Germany were included. For centuries, it was the route by which conquering Central Asian tribes—Huns, Magyars, Tatars, and others—invaded Europe. It was also the path by which Western armies—those of Sweden's Gustavus Adolphus, Napoleon, and Hitler—attacked, attempting to expand east into the center of Russia. This region was an important strategic area called the Euro-Asian *heartland* or *pivot area.* Whoever controlled the territory was said to control the world, which is why CE was repeatedly subject to invasions from east and west. As a result of all these assaults and historical expansions, CE has the most complex ethnic makeup in Europe, peopled in many places by ethnic groups too small to constitute a separate nation-state.

Like Southern and Eastern Europe, CE has been slow, and reluctant, to embrace the Enlightenment as well as the Industrial Revolution; economic development and industrialization evolved more slowly and unevenly than in Western Europe. This may be explained in part by the longtime authority and spiritual power of religion in these countries. Together with other influences, this religious authority contributed to other than economic growth. The specific character of the so-called Slavic mentality generated by the different character of language and cultural heritage is being protected and rescued against the attempts of homogenization resulting from integration with the economically powerful European Union. The problem is very vital in public discussions on the advantages and disadvantages of European integration, and is a strong arguing point for Eurosceptics against Euroenthusiasts. The former oppose treating economic factor as the exclusive criterion of development. They point to the literary tradition of ironic or spiritual distance to terrestrial profits.

Differences between developed countries of the West and the developing CE countries of Poland, the Czech Republic, Slovakia, and Hungary disclose special problems. In the early-twenty-first century, the dynamic growth possible for these countries—sometimes called the Visegrad Group—as a result of having joined the European Economic Community (EEC) as well as the

North Atlantic Treaty Organization (NATO), has opened a new era in their relationship to science, technology, and ethics.

Science

Communist regimes attached great importance to scientific achievements, realizing that such accomplishments constituted a strong position in the Cold War. However the communist ideology, trumpeted in the media and developed in every sphere of public life, that elevated the character of the working class and peasants over that of *elites*, ultimately resulted in scientists being considered parasites, producing nothing of material social value. Such political attitudes fluctuated, being stronger at some times (until Stalin's death in 1953) than at others; nevertheless they had significant impact. Many inventive scholars who did not sign *loyalty declarations* were marginalized or persecuted. After 1968 some university professors, many of Jewish origin, left CE for the West to continue their research in more democratic conditions.

One of them, Leszek Kołakowski, gained international fame as a critic of Marxist theory. His thorough, three-volume monograph of Marxism is outstanding not only due to vast range of materials used in the narration but also due to its clear-sighted style. Kołakowski explains the phenomenon of Marxism and the reasons of its worldwide spreading. He locates the project of radical transformation of social relations on the wide background of history of thought, exposing in it millenarist and eschatological motifs. In such a perspective Marxism is suspected of being one more version of salvation, but a secular one. Kołakowski discusses not only the very conception of Marx but investigates its further history in different European countries, registering meanders of the evolution of the original project, caused by peculiarities of social contexts and processes in different European countries and elsewhere.

Some philosophers nevertheless remained in CE doing work that is directly relevant to the philosophical and ethical understanding of science. Among these are Tadeusz Kotarbinski (Poland) and Jan Patocka (Czech Republic). Their unique achievements, such as the theory of good work called *praxiology* fromulated by Kotarbinski or phenomenological reflection on history and role of technology by Patocka were the exemplary proofs that autonomous and efficient thought has been practiced even under the Soviet Union supremacy. Praxiology was developed in many European countries (such as Norway) as well as in the United States as important contribution to the theories of management and could be also considered as Polish contribution into philosophy of technology within which technology is defined mainly as multi-levels organization. Phenomenological accent on responsibility makes Patocka's considerations actual both at the time he was writing and at the present moment.

After the collapse of communism in 1989, the period of transition and transformation began. The role of science in these countries has been recovering after a totalitarian regime's controlling system. The infrastructure of scientific development is strongly connected to economic growth, which, previously, was not highly advanced. The end of the Cold War opened borders. International cooperation in different fields, especially in the hard sciences (physics, chemistry, and informatics), became more regular and was not ideologically controlled. The Polish Academy of Science became a member of the European Science Foundation in 1991, and scientists had an opportunity to take part in extensive international research programs. International conferences and meetings were organized in all fields, social sciences and the humanities included. The Czech Academy of Science underwent a radical reform in 1994. Social needs fueled research into the economic and legal questions connected to the privatization of socialized property; specialists began to examine critically whether pure liberalism could cope with transition problems, and investigated the role and impact of business ethics in that process.

Social consciousness, which under communism was soothed artificially and often deliberately misdirected, developed rapidly. The problematic character of scientific authority, of science in general, and the issues related to the incorporation of scientific discoveries into society became the substance of public debate as well as of scholarly research. Important works related to these topics discuss issues of science and the search for truth or science and democracy, with the argument that scientific development is central to the future. Nevertheless the growing consciousness of the dilemmas raised by scientific-technological advancement, either generally (for example, spiritual crisis of the contemporary world, technology and civil rights) or more specifically (such as creating quality cultures, or economic and social effects of the lack of adequate technology education) have become vital to the worldview of the CE nations.

Education

Central European University in Budapest, Charles University in Prague, and Warsaw University are representative of the new tendency to liberate education from ideological limits and conditioning. According to some studies, the entire educational system in postcommunist CE countries is undergoing fundamental change. The structure is making a successful transition from the radical-structuralist model that previously dominated to a functional-liberal paradigm.

The communist party ran a hierarchical and strongly centralized educational system. An elementary level of education was easily accessible to all members of society. Education served the needs of the dynamic, industrial society; it also immersed students in *scientific socialism* ideology. In the early-twenty-first century, curricula and teaching methods have undergone serious transformation.

According to some research carried by The World Bank Institute in 1997 on the quality of educational systems in CE countries, the process of decentralizing started and was developed. Comparison of experiences from Hungary, Poland, and the Czech Republic let the researchers expose the main problems and suggest solutions.

Economic underdevelopment makes full educational reform unachievable. In this context, the activity of Hungarian financier and philanthropist, George Soros, appears to be very important. He is a founder and a chairman of Open Society Institute (OSI) in New York and in Budapest and of the Soros foundations network. Promoting a free press and political pluralism in all the postcommunist countries, in spite of being accused and hindered by the authoritarian governments in Eastern European and Post-Soviet countries, he and his foundations are dedicated to building and maintaining the infrastructure and institutions of an open society. Through the global network of nongovernmental organizations (NGOs), he helps to support health programs, to fight discrimination of all sorts, and to promote democracy. In CE countries they help to replace the authoritarian model in education with the civic education style. Different steps and procedures are being introduced to democratize education system, to develop a new way for teachers to relate to their pupils. Financial support providing schools with necessary equipment such as computers, videocassettes, and CDs contributes much to achieving real transition in the education field.

Technology

Under communism, scientific achievements were treated as part of a *scientific-technological revolution* rather than as abstract or pure concepts. At universities and technical schools throughout the CE countries, the philosophy of technology developed first from the Marxist viewpoint (for example, Radovan Richta in Prague, Adam Schaff in Warsaw), and then in a more pragmatic and individualistic way (such as Ladislaus Tondl in Prague; Tadeusz Kotarbiński in Warsaw; Józef Bańka in Katowice, Poland). In general, however, technology has been a subject of systematic philosophical reflection only since the early 1990s.

Apart from comments and attempts to build on the work of established Western thinkers (Karl Jaspers, Hans Jonas, Martin Heidegger, Jacques Ellul, Jose Ortega y Gasset, and others), only a few independent projects for conceiving interrelations between technology and society have appeared in response to contemporary problems and social needs. Ladislav Tondl (Prague) identified different aspects of social control of technology and developed the concept of *delegated intelligence*, which enabled him to investigate the structure of subsystems in technology. Imre Hronszky (Budapest) distinguished technological paradigms and discussed communities in technological change. Józef Bańka (Katowice) studied mutual interactions of modern technology and human personality.

Bańka's research became the basis of a new, individual approach that developed into a philosophical concept called *eutyphronics*. Its main principle was the protection of humankind as it faces the dangers of technological civilization. Andrzej Kiepas and Lech Zacher (Warsaw), editors of the interdisciplinary magazine *Transformacje* (Transformations), which has published numerous articles devoted to that central issue, have been promoting Western European and U.S. traditions of technology assessment and their own original conceptions of it for years.

Ethics

Although CE is increasingly engaged with Western intellectual discussions, such standard fields of applied ethics as environmental ethics, business, computer ethics, and professional ethics in science and engineering have not yet become standard fields for research and discussion. Nevertheless, using such recognized classifications, one can note the following contributions.

Business Ethics

The transformation from planned to market economy in the CE countries is a test bed for applying economic theory and business ethics to an enormous historical transition in the economic and political system. Authors from the Czech Republic, Hungary, Poland,

and Slovakia have analyzed the economic, philosophical and political problems of the transition process. The education and training necessary to combat increasing corruption in public bureaucracies of CE countries are being examined. The transition to democratic institutions must include the participation of all sectors to enhance transparency and build long-term public trust. Anticorruption efforts, including structural and normative approaches to ethical controls, must be aligned with the core values unique to each country's ecology. Key shared values must include honesty, stewardship, respect for human dignity, and concern for others.

Along with the public debate involved in creating a democratic system, social concerns also focus on so-called postmoral spirituality in different areas. For example, Budapest University organized a workshop called Spirituality in Management in Hungary in 2001. Participants discussed spirituality as a search for meaning, which transcends material well-being. The workshop focused on the possible role of spirituality in renewing the contemporary management praxis.

Computer Ethics

Advances in computer and data communication technology have created new ethical issues. Startling advances in biotechnology and genetic engineering offer not only new cures but also open the possibility of modifying existing organisms. Throughout CE, schools dedicated to these technologies have introduced seminars to enhance awareness of the moral implications of working as an engineer or technologist. Engineering ethics, already developed worldwide, is being introduced in CE university curricula and written about in philosophical journals by such authors as Wojciech Gasparski and Andrzej Kiepas (Poland).

Environmental Ethics

Henryk Skolimowski is a leader in the discussion of environmental ethics. His concepts of *cosmocracy* as the next stage of democracy and ecological spirituality constitute an important contribution to the philosophy of technology. Instead of treating the world as a machine, he recommends referring to the world as to the sanctuary. He considers the human race as a guide to realize the eschatological purpose of the universe. The basis of his ethics, which is a practical application of eschatology, is the notion of responsibility of some overreligious, mythological character.

At present, CE focus in this area is on practical problems and their resolution. Technology transfer and technology forecasting make it necessary to consider the expected rate of technological advance and to adjust conditions—material infrastructure and social framework—to various applications in science and technology.

Assessment

In comparison to Western Europe where, as a result of the Enlightenment, the separation of church from government has become the rule, in CE religion retains its importance and influence even in the public sphere. There are political-historical reasons for this situation. In Poland the church was, during the communist period, the center of opposition to the government, shaping opinions and helping to organize resistance to the political regime. Debates among those representing Marxist, atheist, and Roman Catholic views brought ethical problems connected to the scientific-technical revolution to the attention of the public. The vast range of new ethical conflicts and problems are very often still immersed in more general moral worldview religiously or even mythically inspired. Coexistence of these traits with the commonsensical, pragmatic attitude seems to some extent to be politically and socially conditioned. The election of Karol Wojtyła, a Pole, as pope contributed to strengthening the public resolve to reject communism. The great strike organized by the Solidarity movement in Poland in 1980 was the first in a chain of events, which included the fall of the Wall in Berlin in 1989, that culminated in the collapse of the Soviet Union in 1991, ending the communist era. The difficult period of transformation had begun.

MARIA KOSTYSZAK

SEE ALSO *Communism; German Perspectives; Marxism.*

BIBLIOGRAPHY

Bokajło, Wiesław, and Kazimierz Dziubka, eds. (2003). *Unia Europejska: Leksykon integracji* [European union: Lexicon of integration]. Wrocław, Poland: Publishing House "Europa." In each of six chapters there are short encyclopeadic articles devoted to main cultural, legal, social, economic and political issues connected with the process of European integration. Such entries as "National Minority," "Freedom of Belief and Conscience," "Napoleonian Visions of Europe," "Euroregions," "Civil Society," and many others are developed in their conceptual, historical and organizational dimensions.

Davies, Norman. (1997). *Europe: A History*. London: Pimlico. A comprehensive study of genesis, directions of development and the present shape of Europe conceived as a differentiated but integral mixture of nations, interests and complex forces. Some reviewers criticized the book for a somewhat undue emphasis on Poland-Lithuania and Eastern European affairs in general, and for his biased opinions of the Holocaust and the other genocides in modern Eastern Europe. In spite of that, a valuable source.

Gorzelak, Grzegorz; Eva Ehrlich; Lubomir Faltan; and Michal Illmer. (2001). *Central Europe in Transition: Towards EU Membership*. Warsaw: Scholar Wydawnictwo Naukowe [Scientific publishing house "Scholar"]. The transformation process in four countries: Czech Republic, Hungary, Poland and Slovakia in the decade of the 1990s is shown in four major domains: economy; socio-political process; education, science and technology; and the regional dimension of transformation. Prepared by the international team who made the evaluation of development scenarios—political, social and cultural aspects of system changes towards democracy, its institutional and mental obstacles.

Hartman, Jan. (2003). *Short Studies in Bioethics*. Kraków: Jagiellonian University Press. A collection of articles and speeches delivered during European conferences concerning the most vital problems of bioethics in CE countries.

Kaase, Max; Vera Sparschuh; and Agnieszka Wenninger, eds. (2002). *Three Social Science Disciplines in Central and Eastern Europe: Handbook on Economics, Political Science and Sociology (1989–2001)*. Berlin, Bonn, and Budapest: Social Science Information Center and Collegium Budapest. This volume documents the development of economics, political science and sociology in Central and Eastern Europe EU accession countries from 1989 to 2001, with a special emphasis on research. A CD-ROM enclosed in the handbook presents an overview on Central and Eastern European institutions in the respective countries relevant for economics, political sciences, and sociology (about 700 institutions). The CD-ROM enables direct access to the database INEastE in the Internet.

Koczanowicz, Leszek. (1997). "Memory of Politics and Politics of Memory: Reflections on the Construction of the Past in Post-totalitarian Poland." *Studies in East European Thought* 49: 259–270. The article concerns the democratic debate on the communist past in Poland that was carried on the basis of the two models of constructing the past. One was founded on Jaspers's theory of values; the other, dialogical one, conceived the past as the area to be socially negotiated.

Richta, Radovan. (1963). *Člověk a technika v revoluci našich dnů* [Man and technology in the revolution of our day]. Prague: Spolecnost pro sírení politických a vedeckých znalostí. This groundbreaking work did much to bring the concept of technology into the forefront of philosophical thought during the 1960s.

Skolimowski, Henryk. (1981). *Eco-philosophy: Designing New Tactics for Living*. Boston: M. Boyars.

Skolimowski, Henryk. (1992). *Living Philosophy: Eco-philosophy as a Tree of Life*. London and New York: Arkana. This and the preceding title represent two positions presenting basic principles of environmental ethics appealing to the attitude of deep caring about all segments of the surrounding world.

Szkudlarek, Tomasz. (1998). "Television as Adult Pedagogy: Learning Consumerism, Dreaming Democracy." In *Adult Education and Democratic Citizenship*, ed. Agnieszka Bron, John Field, and Ewa Kurantowicz. Kraków: IMPULS. The article provides an analysis of a monthly portion of the news and commercials blocks from Polish television. That case study shows tendencies in informational policy. Viewing contemporary education as the process of defining and re-defining individual and collective identities in constructing and re-constructing the social, the author points to the role media play in the process of production and understanding values.

Tjeldvoll, Arild. (1993). "Edukacja w Europie środkowowschodniej. Przemiany paradygmatu radykalno-strukturalistycznego na funkcjonalno-liberalny" [Education in Central-Eastern Europe. Transformation of a radical-structuralist paradigm into a functional-liberal one), trans A. Dąbrowska and B. Jarzynka. *Edukacja* [Education] 1993, no. 1, 13–33.

INTERNET RESOURCE

Slavonic and East European Review. Available at http://www.ingentaconnect.com/content/mhra/see/2004/00000082/00000003.

CFCS

SEE *Montreal Protocol*.

CHALLENGER ACCIDENT

SEE *Space Shuttle Challenger* and *Columbia* Accidents.

CHANGE AND DEVELOPMENT

• • •

Although it generally is acknowledged that change characterizes many aspects of human life and the larger world and is associated especially closely with science and technology and their influence on society, this phenomenon is not easy to define. One puzzling issue concerns how an object can be one thing, then change, and still remain the same object (that has undergone change). How should such a relationship, which implies both noncontinuity and continuity, be distinguished from replacement? A common response to is to argue that in change there is some development or growth: A thing has immanent within it a feature that over time (through change) is made manifest. The application of this biological notion to scientific, technological, economic, political, or ethical change remains fundamentally problematic and may best be approached through comparisons and in historical terms.

Enlightenment Origins: Change in Science as Progress

Early forms of the interrelated ideas of change and development were expressed in various instances of premodern (European and non-European) thought. Aristotle's *On Coming to Be and Passing Away* is the first systematic discussion of change. However, it was only in association with the scientific revolution of the 1600s and the Enlightenment of the 1700s that change became a theme for systematic articulation and gave rise to a concept of change as progress that has implications for science, technology, and ethics. The scientific revolution was understood by its proponents as a decisive progress in knowledge. Modern science claims as well as strives to represent a truer picture of nature than all previous sciences. In part, this knowledge depends on a more accurate understanding of development and change in the natural world.

The idea that human agency can be understood as social in origin and that all humans have the capacity to change their individual and collective destinies through the deployment of reason to combat tyranny, ignorance, superstition, and material deprivation was an important hallmark of European Enlightenment thinking. The notion that science can explain everything in nature, with the resulting knowledge being available to promote human progress, became the hallmark of modern rationalism and the social sciences. The first systematic compilation of scientific and technological knowledge to this end is contained in Denis Diderot's (1712–1784) *Encyclopédie* (1751–1772).

Armed with their ardent faith in the rationality of scientific methods and their ability to dissect and attack prevailing religious, social, political, and economic practices, many of the followers of the Enlightenment believed in and acted on the possibility of liberty, equality, and the pursuit of happiness for all humanity. Studies of the evolution of human societies gave rise to the notion of modernization as a way to change cultural patterns and social hierarchies and divisions.

The idea of progress through change became ingrained in intellectual thought and social and political action. Imaginative thinkers of modernization such as Auguste Comte (1798–1857), Henri Saint-Simon (1760–1825), and Robert Owen (1771–1858) claimed that the creative application of science and technology in industrial processes could unleash an economy of abundance that could bring an end to the pervasive poverty of the majority of the population in European societies. This progressive vision prevailed despite skepticism on the part of political economists such as Thomas Malthus (1766–1834) that poverty and want could not be eradicated because of unsupportable increases in the human population.

The Nineteenth Century and Beyond

The notion of society as organic in nature and societal progress as an evolutionary process became entrenched in modernization theory in the nineteenth century after the publication of Charles Darwin's (1809–1882) *On the Origin of Species* (1859). Karl Marx's (1818–1883) theorizing of civilizational development in teleological terms so that human history could be read as a dialectical process determined by the specific technological artifacts that are shaped by social forces and relations of production often was associated with Darwin's theory of evolution. Although Marx developed his views well before the *Origin of the Species* was published, Marx certainly thought that his views and Darwin's were compatible. However, the Darwinian theory of natural selection and "survival of the fittest" (a term coined by Herbert Spencer [1830–1903]) was used to justify "both a rugged economic individualism at home and a ruthless collective imperialism abroad" (McNeill 1963, p. 830).

Over the course of the 1800s the idea of progress became the basic ideology of scientific, technological, and economic change in Europe and North America. However, two basic theories about how to promote such progressive change emerged. One was that it was the result of spontaneous order arising from multiple individual sources, none of which has such order consciously in mind (as with Adam Smith's "invisible guiding hand" that operates in market economics); the other was that change requires some kind of central monitor to make sure it serves true human interests and thus becomes progress (by means of planning or some kind of social democratic control).

After World War II: Change as Development

The post–World War II idea of progress through science and technology beyond European and North American shores became the focus of modernization project. The extension of the notion of progress through science and technological industrialization to other nations and later into the colonies of imperial powers came to be known as development. Development as an autonomous practice within what is known as the Third World began in earnest after the World War II with the onset of decolonization. The modernization project implemented through different economic development models began in former colonies at the behest of the United Nations and the World Bank during the 1950s and afterward. The common denominator of those development models was modern technology, the rapid infusion of which was expected to materialize through its transfer from industrialized nations. The modernization project considered foreign aid in capital and technology to be vital for development.

The basic assumption of modernization projects is convergence, an important ontological premise of the Enlightenment: The world is on a Eurocentric path of economic and social change and democratic political dispensation; the West arrived there first, and the rest of the world is expected to catch up eventually. It is axiomatic in modernization theory that "traditional" societies can be transformed through a concerted project of economic development that can be achieved by changing the means of economic production by transforming archaic social structures that lack the incentives for and entrepreneurial spirit of rapid technological innovation. By formulating and implementing the "right" package of policies, the state and other agents of economic power can induce technological change, which is equated to a problem-solving activity. This minimalist, though very effective, model became the heuristic basis for economic development projects. However, this meta-model of modernization and the ensuing universalist narrative of change and development are being challenged by postcolonial and postmodern theorists and deep ecologists for various reasons.

It is important to note that beyond this pervasive notion of economic change and development, there are economists who believe that unleashing the invisible hands of free markets is the "natural" route to economic change and growth. Following this intellectual tradition from Adam Smith to Friedrich Hayek to Peter Bauer, they claim that progress comes from "spontaneous order," not from centrally planned rational design. One of the most influential development theorists of this genre was U.S. presidential adviser Walt W. Rostow (1956, 1960), who distinguished five states of development: (1) traditional society, (2) preconditions for takeoff, (3) takeoff, (4) drive to maturity, and (5) high mass consumption. In this schema, development started in Western Europe and then in North America and Japan and finally the winds of economic change reaches the developing world. It was such orthodox visions of development thinking that became the hallmark of development assistance spearheaded by the World Bank and other aid agencies, until more recently.

Unalloyed faith in scientific and technological knowledge as the most important resource for development was entrenched in all theories of modernization until the 1960s. It generally was agreed that more than capital and labor—the traditional factors of production—it was knowledge manifested as ideas, information, innovation, and technology that would increase productivity, and consequently, the income and wealth of nations.

Criticisms of the Model

However, the unprecedented material progress that the West had experienced as a result of advances in science and technology was challenged when the unintended consequences of controlling and using nature became apparent and problematic. Rachel Carson's (1962) *Silent Spring* brought public attention to the excesses of industrialization in the form of pollution and irreversible environmental changes. The moral qualms that many scientists and intellectuals felt about uncritically pushing the frontiers of scientific knowledge became a matter of serious ethical reflection on the uses and abuses of scientific research. The destruction of Hiroshima and Nagasaki by atom bombs and the invention of recombinant DNA technique added impetus to the notion that the creators of knowledge also bear ethical and moral responsibilities for the application of science and technology, which until that time was thought to be a force for good for all humans. Ulrich Beck (1998), employing a constructivist theoretical framework of self-reflexivity, claims that scientific and technological advances are leading to global risk societies.

The idea that the future of modern industrial civilizations is at risk if the manner and direction of industrialization and economic growth are not reformulated became an important point of discussion among many policy makers, scientists, and public intellectuals after the publication of the *Silent Spring* and the Club of Rome study titled *The Limits to Growth* (Meadows et al. 1972). Through a system-dynamics modeling of global production and consumption patterns, Meadows and associates claimed that the world would run out of food, minerals, and living space as a result of unsustainable population growth, industrialization, and pollution. The alleged inappropriateness of modern technology for the development of the Third World was forcefully argued by E. F. Schumacher (1973) in *Small Is Beautiful*.

Ironically, the advances in science and technology that originally had disproved Malthus's claim that unchecked human procreation would lead to pestilence and famine were presented by many modern neo-Malthusians as the new danger that humans faced. It is a fact that humans are confronted with global environment changes such as global warming, tropical deforestation, industrial pollution, and the proliferation of weapons of mass destruction. However, advances in science and technology are not the reasons for these problems, which are caused by the misuse of science and technology and the domination of the world by unenlightened political, religious, and economic ideologies.

Potential Answers

Advances in science and technology were able to unravel many of the myths of limits to growth and theories concerning unsustainable human population growth. Innovations in agriculture, industry, health, and habitat were shown to be capable of solving many of the problems of food scarcity, disease, and inhospitable living conditions. It became apparent that the difficulties faced by the world's poor are not a production problem any longer but are due to inequitable distribution of resources and denial of access to the opportunities for better living conditions as a result of failed development policies.

New information and communications technologies helped bring about the latest phase of economic, cultural, and political globalization. Recent advances in biotechnology, materials engineering, and communications and information technologies in tandem with globalization are promised to unleash a "new economy" that is predicted to bring prosperity and democracy to all people. However, the benefits of globalization may be a double-edged sword. Although untold wealth is created for a select few connected to the "network society," most people have not yet seen tangible benefits. The globalization of culture and the growth of economic markets are potent forces that threaten to complete the homogenization of cultures and the living patterns of many unique communities and social arrangements.

The ethical consequences of recent advances in molecular biology and genome science are predicted to be much more intractable than all earlier ethical questions concerning science and technology in the industrial age. Cloning, embryonic stem cell research, nanotechnology, biosynthesized and "intelligent" robots, bioengineered organs and tissues, and pervasive computing and human-computer interfaces are going to have a profound effect on the concept of what is "human." The increasingly tenuous divide that has existed between humans and nature will be removed forever. Because humans are now in a position to control their own evolution and because of the tenuous state of the idea of "human nature," the moral challenge will be to construct a collective "human identity" based on political notions such as equality, liberty, and the right to live a dignified life without fear, pain, hunger, and religious and political repression.

The "posthuman future" made possible by the coming biotech revolution (Fukuyama 2002) will allow people to construct the sort of "human essence" they want to preserve. However, questions of what exactly this "essence" is made up of and who can decide these issues and in what manner will be so complex and intractable that no advances in science and technology will be able to answer these questions.

Despite general skepticism in the industrialized countries that further advances in science and technology are the key to continued material well-being, the promise of modern science and technology to improve the material conditions poor people in the developing world is still largely unrealized. Advances in certain domains of science are still needed to conquer deadly diseases and improve the living conditions of billions of people. Unfortunately, funding for recent biomedical and biotechnological advances has been used to improve the dietary practices and treat the diseases that afflict rich people. Diseases such as malaria, tuberculosis, and AIDS that ravage hundreds of millions of poor people in the tropics have not yet received serious attention from the scientific establishment and funding agencies.

In a world riven by unfair social, political, and economic dispensations brought on by untenable religious and nationalistic prejudices in both the East and the West the only hope for a sane world is to rely on the critical rationality of modern science that many believers in the Enlightenment embraced. Science and technology face some crucial ethical dilemmas. Although there is no justification for funding scientific research and technological innovation to enhance the wealth of already rich people, many aspects of existing knowledge and technology could be deployed to liberate billions of people from poverty and deprivation.

Besides playing a direct instrumental role in advancing the material conditions of living, science and technology can be deployed to advance the cause of freedom that humans need to foster development and change. Scientific and technological knowledge is an important resource for advancing the cause of "development as freedom" (Sen 1999). Although scientific knowledge and technological artifacts have bestowed many good things on humanity and have paved the way for progressive change and development, they also have caused serious ethical dilemmas.

The idea behind change and development can be traced to the Enlightenment-driven notion of modernization, which entails two important principles. First, it favors the use of science and technology for human emancipation from wants and regressive social relations as well as inhospitable natural conditions. Second, it offers humans the possibility of becoming autonomous agents so that they can not only take charge of their own destinies but also self-consciously construct and change their identities.

GOVINDAN PARAYIL

SEE ALSO *Cultural Lag; Development Ethics; Progress; Sustainability and Sustainable Development.*

BIBLIOGRAPHY

Beck, Ulrich. (1998). *World Risk Society*. Malden, MA: Polity Press.

Carson, Rachel (1962). *Silent Spring*. Greenwhich, CT: Fawcett Crest Books.

Cowen, Michael P., and Robert W. Shenton. (1996). *Doctrines of Development*. London: Routledge.

Fukuyama, Francis. (2002). *Our Posthuman Future: Consequences of the Biotechnology Revolution*. London: Profile Books.

McNeill, William H. (1963). *The Rise of the West: A History of the Human Community*. New York: New American Library.

Meadows, Donella H.; Dennis I. Meadows; Jorgen Randers; and William W. Behrens III. (1972). *The Limits to Growth: A Report for the Club of Rome's Project on the Predicament of Mankind* London: Earth Island Ltd.

Mokyr, Joel. (1990). *The Lever of Riches: Technological Creativity and Economic Progress*. New York: Oxford University Press.

Parayil, Govindan. (1999). *Conceptualizing Technological Change: Theoretical and Empirical Explorations*. Lanham, MD: Rowman and Littlefield.

Parayil, Govindan. (2003). "Mapping Technological Trajectories of the Green Revolution and the Gene Revolution." *Research Policy* 32(6): 971–990.

Peet, Richard, with Elaine Hartwick. (1999). *Theories of Development*. New York: Guilford Press.

Rostow, Walt W. (1956). "The Take-Off into Self-Sustained Growth." *Economic Journal* 50(4): 150–200.

Rostow, Walt W. (1960). *The Stages of Economic Growth: A Non-Communist Manifesto*. New York: Cambridge University Press.

Rothschild, Emma. (2001). *Economic Sentiments: Adam Smith, Condorcet and the Enlightenment*. Cambridge, MA: Harvard University Press.

Schumacher, Ernst Friedrich. (1973). *Small is Beautiful: Economics as If People Mattered*. London: Blond & Briggs.

Sen, Amartya. (1999). *Development as Freedom*. New York: Anchor Books.

CHAOS

SEE *Complexity and Chaos*.

CHEMICAL WEAPONS

Chemical Weapons (CWs) constitute a major but often under appreciated ethical and political challenge for science and technology. The following entry examines this challenge by describing the character of CWs, the history of their use, and efforts of ethical and political control.

Chemical Weapons: What Are They?

Definitions of *chemical warfare* and *chemical weapon* have changed over time. History is replete with examples of chemicals being employed either to kill individuals, for example, murder or assassination, or larger numbers during warfare, such as the use of *Greek Fire* (a mixture of petroleum, pitch, sulfur, and resins) during at least two sieges of Constantinople (673 and 718 C.E.) However the twenty-first-century understanding of CWs is based on a better scientific appreciation of the underlying chemical and biological processes involved, which began to take shape during the nineteenth century.

Knowledge of how the toxic properties of chemicals could be employed as a method of warfare evolved in conjunction with the industrial and scientific infrastructure that brought about the large-scale production of chemicals. Such an infrastructure provided equipment, production protocols, and analytical techniques from the chemical industry and its research laboratories for CW purposes. Prior to such developments chemical warfare was essentially poisoning by persons who had little or no understanding of how such *weapons* functioned.

The internationally accepted definition of chemical weapons is that contained in the 1993 Chemical Weapons Convention (CWC), which states that such weapons consist of one or more of three elements: (a) toxic chemicals and their precursors (the chemicals used in the synthesis of the toxic chemicals) when intended for warfare; (b) munitions and devices specifically designed to cause harm or death through the use of such toxic chemicals; or (c) any equipment specifically designed to be used directly in connection with such chemicals, munitions, and devices.

Note the presence of a chemical in munitions does not automatically make them CWs. Weapons containing napalm and white phosphorus, for example, are not CWs because their primary effect depends on the incendiary properties of these chemicals and not their toxicity. The CWC definition of CWs contains a *general purpose criterion* (GPC) that bans the production and use of all toxic chemicals except for peaceful purposes.

The GPC is the principal mechanism by which technological and scientific developments can be taken into account by the Organisation for the Prohibition of Chemical Weapons (OPCW), which implements the CWC. The CWC definition of CWs is also phrased to ensure that bulk CW storage containers, and binary or other multi-component systems are covered by the convention.

Finally, toxins—the highly toxic chemical byproducts produced by certain types of living organisms—are covered by both the 1972 Biological and Toxin Weapons Convention (BTWC) and the CWC. Thus the use of a toxin as a *method of warfare* or for *hostile purposes* may be legally defined as both chemical and biological warfare.

Although CWs are, together with nuclear, biological, and radiological weapons, often referred to as *weapons of mass destruction,* they vary widely in terms of effect and lethality. While some CW agents are highly dangerous (i.e., toxic), others were developed to be used as incapacitatants (e.g., BZ or 3-quinuclydinyl benzilate, a hallucinogenic drug).

In terms of killing power, CW agents are not in the same category as nuclear weapons and some biological warfare agents. A fuel-air explosive, or thermobaric device, is generally more lethal (and predictable) than a comparable payload of CW agents. Comparisons are further complicated if one considers low-yield nuclear warheads, such as those being developed for use as part of deep-penetrating munitions, or *bunker busters*. Such weapons could be used in a manner that results in the deaths of only those people located inside a targeted, deeply buried and/or hardened facility.

Finally, LD50 (the amount of agent required to cause 50 percent of those targeted to die) figures do not reflect practical problems associated with the delivery to target of CW agents. The estimated amounts of agent required to effectively contaminate a given area help to illustrate such problems and, therefore, the actual threat posed to individuals by CW agents in the field.

Great attention has been given to the development of firing tables for various types of munitions and agents. For example, a U.S. Army manual estimates that approximately twenty-seven kilograms must be used to achieve a single casualty among protected troops. It has also been estimated that four metric tons of the organophosphorus nerve agent VX would be required to contaminate effectively a six-square-kilometer area. The equivalent figures for CWs employed in enclosed, urban areas, are generally somewhat lower. Additional uncertainties are caused by problems associated with extrapolating research data from various test animals to humans and extrapolating data involving the use of simulants, rather than actual CW agents, particularly in field tests.

CW agents may be divided according to their principal physiological effects: blister or vesicant, blood, choking, incapacitating, nerve, tear gas, and vomiting agents. Vesicants cause skin blisters and can cause severe damage to the eyes, throat, and lungs. Life-threatening infections in the trachea and lungs may result. Lewisite (L), nitrogen mustards (HN-1, HN-2, HN-3), sulfur mustard (H, HD), and phosgene oxime (CX) are examples of blister agents. Their primary purpose is to cause mass casualties requiring intensive, long-term treatment, rather than death. Those exposed may also suffer long-term health problems, such as cancer.

Blood agents, such as arsine (SA), cyanogen chloride (CK), and hydrogen cyanide (AC), inhibit cytochrome oxidase, an enzyme needed to allow oxygen to be transferred from the blood to body tissue and, in the case of significant exposure, rapidly become fatal.

Choking agents, such as chlorine, diphosgene (DP), and phosgene (CG), interfere with breathing. Phosgene and diphosgene interfere with transfer of oxygen via the lung's alveoli sacks. Symptoms of phosgene poisoning do not become apparent for several hours. In addition the chances for survival are a function of physical exertion. The more strenuously victims exert themselves physically after exposure, the more likely they are to die. Complete rest and oxygen treatment are recommended.

Incapacitating agents are designed to induce physical disability or mental disorientation. LSD (a form of lysergic acid) and BZ (3-quinuclidinyl benzilate) are two examples. The United States investigated the potential military uses of LSD. It also weaponized BZ, which can cause constipation, headaches, hallucinations, and a slowing of mental thought processes.

The principal nerve agents, sarin (GB), cyclosarin (GF), soman (GD), tabun (GA) and V-agents, are all organophosphorus compounds that inhibit an enzyme responsible for breaking down acetylcholine, a neurotransmitter. Nerve agents may be inhaled or absorbed through the skin. Symptoms include drooling, dilated pinhead pupils, headache, involuntary defecation, and a runny nose. Death is caused by cardiac arrest or respiratory failure.

Tear gases, such as chloroacetephenone (CN) and O-chlorobenzalmalonitrile (CS), cause irritation of the skin and uncontrolled tearing. Although these are designed to be used as non-lethal, riot control agents, their employment can result in death or injuries if improperly used in enclosed areas or for extended periods of time that results in high levels of exposure.

Although vomiting agents, such as adamsite (DM), diphenylchloroarsine (DA), and diphenylcyanoarsine (DC), have been used for riot control purposes, in the early twenty-first century they are generally considered too toxic for this purpose. All three agents have become obsolete as CWs against an opponent using modern protective equipment. Diphenylchloroarsine and diphenylcyanoarsine, which are in the form of a powder at normal ambient temperatures, were used as *mask breakers* during World War I and by the Japanese in China (1937–1945). The particles were able to penetrate the filters used at the time and could induce a soldier to break the seal of his mask allowing a more toxic agent such as phosgene to take effect. Diphenylchloroarsine and diphenylcyanoarsine were also mixed with sulfur mustard to lower the freezing temperature of the mustard and thus allow the mixture to be used at lower ambient temperatures.

History of Chemical Weapon Use

The first use of a chemical for lethal effect in modern times occurred on April 22, 1915, when the German army released approximately 180 tons of liquid chlorine at Ypres, Belgium, resulting in the deaths of an estimated 5,000 Algerian, Canadian, and French soldiers. The widest variety of chemical compounds developed and used on a large-scale are found among the CW agents produced during this conflict. At least forty different compounds were weaponized. But the most significant development was the production of sulfur mustard. This was first used at the second battle of Ypres in 1917 and, by the end of the war, had become known as the *king of war gases* due to the very large number of casualties resulting from its use. An estimated 1.45 billion shells were fired during the war, of which approximately 66 million contained CW-fill. Approximately 3,500 to 4,000 World War I-era shells were still being recovered annually in Europe during the 1990s, mostly in Belgium and France, of which about 10 to 20 percent are CWs.

Following the widespread use of CWs during World War I, countries with significant military capabilities or security concerns were compelled to consider threats that known or yet-to-be-discovered toxic chemicals might pose, particularly if delivered against vulnerable urban areas by aircraft (or balloons). During World War II, even larger stocks of CWs were produced and stockpiled than in World War I. Despite their widespread availability, however, CWs were, in general, not used during World War II. Most of the stockpiled CWs were either destroyed or disposed of by sea dumping at the end of the war. Their residue is the source of an old CW problem that continues to occur in a number of countries worldwide.

Military establishments have generally been reluctant to embrace chemical weapons, partly out of moral considerations. The use of CWs has generally gone against military codes of conduct. Their use was also generally viewed as an unnecessary complicating factor in military planning and practice operations. This was because of an inability to reliably predict lethal or casualty-causing effects. CW agents may quickly degrade or be dissipated by environmental factors such as rain, heat, and wind. Care must also be taken to ensure that the explosive charge for a CW munition can effectively disperse the agent, without destroying too much of the agent in the process. Aerosol platforms, mainly slow, low-flying aircraft, are also vulnerable to attack. Finally, modern protective clothing, if properly used and maintained, is generally effective against known CW agents.

There have been allegations of the use of CWs during most major armed conflicts in the twentieth century. Many allegations are unproven and appear to be false. This is partly due to deliberate misinformation, information indicating that an opponent possesses CWs or is pursuing a CW program, and the fact that participants may mistake toxic fumes generated during battle as CWs (for instance, fumes generated from the detonation of high explosives). From the early 1980s to the early 1990s, the United Nations Secretary-General investigated allegations of the use of chemical and biological weapon agents in Africa, Armenia, Iran, Iraq, and southeast Asia. The authority of the Secretary-General remains in effect. However if the alleged use were with CWs, the CWC would almost certainly take legal precedence. As previously noted, however, toxins are covered by both the BTWC and the CWC.

CW agents were used by British forces intervening in Russia's Civil War in 1919 (for example, adamsite), by Spain in Morocco in 1924 to 1927 (sulfur mustard), by Italian forces in Abyssinia in 1935 to 1940 (sulfur mustard, phosgene, phenyldichlorarsine), by Japanese forces in Manchuria in 1937 to 1945 (lewisite, diphenyl cyanoarsine, sulfur mustard), by Egypt in the Yemen civil war in 1963 to 1967 (sulfur mustard and phosgene), and by Iraq against Iran in 1982 to 1988 (cyclosarin, sulfur mustard, sarin, and tabun). The use of tear gas by U.S. forces as part of combat operations in Vietnam (to clear tunnel systems, for example) is also generally considered to be an instance of chemical warfare. The CWC forbids the use of riot control agents as a method of warfare. The use of tear gases as part of combat operations is therefore prohibited.

During the Iran-Iraq War (1980–1988), Iraq used CWs, including sulfur mustard and nerve agents (cyclosarin, sarin, tabun) extensively against Iran and its own Kurdish population. Although allegations have been made that Iran used CWs against Iraq, they have not been conclusively proven. By contrast, investigative teams sent to the region during the war by the U.N. Secretary-General conclusively proved Iraqi use of CWs. Iran is a party to the CWC and has declared a past production capability, but has not declared a CW stockpile.

Following the 1991 Persian Gulf War, the U.N. Security Council adopted resolution 687 of 1991 which, *inter alia,* required Iraq to end its CW program and destroy its CW stockpiles. The resolution also established the U.N. Special Commission on Iraq (UNSCOM) to verify the destruction and dismantlement of prohibited weapons and associated programs. (The International Atomic Energy Agency, or IAEA, was given primary responsibility for overseeing the nuclear weapon disarmament of the country.) The principal CW agents produced by Iraq were cyclosarin, sarin, sulfur mustard, and tabun, while the main unresolved CW issue was the nature and extent of Iraq's VX program. Iraq claimed that it had never weaponized VX and had only produced limited, pilot plant-scale quantities of the agent (2–3 metric tons of poor quality material). UNSCOM disputed this claim. UNSCOM inspectors left Iraq in late 1998, as a consequence of a dispute partly based on whether UNSCOM inspectors should be allowed unrestricted access to so-called presidential sites, with the VX issue still unresolved. In December 1999, UNSCOM was replaced by the United Nations Monitoring, Verification and Inspection Commission (UNMOVIC) (U.N. Security Council resolution 1284 of 1999). UNMOVIC conducted its first inspections of Iraq on November 27, 2002, partly under the terms of UN Security Council resolution 1441 of 2002, which deplored Iraq's failure to fully disclose all aspects of its prohibited programs, including with respect to CWs. In describing the nature of Iraqi cooperation with UNMOVIC inspectors, the UNMOVIC Executive Chairman made a distinction between *substance* and *process*. While Iraq did provide immediate access to all requested sites, its active and full cooperation was questioned. Another major unresolved CW issue was the failure by Iraq to account for approximately 6,500 munitions filled with about 1,000 metric tons of chemical agent. As of September 10, 2003, there were no reports of any CWs having been recovered by the U.S.-U.K.-led coalition forces that entered Iraq in March 2003.

The most significant use of CWs by a non-state actor was carried out by the Japanese-based religious cult Aum Shinrikyo. The first major lethal attack occurred in June 1994 when cult members vented sarin vapor from a specially modified van at night in Matsumoto, Japan, outside the homes of three judges who were then involved in a legal case involving the organization. Seven people died and approximately 300 were injured as a result. The incident was not immediately recognized as a CW attack and the police investigation was indecisive and poorly coordinated.

The second attack occurred in March 1995 when group members released sarin in the Tokyo subway. As a result, twelve people died, while approximately 500 people required medical attention or hospitalization. Approximately 5,500 people were examined. In this case, the means of attack and the identity of the perpetrators were quickly determined and the police carried out mass arrests and widespread searches of properties owned by the cult.

At the time, the group had assets worth an estimated 1 billion U.S. dollars. A number of cult members had masters and doctorate degrees in the natural sciences, including chemistry. Despite these factors, Aum Shinrikyo technical ability in creating chemical (and biological) warfare agents was limited. The sarin produced, for example, was unstable and of low purity. Safety precautions during testing and production were poor and a number of cult members were poisoned as a result. In 2004 the cult's founder and head, Chizuo

Matsumoto (a.k.a. "Shoko Asahara" or "bright light"), was sentenced to death.

Attempt at Ethical and Political Control

Agreements regarding CWs include the International Declaration Concerning the Laws and Customs of War (Brussels Conference, 1874); the Acts signed at the First International Peace Conference, Annex to the Convention (The Hague 1899); the Acts signed at the Second International Peace Conference, Annex to the Convention (The Hague 1907); the Treaty of Peace with Germany (also known as the Treaty of Versailles 1919); and the Treaty of Washington of 1922 Relating to the Use of Submarines and Noxious Gases in Warfare (Washington, DC 1922).

A more significant international legal instrument was the Protocol for the Prohibition of the Use in War of Asphyxiating, Poisonous or Other Gases, and of Bacteriological Methods of Warfare, the well-known Geneva Protocol of 1925. The Geneva Protocol did not, however, prevent the stockpiling of CWs and many of the major powers attached conditions to their instruments of ratification (such as, that a state would not consider itself bound by treaty obligations if first attacked with CWs or if involved in a military conflict with states not party to the Protocol or military coalitions which included one or more states not Party to the Protocol).

Since 1993, however, the main international legal instrument dealing with CWs is the CWC. Treaty negotiations began in 1968 within the framework of the U.N. Eighteen-Nation Committee on Disarmament (the present-day Conference on Disarmament).

The CWC is implemented by the OPCW, based in The Hague, Netherlands. The OPCW consists of three parts: The Conference of the States Parties (CSP), the Executive Council (EC), and the Technical Secretariat (TS). The CSP is composed of all member states. It is the highest decision-making body and meets in regular session once each year. The EC is a representative body composed of forty-one members that represent five regional groups (Africa, Asia, Eastern Europe, Latin America and the Caribbean), and Western Europe and other states. Its main task is to oversee operational aspects of treaty implementation. It meets in regular session three to four times each year. Special sessions of the CSP and EC may be convened if a request (made by one or more parties) to convene is supported by at least one-third of the members. A special session of the EC would be convened, for example, if CWs were used. The TS is responsible for the practical implementation of the OPCW, including the processing of annual declarations submitted to the OPCW by the parties and the carrying out of on-site inspections. It has a staff of approximately 485, including about 200 inspectors. The OPCW's budget for 2003 was 68,562,966 euros.

As of July 7, 2003, 153 countries had ratified or acceded to the CWC, while twenty-five countries had signed but not ratified the convention and sixteen countries had neither signed nor ratified the convention. The OPCW's budget for 2004 was 73,153,390 euros. As of March 31, 2004, 161 countries had acceded to the CWC, while twenty countries had signed but not acceded the convention and twelve countries had neither signed nor acceded to the convention. Most of the non-member states are located in Africa or the Middle East, including Iraq, Israel, Egypt, and Syria. Many Arab countries have linked their accession to the CWC to Israel's becoming party to the 1972 Non-Proliferation Treaty (in doing so, Israel would have to demonstrate that it does not possess nuclear weapons). India, Iran, and Pakistan are parties to the CWC.

There are three principal types of inspections under the CWC: routine inspections, challenge inspections, and investigations of alleged use of CWs. CW-related facilities, including CW destruction facilities and facilities that use small quantities of agent for protective purposes, must be declared and are subject to routine on-site inspections. Part of the chemical industry, which processes, produces, or consumes certain chemicals above certain thresholds must also be declared and are subject to inspection. Thus far there have been no challenge inspections or investigations of alleged CW use. The CWC regime has provided a forum in which the parties can consider the contents of each others' declarations and pursue informally further clarification through informal consultations.

The CWC requires that all state parties declare whether they have produced CWs at any time since January 1, 1946. As of March 2004, twelve parties (Bosnia and Herzegovina, China, France, India, Iran, Japan, Libya, South Korea, Russia, the United Kingdom, the United States, and the former Yugoslavia (now Serbia and Montenegro) had declared sixty-four CW production facilities or sites. As of the same date ten parties (Australia, Belgium, Canada, France, Germany, Italy, Japan, Slovenia, the United Kingdom, and the United States) had declared possessing *old* CWs (defined as CWs produced before 1925, or between 1925 and January 1, 1946, and which have been determined not to be usable) and three parties (China, Italy and Panama) have declared having *abandoned* CWs (defined as CWs abandoned by a state on the territory of another state without the permission of the latter).

The CWC is a cooperative regime designed to allow member states to demonstrate their treaty compliance to each other. For such inspections to be completely successful, inspected states must cooperate. If they do not, inspectors should nevertheless be able to acquire some useful information or results. At a minimum, the inspection should serve to provide sufficient information to enable the EC and CSP to formally decide on issues of compliance (for instance, non-cooperation). Under the terms of the CWC, inspected parties may invoke *managed access* provisions to protect sensitive information, including sensitive information about its chemical industry and information sensitive for national security reasons. The burden of satisfying the compliance concern nevertheless lies with the inspected party.

UNSCOM and UNMOVIC, by contrast, were provided mandates that were established as part of an agreement to end military hostilities between Iraq and U.N.-sanctioned, international coalition forces. As such UNSCOM and UNMOVIC were to be provided with unrestricted, immediate access to all requested sites. Their work was also backed by the implicit (or explicit) threat of military action and economic sanctions. If a case of continued, fundamental non-compliance with the CWC were to occur, the OPCW would refer the matter to the U.N. Security Council and U.N. General Assembly for their consideration and action.

Current and Future Trends and Challenges

In the early twenty-first century there is an increased emphasis on ensuring that non-state actors, such as terrorist groups, do not acquire or use CWs. Much of this effort is of a law enforcement or intelligence nature and thus classified or otherwise not openly discussed. There has also been an increased emphasis on harmonizing and strengthening export control regulations and preparing emergency response and management. This is reflected in increased efforts by the OPCW to achieve better uniformity in the collection and reporting of information to the organization, including on the transfers of certain chemicals that appear in the CWC Annex on Chemicals. The OPCW is also implementing a "plan of action" to ensure that the parties have established effective national implementing legislation. The plan has the active political support and engagement of the members.

A number of factors complicate the confirmation or verification of non-production of CWs in chemical industry facilities. In the late twentieth and early twenty-first centuries there was a shift in the size and flexibility of many chemical industry facilities, away from big (e.g., petrochemical) plants that produce large volumes of a limited number of chemicals using a dedicated production method and toward small facilities capable of manufacturing a wide variety of specialized chemicals to order on short notice using smaller, less polluting and more easily reconfigured equipment for different productions routes.

Twenty-first century scientific capabilities also caused a blurring of the distinction between *chemical* and *biological* processes. Many biological substances that could not previously be synthetically manufactured may be *chemically engineered* through such advanced technology. Most biological warfare agents could, in fact, be viewed as chemicals because their action is biochemical in nature and because the derivation of many biological agents involves manufacturing processes—as opposed, for example, to the extraction of substances from naturally occurring organisms. Finally, the manner in which new toxic chemicals are developed and synthesized has been revolutionized through, for example, advances in combinatorial and computational chemistry and microarray processing technologies.

Complete security against CWs will not be achieved. In view of human, financial, and other resource limitations, the approach taken to identify and respond to possible risks posed by CWs should be carefully considered and balanced. The effectiveness of national and international laws against the development and use of CWs is dependent on the amount of attention and resources countries elect to devote to the matter. Any decisions taken with regard to protecting against CWs should be based on the recommendations and experience of CW technical specialists.

JOHN HART

SEE ALSO *Biological Weapons; Just War; Military Ethics; Terrorism; Weapons of Mass Destruction.*

BIBLIOGRAPHY

Dando, Malcolm. (2001). *The New Biological Weapons: Threat, Proliferation, and Control*. Boulder, CO; Lynne Rienner.

Sidell, Frederick R.; Ernest T. Takafuji; and David Franz, eds. (1997). *Medical Aspects of Chemical and Biological Warfare*. Washington, DC: Borden Institute.

Lundin, S. Johan, ed. (1988). *Non-Production by Industry of Chemical-Warfare Agents: Technical Verification under a Chemical Weapons Convention*. Oxford: Oxford University Press.

Lundin, S. Johan, ed. (1991). *Verification of Dual-Use Chemicals under the Chemical Weapons Convention: The Case of Thiodiglycol*. Oxford: Oxford University Press.

Pechura, Constance M., and David P. Rall, eds. (1993). *Veterans at Risk: The Health Effects of Mustard Gas and Lewisite*. Washington, DC: National Academy Press.

Stockholm International Peace Research Institute. (1971–1975). *The Problem of Chemical and Biological Warfare*. 6 vols. Stockholm: Almqvist & Wiksell.

Trapp, Ralf. (1993). *Verification under the Chemical Weapons Convention: On-Site Inspection in Chemical Industry Facilities*. Oxford: Oxford University Press.

Tu, Anthony T. (2002). *Chemical Terrorism: Horrors in Tokyo Subway and Matsumoto City*. Fort Collins, CO: Alaken, Inc.

INTERNET RESOURCES

Organisation for the Prohibition of Chemical Weapons. Available from http://www.opcw.org.

United Nations Monitoring, Verification and Inspection Commission. (2003). "Unresolved Disarmament Issues: Iraq's Proscribed Weapons Programmes." UNMOVIC working document. Available from http://www.unmovic.org under heading "Cluster Document."

CHEMISTRY

• • •

After physics, chemistry is often considered the paradigmatic modern science. The ethical issues associated with chemistry and chemical technologies have nevertheless been more diffuse and less systematically identified than those related to either physics or biology, although the ethical issues associated with chemistry—from worker and consumer safety to environmental pollution, in both public and private contexts, in peace and war—are as broadly present in daily life as those in any other science. The very proliferation of chemistry into analytical chemistry, biochemistry, geochemistry, inorganic and organic chemistry, physical chemistry—not to mention atmospheric, computational, electro-, polymer, and other forms of chemistry—emphasizes the ubiquitousness of this particular science, its technological dimensions, and thus its range of potential ethical and political engagements.

Historical Emergence

The history of chemistry may be divided into three periods: (1) alchemy (from the beginnings of Muslim and Christian knowledge of the subject until the seventeenth century), (2) classical modern chemistry (from the middle of the seventeenth century until the middle of the nineteenth), and (3) theory-based chemistry (twentieth and twenty-first centuries). According to such interpreters as Mircea Eliade and Carl Jung, alchemy was as much a psychological or spiritual practice as a physical one, involving more esoteric religious discipline than a positive science. But at the beginning of the thirteenth century, alchemists such as Roger Bacon, Albertus Magnus, and Ramon Llull, in association with the late medieval desacralization of nature, argued for an ethical shift toward the discovery of new methods and products that had this-worldly value. Thus, the Swiss Theophrastus Bombastus von Hohenheim (known as Paracelsus) dedicated his alchemical labors to the cure of sicknesses. According to him, salt, sulfur, and mercury in adequate proportions were a fountain of health for the human organism (the beginning of medical chemistry).

Standing at the transition from alchemy to chemistry as a positive science is the work of Robert Boyle (1627–1691). In *The Sceptical Chymist* he formulated the modern definition of an element as a substance that cannot be separated into simpler substances, and argued for empirical experimentation as well as the public sharing of scientific knowledge in ways that still define the scientific method. Yet he was a devout if dissenting Christian who saw his scientific studies as an extension of his spiritual life. Boyle also helped found the Royal Society (officially chartered in 1662).

The great positive achievement of the classical modern period in chemistry, and one that became the basis for its transformation into a more theory-based science, was the periodic table. While the Frenchman Antoine-Laurent Lavoisier (1743–1794) advanced the understanding of chemical reactions, the Englishman John Dalton (1766–1844) developed atomic theory, and the Italian Amedeo Avogadro (1776–1856) analyzed relations between molecules and conditions of temperature and pressure (Avogadro's law)—thus creating an international republic of science with a distinctly if unspoken ethical structure. As empirical data accumulated about the properties of various substances, chemists began to consider schemas for classification according to their periodicity. The first was published in 1862, according to which properties repeated with each seven chemicals.

But the initial table mistakenly included some compounds among the elements, and it was the Russian Dmitri Mendeleyev (1834–1907) who created the periodic table as we now know it. Mendeleyev discovered patterns in the properties and atomic weights of halogens and some alkaline metals, similarities in such series as those of chlorine-potassium-calcium (Cl-K-Ca) and iodine-cesium-barium (I-Cs-Ba), and organized the elements according to chemical characteristics and physical properties in order of ascending atomic weight, as published in *On the Relationship of the Properties of the Elements to Their Atomic Weights* (1869).

Nevertheless, no one had yet definitively determined some atomic weights, which caused a few errors. Mendeleyev discovered he had to resituate seventeen elements according to their properties and ignore previously given atomic weights. Furthermore, he left spaces for possible new elements, given that none of those yet identified suited the properties assigned to those spaces. He thus predicted the existence of new elements such as aluminum, boron, and silicon—ten in all, of which seven were eventually confirmed.

The periodic table prepared the way for major advances in both chemical theory and practice. With regard to theory, in the early twentieth century Linus Pauling (1901–1994) employed quantum mechanics to conceptualize subatomic structures at the foundation of the orders reflected in the periodic table. This theoretical achievement at once enhanced the control of chemical processes and increased the ability to design new compounds. With regard to practice, the periodic table effectively predicted the possibility of a whole series of transuranic elements that were experimentally created by Glenn Seaborg (1912–1999). In both instances these newfound powers raised ethical and public policy questions that have been further promoted by the interdisciplinary expansion of chemistry into engineering and biology.

Industrial Chemistry and War

Starting in the eighteenth century—even prior to its theoretical enhancement—chemistry more than any other science contributed to industrial development. Just as Lavoisier is considered a founder of classic modern chemistry as a positive science, his contemporary, Nicolas Leblanc (1742–1806), who developed a process for obtaining soda (sodium carbonate) from sea salt, is credited with founding industrial chemistry. Before Leblanc France depended on foreign imports for the sodium carbonate central to its glass, soap, paper, and related industries. Leblanc's alternative, subsequently improved by the Belgian Ernest Solvay (1838–1922), was thus a major contribution to French industrial independence.

After sodium carbonate, the development of industries to produce nitrogen and fertilizers dominated applied chemical research during the nineteenth century. As contributors to such achievements, the Englishman Humphry Davy (1778–1829) and the German Justus von Liebig (1803–1873) illustrate a special combination of humanitarianism and nationalism. Davy, for instance, along with pioneering work in electrochemistry, invented the miner's safety lamp and promoted improvements in the British agricultural, tanning, and mineralogical industries. Liebig, as a professor of chemistry, pioneered the laboratory as a method of instruction and helped make Germany the world leader in chemical education and research. He also virtually created the field of organic chemistry, which he applied especially to increase German agricultural productivity.

In another contribution to industrial chemistry, the Swedish chemist Alfred Nobel stabilized nitroglycerin in 1866 to make possible the fabrication of new and powerful explosives for military use. Such fabrication, along with the "dye wars" of the late nineteenth and early twentieth centuries, intensified relations between chemistry and national interests, which in turn challenged chemists to reflect on their ethical obligations. It was certainly some such reflection that led Nobel to use the profits from his own chemical industries to establish prizes in honor of "those who, during the preceding year, shall have conferred the greatest benefit to mankind" in the areas of physics, chemistry, physiology or medicine, literature, and peace.

At the Second Battle of Ypres, France (now Belgium), in April 1915, the negative potential of chemistry was nevertheless manifest as never before when chlorine gas was employed for the first time in "chemical warfare." (The term is somewhat anomalous, because gunpowder and all explosives are also chemical products.) In this the physical chemist Fritz Haber (1868–1934) provides a provocative case study. Having previously succeeded in developing a means for synthesizing ammonia from atmospheric nitrogen and hydrogen for industrial and agricultural uses, Haber at the outbreak of World War I placed his laboratory in service of the German government and worked to advance the national cause. One result was his advocacy for the use of chlorine gas at Ypres. But after the war, even though he was awarded the Nobel Prize in chemistry for his prewar work on ammonia synthesis, he remained isolated from the international scientific community. Feeling responsible for the German war debt, he even tried to develop a process to extract gold from seawater. But when Adolf Hitler came to power

Haber's Jewish heritage forced him to flee the country, and he died in exile.

Another feature of industrial chemistry was the creation of large-scale corporations. National efforts to promote self-sufficiency in various chemicals contributed first to overproduction in such basics as fertilizers and dyes, and then to a series of national mergers and consolidations: This produced IG Farben in Germany in 1925 (creating the largest chemical manufacturer in the world), Imperial Chemical Industries (ICI) in England in 1926, and a DuPont–ICI alliance in the United States in 1929. The chemical industry as much as any other anticipated the kind of competition and transnational relations characteristic of the dynamics of globalization—which likewise presents special ethical challenges.

The Chemical World

Despite its contributions to warfare, the primary connotation for chemistry has been, in the words of the long-time DuPont slogan (1939–1999), "Better Things for Better Living ... through Chemistry" (the "through chemistry" was dropped in the 1980s). This vision of chemistry as a primary contributor to better living rests on the creation of a host of substitutes for traditional goods and the creation of new ones. Among substitutes, the most prominent have included first synthetic dyes and then synthetic rubber.

Among new products, plastics and pharmaceutical drugs have played major roles. Synthetic rubber and plastics are outgrowths of the huge development of polymer chemistry and discoveries of ways to use petroleum to create multiple enhancements of or substitutes for traditional materials: Formica (1910s) for wood and stone, Bakelite (patented 1907, but not widely used until the 1920s) for wood and glass, nylon (1930s) for fiber, and more. Complementing a wealth of pharmaceuticals are cosmetics, cleaning compounds, lubricants, and pesticides. From the 1960s there eventually emerged green or environmental chemistry and industrial ecology, with the concept of sustainability coming to play a significant role in chemical research and development. From the 1970s on, research and development also turned toward the design of functional materials, that is, materials fabricated according to the necessities of specific industrial sectors: reinforced plastics for the aerospace, electronics, and automobile industries; silicon for information technology hardware; and more.

In recognition of the chemical world and its pervasive transformation of the world, the American Chemical Society (ACS, founded 1876) undertook in the 1980s to publish a new kind of high school textbook, *Chemistry in the Community* (1988). Through this project professional chemists sought to communicate to those students who were not likely to become science majors some of the lifeworld significance of modern chemistry. The book was thus structured around community issues that had a significant chemical component more than around basic concepts and principles in chemistry itself. It was an effort to exercise professional responsibility in educating the public about the chemical world in which everyone now lived.

Ethical Issues and Responses

Against this historical profile one can identify two distinct chemistry-related ethical issues: those associated with military use and those related to commercial development—that is, the introduction into the world of increasing numbers of chemical compounds not otherwise found there. It is also possible to distinguish two kinds of response: institutional and individual.

With regard to military use, the institutional response has been the practice of military deterrence and development of a chemical weapons convention. The World War I use of chemical weapons was followed by the World War II avoidance of chemical weapons, no doubt in part because possession by all parties led to deterrence. The most dramatic use of chemical weapons since has been by what are sometimes called "rogue states" such as Iraq in the 1980s. The Chemical Weapons Convention (CWC) that entered into force in 1997 is implemented by the Organization for the Prohibition of Chemical Weapons located in The Hague, Netherlands. CWC state party signatories agree to ban the production, acquisition, stockpiling, transfer, and use of chemical weapons.

At the individual level, some activist organizations of scientists such as the Federation of American Scientists or International Pugwash have lobbied for limitations on the development and proliferation of chemical weapons, and in some instances called on chemical scientists and engineers to exercise professional responsibilities by not contributing to related research and development projects. One issue that has not been extensively addressed at either the institutional or individual level, although it has been discussed among scientific professionals concerned with professional responsibility, is the development of nonlethal chemical weapons, that is, weapons that do not kill but only incapacitate.

With regard to the commercial proliferation of chemicals, many governments have development institutional mechanisms for the assessment and regulation of chemicals consumed directly by the public or introduced more generally into the environment. One good example comes from the European Union (EU). According to a regulatory regime established in 1981 (Directive 67/548) all new chemicals manufactured in amounts of 10 kilograms or more must be registered and tested for health and environmental risks, but the more than 100,000 substances on the market at that time were exempted from this process. Because of testing expenses, this meant that innovation and chemical replacement was discouraged, in many instances leaving known dangerous chemicals in place.

In response the EU has proposed a policy reform called the Registration, Evaluation, and Authorisation of Chemicals (REACH) system. Under the new REACH regulatory regime, the manufacture or importation of any chemical in the amount of 1 metric ton or more must be registered in a central database. The registration must include relevant information regarding properties, uses, and safe handling procedures, with a new European Chemicals Agency being charged to review the database and to supplement existing data with other relevant information. No testing is required in the absence of suspected health or environmental dangers. (It may be noted that there is no similar regulatory process in the United States. In fact the U.S. government, along with U.S. chemical producers, have lobbied against REACH, which they argue will negatively affect most goods exported to the EU.)

At the international level, in 2000 negotiations were completed on the Stockholm Convention on Persistent Organic Pollutants (POPs). With 122 negotiating countries represented, the POPs treaty aims to eliminate or severely restrict production and use of nine pesticides, polychlorinated biphenyls (PCBs), and their by-products. The treaty also requires national action plans for its implementation as well as the management and reduction of chemical wastes, while providing funding for the participation of developing countries. According to POPs, trade in the covered chemicals is allowed only for purposes of environmentally sound disposal or in other limited circumstances. The "dirty dozen" substances covered by the treaty are aldrin, chlordane, DDT, dieldrin, endrin, heptachlor, hexachlorobenzene (HCB), mirex, toxaphene, PCBs and their by-products, dioxins, and furans. The treaty includes methods to add new chemicals. Although signed in May 2001, as of 2005 the treaty awaits ratification in the U.S. Senate. Also relevant in the international context is the Globally Harmonized System for chemical classification and labeling that was adopted by Agenda 21 (1992) and is administered by the United Nations Economic Commission for Europe.

At the same time, the pernicious consequences of some chemical substances has led to the creation of a nongovernmental program called Responsible Care, initiated in 1985 by the Canadian Chemical Producers Association and then adopted three years later by the American Chemistry Council (then called the Chemical Manufacturers Association). In 1990 it was also adopted by the Synthetic Organic Chemical Manufacturers Association. Responsible Care is an industry-administered program to certify company compliance with management standards that promote reduced emissions, worker safety, industrial security, product stewardship, public accountability, and research and development. Internationally, Responsible Care is administered by the Brussels-based International Council of Chemical Associations. One stimulus to the creation of the Responsible Care program was no doubt the 1984 chemical accident in Bhopal, India.

One other individual initiative is that of the professional codes of ethics developed by chemists. As a pioneer, the American Chemical Society requires that all professional chemists recognize their obligations to the public, to colleagues, and to science. Building on its federal charter (1937) and "The Chemist's Creed" (1965), the current "Chemist's Code of Conduct" (1994) itemizes nine basic responsibilities to the public, chemistry itself, the profession, employers, employees, students, associates, clients, and the environment. More specifically, with regard to the profession, chemists must strive for the responsible recording and reporting of scientific data, be aware of conflicts of interest and handle them properly, and avoid ethical misconduct defined as fabrication, falsification, and plagiarism. With regard to the public, chemists have obligations both "to serve the public interest and welfare and to further knowledge of science." Indeed, with regard to science, chemists should assure that their work is "thorough, accurate, and . . . unbiased in design, implementation, and presentation."

The STS of Chemistry

The self-presentation of chemistry in its code of conduct and in its work of public education nevertheless raises some more general ethical and public policy issues. Insofar as the chemistry community might take applying chemistry and public science education as the primary ways to serve the public interest, a science, technology, and society (STS) assessment, with chemistry as the leading science, would be appropriate. STS studies in general have highlighted the importance of citizen participation in science and technology decision-making and of public debate appealing to science and technology. One framework that promotes recognition of such interactions is the concept of "post-normal science," defined as issue-driven science in which facts are uncertain, values disputed, but decisions urgent (Funtowicz and Ravetz 1990). Post-normal science calls for broader public education, of a conceptual and philosophical as well as an ethical sort, to manage the science–civil society relationship. In this sense the *Chemistry in the Community* model, with its stress on public problems related to chemistry, is insufficient.

From a philosophical, historical, and chemical education perspective, however, there exists a different but complementary agenda. The philosophy of chemistry, understood as a subdiscipline of the philosophy of science, has been taking shape since the mid-1980s (van Brakel 2000). Its agenda, dominated by the question of whether chemistry can be reduced to physics, has been enlarged to included classic conceptual issues in the philosophy of science (the character of representations and the structure of laws and explanations) as well as debates about ethical, aesthetic, and even sociocultural implications of chemistry. The principal periodicals dealing with such discussions are *Hyle: International Journal for Philosophy of Chemistry* (1995–present) and *Foundations of Chemistry: Philosophical, Historical, Educational, and Interdisciplinary Studies of Chemistry* (1999–present), the latter being the journal of the International Society for the Philosophy of Chemistry. Also relevant are some issues from the early 2000s (for instance, the Vol. 81, numbers 6 and 9 [2004], and the Vol. 82, number 2 [2005]) of the much older issues of the much older *Journal of Chemical Education* (1924–present).

In *Hyle* especially analyses of ethical issues have transcended particular chemical results in order to address questions that underlie all debates about regulation, responsible management, professional codes, or individual conduct. The ethics of chemistry includes questions concerning relations between the chemical community and society—that is, the importance of the particular values of chemists as such and their relation to general social values. This fundamental question can be approached from two directions: one being that of the professional community, the other being that of society. The former treats issues such as the status of the professional codes of conduct of chemical societies, the relation of a putative moral ideal to the specific ethical norms of chemistry, the moral or amoral character of chemical research, and the links that can be found between methodological values and moral values. The latter asks whether chemists have specific kinds of responsibility and duties to the society, or society any responsibility to the science of chemistry. It reflects on what lessons if any might be drawn from the positive and negative effects of chemical research (drugs, increased economic development, weapons, pollution). The responses from both perspectives will, of course, have implications for how the ethics of chemistry should be included within university curricula: as part of the methods of the science, as a technological application, or as a societal framework.

JUAN BAUTISTA BENGOETXEA
TRANSLATED BY JAMES A. LYNCH

SEE ALSO *Chemical Weapons; Environmental Ethics.*

BIBLIOGRAPHY

American Chemical Society. (1988). *Chemistry in the Community*. Dubuque, IA: Kendall/Hunt. 2nd edition, 1993. 3rd edition, 1998. 4th edition, New York: Freeman, 2002. A high-school level textbook structured around community issues related to chemistry rather than chemical concepts.

American Chemical Society. (1994). *Chemistry in Context: Applying Chemistry to Society*. Dubuque, IA: William C. Brown. 2nd edition, 1997. 3rd edition, Boston: McGraw-Hill, 2000. 4th edition, 2003. This is the college-level offspring of *Chemistry in the Community*.

Bensaude-Vincent, Bernadette, and Isabelle Stengers. (1996). *A History of Chemistry*, trans. Deborah van Dam. Cambridge, MA: Harvard University Press. A social and cultural history, originally published, 1993.

Eliade, Mircea. (1971). *The Forge and the Crucible: The Origins and Structures of Alchemy*, trans. Stephen Corrin. New York: Harper and Row. Originally published, 1956.

Funtowicz, Silvio O., and Jerome R. Ravetz. (1990). *Uncertainty and Quality in Science for Policy*. Dordrecht, Netherlands: Kluwer Academic.

Meyer-Thurow, Georg. (1982). "The Industrialization of Invention: A Case Study from the German Chemical Industry." *Isis* 73(268): 363–381.

Schummer, Joachim, ed. (2001). "Ethics of Chemistry." Special issue, *Hyle* 7(2): 83–167. Includes five articles on "Ethics and Science" (Giuseppe Del Re), "Ethics of Chemical Synthesis" (Joachim Schummer), "Handling Proliferation" (Pierre Laszlo), "Gifts and Commodities in Chemistry" (Jeffrey Kovac), and "The Technological Transfer Dilemma" (Brian P. Coppola). A second special issue, from 2002, on the same theme, *Hyle* 8(1): 3–48, includes three more articles: "'Pathological Science' Is Not Scientific Misconduct (Nor Is It Pathological)" (Henry H. Bauer), "Do the Professional Ethics of Chemists and Engineers Differ?" (Michael Davis), and "The Future of Tertiary Chemical Education: A *Bildung* Focus?" (Kathrine K. Eriksen).

van Brakel, Jaap. (2000). *Philosophy of Chemistry*. Leuven, Belgium: Leuven University Press.

INTERNET RESOURCES

American Chemical Society. Available at http://www.chemistry.org.

Globally Harmonized System of Classification and Labeling of Chemicals (GHS). Available at http://www.unece.org/trans/danger/publi/ghs/ghs_welcome_e.html.

International Council of Chemical Associations (ICCA). Available at http://www.icca-chem.org.

HYLE: International Journal for Philosophy of Chemistry. Available at http://www.hyle.org/.

REACH (European Community: Registration, Evaluation and Authorization of Chemicals). Available at http://europa.eu.int/comm/enterprise/reach/.

CHERNOBYL

On April 26, 1986, a catastrophic accident occurred at the Chernobyl-4 reactor near the town of Pripyat, Ukraine, 100 kilometers northwest of Kiev. Figure 1 shows the reactor location and the regions of most intense radioactive contamination. The accident destroyed the reactor and released a large amount of radioactivity into the atmosphere, particularly radioactive iodine (I-131) and radioactive cesium (Cs-137), both of which have the potential to cause cancer. Thirty-one workers at the plant died within a few weeks, most of them from receiving lethal doses of radiation while putting out fires and responding to other emergencies.

Radiation fallout caused significant contamination in parts of Belarus, Russia, and the Ukraine, resulting in the resettlement of more than 350,000 people from 4,300 square kilometers. An approximate five- to ten-fold increase in thyroid cancer has been observed in children from Belarus, Russia, and Ukraine who received a large exposure to I-131. The economic impact has also been large, not only from the direct costs of accident cleanup, decontamination, and entombing the reactor, but also for lost agricultural production from the evacuated areas, and from regions throughout Europe where the radioactive fallout resulted in restrictions on eating certain foods and on limiting imports. Continued health monitoring over many years will be required for citizens who had lived in or are currently living in contaminated areas, and for cleanup workers who received significant doses of radiation.

At the time of the accident, the Chernobyl reactors were owned and operated by the Soviet Ministry of Power and Electrification. The reactor design was a unique Soviet design called an RBMK. A schematic diagram is shown in Figure 2. Like reactors in the United States, RBMK reactors use ordinary water to cool the fuel. Unlike U.S. reactors, which use water to slow down or moderate the neutrons produced in fission, the RBMK uses graphite as the moderator. In this case the water used for the coolant is actually a neutron absorber and reducing the density of the water increases the neutron production. In addition, the ratio of uranium isotopes U-235 to U-238 in the fuel is less in the RBMK than in U.S. reactors. The effect of these differences was that at low power operation, under the right conditions, the power in the RBMK could increase in an uncontrolled manner. Reactor designs that allow power increase in an uncontrolled manner are prohibited by regulation in the United States. The type of accident that occurred was unique to the Soviet-designed RBMK reactor. Another important difference is that Soviet reactors did not have a steel-lined, thick concrete-walled containment building like those in Western Europe, North America, and Asia, using instead an industrial-type building. This final difference had profound consequences.

The Accident

The accident occurred while the operators were conducting a test simulating loss of power at the plant. The goal was to determine if power from the spinning turbines could be used to operate the pumps while backup diesel generators were brought on line. In order to conduct the test, most of the safety systems that would have provided a safe shutdown were disconnected. A test of this type that disconnected the safety systems would never be allowed in the United States, Western Europe, or Asia. The test was to be conducted at about 25 percent power, but when the power level was reduced from 100 percent to 50 percent, the test was delayed for nine hours because the electricity was needed in Kiev. While the operators waited, a strong neutron-absorbing isotope, Xenon-135, built up in the reactor. The operators did not recognize this and did not incorporate the effect into the control computer. When the test resumed, the operators could not control the reduction of power because the Xenon-135 was absorbing neutrons needed for fission and consequent power production. To keep the reactor from shutting itself down, they pulled out most of the neutron-

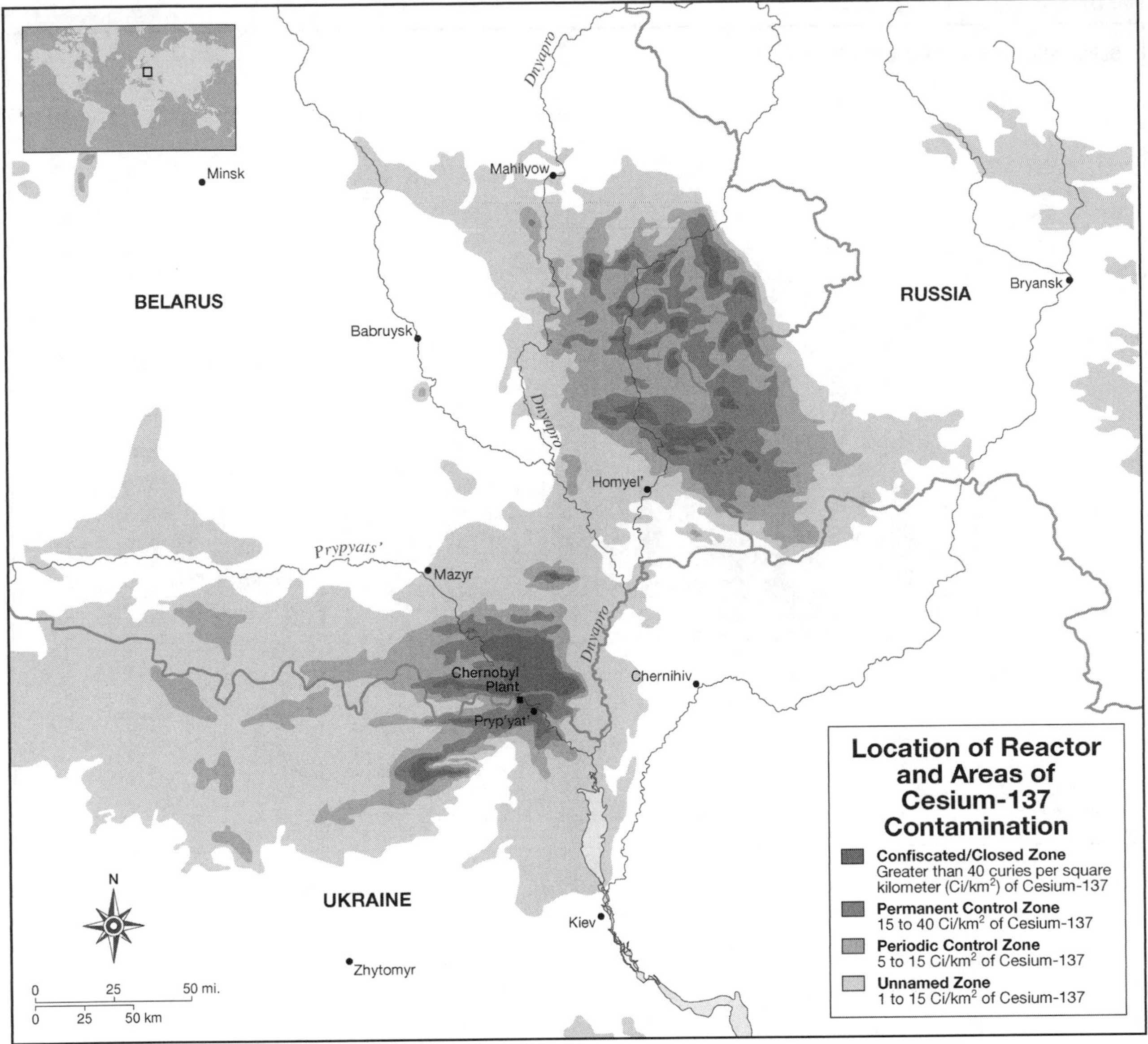

Location of reactor and central spots of Cesium-137 contamination.

absorbing control rods that are used to control the reactor power. This was in violation of Soviet operating procedures. Unknown to the operators, they now found themselves operating under conditions where the reactor could increase in power in an uncontrolled manner.

When the operators continued the test procedure by turning off the water coolant pumps and re-inserting the slow-moving control rods, there was still enough power to cause the water to start boiling, thereby reducing the water density and increasing the neutron production. In addition, there were graphite tips on the bottom of the control rods that added moderator when the control rods were initially inserted, and this further increased neutron production. Instead of the power level decreasing, as the operators expected, it increased rapidly, reaching approximately one hundred times full power in just a few seconds. The increased power resulted in a massive steam buildup inside the reactor leading to an explosion.

A second explosion that followed shortly lifted the large top shield above the reactor, blew off the roof and walls of the building, and dispersed burning fuel and gra-

FIGURE 2

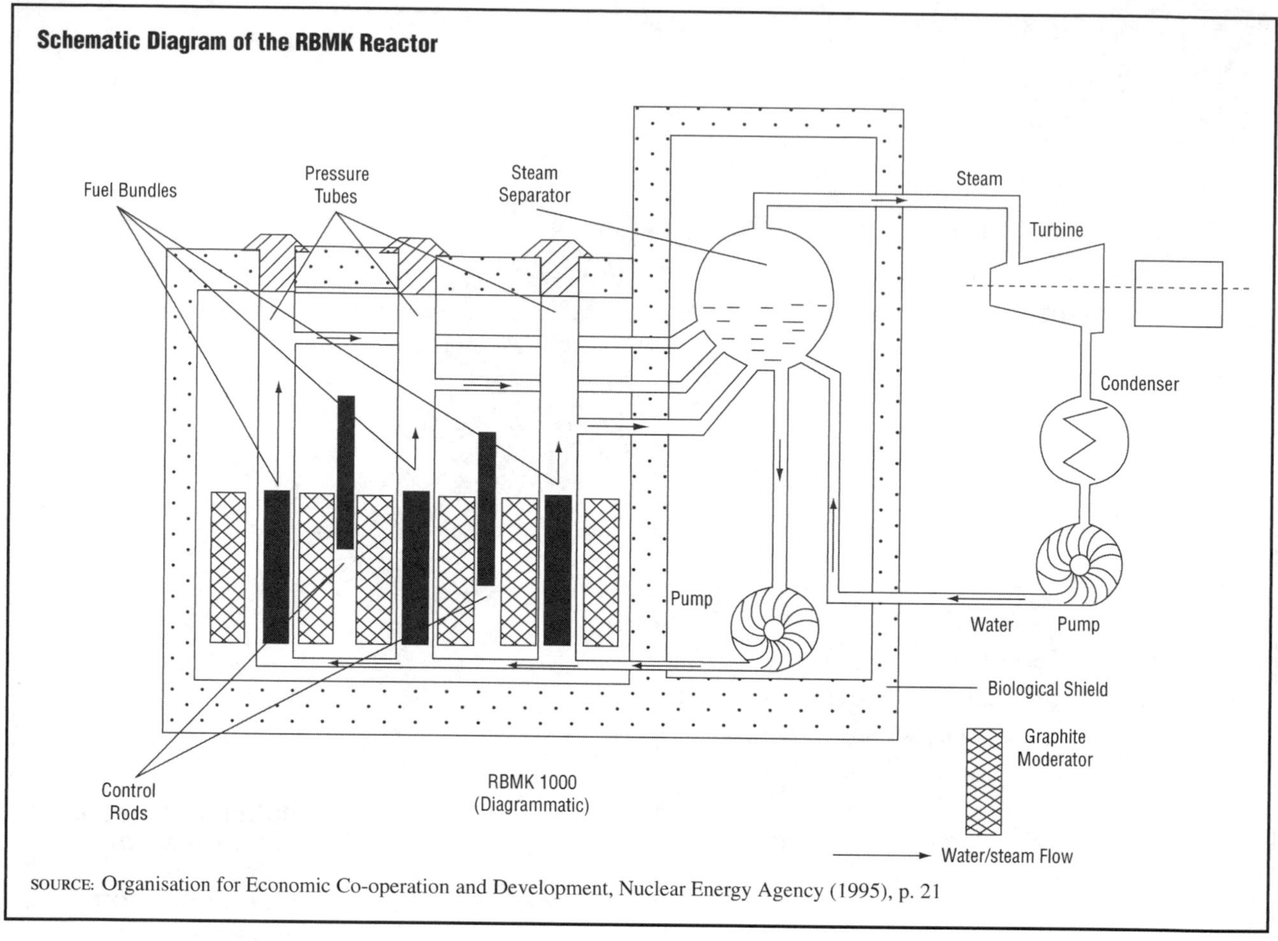

SOURCE: Organisation for Economic Co-operation and Development, Nuclear Energy Agency (1995), p. 21

Schematic diagram of the RBMK reactor.

phite. The steel shield resettled at an angle, allowing air to enter the reactor and the argon gas that normally covers the reactor to escape. Contact with the air caused the hot graphite to ignite, propelling the volatile radioactive materials high up into the atmosphere. Firefighters who went to the room to put out the fires received a lethal dose of radiation. It took ten days to control the fire, and by that time 5 to 10 percent of the radioactive material in the core had been released to the atmosphere.

Evacuation and Health Effects

Evacuation of residents from the nearby town of Pripyat took place the following day, but evacuation from adjacent contaminated areas did not take place for several days, nor did the Soviet government quickly inform the residents or the world what had happened. The radiation release was first made public after the airborne radiation from the accident was detected in Sweden. In 1986 about 116,000 people were relocated from areas surrounding the reactor, and an additional 220,000 people from Belarus, the Russian Federation, and Ukraine were relocated after 1986.

The World Health Organization (WHO) has continually monitored those exposed to radiation, including the residents and 600,000 "liquidators" who came to clean up the accident. Two hundred thousand liquidators built a cooling system under the reactor and a shield building—commonly called the sarcophagus—around the damaged reactor. They received doses of 100 millisieverts (10 rem) or more, with 20,000 receiving doses of at least 250 millisieverts. For comparison, this is five times the U.S. total effective dose limit for radiation workers. Another 400,000 liquidators, who arrived after 1987, received much lower doses.

Extensive analyses of the public health effects from Chernobyl have been conducted by United Nations organizations including WHO. A comprehensive summary is available in a report by the United Nations Scientific Commission on the Effects of Atomic Radiation (UNSCEAR) published in 2000, which concludes that the

FIGURE 3

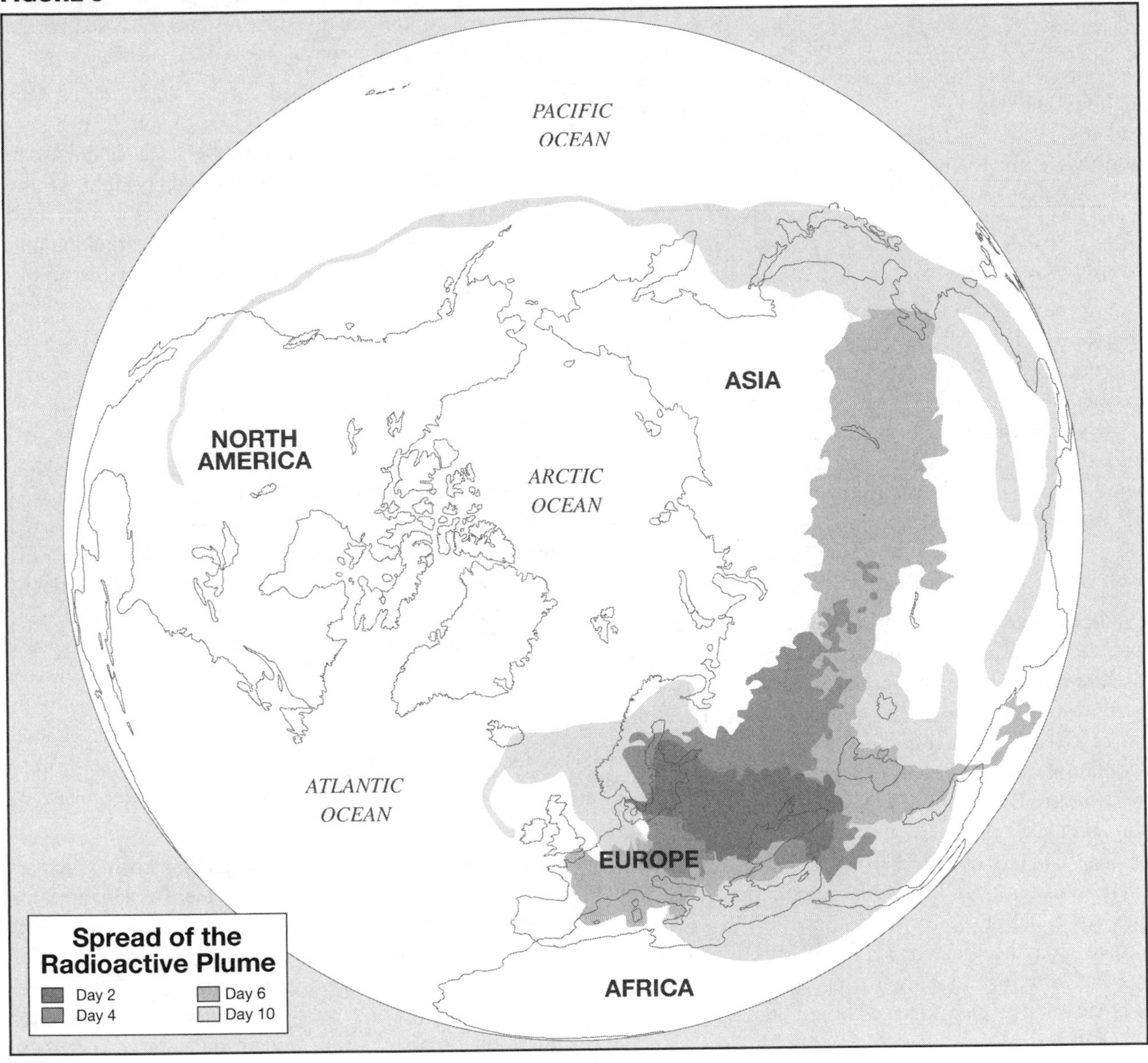

Spread of the radioactive plume over the Northern Hemisphere following the Chenrobyl accident.

only radiation-related effect to that date was an increase in thyroid cancer, largely in children. Through the years 1990–1998 there were about 1,800 cases of thyroid cancer in the contaminated areas of Belarus, Russia, and Ukraine in children 0–17 years old at the time of the accident. The majority of these were related to the accident.

Additional cases of thyroid cancer are expected in the future. Thyroid cancer is generally treatable if caught early. Nonetheless, ten deaths were reported as of 2000. It is not possible to make any accurate prediction about the number of deaths that will ultimately result, because there are no models that accurately predict deaths from low levels of radiation exposure. Many reports indicating much larger numbers of cancers and deaths from the Chernobyl accident were found in the UNSCEAR review to contain misinterpretations of data or use of unsubstantiated data.

While the heaviest radiation doses were received in Belarus, Russia, and Ukraine, the release of radioactivity from the Chernobyl reactor went high into the atmosphere and spread throughout Europe and then around the whole northern hemisphere. Figure 3 shows the dispersion of the radioactive cloud. The fallout in Europe varied considerably among the countries, depending on wind patterns and rain during the ten days the reactor was releasing radioactivity to the atmosphere. The largest doses were received in Poland, followed by Sweden, Germany, Italy, Finland, and Czechoslovakia. While such fallout caused great concern with the governments and the population, the doses received by the populations were relatively low, although there were some localized "hot spots."

Response and Lessons

Engineers from the former Soviet Union have made changes to all RBMK reactors to eliminate the possibility of repeating this type of accident. Nonetheless, in the West the RBMK is considered too unsafe to continue operation. All the Chernobyl units have been shut down. Other RBMK reactors outside Russia are being phased out of operation. However, the Russians still consider it a safe reactor and plan to continue operating all existing units. Plans to build new ones have been nevertheless cancelled.

In addition, the Soviet-designed water-cooled reactors, the VVER, built in the former Soviet Union and its satellite states, have either been shut down for inadequate safety features or modified to enhance safety. The United States and European countries have contributed millions of dollars in equipment and expertise to upgrade the safety of the existing reactors and their operation. The accident also stimulated the creation of the World Association of Nuclear Operations (WANO), whose goal is to improve safety in operations. WANO is an extension of the Institute for Nuclear Power Operations (INPO) that was formed after the accident at Three Mile Island in 1979 and was instrumental in improving safety in the operation of nuclear power plants in the United States.

While the design of the RBMK was flawed, a far greater problem was failures in human performance. How was it possible for the managers to allow the safety systems that would have prevented the accident to be disconnected during the test? How could all the control rods have been removed in violation of fundamental safety procedures? One answer is that written safety and accident response procedures actually did not exist in most RBMK control rooms before the accident. Furthermore, the operators were not trained to respond to different accident scenarios, and surely not to an accident that might occur during an experimental procedure. Importantly, there was no effective safety review of the proposed test. Moreover, the accident occurred in a society where secretiveness rather than openness was standard operating procedure, and this resulted in a lack of communication within the organization and with the public. After the accident, the Soviet government attempted to conceal it and the dangers posed to the local population and to the world.

Western nations learned the lessons of Three Mile Island, but states of the former Soviet Union did not. Specifically, they did not incorporate the fundamental lesson that safety is the most important responsibility of the operators, and that management from the top down must emphasize, encourage, and incorporate this thinking into plant operation. A culture that fosters "safety-first thinking" throughout the organization is necessary if nuclear power is to reach its potential to benefit humanity.

Since Chernobyl, nations of the former Soviet Union (FSU) have made significant improvements in both operations and design. Assistance from the United States and Europe led to the establishment of new training facilities, enhanced operator training, improved procedures for responding to accidents, and upgraded plant equipment. New Russian designs of the VVER type have added safety features and a containment building so they now meet safety standards used elsewhere in the world. There has been a change in management philosophy and an increased emphasis on operations safety. Regulatory agencies are improving in their capabilities. Nonetheless, the culture change needed to reach the safety standards of the United States and Western Europe will be a continuing challenge in the FSU countries.

The Chernobyl accident showed dramatically that an accident anywhere represents an accident everywhere, for it reflects on the ability of nuclear power to serve society as a trustworthy technology. This is a high standard and it raises the question: Can nations throughout the world that desire to use nuclear power maintain this level of attention to safety? Even though future reactors may be designed and built that prevent a catastrophic accident, it is important that an emphasis on a culture of safety be maintained. Ultimately it will reflect on the capacity of the world nuclear industry to serve civilization.

The accident was also global in the sense that radioactive fallout was present throughout the northern hemisphere and caused local contamination in many European countries that were not prepared for such an accident. The reactions of national authorities varied greatly on issues such as restrictions on consumption and marketing of foodstuffs. There was no uniformity in standards for implementation of protective actions. This could be especially disconcerting to the public in the border region when the nations on each side of the border took significantly different actions. National authorities sometimes used interpretations that responded to public fears rather than being based on sound science. This resulted in unnecessarily increasing public confusion and possibly public fears, and caused unnecessary government expense and economic loss. International efforts have been undertaken to produce more uniform regulations and criteria related to radiation accidents, and for emergency management of transnational accidents. Whether these efforts will be effective may not be known unless they are put to the test. In light of the level of terrorism that now exists in the world, and the possibility that biological and chemical agents can cross national boundaries as well as nuclear agents, it has become ever more critical that this type of emergency management be carefully developed and practiced. This could be one of the most important lessons from the Chernobyl accident.

EDWARD H. KLEVANS
DARRYL L. FARBER

SEE ALSO *Nuclear Ethics; Three-Mile Island.*

BIBLIOGRAPHY

Knief, Ronald Allen. (1992). *Nuclear Engineering: Theory and Technology of Commercial Nuclear Power*, 2nd ed. New York: Hemisphere. Nuclear engineering textbook with a good technical discussion of the RBMK reactor, the chronology of the accident, the causes of the accident, the effects of the accident, and comparisons with the Three Mile Island accident.

Medvedev, Grigori. (1991). *The Truth About Chernobyl*, trans. Evelyn Rossiter. New York: Basic Books. Written by a physicist who had been a chief engineer at Chernobyl and was an investigator of the accident; presents both background material and an intimate account of the events leading to the accident and the accident itself.

Organisation for Economic Co-operation and Development, Nuclear Energy Agency. (1995). *Chernobyl—Ten Years On: Radiological and Health Impact*. Paris: Nuclear Energy Agency, OECD. Covers radiation releases, dose estimates, health impacts, agricultural impacts, reactions of national authorities, and other issues.

INTERNET RESOURCES

Nuclear Regulatory Commission. (2000). "Fact Sheet on the Accident at the Chernobyl Nuclear Power Plant." Available from http://www.nrc.gov. Briefly discusses the accident and presents information on the Nuclear Regulatory Commission's post-accident activities to assist the Ukraine in dealing with the accident.

United Nations Scientific Commission on the Effects of Atomic Radiation (UNSCEAR). "Exposures and Effects of the Chernobyl Accident." In Volume II of the *UNSCEAR 2000 Report*. Available from www.unscear.org/chernobyl.html. Comprehensive and authoritative report on the health effects of the Chernobyl accident with data through 1998.

World Nuclear Association. (2001). "Chernobyl." Available from www.world-nuclear.org/info/chernobyl. (Go to Document inf07.) Site has a set of detailed and authoritative documents discussing many facets of the accident and its consequences. There are also maps of the area. Document inf07 describes the accident and its consequences in detail and gives direct access to three short reports on health effects from the accident.

CHINESE PERSPECTIVES

• • •

Overview
Engineering Ethics
Research Ethics

OVERVIEW

China is the oldest continuous civilization in the world and has produced a culture stretching back for millennia. For extended periods of time China led the world in science and technology. Yet traditional Chinese culture focused not on science and technology but on political-ethical issues. Traditional thinkers were more concerned with political morality and developing a centralized bureaucracy to run the country. Ethics was closely associated with politics, and the technical arts were subordinate to political considerations.

The Tradition

Confucianism formed the orthodoxy of premodern Chinese. Confucius (551–479 B.C.E.) himself stressed moral over material goods. He thus considered the technical arts as secondary to the ethical enterprise of living in harmony with the Way of Heaven. According to the Way, people have roles to play in society: the ruler to rule, ministers to administer, fathers to head families, and sons to serve as sons (*Analects*, 12:11). In the Confucian tradition, even acquiring material benefits was subordinate to living in accord with the Way, and technical fields such as agriculture, astronomy, mathematics, and medicine should serve the political needs of the country (Xi Zezong 2001). Though Confucians regarded the technical arts as the lesser way, specialists' use of technology in the service of the people was often held up as an example of moral rectitude.

Daoism was also an important tradition of ancient China. The Daoist view of how the technical arts relate to ethics differed from the Confucian view. While Confucianism subordinated the technical arts to ethics, it did not oppose the special sciences and their results. In contrast, Daoism attributed social disturbances and moral decay in society to science and technology and rejected them outright. For example, Laozi (or Lao-tzu; sixth century B.C.E.), the founder of Daoism, thought that society was already technically too advanced, and that technical invention served only to alienate people from the natural order and destabilize society. He advocated rejecting technical knowledge and skills. According to Laozi, people should not use writing, machines, carriages, or ships. Zhuangzi (or Chuang-tzu; c. 369–c. 286 B.C.E.), another representative of Daoism, presented fables to suggest that the use of machines led individuals to act contrary to their nature.

Among the ancient schools of Chinese thought, only Mohism valued material goods for the benefit of society. This school held that moral value lies in benefiting the people. Mo Di (or Mo-tzu; fl. 476–390 B.C.E.), the founder of Mohism, regarded the technical arts as benefiting human society and proposed to develop them. Mohists even conducted scientific research and made contributions in the fields of geometry, mechanics, and optics. At the same time, Mo Di opposed the use of technology to wage unjust wars and to produce curios for the court (Zhu Yiting 2002).

These ancient philosophies provided the frameworks for traditional ethical thinking in China. A common feature of such philosophies is concern for people, but in a way different from modern ethical thinking: People are conceived in relation to the larger Way. Thus human good is not something that can be pursued scientifically or technologically for individuals in isolation from the cosmos.

For more than 2,000 years such philosophical attitudes predominated in China and influenced science and technology both directly and indirectly, as has been extensively examined not only in China but in the West as well (see Needham 1954–; Sivin 1995). Geoffrey Lloyd and Nathan Sivin (2002) agree that when the Chinese "thought about the universe, what intrigued them was its connection to sociopolitical order" (p. 235). They go on to contrast the emphasis on logical distinction and deductive rigor that separates Greek science from ethics and politics with Chinese efforts "to find and explore correspondences, resonances, interconnections" in ways that "favored the formation of syntheses unifying widely divergent fields of inquiry" (p. 250). Thus from the beginning, and even into the modern period, Chinese scholars pursued what in the West might now be termed synthetic, aesthetic, or interdisciplinary knowledge, which left them vulnerable to more empirical, confrontational ways of knowing and manipulating the world that developed in Europe after the Renaissance.

The Modern Era

The modern period in Chinese history began in the 1500s when European powers established colonies (the first being Macao, founded by Portugal in 1557) for purposes of developing trade. Over the course of the next 300 years, Chinese resources were progressively exploited by Western imperialist forces, culminating in the First Opium War (1839–1842), in which Great Britain fought to deny China the right even to prohibit the importation of an addictive drug that was undermining its social order.

In response to this humiliation and other calamities, there emerged a series of efforts at modernization such as the "self-strengthening movement," which sought to appropriate and adapt "Western learning," especially science and technology, for Chinese benefit. Western models were used to create special schools and factories. Leaders such as Yang Xingfo (1893–1933), who studied engineering and business in the United States and then promoted scientific management in China, and Ren Hongjuan (1886–1961), the founder of the Chinese Society of Science, put forward the idea of "saving the nation by science." Along with such efforts came the eventual overthrow of the Qing dynasty and the creation of the Republic of China in 1911—followed by war with Japan, civil war, and finally the establishment of the People's Republic of China (PRC) in 1949.

In the nearly thirty years from the founding of the PRC to 1978, when China began a policy of reform and opening up to the West, there was little academic research into ethics as related to science and technology. Ethics, as well as science and technology, were viewed through the prism of socialism. On the one hand, scientific socialism held that science and technology were revolutionary forces that drove historical advancement, which was the basis of their social value. On the other, more fundamental was class struggle, to which ethics, along with science and technology, should be subordinated. Intellectuals had to adhere to the party line and to be "both red and expert." During the Cultural Revolution (1966–1976), some aspects of natural science, such as the theory of relativity and modern cosmology, were even viewed as reactionary bourgeois ideas because of supposed antisocialist implications, and scientists in these fields were themselves criticized as reactionaries.

Since 1978, however, China has implemented policies of reform and opening up, and the government and people have come to view science and technology as a primary productive force. The government has implemented strategies for sustainably developing science, education, and the economy to modernize China. In this new intellectual climate, Chinese academics have begun to pay more attention to ethical questions related to science and technology. Their contributions can be broken down into four main categories.

Do Science and Technology Involve Ethical Problems?

One opinion holds that science seeks knowledge or truth and that as such it is a value-neutral cognitive activity devoid of ethical implications (Jin Wulun 2000). The opposite view is that knowledge creation in science and production through technology can involve ethics in any of three ways. First, insofar as scientists and engineers produce objective knowledge and skills, they must follow methodological guidelines, which include professional codes of ethics.

Second, the application of scientific knowledge and the technological manufacture of products may have both positive and negative impacts on the economy, society, and nature in a way that poses ethical problems. But because scientists cannot control how their research results are applied, and engineers cannot determine how their products are used, they are not professionally responsible for the consequences of their work. Only as consumers and citizens are they responsible.

Third, Gan Shaoping (2000) has argued that ethical issues are sometimes inherent in science and technology themselves. Modern science is no longer purely theoretical knowledge, and engineering is not simply design; both are practical activities with built-in purposes oriented toward special applications. Thus researchers cannot pursue science or engineering and ignore the ethical issues implicit in the application of their work.

Justice and Responsibility

The pursuit of science and technology poses ethical issues of justice and responsibility. The problem of justice appears in two forms. The first asks whether the distribution of scientific research resources among scientists, disciplines, and various social needs is just. The second asks whether the application of research results might unfairly favor some and create burdens or harm for others.

The problem of responsibility manifests itself in the human arrangements that science and technology require and make possible. With the ever-increasing power and impact of science and technology in human societies, human arrangements have increasingly replaced natural arrangements. Properly engineering these human arrangements is an ethical concern.

Moreover, with regard to both justice and responsibility, the activities of science and technology have become a global enterprise. The abuse and misuse of science and technology can threaten the entire human species and the habitability of the earth. Scientists and engineers—along with managers, politicians, and the rest of society—are now collectively responsible for how the development of science and technology affects the future of humankind (Zhu Baowei 2000).

Progress

Some scholars maintain that there exists what others have called a "cultural lag" between human ethical standards and scientific-technological progress. On this basis they argue for some limitations in the current uncontrolled growth of science and technology (Lu Feng 2002). Other scholars think that science is superior to ethics, and that ethics should thus conform to developments in science. Most scholars, however, think that there should be an interactive relationship among developments in science, technology, and ethics, and that this constitutes true progress. That is, the correctness of scientific and technological activity should not be judged just from some preconceived ethical standpoint; instead, ethical systems should themselves be rethought, corrected, and developed in light of and in association with science and technology (Li Deshun 2000).

Some scholars have also highlighted dilemmas that arise from interactions between new developments in science and technology and systems of ethical values. On the one hand, new developments in science and technology often bring about new worries in ethics; on the other, if these developments were forbidden, humanity might be deprived of major benefits. In response, it is suggested that a buffer (or soft-landing) mechanism should be introduced between new developments of science and technology and human systems of social values (Liu Dachun et al. 2000).

Ethical Disputes in Particular Hi-Tech Fields

Ethical concerns have come to the fore especially in relation to biotechnology, the environment, and the Internet.

BIOETHICS. In relation to biotechnology, He Zuoxiu, a famous theoretical physicist, argues that no work should be forbidden, not even human cloning (Piao Baoyi 2002) He criticizes bioscientists for caving in to the media and restricting such developments. Zhao Nanyuan, a scholar in the field of automation, further argues that Chinese moralists who simply repeat what foreigners say have become the mouthpiece in China for the antiscientific and antitechnological views of foreign religious zealots. At the same time, most scholars maintain that biotechnology should be pursued prudently because of the risks involved, and that humans should not be cloned because of the ethical and social problems that would arise from human reproductive cloning. The Ministry of Science and Technology and the Ministry of Public Health have firmly opposed human reproductive cloning.

ENVIRONMENTAL ETHICS. Some scholars accept arguments that animals, living things, and indeed the whole ecosystem have inherent value and some rights independent of their instrumental value for humans. Humans should preserve the environment, not only to enhance the well-being of humans and human posterity, but also to preserve the stability, prosperity, and beauty of ecosystems. Most Chinese philosophers, however, still adhere to an anthropocentric view that only humans have moral consciousness and can be morally responsible for their own behavior. Animals do not have rights. Whether holding anthropocentric or nonanthropocentric views, all agree that preserving the environment, reducing pollution, and maintaining biodiversity have long-term benefits.

NETWORK ETHICS. Information transmitted through the Internet may be true or false, healthy or pernicious. These issues have raised the most concern in the field of what is called "network ethics" (also called "computer ethics" or "information ethics" outside China). In addition, some research also focuses on the protection of intellectual property rights and individual privacy. Some scholars suggest that the anonymity of the Internet is the main cause for the ethical problems arising there, and that for this reason maintaining ethical behavior on the Internet ultimately depends on individual moral self-discipline (Wang Lujun 2000). The central government in the PRC also exercises some restrictions over Internet communication in accord with its concerns for social order.

Developments in the Early Twenty-First Century

Generally speaking, traditional Chinese culture, although emphasizing moral issues, has been relatively tolerant of science and technology. There is nothing like the trial of Galileo Galilei (1564–1642) or the rejection of evolution in Chinese history, except during the aberration of the Cultural Revolution.

In the early twenty-first century, China nevertheless lags behind Europe, the United States, Japan, and some other countries in its level of economic and technological development. There thus exists an urgent need to promote science and technology in China. Current studies of ethics in science and technology should thus include promoting the development of science and technology, especially with the aim of benefiting the most people (Chen Ying 2002).

China seeks to promote rapid yet safe and sustainable development of science and technology. This is reflected in an increasing commitment in the PRC to research and development: In 2003 China spent $15.56 billion in this area, an increase of 23.5 percent over that of the previous year. It actually supported more than half again as many researchers. Along with such increases in research support, the Ministry of Science and Technology has promoted efforts to establish ethical systems and adopt ethical codes, and has dealt seriously with issues of scientific misconduct. The China Association for Science and Technology has established a standing committee on morals in science. The Chinese Academy of Sciences and the Chinese Academy of Engineering have likewise adopted codes of behavior for academicians and have established related ethical systems.

In addition, education in the science, technology, and society (STS) studies field has actively cultivated research and teaching on ethics in science and technology. From 1984 to 2004 more than twenty centers or institutes for STS studies, including the Research Center for Science, Technology, and Society, and the Chinese Academy of Social Sciences, have been established. In 2004 Chinese universities have offered more than fifty courses of STS study. Moreover, there have been frequent international and national symposia, and many books and papers in the field have appeared (Yin Dengxiang 1997). STS studies in China seek to promote science and technology in a way that appreciates the ethical dimensions of these activities.

YIN DENGXIANG
LI SHIXIN

SEE ALSO *Buddhist Perspectives; Confucian Perspectives.*

BIBLIOGRAPHY

Chen Ying. (2002). "Keji lunli de genben yuanze: Wei zui da duoshu ren fuwu" [The basic principle for science and technology ethics: Serve the most people]. Paper presented at the Quanguo di si jie yingyong lunli yantaohui [Fourth National Symposium on Applied Ethics].

Gan Shaoping. (2000). "Keji lunli: Yige you zhengyi de keti" [Science and technology ethics: A hotly disputed topic]. *Zhexue dongtai* 10 5–8.

Jin Wulun. (2000). "Kexue yanjiu yu keji lunli" [Scientific research and science and technology ethics]. *Zhexue dongtai* 10 4–5.

Li Deshun. (2000). "Chen si keji lunli de tiaozhan" [Reflections on the challenge of science and technology ethics]. *Zhexue dongtai*, 10, 2–3.

Liu Dachun et al., ed. (2000). *Zai zhen yu shan zhi jian* [Between truth and good]. Beijing: Zhongguo shehui kexue chubanshe.

Lloyd, Geoffrey, and Nathan Sivin. (2002). *The Way and the Word: Science and Medicine in Early China and Greece*. New Haven, CT: Yale University Press. A comparison of the cosmology, science, and medicine of ancient China and Greece that relates conceptual differences to how intellectuals in the two civilizations earned their living, interacted with fellow inquirers, and were involved with structures of authority.

Lu Feng. (2002). "Keji jinbu yu daode jinbu" [Progress in science and technology and progress in morality]. Paper presented at the Quanguo di si jie yingyong lunli yantaohui [Fourth National Symposium on Applied Ethics].

Miller, H. Lyman. (1996). *Science and Dissent in Post-Mao China: The Politics of Knowledge*. Seattle: University of Washington Press. An examination of how scientific dissent led to a renewed liberal voice in China.

Needham, Joseph. (1954–). *Science and Civilisation in China*. 22 vols. to date (2004). Cambridge, UK: Cambridge University Press. A monumental work covering many fields of science in China.

Pourciau, Lester J. (2003). "The Ethics of Electronic Information in China and in Eastern Europe." In *Ethics and Electronic Information*, eds. Barbara Rockenbach and Tom Mendina. Jefferson, NC: McFarland.

Qiu Ren-Zong, ed. (2004). *Bioethics: Asian Perspectives*. Dordrecht, Netherlands: Kluwer Academic. Covers such topics as the intellectual foundation of Asian bioethics, bioethics and Asian culture, life and death, and euthanasia and end-of-life care in Asia.

Sivin, Nathan. (1995). *Science in Ancient China: Researches and Reflections*. Brookfield, VT: Variorum. Examines selected topics in astronomy, medicine, and the scientific revolution in China.

Suttmeier, Richard P. (1974). *Research and Revolution: Science Policy and Societal Change in China*. Lexington, MA: Lexington Books.

Unschuld, Paul U. (1979). *Medical Ethics in Imperial China: A Study in Historical Anthropology*. Berkeley and Los Angeles: University of California Press.

Wang Lujun. (2000). "Wangluo lunli wenti yanjiu zongshu" [Summary of studies on Internet ethical problems]. *Zhexue dongtai* 9.

Xi Zezong, ed. (2001). *Zhongguo kexue jishu sixiang shi: Kexue sixiang juan* [History of Chinese scientific and technical thought: Scientific thought]. Beijing: Kexue chubanshe.

Yin Dengxiang. (1997). *Shidai de huhuan: Kexue jishu yu shehui daolun* [The call of the times: An introduction to science, technology, and society]. Xi'an, China: Shan'xi renmin jiaoyu chubanshe.

Zhu Baowei. (2000). "Kexue jishu lunli: Gongzheng he zeren" [Science and technology ethics: Justice and obligation]. *Zhexue dongtai* 10, 9–11.

Zhu Yiting, ed. (2002). *Lunlixue da cidian* [Dictionary of ethics]. Shanghai: Shanghai cishu chubanshe.

INTERNET RESOURCES

Piao Baoyi. (2002). "Fandui 'fan kelong ren yundong'" [Against 'the movement of against cloning of people']. Available from http://www.people.com.cn/GB/kejiao/42/155/20020607/746960.html.

Zhao Nanyuan. "Minzhu, fazhi, lunli, and kexue: Zai kelong ren xianxiang de beihou" [Democracy, rule of law, ethics, and science: Behind the phenomenon of cloned people]. Available from http://www.oursci.org/magazine/200209/020920.htm.

ENGINEERING ETHICS

China has age-old traditions both in the practical sciences and technology and in ethics, but few studies link the two areas of endeavor. Traditional studies were limited to morality in the practice of medicine and doctrines that promoted harmony between humans and nature. From 1978, when China opened up to the outside world and began a program of reform, scholars in China started studying engineering ethics in the contemporary sense. In China, however, scholars more often talk of science and technology ethics and seldom use the term *engineering ethics*. Since 1978 research on ethics in science and technology has made considerable progress, going through three stages of development: the embryonic stage, the development stage, and the stage of a deepening appreciation of the issues involved.

The Embryonic Stage

During the Cultural Revolution (1966–1976) ethics was a prohibited topic. Scholars started writing about ethics only after 1978, when China began liberalizing. First came the translation of such key works as Rachel Carson's *Silent Spring* (in 1979), J. D. Bernal's *The Social Function of Science* (in 1982), and Donella H. Meadows and colleagues' *The Limits to Growth*. Such specialized volumes were complemented by more general works such as the reprinting of an important anthology of classic Western texts on ethics (Zhou Fucheng 1964) together with a critical biographical study of Western ethical philosophers (Zhou Fucheng 1987). During this period, Chinese scholars focused on the moral practices of Western scientists, using them in the construction of modern science and technology in China (see Xu Shaojin 1995).

Also about this time scholars began studying ethical issues related to specific technologies. One example is interest in environmental problems and ecological ethics, stimulated by the Carson and Meadows translations. As a forum for issues in the field of medical ethics, such as those involving test-tube babies and organ transplants, two new journals were created during the 1980s: *Medicine and Philosophy* (on the philosophy of medicine, published by the Chinese Academy of Medicine) and *Chinese Medical Ethics* (on medical ethics, published by Xinghua University). Relevant monographs include a book on the fundamental principles of medical ethics (Du Zhizheng 1985) that and another on bioethics (Qiu Renzong 1987).

Finally, some general works on ethics in science and technology also appeared: *The Ethics of Science and Technology* (Xu Shaojin 1989), *Essentials of Science and Technology Ethics* (Liu Fengrui 1989), and *Technological Ethics* (Huang and Chen 1989).

The Development Stage

New ethical problems brought about by modern science and technology gave rise to extensive scholarship in China, including frequent academic discussions and numerous publications. Among these was a debate, in the journal *Study of the Dialectics of Nature*, between two opposite views of the relation between humans and the ecosystem: anthropocentrism and ecocentrism. Many works concerned with environmental ethics appeared, among which were four books titled *Ecological Ethics* (Liu Xiangrong 1992, Li and Chen 1993, Ye Ping 1994, Yu Mochang 1999). Other books included *Environmental Ethics* (Li Peichao 1998), and *The Progress of Environmental Ethics* (Xu Songling 1999).

Issues in biomedical ethics also continued to be pursued. Zheng Zhenlu (1992) sought to unify medical and bioethics. Du Zhizheng (2000) undertook a more detailed criticism of the foundations of medical ethics alone.

The Stage of Deepening Appreciation

The beginning of the twenty-first century saw two notable trends in the area of science and technology ethics: Science and technology philosophers turned their attention toward ethics (for example Liu Dachun 2000, Zhou Changzhong 1999), and ethicists focused on science and technology. These two trends converged to form an intellectual climate in which scholars probed more deeply the theoretical and practical problems of science and technology ethics.

The greater attention that philosophers of science and technology gave to ethics aroused concern among scientists, technologists, and the general public about issues of ethical responsibility. Heated disputes about such basic questions as the ethics of human cloning made the study of science and technology ethics ever more important. The beginning of the century also witnessed an increase in exchanges and cooperation between Chinese and foreign scholars in science and technology ethics, especially in medical ethics and bioethics.

Between 2001 and 2003 many works appeared, including two books on general science and technology ethics (Fu Jing 2002, Li Qingzhen et al. 2003), one on engineering ethics (Xiao Ping 2001), a translation on information technology ethics (Spinello 2003), four books on medical ethics or bioethics (Chen and Qiu 2003, Li and Cai 2003, Li and Liu 2003, Qiu and Zhai 2003), and four works on ecological or environmental ethics (Lei Yi 2001, Fu Hua 2002, He Huaihong 2002a, 2002b).

In conclusion, although engineering ethics as such has not become a major theme in Chinese discussions, questions of the ethics of specific types of engineering—such as practiced in relation to the environment, medicine, or the Internet—have been increasingly discussed. In general engineering is seen as simply one aspect of science and technology, and analyzed accordingly. It is worth noting that Chinese perspectives on many of the issues mentioned here have also been increasingly considered in English-language studies, as is illustrated by Ole Döring (1999), Albert R. Jonsen (2000), and Lester J. Pourciau (2003).

LIU ZEYUAN

SEE ALSO *Engineering Ethics*.

BIBLIOGRAPHY

Bernal, J. D. (1982). *Kexue de shehui gongneng*, trans. Chen Tifang. Beijing: Shangwuyin shuguan. Originally published as *The Social Function of Science* (New York: Macmillan, 1939).

Carson, Rachel. (1979). *Jijing de chuntian*, trans. Lü Ruilan and Li Changsheng. Changchun, China: Jilin renmin chubanshe. Originally published as *Silent Spring* (Boston: Houghton Mifflin, 1962).

Chen Yuanfang and Qiu Renzong. (2003). *Shengwuyi yanjiu lunlixue* [Biomedical research ethics]. Beijing: Zhongguo xiehe yike daxue chubanshe.

Döring, Ole, ed. (1999). *Chinese Scientists and Responsibility: Ethical Issues of Human Genetics in Chinese and International Contexts*. Hamburg, Germany: Institut für Asienkunde.

Du Zhizheng. (2000). *Yixue lunlixue tanxin* [Probing medical ethics]. Zhengzhou, China: Henan yike daxue chubanshe.

Fu Hua. (2002). *Shengtai lunlixue tanjiu* [Probing ecological ethics]. Beijing: Huaxia chubanshe.

Fu Jing. (2002). *Keji lunlixue* [Science and technology ethics]. Chengdu, China: Xinan caijing daxue chubanshe.

He Huaihong. (2002a). *Huanjing lunlixue* [Environmental ethics]. Beijing: Gaodeng jiaoyu chubanshe.

He Huaihong. (2002b). *Shengtai lunlixue: Jingsheng ziyuan yu zhexue jichu* [Ecological ethics: Spiritual resources and philosophical foundations]. Shijiazhuang, China: Hebei daxue chubanshe.

Huang Linchu and Chen Aijuan. (1989). *Jishu lunlixue: Lilun yu xhixian* [Technological ethics: Theory and practice]. Xi'an, China: Xi'an jiaotong daxue chubanshe.

Jonsen, Albert R. (2000). "Medical Ethics of India and China." Chap. 3 in *A Short History of Medical Ethics*. New York: Oxford University Press.

Lei Yi. (2001). *Shengtai lunlixue* [Ecological ethics]. Xi'an, China: Shanxi renmin jiaoyu chubanshe.

Li Chunqiu and Chen Chunhua. (1993). *Shengtai lunlixue* [Ecological ethics]. Beijing: Kexue chubanshe.

Li Peichao. (1998). *Huanjing lunli* [Environmental ethics]. Beijing: Zuojia chubanshe.

Li Qingzhen et al. (2003). *Xiandai keji lunlixue* [Modern science and technology ethics]. Jinan, China: Shandong renmin chubanshe.

Li Runhua and Liu Yaoguang. (2003). *Yixue lunlixue* [Medical ethics]. Changsha, China: Zhongnan daxue chubanshe.

Li Yuzhen and Cai Jinhua. (2003). *Yixue yu shengming lunli* [Medicine and bioethics]. Beijing: Kexue chubanshe.

Liu Dachun. (2000). *Zai zhen yu shan zhi jian: Keji shidai de lunli wenti he daode jueze* [Between truth and good: Ethical problems and moral choice in the age of science and technology]. Beijing: Zhongguo shehui kexue chubanshe.

Liu Fengrui, ed. (1989). *Jianming keji lunlixue* [Essentials of science and technology ethics]. Beijing: Hangkong gongye chubanshe.

Liu Xiangrong. (1992). *Shengtai lunlixue* [Ecological ethics]. Changsha, China: Hunan shifan daxue chubanshe.

Pourciau, Lester J. (2003). "The Ethics of Electronic Information in China and in Eastern Europe." In *Ethics and Electronic Information*, ed. Barbara Rockenbach and Tom Mendina. Jefferson, NC: McFarland.

Qiu Renzong. (1987). *Shengming lunlixue* [Bioethics]. Shanghai: Shanghai renmin chubanshe. For an English introduction to this author's work see *Bioethics: Asian Perspectives*, ed. Qiu Renzong (Dordrecht, Netherlands: Kluwer Academic, 2004).

Qiu Renzong and Zhai Xiaomei. (2003). *Shengming lunlixue jiaocheng* [A course on bioethics]. Beijing: Zhongguo xiehe yike daxue chubanshe.

Spinello, Richard A. (2003). *Xinxi he jisuanji lunli anli yanjiu*, trans. Wu Hexin and Zhang De. Beijing: Kexue jishu wenxian chubanshe. Originally published as *Case Studies in Information and Computer Ethics* (Upper Saddle River, NJ: Prentice Hall, 1997).

Xiao Ping. (2001). *Gongcheng lunlixue* [Engineering ethics]. Beijing: Zhongguo tiedao chubanshe.

Xu Shaojin. (1989). *Keji lunlixue* [The ethics of science and technology]. Shanghai: Shanghai renmin chubanshe.

Xu Shaojin. (1995). *Xifang keji lunli xixiang shi* [The history of Western science and technology ethics]. Nanjing, China: Jiangsu jiaoyu chubanshe.

Xu Songling. (1999). *Huanjing lunlixue jinzhan: Pinglun yu chanshu* [The progress of environmental ethics: Comment and exposition]. Beijing: Shehui kexue wenxian chubanshe.

Ye Ping. (1994). *Shengtai lunlixue* [Ecological ethics]. Harbin, China: Dongbei linye daxue chubanshe.

Yu Mochang. (1999). *Shengtai lunlixue* [Ecological ethics]. Beijing: Shoudu shifan daxue chubanshe.

Zheng Zhenlu. (1992). *Yixue daode yu shengming lunli* [Medical morals and bioethics]. Shanghai: Huadong shifan daxue chubanshe.

Zhou Changzhong. (1999). *Puluomixiusi hai shi fushide: Keji shehui de lunlixue* [Prometheus or Faust: Ethics in scientific and technological society]. Wuhan, China: Hubei jiaoyu chubanshe.

Zhou Fucheng. (1964). *Xifang lunlixue mingzhu xuanji* [Selections of famous Western works on ethics]. Beijing: Shangwuyin shuguan. Reissued in 1987.

Zhou Fucheng. (1987). *Xifang zhuming lunlixue jia ping zhuan* [Critical biographies of well-known Western moralists]. Shanghai: Shanghai renmin chubanshe.

RESEARCH ETHICS

In China discussions about research integrity occur in the context of studies of the interaction among science, technology, and society (STS). Such discussions are concerned not only with identifying various types of misconduct in scientific and technological research but also with the institutional reasons for such misconduct in management systems and social culture. In these contexts, scholars suggest measures to counter such misconduct. Their discussions focus mainly on three aspects of STS interactions as follows.

Definition and Prevention of Academic Misconduct

Fan Hongye (1982, 1994), a historian of science, defines "misconduct" according to international standards as fabrication, falsification, or plagiarism to acquire recognition from scientific associations and societies for scientific research. This includes the fabrication or falsification of experimental data, unacknowledged use of others' research, and falsified reports of research results. This definition is generally accepted in academic circles throughout China.

Scholarly work on ethics in science began to develop in the 1980s. Fan's *The Falsification of Scientific Results* appeared in Chinese in 1982. William J. Broad and Nicholas Wade's 1982 book *Betrayers of the Truth* was translated into Chinese in 1988. Xu Shaojin published a monograph in Chinese entitled *The Ethics of Science and Technology* the following year.

As a result of such heightened awareness, since 1990 there have been many reports and criticisms of instances in which researchers, teachers, or graduate students falsified data or plagiarized others' data, such as the Hu Liming (a doctor and professor in Huadong University of Technology) plagiarizing case in 1997, and the Wang Mingming (a professor in Beijing University) plagiarizing case in 2002. Some scholars have pointed out that deficiencies in the system for managing scientific research lead to such misconduct. Others have suggested new laws, regulations, and rules governing scientific and technological research or improvements in systems of research management.

The Social Responsibilities of Scientists and Engineers

A central academic concern in China at the beginning of the twenty-first century is what kind of social responsibilities scientists and engineers should assume. Many scholars have noted that with increased academic freedom in China since the 1980s, scientists have more liberty to determine their research activities. If researchers do not exercise self-discipline and a high sense of responsibility, their research may adversely affect society. Zou Chenglu and Hu Qiheng, members of the Chinese Academy of Sciences, have argued this position, which has attracted much attention.

In 2002 the Chinese Academy of Sciences formulated and published "Self-Disciplining Standards of Scientific Integrity for Members of the Chinese Academy of Sciences," a statement of principles for protecting society, promoting science, and maintaining scientific integrity. Such Chinese works as those by Li Hanlin (1987) and Zhang Huaxia (1999) have analyzed the social responsibilities of scientists and engineers, and proposed measures to guard against weak moral discipline and lack of responsibility. Because of the complex nature of modern science and technology, society has little choice but to rely on technical experts to be responsible in their work.

Dissent as an Ideal in Chinese History

Research is most productive when academic dissent is possible. Academic debate is deeply rooted in Chinese history (though, it should be admitted, so is its opposite, authoritarianism). In the Spring and Autumn period (770–476 B.C.E.) and the Warring States period (475–221 B.C.E.), it was said that a hundred schools of thought contended. (That was before the first emperor of the Qin dynasty, who reigned China from 221 to 210 B.C.E., burned books and unified thought.) At other bright points in history, scholars such as Sima Qian (c. 145–c. 85 B.C.E.), Zhu Xi (1130–1200), and Wang Fuzhi (1619–1692) affirmed the truth, persuaded others by reason, and rejected political suppression of thought.

In 1956 Mao Zedong revived the principle of a hundred schools contending during the hundred flowers campaign. Though in the Soviet Union Trofim Denisovich Lysenko, from 1948 to 1953, successfully led a campaign to repress Mendelian genetics as antisocialist, biologists in China held a symposium on genetics in 1956 in Qingdao, where opposing parties objectively discussed biological research. Unfortunately, by mid-1957 some scholar criticism had been leveled against the leadership of Communist Party, with the result that Mao called a halt to the hundred flowers campaign and suppressed further criticism.

After 1978, when China opened up to the outside world, the pendulum again swung back, and China became an increasingly free and open society. At the beginning of the twenty-first century Chinese researchers enjoy considerable academic freedom. Indeed, the nation has again entered an age when a hundred flowers bloom together and a hundred schools of thought contend.

WANG QIAN

SEE ALSO *Research Ethics.*

BIBLIOGRAPHY

Broad, William J., and Nicholas Wade. (1988). *Beipan zhenli de renmen*, trans. Zhu Jinning and Fang Yuzhen. Beijing: Kexue chubanshe. Originally published as *Betrayers of the Truth—Fraud and Deceit in the Halls of Science* (New York: Simon and Schuster, 1982).

Chinese Academy of Sciences. (2001). "Zhongguo kexue yuan xue bu kexue daode jianshe zuotanhui jiyao" [Summary of a symposium on establishing standards of scientific integrity of Academic Department of the Chinese Academy of Sciences]. *Zhongguo kexue yuan yuankan* 5: 325–326.

Chinese Academy of Sciences. (2002). "Zhongguo kexue yuan yuanshi kexue daode zilü zhunze" [Self-disciplining standards of scientific integrity for members of the Chinese Academy of Sciences]. *Zhongguo kexue yuan yuankan* 1: 2.

Fan Hongye. (1982). *Kexue yeji de bianwei* [The falsification of scientific results]. Shanghai: Shanghai renmin chubanshe.

Fan Hongye. (1994). "Keyan zuowei xingwei ji qi bianshi yu fangfan" [Recognizing and preventing the falsification of scientific research]. *Ziran bianzhengfa tongxun* 1: 25–33.

Hu Qiheng. (1999). "Kexue de zeren yu daode: Yige zhide zhongshi de wenti" [Responsibility and integrity of science: A question worthy of attention]. *Xuehui* 11: 16.

Hu Qiheng. (2000). "Kexue daode bu rong hushi" [Scientific integrity cannot be neglected]. *Zhongguo gaige* 1: 21.

Li Bocong. (2000). "Daode huhuan he daode yueshu" [Moral appeal and moral restraint]. *Kexue dui shehui de yingxiang* 3: 49–50.

Li Hanlin. (1987). *Kexue shehuixue* [Sociology of science]. Beijing: Zhongguo shehui kexue chubanshe.

Li Hongfang. (2000). "Jinnian kexue yuegui wenti yanjiu pingshu" [A review of recent misconduct in science]. *Kexue daobao* 3: 16–18.

Li Peishan. (2000). "Ezhi kexue daode lunluo de jidian qianjian" [Several suggestions for preventing the loss of scientific integrity]. *Ziran bianzhengfa tongxun* 2: 10–11.

Li Peishan, and Xue Pangao. (1996). "Shi Yingwen wenti, haishi kexue daode wenti?" [Is it a problem of understanding English or a problem of scientific integrity?]. *Ziran bianzhengfa tongxun* 4: 74–80.

Li Qingzhen, Su Fuzhong, and An Weifu. (2003). *Xiandai keji lunlixue* [The ethics of modern science and technology]. Jinan, China: Shandong renmin chubanshe.

Liu Fengrui, ed. (1989). *Jianming keji lunlixue* [Essentials of science and technology ethics]. Beijing: Hangkong gongye chubanshe.

Ministry of Science and Technology, Ministry of Education, Chinese Academy of Sciences, Chinese Academy of Engineering and China Association for Science and Technology. (1999). "Guanyu keji gongzuozhe xingwei zhunze de ruogan yijian" [Several opinions on standards of behavior for scientists and technologists]. *Keji yu falü* 4: 3–4.

Shi Yumin. (1994). "Lun kexue huodong zhong de yuegui" [On misconduct in scientific activities]. *Kexue guanli yanjiu* 2: 6–14.

Wang Feng. (1999). "Chaoxi: Bu jinjin shi daode wenti, geng shi weifa xingwei" [Plagiarism: Not just an ethical matter, but an illegal act]. *Keyan guanli* 2: 68–73.

Wang Feng. (2002). "Kexue bu duan xingwei ji qi chengyin pouxi" [Misconduct in science and an analysis of contributing factors]. *Kexue xue yanjiu* 1: 11–16.

Xu Shaojin. (1989). *Keji lunlixue* [The ethics of science and technology]. Shanghai: Shanghai renmin chubanshe.

Yan Bing, and Song Ziliang. (1997). "Kexue zuowei yu daode jiaohua" [Forgery in scientific research and moral instruction]. *Kexue xue yanjiu* 1: 57–62.

Yan Bing, and Song Ziliang. (1999). "Kexue zuowei yu shehui tiaokong" [Forgery in scientific research and social control of it]. *Kexue xue yanjiu* 1: 16–24.

Yang Yusheng. (2002). "Xueshu fubai, xueshu guifan yu xueshu lunli: Guanyu gaoxiao xueshu daode jianshe de ruogan wenti" [Academic misconduct, academic standards, and academic ethics: Several problems for establishing academic standards in higher education]. *Shehui kexue luntan* 6: 28–34.

Yang Yusheng. (2002). "Xueshu piaoqie xianxiang: Women ying you de guanzhu yu fansi" [Academic plagiarism: What we should pay attention to and reflect on]. *Shehui kexue luntan* 4: 24–29.

Zhang Huaxia. (1999). *Xiandai kexue yu lunli shijie* [Modern science and the world of ethics]. Changsha, China: Hunan jiaoyu chubanshe.

Zhou Guangzhao. (1994). "Zai Zhongguo kexue yuan diqici yuanshi dahui shang de gongzuo baogao" [Working report of the seventh conference of members of the Chinese Academy of Sciences]. *Zhongguo kexue yuan yuankan* 3: 196–202.

Zou Chenglu. (2002). "Kexuejia yu kexue yanjiu daode" [Scientists and scientific research integrity]. *Minzhu yu kexue* 1: 10–12.

Zou Chenglu, Shen Sanjiong, Wu Min, and Li Lin. (1991). "Zai lun kexue daode" [On scientific integrity, part 2]. *Zhongguo kexue bao*, 25 October 1991.

CHOICE BEHAVIOR

The ability to make a choice, as opposed to being told what to do, or given only a single option, has been shown to have positive effects (Deci and Ryan 1985). People are more internally motivated and perform better on tasks they have chosen, and they also are more satisfied with their choices and feel more in control. However as decisions become more difficult for decision makers, these benefits begin to disappear. When people face difficult decisions, they experience more anxiety, anticipate potential regret, and are more likely to postpone the decision, relegate it to another person, or avoid making it altogether (Schwartz 2004). In addition after making a difficult decision people are likely to be dissatisfied, and feel less confident that the *right* choice was made. These phenomena have obvious ethical implications for a society in which science and technology are often valued because of their ability to enhance choices.

A number of factors increase the difficulty of a decision. Situations that require decision makers to contrast unattractive options, make large tradeoffs, or compare large numbers of items make decisions difficult, as do those where accountability to others or a lack of information lead to anticipated regret or fear of blame. Increasing the number of options available increases the number of tradeoffs that must be made between desirable attributes of those options. This increases the effort required of the decision maker and induces more severe psychological consequences, which leads decision makers to rely on less of the available information, and to use simplified decision rules, which in turn make mistakes more likely. This result has been found to hold true not only for consumer purchasing decisions, but also for selecting retirement and health insurance plans, and choosing medical treatments (both by patients and doctors) (Schwartz 2004).

In addition to changes in the decision process, researchers have demonstrated effects on decision outcomes. More specifically when the choice involves potential tasks or activities, more options can lead to the decision maker feeling less motivated and performing more poorly on the chosen task. For example, researchers offered students either thirty topic options for an extra credit essay or six options, and found that when students had thirty options to choose from, fewer students chose to write an essay, and the quality of the essays written was worse.

Importantly experts do not appear to be immune to the effects of decision difficulty (Shanteau, Weiss, Thomas, and Pounds 2003). Whereas experts are often able to consider more of the available information, the only experts who appear uniquely equipped to make decisions are those in fields such as physics and mathematics where rules exist for reaching solutions, relative levels of certainty exist, and there are opportunities to learn from feedback. Experts in fields where there are not explicit rules or equations for solving problems (for example, clinical psychologists, legislators, advertising executives) have been found to use simplified decision rules and be affected by the psychological effects of tradeoffs. However the accountability that comes with being an expert has been shown, in many situations, to increase a decision maker's search effort and the complexity of decision strategies (Lerner and Tetlock 2003).

Unfortunately experts and novices alike are commonly unaware of the influences that decision difficulty has on their behavior. People often believe they want more choice options, yet those options make them less happy, and they often want to give such options away once they have them (Schwartz 2004). For example, 65 percent of healthy people say that they would want to choose their own medical treatment if they were to get cancer, whereas among people with cancer only 12 percent want to choose their own treatment. When not actually facing it, people do not realize the difficulty of the decision and the emotional consequences they will face when they have to bear the responsibility of deciding. Likewise as experts make decisions, particularly those concerning outcomes for other individuals, they need to take into consideration both their own cognitive abilities and limitations—in particular, the effects of decision difficulty that they might not be aware of—as well as the abilities and limitations of the individuals who will be affected. For example, legislators deciding not to make changes to an existing program may indicate decision aversion in response to the difficulty that comes from accountability; similarly creating a program that gives more options to the affected citizens (such as giving workers options for investing social security savings) may result from the desire to shift the responsibility of making wise choices to the other party. Whereas the people affected might even think they want the options, if the options leave them with difficult decisions that are undesirable, providing the choice might prove to be a disservice.

ELIZABETH J. MULLIGAN
ERIC D. CLAUS

SEE ALSO *Decision Theory; Psychology.*

BIBLIOGRAPHY

Deci, Edward L., and Richard M. Ryan. (1985). *Intrinsic Motivation and Self-Determination in Human Behavior*. New York: Plenum Press.

Lerner, Jennifer S., and Philip E. Tetlock. (2003). "Bridging Individual, Interpersonal, and Institutional Approaches to Judgment and Decision Making: The Impact of Accountability on Cognitive Bias." In *Emerging Perspectives on Judgment and Decision Research*, ed. Sandra L. Schneider and James Shanteau. Cambridge, UK: Cambridge University Press.

Schwartz, Barry. (2004). *The Paradox of Choice: Why More Is Less*. New York: Ecco. Provides a very thorough, yet accessible, account of research findings on decision making that illuminates the problems that arise when decision makers encounter increasing numbers of options and/or lack information needed to make a decision.

Shanteau, James; David J. Weiss; Rickey P. Thomas; and Julia Pounds. (2003). "How Can You Tell If Someone Is an Expert? Performance-Based Assessment of Expertise." In *Emerging Perspectives on Judgment and Decision Research*, ed. Sandra L. Schneider and James Shanteau. Cambridge, UK: Cambridge University Press.

CHOICE THEORY

SEE *Rational Choice Theory*.

CHRISTIAN PERSPECTIVES

• • •

Contemporary Assessments of Science
Contemporary Assessments of Technology
Historical Traditions

CONTEMPORARY ASSESSMENTS OF SCIENCE

The relationship between science and Christianity is often portrayed as one of perpetual conflict. Although controversies such as that between the science of evolution and claims for religious creationism or intelligent design theory lend credence to this popular perception, the actual relationship is more complex. Indeed, since the Scientific Revolution of the seventeenth and eighteenth centuries and the Industrial Revolution of the nineteenth century, theologians have spent considerable effort just trying to sort out alternatives. Two great contributors to this effort were the German historian of Christianity Ernst Troeltsch (1865–1923) and the American theologian H. Richard Niebuhr (1894–1962). Niebuhr, for instance, distinguishes five basic relationships between Christ and culture: Christ as opposed to culture, as in agreement with culture, as above culture, as paradoxically related to culture, and as transformer of culture. Insofar as science is a kind of culture, these same five types can be found manifest in the Christianity–science relationship. Indeed, in a contemporary adaptation of Niebuhr, Ian G. Barbour (1990) develops a typology of four possible relationships that can serve here as a convenient framework.

Conflict

Barbour contends that this option represents a relatively small group of highly vocal protagonists whom he labels scientific materialists and biblical literalists. Within this schema, materialists use science to discredit religious faith, whereas literalists use religion to dictate the purview and course of scientific investigation.

Scientific materialists assert that science offers the only reliable route to knowledge, and that matter and energy are the fundamental realities of the universe. Drawing heavily on logical positivism, they argue that only verifiable or falsifiable statements have cognitive value. Consequently, religious beliefs are dismissed as meaningless, emotive statements because theological claims can be neither verified nor falsified. Examples of influential scientific materialists include Jacques Monod (1972), Edward O. Wilson (1978), and Richard Dawkins (1986).

Biblical literalists insist that scripture reveals the fundamental truth of the universe as God's creation. Although the Bible does not offer a detailed description of how God brought the universe into existence, it does disclose an underlying intelligent design as unbiased observations of nature confirm. Any so-called scientific evidence to the contrary should be attacked as false, incomplete, or mistaken. Moreover, because God is a supernatural being, divine or miraculous acts are not subject to the principles of verifiability or falsifiability in order to determine their truth. Consequently, given the supernatural origins of the universe, revelation provides the most trustworthy knowledge about ultimate realities.

The difference between materialists and literalists is most pronounced in their conflicting claims on human nature. Materialists argue that human behavior can be best explained as the emergent outcome of a blind evolutionary process. Human values and social mores are thereby the result of adaptive behavioral strategies that gave *Homo sapiens* and predecessors a survival advantage over time. Any moral difference separating humans from other animals is one of degree, not kind. Literalists retort that human beings were specially created by God. Unlike animals, humans possess souls that enable them to have fellowship with God, and the ability to perceive and obey God's moral commands as disclosed in Scripture. Thus human life has a uniquely sacred quality that is fixed rather than malleable. These contending claims over human nature are in turn often reflected in the

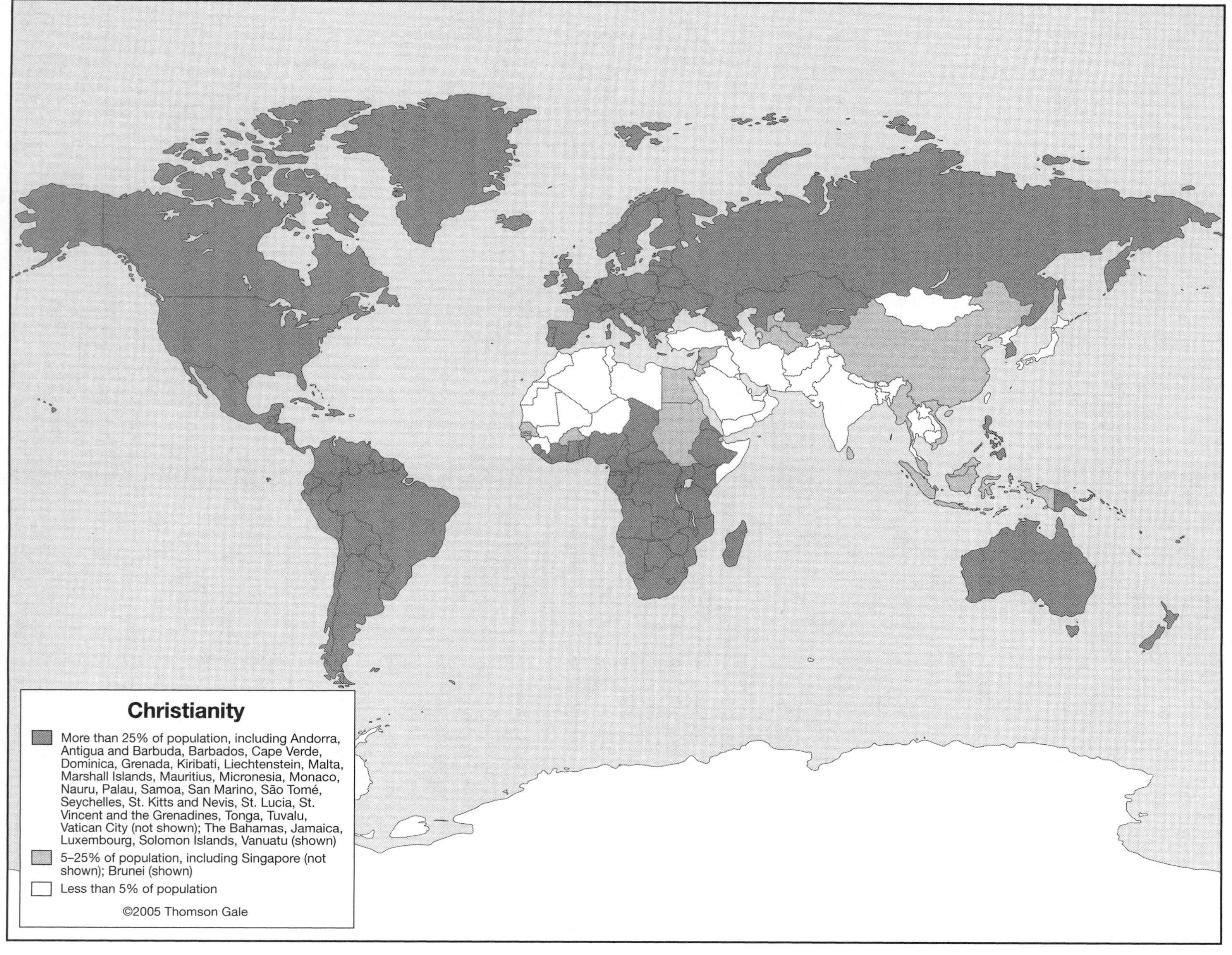
Christianity
More than 25% of population, including Andorra, Antigua and Barbuda, Barbados, Cape Verde, Dominica, Grenada, Kiribati, Liechtenstein, Malta, Marshall Islands, Mauritius, Micronesia, Monaco, Nauru, Palau, Samoa, San Marino, São Tomé, Seychelles, St. Kitts and Nevis, St. Lucia, St. Vincent and the Grenadines, Tonga, Tuvalu, Vatican City (not shown); The Bahamas, Jamaica, Luxembourg, Solomon Islands, Vanuatu (shown)
5–25% of population, including Singapore (not shown); Brunei (shown)
Less than 5% of population
©2005 Thomson Gale

acrimonious "culture wars" fought over such contentious issues as abortion, embryonic stem cell research, and euthanasia.

Independence

One strategy for avoiding conflict between science and religion is to insist on a rigorous separation and mutual honoring of their respective disciplinary boundaries. Science and religion are discrete and autonomous domains of inquiry that do not overlap. Given this détente, science confines itself to questions of *what*, while religion focuses on issues of *why*. Science offers empirical descriptions of physical reality while religion interprets the meaning of human existence by employing theological and moral precepts.

Christian theologians use a variety of methodologies in maintaining their independent sphere of inquiry. Karl Barth and his followers, for example, insist that history rather than nature is the domain of God's activities. It is through God's covenant with Israel, and the life, death, and resurrection of Jesus Christ that God reveals the divine plan for creation. In this respect, the Bible is held in high regard but is not interpreted literally. Consequently, naturalistic explanations of the origin of the universe do not contradict the biblical creation stories because both accounts are offered in delineated and noncompetitive modes of discourse. Moreover, an evolutionary description of human origins is unproblematic because God is revealed in the historical category of human culture rather than the natural category of biology. Similar approaches have been developed by Langdon Gilkey (1965) and Thomas F. Torrance (1969).

Science and theology do not conflict because of their highly disparate objects of inquiry and respective methodologies. They represent differing languages or linguistic constructs that cannot be easily translated into each other's categories. Although mutual independence promotes a peaceful relationship between science and religion, the price is that both appear to be describing two unrelated worlds rather than a common or single reality.

Dialogue

One way of overcoming this artificial division is to promote a dialogue between science and religion. There are two levels at which this dialogue may be pursued. First, both scientists and theologians encounter questions or make discoveries that cannot be easily confined within their respective disciplinary boundaries. Scientific research, for instance, may disclose a natural beauty and elegance inspiring a response of awe and wonder, while theologians are driven to find rational connections between human history and its underlying natural foundations. These transdisciplinary insights raise the prospect that although science and religion invoke two incompatible languages, they are nonetheless making correlative claims. Ernan McMullin (1998), for example, contends that although the big bang theory does not prove the Christian doctrine of creation, there is an implicit consonance suggesting that the universe is dependent upon God. More explicitly, Karl Rahner (1978) contends that Christian anthropology is compatible with evolutionary theory because it is through the emergence of spirit within matter that God has brought into being a creature with the capabilities of self-transcendence and divine fellowship.

The second level of dialogue focuses on the methodological parallels between science and religion. Wolfhart Pannenberg (1976), for instance, contends that theological doctrines are equivalent to scientific hypotheses that can be tested against universal rational criteria. The principal difference between science and theology is that the latter is concerned about reality as a whole, and given its unfinished and unpredictable character is not subject to as rigorous disciplinary scrutiny. In a similar vein, Alister E. McGrath (2003) argues that theological doctrines should be thought of as theories about nature and reality, whose truthfulness should be tested by rigorous theological and philosophical criteria. Other writers, such as Janice Martin Soskice (1985), Barbour (1990), and Mary Gerhart and Allan Russell (1984) insist that the dichotomy between "objective" science and "subjective" religion is false and misleading. Scientific research is itself theory-laden rather than neutral, and scientists often resort to intuition and analogies in constructing their theories. Similarly, theologians use theory-laden models and metaphors to investigate and describe religious experience. The work of both scientists and theologians may therefore be assessed in terms of coherence, comprehensiveness, and fruitfulness, thereby acquiring a common form of knowledge that Michael Polyani asserts is personal but not merely subjective. Consequently, the models and metaphors employed respectively by science and theology may prove mutually enriching in investigating the origin and nature of the universe in general and those of human beings in particular.

Integration

Although the dialogue approach promotes a closer relationship between science and religion than that offered by the independence model, the resulting conversation tends to be cursory given the focus on methodological issues. A number of writers assert that in order to correct the incomplete character of this dialogue, the content of science and religion needs to be integrated. Following Barbour, there are three prominent ways for pursuing such integration, which he identifies as natural theology, theology of nature, and systematic synthesis.

Natural theology is based on the premise that the order and intelligibility of the universe suggests an underlying purpose or design. This is especially the case with respect to the emergence of life, which proponents claim implies a natural teleology; that is, the evolution of the universe is itself oriented toward an emergent intelligence. Religious experience and revelation confirm this basic scientific insight. Consequently, natural theological arguments often begin with science in order to construct subsequent religious claims. Richard Swinburne (2004), for instance, contends that given all the available scientific evidence, it is more probable than not that a deity or creator exists. A variety of authors have also invoked the anthropic principle, claiming that the universe appears to be "fine-tuned" for the emergence of life. Freeman Dyson (1979) claims that although the anthropic principle does not prove God's existence, the universe's architecture is consistent with a structure in which something like a mind plays a dominant role. More expansively, Simon Conway Morris (2003) contends that evolution is not a random process, and that the emergence of human life was inevitable given the rare physical conditions of planet Earth.

A theology of nature approach starts with traditional religious claims and reformulates them in light of contemporary science. Arthur Peacocke (1993), for example, explicates a pantheistic understanding of God to account for the necessity of randomness and chance in God's created order. It is through natural processes as disclosed by science that God participates in the ongoing creation of the universe. In this respect, Peacocke asserts that God is in the world but the world in also in God, and he uses the analogy of the universe as God's body and God as the universe's mind or soul to illustrate his argument. John Polkinghorne (1994) and Ted Peters (2000) have also undertaken similar reformulations, though with differing doctrinal emphases. More radically, Pierre Teilhard de Chardin (1964) offers a reinterpretation of Christian eschatology in which the evolution of self-conscious and intelligent life is being drawn toward an "Omega Point" of a single, universal consciousness.

Other writers advocate a systematic synthesis of science and religion resulting in an all-embracing metaphysics. The process philosophers Alfred North Whitehead (1978/1929) and Charles Hartshorne (1967) are leading examples of this approach. Both reject traditional doctrines of divine omnipotence in favor of a persuasive God, thereby accounting for the necessity of freedom, chance, and suffering in the world. Creation is an incomplete process, and God encourages its self-creation and completion, thereby allowing humans to exhibit genuine freedom and novelty within malleable natural structures. More modestly, James Gustafson (1981) and Charles Birch and John B. Cobb Jr. (1981) use scientific and religious principles to develop a non-anthropocentric ethic in which nature and nonhuman life-forms are valued in respect to God rather than for their usefulness to humans. In formulating their respective ethics, they draw heavily on the biological and environmental sciences. Philip Hefner (1993) has also used a variety of sciences in pursuing a thorough and systematic recasting of theological anthropology. Humans are the products of genetic and cultural information to such a degree that technological civilization has become their natural habitat. Humans have therefore emerged as created cocreators, who in partnership with God are responsible for the eventual fate of creation.

Assessment

There is thus no such thing as *the* Christian assessment of contemporary science. Rather, there is a wide range of assessments reflecting denominational and doctrinal differences, as well as the diversities of contemporary culture. Moreover, the typologies employed should not be construed as rigid categories but as markers within a highly fluid range of options. This is in keeping with the fact that the various relationships between science and religion are themselves subject to frequent reevaluation and revision in response to rapid developments in scientific, theological, and philosophical inquires.

It might also be noted that Barbour's typology has been criticized for a failure to take revelation seriously enough or as containing a built-in bias toward integration. Certainly there is a sense in which, from Barbour's perspective, integration appears to be the highest type of relationship between science and Christian theology.

BRENT WATERS

SEE ALSO *Anglo-Catholic Cultural Criticism; Evolution-Creationism Debate; Natural Law.*

BIBLIOGRAPHY

Barbour, Ian G. (1990). *Religion in an Age of Science*. San Francisco: Harper and Row. An extensive summary and overview of various relationships between modern science and religion.

Birch, Charles, and John B. Cobb Jr. (1981). *The Liberation of Life*. Cambridge, UK: Cambridge University Press. A theistic interpretation of human evolution from the perspectives of a leading ecologist and process theologian.

Dawkins, Richard. (1986). *The Blind Watchmaker*. New York: Norton. A materialist and contemporary Darwinian account of nature.

Dyson, Freeman. (1979). *Disturbing the Universe*. New York: Harper and Row. A philosophical reflection by a leading scientist on the meaning of recent and anticipated scientific and technological advances.

Gerhart, Mary, and Russell, Allan. (1984). *Metaphoric Process: The Creation of Scientific and Religious Meaning*. Fort Worth: Texas Christian University Press. An inquiry into the similarities and differences between religious and scientific ways of knowing.

Gilkey, Langdon. (1965 [1959]). Maker of Heaven and Earth: A Study of the Christian Doctrine of Creation. Garden City, NY: Anchor Books. A contemporary Protestant theological treatise on the Christian doctrine of creation in light of modern science.

Gustafson, James M. (1981). *Ethics From A Theocentric Perspective*, Vol. 1: *Theology and Ethics*. Chicago and London: University of Chicago Press. A leading Protestant theologian argues that ethics should be theopocentric (God-centered) rather than anthropocentric (human-centered).

Hartshone, Charles. (1967). A Natural Theology for Our Time. La Salle, IL: Open Court. A concise exposition on the primary themes of process philosophy and theology.

Hefner, Philip. (1993). *The Human Factor: Evolution, Culture, and Religion*. Minneapolis, MN: Fortress Press. Argues that advances in biological cultural evolution disclose humans as god's created co-creators.

McGrath, Alister E. (2001). *A Scientific Theology*, Vol. 1: *Nature*. Grand Rapids, MI: Eerdmans. Contends that from the perspective of Christian theology, nature should be regarded as creation.

McGrath, Alister E. (2002). *A Scientific Theology*, Vol. 2: *Reality*. Grand Rapids, MI: Eerdmans. Argues that creation can only be properly understood from a critical realist vantage point.

McGrath, Alister E. (2003). *A Scientific Theology*, Vol. 3: *Theory*. Grand Rapids, MI: Eerdmans. Contends that theological doctrines constructed from a critical realist perspective are similar to scientific theories that can be tested through the employment of various criteria.

Mcmullin, Ernan. (1998). "Cosmic Purpose and the Contingency of Human Evolution." *Theology Today* 55(3): 389–414. Examines the relevance of recent advances in cosmology for Christian theology.

Monod, Jacques. (1972). *Chance and Necessity*, trans. Austryn Wainhouse. New York: Vintage Books. A philosophical interpretation of biology from an evolutionary and materialist perspective.

Morris, Simon Conway. (2003). *Life's Solution: Inevitable Humans in a Lonely Universe*. Cambridge, UK: Cambridge University Press. Argues that the evolution of life is not radically open-ended but is convergent, and thereby limited to a restricted range of outcomes.

Pannenberg, Wolfhart. (1976). *Theology and the Philosophy of Science*, trans. Francis McDonagh. Philadelphia: Westminster Press. A highly detailed investigation into the relation between systematic theology and philosophy of science.

Peacocke, Arthur. (1993). *Theology for a Scientific Age*, rev. edition. Minneapolis, MN: Fortress Press. Crafts a concept of God that is intended to be compatible with contemporary scientific perspectives.

Peters, Ted. (2000). *God—the World's Future: Systematic Theology for a New Era*. Minneapolis, MN: Fortress Press. The author reconstructs a wide range of traditional Christian doctrines in response to new scientific theories and developments in postmodern philosophy.

Polkinghorne, John. (1994). *The Faith of a Physicist: Reflections of a Bottom-Up Thinker: The Gifford Lectures for 1993–4*. Princeton, NJ: Princeton University Press. An interpretation of Christian belief in light of new developments in physics, with an emphasis on natural theology, and the relationship between faith and reason.

Price, Daniel J. (2002). *Karl Barth's Anthropology in Light of Modern Thought*. Grand Rapids, MI: Eerdmans. A critical and detailed assessment of Barth's theological anthropology.

Rahner, Karl. (1978). *Foundations of Christian Faith*, trans. William V. Dych. New York: Seabury Press. General reflections on the relationship between science and religion by one of the twentieth century's leading Catholic theologians.

Soskice, Janet Martin. (1985). *Metaphor and Religious Language*. Oxford: Clarendon Press. A comparative study of religious and scientific forms of discourse.

Swinburne, Richard. (2004). *The Existence of God*, 2nd edition. Oxford: Clarendon Press. A highly detailed philosophical argument for the existence of God.

Teilhard de Chardin, Pierre. (1964). *The Future of Man*, trans. Norman Denny. London: Collins. A collection of highly speculative essays on the future course of human evolution.

Torrance, Thomas F. (1969). *Theological Science*. London: Oxford University Press. A philosophical and theological inquiry into the relationship between science and religion by a leading Protestant theologian.

Whitehead, Alfred North. (1978 [1929]). *Process and Reality*. New York: Free Press. A highly detailed account of the principal themes of process philosophy by its leading proponent.

Wilson, Edward O. (1978). *On Human Nature*. Cambridge, MA: Harvard University Press. An overview of the principal claims and methodologies of sociobiology.

CONTEMPORARY ASSESSMENTS OF TECHNOLOGY

Insofar as Christianity, like any religion, is a way of life as much or more than a system of thinking, its relations to modern technology are even more problematic than those with science. The Christian life aspires to provide guidance for daily behavior, from the saying of prayers to charitable care for others. When Jesus of Nazareth was asked about the most fundamental commandments (not ideas), he answered that they were "to love the Lord your God . . . and your neighbor as yourself" (*Luke* 10:27). And when asked who is the neighbor, he answered not with a theoretical discourse but the parable of the Good Samaritan (*Luke* 10:30–37). The most fundamental question for Christianity is the degree to which technology is or is not a way to practice love of one's neighbor.

The Origins of Technology

The historical fact is that modern technology arose within the context of a Christian culture. This has led to numerous debates about the degree to which Christianity has itself contributed to this origin. The most radical position is that of the historian Lynn White Jr. (1967) who has argued at length that the roots of technology in its distinctly modern form lie in Christian theology as it developed in the Latin West.

White's chief contention is that Christian theology, particularly the teaching of human dominion over creation, is the primary culprit underlying the environmental crises of the late twentieth centuries. In exercising this dominion humans have developed and deployed various technologies in an irresponsible manner, leading to ecological instability.

Although White's thesis has been subjected to subsequent criticism noting his failure to take into account the attending biblical emphasis on stewardship, which blunts the more egregious forms of exploitation he deplores, he nonetheless identifies a dilemma regarding a Christian moral assessment of technology per se. If on the one hand, technology is a valuable instrument humans use in exercising their dominion and stewardship, then it is inherently good. If on the other hand, technology is used in an exploitive and environmentally destructive manner thereby distorting human dominion and stewardship, then it is inherently evil. Various Christian theologians have adopted one or the other of these options, as well as a range of alternative assessments between these two extremes.

This historico-theological debate easily invites further analysis of the spectrum of theological attitudes toward technology. Drawing on a typology developed by Ian G. Barbour (1993), it is convenient to classify these basic attitudes as those of optimism, pessimism, and contextualism.

Christian Optimism

The first approach, optimism, perceives technology as a liberating force. Optimists contend that technology has been a singularly effective means for improving the quality of human life by overcoming a series of natural, social, and psychological constraints. This impressive accomplishment has been achieved by enabling higher living standards, improved health care, an expanded range of individual lifestyles, greater leisure, and rapid communication. Moreover, there is no compelling reason to believe that technological development will not continue this progressive trend in the foreseeable future.

Pierre Teilhard de Chardin (Teilhard, 1964) offers an expansive vision in which technology is used by humans to determine their own destiny as a species. Current technical interventions are prompting the evolution of a global spiritual consciousness, and Teilhard foresees the day when humans will no longer be discrete organisms. Subsequent theologians in this category draw heavily on the works of such futurists as Daniel Bell, R. Buckminster Fuller, Herman Kahn, and Alvin Toffler. Harvey Cox (1965), for instance, praises technology for rescuing humans from the tyranny of tradition, thereby expanding the range of their freedom and creativity. Philip Hefner (1993, 2003) portrays it as a principal mechanism for humans to fulfill their calling as God's created co-creators. In general, optimists tend to regard technology as a means for humans to better display the divine image they bear, or to more effectively express a love of neighbor.

Critics charge that optimists too easily disregard the costs and risks of technological development; unintended consequences and catastrophic accidents can and do occur. In addition, the large-scale technologies advocated by most optimists concentrate economic and political power in the hands of the few, which is inherently antidemocratic. Most importantly, the emergence of a technological age has alienated humans from nature, and is unsustainable because it is consuming natural resources and destroying ecologies at voracious rates that will eventually threaten human welfare. Confidence in unlimited technological development is more an unorthodox than an orthodox leap of faith. In the words of Jacques Ellul (1964), it is a non-Pascalian wager on human power, not the existence of God.

In reply, optimists contend their reliance on technology is justified. History is a relatively accurate indicator of future trends, and the path of technological development has an impressive track record. Whatever problems may exist presently or in the future can be solved through rational policies governing further technological development.

Christian Pessimism

The second type, pessimism, is the polar opposite of the first because it views technology as a grave threat, a consequence of living in a fallen creation. Pessimists perceive the emerging technological society as a place of unrelenting uniformity and conformity that undermines individual freedom. They decry a narrow understanding of efficiency leading to numbing specialization and social fragmentation. Moreover, the process of developing and maintaining various technological projects is inherently alienating; genuine communities are displaced by functional and manipulative relationships. More menacingly, technology takes on a life of its own that is not easily subjected to human control.

Ellul is the dominant figure here. His principal thesis is that society now comprises a series of interdependent ensembles of economic, political, and psychological techniques. More troubling, these ensembles are merging into a singular, comprehensive, and autonomous technique that resists, if not defies, meaningful human participation or control. In short, modern technological development is totalitarian and dehumanizing. A number of other writers have either expanded on or formulated similar arguments. George Grant (1986), for example, contends that modern technology embodies Friedrich Nietzsche's will to power, resulting in an unrelenting desire to master nature and human nature.

This fixation on mastery creates two related moral problems: First, technology is the means of the powerful to assert their will over the weak, and second, rather than enabling human flourishing, technical efficiency becomes a standard to which human behavior must conform. As a consequence, basic notions of truth, beauty, goodness, and justice become profoundly disfigured and corrupted in a technological age. Albert Borgmann (2003), for instance, argues that the principal values underlying technological development distort normative patterns of human interaction. The fast-food industry has transformed the art of dining into a quick meal on the run. What is lost in the process is a rich set of cooperative practices involving the careful preparation of the meal, its leisurely consumption, and accompanying conversation. This loss in turn has a detrimental effect on the quality of life for individuals, families, and communities. It should be noted that although pessimists are certainly not sanguine about the future, neither are they without hope. For instance, Grant and Borgmann assert, respectively, that a recovery of Platonic principles and Christian moral convictions, and the employment of key focal practices can at least mitigate the ill effects of a technological society.

Critics charge pessimists with such a high level of abstraction that their ensuing analysis diverts rather than focuses attention on the ethical issues at stake. They grant technology a deterministic power that cannot be challenged; the outcomes of technological development will, by definition, always be evil or at least menacing. This conclusion is unwarranted because pessimists have concocted a self-fulfilling prophecy instead of demonstrating an inherent inevitability. This is reflected in their failure to make any discrimination among discrete technologies, and how their development has varied within different cultural settings. More importantly, pessimists refuse to entertain the possibility that technology can be redirected in ways that strengthen rather than corrode the values they commend. More control over the direction of technological development can be exerted than they are willing to admit.

In response, pessimists insist that their level of abstraction is no less than that employed by optimists. Consequently, the resulting analysis in behalf of progressive technological development serves to confuse rather than clarify the ethical issues in question. Moreover, the contention that technology can be easily redirected to serve human values is naive, because it fails to recognize the extent to which the purported values have been deformed by a pervasive technical rationality, thereby rendering them unsuitable as a moral rudder.

Contextualism

The third type, contextualism, occupies the middle space between the previous two. Rejecting the generalizations of both optimists and pessimists are those who claim that technology is an ambivalent instrument of power that can be used for good or evil purposes in varying socioeconomic contexts. Consequently, contextualists contend that through a combination of social, political, and economic reforms, technological development can be redirected toward more just and humane goals.

Given their heavy emphasis on reform, contextualists devote a great deal of attention to issues involving regulatory policies. Victor C. Ferkiss (1969, 1974), for instance, argues that existing political structures can redirect technological development, but that this requires two prior steps. First, technology must be directed away from generating private wealth (for example, corporate profit) and toward promoting the common good (such as the environment). Second, a rampant individualism that diminishes the common good must be tempered with more decentralized, inclusive, and participatory decision-making processes. Roger Shinn (1982) agrees with the pessimists that various technologies form an interlocking structure that tends to concentrate and centralize economic and political power, but he argues that citizens can marshal sufficient pressure to garner greater democratic control.

Barbour places himself in the contextual camp because he believes it embodies a biblical perspective that combines the ideal of social justice with a realistic assessment of self-interested power. Contextualists seek the practical application of moral convictions that direct technology toward meeting basic human needs, and this goal is best accomplished by creating more distributive economic systems, implementing widely participatory and democratic regulation, and developing appropriately scaled and sustainable technologies.

As might be expected, optimists and pessimists offer differing criticisms of this middle position. Optimists contend that the reforms envisioned by contextualists would serve only to retard economic growth. Without sufficient incentives for return on investment little innovation or technical progress will be achieved, even on the modest scale envisioned. The net effect would be to amplify the very injustice and suffering of the disadvantaged groups the contextualists purportedly wish to serve. Pessimists dismiss reform as little more than a rearguard action that may slow the pace but will not change the direction of technological development. Once enacted, reforms will be subsumed within a more encompassing framework of techniques, thereby rendering them ineffectual. There is scant evidence that the course of modern technological development has been redirected once it has achieved sufficient momentum. In reply, contextualists argue that the dire predictions of optimists and pessimists cannot be known in advance. The only way to test the validity of reform is its implementation in order to judge the failure or efficacy of actual results.

Illustrative Issue: Energy

Although this typology identifies three basic approaches for assessing technology, the question remains: What difference do these approaches make in respect to specific ethical issues and religious life? Consider two illustrative case studies. First, since the 1960s environmental issues have commanded public attention. Focusing on the related issue of energy allows for a more clear focus on the arguments originating in the categories outlined above. In each instance a dominant theological doctrine or theme underlying these arguments is also identified.

Optimists assert that the so-called energy crisis is greatly exaggerated. There is admittedly a finite limit to fossil fuels, but new and more plentiful sources, such as hydrogen and nuclear power, can be developed. The adverse impact on the environment caused by steadily increasing energy consumption has also been overstated. Automobile and power plant emissions have already been reduced through the use of more efficient technologies, and the development of new fuels promises even cleaner sources of energy. Individuals do not need to forsake their affluent lifestyles as claimed by many environmentalists. Rather, what is needed are economic incentives and investment opportunities that promote rapid technological development to ensure plentiful and relatively cheap sources of energy.

The principal theological justification of this position is an underlying *anthropocentrism*. Human benefit is the measure for determining whether certain acts are good or evil, a belief stemming from the biblical mandate that humans have been given dominion over creation. Consequently, humans may exploit natural resources to improve the quality of their lives, and the standard used to evaluate this improvement is predominantly materialistic.

The optimists' energy manifesto merely confirms the worse fears of the pessimists. On the one hand, hope is being placed largely on unproven technologies with unknown risks. The entire enterprise could prove disastrous. On the other, even if successful the envisioned programs would centralize political and economic power even more, thereby exacerbating the gap between rich and poor, and further eroding the already fragile bonds of various communities. This is but another ploy for tightening the grip of an autonomous technological system already beyond democratic control.

The primary religious imagery informing this perspective may be described as *theopocentric*. The morality of certain acts is judged in relation to God's will or commands. Moreover, nature is not a storehouse of raw material waiting to be exploited, but part of God's creation, and should be honored as such. Consequently, natural limits should shape normative patterns of both individual lives and communal life. This may require adopting far simpler lives of restricted mobility and reduced consumption of material goods, but such is the price, as well as the joy, of being God's faithful and obedient servants.

Contextualists claim that pessimists and optimists proffer, respectively, a mistaken diagnosis and remedy. Technology per se is neither the problem nor the solution. The real issue at stake is the purposes that various technologies serve. The generation and delivery of energy should be directed primarily toward meeting needs rather than wants. This means that a combination of renewable and nonrenewable sources of energy should be developed, and the delivery mechanisms scaled down, decentralized, and subjected to participatory and democratic control. These reforms admittedly require adopting less mobile and consumptive lifestyles, but not a wholesale rejection of technology as feared by the optimists. In addition, greater democratic participation and less hectic lives may also promote the kind of human relationships and communities advocated by the pessimists.

The principal theological theme informing the contextualist approach is *stewardship*. Humans do not own the earth and may not do with it what they wish. They are instead entrusted by God to oversee its care. Because humans are accountable to God, there are certain normative convictions inherent to the role they have been called to perform. Consequently, there are limits to the extent to which natural resources should be exploited, but this does not mean that technology should be rejected because its appropriate use can assist humans to be good and faithful stewards.

Illustrative Issue: Biotechnology

Although Barbour's typology helps to identify differing ethical assessments of and theological perspectives on technology, the analysis is confined principally to mediating a perceived dualistic relationship between nature and human culture. But are the three approaches still illuminating when technology is used to bridge or even eliminate the nature–culture distinction? This question is prompted by anticipated developments in biotechnology, artificial intelligence, robotics, and nanotechnology. The most promising advances presumably involve the complementary approaches of designing sophisticated machines that emulate biological processes, while at the same time engineering biological organisms. Such an approach blurs the line separating the natural from the artificial. In practical terms, this implies a gradual merging of humans with their technology. Presumably this will occur initially through the introduction of more effective prosthetics (for example, optical implants to relieve blindness), but these therapeutic interventions could be used to enhance normal functions (such as telescopic or night vision). Some writers, such as Rodney A. Brooks (2002), Hans Moravec (1988, 1999), and Ray Kurzweil (1999), predict that this merging will prove so beneficial and complete that someday humans will be more like software than hardware. Minds will be uploaded into computers and then downloaded into organically engineered, robotic or virtual substrata. Yet how would Christians assess the prospect of an emerging technoculture populated by a new species of "technosapiens"?

Technological optimists and pessimists have an apparently easy time answering this question. Optimists presumably support these envisioned advances. Alleviating suffering and extending longevity, to say nothing of the virtual immortality predicted by bold visionaries, would certainly benefit humankind. Against the assertion that developing technosapiens negates the anthropocentric base of the optimists' moral stance, it can be maintained that the possibility that humans might evolve into a superior species is not ruled out in principle. Natural selection, which is slow paced and indifferent to human well-being, is being replaced by a more efficient and purposeful form of selection that favors human flourishing. Moreover, the quintessential characteristic of the human mind will be preserved and amplified in technosapiens. This emphasis upon a technologically enhanced human could in turn enable the emergence of the kind of global and spiritual consciousness envisioned by Teilhard de Chardin.

Pessimists are appalled by the prospect of a technoculture because it is little more than thin veneer disguising a death wish for the human species. On the one hand, no one can foresee the potentially lethal consequences of the proposed technological developments. Pessimists echo the concerns of Bill Joy (2000) and others, who contend that these new technologies could very easily run amok, leading to the extinction of *Homo sapiens*. On the other hand, if the project proves successful, the emergence of posthumans nonetheless signals the end of human life. Individuals are formed within a series of relationships that are experienced in and mediated through organic bodies. To ignore this embodied quality is also to reject what it means to be human. Asserting their underlying theopocentric stance, the pessimists contend that humankind is a unique creature bearing the image of God. Bearing that image faithfully requires that the vulnerable and mortal nature of embodied existence be accepted and honored as a gift instead of despised as a burden to be escaped. Any presumption that humans can improve or perfect themselves is an idolatry predicated upon and ending in death.

It is difficult to determine how contextualists might assess the emergence of a technoculture. First, contextualists tend to use conceptual frameworks that may not be applicable in an emerging technoculture. How, for instance, are concepts of scale, sustainability, participation, and identifying risks and benefits applicable to the interests of posthumans? The reformist agenda promotes a responsive rather than proactive ethic, one more suited to redirecting rather than charting a new course of technological development.

Second, the dualism presupposed in the underlying theological rationale of stewardship is severely eroded if not rendered unintelligible. The role of the steward is to somehow protect nature or creation from what are judged to be unwarranted intrusions by human culture. Yet the force driving the technology in question itself collapses the boundaries separating these categories. Recovering a role for the steward in the context of an emerging technoculture would require making normative claims about nature or humankind. Such a maneuver, however, would also presumably entail moving closer to either the optimist or pessimist camp, thereby forsaking the middle ground.

Assessment

To ponder the prospect of an emerging technoculture populated by technosapiens is admittedly highly speculative. If history is a reliable guide, many, if not most, prognostications about this future will prove mistaken. Moreover, the immodest predictions about digitized beings enjoying their immortality within the friendly confines of virtual reality can be easily dismissed as science fiction posing as science. Such a casual dismissal, however, should be resisted. Again, if the past is any guide, the wildest dreams of many scientists and inventors that never came true, nonetheless sparked the imagination of previous generations to form a culture, for good or ill, intricately dependent upon an evolving technology. Even if none of the predictions about a technoculture and technosapiens prove true, the speculation itself reveals how humans are coming to perceive themselves and their future. This imaginative enterprise in turn poses a crucial question: In light of humankind's technological potential, what does it mean to be human? And more importantly, should the question be answered in terms of an essential feature (mind or body), or function (stewardship), or some combination? Answering these questions requires both critical *and* constructive engagement, and given the unprecedented transformative power these new technologies embody this will also require creating new categories which go beyond either optimism or pessimism. The Christian theological tradition can offer both critical constructive resources for answering these questions, and hopefully its contribution will help forge an ethic to guide the future course of technological development.

BRENT WATERS

SEE ALSO *Ellul, Jacques; Kierkegaard, Søren; Nietzsche, Friedrich.*

BIBLIOGRAPHY

Barbour, Ian G. (1993). *Ethics in an Age of Technology*. San Francisco: HarperSanFrancisco. An overview of historical and contemporary ethical assessments of technologies from various religious perspectives.

Borgmann, Albert. (1984). *Technology and the Character of Contemporary Life: A Philosophical Inquiry*. Chicago: University of Chicago Press. A detailed and in-depth analysis of how technology shapes modern social structures and institutions.

Borgmann, Albert. (2003). *Power Failure: Christianity in the Culture of Technology*. Grand Rapids, MI: Brazos Press. A critical and constructive assessment of the emerging technoculture from an overtly Christian philosophical perspective.

Brooks, Rodney A. (2002). *Flesh and Machines: How Robots Will Change Us*. New York: Pantheon. A leader in the field of robotics argues that human nature contains key features that are essentially mechanistic, especially in respect to artificial intelligence.

Cole-Turner, Ronald. (1993). *The New Genesis: Theology and the Genetic Revolution*. Louisville, KY: Westminster/John Knox Press. Argues that Christian theology needs to be reformulated in light of recent advances in genetics and biotechnology.

Cox, Harvey. (1965). *The Secular City: Secularization and Urbanization in Theological Perspective*. New York: Macmillan.

Deane-Drummond, Celia. (1997). *Theology and Biotechnology: Implications for a New Science*. London: Geoffrey Chapman. An introductory overview of issues theological issues raised by recent developments in biotechnology.

Deane-Drummond, Celia, and Bronislaw Szerszynski, eds. (2003). *Re-Ordering Nature: Theology, Society, and the New Genetics*. London: T and T Clark. A collection of essays by theologians and scientists examining a range of ethical issues related to genetic research.

Ellul, Jacques. (1964). *The Technological Society*, trans. John Wilkinson. New York: Knopf. A wide ranging critique of the social and political influence of modern technology.

Ferkiss, Victor C. (1969). *Technological Man: The Myth and the Reality*. New York: Braziller. Argues that modern political philosophy must be formulated in light of technology as the most influential formative social and economic force.

Ferkiss, Victor C. (1974). *The Future of Technological Civilization*. New York: Braziller. A philosophical argument in behalf of "ecological humanism."

Fern, Richard L. (2002). *Nature, God, and Humanity: Envisioning an Ethics of Nature*. Cambridge, UK: Cambridge University Press. Argues that a Christian solution that places a higher premium on the inherent value of nature needs to be developed in response to the environmental crisis.

Grant, George. (1986). *Technology and Justice*. Notre Dame, IN: University of Notre Dame Press. A highly critical assessment of the destructive nature of modern technological development.

Gustafson, James M. (1996). *Intersections: Science, Theology, and Ethics*. Cleveland, OH: Pilgrim Press. A series of theological reflections on the cultural and religious meaning of modern technological development.

Hefner, Philip. (1993). *The Human Factor: Evolution, Culture, and Religion*. Minneapolis, MN: Fortress Press. A theological apology for the human as God's created co-creator.

Hefner, Philip. (2003). *Technology and Human Becoming*. Minneapolis, MN: Fortress Press. A religious reflection on the prospect and potential benefit of humans becoming cyborgs.

Heidegger, Martin. (1977). *The Question Concerning Technology, and Other Essays*, trans. William Lovitt. New York: Harper and Row. Argues that technology is enfolding humanity within its own rationality and destiny.

Herzfeld, Noreen L. (2002). *In Our Image: Artificial Intelligence and the Human Spirit*. Minneapolis, MN: Fortress Press. Written by a computer scientist; the central contention is that various approaches to research in artificial intelligence reveals corresponding assumptions regarding the central aspects of human nature.

Houston, Graham. (1998). *Virtual Morality: Christian Ethics in the Computer Age*. Leicester, UK: Apollos. Proposes that Christian theology can provide basic moral principles for governing behavior within the growing virtual domain of cyberspace.

Kurzweil, Ray. (1999). *The Age of Spiritual Machines: When Computers Exceed Human Intelligence*. New York: Viking. Argues that in the twenty-first century computer intelligence will become superior to that of humans, and that humans in turn will be forced to merge with computers in order to survive.

Mitcham, Carl, and Jim Grote, eds. (1984). *Theology and Technology: Essays in Christian Analysis and Exegesis*. Lanham, MD: University Press of America. A collection of essays assessing the theological and religious significance of modern technology.

Moravec, Hans. (1988). *Mind Children: The Future of Robot and Human Intelligence*. Cambridge, MA: Harvard University Press. A highly speculative essay on future developments in which artificial life will emerge as earth's most highly evolved species.

Moravec, Hans. (1999). *Robot: Mere Machine to Transcendent Mind*. New York: Oxford University Press. A description of anticipated developments in robotics and artificial intelligence that will pave the way for humankind's "mind children" described in his previous book.

Pullinger, David. (2001). *Information Technology and Cyberspace: Extra-Connected Living*. London: Darton Longman and Todd. Examines such issues as privacy and community in light of the Christian teaching on love of neighbor in a world being shaped by communication technology.

Schultze, Quentin J. (2002). *Habits of the High-Tech Heart: Living Virtuously in the Information Age*. Grand Rapids, MI: Baker Books. Discusses, from a Christian perspective, how people can be virtuous in a high-tech world.

Schuurman, Egbert. (1995). *Perspectives on Technology and Culture*, trans. John H. Kok. Sioux Center, IA: Dordt College Press. A critical analysis of technology's historic and scientific background in contemporary culture.

Shinn, Roger. (1982). *Forced Options: Social Decisions for the 21st Century*. San Francisco: Harper & Row. A leading Christian social ethicist examines a wide range of issues related to science, technology and society.

Teilhard de Chardin, Pierre. (2004 [1964]). *The Future of Man*, trans. Norman Denny. London: Collins; New York: Image Books/Doubleday. A collection of highly speculative essays on the future course of human evolution.

White, Lynn, Jr. (1967). "The Historical Roots of our Ecological Crisis." *Science* 155(3767): 1203–1207.

INTERNET RESOURCE

Joy, Bill. (2000). "Why The Future Doesn't Need Us." *Wired* 8(4): April. Available from http://www.wired.com/wired/archive/8.04/joy.html. Argues that research in biotechnology, artificial intelligence, robotics and nanotechnology should be regulated, and in some instances prohibited, because they pose a threat to human survival.

HISTORICAL TRADITIONS

The relation between science, technology, and Christianity has been subjected to varying interpretations. A popular impression inherited from the Enlightenment of the eighteenth century is that the relationship is one of perpetual conflict. Science is opposed to religious belief and, by focusing on material phenomena, diverts attention from spiritual concerns. In reaction, some scholars contend that Christian theology provided the intellectual foundations for modern science and technology. Because nature was not sacred it was open to investigation and manipulation, activities that improved the human condition and were therefore compatible with Christian convictions. In distinction from both these analyses, other theologians contend that the relationship is characterized by neither hostility nor affinity. Science and religion represent two different forms of inquiry and discourse, and technology consists of neutral instruments that can be used for either good or evil purposes. These varying interpretations are reflected in the historical development of this relationship, which in turn informs contemporary assessments.

Premodern Christian Attitudes toward Nature, Science, and Technology

Early Christian interpretation of Scripture reflects ambiguous appraisals of activities associated with science and technology. Work, for example, is both extolled as a sacred vocation and portrayed as punishment for Adam's original sin. The grandeur and beauty of creation that humans cannot fully understand and master is juxtaposed with a dominion that they are called by God to exert over a world that is often inimical to their welfare, a mandate that can be accomplished only with the aid of tools and artifacts. There is, in short, no obvious endorsement or condemnation of what is now called science (*episteme* and *logos*) and technology (*techne*) in the Bible.

As recorded in the Gospels, Jesus of Nazareth often alluded to nature in his parables and, except for the last week in Jerusalem before his crucifixion, confined his ministry largely to the countryside. Care should be taken, however, not to read too much into such general observations. It is not clear if these allusions imply a positive view of nature or if Jesus employed familiar scenes for his predominantly rural audience; nor should his death be construed as a blanket condemnation of urban life. Jesus, after all, was also a carpenter (*tekton*).

In contrast, Paul makes few references to nature, and spent his ministry almost entirely in cities along the Mediterranean. Moreover, he thanked God for the mobility and safety afforded by Roman roads and ships that assisted his missionary work. Caution dictates, however, against concluding that Paul valued human artifacts more highly than nature. Although he appreciates the ability to transform natural resources into useful tools, Paul also offers the enigmatic vision of creation groaning in futility awaiting its salvation, implying that nature has an intrinsic value and will be included in God's final redemptive act (*Romans* 8:18–25).

This ambiguity extends into the patristic period (the first few centuries C.E.). Although Tertullian (c. 155 or 160–after 220 C.E.), for instance, admits that natural philosophy may disclose some of the workings of creation, he insists that the knowledge revealed in Scripture is of far greater importance. The former deals only with temporal matters, whereas the latter is focused on eternity. The science of Athens has nothing significant to add to the faith of Jerusalem. Gregory of Nazianzus (c. 330–c. 389) was more open to what Athens offered, using the science of his day to expound the creation stories found in Genesis. But he too concludes that mystical experience is superior to natural knowledge.

Augustine of Hippo (354–430) likewise insisted that revelation was superior to unaided reason, but he exhibited, to a greater extent than previous theologians, an appreciation of natural philosophy. He rebuked fellow Christians who were uninformed about the natural workings of the world that were well known to educated unbelievers, complaining that their ignorance brought the faith into disrepute. Moreover, Augustine argued that because the world is God's good creation, the material aspects of life should not be despised. In contrast to Greek philosophy, the physical world is not a place of vulgar necessity in which the craft of artisans is inherently inferior to contemplative pursuits. Augustine praised human intellect and ingenuity, singling out achievements in such areas as agriculture, architecture, navigation, communication, medicine, military weaponry, and the arts (*City of God*, Book XXII, Chapter 24).

Interest in science and technology waxed and waned among subsequent generations of theologians. It was with the recovery of Aristotle (384–322 B.C.E.) by scholastic theologians that attention gathered momentum. This is particularly the case in the synthesis of Augustinian and Aristotelian themes by Thomas Aquinas (1225–1274). Thomas argued that reason and revelation do not contradict each other, and grace perfects rather than negates nature. Knowledge about the world complements and amplifies religious belief.

The recovery of Aristotle also transformed the medieval university. Alongside the faculties of theology, law, and medicine, the arts and sciences grew in prestige and intellectual rigor. Inevitable tensions arose as rediscovered texts in Greek mathematics, physics, and astronomy were refined and elaborated upon, but a great deal of latitude was given to scientific inquiry so long as it did not challenge directly the church's core theological teachings.

Modern Christian Attitudes toward Science and Technology

Tension nevertheless grew more intense as scientists gained greater confidence in their methods of investigation. Galileo Galilei (1564–1642), for example, was tried and convicted of heresy because his defense of a heliocentric universe displaced the earth from its central position. More importantly, this shift from the center to the periphery implied that humankind could no longer regard itself as the apex of creation. The case of Galileo, however, is not representative of the relation between Catholicism and science throughout the sixteenth and seventeenth centuries. Many Catholics, such as Marin Mersenne (1588–1648), René Descartes (1596–1650), and Pierre Gassendi (1592–1655), made important contributions to science during this period.

Protestants, however, tended to view science and technology in a more accommodating manner. The ordering of creation was subject to God's providential governance, which though at times inscrutable was ultimately intelligible. Scientific inquiry could disclose the workings of divine providence, and scientists were thereby encouraged to explore the created order. Many Protestants, for example, were influential members of the Royal Society. This framework led scientists such as Johannes Kepler (1571–1630), Isaac Newton (1642–1727), and Robert Boyle (1627–1691) to investigate nature with relative freedom, leading to numerous important discoveries. More significantly, many discoveries contributed to inventive developments in commerce and industry.

The Enlightenment and its aftermath placed severe strains on this Protestant framework. The problem was primarily philosophical. A number of philosophers claimed that the physical world could be described in naturalistic terms independently from theistic beliefs. Initially, many theologians invoked science as an ally in defending traditional doctrines against deist and atheist attacks. Natural theology in particular drew heavily upon science to argue that nature had been designed by a creator. The image of a watch and watchmaker was often used as a popular analogy. Yet the analogy required appeal to consistent laws of nature rather than an inscrutable divine providence to account for the rational ordering of the universe. Significantly, scientists could appeal to these same laws without attributing their legislation to the God of the Bible and Christian dogma.

Both theologians and scientists referred to nature in increasingly mechanistic terms. This in part reflected the rapid proliferation of inventions and other technological innovations associated with scientific discoveries. A growing knowledge of natural laws could be applied to improving the quality of human life by constructing more effective tools and artifacts. Progress thus displaced providence as the dominant conceptual framework for charting the course and destiny of human history. This progressive ideology introduced a tacit division of labor in which nature was a realm studied by science, whereas spiritual and moral concerns fell within ecclesiastical purview. Conflict was avoided so long as neither party crossed these jurisdictional boundaries.

In the nineteenth century this tacit division began to unravel. Charles Darwin's *The Descent of Man* (1871) implied that even human nature could be explicated in naturalistic categories. Natural selection and not the presence of a soul shaped human behavior. In short, there was no longer a unique sphere that Christianity could claim as its own. It should not be assumed, however, that the ensuing battle lines were drawn evenly or predictably. Darwin had both his scientific critics and religious defenders, and it is arguable that new forms of biblical criticism (before Darwin) and Freudian psychology (after him) presented more severe challenges to traditional Christian beliefs than evolution.

Christianity in the Industrial Revolution

Nevertheless, Darwinian evolution influenced later developments in ethics and social theory related to the rapid industrialization of the late nineteenth and early twentieth centuries. The image of nature red in tooth and claw captured both public and intellectual attention. Social Darwinists, such as Herbert Spencer (1820–1903) and William Graham Sumner (1840–1910), contended that what was true in nature was also true in society, namely, that competition over scarce resources promoted a strong and vibrant human race. Moreover, science and technology were key factors in ensuring the survival of the fittest. This was readily apparent in the economic realm, where the rapid development of new industrial, transportation, and communication technologies offered competitive advantages.

Although the Industrial Revolution generated unprecedented wealth and created new markets and employment opportunities, the ensuing economic benefits were unevenly distributed. Factory workers were usually underpaid and overworked, and endured dangerous working conditions. Rapidly growing cities suffered from overcrowded tenements, inadequate sanitation, stifling pollution, widespread poverty, and violent crime. These deplorable conditions inspired mounting social unrest. In defense of industrialization it was often argued that these conditions were regrettable but necessary in the short term, and would eventually be remedied through greater economic growth driven by technological innovation. Workers must be patient, for any attempt to redistribute wealth along socialistic lines would serve only to derail the necessary competition that would eventually provide greater material comfort to a wider range of people, especially those devoted to thrift and hard work.

Religious responses to industrialization and its accompanying ethical issues were far from uniform. Proponents of the gospel of wealth maintained that economic competition was not incompatible with biblical and Christian teaching. Indeed, the accumulation of wealth promoted a philanthropic spirit as demonstrated in the largesse of such industrialists as Andrew Carnegie (1835–1919) and John D. Rockefeller (1839–1937). Critics countered that the plight of workers was patently unjust and dehumanizing. Laborers were little more than commodities exploited by owners driven by monopolistic greed instead of genuine competition. In response, the Social Gospel movement, drawing especially on the works of Walter Rauschenbusch (1861–1918), advocated workers' rights, the formation of labor unions, large public expenditures to improve urban life, antitrust legislation, and at times more radical proposals for public ownership of various industries.

What was at stake in these disputes was purportedly the progressive trajectory and destiny of history. Although various protagonists tried to wrap themselves in the mantel of progress, the perception of science and technology as the twin engines driving the steady improvement of human life made a powerful public impression. This impression was reinforced by the publication of John William Draper's *The History of the Conflict between Religion and Science* (1874) and Andrew Dickson White's *A History of the Warfare of Science with Theology in Christendom* (1896), both of which portrayed a perpetual battle between science and religion. The popularity of these books helped create a public perception that the progressive forces of science and technology were once again struggling against their old foes of religion and superstition. Although the myth of perpetual warfare is a modern invention, it continues to influence popular perceptions. As other entries demonstrate, however, contemporary Christian assessments of science and technology are more varied and nuanced than the myth admits.

BRENT WATERS

SEE ALSO *Augustine; Natural Law; Thomas Aquinas.*

BIBLIOGRAPHY

Augustine of Hippo. (1984). *Concerning the City of God against the Pagans*, trans. Henry Bettenson. London: Penguin Books. A classic theological interpretation of human history and destiny in light of the Christian doctrines of providence and eschatology.

Basalla, George. (1988). *The Evolution of Technology*. Cambridge, UK: Cambridge University Press. A historical and philosophical overview of technological innovation and its influence on cultural development.

Brooke, John Hedley. (1991). *Science and Religion: Some Historical Perspectives*. Cambridge, UK: Cambridge University Press. A series of in-depth essays examining a broad range of historical events shaping the relationship between science and religion.

Kasson, John F. (1976). *Civilizing the Machine: Technology and Republican Values in America, 1776–1900*. New York: Penguin. A critical overview of industrialization in nineteenth century America.

Manuel, Frank E. (1983). *The Changing of the Gods*. Hanover, NH: University Press of New England. A historical overview of the rise of deism and atheism in the enlightenment.

Nisbet, Robert. (1980). *History of the Idea of Progress*. New York: Basic. A philosophical inquiry tracing the idea of progress through its stages of historical development.

Rauschenbusch, Walter. (1997 [1917]). *A Theology for the Social Gospel*. Louisville, KY: Westminster John Knox Press. A treatise offering a biblical and theological foundation for the social gospel by the movement's leading proponent.

Tennant, F. R. (1928–1930). *Philosophical Theology*. 2 vols. Cambridge, UK: Cambridge University Press. A systematic attempt to reconcile key Christian doctrines with leading scientific theories of the late nineteenth and early twentieth centuries.

Turner, James. (1985). *Without God, without Creed: The Origins of Unbelief in America*. Baltimore: Johns Hopkins University Press. An in-depth analysis of the rise of atheism in the nineteenth and early twentieth centuries.

CICERO'S CREED

Lawyer, author, statesman, and scholar Marcus Tullius Cicero (106–43 B.C.E.) is considered Rome's greatest orator. His philosophical writings are impressive. In vocabulary alone, Cicero gave Rome the words *quality*, *individual*, *moral*, *definition*, *comprehension*, and *infinity* (Everitt 2001). Also attributed to him is Cicero's Creed, called the oldest statement of engineering ethics specifically, "Salus populi suprema est lex," or "the safety of the public shall be the[ir] highest law" (Broome 1986), which is comparable in stature to medicine's "primum non nocere" ("first, do no harm," attributed to Hippocrates, but found in his *Epidemics* rather than his *Oath*). Varying versions of Cicero's Creed have been incorporated into each of the major engineering professional organizations' codes (Martin and Schinzinger 2005). As such, it has served as a common reference point for contemporary engineers navigating the moral boundaries of their work.

As with "first, do no harm," however, the practicality of applying Cicero's Creed came into question during the 1980s. Just as the new field of bioethics scrutinized how physicians made ethical decisions and asked what role (if any) the public had in this process (Veatch 1981), three contending criticisms challenged Cicero's Creed. The contractarian code denied any implied or explicit contract between engineers and the public and posited that social contracts were "abstract, arbitrary, and absent of authority." The only operative contract was one between professional engineers and their employers. The personal-judgment imperative maintained that the interests of business and government never conflict with the interests of the public. Engineers, de facto, then represent the public in their safety decisions. The third criticism defined engineering as consisting of "theories for changing the physical world before all relevant scientific facts are in." Hence, engineering could never be totally risk-free or absolutely safe (Broome 1986).

Rosa Pinkus, et al. (1997) incorporated these disparate views into a framework for gauging the ethical practice of both the individual and the organization. It consists of three principles: competence, responsibility, and Cicero's Creed II. Adding specificity to the historic code, Cicero's Creed II suggests that the "ethical engineer should be cognizant of, sensitive to, and strive to avoid the potential for harm and opt for doing good." Operationalizing this implies understanding the risk and failure characteristics of the product or process at hand. Further, "the ethical organization manages technology so as not to betray the public trust," thus introducing the concept of stewardship for public resources that embodied the intent of Cicero's original ethic. Hence, the ethical engineer must have the "competence" to assess risk and should exercise the "responsibility" to communicate it when it is known.

The longevity of Cicero's Creed is a tribute to the rhetorical power and wisdom of its originator. When Cicero coined the phrase, "the safety of the people shall be their highest law," rather than engineers, he was referring to newly appointed "praetors, judges, and consuls" who were, in turn, directed to decide civil cases in the Roman Empire. However, as noted by Harris, Pritchard, and Rabins (2004, p. 12), it was not until 1947, when the engineers' council for professional development issued the first major code proclaiming that engineers "will have due regard for the safety and health of the public." Until then, engineers were to consider the protection of their clients or employers interests as their highest professional obligation.

Hence one can conjecture that around this time some engineers began to refer to the safety of the public as "Cicero's Creed." Perhaps it was first used in a popular speech or article and caught on as a professional ethic. Mistaken context aside, when balanced within the cost and schedule of completing a project, Cicero's Creed can provide direction for weighing the competing ethical demands that are built into the profession of engineering.

ROSA L. PINKUS
LARRY J. SHUMAN
HARVEY WOLFE

SEE ALSO *Codes of Ethics; Engineering Ethics*.

BIBLIOGRAPHY

Broome, Taft H., Jr. (1986). "Ethics: The Slippery Ethics of Engineering." *Washington Post*, December 28, p. D3.

Cicero, M. T. *De Legibus* III, trans. C.W. Keyes. Cambridge, MA: Harvard University Press.

Everitt, Anthony. (2001). *Cicero: The Life and Times of Rome's Greatest Politician*. New York: Random House.

Harris, Charles E., Jr.; Michael S. Pritchard; and Michael J. Rabins. (2004). *Engineering Ethics: Concepts and Cases*. Belmont, CA: Thomson Wadsworth.

Martin, Mike W., and Roland Schinzinger. (2005). *Ethics in Engineering*, 4th ed. Boston: McGraw-Hill.

Pinkus, Rosa Lynn; Larry J. Shuman; Norman P. Hummon; and Harvey Wolfe. (1997). *Engineering Ethics: Balancing Cost, Schedule, and Risk—Lessons Learned from the Space Shuttle*. Cambridge, UK: Cambridge University Press.

Veatch, Robert A. (1981). *A Theory of Medical Ethics*. New York: Basic Books.

INTERNET RESOURCE

Augustine, Norman R. (2002). "Ethics and the Second Law of Thermodynamics." *The Bridge* 32(3). Available from http://www.nae.edu.

CITIZENSHIP

• • •

Citizenship is the status of being a legally recognized member of a nation-state or other political community, possessing rights such as voting and owing duties such as jury service. In democratic thought, citizens generally are expected to be more actively involved and influential than citizens of authoritarian political systems. By joining environmental organizations, writing letters to government officials, working as volunteers, and otherwise affecting civic life, millions of citizens have helped bring about improvements in environmental policy, AIDS-HIV treatment, civilian nuclear power, genetically modified foods, and other technological endeavors.

In the city-state of ancient Athens, members of the *demos* participated directly in public debates and governmental choices, a time-consuming responsibility and honor—but only for the minority of the adult population who were not females, slaves, or otherwise excluded. When democracy was reinvented on the scale of the nation-state in Western Europe and the United States, citizenship extended only to property-owning males. Although such legal constraints have been abolished, the affluent and well educated continue to participate at higher rates, donate more money to candidates, and speak and write more persuasively. Women are underrepresented in political life due to the legacy of being hindered in "their access to full citizenship (including their capacity to speak and write freely, to acquire education, or to run for political office)" (Kessler-Harris 2001, p. 3–4). Ethnic minorities are disadvantaged almost everywhere.

New Citizenship Problematics

Challenges for citizenship now arise from globalization and the erosion of national sovereignty. The governmental unit one should identify with—the city of Paris, the nation of France, the European Union, or humanity most generally—is no longer clear (Balibar 2004). Because technological innovation emerges primarily in the affluent nations, moreover, those who reside elsewhere—a majority of humanity—in some respects are not citizens of the technological world order. Transnational citizenship seems increasingly sensible, therefore, yet institutions for it are weak.

Citizenship also becomes less salient when technological choices occur via the economy more than via government. Business executives exercise primary discretion over job creation, quality of work life, and new technological products, and computerized transactions in a few financial centers such as London affect monetary matters worldwide (Dean 2003). The privileged position of business extends to ordinary politics, where industry executives marshal unrivaled expertise, enjoy easy access to public officials, and have ample funds for lobbying and for legal challenges to government regulations (Lindblom and Woodhouse 1993).

In contrast, most adults work in semiauthoritarian organizations and exert little influence over whether technological innovations are used to make jobs more interesting, or to displace and down-skill those affected. Workers may learn a more general lesson: Don't expect to be full citizens whose opinions are valued and influential. Industrial democracy in the former Yugoslavia, codetermination laws in Scandinavia, and other experiments in economic democracy have not been widely emulated (Dahl 1985).

To the extent that ordinary people do participate in economic-technological choices, it is via consumer purchasing or *market voting*. Thus new homes in the United States grew from 800 to 2,300 square feet from 1950 to 2000, affecting energy usage, environmental despoliation, and even the level of envy. Consumer-citizens catalyzed global proliferation of a high-consumption lifestyle including air conditioning, television, and leisure travel—thereby distributing endocrine-disrupting chemicals throughout the biosphere, causing the extinction of several thousand languages and traditional cultures, endangering myriad species, and increasing rates of psychological depression.

The Challenge of Technoscientific Expertise

Another difficulty confronting citizenship is that technical knowledge increasingly required for informed discussion. When a U.S. congressional committee considered tax credits to help professional cleaners switch away from the dangerous solvent perchloroethylene in 1999, not a single citizen or public interest group wrote, phoned, or visited: Hardly anyone understood the problem of toxic air pollution from professional cleaning. Technologists do not themselves control governments, but expertise complexifies and effectively restricts participation in governance (Laird 1993).

A subtle way this occurs is that technoscientists accelerate innovation to a pace that government regulators, interest groups, and the attentive public cannot match. Roboticists, developers of esoteric weapons, biomedical researchers, nanotechnologists, and others ride a juggernaut fundamentally altering everyday life worldwide. If representative processes do not apply to technologists—most of whom are upper-middle-class males from the European Union, Japan, and the United States—and if there is insufficient time for deliberation, what meaning does citizenship have?

For all the shortcomings of traditional democratic procedures, that realm at least has competing parties, electoral campaigns, interest groups, and other forms of public inquiry, advocacy, deliberation, and dissent. Consumer-citizens enjoy none of these advantages—for example, shoppers rarely hear informed, conflicting views about environmental and other public consequences of products they purchase. Should citizenship be extended to the technological-economic sphere? To do so might require a set of citizen rights and obligations to "reconcile democracy . . . with the right of innovators to innovate . . . (and) to reconcile technology's unlimited potentials for human benefit and ennoblement with its unlimited potentials for human injury, tyrannization, and degradation" (Frankenfeld 1992, p. 462). Citizens arguably deserve relevant information, informed consent, and a limit on endangerment; and they presumably should embrace a corresponding duty to learn enough to exercise informed judgment.

In the early twenty-first century, technoscientists often proceed without obtaining informed consent, publics are mostly quiescent, and decision-making processes are not designed for timely deliberation. Extensive political research and development would be required to develop new mechanisms for holding technoscientific-economic *representatives* accountable, while organizing intermediary institutions to assist citizens in gaining requisite knowledge and shouldering other burdens of responsible participation.

There are a few encouraging signs: Some European political parties now require that women occupy 50 percent of elected offices, international norms and governance mechanisms may be emerging, and small-scale experiments with consensus conferences and other participatory innovations are gaining credibility. Nevertheless *no innovation without representation* is a long way from becoming the twenty-first-century equivalent of American colonists' cries against taxation without representation; there are formidable obstacles to an ethically defensible citizenship for wisely governing technoscientific trajectories and for fairly distributing rights and duties in a technological civilization.

E. J. WOODHOUSE

SEE ALSO *Civil Society; Consensus Conferences; Democracy; Expertise.*

BIBLIOGRAPHY

Balibar, Etienne. (2004). *We, the People of Europe?: Reflections on Transnational Citizenship*. Princeton, NJ: Princeton University Press.

Beiner, Ronald, ed. (1994). *Theorizing Citizenship*. Albany: State University of New York Press. Widely cited text that nevertheless fails to theorize the problems that science and technology pose for citizenship.

Dahl, Robert A. (1985). *A Preface to Economic Democracy*. Berkeley: University of California Press.

Dean, Kathryn. (2003). *Capitalism and Citizenship: The Impossible Partnership*. New York: Routledge.

Frankenfeld, Phillip J. (1992). "Technological Citizenship: A Normative Framework for Risk Studies." *Science, Technology, & Human Values* 17: 459–484.

Kessler-Harris, Alice. (2001). *In Pursuit of Equity: Women, Men, and the Quest for Economic Citizenship in 20th-Century America*. New York: Oxford University Press.

Laird, Frank N. (1990). "Technocracy Revisited." *Industrial Crisis Quarterly* 4: 49–61.

Lindblom, Charles E., and Edward J. Woodhouse. (1993). *The Policy-Making Process*, 3rd edition. Englewood Cliffs, NJ: Prentice Hall.

Lister, Ruth. (2003). *Citizenship: Feminist Perspectives*, 2nd edition. New York: New York University Press. Analyzes gender and citizenship in terms going beyond women's participation.

CIVIL SOCIETY

Civil society refers to the sphere of human activity outside government, the market economy, and the family. It includes communities, churches, voluntary associations, philanthropic organizations, and social movements. Civil society potentially constitutes a venue for reasoned discussion that bridges social differences, empowers participation in public life, and encourages deliberation concerning ethical issues pertaining to science and technology.

Development and Problems

Derived from Aristotle and applied to the modern nation-state by eighteenth-century liberal reformers, the concept of civil society came to be so closely associated with bourgeois economic and political life that Karl Marx distrusted the idea. Neo-Marxists came to endorse a public arena independent of state- or party-controlled communication, however, and contemporary social scientists generally view intermediary associations as conducive to stable democracy. As civic disengagement became widespread in the 1970s and thereafter, coupled with globalization, deregulation of industry, and the rise of new social movements, the idea of *building social capital* by strengthening nongovernmental organizations (NGOs) and other social institutions that *make democ-*

racy work seemed attractive to many social thinkers and activists, especially in the former Soviet sphere and in Latin America.

Defining the boundaries of civil society proves difficult, however. Publicly funded educational institutions catalyze research and discussion, yet are part of government. Most mass media are profit-making businesses; yet civic life depends on these institutions for informed inquiry. Conversely, some not-for-profit organizations such as hospitals are hard to distinguish from private businesses. Quakers and Unitarians may think deeply about social justice, but other religious groups turn away from social problems. So where exactly *is* civil society?

Also problematic is the idea of a venue/network where people with public-regarding values interact to produce outcomes endorsed by progressive social forces—saving the Mediterranean, stopping abusive labor practices, bringing AIDS drugs to Africa. However the Heritage Foundation and the Hoover Institution helped conservative Republicans create reform agendas that progressives perceive as exacerbating social differences and disempowering non-elites. Yet those research institutions clearly belong to the system of organized social inquiry and discourse. Perhaps, then, civil society belongs to no particular ideological camp, but can be mobilized by one's allies or opponents in the service of both good and ill.

A third difficulty is that most nongovernmental organizations are not altogether *public*. The American Association for the Advancement of Science (AAAS) lobbies government for taxpayer subsidies for well-paid scientists, with much research arguably serving scientists' hobbies more than the public good. Auto and chemical workers' unions focus on higher wages for current members rather than on fairer income distribution or on innovating technologically to improve the quality of work life for all. And if admission to a not-for-profit science museum costs more than seeing a Hollywood film, in what sense is the museum a *public* institution?

Fourth, governments and corporations dominate technological decisions, relegating civil society to the periphery of innovations in robotics, nanotechnology, weaponry, computers, pharmaceuticals, electronics, transport, chemicals, and agriculture. There are too many businesses for the few NGOs to watch, and government officials usually side with business. Thus, although Consumers Union and Mothers Against Drunk Driving (MADD) make modest contributions to transportation safety, they are no match for investment tax credits to industry, trust funds for building highways, and billions spent marketing new cars.

Achievements and Limitations

Nevertheless, NGOs have been influential on aspects of environmental policy, including technological changes such as catalytic converters on cars, scrubbers on electric power plants, and support for renewable energy. The environmental movement has enrolled millions of people in opposing hazardous waste dumping, fighting installation of polluting facilities, and lobbying for tighter regulations. Health social movements have tilted medical care toward AIDS prevention and treatment. Although quite important, these are exceptions to the rule, and the rule is that civil society organizations participate in only a small fraction of technoscientific choices, rarely winning a large fraction of what they seek.

Such inherent disadvantages are magnified by elite dominance over fundamental ideas circulating within civil society. From clergy and nobles of centuries past to contemporary scientific spokespersons, government officials, and business executives, elites sometimes reinforce myths that limit critical inquiry and thoughtful deliberation concerning science, technology, and ethics. Such myths include, among many others:

- That technoscience benefits all more or less equally, even though poorer persons and countries obviously are less able to purchase innovations;
- That research and development should proceed quite rapidly, despite the fact that humans learn and react rather slowly to the many unintended consequences of technology;
- That inherited economic and political institutions need not be fundamentally reconsidered, despite new organizational challenges involved in governing technological civilization.

It is of course rare to find societies where the dominant myths do not serve the interests of powerful organizations, affluent people, and experts themselves (Lindblom and Woodhouse 1993).

Perhaps the clearest connection between technological innovation and civil society is that television has displaced political conversation and other leisure activities, because "more television watching means less of virtually every form of civic participation and social involvement" (Putnam 2000, p. 228). Television maximalists lack time for civic engagement; the medium encourages individuation—as epitomized by the ubiquity of television sets in children's bedrooms; and an emphasis on individual rather than collective failings discourages viewers from trying to ameliorate social problems. Cell phones and email have been used in organizing public protests and even toppling a few governments, but cyberspace generally has not lived up to the hopes of early advocates as a space for public inquiry.

Capacities for public thought and action would be stronger in a commendable technological civilization, where civil society might function closer to the *ideal speech situation* envisioned by Jürgen Habermas. One of the most important changes would be to reduce the domination of public discourse by those with governmental, business, media, religious, and scientific authority; this would allow organizations and spokespersons to champion many more facets of many more issues than now occurs. Another important change, now partially under way, would be the evolution of an international civil society capable of reining in the worst practices of national governments, multinational corporations, and the global communities of technoscientists. Third, civil society participants would need to pay far more attention to ethical and policy issues pertaining to science and technology.

Overall, then, civil society advocates from Alexis de Toqueville to Michael Walzer surely are correct in recognizing that *social capital* plays an important role in building a society worth living in. Civil society plays an indispensable role in focusing, channeling, and helping to improve the quality of public thought: When anti-environmentalists win public office, for example, they cannot reverse most policies because pro-environmental discourse has become so widespread. Advocacy organizations play important roles in raising questions about the conduct of science and technology, and strengthening civil society probably is a necessary condition for a wiser, fairer technological civilization. However a balanced understanding of civil society must include recognition that it is difficult to conceptualize, is relatively weak compared with market and state, and possibly has been undermined as much as strengthened by the rise of global science and by recent technological developments.

E. J. WOODHOUSE

SEE ALSO *Citizenship; Liberalism; Nongovernmental Organizations.*

BIBLIOGRAPHY

Breyman, Steve. (2001). *Why Movements Matter: The West German Peace Movement and U.S. Arms Control Policy*. Albany: State University of New York Press. Argues that social movements are responsible for much of what works well in technological civilization.

Cohen, Jean L., and Andrew Arato. (1992). *Civil Society and Political Theory*. Cambridge, MA: MIT Press. One of the most sophisticated analyses of the relationship between civil society and democratic government.

Ehrenberg, John. (1999). *Civil Society: The Critical History of an Idea*. New York: New York University Press. Places contemporary discussions in a historical context going back to Plato's concern that ambition, greed, and other private concerns would undermine public order.

Habermas, Jürgen. (1981). *The Theory of Communicative Action*. Boston, MA: Beacon Press. A classic perspective from a later version of critical theory.

Kaldor, Mary. (2003). *Global Civil Society: An Answer to War*. Cambridge, UK: Polity Press. Proposes that civil society could become an antidote to the war-making propensities of nation states.

Kaviraj, Sudipta, and Sunil Khilnani, eds. (2001). *Civil Society: History and Possibilities*. Cambridge, UK: Cambridge University Press, 2001. Learned essays on European civil society and its possible futures, but under-emphasizing science and technology.

Keane, John. (2003). *Global Civil Society?* Cambridge, UK: Cambridge University Press. Argues that new democratic ways of living in a stronger global civil society are the key combating "turbocapitalism."

Lindblom, Charles E., and Edward J. Woodhouse. (1993). *The Policy-Making Process*, 3rd edition. Englewood Cliffs, NJ: Prentice Hall. Traces shortcomings in governmental policy to weaknesses in civil society, including "circularity" in public discourse whereby citizens tend to ask for no more than what business and governmental elites find it convenient to provide.

Nyhart, Lynn K., and Thomas H. Broman, eds. (2002). *Osiris*, Vol. 17: *Science and Civil Society*. Chicago: University of Chicago Press.

Putnam, Robert D. (1993). *Making Democracy Work: Civic Traditions in Modern Italy*. Princeton, NJ: Princeton University Press. Demonstrates the importance of civic life to democracy via careful study of the public sphere in many Italian communities.

Putnam, Robert D. (2000). *Bowling Alone: The Collapse and Revival of American Community*. New York: Simon and Schuster. One of the most thorough empirical studies of declining participation in community life.

Walzer, Michael, ed. (1997). *Toward a Global Civil Society*. New York: Berghahn Books. A leading political philosopher's interpretation of globalization and the prospects for a planet-wide civil society.

CLASS

Social inequalities are ancient, but the concept of class evolved only in the nineteenth century with the increasing division of labor accompanying industrialization. Karl Marx (1818–1883) sometimes wrote as if there were just two distinct, inherently antagonistic classes—the bourgeoisie owning the means of production, and the proletariat working for them. Class structures actually were more complex than that even in prerevolutionary Europe, and are all the more so in a global technological civilization. Class is thus a special form of inequality tied to the development of modern science and technology; it is of ethical significance because the costs and benefits of innovation tend to be distributed along class lines.

Sociology of Class

Sociologists studying class tend to categorize households by the male breadwinner's occupation. John H. Goldthorpe (1987) uses eleven categories ranging from professionals, administrators, and corporate managers at the top, to small proprietors, farmers, and personal service workers in the middle, to unskilled manual and agricultural workers near the bottom. Using this and other measures, it becomes apparent that there is substantial variation in distribution of what Max Weber called life chances: The United States has much greater income inequality than most other affluent nations, and low-income American families have worse access to health care, education, and other desired social outcomes (Lareau 2003, Hofrichter 2003). Intergenerational social mobility turns out to be poor just about everywhere, however, with the odds of a middle-class child remaining in that class as an adult about fifteen times greater than the chances of a working-class child moving into the middle class (Marshall et al. 1997).

Increasing participation of women in the workforce means that spouses may work in different job categories, which makes the above classification scheme harder to apply. Two-income households can afford different lifestyles than single-income households, moreover, so categorizing by occupation has become less meaningful. Parents' education may matter more than their occupations in determining a family's Internet usage, leisure activities, nutrition and health, and aspirations for children's futures. More fundamentally, conventional depictions of social class capture rather poorly the creative destruction that technological innovation brings, creating new types of careers while undermining older occupations. The winners celebrate, but several hundred million worldwide have been displaced from farms, factories, and other workplaces in the past generation at considerable personal and social cost.

Likewise being reconstructed over time are the everyday lives of various social strata. At the beginning of the twenty-first century, the affluent enjoy transportation, communication, medical care, food, and leisure opportunities superior to what has previously been available to anyone. Even persons of comparatively modest means have access to television, refrigeration, T-shirts, plastic bags, and other manufactured artifacts. Their shared participation in a *consumer class* may be a more salient social fact than their occupational or even income differences. Because people's realities are substantially structured in relationship with material things, class warfare arguably has become less a conflict among classes than one between the consumer class and the planet.

International Dimensions

Older understandings of class are challenged as well by international stratification. Most of the affluent live in the northern hemisphere, and a *working-class* household in Europe or Japan is well above average for the world as a whole—and may include a comfortable dwelling, reliable electricity, convenient mass transit or automobile, and government-funded medical care. Peasant farmers and stably employed urban dwellers in poor countries have far less access to technological benefits, and yet they are well above the billion or more persons living in absolute poverty.

Possibly on the lowest rung of the ladder are those who speak one of the 3,000 languages likely to become extinct in the twenty-first century. For example, in 2003 the Danish Supreme Court turned down the final appeal of 150,000 indigenous peoples forcibly expelled from their ancestral lands in northern Greenland during the Cold War to make way for a U.S. missile base. "The Inuit will, in all likelihood, join other indigenous peoples globally whose language, culture, and presence are no longer with us" (Lynge 2002, p. 103).

Thus conventional depictions of social class are too *nice*, and fail to convey the raw power and powerlessness that often accompany technologies deployed in contexts of socioeconomic inequality. Large dams that flooded villages while failing to deliver the promised irrigation benefits displaced millions. Millions more have been dislocated, maimed, or killed in civil wars fought with helicopter gun ships and automatic weapons. Subsistence farming was undermined by the imposition of export-oriented monocultures and European and North American *scientific* agricultural methods. International financial markets enabled by computerized data processing have caused ruinous fluctuations in local currencies. Toxic wastes and environmentally hazardous manufacturing processes have been transferred to poor countries (Clapp 2001).

Even within affluent societies, technological bads tend to follow class lines. As environmental justice advocates point out, those with less capacity to buy their way out or to organize politically often get stuck living near noisy factories, polluted waterways, traffic noise and exhaust fumes, hazardous waste dumps, landfills, and other noxious facilities (Bullard 2000). Those with less power in the labor market often find themselves disadvantaged by technological changes in the workplace (Wyatt et al. 2000)

The Future of Class

In sum, an adequate understanding of social class requires dealing with the ugly realities of power, gross international inequalities, post-industrial socioeconomic issues going well beyond occupational stratification, consumers as a new kind of class, and upheavals in work roles and lifestyles associated with technological innovation. The technoscientists' predicament is that their findings and innovations enter a highly stratified world; although few technologists might be comfortable acknowledging it, in effect they work for some social classes much more than for others. Class consciousness has long been weak in the United States, and has diminished even in European social democracies; many social observers speak as if inequality were unimportant. Yet the pervasive, harmful effects of inequalities are well documented, and one need not return to simplistic notions of a ruling class in order to think that ethically charged questions about who gets what deserve the same careful attention accorded to technical aspects of innovation.

E. J. WOODHOUSE

SEE ALSO *Affluence; Citizenship; Marxism; Money; Race; Work.*

BIBLIOGRAPHY

Bullard, Robert D. (2000). *Dumping in Dixie: Race, Class, and Environmental Quality*, 3rd edition. Boulder, CO: Westview Press.

Clapp, Jennifer. (2001). *Toxic Exports: The Transfer of Hazardous Wastes from Rich to Poor Countries*. Ithaca, NY: Cornell University Press.

Gibson-Graham, J. K.; Stephen A. Resnick; and Richard D. Wolff, eds. (2000). *Class and Its Others*. Minneapolis: University of Minnesota Press. A feminist, cultural studies approach to the subject of class.

Goldthorpe, John H. (1987). *Social Mobility and Class Structure in Modern Britain*, 2nd edition. Oxford: Clarendon Press.

Hofrichter, Richard, ed. (2003). *Health and Social Justice: Politics, Ideology, and Inequity in the Distribution of Disease*. San Francisco: Jossey-Bass.

Lareau, Annette. (2003). *Unequal Childhoods: Class, Race, and Family Life*. Berkeley: University of California Press.

Lynge, Aqqaluk. (2002). *The Right to Return: 50 Years of Struggle by Relocated Inughuit in Greenland*. Nuuk, Greenland: Forgalet Atuagkat.

Marshall, Gordon; Adam Swift; and Stephen Roberts. (1997). *Against the Odds?: Social Class and Social Justice in Industrial Societies*. New York: Oxford University Press.

Wyatt, Sally; Flis Henwood; Nod Miller; and Peter Senker, eds. (2000). *Technology and In/equality: Questioning the Information Society*. New York: Routledge.

CLINICAL TRIALS

• • •

Clinical trials are systematic investigations on human subjects testing the safety and efficacy of novel medical interventions, including drug, surgical, or behavioral treatments. Conventionally clinical trials are divided into four types or *phases*. In a phase I clinical trial, typically involving tens of subjects, a novel procedure is tested for the first time in human beings and data is collected on safety. In a phase II trial, which may involve hundreds of patients, evidence is sought that a novel intervention has a therapeutic effect on the disease of interest. In a phase III clinical trial, often involving thousands of patients, the novel intervention is compared to a standard intervention or placebo. In a phase IV trial, called a post-marketing study, information is collected on the long-term safety and efficacy of the intervention from patients receiving the intervention in clinical practice and measured against a control treatment. The rigorous evaluation of novel medical interventions in clinical trials is a foundation of evidence-based medicine.

Historical Development

The randomized clinical trial is one of the most important advances in medicine in the twentieth century. Prior to its development, treatments were adopted on the basis of the publication of a series of cases in which their use had proved helpful. Due to numerous sources of potential bias, including variation in expertise from one clinician to the next, and the selection of patients more likely to recover for inclusion in the study, case series often led to misleading results. Clinicians were faced with numerous treatments from which to choose, and little evidentiary basis upon which to ground a choice. For example, Richard Doll, a well-known British clinical trialist, described how at the start of his research into the treatment of peptic ulcer in 1948, he was able to list purported treatments beginning with each letter of the alphabet.

In clinical trials until mid-twentieth century, two treatments for comparison were allocated to alternating patients. This method was flawed by the fact that physicians could anticipate the treatment assignment, and thereby select which treatment a particular patient would receive by changing the patient's position in the queue. It was not until mid-century that R.A. Fisher's allocation strategy using random numbers, developed in 1926 for agricultural experiments, was used in clinical trials. Allocation using random numbers countered bias in selection and provided statisticians with an estimate of random error, a key component of modern statistical analysis. The first clinical trial to randomly allocate treatments to patients using random numbers was the United Kingdom Medical Research Council (MRC) whooping cough immunization trial initiated in 1946. The better-known MRC streptomycin trial in tuberculosis started a few months later, but published its results before the whooping cough trial in 1948.

Since the mid-twentieth century, the clinical trial has undergone a dramatic increase in use for the evaluation of the safety and efficacy of novel medical interventions. A variety of social and political factors supported this trend. The period following the Second World War witnessed an unprecedented public investment in health research. In 1945, the United States National Institutes of Health budgetary appropriation was $700,000; in 1970 its appropriation was $1.5 billion. Drugs regulation underwent significant changes in this period as well. From 1938 to 1962, the U.S. Food and Drug Administration (FDA) was only empowered to require that new drugs be tested for safety. Following on the heels of the thalidomide tragedy, in which hundreds of infants were born with congenital malformations after exposure to thalidomide in utero, legislative reform dramatically increased the FDA's power. The 1962 Kefauver-Harris Act expanded the FDA's mandate to test new drugs for both safety and efficacy.

The testing of new drugs for safety and efficacy in clinical trials occurs in an increasingly international environment. Cooperation among drugs regulators and manufacturers seeks to standardize the conduct of clinical trials and their review by drugs regulators. The International Conference on Harmonization *Good Clinical Practice* guidelines are a key instantiation of this effort. The protection of human subjects in research is similarly seen as a matter of global concern. Perhaps the most influential ethics document in the international forum is the World Medical Association's *Declaration of Helsinki*. The declaration requires that clinical trials be reviewed by appropriately constituted research ethics committees; research be free of misconduct; the consent of human subjects be obtained; study participation pose a favorable balance of benefits to harms; and subjects be selected equitably.

Ethical Issues

Some of the most important ethical challenges of clinical trials stem from conflicting duties of the physician-researcher. Physicians have fiduciary obligations to patients, including a duty to provide competent personal care. Researchers, by contrast, have obligations to science and society, including duties to provide treatment as prescribed in the trial protocol, ensure that patients comply with treatment, and encourage them to stay in the study. Given that these duties may conflict, the central moral question of the clinical trial is: When may physicians legitimately offer patients enrollment in a clinical trial? While a variety of answers have been provided to this question, the most widely accepted is that of clinical equipoise. According to clinical equipoise, physicians may legitimately offer patients enrollment in a clinical trial only if the medical interventions within the study are consistent with competent medical care. More formally, it requires that at the start of the study there exists a state of honest, professional disagreement as to the preferred treatment. The consequences of clinical equipoise for the design of clinical trials are far reaching.

Two issues in respect to the design of clinical trials have dominated research ethics literature since the 1990s. The first is the proper role of placebo controls in the new drug approval process in developed countries. Drug regulatory agencies in developed countries, such as the FDA, have long required that new drugs prove superior to placebos in at least two clinical trials before licensure. The practice in the United States is rooted in legislation that requires the FDA to ensure new drugs are efficacious, that is, that they have some effect in treating the condition of interest, but generally restricts its ability to demand evidence of comparative effectiveness. According to clinical equipoise, placebo-controlled clinical trials are unproblematic when there is no proven treatment for the condition of interest.

Criticism has focused on the use of placebo controls in clinical trials testing novel interventions for treatable medical conditions, such as severe depression and schizophrenia. The use of placebos in these cases is impermissible, because no competent physician would fail to offer a patient treatment and, accordingly, clinical equipoise is violated.

The 2002 revision of the Declaration of Helsinki sets aside this fundamental moral requirement, and for the first time permits the use of a placebo control when "compelling and scientifically sound methodological reasons" exist. This change seems to violate a core provision of the declaration requiring that "[i]n medical research on human subjects, considerations related to the well-being of the human subject should take precedence over the interests of science and society." Whether there are in fact *scientifically sound methodological reasons* to prefer a placebo control over a standard treatment control remains an open question.

The second clinical trial design issue to receive considerable attention in the literature is the choice of control treatment in clinical trials of new and affordable treatments for developing countries. Disagreement was originally sparked by clinical trials testing the efficacy of short-course zidovudine against placebos for the prevention of transmission of HIV from mother to child. Critics of the clinical trials pointed to the existence of an effective prevention regimen called ACTG 076 used in developed countries. Denying subjects in the clinical trials conducted in developing countries access to this prevention regimen, they claimed, constitutes an ethical double standard between developed and developing countries.

Proponents of the clinical trials countered that the ACTG 076 regimen is not suited to administration in many developing countries and the cost is prohibitive. Changes in international regulation have tended to entrench rather than resolve the dispute. The Declaration of Helsinki proscribes placebo controlled trials in developing countries when effective treatment exists in developed countries saying that "[t]he benefits, risks, burdens and effectiveness of a new method should be tested against those of the best current prophylactic, diagnostic, and therapeutic methods." Yet the "International Ethical Guidelines for Biomedical Research Involving Human Subjects" permit placebo controlled trials under these circumstances provided the clinical trial is "responsive to the health needs of the population from which the research subjects are recruited and there [is] assurance that, if it proves to be safe and effective, it will be made reasonably available to that population."

OPEN QUESTIONS. The interface between the ethics and science of clinical trials is replete with challenging questions yet to be addressed adequately. What ought the role be for adaptive designs, for instance, clinical trials in which the probability of being assigned to one treatment or another is dynamic in an attempt to minimize the number of subjects who receive the treatment that turns out to be inferior? Can alternative medical treatments be evaluated rigorously in clinical trials? Alternative practitioners may claim that alternative treatments cannot be removed from a holistic treatment context, a substantial obstacle to the rigorous assessment of the treatment's efficacy. How will pharmacogenetic testing impact the conduct of clinical trials? Proponents of pharmacogenetics suggest that identification by genetic testing of those likely to respond to treatments and those likely to suffer adverse events would increase the efficiency and safety of clinical trials. Critics wonder if the gains from such testing will be as large as promised and what impact it will have on the generalizability of clinical trial results.

While ethical issues in the design of clinical trials are the subject of ongoing scholarship, ethical aspects of the conduct and reporting of clinical trials are relatively ignored. As clinical trials accumulate data on outcomes, disparities may emerge between the treatments in the clinical trial raising questions as to whether the trial ought to be stopped early. It is generally agreed that when clinical trials use outcome measures of mortality or serious morbidity an independent data monitoring committee should be established to periodically review accumulating data. A satisfactory moral framework to guide the decisions of data monitoring committees has yet to be developed.

Ethical issues in the reporting of clinical trial results also deserve attention. If researchers fail to report the results of a negative clinical trial, subjects in the trial were exposed to risk for naught and the problem of publication bias is compounded. While this seems problematic intuitively, a moral basis for an obligation to publicize clinical trial results has yet to be articulated.

CHARLES WEIJER

SEE ALSO *Complementary and Alternative Medicine; Drugs; Human Subjects Research.*

BIBLIOGRAPHY

Angell, Marcia. (1997). "The Ethics of Clinical Research in the Third World." *New England Journal of Medicine* 337: 847–849. A stinging critique of placebo controlled clinical trials testing the efficacy of short course zidovudine in the prevention of perinatal transmission of HIV in developing countries.

Crouch, Robert, and John Arras. (1998). "AZT Trials and Tribulations." *Hastings Center Report* 28(6): 26–34. A sober response to Marcia Angell's critique.

Doll, Richard. (1998). "Controlled trials: The 1948 Watershed." *British Medical Journal* 317: 1217-1220. A historical review of changes to clinical trials at mid-century by one of the central figures in its development.

Emanuel, Ezekiel; Robert Crouch; John Arras; et al., eds. (2003). *Ethical and Regulatory Aspects of Clinical Research: Readings and Commentary*. Baltimore, MD: Johns Hopkins University Press. The definitive collection of readings in the ethics of human experimentation.

Miller, Paul, and Charles Weijer. (2003). "Rehabilitating Equipoise." *Kennedy Institute of Ethics Journal* 13: 93–118. A philosophical piece examining the conflicting duties of physician and researcher, and how these conflicts might be resolved.

Piantadosi, Steve. (1997). *Clinical Trials: A Methodologic Perspective*. New York: John Wiley & Sons. An authoritative methodological resource for the design of clinical trials.

INTERNET RESOURCES

Council for International Organizations of Medical Sciences. "International Ethical Guidelines for Biomedical Research Involving Human Subjects." Available from http://www.cioms.ch/frame_guidelines_nov_2002.htm. Guidelines for the conduct of research in resource-poor nations.

National Placebo Working Committee. "National Placebo Initiative: Draft Report of the National Placebo Working Committee." Canadian Institutes of Health Research. Available from http://www.cihr-irsc.gc.ca/e/services/pdf_19320.htm. A thorough document setting out the scientific, ethical, legal, and policy case for and against the widespread use of placebo controls in clinical trials.

World Medical Association. "Declaration of Helsinki." Available from http://www.wma.net/e/policy/b3.htm. The most influential international statement of moral standards for the conduct of clinical research.

CODES OF ETHICS

A code of ethics may appear in disciplines such as engineering, science, and technology under several other names: professional principles, rules of conduct, ethical guidelines, and so on. However denominated, a code of ethics can be placed in one of three categories: (1) *professional*, such as the Chemist's Code of Conduct of the American Chemical Society, applying to all the members of a certain profession (chemists) and only to them; (2) *organizational*, such as the Code of Ethics of the Institute of Electrical and Electronic Engineers, applying to members of the technical or scientific society that has enacted it and only to them or, in the case of the code of ethics of a university or industrial laboratory, only to a certain class of the enacting organization's employees; (3) *institutional*, such as the Computer Ethics Institute's Ten Commandments of Computer Ethics, applying to anyone involved in a certain activity (in this case, using a computer).

Codes of ethics may include ordinary moral rules ("Do not steal" or "Treat others fairly"). Codes of ethics may be enacted into law. For example, some codes (such as the engineer's code of ethics in Chile) have the status of domestic administrative law. Other codes, such as the "Nuremberg Code" on human experimentation, have become part of both international law and the general domestic law of many countries. Nonetheless, a code of ethics is never simply a matter of law or ordinary morality. To call a document a code of *ethics* is to make a claim for it one does not make when one claims that the document in question is a statute or statement of ordinary morality.

The Meaning of Codes

The word *code* comes from Latin. Originally it referred to any wooden board, then to boards covered with wax that were used to write on, and then to any book (*codex*). That was the sense it had when first applied to the book-length systemization of Roman statutes that the Emperor Justinian enacted in 529 C.E. Justinian's *Code* differed from an ordinary compilation of law in one important respect: He had the legal authority to make his compilation law, replacing all that preceded it.

Since that time, any document similar to Justinian's *Code* could be called a code. Sometimes the analogy with Justinian's *Code* is quite close (as it is, for example, for the *Code Napoleon*). Sometimes it is not. For example, computer code is code in a rather distant sense: Although the rules are presented systematically, computer code is written for machines, not for humans.

An important feature of Justinian's compilation is that it was written. Could a code be *unwritten*? Certainly there are unwritten laws. However, because the point of codification is to give law (and by analogy any similar system of guidance) an explicit and authoritative formulation, an unwritten code would seem not to be a code at all. There are nonetheless at least two ways in which codes can be unwritten. First, a code that is not in writing may have an authoritative *oral* formulation. Second, an unformulated code may be so obvious to those familiar with the practice that the code need only be formulated to be accepted. Although some parts of engineering or science may have a few rules unwritten in one or both of these senses, no large discipline or organization seems to have enough of those rules to constitute an unwritten code. In a world in which so much changes so quickly, how can individuals separated by education, experience, and distance reach agreement on much without putting that agreement in writing?

The Meaning of Ethics

The term *ethics* has at least four senses. In one, it is a synonym for ordinary morality, the universal standards of conduct that apply to moral agents simply because they are moral agents. Etymology justifies this sense. The root of the word *ethics* (*ēthos*) is the Greek word for "habit" (or "character"), just as the root of the word *morality* (*mores*) is the Latin word for that concept. Etymologically, *ethics* and *morality* are twins (as are *ethic* and *morale*). In this sense of the term, codes of ethics are systematic statements of ordinary morality; there is no point in speaking of ethics rather than morality.

In at least three other senses, however, ethics differs from morality. In one, ethics consists of the standards of conduct that moral agents *should* follow (*critical morality*); morality, in contrast, consists of the standards that moral agents generally do follow (*positive morality*). Ethics in this sense is very close to its root *mores*; it can refer to unethical acts in the first sense of *ethics*. What some believe is morally right (slavery, forced female circumcision, and the like) can be morally wrong. Morality in this sense has a plural: There can be as many moral*ities* as there are moral agents. Nonetheless, ethics in this sense can be a standard that is common to everyone. This second sense of ethics is, then, as irrelevant to the purposes here as is the first. Codes of ethics generally contain some rules ordinary morality does not.

Sometimes ethics is contrasted with morality in another way: Morality consists of the standards that every moral agent should follow. Morality is a universal minimum, the standard of moral right and wrong. Ethics, in contrast, is concerned with moral good, with whatever is beyond the moral minimum. This is another sense that seems not to fit codes of ethics. First, this ethics of the good is still universal, applying outside professions, technical societies, and institutions as well as within them. Second, codes of ethics in fact consist largely of *requirements*, the right way to conduct oneself rather than just a good way. Any sense of ethics that excludes requirements cannot be the sense relevant to codes of ethics.

The term *ethics* can be used in a fourth sense to refer to the morally permissible standards of conduct that govern the members of a group simply because they are members of that group. In this sense, research ethics is for people in research and no one else, engineering ethics is for engineers and no one else, and so on. Ethics in this sense is relative even though morality is not; like law and custom, it can vary from place to place, group to group, and time to time.

Though relative, ethics (in this sense) is not mere mores. It must (by definition) set a standard that is at least morally permissible. There can be no thieves' ethics or Nazi ethics, except with quotes around the word to signal an analogical or perverted use. Because ethics in this fourth sense must both be morally permissible and apply to members of a group simply because of their membership, it must demand more than law, market, and ordinary morality otherwise would. It must set a "higher" or "special" standard.

The Meaning of Codes of Ethics

A code of ethics, though not a mere restatement or application of ordinary morality, can be morally binding on those to whom it applies; that is, it can impose new moral obligations or requirements. How is this possible? Some codes of ethics are morally binding in part because they require an oath, a promise, or other "external sanction" (for example, one's signature on a contract that makes accepting an employer's code of ethics a condition of one's employment). In general, though, codes of ethics are binding in the way the rules of a morally permissible game are binding on those who voluntarily participate. The sanction is "internal" to the practice. When a person voluntarily claims the benefits of a code of ethics—for example, the special trust others place in those whom the code binds—by claiming to be a member of the relevant group ("I am an engineer"), that person has a moral obligation, an obligation of fairness, to do what the code says. Because law applies to its subjects whether they wish it to or not, law cannot bind in the way a code of ethics (a voluntary practice) can. Because a code of ethics applies only to voluntary participants in a special practice, not everyone, a code, if it is generally followed, can create trust beyond what ordinary moral conduct can. It can create a special moral environment. So, for example, if engineers generally "issue public statements only in an objective and truthful manner [including] all relevant and pertinent information" (as the Code of Ethics of the National Society of Professional Engineers requires), their public statements will generally (and justifiably) be trusted in a way those of politicians, lobbyists, and even ordinary private citizens would not be. Engineers will therefore have a moral obligation to do as required to preserve that trust. They will have a special moral obligation to provide all relevant and pertinent information even when others do not have such an obligation.

Attempts have been made to distinguish between short, general, or uncontroversial codes (code of ethics) and longer, more detailed, or more controversial ones (code of conduct, guidelines, and the like). Although this type of distinction may occasionally be useful in practice, it is hard to defend in theory. A typical code of conduct is as much a special standard as a typical code of ethics is, except when the code of ethics, being a mere restatement of morality, is just a moral code. Codes of conduct are also generally as morally binding as other codes of ethics. Sometimes, as in the Code of Ethics and Professional Conduct of the Association of Computing Machinery, the code does not even distinguish between the two.

Attempts have also been made to distinguish between (hard and fast) "rules" and mere "guidelines". Rules are then said to be typical of law, to allow only for submission or defiance, and therefore to interfere with moral autonomy. Guidelines, in contrast, are said to be typical of ethics, to require interpretation rather than "mindless submission", and therefore to preserve moral autonomy. In fact, all rules, including statutes, require interpretation (rather than mindless submission). In this respect, all rules are mere guidelines. There is, then, no reason why a code of ethics, understood as rules, should interfere with moral autonomy—or, at least, no reason why it should interfere any more than a promise or obligation of fairness does. On the other hand, "guidelines" such as those in ACM's Code often have the same mandatory form as other rules. They function as a commentary on the code rather than as a distinct document.

Uses and Design of Codes of Ethics

Codes of ethics have at least five uses: First and most important, a code of ethics can establish special standards of conduct in cases in which experience has shown that common sense is not adequate. Second, a code of ethics, being an authoritative formulation of the rules that govern a practice, can help those new to the practice learn how to act. Third, a code can remind those with considerable experience of what they might otherwise forget. Fourth, a code can provide a framework for settling disputes even among persons with considerable experience. Fifth, a code can help those outside the group ("the public") understand what they may justifiably expect of those in the group.

A code of ethics can also be used to justify discipline, legal liability, or other forms of external accountability, but such uses threaten to turn the code into something like law. Even when a code of ethics has been enacted into law, obedience to it must rely in large part on conscience or there is no point in describing it as a code of *ethics* (rather than just another legal requirement).Therefore, to object to a code of ethics that it cannot be enforced in the way laws generally are is to confuse ethics with law.

Some writers have claimed that a code of ethics must have a certain content (something more specific than "a higher standard"), for example, that any "true professional code" must have a provision giving special prominence to the public interest. For some professions, such as engineering, the claim is plausible. Engineers have long agreed that the public health, safety, and welfare should be "paramount" in their professional work. But for other professions, such as mathematics, the claim is much less plausible. The Ethical Guidelines of the American Mathematical Society commit mathematicians to mathematical truth, whether in the public interest or not. Many other scientific professions have a similar commitment to truth rather than the public interest as such. There can be no moral objection to such a failure to emphasize the public interest so long as the code does not require or allow anything ordinary morality forbids.

Because codes of ethics have no necessary content, they have no necessary structure or design. So, for example, the Software Engineering Code of Ethics divides its requirements into eight major categories (Public, Client and Employer, Product, Judgment, Management, Profession, Colleagues, and Self); the Codes of Ethics of the Australian Computer Society divides its requirements into six (Priorities, Competence, Honesty, Social Implications, Professional Development, and Computing Profession); and other codes have adopted other divisions, some similar to these and some quite different. About all that can usefully be said about the structure of codes of ethics generally, is that the structure should help ordinary users understand the code as a whole and to find what in particular they need.

MICHAEL DAVIS

SEE ALSO *Accountability in Research; Cicero's Creed; Engineering Ethics; Profession and Professionalism; Sociological Ethics.*

BIBLIOGRAPHY

Anderson, Ronald E. (1994). "The ACM Code of Ethics: History, Process, and Implications." In *Social Issues in Computing: Putting Computing in Its Place*, ed. Chuck Huff and Thomas Finhold. New York: McGraw-Hill. An excellent (and rare) description of the writing of a code of ethics in a major technological organization (with lots of details about the give-and-take involved).

Baker, Robert B.; Linda L. Manuel; Arthur L. Caplan; and Stephen R. Latham, eds. (1999). *The American Medical Ethics Revolution: How the AMA's Code of Ethics Has Transformed Physicians' Relationship to Patients, Professionals, and Society*. Baltimore: Johns Hopkins University Press. An excellent resource for understanding the writing of what is (arguably) the first code of ethics governing a technological profession.

Berleur, Jacques, and Marie d'Udekem-Gevers. (2001). "Codes of Ethics: Conduct for Computer Societies: The Experience of IFIP". In *Technology and Ethics: A European Quest for Responsible Engineering*, ed. Philippe Goujon and Bertrand Hériard Dubreuil. Leuven, Belgium: Peeters. Describes the failure of a major international organization to adopt a code of ethics, apparently because of worries about enforceability.

Coady, Margaret, and Sidney Block, eds. (1996). *Codes of Ethics in the Professions*. Carlton South, Victoria, Australia: Melbourne University Press. A good collection of essays providing a benchmark for current philosophical understanding of codes of ethics in professions.

Davis, Michael. (2002). *Profession, Code, and Ethics*. Aldershot, UK: Ashgate. A major challenge to the current philosophical (and sociological) understanding of professions, with engineering (along with law and police) as one of the three major professions studied.

Ladd, John. (1980). "The Quest for a Code of Professional Ethics: An Intellectual and Moral Confusion." In *AAAS Professional Ethics Project: Professional Ethics Activities in the Scientific and Engineering Societies*, ed. Rosemary Chalk, Mark S. Frankel, and Sallie B. Chafer. Washington, DC: American Association for the Advancement of Science. A classic challenge to the very idea of a code of ethics.

Luegenbiehl, Heinz. (1983). "Codes of Ethics and the Moral Education of Engineers." *Business and Professional Ethics Journal* 2: 41–61. An important work on the educational uses of codes of ethics, as well as the classic statement of the distinction between rules and guidelines.

INTERNET RESOURCE

Center for the Study of Ethics in the Professions, Illinois Institute of Technology. "Codes of Ethics Online." Available at http://www.iit.edu/departments/csep/PublicWWW/codes. An enormous collection of codes of ethics, by far the largest available online. All codes cited here can be found at this site.

COLONIALISM AND POSTCOLONIALISM

Colonialism, understood provisionally as the European annexation and administration of lands and populations in the Americas, Africa, and Asia, has been intertwined with science, technology, and ethics since the Renaissance. Certainly one prelude to colonial expansion was the European acquisition of military and navigational technologies superior to those found in other continents. But the colonial experience also had a formative impact on the nascent European science, because it permitted the region's scholars to come into contact with new environments and data and provided access to alternative systems of knowledge developed by other cultures. In fact, the requirement of controlling and cataloging colonial populations and resources led to the creation of new disciplines in the social sciences, such as ethnography, linguistics, and archaeology. Moreover, this impact has continued to the early twenty-first century, as a new scientific discipline, ecology, has found inspiration in the practices of non-western precolonial cultures and on the nineteenth century British and French "colonial conservationism" that attempted to deal with the degradation caused by the exploitation of recently acquired environments was "able to foresee, with remarkable precision, the apparently unmanageable environmental problems of today"(Grove 1995, p. 12).

Indeed, colonialism had an indirect, though profound, impact on European culture. In reaction to the frequently genocidal military tactics used by Europeans and the exploitation of indigenous populations that characterized the administration of colonies, few, if any, other historical events did more to promote the extension of ethics into the political, social, and legal spheres. In politics, such central contemporary concepts as human rights, representative democracy, and socialism developed, at least in part, as reactions to the brutality of the process of colonization and to the contact with non-European cultures and their political systems. Moreover, colonialism, by transferring enormous amounts of gold and silver from the Americas to Europe during the sixteenth and seventeenth centuries, thereby permitting the development of a money economy, may be seen as a factor that contributed to the development of capitalism, and the science that studies it, economics. The European colonization of Africa, the Americas, and Asia is thus one of the founding experiences of modernity, its impact felt on every aspect of contemporary life, even in countries that did not embark on colonial adventures.

Conceptual Issues

Despite its importance, however, any attempt to define colonialism in a manner that goes beyond the mere recounting of a set of historical facts runs into a series of conceptual problems. The difficulty in defining colonialism and related concepts—such as imperialism, anticolonialism, neocolonialism, or even postcolonialism—is that they can be interpreted as linked to social phenomena existing since antiquity throughout the world. Yet, it is customary to see colonialism as bounded, on the one hand, by a European expansion that began in the fifteenth century with the Portuguese and Spanish forays into Africa and the Americas, and, on the other, by the decolonization of Asia and Africa, a process that concluded in 1975 with the independence of the last Portuguese dominions, Mozambique and Angola. Although the United Nations reported that, as of 2003, there were still sixteen "non-self-governing territories," colonialism, as customarily defined, is no longer at the core of the world economy, and the impetus for self-governance, while not fully realized, concerns smaller populations and areas.

These temporal boundaries are justified by a central difference between classical and modern empires. In the latter, colonization was characterized not only by the conquest of a territory and its population, or by the extraction of monetary, human, or material resources, as was the case in antiquity, but also by a thorough restructuring of the colonial economy for the benefit of the economic interests of the metropolis. The securing of raw materials to be used exclusively by imperial industries or the restrictions placed on the production of goods in the colonies in order to transform them into exclusive markets for metropolitan products are examples of such restructuring.

In addition to reshaping economic structures, modern colonialism also attempted to change the cultures of the populations conquered. The successful catechization of Latin America in the sixteenth century, despite the frequently syncretic character of the resulting religion (that is, its being a combination of originally Amerindian and European beliefs), is a case in point. In fact, this cultural change was often a prerequisite for the economic exploitation of the acquired territories, because traditional labor patterns and economic structures had to be transformed according to the economic requirements of European industries and settlers. Colonialism's practical emphasis on the modification of the cultures of the conquered populations and the concomitant resistance of the latter, as well as the unavoidable hybrid identities generated by this encounter, have become key objects of study for contemporary theorists.

But the difficulties to be found in conceptually delimiting colonialism remain implicit in such a description. The most obvious problem is that processes of colonization and decolonization are not discrete and chronological. In fact, the first postcolonial societies in the Americas arose before the second wave of European imperialist expansion crested in the nineteenth century. Furthermore, as José Carlos Mariátegui (1894–1930) noted in the 1920s, colonial practices, institutions, and ideologies did not disappear with formal independence, but frequently constituted the bases on which the new nations were built. Thus it becomes possible to talk of an internal colonialism present in politically independent nations in which cultural, racial, ethnic, religious, linguistic, or caste differences form the basis for the institutionalized economic exploitation of one group by another. Then, moreover, there is the unique case of the United States: a postcolonial society that itself became a full-fledged colonial power in the second half of the nineteenth century through the annexation of Puerto Rico, the Philippines, and Hawaii, and that in the twentieth century helped establish new patterns of international domination and unequal resource flows. Given this inequality, it is possible to argue that current international economic structures and relationships among different national and regional economies constitute a continuation and development of colonialism rather than its abolition.

Imperial Differences

Critics have questioned the validity of the chronology proposed above by distinguishing Spanish and Portuguese colonialism, on one side, and the later French and British empires, on another. Unlike the more fully capitalist British or French colonial regimes, the earlier Iberian empires were frequently mercantilist and precapitalist, even medieval. While the former restructured the new colonies' economies so as to propel metropolitan capitalist growth, the latter colonial enterprises were based mainly on the acquisition or extraction of directly marketable resources, such as gold or spices, and on the taxation of native and settler populations as direct sources of income. From this perspective, colonialism as a fully modern capitalist undertaking must be differentiated from earlier Iberian empire building. In fact, critics have argued that terms such as colonialism, imperialism, or postcolonial "evince the history of British colonial/imperial involvement with Ireland, India, and South Africa" and that their use leads to the "(mis)-understanding and (mis)labeling of the so-called colonial American situation" (Klor de Alva 1995, p. 264). Thus mainstream analyses of colonialism would be applicable only to the European empires built in Asia and Africa during the eighteenth and particularly the nineteenth centuries.

A concept frequently used to separate earlier Iberian and later colonialisms is that of imperialism. In 1917 V. I. Lenin (1870–1924), arguably the most influential critic of imperialism, claimed that it constituted "the monopoly stage of capitalism." For him, colonial expansion responded to the needs of monopolistic finance capital, which he believed to be the hegemonic sector in a modern economy, to find a "guarantee against all contingencies in the struggle against competitors" by ensuring access to markets and resources (Lenin 1977, p. 260). Because Lenin saw finance capital as firmly national, imperialism necessarily led to war as the colonial powers attempted to acquire "precapitalist" areas, to forcibly take over each other's colonies, or even to try to gain access to the natural resources located in Europe. (World War I was Lenin's prime example of how the hegemony of financial monopoly capital invariably led to war.)

Critics have noted, however, that one can free Lenin's arguments from his national, political, and military framework. In this way it becomes possible to speak of a U.S. imperialism that is no longer based on the formal possession of colonies, as Harry Magdoff (1969) first argued; or of a neocolonialism in which "First World" nations use international economic, political, and cultural structures and institutions to maintain their political and economic control over nominally independent nations, as the Ghanaian independence leader Kwame Nkrumah (1909–1972) proposed in 1965. In their 2000 book, *Empire*, Michael Hardt and Antonio Negri have taken this loosening of the ties between economic relations and the national sphere to its ultimate conclusion. For them, globalization has led to the creation of a true empire of capital in which unequal flows of resources are organized by means of a "decentered and deterritorializing apparatus of rule" that no longer has a geographically defined direction (Hardt and Negri 2000, p. xii). While inequality is seen as probably growing, the concept of imperialism, based on notions of metropolises and colonies, and its dependency theory derivation of center and periphery, is, therefore, obsolete.

Paradoxically, this postmodern interpretation of empire has been proposed at precisely the moment when the United States has acquired unparalleled economic, military, and technological superiority, and has claimed the right to use military force to achieve its goals, exercising this "right" first in Afghanistan (2001) and then in Iraq (2003). Indeed, critics as well as supporters of contemporary U.S. foreign policy frequently describe it as imperial. Thus current discussions of imperialism and empire frequently attempt to elucidate the role played by the United States in international economic inequalities. For instance, Aijaz Ahmad argues "what we actually have is, finally, for the first time in history, a globalised empire of capital itself, in all its nakedness, in which the United States *imperium* plays the dominant role, financially, militarily, institutionally, ideologically" (Ahmad 2000, Internet page). Whether this new globalized capitalism is a dramatically new stage in capitalism that invalidates earlier analyses whether Marxist or not, as Hardt and Negri argue, or simply an intensification and elaboration of the basic traits of capitalism and imperialism, as analyzed by Marx and Lenin, as Ahmad and others propose, is a matter of disagreement.

The standard chronology of colonialism has also been put into question by arguments that in order to understand European colonization it is necessary to analyze its underlying discursive and ideological underpinnings. Thus in his 1978 book, *Orientalism*, arguably the foundational text of postcolonial studies, Edward Said (1935–2003) traces the construction of the "Orient" back to early modern and even Greek sources, analyzes its influence on the self-construction of the "West," and notes how this European production of knowledge affected colonialist practice in the region. From a related perspective, Nelson Manrique (1993) has emphasized the manner in which the mind-set formed by 700 years of contradictory interaction among Christians, Muslims, and Jews was transplanted by the Spanish conquistadors to very different American realities. According to these and related studies, the conventional chronology of European colonialism leads only to the distortion, even the mutilation, of history.

Given these difficulties in establishing a clearly bounded definition of colonialism and related terms, these must be seen as constituting a semantic field in which conceptual boundaries blur into one other, and in which historical frameworks, though necessary, necessarily break down.. But underlying the semantic field there exists a continuum of unequal and exploitative economic, social, and political phenomena that impacts directly on the relationships among science and technology, and has ethical consequences that have yet to be fully explored.

Colonialism as Turning Point

Iberian colonialism nevertheless signaled a turning point in world history. Not only did European power and culture begin its process of expansion and imposition throughout lands and populations unknown by the West, but also new unequal flows of resources favoring colonial powers were for the first time established on a planetary scale. British and French colonialism, even contemporary international trade relations, are subsequent, capitalist developments within this unequal planetary framework. Furthermore, the pivotal role played by the Iberian empires is evidenced by the way they developed two of the central institutions characteristic of eighteenth- and nineteenth-century colonialism and beyond, slavery and the plantation system, as well as the ultimate ideological basis on which colonialism would be built: racism. As the Spanish philosopher Juan Ginés de Sepúlveda (1490?–1572 or 1573) argued, the colonization of the Americas and the exploitation of the Amerindians was justified by the fact that these were "as inferior to Spaniards as children are to adults and women to men ... and there being between them [Amerindians and Spaniards] as much difference as there is between ... monkeys and men" (Sepúlveda 1951 [1547], p. 33). Although miscegenation (the mixing of races) was more frequent in Iberian colonies than in those of France or England, it was the product of necessity, given the limited number of women who traveled with the conquistadors, and was not incompatible with the development of intricate racial hierarchies that became legacies of the Spanish and Portuguese empires. Indeed, the scientific racialism of the nineteenth century would ground a similar discourse, not on philosophical and religious reasons, as Sepúlveda did, but on (pseudo)scientific ones.

Colonialism is thus more than a set of institutions or practices that permit the establishment and maintenance of unequal economic exchanges among regions or countries. Underlying colonial economic relations and institutions are evolving beliefs or ideologies that make possible the permanence and reproduction of colonialism. For instance, the Spanish conquistadors saw even their most brutal actions justified by their role in spreading the Catholic religion. It is reported that Hernán Cortés (1485–1547), the conqueror of Mexico, claimed that "the main reason why we came ... is to praise and preach the faith of Christ, even if together with this we can achieve honor and profit" (Zavala 1972, p. 25). In a similar vein, the British and French empires found their justification in supposedly bringing civilization to "primitive" regions of the world.

Western culture is thus permeated by pseudorational justifications of racial hierarchies, which would seem to ground colonialism on nature. Even the usually skeptical David Hume (1711–1776) accepts colonial racial hierarchies when he states "the Negroes and in general all other species of men (for there are four or five different kinds) to be naturally inferior to the whites. There never was a civilized nation of any other complexion than white, nor even any individual eminent either in action or speculation" ("Of National Characters," *Philosophical Works III*, p. 228). Writing about "Locke, Hume, and empiricism," Said has argued "that there is an explicit connection in these classic writers between their philosophic doctrines [and] racial theory, justifications of slavery [and] arguments for colonial exploitation" (Said 1978, p. 13). Other canonic names are easily added to that of Hume, and many other disciplines to that of philosophy, from evolutionary biology—which, despite the misgivings of Charles Darwin (1809–1882), ended up applying its notions of competition to humanity—to historical linguistics, which helped provide a pseudoscientific basis for racist celebration of the so-called Aryan race.

Anticolonialism

Yet just as colonialism found occasional supporters among its subjects in the Americas, Africa, and Asia, European reaction to colonialism was not homogeneous. There was an important streak of anticolonial thought and action in Europe as long as colonies existed, and this too left an imprint on Western thought. Indeed, colonialism not only permeated Western culture, it also established the framework within which anticolonialist thought and action frequently developed. Because of the central role played by Catholicism in the justification of Spanish expansion, the anticolonialist reaction in sixteenth-century Spain used the intellectual tools provided by the church. Thus Bartolomé de Las Casas (1474–1566), the greatest critic of the Spanish conquest, used Biblical exegesis, scholastic philosophy, canonic law, historiography, and his own and others' eyewitness accounts to convince the Spanish court and the church of the humanity of the Native American populations and to achieve partial recognition of their rights. In fact, the arguments of Las Casas and other like-minded contemporary critics of colonialism, such as Francisco de Vitoria (c. 1486–1546), are the seeds from which contemporary notions of human rights and international law have sprung. But Las Casas did not deny the need to evangelize Native Americans or fail to acknowledge the sovereignty of the Spanish monarchy over them, even as he vindicated their right to self-government and to be treated as human beings.

Even texts produced in the Americas that are generally taken to be expressions of indigenous cultures, such as the anonymous seventeenth-century compilation of Meso-American myths, the *Popol Vuh*, or the Andean chronicler Felipe Guaman Poma de Ayala's *El primer nueva corónica y buen gobierno* (The first new chronicle and good government), also finished in the early seventeenth century, were intellectually framed by Catholicism. While the *Popol Vuh* uses Latin script to reconstruct the Mayan hieroglyphic books destroyed during the Spanish catechization, and can, therefore, be considered an act of absolute resistance to the Spanish conquest, its anonymous author describes the text as written "in Christendom." Although Guaman Poma de Ayala's very title implies criticism of Spanish rule, it is a hybrid text in which traditional Andean structures, such as the *hanan/hurin* (upper/masculine–lower/feminine) binary, are maintained while acknowledging Catholicism and incorporating into its narrative idiosyncratic versions of biblical stories.

This dependence on European thought, even on some of the basic presuppositions of colonialism itself, will be continued by most oppositional movements and texts produced after the first moment of resistance to European invasion. For instance, while for Lenin imperialism is rooted in the nation and in national capital, anti-imperial movements will likewise be national movements struggling to achieve independence. If the spread of "civilization" is seen in the nineteenth century as validating colonial expansion, the Cuban anticolonial activist, revolutionary, and scholar José Martí (1853–1895), in his classic essay, "Our America," proposed the establishment of the "American University," in which a decolonized curriculum would, for example, privilege the Incas and not the Greeks as the foundation of culture. Even the appeal of Mahatma Gandhi (1869–1948) to nonviolence as the basis of the struggle against colonial oppression, while rooted in his reading of the *Bhagavad Gita*, is also a reinterpretation of principles first proposed by David Henry Thoreau (1817–1862) and developed by Leo Tolstoy (1828–1910), with whom the great Indian leader corresponded.

A similar appropriation and modification of Western discourse can be found in twentieth-century anticolonialism's relationship with Marxism, even if in this case, as in that of nonviolence, it is an oppositional rather than a hegemonic one that is being used. Thus Mariátegui argued: "[Socialism] must be a heroic creation. We must give life to an Indo-American socialism reflecting our own reality and in our own language" (Mariátegui 1996, p. 89). And this attempt at translating Marxism into local cultural traditions was replicated throughout most of the colonial and neocolonial world, as authors as diverse as Ernesto "Che" Guevara (1928–1967), Amilcar Cabral (1921–1973), and Mao Zedong (1893–1976) attempted to create "socialisms" not only compatible with the social and cultural conditions of Latin America, Lusophone (Portuguese-speaking) Africa, and China, but also rooted in them. Precisely because of the importance given to local conditions, this anticolonial and nationalist Marxism was characterized by an emphasis on the cultural effects of political actions, and vice versa. Although not completely ignored, culture and nation did not play prominent positive roles in classic European revolutionary authors such as Karl Marx (1818–1883), Friedrich Engels (1820–1895), and Lenin. The subsequent preoccupation with culture is a link between anticolonial Marxism and postcolonialism, understood as a cultural and political critique of the surviving colonial and developing neocolonial structures and discourses.

Postcolonialism

But questions remain regarding postcolonialism. Is the *post* in *postcolonialism* merely a temporal marker? If so, all postindependence literary and critical production in all former colonies, regardless of whether they deal with or promote cultural and structural decolonization, would be postcolonial. Or is it a reference to those writings that attempt to deal with the aftermath of colonialism, with the social and cultural restructuring and healing necessary after the expulsion of the European colonists? In this case the novels of James Fenimore Cooper (1789–1851) and even those of Henry James (1843–1916), all of which, in one way or another, deal with the problem of establishing a U.S. identity distinct from those of England and Europe, could be classified as "postcolonial." In Latin America, several figures would qualify as postcolonial thinkers: the nineteenth-century polymath Andrés Bello (1781–1865), with his didactic poetry praising and, therefore, promoting "tropical agriculture," and his attempt at modifying Spanish orthography so as to reflect Spanish-American pronunciation; the Cuban scholar Fernando Ortiz (1881–1969), producer of pioneering studies of the cultural hybridity characteristic of the colonial and postcolonial experiences for which he coined the term *transculturation*; and, as well, the aforementioned Martí and Mariátegui, who among others, initiated in the region the systematic criticism of neocolonialism, internal colonialism, racism, and cultural dependence.

Or is the *post* in the term a not-so-implicit alignment with poststructuralism and postmodernism, that is with the antifoundational philosophies developed by, among others, Jacques Derrida (1930–2004), Gilles Deleuze (1925–1995) and Félix Guatari (1930–1992) and Michel Foucault (1926–1984)? If so, despite the existence of transitional figures such as Frantz Fanon, whose writings combine anti-colonial agitation, Marxism, French philosophy and psychoanalysis, postcolonialism could be seen as opposed to Marxist and non-Marxist anticolonialism and to mainstream attempts at understanding and undermining neocolonialism. From this antifoundational perspective, if the stress on cultural topics characteristic of anticolonial and postindependence fictional and theoretical texts establishes a connection with postcolonialism, their frequent essentialism, occasional blindness toward gender hierarchies, emphasis on politics and economics over constructions of subjectivity, make them at best flawed precursors. And from the point of view of scholars who claim to be developing the perspectives proposed by anticolonial theorists—Marxist or otherwise—postcolonialism can be interpreted as the direct application of theories developed in Europe and the United States that disregard earlier local theorizations and mediations.

Regardless of how one understands its relationship with anticolonial thought, this postcolonialism as exemplified by the works of Said, Homi K. Bhabha, and Gayatri Chakravorty Spivak, among others, has generated challenging analyses of the role of gender within colonial and postcolonial institutions, of the political implications of hybridity and diaspora, of racism, and of the importance of constructions of identity within colonial, neocolonial, and postcolonial situations. Moreover, it has permitted the extension of its analyses of subjectivity and of heterogeneous social groupings to the colonial archive, permitting the elaboration of innovative historical reconstructions that go beyond the obsession with facts and events of conventional historiography, or the frequently exclusive preoccupation with classes and economic structures characteristic of Marxism.

Assessment

The importance of the study of colonial and postcolonial structures and ideologies resides in the fact that contemporary international economic and cultural relations and realities, rather than being their negation, can be read as their continuation. In fact, contemporary American, African, and Asian national boundaries are part of the colonial inheritance. These borders, drawn according to purely administrative and political criteria by the imperial powers without taking into account cultural, ethnic, linguistic, or historical differences among the diverse populations thus brought together, have been a contributing factor to the ethnic and national violence that have plagued postcolonial areas.

But international economic inequality is the most egregious legacy of empire. The depth of this continuing disparity is such that, according to the Food and Agriculture Organization (FAO) of the United Nations, of the 842 million people classified as undernourished between 1999 and 2001, 798 million lived in postcolonial areas (FAO 2003). A similar inequality, though undeniably less dramatic in its immediate consequences, is present in the field of science and technology. For instance, Latin America holds only 0.2 percent of all patents (Castro Díaz-Balart and Rojas Pérez 2002, p. 331). While this is the direct result of the countries of the so-called developing world investing only 0.3 to 0.5 percent of their gross domestic product in the fields of science and technology—in contrast, "First World" countries set aside 2 to 5 percent for the same purpose (Castro Díaz-Balart and Rojas Pérez 2002)—it is also a consequence of the unequal manner in which the contemporary global economy is structured, which transforms scientific and technological research into a luxury. Moreover, this low investment in science and technology constitutes a contributing factor to the perpetuation of this international inequality (Castro Díaz-Balart and Rojas Pérez 2002). Furthermore, colonialism and the continuing global inequality it created can be seen as determining patterns of consumption of natural resources that have played a central role in past and current exploitation and destruction of colonial and postcolonial environments. For instance, Richard Tucker (2000) has noted that the United States, as a neocolonial power, has come "to be inseparably linked to the worldwide degradation of the biosphere" (p. 2). Thus the inheritance of colonialism, described by the constellation of heterogeneous terms postcolonialism, neocolonialism, or imperialism—in both its territorialized and deterritorialized conceptualizations—not only constitutes a central problematic in the fields of science and technology but also is at the core of the major ethical dilemmas faced by humanity in the early twenty-first century.

JUAN E. DE CASTRO

SEE ALSO *African Perspectives; Development Ethics; Globalism and Globalization; Industrial Revolution; Scientific Revolution.*

BIBLIOGRAPHY

Adas, Michael. (1989). *Machines as the Measure of Men: Science, Technology, and Ideologies of Western Dominance*. Ithaca, NY: Cornell University Press. A broad, comparative study of the history of European responses to the cultures of sub-Saharan Africa, India, and China that emphasizes the role played by western evaluations of technological differences.

Ahmad, Aijaz. (1992). *In Theory: Classes, Nations, Literatures*. London: Verso. Ahmad, an Indian poet and literary critic, provides a stringent critique of postcolonial theory and a defense of the continuing relevance of Marxist analysis to the understanding of colonial and neocolonial literatures, cultures, and politics.

Ahmad, Aijaz. (2000). "Globalisation: A Society of Aliens?" *Frontline* 17(20). Also available from http://www.frontlineonnet.com/fl1720/17200490.htm. In this article Ahmad provides a Marxist analysis of globalization.

Bello, Andrés. (1997). *Selected Writings of Andrés Bello*, trans. Frances M. López-Morillas. New York: Oxford University Press. This is a selection of essays and didactic poetry by the most influential nineteenth-century Spanish American intellectual of the independence and post-independence periods.

Bhabba, Homi K. (1994). *The Location of Culture*. London: Routledge. One of the key sources for the concept of hybridity in the humanities and social sciences. Bhabba analyzes the manner in which cultural mixture undermines colonial and postcolonial projects.

Cabral, Amilcar. (1979). *Unity and Struggle: Speeches and Writings*, trans. Michael Wolfers. New York: Monthly Review Press. In this collection of his writings, Cabral, the main leader of the struggle for the independence of Guinea-Bissau, provides sophisticated Marxist analyses of the roles played by culture and nationalism in anticolonial revolutions.

Brockway, Lucile H. (1979). *Science and Colonial Expansion: The Role of the British Royal Botanic Gardens*. New York: Academic Press. Arguably the foundational text in the new literature of case studies on the relationship between Western science and colonialism.

Castro Díaz-Balart, Fidel, and Hugo Rojas Pérez. (2002). "Globalization, Science and Development." *Perspectives on Global Development and Technology* 1(3/4): 322–339. A study of contemporary international inequalities in scientific research and their impact on development.

Fanon, Frantz. (1963). *The Wretched of the Earth*, trans. Constance Farrington. New York: Grove Press. Fanon, at the time deeply involved in the struggle for Algerian independence, applies insights from psychoanalysis, Sartrean existentialism, and Marxism to the colonial situation in Africa. Originally published in 1961.

Fanon, Frantz. (1967). *Black Skin, White Masks*, trans. Charles Lamm Markmann. New York: Grove Press. In this book, Fanon, a trained psychoanalyst, studies the influence of colonialism on black colonial subjects. Originally published in 1954.

Fishlock, Trevor. (2004). *Conquerors of Time: Exploration and Invention in the Age of Daring*. London: John Murray. Extended narrative stressing how technological inventiveness was often stimulated by problems encountered in colonial settings.

Food and Agriculture Organization (FAO) of the United Nations. (2003). *The State of Food Insecurity in The World*, 5th edition. Rome: Author.

Gandhi, Mahatma. (1983). *The Essential Gandhi*, ed. Louis Fischer. New York: Vintage Books. Useful collection of essays, speeches, and interviews by the spiritual leader of India's independence movement.

Grove, Richard H. (1995). *Green Imperialism: Colonial Expansion, Tropical Island Edens, and the Origins of Environmentalism, 1600–1860*. Cambridge, UK: Cambridge University Press. Argues that modern notions of ecology are heavily dependent on the colonial experience, and especially professional scientists working in the Dutch, French, and British colonies.

Guaman Poma De Ayala, Felipe. (1978 [1613]). *Letter to a King: A Peruvian Chief's Account of Life under the Incas and under Spanish Rule*, ed. and trans. Christopher Wentworth Dilke. New York: Dutton. Originally published in Spanish as El primer nueva corónica y buen gobierno [the first new chronicle and good government]. Written by a regional noble, this early seventeenth-century illustrated chronicle is not only the most thorough expression of the indigenous perspective on the conquest of Peru, but its 398 drawings provide an invaluable visual record of life during the first years of the colony.

Guevara, Ernesto "Che." (1997). *Che Guevara Reader*, ed. David Deutschmann. Melbourne: Ocean Press. Comprehensive selection of essays and speeches by the Argentine-born leader of the Cuban revolution who became the international symbol for anti-imperialist and anti-capitalist struggles.

Hardt, Michael, and Antonio Negri. (2000). *Empire*. Cambridge, MA: Harvard University Press. Influential attempt to apply poststructuralist theories to the analysis of contemporary international economic reality.

Johansen, Bruce E. (1982). *Forgotten Founders: Benjamin Franklin, the Iroquois, and the Rationale for the American Revolution*. Ipswich, MA: Gambit. Study of the influence of Iroquois political institutions on those of the United states.

Klor de Alva, Jorge. (1995). "The Postcolonization of the (Latin) American Experience: A Reconsideration of 'Colonialism,' 'Postcolonialism,' and 'Mestizaje.'" In *After Colonialism: Imperial Histories and Postcolonial Displacements*, ed. Gyan Prakash. Princeton, NJ: Princeton University Press. Analysis of the difficulties found in the application of postcolonial theory to the Latin American colonial experience.

Las Casas, Bartolomé de. (1992). *A Short Account of the Destruction of the Indies*, trans. Nigel Griffin. London: Penguin. Originally published in Spanish as *Brevísima relación de la destrucción de las Indias*, 1552. A devastating account of the genocide of the Amerindian by the Spanish conquistadors written by the foremost sixteenth-century defender of indigenous rights.

Lenin, Vladimir Illich. (1977). "Imperialism, the Highest Stage of Capitalism." In *Lenin: Collected Works*, Vol. 22.

Moscow: Progress Publishers. Originally published in German, 1917. Classic Marxist analysis of nineteenth-century and early-twentieth-century imperialism.

Loomba, Ania. (1998). *Colonialism-Postcolonialism*. London: Routledge. Careful introduction to the history of colonialism and its contemporary incarnations and to the different theoretical approaches to the topic.

Magdoff, Harry (1969). *The Age of Imperialism: How US Foreign Policy Works*. New York: Monthly Review Press. Collection of essays which analyze from an economic perspective how U.S. imperialism works.

Manrique, Nelson. (1993). *Vinieron los sarracenos: El universo mental de la conquista de América* [The Sarracens arrived: the mental universe of the conquest of America]. Lima: Desco. Well-documented reconstruction of the evolution of the "mental world" of medieval Spain that helped determine the attitudes and actions of the Spanish conquistadors in the New World.

Mao Zedong. (1970). *Mao Papers, Anthology and Bibliography*, ed. Jerome. Ch'en. London: Oxford University Press. Collection of some of the Chinese revolutionary's most influential essays and speeches.

Mariátegui, José Carlos. (1971). *Seven Interpretive Essays on Peruvian Reality*, trans. Marjory Urquidi. Austin: University of Texas Press. Originally published in Spanish, 1928. Foundational text of Latin American Marxism. It studies Peru's history, social and economic structures, and cultural topics from a heterodox perspective; also influenced by Georges Sorel, Friedrich Nietzsche, and Sigmund Freud.

Mariátegui, José Carlos. (1996). *The Heroic and Creative Meaning of Socialism*, ed. and trans. Michael Pearlman. Atlantic Highlands, NJ: Humanities Press. A collection of essays on cultural and political topics written by the Peruvian Marxist in the 1920s.

Martí, José. "Our America." (2002). In *José Martí: Selected Writings*, ed. and trans. Esther Allen. New York: Penguin. Originally published in Spanish as "Nuestra América," 1881. Influential anti-imperialist tract written by the great poet and martyr of Cuba's independence.

Nkrumah, Kwame. (1965). *Neo-Colonialism: The Last Stage of Imperialism*. London: Heinemann. Analysis of the manner in which political independence does not necessarily lead to economic independence.

Ortiz, Fernando. (1995). *Cuban Counterpoint: Tobacco and Sugar*, trans. Harriet de Onís. Durham, NC: Duke University Press. Originally published in Spanish in 1940. By means of a study of the histories of the tobacco and sugar industries in Cuba, Ortiz provides the foundational study of cultural contact in Cuban and the Caribbean.

Reingold, Nathan, and Marc Rothenberg, eds. (1987). *Scientific Colonialism: A Cross-Cultural Comparison*. Washington, DC: Smithsonian Institution Press. The proceedings of an Australian conference that studies the role played by colonialism in stimulating the development of Western science.

Said, Edward. (1978). *Orientalism*. New York: Pantheon. One of the foundational texts in postcolonial theory that studies the discursive formation of the notion of the Orient and its distorting effect on colonial and contemporary policies and analyses.

Sepúlveda, Juan Ginés de. (1951 1547]). *Demócrates segundo, o, De las justas causas de la guerra contra los indios* [The second democrates, or, of the just causes for the war against the Indians], ed. Ángel Losada. Madrid: Instituto Francisco de Vitoria. Justification of the conquest written by Las Casas's main ideological rival and one of Spain's foremost Rennaisance Aristotelians.

Spivak, Gayatri Chakravorty. (1996). *The Spivak Reader: Selected Works by Gayatri Chakravorty Spivak*, ed. Donna Landry and Gerald Mac Lean. New York: Routdledge. A collection of essays informed by postructuralism and feminism on postcolonial literature and history, as well as on other theoretical and cultural topics.

Tucker, Richard P. (2000). *Insatiable Appetite: The United States and the Ecological Degradation of the Tropical World*. Berkeley and Los Angeles: University of California Press. An extended analysis of the negative impact of the United States as a neocolonial power on the world's environment.

Tedlock, Dennis, trans. (1996). *Popol Vuh: The Mayan Book of the Dawn of Life*, rev. edition. New York: Simon and Schuster. A collection of Maya creation myths written shortly after the conquest by an anonymous indigenous author.

Young, Robert J. C. (2001). *Postcolonialism: An Historical Introduction*. Oxford: Blackwell. Good overview that emphasizes the links between anti-colonial and postcolonial theories.

Zavala, Silvio. (1972). *La filosofía política en la conquista de América* [Political philosophy in the conquest of America]. Mexico City: Fondo de Cultura Económica. Originally published in 1947. Analysis of the philosophical and political debates in Spain and the Americas on the rights of Amerindians that emphasizes their links to the development of modern notions of democracy and human rights.

INTERNET RESOURCES

Ahma, Aijaz. "Globalisation: A Society of Aliens." Available from http://www.flonnet.com/fl1720/17200490.htm.

Food and Agriculture Organization of the United Nations. "The State of Food Insecurity in the World 2003." Available from http://www.fao.org/DOCREP/006/J0083E/J0083E00.HTM/.

Guamán Poma de Ayala, Felipe. *El primer nueva corónica y buen gobierno* [The first new chronicle and good government]. Available from http://www.kb.dk/elib/mss/poma/.

United Nations. "The United Nations and Decolonization." Available from http://www.un.org/Depts/dpi/decolonization/main.htm.

COLUMBIA ACCIDENT

SEE *Space Shuttle Challenger* and *Columbia* Accidents.

COMMON HERITAGE OF MANKIND PRINCIPLE

The Common Heritage of Mankind Principle (CHP) as it was presented to the United Nations General Assembly in various declarations and treaties, and as it is understood in the early-twenty-first century, affirms that the natural resources of the deep seabed and of outer space are held in common by all nations, and should be distributed equitably for the benefit of all humankind. Specifically the CHP of the 1979 Treaty Governing the Activities of States on the Moon and Other Celestial Bodies (The Moon Treaty), refers to: the equitable sharing of outer space resources; the nonappropriation of *in-place* resources particularly with regard to outer space mining activities; and the institution of an *international regime* to supervise commercial activities in space.

The CHP was presented with the understanding that it was crucial to plan for future exploration and uses of these important regions in order to insure not only an equitable distribution of their natural resources, but to prevent conflicts among nations as have occurred during earlier eras of exploration. Proponents of the CHP believe the principle confers on a region the designation of *domino util* or beneficial domain that should be legally defined as *res communis humanitatis*, a common heritage that is not owned by any nation, but from which all nations may garner profits and benefits.

Early Usage

Notions designating global resources as the common property of humankind (*res communis*) are not new, particularly in relation to the oceans, but date back more than 400 years. During the great age of discovery in the fifteenth century, Spain and Portugal claimed sovereignty over the high seas in accordance with the Papal Bull of 1493. This Bull established the border between Portuguese and Spanish waters "by a meridian line running 100 leagues west of the Azores, through both poles." In the late 1500s, however, the Protestant, seafaring nations of England and Holland challenged these claims of exclusive sovereignty over the oceans. Elizabeth I, in 1577, specifically dismissed Spanish claims of sovereignty over the high seas by "declaring that the sea, like the air, was common to all mankind and that no nation could have title to it" (Schachter 1959, p. 10). This began the establishment of the principle of *freedom of the seas*, or open access and nonappropriation in maritime law, which later was seen as a positive-sum game that encourages the usage and development of ocean resources as well as international trade for the common interest of nations (DeSaussure 1989, p. 29).

Modern Applications

The International Geophysical Year (IGY) was a main motivating factor behind the development of contemporary legal notions concerning open access and common property as applied to new territories such as Antarctica, the deep seabed, and outer space. The international scientific investigations conducted during 1957 and 1958 were enormously successful, and created a new paradigm for international prestige through cooperation in quality scientific research. In fact, the collaborations forged during the IGY fostered the formation of a number of new international committees and agreements including the 1958 United Nations General Assembly Conference in Geneva on the Law of the Sea, which reaffirmed the freedom of the high seas and began negotiations concerning the natural resources of the continental shelf and deep seabed; the 1959 Antarctic Treaty; the United Nations Committee on the Peaceful Uses of Outer Space (COPUOS); and ultimately the 1967 Outer Space Treaty containing the Common Benefit Principle (a modified *res communis*), which mandates that space exploration and the utilization of its resources be "for the benefit and in the interests of all countries."

The Law of the Sea and the Moon Treaty

During the late 1960s, the development of new technologies capable of taking commercial advantage of natural resources in the deep seabed and outer space, rendered the common benefit and nonappropriation clauses of earlier treaties obsolete. Ambassador Arvid Pardo of Malta introduced to the United Nations in 1967 a declaration related to the peaceful uses of the seabed and ocean floor that referred to these areas as a common heritage of humankind (Gorove 1972). According to Pardo, the CHP would establish "an administrative process whereby benefits derived from the resources of the [ocean] would be used for the common advantage of all peoples without regard to conditions of poverty or of wealth"; require supplementary programs of environmental protection to insure that the ocean's resources would be "passed on to succeeding generations"; and imply that the ocean and its resources "will be used exclusively for peaceful purposes" (Christol 1976, p. 44.) This declaration was accepted by the United Nations General Assembly without major criticism, and work began on the Declaration of Principles Governing the Sea-bed, the Ocean Floor, and the Subsoil Thereof,

Beyond the Limits of National Jurisdiction, which was presented to the General Assembly in 1970.

The opening of outer space territories and resources to the possibility of commercial ventures also raised new questions with regard to the activities of states and private entities in outer space. The Common Benefit Principle of the 1967 Outer Space Treaty in combination with its nonappropriation clause in Article 2 left open certain questions concerning sovereignty and property rights in relation to permanent space stations, lunar stations, and astral and lunar mineral resources. The CHP was offered as a complementary principle that would fill these legal gaps by defining the nature and use status of outer space and its resources; clarifying the rights and obligations of states and private entities in relation to these resources; and providing regulatory guidelines that would reduce the monetary risks of commercial space ventures. In 1972 the United States made a formal presentation to the COPUOS committee working on the Moon Treaty draft advocating inclusion of the CHP into the treaty text.

Implications for Science and Technology

The possible implications of the CHP for advancement of science and technology can be found in the debate between First and Third World nations as to the effects the implementation of this principle might have on the commercial development of space resources and the technologies that access them.

As committee work on the Moon Treaty continued, controversy grew in the United States concerning the CHP and its implications for the development, use, and allocation of outer space resources. There was considerable debate on both the definition of what the equitable sharing of resources meant under the CHP, and whether or not that sharing included access to space technology. In particular, a swarm of small but powerful U.S. space interests, especially the L-5 Society, began to publicly protest against the treaty, and managed to challenge the original U.S. position in several important areas.

Consequently U.S. representatives began arguing that the implementation of the CHP, with its mandate for profit sharing through an international regime, would be a disincentive to capital investment by private enterprise in the development of space resources and technologies. In addition, the principle's affirmation of equitable sharing and open access to space resources and technologies would bring about static inefficiency in the development of these resources, resulting in fewer benefits being produced for all concerned. Finally the equitable sharing of space technologies would be a threat to national security, both undermining the economic base of the United States and supplying potentially unstable nations with technology that had possible dual-use military applications.

Third World nations argued that the CHP did not constitute a disincentive to space resource development because its provisions were designed to grant positive rights that would allow humankind to exploit the benefits of space resources for the first time (Cocca 1973). This was a clear improvement to the 1967 Outer Space Treaty that specifically excluded the possibility of appropriating these resources. In addition, the CHP authorizes an equitable, not an equal, sharing of profits, and contains a compromise clause that balances the distribution of benefits by taking into consideration both the needs of Third World countries, and the efforts put forth by the nations or entities developing these specific resources.

Third World nations also argued that the international regime, rather than obstructing the development of space resources, actually furnishes a system capable of facilitating cooperative space ventures between nations for the accessing of space resources. Moreover the mitigation of Third World underdevelopment and external dependency on the First World through the equitable sharing of outer space resources would in reality further international cooperation and reap greater economic benefits for all nations. In fact, economic research studies have recommended that "for the sake of American commercial competitiveness in space," the United States should maintain lenient policies in relation to international technology transfers and encourage the cooperative exchange of information among scientists from all nations as a means of accelerating technological innovation (Corson 1982, pp. 59–61).

Status and Assessment

The Moon Treaty, with its common heritage language, spent seven years in the COPUOS working committee before it was finally passed by consensus and sent to the UN General Assembly in 1979 for a vote, where it was adopted by all 152 member nations. However the Moon Treaty was subsequently ratified by only thirteen nations, and while it is technically in force in 2004, the lack of support by First World, spacefaring nations has undermined the treaty's inherent authority, and ultimately created a large and growing gap between the uses of space resources and technologies and the adequacy of the laws regulating them.

In the absence of an accepted system of international space law, nations have been turning to the formation of their own domestic law to furnish at least some legal guidance and security for the conduct of space activities (Goldman 1988, p. 85). Domestic space law, however, generates even more complex issues of compliance, particularly given the international nature of outer space and space activities. Questions regarding whose law will apply for joint space ventures such as the international space station, or in areas of liability for space accidents occurring between states, will be extremely troublesome to answer.

Yet the compromises that occurred during the laborious process of consensus in developing the Moon Treaty and the CHP were made to "assure developed and developing nations the opportunity to benefit from space activities" taking place within a commonly held region beyond national territorial boundaries (Jasentuliyana 1984, p. 4). The Moon Treaty offered an indispensable legal framework for maintaining international stability and clarifying the expectations of the international community, thereby reducing the potential for conflict, creating a safer investment climate for both government and private entities, and furnishing an organizational mechanism for cooperative commercial ventures in outer space (Jasentuliyana 1980, pp. 6–7; Goldman 1985, p. 85).

Consequently First World suspicions regarding the CHP and its mandate for the equitable sharing of space resources and technologies, along with the belief that open access, and/or cooperative ventures with *less qualified* Third World nations would lead to the inefficient development of these resources, ended an unprecedented era of international collaboration in scientific exploration, technological advancement, and the development of positive international law.

KIM ALAINE RATHMAN

SEE ALSO *Development Ethics; Space Exploration.*

BIBLIOGRAPHY

Christol, Carl. (1976). "The Legal Heritage of Mankind: Capturing an Illusive Concept and Applying it to World Needs." In *Proceedings of the XVIII Colloquium on the Law of Outer Space*. Paris: International Institute for Space Law.

Cocca, Aldo A. (1973). "The Principle of the *Common Heritage of All Mankind* as Applied to Natural Resources form Outer Space and Celestial Bodies." In *Proceedings of the XVIth Colloquium on the Law of Outer Space*. Paris: International Institute for Space Law.

Corson, Dale. (1982). *Scientific Communication and National Security*. Panel on Scientific Communication and National Security. Committee on Science, Engineering, and Public Policy. Washington, DC: National Academy Press.

DeSaussure, Hamilton. (1989). "Maritime and Space Law, Comparisons and Contrasts: An Oceanic View of Space Transport." In *Outer Space: Problems of Law and Policy*, ed. Glenn Reynolds and Robert Merges. Boulder, CO: Westview Press.

Goldman, Nathan. (1985). "The Moon Treaty: Reflections on the Proposed Moon Treaty, Space Law, and the Future." In *People in Pace: Policy Perspectives for a Star Wars Century*, ed. James Katz. New Brunswick, NJ: Transaction Books.

Goldman, Nathan. (1987). "Transition of Confusion in the Law of Outer Space." In *International Space Policy: Legal, Economic, and Strategic Options for the Twentieth Century and Beyond*, ed. Daniel Papp and John McIntyre. New York: Quorum Books.

Goldman, Nathan. (1988). *American Space Law, International and Domestic*. Ames: Iowa State University Press.

Gorove, Stephen. (1972). "The Concept of the *Common Heritage* of Mankind: A Political, Moral or Legal Innovation?" *San Diego Law Review* 9: 390–403.

Jasentuliyana, Nandasiri. (1980). "A United Nations Perspective of the Moon Agreement." Paper presented at the Seminar on the United Nations Moon Treaty, Center for Strategic and International Studies, Georgetown University, Washington, DC, February 19.

Jasentuliyana, Nandasiri. (1984). "Conditions Essential for the Peaceful Uses of Outer Space: The Moon Treaty," Paper presented at the symposium cosponsored by the United Nations and the International Institute of Space Law, Peace Palace, The Hague, Netherlands, March 12–15.

Jasentuliyana, Nandasiri, ed. (1992). *Space Law: Development and Scope*. Westport, CT: Praeger.

McDougall, Walter A. (1985). *. . . The Heavens and the Earth: A Political History of the Space Age*. New York: Basic Books.

Reynolds, Glenn, and Robert Merges, eds. (1989). *Outer Space: Problems of Law and Policy*. Boulder, CO: Westview Press.

Schachter, Oscar. (1959). "Who Owns the Universe?" In *Space Law: A Symposium*. Washington, DC: Government Printing Office. *Space Law: A Symposium* was prepared at the request of Honorable Lyndon B. Johnson, Chairman Special Committee on Space and Astronautics, U.S. Senate, 85th Congress, 2nd Session.

Vicas, Alex. (1980). "Efficiency, Equity and the Optimum Utilization of Outer Space as a Common Resource. *Annals of Air and Space Law* 5: 589–609.

COMMUNICATION ETHICS

Communication ethics is concerned primarily with human communication mediated by communications technologies, from print to radio, television, and other advanced electronic media. As such it assumes the importance of ethical responsibilities in direct or immediate communication, such as the obligation to speak truthfully, and seeks to reflect on how these carry over into the complex circumstances that arise with the development of communications science and technology. Because of the historical role played by reflection on ethics in relation to mass circulation print technologies in the form of newspapers during the first half of the twentieth century, communication ethics has its roots in journalism ethics. Because of the multiplicity of communications media during the last half of the twentieth century, the term *media ethics* is sometimes used as a synonym for communication ethics.

Contemporary Context

The communication technologies that produce and distribute information are an economic paradise. Massive multimedia conglomerates are at war for the trillions of dollars at stake—Pearson PLC in England, Bertlesmann in Germany, Microsoft and Disney in the United States, the Rupert Murdoch empire, and Sony of Japan. The business tycoons of these global companies do not specialize in hard goods, but control images, data, software, and ideas. Clusters of high-tech communication firms are re-mapping the planet. Previous geographical alignments organized by political power are being reordered in terms of electronic megasystems.

The revolution is not taking place in abstraction, outside of everyday affairs. Banking, the stock market, entertainment media, and the military represent the most advanced electronic communication systems. However the menagerie of fiber optics, supercomputer data, and satellite technology, although inescapably global, is local and personal as well. Television, CDs and CD-ROMs, DVDs and VCRs, online databases, rock music channels, PCs, video games, cellular telephones, and virtual reality—the electronic highway has become the everyday world of advanced industrial societies

Public life in the twenty-first century is being altered in complex ways through ubiquitous multimedia technologies, and ethics is essential for coming to grips with them. Language is indispensable to humanness and to the social order; therefore when human communication capacity is mediated in fundamentally different ways than before, the impact is substantial and far-reaching. Accounting for the social influence of media technologies is an historical and empirical task, but clearly the domain of communication ethics as well.

Communication as Symbol Making

The mainstream view in communication studies has been a mechanistic stimulus-response model rooted in empiricist science. However since the 1990s, communication theory has been complemented with an interpretive turn. From this perspective, human discourse and culture become fundamental, and language is the public agent through which identity is realized. Individuals are integrated into social units through symbol, myth, and metaphor. Communication is the creative process of building and reaffirming through symbols, with cultures the constructions that result. In a symbolic approach to communications, concepts are not isolated from their representations. The social and individual dimensions of language are a unified whole. Through the social nature of language, human beings integrate specific messages with the larger project of cultural formation.

Although not identical to that which they symbolize, symbols participate in their meaning. They share the significance of that to which they point. Symbols create what human beings call reality. Human identity embedded in representations matters to people. Thus worries about racism, sexism, and age discrimination in language are not marginal but central to socially responsible communication. The manner in which race, age, gender, class, disabilities, economic status, and ethnicity are represented symbolically influences the possibilities for a just sociopolitical order.

From a symbolic perspective, when symbols are mediated technologically, the changes in human life and culture must be understood historically and evaluated morally. Walter Ong (2002) calls this *technologizing the word*. Symbolic theory presumes that the history of communications is central to the history of civilization, that social change results from media transformations, that changes in symbolic forms alter the structure of consciousness.

The Canadian scholar Harold Innis (1951), for instance, studied the introduction of papyrus, the printing press, radio, and the telegraph—and documented a bias regarding space and time. Oral communication systems, he argued, are biased toward time, making time continuous while rendering space discontinuous. Print systems, by contrast, are biased toward space, making geography continuous and breaking time into distinct units.

Thus from the introduction of cuneiform writing to contemporary communication satellites and fiber optics, media technologies have attracted considerable attention—scholars in the symbolic tradition examining all significant shifts in technological form, associating with them alternations in culture and in perception. Within this paradigm of bias in communication systems, the intellectual challenge is to identify the distinguishing properties of particular media technologies such as books, cinema, and the Internet. As the physicist steps inside the world of atoms to understand them from the inside, so communications scholars, regarding television or magazines or billboards, must delve into their aesthetic properties in order to know them fundamentally and distinctively (McLuhan 1966).

As a minor premise, Innis (1952) argued that one form of communication tends to monopolize human knowledge and render other forms residual. Communications media never exist innocently and equally alongside one another. Elizabeth Eisenstein (1979), for example, documents the overriding significance of symbolic formation in her definitive work on the invention of printing. The printing press reformulated symbols at a historical watershed, fostering prescriptive truth and decentering papal authority by empowering the home and countryside with vernacular Bibles and Martin Luther's pamphlets. The ninth-century Carolingian and twelfth-century Gothic renascences were limited and transitory. The preservative power of Johannes Gutenberg's invention made the Renaissance permanent and total.

If oral cultures make time stand still, and print cultures foster empire and objectivism, the ongoing shift, from invention of the telegraph to early-twenty-first-century electronic culture, dislocates individuals from both space and history. It ruptures historical consciousness and pushes people into world citizenship, ill-equipped as they may be to accept that role. Without specific anchors in time and space, humans are ripe for electronic picking. Linear rationality facilitated by print is co-opted by mass media images. In sociological terms, the large-scale electronic media radically disconnect human beings from the mediating structures that serve as their everyday habitat—family, school, church, neighborhoods, and voluntary associations. Such primary groups lose their resonance.

The development of Internet technology marks another era of rapid growth and change in the media. Mass media technologies are converging into digital formats. Internet chat rooms, e-mail, multi-user domains (MUD), web-based publications, and the ability to hyperlink are producing new forms of human interaction. The 3-D virtual world is the innovative edge of these online technologies. In principle, interactive Internet technology gives people a voice and connects users directly without professionals or gatekeepers in between. Internet technologies can be democratic tools that serve people's everyday needs rather than those of special interest groups or the market.

Jacques Ellul developed the argument that technology is decisive in defining contemporary culture. Indeed not only productivity, but also economics, politics, and symbolic formations are dominated by the technological. In Ellul's (1969) framework, communications media represent the world of meaning in the technological system at large, the arena where the latter's character is most clearly exposed. Though exhibiting the structural elements of all technical artifacts, their particular identity as a technology inheres in their function as bearers of symbols. Information technologies thus incarnate the properties of technology while serving as agents for interpreting the meaning of the very phenomenon they embody.

Ellul calls communication systems the "innermost, and most elusive manifestation" of human technological activity (Ellul 1978, p. 216). All artifacts communicate meaning in some sense, but media instruments play this role exclusively. As the media sketch out the world, organize conversations, influence decisions, and impact self-identity, they do so with a technological cadence, massaging a technological rhythm and disposition into the human soul. With moral and social values disrupted and reoriented in the process, the ethics of communications technologies are an important arena for examining life in technological societies at present.

History of Communication Ethics

Historically communication ethics arose in conjunction with concerns related to print media, so that it requires work to extend the original developments to the more prominent digital technologies. Print news and the ethical standards for newspaper reporters were the first concerns of anything that could be called communication ethics. The harm that an unregulated press could do to society was first explicitly linked to ethical principles in North America and Europe during the 1890s, when critics began assessing journalism philosophically. These initial forays blossomed into the first systematic work in communication ethics during the 1920s in the United States. Four major books emerged from

America's heartland during that decade, their authors among a *Who's Who* of journalism luminaries: Nelson Crawford's *Ethics of Journalism* (1924), Leon Flint's *The Conscience of the Newspaper* (1925), William Gibbons's *Newspaper Ethics* (1926), and Albert Henning's *Ethics and Practices in Journalism* (1932). These authors understood ethics as a scholarly enterprise and left a permanent legacy. In Europe also several ethical issues emerged during the early-twentieth-century. Sensationalism was considered contrary to the public service role of the newspaper. Freebies and junkets, scourged by media critics as early as 1870, were treated more systematically in the context of rising business competition. Truthfulness as a moral principle was abstracted for the first time from the practice of accurately reporting facts. During this period, a platform for the free press/fair trial debate was created, though it was one-sided in promoting the rights of the press. Together they carved out much of the structure that dominates journalism ethics across Europe and North America in the early-twenty-first century, and with some nuances, in various regions around the world.

The intellectual roots of the democratic press were formed when print technology was the exclusive option. Most of the heavyweights in communication ethics in industrialized democracies demonstrate like predilections for news, and news in its literary rather than electronic broadcast form. Yet extensive research remains to be done on various aspects of the news business: declining readership among youth and in urban cultures, production practices, multiculturalism, the problematic status of objectivity, technological innovation, newspaper credibility, hiring practices, and others. Most of the perpetual issues in media ethics—invasion of privacy, conflict of interest, sensationalism, confidentiality of sources, and stereotyping—get their sharpest focus in a print context. Meanwhile newspapers outside the mainstream have scarcely been considered.

But the context has changed. Television is the primary source of news for most people and information radio remains vital. Even research that emphasizes the news function tackles cases and problems from broadcasting, the wire service agencies, and documentaries, in addition to everyday reporting. And beyond the daily paper, magazines and instant books are increasingly prominent. In a more dramatic trend, reporting is being removed from its pedestal and treated in the same way as other mass media functions. News is now being integrated with other aspects of the information system, that is, to persuade, to entertain, and to serve bureaucracy. In fact, practitioners of journalism, advertising, entertainment, and data management are often part of the same institutions and encounter other media functions directly in their work.

Arguably heads of media corporations should ideally come from a news background, and clearly the demands on news operations have never been more intense. But it is empirically true that the media's role in persuasion, entertainment, and digital transmission has also become pervasive, socially significant, and ethically charged—thus the burgeoning research in the ethics of public relations, organizations, face-to-face encounters, the music business and cinema, libraries, book publishing, confidentiality in computer storage, fiction, new media technologies, the mass-mediated sports industry, and more.

The dark side of ethical research into this expanding field is faddishness and fragmentation. However there is hope that the widening spectrum will open new insights and fresh approaches to the substantive issues. Deception and economic temptation are common in all mass-mediated communication. Sexism and racism are deep-seated everywhere. Reporters often fail to recognize sensationalism in the news until they confront the difference between gratuitous violence and realism in entertainment media. Invasion of privacy, easily excused in news, becomes an insufferable evil when government agencies access confidential information from data banks without permission. The challenge is to demonstrate how ongoing ethical quandaries can be fruitfully examined across a diverse range of media technologies and functions.

Ethical Issues

In outlining an agenda for communication ethics in terms of global media technologies rather than print journalism alone, several issues emerge as primary. Each can profit from the past, though several are new or have such dramatic intensity in the early twenty-first century that thinking rooted in the communication ethics of the first half of the twentieth century is no longer directly relevant. Meanwhile the electronic media have achieved some important successes. The Internet makes it possible for people who disagree with government policies to unite and protest against them. The Montreal Protocol and the Landmine Ban Treaty, for example, could not have happened without new media technologies. Television was the stimulus for humanitarian intervention in Somalia and prison reform in the U.S. military. Strengthening the media's role in democracy is important for communication ethics, while identifying the negative dimensions that are already obvious.

DISTRIBUTIVE JUSTICE. An ethics of distributive or social justice is mandatory for understanding the communications revolution. The mainstream view of social justice centers on fairness. As a formal concept, justice means "the consistent application of the same norms and rules to each and every member of the social cluster to which the norms and rules apply" (Heller 1987, p. 6). But in the more dynamic and multidimensional terms of distributive justice, the overriding question is accessibility. Just distribution of products and services means that media access ought to be allocated to everyone according to essential needs, regardless of income or geographical location. Comprehensive information ought to be ensured to all parties without discrimination.

In contrast, the standard conception among privately owned media is allocating to each according to ability to pay. The open marketplace of supply and demand determines who obtains the service. Consumers are considered at liberty to express their preferences and to select freely from a variety of competing goods and services. The assumption is that decisions about allocating the consumer's money belong to the consumer alone as a logical consequence of the right to exercise social values and property rights without coercion from others.

An ethics of justice where distribution is based on need offers a radical alternative to the conventional view. Fundamental human needs are related to survival or subsistence. They are not frivolous wants or individual whims or deserts. Agreement is rather uniform on a list of most human necessities: food, housing, clothing, safety, and medical care. Everyone is entitled without regard for individual success to that which permits them to live humanely.

The electronic superhighway is swiftly becoming indispensable. Communications networks make the global economy run, they provide access to agricultural and health care information, they organize world trade, they are the channels through which international and domestic political discussions flow, and through them people monitor war and peace. Therefore as a necessity of life in a global order, communication systems ought to be distributed impartially, regardless of income, race, religion, or merit.

What is most important about Internet technology is not so much the availability of the computing device or the Internet line, but rather the ability to make use of the device and conduit for meaningful social practices. Those who cannot read, who have never learned to use a computer, and who do not know the major languages of software and Internet content will have difficulty getting online, much less using the Internet productively.

There is no reasonable likelihood that need-based distribution will ever be fulfilled by the marketplace itself. Technological societies have high levels of computer penetration, and nonindustrial societies do not. Digital technology is disproportionately concentrated in the developed world, and under the principle of supply and demand there are no structural reasons for changing those disproportions. Even in wired societies, the existence of Internet technology does not guarantee it will reach its potential as a democratic medium. There is a direct correlation between per capita gross domestic product (GDP) and Internet distribution. The geography of the digital world is not fundamentally different from that of the off-line world. The history of the communications media indicates that existing political and economic patterns will prevail; inequities in society lead to inequities in technology.

In the digital age—rooted in computers, the Internet, fiber optics, and communication satellites—ideally all types of persons will use all types of media services for all types of audiences. Therefore the normative guideline ought to be universal access, based on need. And universal service is the Achilles' heel of new technologies driven by engineering and markets. As the economic disparity between rich and poor countries grows, an information underclass exacerbates the problem because information is an important pathway to equality. An ethics of justice requires that the approach to media institutions should be modeled after schools, which citizens in democracies accept as their common responsibility. Without intervention into the commercial system on behalf of distributive justice, the world will continue to be divided into the technologically elite and those without adequate means to participate.

CULTURAL DIVERSITY. Indigenous languages and ethnicity have come into their own in the early-twenty-first century. Sects and religious fundamentalists insist on recognition. Culture is more salient at present than countries. Muslim immigrants are the fastest-growing segment of the population in France and longstanding policies of assimilation are no longer credible. Thirty thousand Navajos live in Los Angeles isolated from their native nation and culture. The nomadic Fulani, searching for good pasture throughout sub-Saharan West Africa, are held together by clan fidelity, but their political future hangs in the balance. More than 30 percent of the information technicians working for the Microsoft Corporation in the United States come from India. In the early 1900s, 80 percent of immigrants to the United States were from Europe. Since the 1960s, the majority has come from Asia, Latin America, and developing countries in Africa. Rather than the melting pot of the last century, immigrants to the United States in the early-twenty-first century insist on maintaining their own cultures, religions, and languages. Identity politics has become dominant in world affairs since the Cold War, and ethnic self-consciousness is now considered essential to cultural vitality. As a result, social institutions such as the mass media are challenged to develop a healthy cultural pluralism instead of strident tribalism.

In order to integrate the new demands of cultural diversity into media practices and policies, an individualistic morality of rights must be modified by a social ethics of the common good. A commitment to cultural pluralism makes sense when the community is understood to be axiologically and ontologically superior to the individual. Human beings in this communitarian perspective do not disappear into the tribe, but their identity is constituted organically. Persons depend on and live through the social realm. Human beings are born into a sociocultural universe where values, moral commitments, and existential meanings are both presumed and negotiated. Thus in communitarian ethics, morally appropriate action intends community. Unless a person's freedom is used to help others flourish, that individual's well being is itself diminished.

Communitarianism as the basis for ethnic plurality moves media programming and organizations away from melting pot homogeneity and replaces it with the politics of recognition. The basic issue is whether democracies discriminate against their citizens in an unethical manner when major institutions fail to account for the identities of their members (Taylor et al. 1994). In what sense should the specific cultural and social features of African Americans, Asian Americans, Native Americans, Buddhists, Jews, the physically disabled, or children publicly matter? Should not public institutions insure only that democratic citizens share an equal right to political liberties and due process without regard to race, gender, or religion? Charles Taylor considers the issue of recognizing multicultural groups politically as among the most urgent and vexing on the democratic agenda. Beneath the rhetoric is a fundamental philosophical dispute that Taylor calls the *politics of recognition*. As he puts it, "Nonrecognition or miscrecognition can inflict harm, can be a form of oppression, imprisoning someone in a false, distorted, and reduced mode of being. Due recognition is not just a courtesy we own people. It is a vital human need" (Taylor et al. 1994, p. 26). This foundational issue regarding the character of cultural identity needs resolution for cultural pluralism to come into its own.

As one illustration of this framework, Robert Entman and Andrew Rojecki (2000) indicate how the race dimension of cultural pluralism ought to move forward in the media. Race in the early-twenty-first-century United States remains a preeminent issue, and Entman and Rojeck's research indicates a broad array of white racial sentiments toward African Americans as a group. They emphasize not the minority of outright racists but the perplexed majority. On a continuum from comity (acceptance) to ambivalence to animosity and finally racism, a complex ambivalence most frequently characterizes the majority. "Whites bring complicated combinations of assumptions, misinformation, emotional needs, experiences, and personality traits to their thinking about race" (Entman and Rojecki 2000, p. 21). They may believe, for example, that blacks face discrimination and merit aid, but argue against welfare spending out of a suspicion of government programs. Ambivalence means that the majority of whites do not necessarily harbor deep-seated fears or resentment, but become conflicted about the best strategies to follow and sometimes lose their patience with the slow progress of change.

Correcting white ignorance and dealing with ambiguities hold the most promise for the media. The reality is, however, that the media serve as resources for shading ambivalence off into animosity. There is little evidence that television or other popular media pull their viewers toward comity. The white majority mostly experiences "media images of Blacks on welfare, of Black violence on local news, and of crude behavior—open sexuality and insolence—in entertainment television.... The habits of local news—for example, the rituals in covering urban crime—facilitate the construction of menacing imagery" (Entmann and Rojecki 2000, p. 34). Thus the media do little to enhance racial understanding among the ambivalent majority most open to it. Unfortunately the media do not provide the information that this important swing group needs to move policy and institutions toward cultural pluralism.

VIOLENCE. Violence in television and film has been a major ethical issue for decades. Internet technology has complicated the problem with hate speech and cyberterrorism.

In the United States, for example, studies have shown that by high school graduation the average seventeen-year-old will have seen 18,000 murders in the movies and on television. From the horrific shootings at Columbine High School in 1999 to similar tragedies in other states and countries before and since, teenagers who slaughter their classmates and teachers, and then kill themselves, are linked by debate or research to the culture of violence in which they live. While the United States leads the world in the amount of violence on television, television programming in all parts of the globe contains a great deal of violence, including a high percentage of guns as weapons and indifference to brutality, with the terrible consequences only hinted at or not depicted at all (Potter 1999). Gun-related deaths in the United States have reached the level of a public health epidemic.

Meanwhile media industries and civil libertarians opposed to censorship claim that no direct effects from violent programming have been documented or proved. In fact, this argument against curtailing violence in the media has long been the most persistent and persuasive. However the no-effects conclusion is no longer credible. Evidence of a positive association between media violence and real violence has been accumulating for at least forty years. Analyses during the 1990s of literally hundreds of studies on media violence verify a causal link between televised violence and real-life aggression with some of the strongest effects among young children. Research conducted for the American Medical Association (AMA) and the National Centers for Disease Control and Prevention, and the results of the exhaustive National Television Violence Study (1994–1998) support the same conclusion (Wilson et al. 2002).

Based on a review of the research, James Potter (1999) concludes that there exist both immediate and extended consequences from televised violence—with the caveat that the effects process is highly complex. In the short term, fear and habituation occur, but increased aggressiveness toward others is strongly supported also. The same is true for effects over a longer period: Research shows that exposure to violence in the media is linked to long-term negative effects such as increased aggression, a worldview based on fear, and desensitization to violence.

Violence is a serious ethical issue because it violates the persons-as-ends principle. In Immanuel Kant's standard formulation, people must treat all other people as ends-in-themselves and never as means only. In Judeo-Christian agape and feminist relational ethics, violence contradicts Other-regarding care. On multiple grounds, the gratuitous cheapening of human life to expand ratings is a reprehensible mistreatment of human beings. From the persons-as-ends perspective, there is a special interest in the sexual violence so common in music video, horror movies (especially slasher films), pornographic literature, and video games. Sadistic, bloodthirsty torture in a sexual context is a particularly offensive form of dehumanization.

A new dimension of violence has emerged with hate speech on the Internet. In 1995, former Ku Klux Klan (KKK) leader Don Black established Stormfront, the first white supremacist Internet site. As access to the Internet became less expensive and creating web pages much simpler, the number of Internet sites and people visiting them grew exponentially. Mirroring this growth, Internet sites espousing various kinds of bigotry have multiplied dramatically, now numbering in the thousands. In the past, hate was promoted through crude graffiti and low quality pamphlets. Bulk mailings to even a few hundred people were difficult. But with the Internet, slick web sites devoted to hate are available to a potential audience of millions.

In the early-twenty-first century, though the KKK is more fragmented than at any time since World War II, its factions are using the Internet to revitalize the organization. The KKK sites maintain and defend the superiority of the white race, and warn against interracial marriage. Jews are vilified as Satan's people, and immigration is condemned as an uncontrolled plague. In addition, the number of Internet sites for the National Association for the Advancement of White People, founded by former KKK leader David Duke, has mushroomed and energized the so-called *Klan without robes*.

Numerous neo-Nazi Internet sites promote the anti-Semitic racism of Adolf Hitler, with the National Alliance being the most prominent Hitlerian organization in the United States. Jews are blamed for inflation, media brainwashing, and government corruption, with blacks depicted as criminals and rioters. A host of sites are devoted to Holocaust revisionism, denying the murder of Jews in World War II.

Internet sites of hate groups that claim religious legitimacy are flourishing as well. The Christian Identity site is virulently racist and anti-Semitic. The World Church of the Creator calls nonwhites physiologically subhuman. The site for White Aryan Resistance rails against the nonwhite birthrate. Other sites are anti-Catholic and anti-Muslim, or militantly anti-abortion.

Most organizations that monitor Internet hate activity do not advocate censorship. Education is seen as more effective than trying to silence bigots. With many moral problems in the media, some ethical theories are more appropriate than others, but hate speech on the Internet is contradicted by all major theories without exception. This across-the-board condemnation suggests that all personal, educational, and policy efforts to combat Internet hate speech are permissible, even mandatory, but obviously without the revenge and aggressiveness that contradict good ends.

Another kind of violence made possible by digital technology is cyberterrorism, that is, attacks on human targets abetted by machines and direct attacks on the telecommunications infrastructure. Financial transaction systems, electrical supply networks, military operations, police and emergency electronic devices, water purity management, air traffic control, and other essential services are vulnerable to computerized sabotage. All attempts at protecting societies through cybersecurity have tended to lead to increased surveillance, intrusions upon private data, and centralized government authority. High-level encryption technology is essential for protecting civil liberties and societies from terrorist attacks. Many security issues in advanced societies are still unclear and their resolution ill-defined. Should diagrams of nuclear power plants or city water systems, for example, be easily available to the public as they were before September 11, 2001? Resolving the conundrums requires as much open communication as possible, but the profusion of communication itself is sometimes counterproductive. In all aspects of cyberterrorism, a proactive citizenry and enlightened legislation are indispensable.

INVASION OF PRIVACY. Public opinion polls indicate that privacy is the premier issue in media ethics, at least in European and North American cultures. Intruding on privacy creates resentment and damages the credibility of the news media. But for all of the advances in privacy and tort law, ethicists consider legal definitions an inadequate foundation. How can the legally crucial difference between newsworthy material and gossip or voyeurism be reasonably determined?

Therefore while acknowledging legal distinctions and boundaries, the ethics of privacy is constructed from such moral principles as the dignity of persons and the redeeming social value of the information disclosed. Privacy is a moral good because it is a condition for developing a healthy sense of personhood. Violating it, therefore, violates human dignity. But privacy cannot be made absolute because people are cultural beings with responsibility in the social and political arena. People are individuals and therefore need privacy; people are social beings and therefore need public information about others. Because people are individuals, eliminating privacy would eliminate human existence as they know it; because people are social, elevating privacy to absolute status would likewise render human existence impossible. These considerations lead to the formal criterion that the intimate life space of individuals cannot be invaded without permission unless the revelation averts a public crisis or is of overriding public significance and all other means to deal with the issue have been exhausted.

From an ethical perspective, legal definitions of privacy beg several questions about the relationship between self and society. A legal right to privacy presumes a sharp line dividing an individual from the collective. An ethics of privacy prefers the richer connections between public and private advocated by social theorists since Alexis de Tocqueville, who have centered their analysis on a viable public life. While participating in theoretical debates over the nature of community, media ethicists have been applying moral principles to three areas: (a) the reporting of personal data on various social groups from innocent victims of tragedy to public officials to criminals; (b) protecting confidential information stored in computer data banks—medical, financial, library, educational, and personal records, for example, and (c) ubiquitous advertising that intrudes on our everyday activities.

Conclusion

The cosmopolitan reach of high-speed electronic technologies has made communication systems and institutions of global scope possible. Dealing with these new entities requires a technologically sophisticated, cross-cultural ethics commensurate with the worldwide reach of the media. In the process of identifying and responding to specific issues, communication and media ethics must make the questions raised by technology the central focus while repositioning them internationally. As true of professional ethics generally, communication ethics ought to become comparative in character. In place of its largely European and North American, gender-biased, and monocultural canon, media ethics of the future must be ecumenical, gender-inclusive, and multicultural.

A diversified comparative ethics, with a level playing field rooted in equal respect for all cultures, is by no means unproblematic and involves an act of faith. The claim that all cultures have something important to say to all human beings is an hypothesis that cannot be validated concretely. Yet it serves as an open horizon for moving comparative, transnational study forward in an interactive mode. Of the various types of applied and professional ethics, communication ethics has its roots most deeply in language, culture, and dialogue. In that sense, a multicultural style is required for its own authenticity.

CLIFFORD G. CHRISTIANS

SEE ALSO *Communications Regulatory Agencies; Computer Ethics; Computer Viruses/Infections; Ellul, Jacques; Journalism Ethics; Information Society; Internet; Networks; Rhetoric of Science and Technology; Science, Technology, and Literature.*

BIBLIOGRAPHY

Bertrand, Claude-Jean. (2000). *Media Ethics and Accountability Systems*. New Brunswick, NJ: Transaction.

Christians, Clifford. (2000). "An Intellectual History of Media Ethics." In *Media Ethics: Opening Social Dialogue*, ed. Bart Pattyn. Leeuven, Belgium: Peeters. Gives a detailed account of the development of media ethics during the twentieth century in North America and Europe.

Christians, Clifford, and Michael Traber. (1977). *Communication Ethics and Universal Values*. Thousand Oaks, CA: Sage. A common commitment to truth-telling, human dignity, and nonviolence is documented from thirteen countries on four continents.

Crawford, Nelson. (1924). *The Ethics of Journalism*. New York: Knopf.

Eisenstein, Elizabeth. (1979). *The Printing Press as an Agent of Change: Communications and Structural Transformations in Early-Modern Europe*, 2 vols. New York: Cambridge University Press.

Ellul, Jacques. (1969). *Propaganda: The Formation of Men's Attitudes*, trans. Konrad Kellen, and Jean Lerner. New York: Alfred A. Knopf. The mass media are described as agents of covert, sociological propaganda in contrast to overt, political propaganda.

Ellul, Jacques. (1978). "Symbolic Function, Technology and Society." *Journal of Social and Biological Structure* October: 207–218.

Entman, Robert, and Andrew Rojecki. (2000). *The Black Image in the White Mind: Media and Race in America*. Chicago: University of Chicago Press.

Flint, Leon Nelson. (1925). *The Conscience of the Newspaper*. New York: Appleton.

Gibbons, William Futhey. (1926). *Newspaper Ethics: A Discussion of Good Practice for Journalists*. Ann Arbor, MI: Edwards Bros.

Hamelink, Cees J. (2000). *The Ethics of Cyberspace*. Thousand Oaks, CA: Sage. Universal human rights as defined by the United Nations are presented as the standard for evaluating cyber technologies.

Heller, Agnes. (1987). *Beyond Justice*. Oxford: Basil Blackwell.

Henning, Albert F. (1932). *Ethics and Practices in Journalism*. New York: Long and Smith.

Innis, Harold. (1951). *The Bias of Communication*. Toronto: University of Toronto Press.

Innis, Harold. (1952). *Empire and Communication*. Toronto: University of Toronto Press.

Lester, Paul M., and Susan D. Ross, eds. (2003). *Images That Injure: Pictorial Stereotypes in the Media*, 2nd edition. Westport, CT: Greenwood.

Lyon, David. (2001). *Surveillance Society: Monitoring Everyday Life*. Philadelphia: Open University Press.

McLuhan, Marshall (1966). *Understanding Media: The Extensions of Man*. New York: McGraw Hill.

Ong, Walter J. (2002). *Orality and Literacy*. Florence, KY: Taylor and Francis Books. The psychological and sociological differences between oral and print societies are outlined historically.

Potter, W. James. (1999). *On Media Violence*. Thousand Oaks, CA: Sage Publications.

Taylor, Charles K.; Anthony Appiah; Jürgen Habermas, et al. (1994). *Multiculturalism: Examining the Politics of Recognition*. Princeton, NJ: Princeton University Press. Taylor argues that the need for a voice from various ethnic groups is one of democracy's most difficult challenges at present.

Wilson, Barbara; Stacy L. Smith; W. James Potter; et al. (2002). "Violence in Children's Programming: Assessing the Risks." *Journal of Communication* 52: 5–35.

COMMUNICATIONS REGULATORY AGENCIES

• • •

Human beings are animals that communicate intensively, and all communication systems, beginning with spoken and written languages, are regulated in at least informal ways. Most people feel that there are certain things that should not be said or written and that certain forms of speech and writing are appropriate for different contexts. However, with the development of physical communication systems such as the postal system and even more with that of the telegraph, telephone, radio, and television, regulation guided by ethical principles has become an increasingly prominent feature of those technologies. Ethical principles concerning content and access have created the foundation for regulation of communication systems. Concerns about content include privacy and anonymity, copyright, defamation, censorship, and profanity. Ethical issues relating to access include concerns about the availability of communication systems and control of content production.

Speech and Postal Systems

Law has been used to regulate the content of speech and writing since at least Roman jurisprudence, in which speakers were held liable for defamatory communication that caused injury to another party. The United Kingdom began regulation of defamation during the sixteenth and seventeenth centuries; this area of law was established in the United States after independence and is mirrored in other countries. Defamation law protects individuals from falsehoods that may cause economic or emotional harm and covers utterances in both speech (slander) and writing (libel). This principle of constrained communication extends to all media and their respective systems and has a lengthy judicial history in the United States and the United Kingdom (Jones 2003).

A number of ancient civilizations created courier services to deliver official documents and messages, with the earliest evidence of an organized infrastructure appearing in Egypt in 2000 B.C.E. From initially serving the government, a number of those systems were expanded to include public and private correspondence; that led to the almost complete control of postal services by nations by 1875. Regulation of those entities focused primarily on efficient administration.

However, by the middle of the twentieth century an expectation of privacy had made its way into many legal systems, including the Mexican constitution, U.S. and British law, and the European Convention on Human Rights. This principle restricts readership of mail to the addressee but generally is qualified to give states the ability to censor materials in the name of security; this explains the still widespread censorship of mail within military forces (Scheele 1970).

Toward the end of the twentieth century government-run monopolies on postal services began to compete again with private courier services. In response a number of government services, including the U.S. Postal Service, began to operate with more independence from the government. Ensuring complete access to the global postal network remains a key factor supporting government-run services as many small or hard to reach communities fear complete isolation in an entirely privately run system.

Telegraph and Telephone

The wire telegraph was invented by Samuel F. B. Morse (1791–1872) in 1835 and saw widespread deployment within ten years of its invention. Alexander Graham Bell (1847–1922) was granted a patent on the telephone in 1876, but that technology grew somewhat more slowly than did the telegraph, with the first transcontinental line in North America not being finished until 1915. In 1865 the International Telegraph Union was founded to support international interoperability of the telegraph system. That union was the first international body to regulate communications and attempted to allow easier communication across national boundaries. The union has expanded to include all telecommunications activities but does not address ethical issues involving content or access directly.

The telegraph initially was regulated in the United States through the Post Roads Act of 1866, which gave authority to the postmaster general to fix rates for telegrams sent by the government. Greater government involvement in the industry did not come until twenty-one years later, when the U.S. Congress passed the Interstate Commerce Act of 1887 to regulate railroads and laid the foundation for the regulation of common carriers within the United States. A common carrier is any transporter that offers services to the general public to transport goods. Court interpretation of common carriers to include communication services provided the legal authority for government to become more actively involved in the communications industry. The 1887 act was amended explicitly to include that extension of government jurisdiction to regulate telephone and telegraph companies by the Mann-Elkins Act of 1910. Regulation of telephone and telegraph was taken over by the Federal Communications Commission (FCC) at its inception in 1934.

Although some nations tried to regulate privately owned telephone and telegraph companies as the United States did, a number of others followed a model closer to that of the postal service and created nationalized phone utilities. In many countries, including the United States, regulators oversaw private companies with full or partial monopolies. Regardless of the details of the regulatory structure or the preference for government involvement or free market competition, each nation faced similar ethical questions.

Early regulation of the telegraph and telephone industries focused on improving interoperability between competing networks, allowing consumers to send messages to any recipient regardless of the network to which they subscribed. Regulators also attempted to ensure that telephone and telegraph companies charged consumers equally for the same service, thus supporting equal access to the system. Each regulatory regime also grappled with questions concerning the privacy and content of those transmissions. When can the state or an individual record or intercept those messages? In the United States third-party taping of conversations requires a court order, whereas rules for recording by parties to a conversation vary from state to state. As with the postal service, most nations have formulated some expectation for the privacy of telephone and telegraph messages.

Radio and Television

Concurrent with the initial development of the telephone and telegraph, research into wireless communications systems led to the creation of the first wireless telegraph by Guglielmo Marconi (1874–1937) in 1895. Maritime adoption of that technology for ship-to-ship and ship-to-shore communication spread rapidly and led to the Berlin International Radiotelegraphic Convention, a series of international conferences in Berlin in 1903 and 1906 and in London in 1912 to discuss radio telegraphy. Beyond determining SOS as the standard distress signal, the 1912 conference led directly to the U.S. Radio Act of 1912, which, along with the Mann-Elkins Act, became the foundation for the regulation of communication systems by the U.S. government.

Radio transmission of voice developed slowly during that period and remained closely tied to telephony. However, by 1920 radio broadcasting had begun in earnest with the November 2 broadcast of election returns by the Pittsburgh station KDKA. The early years of broadcast radio were marked by turmoil. Stations went on and off air, using a frequency and power of their choosing, resulting in widespread interference and confusion. The Radio Act of 1912 required stations to obtain a license from the U.S. Department of Commerce, although the department had no enforcement authority and issued licenses with little oversight. As a result Congress passed the Dill-White Radio Act of 1927, which established the Federal Radio Commission and granted it authority to assign and revoke broadcast licenses at particular powers and frequencies. The act also included provisions for the regulation of programs that exploited or misled the public; that allowed the commission to end broadcasts of fraudulent drug claims or religious scams.

Faced with a growing number of regulatory bodies responsible for communication, Congress created the Federal Communications Commission in the Communications Act of 1934 to take over all communication regulatory activities of the U.S. government. The U.S. regulatory structure stayed largely unchanged until the passage of the Telecommunications Act of 1996.

By the time of the creation of the FCC the television pioneers Vladimir Zworykin (1889–1982) and Philo Taylor Farnsworth (1906–1971) had succeeded in designing and producing all-electronic televisions and television broadcasting was beginning. By the mid-1930s over a dozen stations were broadcasting within the United States. As with radio, the FCC regulated the licensing, power, and frequency of new broadcasters to limit or eliminate interference and ensure that airways were used in the public interest. Television began to grow rapidly in the 1950s and 1960s, quickly reaching a large majority of the public. In the United States satellite and cable television entered the market in the late1970s, but its development was largely unregulated after the Cable Communications Act of 1984 removed much of FCC jurisdiction over those industries.

Although the telegraph and the telephone were accessible by a wide range of the public, broadcast radio and television were limited to a few stations that could broadcast without interference. As a result of the limited nature of broadcasting, governments created various methods to ensure programming in the public interest. In the United Kingdom owners of television sets are required to pay a license fee for partial funding of the government-sponsored British Broadcasting Corporation. In the United States the FCC requires broadcast stations to meet public interest requirements as terms for receiving a broadcasting license. In 1967 the Public Broadcasting Act created public television and radio stations in the United States and partially excluded them from FCC regulation. However, the FCC did act to revoke the license of the Alabama Educational Television network in 1975 because of its racist programming and hiring practices.

From the creation of the Federal Radio Commission to 1987 the FCC enforced a regulatory principal known as the Fairness Doctrine, which holds that stations are obligated to seek out issues of public importance and present contrasting points of view. During the presidency of Ronald Reagan (1980–1988) the FCC began to deregulate all the industries in its jurisdiction. Court cases in 1987 held that the Fairness Doctrine was not required by an act of Congress, allowing the FCC to rescind the policy. Two related rules requiring equal time for targets of personal attacks or political editorials to respond were removed in 2000. Advocates of the change argued that the growth in media outlets negated the need for the doctrine; opponents argued that broadcasters would attempt to further specific political and/or economic agendas to the detriment of the public at large. The Fairness Doctrine was a prime example of regulation of communication that was intended to benefit the public by influencing the content of broadcasts.

Regulators also grappled with control of the limited means of production in the broadcast industry. In light of the limited number of voices that can be brought to air, the distribution of those voices is an important ethical question. A poignant example of the perceived power of broadcasting was the capitulation of broadcasting companies in 1950s to the blacklisting of performers, writers, and directors for alleged leftist political leanings by the organization aware. In that case regulators at the FCC took no action, as they would later do in cases of race or gender discrimination.

Regulators often have attempted to limit ownership of multiple media outlets by single companies to maintain diversity, seeking a balance between preserving independent ownership and allowing free competition. Advances in technology also have changed the availability of the broadcast spectrum by decreasing the amount of interference between nearby stations. At the beginning of the twenty-first century the FCC examined the viability of low-power television and radio stations that would serve small areas and determined that those neighborhood broadcasters did not pose a significant risk of interference with established stations. However, legislation to grant the FCC authority to license those stations has not gotten support from the U.S. Congress.

Internet, Convergence, and the Information Society

The last two decades of the twentieth century saw tremendous growth in a number of new telecommunications fields, especially the worldwide network of computers now known as the Internet. The potential movement of traditional telephone, radio, and television communication to the Internet is known as convergence. The technological underpinning of the Internet makes no distinction between data as the data travel. E-mail, pirated video, and Internet telephony all move equally and without distinction. Data can be identified only by destination or origin. The transformation of all types of data (writing, speech and audio, pictures and video) to computer-based digital data has profound implications for all previous systems: Anyone with access to the Internet can transfer text, audio, or video around the globe and can compete with or avoid traditional communication systems.

Regulation in the new media has been minimal for the most part, with China being a striking exception. The easy accessibility of information on the Internet has led to concerns about the content being provided. The Chinese government regularly blocks content from outside the country and exerts strong control of the information posted within the country. In the United States some have found the availability of pornography to be repugnant and have pushed for greater control over content. In 1996 that desire led to the Communications Decency Act, which created stiff penalties for the distribution of pornographic works to minors; however, the act was struck down by courts as a violation of First Amendment freedom of speech rights.

In light of the growing importance of the Internet, access has become a vitally important question. Disparities between rich and poor individuals and nations in computer access have created a digital divide that has implications for the future growth and equality of those groups.

Assessment

Communication regulation helps define the limits of freedom of speech. Regulators set limits on the content of communication for a variety of reasons, including the protection of personal or secret information, a desire to limit false and misleading claims, and the encouragement of debate. Communication may have negative consequences for individuals, groups, or entire societies. Lessening these harms, however, can require sacrifices in terms of the privacy, anonymity, and freedom of individuals. Modern regulatory agencies must balance the rights of the individual broadcaster with the interests of society as a whole.

Coupled with the regulation of communication content, regulatory agencies also try to control access to communication technologies. Some technologies have a limited capacity for public use, such as radio and over-the-air television. Thus, access to those means of communication is a unique benefit that government has seen fit to control. Other technologies may have limited access because of economic inequities or limits to the physical interconnection of communication networks. Here too regulatory agencies have interfered with the market to promote access to the widest possible set of consumers.

The great power of communications systems as a persuasive force makes these determinations of appropriate content and access disputed issues. Changes in these systems affect millions of consumers and billions of dollars of economic activity. Regulatory agencies sit at the center of political and ethical debate over the appropriate use of these rapidly evolving technologies.

TIND SHEPPER RYEN

SEE ALSO *Communication Ethics; Internet; Radio; Science, Technology, and Law; Television.*

BIBLIOGRAPHY

Brennen, Linda L., and Victoria E. Johnson, eds. (2004). *Social, Ethical and Policy Implications of Information Technology*. Hershey, PA: Idea Group. A social science text on the ethical dimensions of IT communication

Hilliard, Robert L. (1991). *The Federal Communications Commission: A Primer*. Boston: Focal Press. A general guide to the commission's organization, decisions, and history

Jones, William K. (2003). *Insult to Injury: Libel, Slander, and Invasions of Privacy*. Boulder: University Press of Colorado. Provides a detailed history of libel cases within the United States

Jordana, Jacint, ed. (2002). *Governing Telecommunications and the New Information Society in Europe*. Cheltenham, UK: Edward Elgar. Detailed work on the structure of IT governance within the European Union

Levy, Brian, and Pablo T. Spiller, eds. (1996). *Regulations, Institutions, and Commitment: Comparative Studies of Telecommunications*. New York: Cambridge University Press. Describes telecommunication industries in a number of industrialized and developing countries, particularly in South America

Scheele, Carl H. (1970). *A Short History of the Mail Service*. Washington, DC: Smithsonian Institute Press. A detailed history of mail service, especially with in the United States

INTERNET RESOURCES

Federal Communications Commission History Project. (2004). Available from http://www.fcc.gov/omd/history.

COMMUNISM

The word Communism has been used in different senses by different authors, but from 1917 onward it was most readily associated with the type of political and economic system established in Russia and the other lands that became the Union of Soviet Socialist Republics (USSR). By the 1970s Communism in this sense of the term prevailed in Latvia, Lithuania, Estonia, and parts of Bessarabia, all of which were incorporated directly into the USSR, as well as in Mongolia, Poland, Hungary, Czechoslovakia, Bulgaria, Romania, Yugoslavia, Albania, East Germany, North Korea, China, Tibet, Cuba, Vietnam, Laos, and Cambodia. A number of other states, including Nicaragua, Granada, Afghanistan, Angola, Mozambique, and Ethiopia, were ruled by parties closely allied with the USSR, but whether they were full-fledged Communist states is open to debate. In addition, parties advocating the Soviet model of government formed in most other countries. These states and parties, although they used various names—workers, people's, democratic—were commonly referred to as Communist.

From its earliest period of development Communism made two important claims about its relation to science. The first was that it was itself a scientific theory. The second was that it put science and technology to greater benefit than any competitor political practice. Both claims were disputed by non-Communists.

Marxism-Leninism as Science

Although Communism was not alone among the various schools of socialism in tracing its roots to Marxism, Communists were the most emphatic in asserting the absolute validity of that doctrine. Under Vladimir I. Lenin (1870–1924), the founder of Communism, the scientific claims of Marxism were treated as undeniable dogma. Lenin wrote, "From the philosophy of Marxism, cast of one piece of steel, it is impossible to expunge a single basic premise, a single essential part, without deviating from objective truth, without falling into the arms of bourgeois-reactionary falsehood" (Lenin 1977, p. 326).

Lenin wrote these words years before he came to rule Russia. Once he took power, the dogmatic spirit they reflect was reinforced by the exigencies of revolutionary government. Lenin's party, first called Bolshevik and later Communist, was a small, elite group. In order to hold onto power its members saw they would have to suppress the opposition, and Lenin made no bones about this. Since he was confident that his party was the authentic representative of the proletariat, any opposition would inevitably reflect hostile class interests that deserved to be suppressed for the sake of human progress.

Thus did Lenin introduce the practice of silencing criticism or dissent. Under his heir, Joseph Stalin (1879–1953), and their disciples in other countries, such as China's Mao Zedong (1893–1976), Communism assiduously policed the expression of opinion, exacting draconian penalties against any deviation from party policy. All of this was accompanied by sweeping assertions of the scientific character of Marxist and Commu-

Banner depicting (left to right) Vladimir Lenin, Karl Marx, and Friedrich Engels. The three men can be thought of as the "fathers" of Communism. *(© Brian A. Vikander/Corbis.)*

nist doctrine, evidenced by the fact that almost any speech, book, essay, or paper required numerous citations from the texts of Marx, Engels, and Lenin. This semblance of scientific procedure was topped off by the claim that Communism was possessed of a unique form of philosophical reasoning called *dialectic* or *dialectical materialism* that somehow offered more penetrating insights than did conventional logic.

In the usage of Lenin and subsequent Communists, calling something science or scientific meant that something was true. To real scientists, the term has nearly the opposite meaning, connoting a search for truth in which all conclusions are provisional.

In the end even Communist leaders themselves acknowledged that the legacy they inherited was less one of science than dogma. In the USSR Mikhail Gorbachev, under the rubric *glasnost,* reversed the tradition initiated by Lenin and opened the way to freedom of speech. And in China Deng Xiaoping (1904–1997) sought to undo Mao's worship of doctrine, coining the slogan: "It does not matter if a cat is black or white as long as it catches mice."

If free inquiry and acceptance of the notion that all conclusions are subject to revision in the face of new evidence are the touchstones of science, then Communism presented an environment that was inimical to science. This went even a step further in China, where for a time Mao actively discouraged the reading of books and education other than practical training. Peasants, although harshly exploited in collective farms, were nonetheless held by Mao to be the repositories of revolutionary virtue, and urbanites that fell afoul of the regime were often exiled to the countryside to "learn from the peasants." During the Great Proletarian Cultural Revolution (1966–1977), schools were closed for years as teenagers were mobilized into perpetual street mobs in byzantine power struggles between rival party factions. All of this fierce anti-intellectualism, so at odds with traditional Chinese reverence for education, was justified as being egalitarian and antielitist. But such contempt for formal learning was also antiscientific.

Communist Achievements in Science and Technology

This is not to say that Communist societies were without their accomplishments in scientific fields. There were some, particularly in engineering and applied research. Great investments were made in military equipment and in other technologies such as space exploration that were part of a symbolic competition with the capitalist world. Moreover Communist regimes were disdainful or indifferent to *soft* fields of scholarship—the arts, humanities, and social sciences—so that the finest minds of these societies almost necessarily found their outlets in hard science or engineering.

In addition to discouraging free inquiry, Communist regimes sometimes intervened directly in scientific questions, most famously when Stalin directed Soviet biology to embrace the tenets of Trofim Denisovich Lysenko (1898–1976). Ironically, in light of Marx and Engels's belief that their theories were analogs to Darwin's, Lysenko was a Soviet scientist who dissented from a key tenet of Darwin's principle of natural selection. Lysenko believed that acquired, as opposed to inherited, traits could be passed on genetically. Because Stalin had a deep fondness for great projects of social engineering, the idea that one might alter life itself in this manner appealed greatly. For some years genetic research in the USSR was forced to devote itself to Lysenko's eventually discredited theories.

Through the concentration of material and human capital, Communist regimes competed effectively, albeit usually coming in second, in the fields of weaponry and space exploration. Sometimes what these endeavors lacked in fine-tuning they made up for in size—for example, less accurate missiles armed with larger warheads. Usually they competed a lot less well in technologies devoted to consumer goods. The lack of marketplace incentives to maintain or improve the quality of products, combined with the general dampening of innovation and the low priority given to economic planning involving consumer goods, resulted in a generally shoddy quality of merchandise. Popular discontent on this score was an important factor that eventually resulted in pressure for political change in China and the USSR.

The most singular episode in the history of technology under Communism was the Great Leap Forward (1957–1960), a program guided by Mao's conviction that a collective farm could produce industrial as well as agricultural goods and thereby become completely self-sufficient. In a fervent national campaign from which dissent was not tolerated, collectives began trying to produce industrial goods including that sine quo non of industry, steel. Mao announced that small *backyard* smelters could replace large steel mills. One of the many flaws in this theory was the absence of thought given to the question of material inputs for these smelters. Egged on and intimidated, peasants felt compelled to contribute not only scrap but whatever was available in existing tools and utensils, so that these might be melted down to make new steel. Little real steel was produced by this method, but many small tools and even cooking woks were sacrificed. Add to this the sacrifice of peasant labor diverted from the fields, and the result was a mass famine during the years 1959–1962 that most sinologists estimate took some 30 million or more lives.

Ethics: New Ends Justify Any Means

The large-scale loss of life under Communism in China, the USSR, and a few other places, notably Cambodia and North Korea, highlights the ethical issues raised by Communism. Although the facts of these cases were once hotly disputed, for the most part disputes ended when successor Communist rulers acknowledged the respective tragedies. That is, the deaths caused by Stalin's regime in the USSR were decried first by Nikita Khrushchev (1894–1971), then more fully by Gorbachev. The depredations of Pol Pot (1926–1998) were roundly denounced by the Communists who threw him out of power in Cambodia. And some of the carnage caused by Mao—that associated with the Great Proletarian Cultural Revolution—was recognized at least implicitly after Deng Xiaoping took the helm in China in 1978, although Mao was not directly blamed.

The needless deaths of large numbers of human beings would in itself seem to constitute a moral transgression of the highest order. And yet under Communism this was not deemed axiomatic. Communists asserted the moral standards that were traditional to Christianity (and in the East to Confucianism or other longstanding codes) were themselves expressive of the domination of the wealthy classes. As Lenin put it: "People always have been the foolish victims of deception and self-deception in politics, and they always will be until they have learned to seek out the *interests* of some class or other behind all moral, religious, political and social phrases, declarations and promises." (Lenin 1969).

Mao Zedong waves to the cheering crowd at Tiananmen Square in Beijing as they celebrate May Day, 1967. Mao was influenced by the writings of Marx and Lenin, but was also inescapably a Chinese nationalist. He believed that the communist revolution in China was distinct from all others because of the weight of its history and culture. *(Getty Images.)*

Therefore the proletariat would embody its own ethical standards. And these would be closely tied to the fulfillment of its mission to overthrow capitalism and usher in a new historical age. Communism would provide a fulfilling life for all people, and since society would no longer be divided by classes, it would make possible for the first time the emergence of truly universal moral principles.

Since so much is at stake in the triumph of the socialist revolution—nothing short of the achievement of humankind's ultimate destiny—everything must be put at the service of this goal. As Lenin wrote:

> "Our morality is entirely subordinated to the interests of the proletariat's class struggle. When people tell us about morality, we say: to a Communist all morality lies in conscious mass struggle against the exploiters. We do not believe in an eternal morality Communist morality is based on the struggle for the consolidation and completion of communism." (Lenin 1968).

In taking this approach, Lenin rested on a strong but nonetheless ambiguous tenet in Marxist theory. Marx and Engels asserted that all ideas spring from class roots, which suggests that no objective ethical standards exist. Yet their condemnation of capitalism drew its power from its implied moral terms. Marx and Engels often claimed that they had done no more than lay bare the laws of history, showing that capitalism was destined to be replaced by socialism. But if so, there was no reason to work for the advancement of socialism. In practice Marx and Engels worked with all the energy they could muster. They were as much activists as philosophers, and the only explanation for this, even if implicit, was that socialism was not only inevitable but also highly desirable—which implies some standard of good and bad.

At the same time, Marx also proclaimed that "Communism is the riddle of history solved." If indeed this is the case, then it is hard to take exception to Lenin's very instrumental approach to ethics, for nothing else could possibly take priority. The achievement of Communism would be the measure of all things.

A companion aspect of the view that all else must be subordinated to the fulfillment of the destiny of humankind as a whole is that any given individual's well being might be subordinated to this higher, collective good. As explained by Aleksandr F. Shishkin, author of the leading Soviet text on ethics, Communist morality teaches the individual "not to look upon himself as an end in himself." Rather "the new society cultivates the individual in such fashion as to cause him to see the fullness of human existence to lie in struggle for a common cause and to be able to resolve in favor of society any contradiction arising between the needs of society and his personal ambition" (Shishkin 1978, p. 88).

JOSHUA MURAVCHIK

SEE ALSO *Chinese Perspectives; Lysenko Case; Marxism; Marx, Karl; Russian Perspectives; Socialism.*

BIBLIOGRAPHY

DeGeorge, Richard T. (1969). *Soviet Ethics and Morality*. Ann Arbor: University of Michigan Press.

Graham, Loren R. (1972). *Science and Philosophy in the Soviet Union*. New York: Knopf.

Graham, Loren R. (1993). *Science in Russia and the Soviet Union: A Short History*. New York: Cambridge University Press.

Graham, Loren R., ed. (1990). *Science and the Soviet Social Order*. Cambridge, MA: Harvard University Press.

Heller, Mikhail, and Aleksandr M. Nekrich. (1986). *Utopia in Power: The History of the Soviet Union from 1917 to the Present*. New York: Summit.

Kamenka, Eugene. (1969). *Marxism and Ethics*. New York: St. Martin's Press.

Kolakowski, Leszek. (1979). *Main Currents of Marxism*. Oxford: Oxford University Press. The work consists of three volumes; see especially volumes I and III.

Lenin, Vladimir I. (1969). *Three Sources and Three Component Parts of Marxism*. Moscow: Progress Publishers. Originally published in 1913.

Lenin, Vladimir I. (1968). *The Tasks of the Youth Leagues*. Moscow: Progress Publishers.

Lenin, Vladimir I. (1977). "Materialism and Empirico-Criticism." In *Collected Works*, Vol. 14. Moscow: Progress Publishers.

Marx, Karl. (1978). "Economic and Philosophical Manuscripts of 1844." In *The Marx-Engels Reader*, ed. Robert C. Tucker. Text originally published in 1844.

Medvedev, Zhores A. (1978). *Soviet Science*. New York: Norton.

Pipes, Richard. (1993). *Russia under the Bolshevik Regime*. New York: Knopf.

Shishkin, Aleksandr F. (1978). "On Moral Values in the Contemporary World." *Soviet Studies in Philosophy* 17(1): 71–99.

COMMUNITARIANISM

Communitarianism is part of the neo romantic reaction to rationalism. It emphasizes moral and social values and the societal institutions that support them, especially community and its traditions, passions and beliefs, religion, and the habits of the heart. Communitarianism is not blind to facts and logic, the cool calculations of the rational mind, or the importance of science, technology, and economic progress. Nevertheless, it is concerned that such perspectives may override, if not ignore, other *human* considerations, to which communitarianism is attentive. For the same reasons, communitarianism seeks to balance concern for individual rights and liberty with concerns for the common good and community.

Definition and History

The term *communitarian* was first introduced in 1841, to mean "of, pertaining to, or characteristic of a community or communistic system; communitive." It was infrequently employed from then until the mid-twentieth century.

Several critics have argued that the concept of the community is of questionable value because it is so ill-defined. In *The Myth of Community Studies*, Margaret Stacey (1974) argues that the solution to this problem is to avoid the term altogether. In the same publication, Colin Bell and Howard Newby similarly point out, "There has never been a theory of community, nor even a satisfactory definition of what community is" (p. xliii).

Amitai Etzioni (1996) has nevertheless argued that community can be defined with reasonable precision. Community has two characteristics: first, a web of affect-laden relationships among a group of individuals, relationships that often crisscross and reinforce one another (as opposed to one-on-one relationships); and second, a measure of commitment to a set of shared history and identity—in short, a particular culture. David E. Pearson stated, "To earn the appellation 'community,' it seems to me, groups must be able to exert moral suasion and extract a measure of compliance from their members. That is, communities are necessarily, indeed by definition, coercive as well as moral, threatening their members with the stick of sanctions if they stray, offering them the carrot of certainty and stability if they don't" (Pearson 1995, p. 47)

Among early sociologists whose work is focused on communitarian issues (though they did not draw on the term) are Ferdinand Tönnies (1855–1936), especially his comparison of the *Gemeinschaft* and *Gessellschaft*; Emile Durkheim (1858–1917), particularly his studies of the socially integrating role of values and the relations between the society and the person; and George Herbert Mead (1863–1931) in his work on the self. Other early relevant sociological works are those of Robert E. Park, William Kornhauser, and Robert Nisbet.

While the term *communitarian* was coined in the mid-nineteenth century, ideas that are essentially communitarian appear much earlier. They are found in the Old and New Testaments, Catholic theology (for example, the emphasis on the Church as a community), more recently in socialist doctrine (for example, writing about early communes and workers' solidarity), and finally *subsidiarity*—the principle that the lowest level of authority capable of addressing an issue is the one best able to handle it. In essence, moral judgments are best made at the community level rather than from the higher governing bodies.

Balancing Liberty with the Common Good

In the 1980s, communitarianism was largely advanced by political theorists Charles Taylor, Michael Sandel, and Michael Walzer. They criticized liberalism for overlooking that people can have a strong attachment to their societies. They lamented liberalism's focus on individualistic self-interest.

Since that time, two main forms of communitarianism have emerged. Authoritarian communitarians, who typically concern themselves with Asian culture, argue that to maintain social harmony, individual rights and political liberties must be curtailed. Some emphasize the importance of the state to maintain social order (for instance, leaders and champions of the regimes in Singapore and Malaysia), and some focus on strong social bonds, morality, and traditional culture (as in Japan). Some Asian communitarians also hold that the West's notion of liberty actually amounts to anarchy, that strong economic growth requires limiting freedoms, and that the West uses its idea of legal and political rights to chastise other cultures.

In 1990 a new school of communitarianism developed. Among its leading scholars are political theorist William A. Galston, legal scholar Mary Ann Glendon, political scientist Thomas Spragens, Jr., writer Alan Ehrenhalt, and sociologists Philip Selznick, Robert Bellah and his associates, and Amitai Etzioni. The work of these authors laid the foundations in 1990 for the second form of communitarianism: responsive (democratic) communitarianism.

Responsive communitarianism assumes that societies have multiple and not wholly compatible needs, in contrast to philosophies built on one core principle, such as liberty. In communities, there is an irrepressible tension between exclusion and inclusion, and between civility and piety. Thus community is not a restful idea, a realm of peace and harmony. On the contrary, community members must recognize and deal with competing principles. Responsive communitarianism assumes that a good society is based on a balance between liberty and social order, and between particularistic (communal) and society-wide values and bonds. This school stresses the responsibilities that people have to their families, kin, communities, and societies. These exist above and beyond the universal rights that all individuals command, which is the main focus of liberalism.

While a carefully crafted balance between liberty and social order defines a generic concept of the good society, communitarians point out that the historical-social conditions of specific societies determine the rather different ways that a given society in a given era may need to change to attain the same balance. Thus, contemporary Japan requires much greater tolerance for individual rights, while in the American society excessive individualism needs to be curbed.

To achieve this balance, unlike laissez faire conservatives and welfare liberals who differ mainly with regard to the respective roles of the private sector and that of the state, communitarians are especially concerned with the third sector, that of civil society. They pay special attention to the ways that informal communal processes of persuasion and peer pressure foster social responsibilities for the common good.

Communitarians are also concerned with the relationship between the self and the community. Political theorists depict the self as "embedded," implying that the self is constrained by the community. Responsive communitarians stress that individuals who are well integrated into communities are better able to reason and act in responsible ways than are isolated individuals, but if social pressure to conform rises to high levels, it will undermine the individual self and therefore disrupt the balance.

This issue is reflected in questions that arise when associations of scientists and professions such as engineering address ethical and policy issues relevant to their work. Should the decisions involved, say whether or not to proceed with human cloning, be made by each scientist or by their informal communities or associations? And what role, if any, should the public and its elected representatives have in making these decisions? Closely related are similar questions such as to how to deal—and above all, who should deal—with instances of fraud in research, misappropriation of funds, and violations of security.

Communitarianism's Critics

Critics generally suggest that those who long for communities ignore the darker side of traditional communities. "In the new communitarian appeal to tradition, communities of 'mutual aid and memory,'" writes Linda McClain (1994), "there is a problematic inattention to the less attractive, unjust features of tradition" (p. 1029). Amy Gutmann (1985) pointedly remarks that communitarians "want us to live in Salem" (p. 319), a community of strong shared values that went so far as to accuse nonconformist members of witchcraft during the seventeenth century.

Communitarians counter that behind many of these criticisms lies an image of old, or total, communities, that are neither typical of modern society nor necessary for, or even compatible with, a communitarian society. Old communities (traditional villages) were geographically bounded and the only communities of which people were members. In effect, other than escaping into no-man's-land, often bandit territories, individuals had few opportunities for choosing their social attachments. In short, old communities had monopolistic power over their members.

New communities are often limited in scope and reach. Members of one residential community are often also members of other communities, for example work, ethnic, or religious ones. As a result, community members have multiple sources of attachments; if one community threatens to become overwhelming, individuals will tend to pull back and turn to another for their attachments. Thus, for example, if a person finds herself under high moral pressure at work to contribute to the United Way, to give blood, or to serve at a soup kitchen for the homeless, and these are lines of action she is not keen to follow, she may end up investing more of her energy in other communities—her writers' group, for instance, or her church. This multi-community membership protects the individuals from both moral oppression and ostracism.

Another criticism is that communities are authoritarian. Derek Phillips (1993), for instance, remarks, "[C]ommunitarian thinking ... obliterates individual autonomy entirely and dissolves the self into whatever roles are imposed by one's position in society" (p. 183). As the political scientist Robert Booth Fowler (1991) puts it, critics "see talk of community as interfering with the necessary breaking down of dominant forces and cultures" (p. 142). Some critics mean by this that communities are totalistic, a point already covered. Others mean that they are dominated by power elites or have one group that forces others to abide by the values of those in power.

Communitarians find that this criticism has merit but is misdirected. There are communities both past and present that have been or still are authoritarian. The medieval phrase *Stadt Luft macht frei* ("the air of the cities frees") captures what the farmers of traditional villages must have felt when they first moved into cities at the beginning of the industrial era. (Poor working conditions and slums aside, being away from the stricter social codes of their families and villages seems to have given them a sense of freedom, which in some cases led to anarchic behavior.) Totalitarian communities exist in contemporary societies, such as North Korea. However, most contemporary communities, especially in communitarian societies, are not authoritarian even when they are defined by geography. Also, the relative ease of mobility means that people often choose which community to join and within which to live. Agnostics will not move into a Hasidic community in Brooklyn, and prejudiced whites will not move into a neighborhood dominated by the Nation of Islam.

Science and technology help open up societies and they promote relatively empirical, rational approaches to the world. New communications technologies, such as the Internet and satellite dishes, help undermine authoritarian regimes. However, no one should assume that on their own, these devices are capable of delivering a truly democratic state—especially when such technological advances are not accompanied by a proper change in values, as has been seen in Russia, Singapore, and China in the early twenty-first century.

Contemporary Issues

Communitarians have developed several specific concepts and policies that draw on their philosophy. They favor shoring up families, not traditional-authoritarian ones but peer marriages (in which mothers and fathers have equal rights and responsibilities). They fostered schools that provide character education rather than merely teach, but avoid religious indoctrinization. They developed notions of community justice, in which offenders, victims, and members of the community work together to find appropriate punishments and meaningful reconciliation. Communitarians favored devolution of state power, and the formation of communities of communities (within national societies and among nations), among many other policies.

Following the growing popularity of the concept of civic society, Etzioni (1999) argues that contemporary civic society is insufficient because it tends to be morally neutral on all matters other than the attributes that citizens need to make themselves into effective members of a civic society, for instance, the ability to think critically. In contrast, a good society seeks to promote a core of substantive values, and thus views some voluntary associations and social activities as more virtuous than others.

In the same vein, communitarians argue that while everyone's right to free speech should be respected, some speech—seen from the community's viewpoint is morally highly offensive and when children are exposed, damaging. For instance, the (legal) right to speak does not render verbal expressions of hate (morally) right.

Science has long been associated with rational thinking and in turn with secularism. Indeed, historically, science has often been considered antithetical to religion. However, communitarians are concerned with the moral fabric of society and they find religion one source of moral values. A communitarian may prefer to divide the issues people face among those that are subject to rational or scientific analysis and those that belong to a different sphere, reserved for belief. These include questions such as is there a god, why people are cast in this world born to die, what people owe their children and members of their community, among others.

Closely related is the question of a proper balance between the two *sectors*. Since the enlightenment, the sector of rationality (and within it science and technology) has increased dramatically in western societies. Communitarians ask whether in the process resources and time dedicated to the family, social and public life, culture, and spiritual and religious activities have been neglected.

While sociologists made numerous contributions to altered communitarian thinking, in turn communitarian philosophy has challenged sociology to face issues raised by cross-cultural moral judgments. Sociologists tend to treat all values as conceptually equal; thus, sociologists refer to racist Nazi beliefs and those of free societies by the same "neutral" term, calling both *values*. Communitarians instead use the term *virtue* to indicate that some values have a high moral standing because they are compatible with the good society, while other values are not and hence they are "aberrant" rather than virtuous.

AMITAI ETZIONI

SEE ALSO *Community; Durkheim, Émile; Liberalism; Libertarianism; Neoliberalism.*

BIBLIOGRAPHY

Bell, Daniel A. (2000). *East Meets West: Human Rights and Democracy in East Asia*. Princeton, NJ: Princeton University Press. Is liberal democracy universal? Daniel Bell explores this question with a look at Asian societies. Bell criticizes the "Asian values" perspectives, but finds that drawing on East Asian cultural traditions can create a model for political institutions that differs substantially from western-style democracy.

Bellah, Robert Neelly, ed. (1996). *Habits of the Heart: Individualism and Commitment in American Life*. Berkeley: University of California Press. Argues that democratic communities must focus on forces that humanize markets and government administrative powers.

Etzioni, Amitai. (1996). *The New Golden Rule*. New York: Basic Books. A foundational work of new communitarian theory. This book argues for the need to balance freedom with morality, and automony with community. It paves the way for a social order largely based on moral suasion rather than coercion.

Etzioni, Amitai. (1999). *The Limits of Privacy*. New York: Basic Books. An exploration of the right to privacy and the potentially negative impact it can have on public health and safety. Suggests criteria when privacy ought to yield and when it needs to be further extended. Case studies include sex offenders; HIV testing; medical records; ID cards; and encrypted communications.

Fowler, Robert Booth. (1991). *The Dance with Community: The Contemporary Debate in American Political Thought*. Lawrence, KS: University Press of Kansas.

Glendon, Mary Ann. (1993). *Rights Talk: The Impoverishment of Political Discourse*. New York: Free Press. Contends that the rise of the language of rights in Western discourse has stifled real discussion, debate, and even the liberties that it seeks to protect.

Gutmann, Amy. (1985). "Communitarian Critics of Liberalism" *Philosophy And Public Affairs* 14 (3): 308–322.

Kornhauser, William. (1959). *The Politics of Mass Society*. Glencoe, IL: The Free Press.

McClain, Linda C. (1994). "Rights and Irresponsibility" *Duke Law Journal* 43 (5): 989.

Nisbet, Robert. (1962). *Community and Power*. New York: Oxford University Press.

Pearson, David E. (1995). "Community and Sociology" *Society* 32 (5): 44–50.

Phillips, Derek L. (1993). *Looking Backward: A Critical Appraisal of Communitarian Thought*. Princeton, NJ: Princeton University Press.

Sandel, Michael. (1998). *Liberalism and the Limits of Justice*. Cambridge, UK: Cambridge University Press. Liberal societies try to avoid imposing value judgements on their members so that each individual is free to pursue the principles that meet with their interpretation of the good life. But do such principles exist? Sandel critiques this form of democratic liberalism.

Stacey, Margaret. (1974). "The Myth of Community Studies" In *Sociology of Community: A Collection of Readings*, ed. Colin Bell and Howard Newby. London: Frank Cass & Co.

INTERNET RESOURCES

The Communitarian Network. Available from http://www.gwu.edu/~ccps/index.html.

COMMUNITY

Community is a term with widely varying historical and current meanings in both specialized and everyday discourse. It also possesses several dimensions ethical, political, social, ontological, psychological, and epistemological—many of which are relevant to discussions of science and technology.

Theorists generally consider community to be a good that, carried too far, may undermine its own moral and political values for those both within and outside it. Community is an important source of meaning in human lives, and it encompasses the sets of values, beliefs, and interpretative frameworks by which the world takes on meaning. Indeed, the scientific and technological enterprise is often described as dependent on the special values of a scientific or technical community (Merton 1942). Community, however, may also manifest itself in oppressive political forms that defy universal values, shared rights, or basic forms of well-being of certain members of a society in the name of community. Political forms of community such as nationalism or populist fascism, or Thomas Hobbes's or Jean-Jacques Rousseau's different versions of collective identity, may belie other human values such as individual liberty. Members of the scientific community have also sometimes ignored the rights of nonscientists or the larger social orders of which science is a part.

At a minimum, community is a set of shared goals or values perceived as good by those who participate in their formation or by those who belong to the heritage these shared values define. A qualitative sense of belonging therefore attends community, and a broader notion of community also includes common language, rituals, geographical territory, religion, historical memory, and ethnic identification.

Community versus Society

In 1887 the German sociologist Ferdinand Tönnies developed the distinction between community and society in terms of the informal, moral, familial bonds of traditional communal life and the formalized and impersonal, amoral, juridical, and administrative relations of industrial society. "Community" was taken to have an organic quality, whereas society was mechanistic in nature. The importance of this distinction relates to the sense of belonging in social relations and has applications for notions of citizenship, the legitimacy of political representation, ideas of the common good, and the meaning of public participation. "Society" corresponds to the "neutral" structural conditions of modern life. For better or worse, Tönnies's distinction has served to circumscribe much of the sociological, political, and moral meaning of community to this day. In contemporary political thought, for example, the distinction is manifested in terms of holistic communitarianism versus individualistic liberalism. While complex, these latter terms highlight the relative importance of participation in political life, whether the good is best articulated collectively or individually, and the extent to which institutions should choose between a fully embodied moral community and a minimally protective framework for individual liberties.

If one assumes that shared values and frameworks of belief are paramount in the legitimate governance of societies, Tönnies's distinction between community and society has also influenced modern science and technology in important ways. Twentieth-century critics of technology as varied as Martin Heidegger, Herbert Marcuse, Jacques Ellul, Ivan Illich, and early Jürgen Habermas maintain that the intrinsic qualities of communal life were slowly eroded by a postindustrial society of "technoscientific," instrumental emphasis on values of use and efficiency. Langdon Winner (1986) further argues that technological choices determine broader social and administrative structures and reframe the conditions of moral and political life, even though these choices remain beyond the scope of communities. In such views, modern society is an "organizational" society in which rationalizing "technoscientific" approaches to social organization root out the affective (emotional) characteristics of sociality and the bonds of community that Tönnies and others ascribe to community. Expert management replaces participation and communal frameworks of value as the main force of modern social and value formation. If moral value is rooted in community, then technical social management entails institutions that express a small set of values disguised as socially neutral instruments. Others argue, more specifically, that the global spread of modern technologies has served to destroy traditional cultures, communities, and economies and to undermine modern values such as sustainability (see, for example, Helena Norberg-Hodge's studies of Ladakh [1991]).

Scientific and Democratic Community

In contrast, John Dewey (1954 [1927]) and others argue that technological society, especially through new communications technologies, harbors the potential to revive local community. Similarly, Thomas C. Hilde (2004) suggests that the integration of modern technologies and science into the global formation of norms presents not only risks to traditional notions of community but also new possibilities. For Hilde this framework constitutes an "epistemic cosmopolitanism" capable of facilitating new forms of community.

The scientific sense of communal inquiry and of the production of knowledge is further developed by Peter M. Haas (1990) and others as "epistemic community." Epistemic communities are, according to Haas, scientists and others united by both causal explanations (of, for example, ecological damage) and shared values regarding which policies should emerge from scientific evidence.

If scientific inquiry always harbors the preferences of a broader community and social organization in which scientists work (Kuhn 1962, Longino 2002, Harding 1998), then the currently dominant utilitarian values enframe the broader technological/scientific project. This, in turn, constricts the range of values embodied in technical decisions that influence the shape of society and its future policy outcomes. If this basic thesis regarding the importation of value is correct, then both community and scientific inquiry merit further discernment of beneficial preferences from damaging ones, and in such cases deliberation may be better sought through the broader community. Scientific inquiry might then better serve to advance not only the knowledge of the broader community, but also its methods of inquiry.

THOMAS C. HILDE

SEE ALSO *Civil Society; Communitarianism.*

BIBLIOGRAPHY

Bauman, Zygmunt. (2001). *Community: Seeking Safety in an Insecure World.* Cambridge, UK: Polity Press.

Dewey, John. (1954 [1927]). *The Public and Its Problems.* Athens, OH: Swallow Press.

Feyerabend, Paul. (1993). *Against Method,* 3rd edition. London: Verso.

Haas, Peter M. (1990). *Saving the Mediterranean: The Politics of International Environmental Cooperation.* New York: Columbia University Press.

Habermas, Jürgen. (1987). *The Theory of Communicative Action,* Vol. 2: *Lifeworld and System,* trans. Thomas McCarthy. Boston: Beacon Press.

Harding, Sandra. (1998). *Is Science Multicultural? Postcolonialisms, Feminisms, and Epistemologies.* Bloomington: Indiana University Press.

Hilde, Thomas C. (2004). "The Cosmopolitan Project: Does the Internet Have a Global Public Face?" In *The Internet in Public Life,* ed. Verna V. Gehring. Lanham, MD: Rowman and Littlefield.

Kuhn, Thomas. (1962). *The Structure of Scientific Revolutions.* Chicago: University of Chicago Press.

Longino, Helen E. (2002). *The Fate of Knowledge.* Princeton, NJ: Princeton University Press.

Marcuse, Herbert. (1964). *One-Dimensional Man.* Boston: Beacon Press.

Merton, Robert K. (1942). "Science and Technology in a Democratic Order." *Journal of Legal and Political Sociology* 1: 115–126. Reprinted as "The Normative Structure of Science." In *The Sociology of Science: Theoretical and Empirical Investigations,* ed. Norman W. Storer. Chicago: University of Chicago Press, 1973.

Norberg-Hodge, Helena. (1991). *Ancient Futures: Learning from Ladakh.* San Francisco: Sierra Club Books.

Peirce, Charles S. 1992 (1868). "Some Consequences of Four Incapacities." In *The Essential Peirce,* Vol. 1, ed. Nathan Houser and Christian Kloesel. Bloomington: Indiana University Press.

Taylor, Charles. (1985). *Philosophical Papers,* Vol. 2: *Philosophy and the Human Sciences.* Cambridge, UK: Cambridge University Press.

Tönnies, Ferdinand. (1957 [1887]). *Community and Society.* New York: Harper and Row.

Winner, Langdon. (1986). *The Whale and the Reactor: A Search for Limits in an Age of High Technology.* Chicago: University of Chicago Press.

COMPLEMENTARY AND ALTERNATIVE MEDICINE

The term *alternative medicine* refers to therapies and diagnostic procedures that are used instead of those of conventional medicine, whereas *complementary medicine* refers to therapies and diagnostic procedures that are used in addition to those of conventional medicine. The same therapy can be alternative or complementary, depending on its use. For example, a dietary program for treating cancer sometimes is used as a complement to surgery but also may be employed as an alternative to chemotherapy. The term *complementary and alternative medicine* (CAM) is a standard way of referring to both, whereas *integrative medicine* refers to medical practices that bring together conventional medicine and CAM. *Experimental medicine* refers to therapies, usually drugs, that are undergoing testing for regulatory approval.

Classification of CAM

The National Center for Complementary and Alternative Medicine (2003) of the U.S. National Institutes of Health classifies CAM into the following subcategories: alternative medical systems such as Chinese medicine and naturopathic (a type of nutritional and dietary) medicine; mind-body interventions that are not mainstream, such as prayer, meditation, and mental healing; biologically based therapies such as dietary supplements and herbs (one also would include here immunological therapies that are not in clinical trials); manipulative therapies such as chiropractic; and therapies based on electromagnetic energy or forms of energy that are not accepted by contemporary science.

Applying Bioethics Principles

Most of the literature on medical ethics and CAM is based on the mainstream bioethical principles of beneficence (guiding and helping a patient), nonmaleficence (avoiding harmful and futile treatments), autonomy (protecting a patient's informed consent to choose treatments), and justice (fairness in terms of the right of access). There have been attempts to introduce other principles that are relevant to CAM (Guinn 2001), but the vast majority of the discussions take place in the context of conventional bioethical principles (Sugarman and Burk 1998). Furthermore, most ethics discussions related to CAM focus on the relationship between the health-care provider and the patient. The principle of justice gets much less attention.

Patients' use of CAM therapies has grown since the 1980s, and in the United States CAM increasingly is offered through licensed alternative professions such as naturopathy, chiropractic, and acupuncture/Chinese medicine. Other countries also provide legal recognition of various types of CAM providers. In the United States the Federation of State Medical Boards (2002) developed guidelines for physicians who use CAM or work with licensed CAM providers. The statement outlines three types of possible harm from CAM: economic harm from spending money on futile therapies, indirect harm caused by avoiding efficacious conventional therapies or having hopes raised falsely, and direct harm caused by negative side effects of CAM therapies. The statement also specifies four ethically relevant categories of CAM: documented as effective and safe, documented as effective but with side effects and risk, inadequately studied but safe, and ineffective and dangerous. The federation suggests that physicians recommend CAM treatment when there is a favorable risk-benefit ratio, a favorable expected outcome, and a greater benefit with CAM than with no treatment. Under those conditions physicians should not lose their licenses for recommending CAM and should respect a patient's right to choose CAM (the principle of autonomy).

The guidelines provide some help in answering two of the most frequently discussed ethical issues regarding CAM: obligation to inform and obligation to treat. Is a physician obligated to inform a patient of an available CAM option? Failure to do so for conventional therapies generally is considered a violation of informed consent, and this principle is being extended to CAM therapies, but only if they meet fairly high standards for efficacy and/or safety. Is a physician obligated to treat a patient if the patient has full informed consent with regard to the risk-benefit ratio yet opts instead for a CAM therapy that the physician considers dangerous? Here there is a conflict between the principles of beneficence (the physician's assumed superior knowledge) and maleficence (the purported danger of the CAM therapy) and the principle of autonomy (the patient's right to choose). Frequently in this situation physicians will refuse additional treatment or comanagement of the case. In addition to their ethical defense based on concern for the patient's well-being physicians may cite their personal risk of malpractice litigation or loss of license (Studdert et al. 1998).

The guidelines represent a significant shift from older medical approaches to CAM, which dismissed it as quackery and considered recommendations to consider CAM therapies to be unethical. However, ethical ambiguities and questions remain.

First, frequently patients opt to receive a CAM therapy from a nonlicensed provider, such as a noninvasive spiritual or mind-body therapy that often is associated with a patient's religious belief in shamanism, spiritualism, or evangelical faith healing. How should physicians or CAM professionals answer questions about healing services offered by religious groups or other nonlicensed providers? Should absence of indirect and direct harm suffice to warrant discussions or even referrals?

A second ambiguity is the question of what constitutes adequate evidence for evaluations of safety and efficacy. In contemporary medical science evidence usually is organized in a hierarchy of credibility. At the top is the gold standard: the controlled clinical trial. In this form of research patients are divided on a random basis into two or more groups, one with the test therapy (in this case a CAM therapy) and others with placebos or conventional therapies. The following alternative methods often are viewed in a descending order of evidential value: a retrospective form of data analysis that takes existing cases, such as patients who used a CAM therapy, and compares them with a control group; the best case series, which shows promising results in a series of patients but lacks a statistical analysis with a comparison group; subclinical research such as experiments that test CAM substances on animals or cell cultures; and a lower level of subclinical research that provides biochemical analyses of a CAM substance (such as an herb) to determine if it has any known pharmacologically active agents. A significant debate has emerged regarding the value of clinical trials versus other methods for the evaluation of CAM (Hess 1999).

Most CAM therapies lack a body of consistent clinical trials with supporting evidence at the other levels. If the evidence were complete, consistent, and highly positive, the therapy probably would be considered conventional, not CAM. As a result both conventional and CAM providers face the dilemma of making recommendations in the absence of complete evidence. In many cases there is only some, often mixed, evidence for efficacy, but there is a long record of use with few or no risks, side effects, or negative interactions with other therapies. In such cases physicians who practice integrative medicine sometimes will add CAM therapies, but only as a complementary modality.

To understand some of the complexities one can consider the case of a patient whose tumor has metastasized, or spread, to other organs. Surgery was only partially successful, and the oncologist recommends additional chemotherapy. The chemotherapy for this tumor type has serious side effects and is not curative; it prolongs life for a few weeks or months at the cost of highly reduced quality of life. There are some CAM treatments with claims of long-term survival, but those treatments are expensive. There have not been clinical trials yet, but there are a few case study series that show impressive remissions, and there is a good biological rationale with some supportive subclinical data. If the patient opts for the alternative therapy instead of chemotherapy, is the oncologist's decision to abandon the patient ethically justified?

Justice Issues

A broader set of ethical issues involves the principle of justice. Conventional providers often place the responsibility for gathering evidence with CAM providers. They argue that it is unethical for CAM providers to offer therapies to patients without providing adequate evidence in support of their claims of therapeutic benefit; that CAM providers should enroll patients in clinical trials or other forms of clinical evaluation; that by failing to do so those providers put personal gain ahead of potential economic, indirect, or direct harm to patients; and that it would be legitimate for the government and medical associations to close down such providers.

From the CAM perspective the same argument applies in reverse. CAM practitioners charge that the pharmaceutical industry and the members of many medical specialties are economically threatened by the potential of alternative (rather than complementary) therapies. For example, if chelation therapy (the use of mineral ions to remove cardiovascular blockages) and dietary/lifestyle programs were to replace bypass surgery, hospitals and surgeons would lose revenue. Similarly, if dietary programs were to replace chemotherapy as follow-up to surgery for solid tumors, oncologists and pharmaceutical companies also would lose money. Consequently, by failing to investigate promising CAM therapies developed by credentialed researchers or clinicians, the medical profession and affiliated industries put their own financial gain ahead of potential benefits to patients.

CAM advocates argue that the lack of ethics lies not in their failure to provide extensive positive evidence but in the long history of suppression of CAM research and therapies. They argue that clinical trials are very expensive and that their applications for research support go unfunded. Even worse, applications for research support often trigger investigations that lead to the loss of licenses or clinic closures. CAM advocates further argue that in the few cases in which public pressure has led to government-supported clinical trials (e.g., laetrile, hydrazine sulfate, and vitamin C) studies of CAM by conventional researchers have been weakened by exclusion of CAM advocates from research teams, protocol modifications that introduce biases against CAM, biased interpretation of equivocal data, and follow-up media campaigns intended to discredit CAM.

Historical research has documented suppression of CAM research and therapies (Hess 1997, Moss 1996, Richards 1981). Researchers and clinicians who have attempted to investigate CAM have faced denial of

U.S. Food and Drug Administration (FDA) investigational drug permits, dismissal from universities or other organizations, bias and blockage of publication in peer-reviewed journals, media campaigns against CAM, and loss of funding. Clinicians who use alternative therapies, particularly for cancer, have faced restraining orders, raids on clinics, warnings and denial of drug permits from the FDA, hostile tax audits, revocation of licenses and hospital privileges, and criminal charges (fraud, manslaughter, etc.) and civil lawsuits by CAM opponents.

Where cases have ended up in court, in some cases the rulings have favored CAM practitioners and in other cases the medical profession and state. Whether the historical cases represent unethical suppression of potentially beneficial therapies or an ethically legitimate watchdog function of the medical profession and state depends on one's assessment of the potential of CAM. If CAM is viewed as largely the product of quacks who want to make money from suffering patients, an ethical public policy would emphasize paternalism (protection from maleficence), suppress those alternatives, and limit the range of therapeutic options available to patients. If the promise of CAM is viewed in a more favorable light, an ethical public policy would emphasize autonomy, favor a more tolerant approach to alternatives, and increase both research funding and clinical access for CAM.

To some extent the older patterns of suppression have subsided as the medical profession has called for limited acceptance of CAM on the basis of evidence. However, although surveys continue to document high levels of patient utilization, federal government funding for CAM research amounts to less than 1 percent of funding for conventional medicine. Furthermore, the pattern of integration tends to favor complementary usage of CAM over alternative usage (Hess 2002). For example, in cancer research nutritional programs are being incorporated as complements to conventional therapies rather than as alternatives to them.

Does CAM offer the possibility of more than complementary, palliative care for chronic disease? Does it offer the potential for less toxic, less expensive, and more efficacious alternative therapies for a significant range of chronic diseases? Although the framework of evidence-based medicine can answer those questions, the lack of funding and the channeling of existing funding toward complementary therapies suggest that the answer will be deferred for many years.

DAVID J. HESS

SEE ALSO *Acupuncture; Alternative Technology; Bioethics; Clinical Trials; Drugs; Health and Disease; Medical Ethics; National Institutes of Health.*

BIBLIOGRAPHY

Federation of State Medical Boards. (2002). "New Model Guidelines for the Use of Complementary and Alternative Therapies in Medical Practice." *Alternative Therapies in Health and Medicine* 8(4): 44–47.

Guinn, David E. (2001). "Ethics and Integrative Medicine: Moving beyond the Biomedical Model." *Alternative Therapies in Health and Medicine* 7(6): 68–72.

Hess, David. (1997). *Can Bacteria Cause Cancer?* New York: New York University Press. Historical survey of research and therapy based on the assumption that cancer is an infectious disease.

Hess, David. (1999). *Evaluating Alternative Cancer Therapies*. New Brunswick, NJ: Rutgers University Press. A survey of methodological and evaluation issues as seen by leaders of the CAM cancer therapy movement.

Hess, David. (2002). "Stronger Versus Weaker Integration Policies." *American Journal of Public Health* 92(10): 1579–1581.

Moss, Ralph. (1996). *The Cancer Industry*. Brooklyn, NY: Equinox Press. The classic source on the suppression of alternative cancer therapies by a leader of the CAM cancer therapy movement.

Richards, Evelleen. (1981). *Vitamin C and Cancer*. New York: St. Martin's Press. Study of clinical trials controversy involving Linus Pauling.

Studdert, David; Eisenberg, David; Miller, Frances; Curto, Daniel; Kaptchuk, Ted; and Brennan, Troyen. (1998). "Medical Malpractice Implications of Alternative Medicine." *Journal of the American Medial Association* 280(18): 1610–1615.

Sugarman, Jeremy, and Larry Burk. (1998). "Physicians' Ethical Obligations Regarding Alternative Medicine." *Journal of the American Medical Association* 280(18): 1623–1625. Introduction to ethical aspects of CAM from the perspective of the medical profession and conventional medical ethics.

INTERNET RESOURCES

National Center for Complementary and Alternative Medicine. (2003). "What Is Complementary and Alternative Medicine?" Available from http://nccam.nih.gov/health/whatiscam/index.htm. Official definitions by the U.S. government organization.

COMPLEXITY AND CHAOS

Complexity and chaos are intuitive notions not easily rendered into formal definitions, and yet they have become increasingly important to both science and technology—and thereby to ethics. One useful way to approach complexity is through the analysis of dynamic systems.

Dynamic or changing systems are of two types: those in which knowledge of current states enables the prediction of future states, and those in which knowledge of current states does not enable the prediction of future states. In general ethics has attributed the first type of system to the world (because this appears to reflect a large part of reality, and in the absence of such a system it would be hard to hold human beings responsible for the consequences of their actions), and the second type to human beings (again because this appears confirmed by some aspects of human behavior, and without it humans could not be held accountable for voluntarily choosing to perform one action rather than another). Only since the last third of the twentieth century has scientific understanding of dynamic systems been advanced enough to explain the intellectual framework behind these two attributions.

Linear Dynamics and Its Limits

In the wake of the scientific revolution of the sixteenth and seventeenth centuries, science projected that all natural phenomena, including human actions, could be fully explained with the same logic used to predict planetary motion. According to this view, events are fully explained only when their occurrence is inferred from a covering law together with initial condition statements. The following assumptions framed this approach to explanation: (a) All phenomena are essentially atemporal or, in the case of near-equilibrium thermodynamics, independent of their history; that is, only the future, not the past, is packed into the present; (b) All phenomena are linear, that is, similar causes, under similar conditions, always produce similar results; and (c) Wholes are epiphenomenal by-products no different from aggregates and can therefore be functionally decomposed into their component parts. Insofar as these assumptions hold, all phenomena were taken to be reducible and decomposable in a way that made them tractable to deductive explanation, and thus predictable. For many centuries, in short, a deterministic, clockwork universe served as the ontological underpinning for Western epistemology and ethics.

Because free will is commonly viewed as a precondition of normative behavior (if everything is fully determined and predictable, responsibility and agency go by the board), a mechanistic worldview makes it necessary either to conclude that human beings are as determined as the rest of the universe (which yielded Calvinist ethics as its axiological counterpart), or to imagine free will as a nonnatural faculty itself uncaused but with the ability to exercise causal power (a view espoused by Immanuel Kant). For ascriptions of moral responsibility to be possible, behavior must be voluntary and caused or controlled by a *meaningful* intention, reason, or purpose (and not just triggered by a forceful Newtonian cause). Postulating free will as a nonnatural trait in human beings allowed theorists to account for the philosophical concepts of moral value and responsibility. Grounding moral responsibility on an uncaused act of will is, however, as problematic a tactic for ethical theory as the determinism it was supposed to correct: If intentions are caused by external events, then they are not freely formed; if intentions just pop into existence for no reason whatsoever, ascriptions of moral responsibility are as arbitrary as their causal origins. In any case, according to the received worldview (and paralleling the received logic of explanation), moral education consists in learning a set of universal moral principles and then exercising free will to implement the specific normative prescriptions that follow from those principles in particular circumstances.

In the nineteenth century, the mechanistic framework was challenged by the appearance of two new scientific theories: thermodynamics and evolution. Unlike the time-reversible equations of Newtonian mechanics, the second law of thermodynamics postulates an arrow of time. For near-equilibrium thermodynamics, usable energy decreases inexorably over time, a death march that will ultimately end in a state characterized by a complete lack of energy potential. Since usable energy is associated with order, and unusable energy is associated with disorder, Victorians worried about the ethical implications of thermodynamics.

Charles Darwin's theory of evolution, by contrast, appeared to identify the mechanism responsible for the increasing complexity and order characteristic of ontogeny and phylogeny. Nineteenth-century moralists did not quite know what to make of Darwin's ideas. On one hand, they were welcomed because the sequence of creation described by Genesis—from simple organisms to the most complex human beings—seemed to find support in the trajectory of evolution. On the other, his ideas were uncomfortable insofar as evolution suggested that nature was *red in tooth and claw*, removed altruism and agape from the natural realm, and called into question the origin and ontological status of the human mind and soul. Finally, because of the role of random mutations in evolution, its trajectory was shown not to be predictable and determinable, even in principle, an obstacle that made Darwin (who subscribed to the deductive logic of explanation) doubt that evolution was even explicable.

Attempts to force evolution (and biology in general) to fit the mechanistic view met with failure time and again; it became clear that organisms are not clockwork-like. Because complex systems (including biological organisms) are described by second order, nonlinear differential equations that are not formally solvable, they were for centuries considered intractable.

Nonlinear Dynamics and Its Achievements

The advent of computer simulation changed all this. Computer simulation research during the last quarter of the twentieth century demonstrated that turbulent flow and other seemingly *chaotic* processes in fact exhibit a very sophisticated form of order that is nevertheless unpredictable in detail. In the early 1960s, Edward Lorenz of the Massachusetts Institute of Technology (MIT) discovered the underlying mechanism responsible for deterministic chaos. Working with meteorological models, Lorenz showed that systems with only a few variables, even though deterministic, display highly complex behavior that is unpredictable in fact because slight differences in one variable produce dramatic effects on the overall system. This feature of complex and chaotic systems has come to be called *sensitivity to initial conditions*.

In 1977 the Russian-born Belgian scientist Ilya Prigogine received the Nobel Prize in chemistry for his formulation of the theory of *dissipative structures*, whose fundamental insight is that nonequilibrium is a source of order and complexity. Prigogine demonstrated that open systems (which include organisms) that exchange matter and energy with their environments can show a reduction of local or internal entropy; that is, they are able to self-organize and complexify. Complex systems are dynamical systems whose cooperating and interacting parts display spontaneous, self-organized pattern formation with emergent properties that are not reducible to the sum of their constituent parts. Early-twenty-first-century proponents of a complex dynamical systems approach to the mind (Scott Kelso, Francisco Varela) maintain that mental and axiological properties are high-level dynamical neurological patterns.

For a dynamical system to show structure formation, the process must take place far from equilibrium; it must be nonlinear; and the system must be open to exchanges with its environment. Nonlinearity appears whenever there is interaction among components, whenever the *organizational relationships* among parts determine the overall systemic behavior. Such nonlinear dynamical systems are typically characterized by feedback loops that embed the systems in their environment and history in such a way that their trajectory history is inscribed in their very structure. Thus the dynamical systems become deeply contextual and extremely sensitive to initial conditions. After a few iterations, the trajectory of two initially close nonlinear dynamical systems will diverge exponentially, and long-term predictions become impossible.

Phenomenologically, however, it was evident that some systems eventually settle down to an oscillatory pattern. Others, such as the Belousov-Zhabotinsky reaction (B-Z reaction), trace complexly patterned trajectories. Yet others, such as turbulent flow, become chaotic, displaying (not no order at all, as had initially been thought) a highly complex form of order. These complex and chaotic systems are described by second order nonlinear differential equations and, as noted, had previously been considered intractable.

The B-Z reaction sequence is an illustration of the abrupt self-organization of hidden order that occurs in open systems far from equilibrium. It shows what can happen when potassium bromate, malonic acid, and manganese sulfate are heated in a bath of sulfuric acid. The first three reactions of the sequence are not remarkable, but the fourth has the unusual feature of being autocatalytic: The product of the process is necessary for the activation of the process itself. Instead of damping oscillations, positive feedback loops around autocatalytic cycles increase system fluctuations around a reference value.

With the system driven far from equilibrium by this runaway process, at a certain critical distance an instability occurs: a threshold point at which small, randomly occurring fluctuations can no longer be damped. Instead the internal dynamics of the autocatalytic cycle amplify a fluctuation, driving the reaction to a new mode of organization. The new system is characterized by the *coherent behavior* of an amazingly large number of molecules that synchronize to form a chemical wave that oscillates from blue to red. A colorful macroscopic structure (the visible evidence of a phase change) appears. True self-organization has taken place because the internally driven dynamics of autocatalysis precipitate the sudden change.

Biological complex systems are adaptive: As a result of feedback, they change their internal structure to respond to a changing environment. Virus mutations are a good illustration. Fundamentally rooted in their environment and history through context-dependent constraints, complex adaptive systems are thus deeply enmeshed in their surroundings. Nor do they start from scratch; they are fundamentally historical entities that embody in their structure the very conditions under which they were created and the trajectory they followed. Snowflakes are examples of such systems. Not only is each unique; its very structure *carries its history on its back* by embodying the pressure and temperature conditions in which it formed. At the same time, self-organizing systems such as slime molds display an autonomy that effectively decouples them from their environment.

Such complex adaptive systems, a category that includes people and their actions, are not isolated atoms. They are always already networked and entangled in both time and space. Their relationships create an interdependent whole that is ontologically new. Thus the environment coevolves with human beings; niches change in response to the organisms that occupy them, every bit as much as the organisms are selected by the niches. And both ontogenetically and phylogenetically, they become increasingly individuated over time.

Ethics in and of Nonlinear Dynamics

The dynamical systems approach suggests an interesting new ethical discussion (Dupre 1993, Juarrero 1999). From the perspective of this new science, the prerequisite for moral action known as free will is not the absence of external determining (Newtonian) causes, but the human capacity to impose order on a progressively disordered world. Because all self-organizing systems select the stimuli to which they respond, their behavior is constrained top-down and becomes increasingly autonomous from environmental impact. More complex systems are more autonomous. Self-organized processes, in other words, act from their own point of view. Furthermore the more complexly structured the entity, the more varied its organization and its behavior, and the more decoupled from and independent of its environment—the more autonomous and authentic, in short.

In another sense, the more complex a nonlinear dynamical system is, the freer it is because increasing complexity corresponds to an increase in state space: The system has new, different, and more varied states to access. Intentional human action is free to the degree and extent that the behavior is controlled by higher-level neurological contextual constraints, those with the emergent properties of meaning, value, and even awareness to a certain degree. Insofar as a wink is an action for which an agent can be held morally responsible precisely because the behavior is caused and controlled by a meaningful intention, and the agent is aware of so acting, a wink is freer than a blink because the latter originates in less complex neurological structures that do not embody meaning and value, and may occur as a reflex reaction.

The atoms of a Newtonian universe are independent of one another. So too are moral agents in a Kantian world. Because they are essentially relational entities, however, complex adaptive systems show how interdependence can create an ontologically distinct phenomenon, an organic whole greater than its parts. This is a fundamental axiological lesson of nonlinear dynamics.

Beginning with Plato's utopia, *The Republic*, Western philosophers have attempted to design fail-safe social systems (whether legal, educational, penal, or other) that are perfect and so never go wrong, morally or otherwise. Complex systems theory shows this is a hopeless task. First, since people carry their history on their backs, they can never begin from scratch, either personally or as societies. Second, perfection allows no room for improvement. Plato was one of the few thinkers who understood that if a utopia were ever successfully established, the only way it could change would be for the worse. Stasis and isolation are therefore essential to maintaining the alleged perfection, not only of Plato's *Republic*, but of most other utopias as well. The noumenal self that Kant postulates as the seat of moral choice and free will is likewise not part of this world. The possibility of perfection requires isolation.

The only choice, from an evolutionary perspective, is to cobble together safe-fail family and social organizations, structures flexible and resilient enough to minimize damage when things go wrong as they inevitably will. But to do so, human beings must recognize the potential of interdependence to create an ontologically distinct, metastable entity. Society needs to reintegrate those pieces torn apart by the old Newtonian framework, whether personally or socially, in both its means of communication and its advocacy of public policy. "Personal ethics must now be augmented by policy making" (Mitcham 2003, p. 159).

The downside of historical and environmental embeddedness is that, as members of a community, human beings do lose some of their freedom. Living in society can and often does cramp one's style. By contrast, components in a system acquire characteristics and identities they previously lacked (and could never acquire on their own): They become nodes in a network of relationships that permits new forms of life and act-types unavailable either to the hermit or to Kant's noumenal self: Only as members of complex social systems can humans be citizens and senators, teachers and wives, scientists and philosophers. The more complex the entity, the more meaningful the choices as well: As citizens and teachers, senators and wives, whatever roles they choose, people can be responsible or irresponsible, conscientious or careless, virtuous or not.

Because of their sensitivity to initial conditions, complex dynamical systems are not only unpredictable, they also become increasingly individuated over time making each developmental or ontogenetic trajectory unique. In contrast to the science of both Aristotle and Newton, non-linear dynamical systems theory incorporates individuation and concreteness into its conceptual framework. Knowing that each complex system's trajectory is unique raises questions about the universality at the heart of Kant's famous moral command, the categorical imperative. Human individuality, historicity, and contextuality are forced into a one-size-fits-all mold. Unacknowledged recognition of the inevitable interdependence and entanglement highlighted by both complexity and quantum theories might well be behind the more recent emphasis on Kant's second formulation of the categorical imperative: Always treat people as ends, never merely as means.

In a world with room enough for both societies and unique individuals, and the creativity and novelty they promote, precise prediction is impossible. Accordingly, dynamical systems theory calls into question the morality of consequentialism, whether in the utilitarianism of John Stuart Mill or elsewhere. In a world where precise consequences cannot be predicted, and where phenomena are intertwined and entangled in their own histories, basing morality on the *actual* outcome of individual behavior is a poor foundation for moral decisions and judgments.

Both consequentialism and Kantian formalism reduce morality and ethics to a set of formal rules. The highly contextual nature of complex systems suggests, in contrast, a different approach to moral education, one that references the virtue ethics of Aristotle and the ancients. Instead of memorizing a set of moral principles, which the agent is then suppose to implement moral education would consist of a gradual shaping of character through feedback and habituation. Moral education under this approach is the process of molding certain desires and character traits that are activated in appropriate contexts.

Nonlinear dynamical systems theory also calls for an ethics appropriate to a universe of interdependence and uncertainty. The recent renewal of interest in virtue ethics seems to implicitly recognize this. By contrast, as Carl Rubino notes, because of the ruling mechanistic paradigm's continuing influence on axiology, uncertainty still carries negative connotations. It should not. Complex dynamical systems teach that "change, novelty, creativity and spontaneity are the real laws of nature, which makes up the rules as it goes along. This is good news, cause for rejoicing; we should lift up our voices, as the prophet says, and not be afraid" (Rubino 1990, p. 210).

ALICIA JUARRERO

SEE ALSO *Free Will; Incrementalism; Systems.*

BIBLIOGRAPHY

Dupre, John. (1993). *The Disorder of Things*. Cambridge, MA: Harvard University Press. In a fundamentally disordered world in which different perspectives reveal distinct domains of partial order, our choices about which scientific project to pursue determines not only what kinds of order we observe in nature but also what kinds of order we impose on the world we observe.

Gleick, James. (1987). *Chaos*. New York: Viking. An excellent popular introdution to chaos theory by an award-winning science writer.

Juarrero, Alicia. (1999). *Dynamics in Action: Intentional Behavior as a Complex System*. Cambridge, MA: MIT Press. A rethinking of the concept of causality in terms of context-sensitive constraints, and applied to intentional action.

Juarrero-Roque, Alicia. (1991). Fail-safe versus safe-fail: Suggestions toward an evolutionary model of justice. *Texas Law Review* 69(7):1745–1777. A diagnosis of the failures of utopian literature, and pointers toward a new model of justice, both formulated from the perspective of complex adaptive systems theory.

Kant, Immanuel. (1952). "The Critique of Teleological Judgement." In *The Critique of Judgement*, trans. J. C. Meredith. Oxford: Clarendon Press.

Kauffman, Stuart. (1993). *The Origins of Order*. Oxford: Oxford University Press. Centered on the debate on the origins of life and the maintenance of order in complex biological systems, the book focuses on how self-organization can be incorporated into evolutionary theory.

Kelso, J. A. S. (1995). *Dynamic Patterns: The Self Organization of Brain and Behavior*.Cambridge, MA: MIT Press. A new general framework within which to connect brain, mind and behavior, from the perspective of self-organization dynamic pattern formation.

McIntyre, Alasdair. (1981). *After Virtue*. Notre Dame, IN: Notre Dame University Press. With a thorough analysis of the failure of the ethical theories of the enlightenment, this work singlehandedly revived the ancient classical approach to values known at virtue ethics.

Mitcham, Carl. (2003). "Interacciones de complejidades: Ciencia, tecnología, sociedad y ética" ("Interactions of Complexities: Science, Technology, Society, and Ethics"), *Paradoxa* 10 (137–163).

Prigogine, I., and I. Stengers. (1984). *Order out of Chaos: Man's New Dialogue with Nature*. Princeton, NJ: Princeton University Press. One of the classics of the breakthrough period of chaos theory, complex systems, and self-organization theories, the work explores the history of the relations of the human and natural sciences, to the present from the perspective of the theory of dissipative structures.

Prigogine, Ilya. (1996). *The End of Certainty: Time, Chaos, and the New Laws of Nature*. New York: Free Press. An account of how the end of certainty and determinism implied by theory of evolving self-organizing structures is the birth of a new formulation of the natural laws of both science and culture.

Rubino, Carl. (1990). "The Evolution of Our Choices: Notes toward an Ethic of Uncertainty." In *Toward a Just Society for Future Generations: Proceedings of the 34th Annual Meeting of the International Society for Systems Sciences*, 1: 205–212. A thoughtful analysis of the axiological implications of nonlinear dynamical systems theory.

Waldrop, Mitchell. (1992). *Complexity: The Emerging Science at the Edge of Order and Chaos*. New York: Simon and Schuster. An excellent if popular introduction to complexity theory as seen through the eyes of five of the major scientists responsible for the new science.

Yates, F. E. ed. (1987). *Self-Organizing Systems: The Emergence of Order*. New York: Plenum. The best anthology on self-organization, it includes articles from leading researchers in the field, on a wide range of topics including neural organization, morphogenesis, and the genesis and evolution of life.

INTERNET RESOURCE

Peterson, Gabriel. "The Belousov-Zhabotinsky Reaction." College of the Redwoods. Available from http://online.redwoods.cc.ca.us/instruct/darnold/DEProj/Sp98/Gabe/intro.htm. A cinematic demonstration of the reaction.

COMPUTER ETHICS

The field of study referred to as *computer ethics* addresses ethical issues arising around the development and use of computers and related technology. Computer ethics can be thought of as the field of study that examines ethical issues distinctive to an *information society*. Information society is the term often used (especially by economists and sociologists) to characterize societies in which human activity and social institutions have been significantly transformed by computer and information technology (Webster 2002). The focus of attention in this field has varied over its twenty-five- to thirty-year history as the technology has evolved. Because the field is relatively new and computer technology is continually changing and being used in new domains, computer ethics overlaps with other fields of study such as information ethics, media ethics, and communication ethics, as well as domain-specific ethics such as medical ethics, business ethics, environmental ethics, and legal ethics. Computer ethics is centrally focused on understanding the interactions among science, technology, and ethics and, arguably, it is one of the most developed fields with such a focus.

A Short History of Computer Ethics

From the moment of their invention, computers raised complex social, ethical, and value concerns. While computers are not the first technology to raise ethical issues, they have been especially fascinating to scholars, science fiction writers, and the public. The origin of this fascination may well be related to computers having been initially perceived and characterized as *thinking machines*. As such, they were thought to challenge the distinguishing feature of humankind. For centuries, human beings had been thought of as unique because they were able to *reason* and had the capacity for *rational thinking*. When computers were first developed and used, they seemed capable of being programmed to think in some of the ways that humans think; some believed they had the potential to become even more sophisticated and eventually reach or even surpass human intelligence. In that context, it was thought that computers would revolutionize the way humans think about themselves and what it means to be human. While many of the original hopes and promises of artificial intelligence (AI) researchers have not come to fruition, computers have changed the way scientists think about human cognition and brain functions. Computer technology continues to be a fascination for scientists, science fiction writers, and humanities and social science scholars as well as ethicists.

From a historical perspective, the ethical issues identified in relation to computers seem to follow the sequence of development of the technology. In addition to the threat to notions of what it means to be human, in the very early days of computing the first ethical issues arose in relation to the enormous power that computers might give to government and large bureaucratic organizations. By the late 1970s, the first books on this topic were published. Joseph Weizenbaum's *Computer Power and Human Reason* (1976) and Abbe Mowshowitz's *Conquest of Will* (1976) were, perhaps, the most notable. In this period, the record-keeping capabilities of computers were a key focus, especially the privacy issues raised by this record keeping. Several major government reports were issued including: in 1972, *Databanks in a Free Society: Computers, Record-Keeping and Privacy* by Alan F. Westin and Michael A. Baker, a report of the National Academy of Sciences; in 1973, *Records, Computers, and the Rights of Citizens*, a report of the U.S. Department of Health, Education, and Welfare from the Secretary's Advisory Committee on Automated Personal Data Systems; and in 1977, *Personal Privacy in an Information Society: The Report of the Privacy Protection Study Commission*. The issues that took shape in this period were largely issues of privacy and the power of centralized government was often characterized as the threat of Big Brother. In the aftermath of World War II and the fight against totalitarianism, it was feared that computers would give government unprecedented power and reach.

In hindsight this concern was the result in part of the size of computers. At that time, they were huge mainframe systems that cost a lot, took up a lot of space, and were labor-intensive; hence large organizations were the only viable users. Moreover, in those early days of computing, mainframes were used for large-scale calculations and to create and maintain huge databases. Such calculations made weapons development, space travel, and census tracking possible on a broader scale than ever before. The databases mostly contained personal information. In any event, large organizations were the likely users and hence the concerns about centralization of power and privacy.

The next major technological shift was the development of small computers referred to initially as *microcomputers* and later *personal computers*. Public interest, for a time at least, turned to the democratizing aspects of computers. Computer enthusiasts saw in these small machines the potential for a major social revolution. With visions of computers in every home and shifts in power from large organizations to small businesses and individuals, the fear of Big Brother dissipated somewhat.

As microcomputers were being developed and taking hold in the marketplace, remote access became possible, first to contact large mainframes and later as a component of a network of telecommunications connections between large and small computers. That network eventually became the Internet. However, long before the advent of the Internet, attention turned to software. Microcomputers were less expensive and easier to use; this meant a much broader range of users and, in turn, a broad range of uses. During this phase in the development of computers, software became extremely important both for the development of the technology but also, in parallel, for computer ethics.

To make computers effective tools for the wide range of activities that seemed possible, user-friendly software was critical. Companies and individuals began developing software with a fury, and with that development came a new set of ethical issues. Issues having to do with property rights and platform dominance in software were particularly important in this era. Software was recognized as something with enormous market value; hence, the questions: Should software be owned? If so, how? Would existing intellectual property law—copyright, patents, trade secrecy—be adequate protection for software developers? Ownership rights in programs used to create computer or video games were the first kinds of software cases brought before the courts; the market value of owning these programs was significant.

Along with property rights issues came issues of liability and responsibility. Consumers who buy and use computers and software want to be able to rely on these tools, and when something goes wrong, they want to know whom to blame or they want to be compensated for their losses. Computer ethicists as well as lawyers and computing professionals rose to the challenge and questions of property rights and liability were debated in print as well as in courts.

In the 1980s, more attention began to focus on hackers. Hackers did not like the idea of property rights in software. However those who were acquiring such property rights or making a business of computing were threatened by hackers not only because the latter were breaking into systems but also because they had a different vision of how the system of computers, software, and telecommunications connections should be set up and how software should be distributed. At that time, there were no laws against breaking into computer systems or duplicating software. Hackers argued for open systems with fewer controls on access to information. Perhaps the best illustration of this movement is Richard Stallman's work and the development of the Free Software Association (Stallman 1995).

By the 1990s, the development of the Internet was well underway and seen as a revolutionary event. The coming together of computers, telecommunications, and media and the global scale of the Internet produced a seemingly endless array of ethical issues. The

Internet was being used in many different ways, in many different domains of life. In effect the Internet recreated much of the world in a new medium. Property rights, freedom of speech, trust, liability, and privacy had to be rethought for a medium in which instantaneous communication was the norm; the reproduction of information, documents, or programs was almost effortless; and anonymity was favored. Moreover the new medium facilitated interaction on a global scale, raising issues regarding what laws and conventions applied in *cyberspace*.

During the 1980s and 1990s, computer technology also began to be used for a wide variety of visualization activities. Computer graphics and gaming were part of this, but equally if not more important was the development of many simulation applications including medical imagining and graphical dynamic models of the natural world. The power and reliability of these technologies raised ethical concern. An offshoot of these developments was a focus on *virtual reality* and what it might mean to human experience. Would human beings become addicted to living in fantasy worlds? Would experiences in violent, virtual computer games make individuals more violent than they would otherwise be? These concerns continue in the early-twenty-first century as new applications are developed. For example, important ethical issues are being raised about *tele-medicine*. Computing together with the Internet makes it possible for many aspects of medical treatment to be performed electronically. Issues of responsibility and liability are diffused when doctors do surgery remotely. A doctor in one location can manipulate machines that are electronically connected to machines in a second location where the surgical procedure actually occurs. Should doctors be allowed to do this? That is, is it appropriate? Is it safe? Who is responsible if something goes wrong?

Ethical issues surrounding computer technology continue to arise as new developments in the technology occur. Many of these involve computing applications. For example, new areas of concern include surveillance technologies that result from using geographic information systems and digital imagining to keep track of individuals via digital cameras and satellites. There are projections about the use of tiny, biological computers that might be deployed in human bodies to seek out poorly functioning cells and fix them. Computer technology makes possible human behavior and social arrangements that have a moral character. Hence activities involving computers will continue to be a focus for computer ethics.

Persistent Issues

As computer technology evolves and is deployed in new ways, ethical issues proliferate. To illustrate the kinds of concerns that arise, issues of professional ethics, privacy, hacking and cracking, and the Internet will be briefly described.

PROFESSIONAL ETHICS. In an information society, a large number of individuals are educated for, and employed in, jobs that involve development, maintenance, buying and selling, and use of computer and information technology. Indeed an information society is dependent on such individuals—dependent on their special knowledge and expertise and on them fulfilling social and professional responsibilities. Expertise in computing can be used recklessly or cautiously, for good or ill, and the organization of information technology experts into occupations and professions is an important social means of ensuring that the expertise is used in ways that serve human well-being.

The social responsibilities of computer experts are connected to more general notions of duty and responsibility and computer ethicists have drawn on a variety of traditional philosophical concepts and theories to understand them. Computing professional associations have developed codes of ethical and professional conduct that represent what computer professionals believe to be their duties and the ideals to which they should aspire. However it is important to note that computing is not a single, homogenous profession. The responsibilities and likely areas of ethical concern vary widely with the computer professional's particular job and employment context. Consider, for example, the differences between academic computer scientists, software engineers working in industry, programmers, managers of information technology units in organizations, and computer and software marketers.

The largest and most visible organization of computer professionals is the Association for Computer Machinery (ACM). The ACM has a code of ethics and professional conduct and, with the Institute for Electrical and Electronic Engineers (IEEE), also has developed a code for software engineers, the ACM/IEEE Code of Ethics for Software Engineers. The key elements in both codes are very general edicts to contribute to society and human well-being; avoid harm; be honest and trustworthy; and act in a manner that is consistent with the interests of client, employer, and public. Yet both codes go beyond these general principles and give content and meaning to the principles. While one can argue that codes of conduct are not a very effective mechanism for regulating behavior, they are an important component in constituting a responsible profession. The codes are statements to the public as to what to expect; they articulate standards for the field and make clear that members are professionals. Codes can be used in relation to employers and others to emphasize that computer professionals must adhere to standards independent of the orders they receive at work.

PRIVACY. In an information society, privacy is a major concern in that much (though by no means all) of the information gathered and processed is information about individuals. Computer technology makes possible a magnitude of data collection, storage, retention, and exchange unimaginable before computers. Indeed computer technology has made information collection a built-in feature of many activities, for example, using a credit card, making a phone call, and browsing the Worldwide Web (WWW). Such information is often referred to as transaction-generated information (TGI).

Computer ethicists often draw on prior philosophical and legal analyses of privacy and focus on two fundamental questions, What is privacy? and Why is it of value? These questions have been contentious and privacy often appears to be an elusive concept. Some argue that privacy can be reduced to other concepts such as property or liberty; some argue that privacy is something in its own right and that it is intrinsically valuable; yet others argue that while not intrinsically valuable, privacy is instrumental to other values such as friendship, intimacy, and democracy.

Computer ethicists have taken up privacy issues in parallel with more popular public concerns about the social effects of so much personal information being gathered and exchanged. The fear is that an information society can easily become a *surveillance society*. Computer ethicists have drawn on the work of Jeremy Bentham and Michel Foucault suggesting that all the data being gathered about individuals may create a world in which people effectively live their daily lives in a *panopticon* (Reiman 1995). Panopticon is a term that describes the shape of a structure that Bentham designed for prisons. In a panopticon, prison cells are arranged in a circle with the inside wall of each cell made of glass so that a guard, sitting in a guard tower situated in the center of the circle, can see everything that happens in every cell. The effect is not two-way; that is, the prisoners cannot see the guard in the tower. In fact, a prison guard need not be in the guard tower for the panopticon to have its effect; it is enough that prisoners believe they are being watched. When individuals believe they are being watched, they adjust their behavior accordingly; they take into account how the watcher will perceive their behavior. This influences individual behavior and how individuals see themselves.

While computerized information gathering does not physically create the structure of a panopticon, it does something similar insofar as it makes much individual behavior available for observation. Thus the data collection activities of an information society could have a panoptic effect. Individuals know that most of what they do can be observed and that knowledge could influence how they behave. When human behavior is monitored, recorded, and tracked, individuals may become intent on conforming to norms for fear of negative consequences. If this were to happen to a significant extent, the ability of individuals to act freely and think critically—capacities necessary to realize democracy—may be compromised. In this respect, the privacy issues around computer technology go to the heart of freedom and democracy.

A good illustration of the panoptic environment is the use of cookies at web sites. A cookie is a file placed on a user's computer when the user visits a web site. The file allows the web site to keep track of subsequent visits by the user. Thus, the web site maintains a record of the user's visits. While this can help the web site provide better service to the user—based on information about use—users are being watched, records are being created and the panoptic effect may occur. Moreover, the records created can be matched with information from other web sites and domains.

It might be argued that the panoptic effect will not occur in information societies because data collection is invisible; individuals are unaware they are being watched. This is a possibility, but it is also possible that as individuals become more and more accustomed to information societies, they will become more aware of the extent to which they are being watched. They will see how information gathered in various places is put together and used to make decisions that affect their interactions with government agencies, credit bureaus, insurance companies, educational institutions, and employers, among others.

Concerns about privacy have been taken up in the policy arena with the passage of legislation to control and limit the collection and use of personal data. An important focus is comparative analyses of policies in different countries. The U.S. approach has been piecemeal with separate legislation for different kinds of records, for instance, medical records, employment histories, and credit records. By contrast, several European countries have comprehensive policies that specify what kind of information can be collected under what conditions in *all* domains. The growing importance of global business influences policy debates. Information-gathering organizations promise that they will use information only in certain ways; yet, in a global economy, data collected in one country—with a certain kind of data protection—can flow to another country where there is no protection, or where such protection differs from that of the original country. To assure that this does not happen, a good deal of attention is focused on working out international arrangements and agreements to protect data internationally.

HACKERS AND CRACKERS. While the threats to privacy described above arise from *uses* of computer and information technology, other threats arise from *abuses*. As individuals and companies do more and more electronically, their privacy and property rights become increasingly important. Individuals who defy the law or test its limits can threaten these rights. Such individuals, often called *hackers* or *crackers*, may seek personal gain or may just enjoy the challenge of figuring out how to *crack* security mechanisms. The term hacker originally referred to individuals who simply loved the challenge of working on programs and figuring out how to do complex things with computers, but who did not necessarily break the law. Crackers referred to individuals who did. However, in the early-twenty-first century, the terms are used somewhat interchangeably to refer to those who engage in criminal activity.

Distinguishing the terms, however, reveals two streams of development in computing and two streams of analysis in computer ethics. Hackers are not only individuals who love computing and are very knowledgeable about it, but in particular are those who advocate an alternative vision of how computer technology might be developed and used. Hackers are interested in a computing environment that has more sharing and less ownership. For many hackers, this is not just talk. They are involved in what is sometimes called the open source movement, which involves the development of software that is available for free and can be modified by the user. Over the years, through various organizations, a good deal of open source software has been developed including, notably, the Linux operating system.

Because hackers represent an alternative vision of software, they are seen as part of a social and political movement, a kind of counterculture. A strand of this movement goes beyond the development of open source software and engages in political activism, using computing expertise to make political statements. The term *hacktivism* refers to on-line political activism. Whether such behavior is legal or illegal remains ambiguous.

Another stream of analysis centers around crackers. Cracker refers, simply, to an online criminal. Crackers break into systems or disrupt activities on the Internet by launching viruses or worms or by engaging in a host of other kinds of disruptive behavior, including pinging, and taking control of websites. The ethical issues are not particularly deep. Cracking behavior interferes with innocent users who are trying to do what they have legal rights to do; the behavior of crackers may violate property rights or privacy, involve harassment, and more. Computer ethics literature examines this behavior for its ethical content but also to try to understand whether there is anything unique or special about cracking behavior and computer crime.

Law often lags behind technology and, in the early days of computing, there were no prohibitions against the disruptive behavior of crackers. In the early-twenty-first century, however, there are many laws regulating behavior on the Internet. Yet issues and problems persist. New technologies facilitate crackers and there are serious questions regarding harmonization of laws globally. Anonymity makes it difficult to catch computer criminals.

INTERNET ISSUES. Arguably the Internet is the most powerful technological development of the late-twentieth century. The Internet brings together many industries but especially the computer, telecommunications, and media enterprises. It provides a forum for millions of individuals and businesses around the world. It is not surprising, then, that the Internet is a major focus of attention for computer ethicists. The development of the Internet has involved moving many basic social institutions from a paper and ink environment to an electronic environment. The change in environment changes the features of activities. Thus a number of ethical issues arise as regards the behavior of individuals and organizations on the Internet.

The Internet has at least three features that make it unique. First, it has unusual scope in that it provides many-to-many communication on a global scale. Of course, television and radio, as well as the telephone, are global in scale, but television and radio are one-to-many forms of communication, and the telephone, which is many-to-many, is expensive and more difficult to use. Individuals and companies can communicate with one another on the Internet frequently, in real time, at relatively low cost, with ease, and with visual as well as sound components. Second, the Internet facilitates a certain kind of anonymity. One can communicate with individuals across the globe (with ease and minimal cost), using pseudonyms or real identities, and yet never actually meet those people. This type of anonymity affects the content and nature of the communication. The third special feature of the Internet is its reproducibility. Text, software programs, music, and video on the Internet can be duplicated ad infinitum and altered with ease. The reproducibility of the medium means that all activity on the Internet is recorded and can be traced.

These three features—global, many-to-many scope; anonymity; and reproducibility—have enormous positive, as well as negative, potential. The global, many-to-many capacity can bring people closer together, relegating geographic distance to insignificance. This feature is especially liberating to those for whom travel is physically challenging or prohibitively expensive. However these benefits come with drawbacks; one is that such capabilities are also available to those who use them for heinous purposes. Individuals can—while sitting anywhere in the world, with very little effort—launch viruses and disrupt communication. They can misrepresent themselves and dupe others on a much larger scale than was possible before the Internet.

Similarly anonymity has both benefits and dangers. The kind of anonymity available on the Internet frees some individuals by removing barriers based on physical appearance. For example, in contexts in which race and gender may get in the way of fair treatment, the anonymity provided by the Internet can eliminate bias (for example, in online education, race, gender, and physical appearance are removed as factors affecting student-to-student interactions as well as teacher evaluations of students). Anonymity may also facilitate participation in beneficial activities such as discussions among rape victims, battered wives, or criminal offenders, in which individuals might be reluctant to participate unless they had anonymity.

Nevertheless anonymity leads to serious problems of accountability and integrity of information. Perhaps the best illustration of this is information acquired in chat rooms on the Internet. It is difficult (though not impossible) to be certain of the identities of people with whom one is chatting. One person may participate under multiple identities; a number of individuals may use the same identity; or participants may have vested interests in the information being discussed (for instance, a participant may be an employee of the company or product being discussed). When one cannot determine the true source of information or develop a history of experiences with a particular source, it is impossible to gauge the reliability of the information.

Like global scope and anonymity, reproducibility also has benefits and dangers. Reproducibility facilitates access to information and communication; it allows words and documents to be forwarded (and downloaded) to an almost infinite number of sites. It also helps in tracing cybercriminals. At the same time, however, reproducibility threatens privacy and property rights. It adds to problems of accountability and integrity of information arising from anonymity. For example, students can send their assignments to teachers electronically. This saves time, is convenient, and saves paper. However the reproducibility of the medium raises questions about the integrity of the students' product. How can a teacher be sure a student actually wrote the submitted paper and did not download it from a web site?

As the daily activities of individuals and businesses have moved online, distinctive ethical questions and issues have been identified; some of these issues have been addressed by adopting or modifying relevant laws; others have been addressed by new technology; yet others persist as nagging problems without solution or only with solutions that are worse than the problem. Plagiarism is an example of a problem that can be at least partially addressed via new technology; that is, there are tools available for teachers and professors to use to detect student work that has been copied from the Internet or copied from other students. On the other hand, pornography is an example of an issue that defies solution. An incredibly large proportion of the traffic on the internet involves distributing, advertising, and accessing pornography. This seems an unworthy use of one of the most important, if not the most important, inventions of the twentieth century. Yet, eliminating or reducing pornography on the Internet would seem to require censorship and policing of a kind that would undermine the freedom of expression that is the bedrock of democratic societies. Hence, pornography on the internet persists.

Conclusion

Perhaps the deepest philosophical thinking on computer-ethical issues has been reflection on the field itself—its appropriate subject matter, its relationship to other fields, and its methodology. In a seminal piece titled "What is Computer Ethics?" James Moor (1985) recognized that when computers are first introduced into an environment, they make it possible for human beings (as individuals and through institutions) to do things they could not do before and that this creates *policy vacuums*. People do not have rules, policies, and conventions on how to behave with regard to the new possibilities. Should employers monitor employees with computer software? Should doctors perform surgery remotely? Is there any harm in taking on a pseudoidentity in an on-line chat room? Should companies doing business online be allowed to sell the TGI they collect? These are examples of policy vacuums created by computer technology.

Moor's account of computer ethics has shaped the field. Many computer ethicists see their role as that of filling policy vacuums. Indeed one topic of interest in computer ethics is defining the activity of filling policy vacuums.

Because computers and information technology will continue to evolve and become further integrated into human life, new ethical issues will certainly arise. However, as human beings become more and more accustomed to interacting with and through computer technology, the difference between ethics and computer ethics may well disappear.

DEBORAH G. JOHNSON

SEE ALSO *Artificial Intelligence; Communication Ethics; Computer Viruses/Infections; Engineering Ethics; Gates, Bill; Geographic Information Systems; Hardware and Software; Hypertext; Internet; Networks; Security; Special Effects; Turing, Alan; Video Games.*

BIBLIOGRAPHY

Advisory Committee on Automated Personal Data Systems. *Records, Computers, and the Rights of Citizens*. U. S. Department of Health, Education, and Welfare Publication No. (OS) 73–94, July 1973.

Johnson, Deborah G. (2001). *Computer ethics*, 3rd edition. Upper Saddle River, NJ: Prentice Hall. A philosophical survey of the ethical issues arising around computer technology.

Moor, James H. (1985). "What Is Computer Ethics?" *Metaphilosophy* 16(4): 266–275. Classic piece on why the study of computer ethics is needed.

Mowshowitz, Abbe. (1976). *The Conquest of Will: Information Processing in Human Affairs*. Reading, MA: Addison-Wesley.

Privacy Protection Study Commission. (1977). *Personal Privacy in an Information Society: The Report of the Privacy Protection Study Commission*. Washington, DC: Author. For sale by the U.S. Government Printing Office.

Reiman, Jeffrey H. (1995). "Driving to the Panopticon: A Philosophical Exploration of the Risks to Privacy Posed by the Highway Technology of The Future." *Computer and High Technology Law Journal* 11: 27–44. Provides an argument for the importance of privacy incorporating Foucault's analysis of the panoptic effect.

Stallman, Richard. (1995). "Are Computer Property Rights Absolute?" In *Computers, Ethics, and Social Values*, ed. Deborah G. Johnson and Helen Nissenbaum. Upper Saddle River, NJ: Prentice-Hall. Articulates a rationale for why software should not be private property.

Webster, Frank. (2002). *Theories of the Information Society*. New York: Routledge. Provides an account of the major theories describing information societies, their key features, and how they work.

Weizenbaum, Joseph. (1976). *Computer Power and Human Reason*. San Francisco: W. H. Freeman. A classic early work on the social implications of computers.

Westin, Alan F., and Michael A. Baker. (1972). *Databanks in a Free Society: Computers, Record-Keeping and Privacy*. New York: Quadrangel Books. Report of the Project on Computer Databanks of the Computer Science and Engineering Board of the National Academy of Sciences.

COMPUTER VIRUSES/ INFECTIONS

A computer virus is a piece of software that "invades" a computer. As such, a computer virus is one of several kinds of infections, including Trojan horses and worms. Infections are themselves a subset of possible attacks on computers and networks; other attacks include probes, unauthorized access, denial of service, Internet sniffers, and large-scale scanning. This entry focuses on viruses, worms, and Trojan horses—collectively termed *electronics infections*—the three most common kinds of attacks and the ones best known by the public (Carnegie Mellon University Internet site). All such infections constitute multiple ethical and political issues: the responsibilities to protect against them, determining consequences for those responsible for attacks, and how to educate users about their vulnerabilities.

Technical Features

A virus is a piece of software that is hidden inside a larger program. When the larger program is executed, the virus is executed as well. During that execution, the virus can try to fulfill its purpose, often to replicate (that is, copy) itself in other programs on its host machine or (via the Internet) to new host machines. This copying and sending takes up resources on the original machine, on the Internet's communications capacity, and on any new machines infected. For a major virus attack, the loss of resources can cost billions of dollars.

One variation on the more traditional application-borne computer virus is the e-mail virus. An e-mail virus attaches itself to a piece of e-mail instead of to a program. Another subspecies of computer virus is the "logic bomb." A logic bomb is a virus because it resides inside the operating system or an application; the variation is that a logic bomb executes its harmful side effect only when certain conditions are met, typically when the system clock reaches a particular date. At the appointed time, the virus can do something relatively harmless, like flashing a provocative text message on the screen; but it could also do something far more serious, such as erasing significant portions of the resident host's hard drive.

A virus requires a program or e-mail to hide in. But a computer worm works independently. A computer worm uses computer networks and security flaws to replicate itself on different networked computers. Each copy of the worm scans the network for an opening on another machine and tries to make a new copy on that machine. As this process is repeated over many generations, the computer worm spreads. As with viruses, both the propagation and any other side effects can be frivolous or draconian.

A Trojan horse is a complete computer program that masquerades as something different. For example, a web site might advertise a freeware computer game called Y. But when someone downloads and runs a copy of Y, Y erases the hard drive of the host machine. Unlike viruses, a Trojan horse does not include a self-replication mechanism inside its program.

Ethical Issues

Early in the history of computer development, some people thought of electronic infections as relatively harmless, high-tech pranks. But once these infections began to cost the public enormous amounts of time, energy, and money, they ceased to be laughing matters. The technical details that separate viruses, worms, and Trojan horses are useful distinctions when understanding the different techniques, but all infections share a common feature: They enter someone's computer without permission. Although different infections have different effects (and some claim to be benign), all of them take unauthorized control of another machine and/or memory.

In the early 1990s, there was actually some controversy about whether or not computer infections and other "hacking" activities were always unethical. In some instances benign infections simply used underutilized computer power in ways that did not compromise the owner's uses (Spafford 1992). But the reflective consensus in the early twenty-first century is that all infections and break-ins are wrong. Reasons for this consensus include the view that it causes real harms, it violates legitimate rights to non-intrusion, it steals resources that could be put to better use, and it encourages otherwise unnecessary spending on security that could be spent on better things (Johnson 2001).

Even when it is agreed that all computer infections are unethical, important questions remain. For example, most computer infections now known are aimed at Microsoft Corporation operating systems and applications. That may be a consequence of Microsoft's market share, of technical details about Microsoft's software, of hackers' attitudes toward Microsoft, or a combination of these. Each has ethical dimensions. When one condemns the creator of a harmful infection, should some of the blame for the damages not be shared by vendors who release software with security holes that are easily exploited? Are users who fail to install security updates or adopt easily broken passwords not partially responsible? Such questions are part of an ongoing discussion of responsibility that can be found in analyses of the degrees of victim contributions and extenuating circumstances with regard to a wide range or crimes, from fraud to theft and assault and battery.

Education

Consider also questions raised by teaching students about computer infections. Those offering such classes defend their actions as helping students learn how to defend against such infections; critics have argued that such classes may actually encourage students to write and propagate new infections.

Both the defenders and the critics of academic work on computer infections raise legitimate issues. Considering their positions consequentially, if such classes reduce the number and severity of infections, then they are morally justified; conversely, if they increase the number or severity of infections, then they are not justified. But it seems unlikely that enough information about consequences can be easily gathered to settle the question.

Another approach is to analyze classes that teach about computer infections in terms of course content. Surely it would be noncontroversial to teach historical facts about the occurance and severity of computer infections. Furthermore, discussing the ethics of computer infections and other attacks are also unlikely to raise objections. The content most likely to prove objectionable would be teaching the technical details of how to construct computer infections, with assignments that require students to design new infections.

Is it ethical to teach the technical details of computer infections? Consider an analogy: Is it ethical to teach accounting students the details of accounting fraud? Such classes exist and have not elicited the same kind of criticism that has been leveled against computer infection classes. It seems reasonable in both cases that professionals in the field should know how people have conducted "attacks" in order to detect and defend against them in the future.

Yet there are ethically significant differences between accounting and computing—the rules of proper accounting are more explicitly spelled out than the rules of "proper computing." Accountants are held to more formal, legal, and professional standards than computing professionals. Furthermore, it takes very little advanced skill to launch a computer attack, but it requires some sophistication (and often a high position in a company) to launch a major accounting fraud. Finally, although an accounting class might include the study of strategies to defraud a company, it seems unlikely that a student could actually implement a fraud during the class, whereas computer science students can indeed launch a computer virus (and some have).

The analogy suggests that the notion of teaching computer science students about computer infections seems reasonable, but that some cautions about what is taught and how it is taught may be necessary. There is no airtight case for or against classes that include details of computer infections, but there are two important perspectives to consider: consequentialist arguments and arguments from analogy. Other perspectives might include deontological obligations to share knowledge or to recognize traditions of forbidden knowledge, and the character or virtue implications for both teachers and students in such classes.

Preliminary explorations nevertheless suggest that the content of the courses and the context in which technical details are presented will determine whether or not such courses are ethical. One can envision a course in which a professor does not emphasize the responsibilities of a programmer and does not discuss the negative impact of computer infections; in such a class, the presentation of technical details of computer infections are likely inappropriate. One can also envision a course in which professional responsibilities and public safety are central themes; in such a course, details of computer infections might be entirely appropriate.

TABLE 1

Possible Sanctions Against Those Who Create and Launch Computer Infections

	Minor consequences to others	Major consequences to others
Unintended	Education	Education plus minor punishment
Intended	Education plus minor punishment	Education plus major punishment

SOURCE: Courtesy of Keith W. Miller and Carl Mitcham.

Sanctions

If infecting systems that don't belong to you is wrong, it is necessary to consider appropriate sanctions against those who create and launch computer infections. In general, punishments for any unethical behavior should take into account both consequences of the act and intentions of the actor. Table 1 shows a broad view of how sanctions could be applied using considerations of intent and consequences.

Unintended minor consequences (as when a person experimenting designs a virus to see how it works and accidentally lets it get away, but it does very little damage) surely deserves little in the way of punishment, although some acknowledgement of the damage done seems appropriate. Unintended major consequences and intended minor consequences both deserve education plus some form of punishment, although probably not the same in each case. But intended major consequences could be assigned significant punishments, including jail and restrictions on future computer use.

The computer software community of hackers also has responsibilities to exercise social pressure and the punishment of ostracism on intentional offenders. Indeed, to some extent it seems to do this by reserving the pejorative term *crackers* for such persons. But professional organizations such as the Association for Computing Machinery might also instigate formal forms of ostracism. Codes of ethics for computing professionals such already include explicit prohibitions against computer attacks. For example, section 2.8 of the ACM Code of Ethics states: "Access computing and communication resources only when authorized to do so" (ACM Internet site). However, the ACM rarely disciplines members, and removal from the ACM is not seen as a significant threat to most hackers.

KEITH W. MILLER

SEE ALSO *Communication Ethics; Computer Ethics; Internet; Security.*

BIBLIOGRAPHY

Johnson, Deborah G. (2001). *Computer Ethics*, 3rd edition. Upper Saddle River, NJ: Prentice Hall. A groundbreaking computer ethics textbook.

Spafford, Eugene H. (1992). "Are Computer Break-ins Ethical?" *Journal of Systems and Software* 17(1): 41–48. An relatively early work on the ethics of hacking by an expert in Internet security.

INTERNET RESOURCES

Association for Computing Machinery (ACM). "ACM: Code of Ethics." Available from http://www.acm.org/constitution/code.html#sect2. An influential code of ethics in the United States.

Brain, Marshall. "How Computer Viruses Work." HowStuffWorks. Available from http://computer.howstuffworks.-com/virus1.htm.

Carnegie Mellon University. CERT Coordination Center. "CERT/CC Overview Incident and Vulnerability Trends," Module 4: "Types of Intruder Attacks." Available from http://www.cert.org/present/cert-overview-trends/. CERT is an acknowledged authority on computer security, and is often involved in quick reactions to Cyberspace threats.

COMTE, AUGUSTE

One of the French founders of modern sociology, Isidore-Auguste-Marie-François-Xavier Comte, better known simply as Auguste Comte (1798–1857), was born in Montpellier on January 19 (30 *Nivose* Year VI in the revolutionary calendar) and tried to reconcile the ideals of the Revolution of 1789 with early nineteenth century society. Comte's higher education began at the École Polytechnique in Paris, although he was expelled after two years following a quarrel with one of his mathematics professor. He then briefly studied biology at the École de Médecine in Montpellier before returning to Paris. Among his early influences, the philosophy of the Marquis de Condorcet (1743–1794) had the greatest impact. In 1817, Comte began his close association with Claude-Henri de Saint-Simon (1760–1825), one of the founders of French socialist thought who envisaged the reorganization of society by an elite of philosophers, engineers, and scientists. After an angry break between the two in 1824, Comte spent the next twenty years delivering lectures on "social physics." He suffered periods of intense mental collapse and died isolated and bitter on September 5 in Paris.

Positive Philosophy

Building on Condorcet's theory of human progress, Comte constructed what he called a "positive philosophy." Central to his philosophy was the "law of the three stages" between theological (mythological or fictitious), metaphysical (abstract), and positive (empirical and descriptive) knowledge. Over the course of history and across a broad range of disciplines and dimensions of human culture, the myths of theology have been gradually replaced by the general principles of metaphysics that were, in Comte's own time, being superseded by positive or empirical scientific knowledge. The positive stage constitutes the highest stage of human history because it is only when science has become "positive" that human beings will truly understand the world. For Comte, astronomy was the first science to become positive, because its phenomena are universal and affect other sciences without itself being affected. Because it is so complex, the last science to become positive is "social physics" or sociology.

Comte divided social physics into statics and dynamics, order and progress. The idea of order appears in society when there is stability because all members hold the same beliefs, a stage that occurred with the triumph of medieval Christianity. The idea of progress appeared with the Protestant Reformation and the French Revolution. For Comte, the contemporary challenge was to reconcile or synthesize order and progress, because revolution had destroyed the medieval sense of order but not yet created a new one to take its place. According to Comte, this new order required not only science but religion, with a new clergy to preach the laws of society. Comte eventually proposed himself as the high priest of this new scientific religion, and from 1844 signed his works, "The Founder of Universal Religion, Great Priest of Humanity."

Comte's Influence

Comte has been severely criticized for proposing that a technocratic elite was needed to educate and discipline society (see, for instance, the remarks on Comte in his contemporary John Stuart Mill's book *On Liberty*, 1859). But Comte was also interested in the moral improvement of humanity as a whole, and a social order in which self-interest is restrained within the bounds of an appreciation of the good of others as well as oneself. Morality for him was constituted by devotion to the whole of society. Such an idea clearly represented a critique of the unqualified competitiveness characteristic of the Industrial Revolution. Indeed, the need for some authoritarian, technocratic guidance—perhaps imbued

Auguste Comte, 1798–1857. Comte developed a system of positive philosophy. He held that science and history culminate in a new science of humanity, to which he gave the name "sociology." *(The Library of Congress.)*

to some degree with a religious sensibility—to facilitate the creation of a legal framework that supports qualified capitalist competition is not easily dismissed.

The importance of Comte must be placed in the historical context of a century in which vast systems of ideas were being fashioned in response to the forces unleashed by the French and Industrial Revolutions. Although the law of the three stages sounds contrived, and his plans for a new positive religion utterly fantastic, Comte succeeded in introducing the scientific study of society into nineteenth century intellectual discourse. His vision of a science of society to complement the emerging science of nature remains of fundamental importance to the relationship between science, technology, and ethics.

STEVEN KREIS

SEE ALSO *Enlightenment Social Theory; Secularization.*

BIBLIOGRAPHY

Comte, Auguste. (1853). *The Positive Philosophy of Auguste Comte*, trans. Harriet Martineau. London: J. Chapman.

Comte, Auguste. (1865). *A General View of Positivism*, trans. J. H. Bridges. London: Trübner.

Comte, Auguste. (1875–1877). *The System of Positive Polity*, trans. J. H. Bridges, et al. London: Longmans, Green.

Comte, Auguste. (1958). *The Catechism of Positive Religion*, trans. Richard Congreve. London: Kegan Paul.

Mill, John Stuart. (1866). *Auguste Comte and Positivism*. Philadelphia: Lippencott. One of the best introductions to Comte's work, by one of his contemporary admirers and critics.

Pickering, Mary. (1993). *Auguste Comte: An Intellectual Biography*. Cambridge, UK: Cambridge University Press. Excellent study of Comte's early life and difficulties; places his ideas within the context of the intellectual history of early nineteenth century France.

Scharff, Robert C. (1995). *Comte after Positivism*. Cambridge, UK: Cambridge University Press. A detailed consideration of Comte's ideas and of their applicability in the early twenty-first century.

CONDORCET, MARQUIS DE

SEE *Progress*.

CONFLICT OF INTEREST

• • •

A conflict of interest is a situation in which some person (whether an individual or corporate body) stands in a certain relation to one or more decisions. Often such persons are engineers, scientists, or organizations of engineers or scientists. On the standard view, *a person has a conflict of interest if, and only if, that person (a) is in a relationship with another requiring the exercise of judgment in the other's behalf and (b) has a (special) interest tending to interfere with the proper exercise of such judgment*.

Key Features of Conflict of Interest

The crucial terms here are *relationship*, *judgment*, *interest*, and *proper exercise*. Relationship is quite general, including any connection between persons or organizations justifying one's reliance on the other for a certain purpose. A relationship may be formal (as is that between the Academy of Science and the government it advises) or informal (as when an engineer responds to a neighbor's question about the best bicycle to buy). A relationship can last years (as the relationship between colleagues in a lab often does) or only a minute (as when one answers a stranger's question at a talk). The relationship required must, however, be fiduciary, that is, involve one person justifiably trusting (or, at least, being entitled to trust) another—to exercise judgment in the other's service.

Judgment refers to the ability to make certain kinds of decision correctly more often than would a simple clerk with a book of rules and all, and only, the same information. Insofar as decisions do not require judgment, they are *routine, ministerial, mechanical,* or *something a technician could do*; they have (something like) an algorithm. The decision maker contributes nothing special. Any difference between the decision maker's decision and that of someone equally well trained would mean that (at least) one of them erred (something easily shown by examining what they did). Ordinary math problems are routine in this way; so is the taking of readings from a gauge.

Where judgment is required, the decision is no longer routine. Judgment brings knowledge, skill, and insight to bear in unpredictable ways. Where judgment is necessary, different decision makers, however skilled, may disagree without either being obviously wrong. Over time, observers should be able to tell that some decision makers are better than others (indeed, that some are incompetent). But, except in extraordinary circumstances, an observer will not be able to do that *decision by decision;* nor will an observer be able to explain differences in outcomes in individual decisions merely by error—or even be able to establish decisively that one decision maker's judgment is better than another's in this or that case. Even if one decision maker is successful this time when another is not, the difference might as easily be the result of *dumb luck* as *insight*. Good judgment lasts. What makes a good scientist or a good engineer is good scientific or engineering judgment. Judgment is less general than *expertise*. Some of what is expected from experts is not judgment but merely special knowledge or routine application of a special skill.

Not every relationship, not even every relationship of trust or responsibility, requires judgment. A person may, for example, be asked to keep safe—but not look at—important lab notebooks until the owner returns. That person has been charged with a great trust as a fiduciary upon whom the owner may be relying to protect an important discovery. But the person need not exercise judgment to carry out the task. The task is entirely routine, however much the ability to behave as required is strained by a desire to peek. The notebooks need only be placed in a desk and left there until the owner returns and asks for them. The holder of the notebooks is a mere trustee, lacking the permissible options that make conflict of interest possible. Not all temptations to misbehave constitute conflict of interest in the strict sense.

Interest refers to any influence, loyalty, concern, emotion, or other feature of a situation tending to make a person's judgment (in that situation) less reliable than it would normally be, without rendering that person incompetent. Financial interests and family connections are the most common sources of conflict of interest, but love, prior statements, gratitude, and other subjective tugs on judgment can also be interests. For example, a biologist hired by a drug company to test some drug for efficacy has an interest (in the relevant sense) if the drug's inventor is a friend or enemy (just as if the biologist were paid with stock in the drug company).

What constitutes proper exercise of judgment is a *social fact*, that is, something decided by what people ordinarily expect, what the person exercising judgment or the group to which that person belongs invites others to expect, what that person has expressly contracted to do, and what various laws, professional codes, or other regulations require. Because what is proper exercise of judgment is so constituted, it changes over time and, at any time, may have a disputed boundary. For example, civil engineers in the United States today are expected to give substantial weight to considerations of environmental harm when deciding what to recommend, something (probably) not within the proper exercise of their judgment until the second half of the twentieth century.

The Problem with Conflict of Interest

What is wrong with conflict of interest? Having a conflict of interest is not wrong. However what one does about the conflict may be—for one of three reasons.

First, the person exercising judgment may be negligent in not responding to the conflict of interest. Society expects those who undertake to act in another's behalf to know the limits of their judgment when the limits are obvious. Conflict of interest *is* obvious. One cannot have an interest without knowing it—though one can easily fail to take notice of it or misjudge how much it might affect one's judgment. Insofar as the person exercising judgment is unaware of the conflict of interest, that person has failed to exercise reasonable care in acting in another's behalf. Failing to exercise reasonable care is negligent, and therefore the conduct is morally objectionable.

Second, if those justifiably relying on a person for a certain judgment do not know of the conflict of interest but the person knows (or should know) that they do not, then the person is allowing them to believe that the judgment in question is more reliable than it is—in effect, deceiving them. That deception is a betrayal of their (properly-placed) trust and therefore morally objectionable.

Third, even if the person exercising judgment informs those justifiably relying on that judgment that a conflict of interest exists, the judgment will still be less reliable than it ordinarily is. The person will still be less competent than usual—and perhaps appear less competent than members of the profession, occupation, or discipline in question should be. Conflict of interest can remain a technical problem, affecting reputation, even after it has ceased to be a moral problem.

How to Respond to Conflict of Interest

What can be done about conflict of interest? One common answer, one still enshrined in many codes, is: Avoid all conflicts of interest. That answer probably rests on at least one of two possible mistakes. One is assuming that all conflicts of interest can, as a practical matter, be avoided. Some certainly can be. For example, a journal editor can avoid most conflicts of interest by making sure all reviewing is *blind*. Reviewers would then (generally) not know what effect their official recommendations had on friends or enemies. An editor cannot, however, avoid all conflicts of interest in this way. Sometimes a reviewer will know enough to recognize that the author of a submission is an old friend (or enemy).

The other mistake is to assume that avoidance is the only proper response to conflict of interest. In fact, there are at least three others: escape, disclosure, and management.

Escape ends the conflict. So, for example, a reviewer who discovers that he or she is reviewing a friend's submission can stop reading, send the submission back to the editor with an explanation, and recommend a replacement.

Disclosure, even if sufficiently complete (and understood), merely gives those relying on a person's judgment the opportunity to give informed consent to the conflict of interest, to replace that person with another, or to adjust reliance in some less radical way (for example, by seeking a second opinion). Unlike escape, disclosure as such does not end the conflict of interest; it merely avoids the betrayal of trust.

Managing, though often the resolution reached after disclosure (as illustrated above), need not follow disclosure. Where disclosure is improper (because it would violate some rule of confidentiality) or impossible (because the person to whom disclosure should be made is absent, incompetent, or unable to respond in time), managing may still be a legitimate option.

Conclusion

Too frequently discussions of conflict of interest start with the biblical quotation, "Can a man have two masters?" This seems to be the wrong way to begin. The reason one cannot have two masters is that a master is someone to whom one owes complete loyalty, and complete loyalty to one excludes any loyalty to another. Having only one master *is* a strategy for avoiding conflict of interest, but a strategy making the concept uninteresting. Society must worry about conflict of interest only when avoiding all conflicts of interest is virtually impossible or so socially inefficient that there is general agreement that avoidance is often undesirable. Conflict of interest is an interesting concept only when loyalties are regularly and legitimately divided.

The term *conflict of interest* seems to have separated off from the related terms *conflicting interests* and *conflict of interests*," taking on the meaning given here, only in the middle of the twentieth century, a period in which two related trends seem to have accelerated. First, society has become more complex, making people increasingly dependent on experts. Second, society has become increasingly unsettled, making people increasingly reliant on strangers rather than on people they have known well for many years. People cannot manage the conflict of interest of those relied upon when they do not know enough about them. Society cannot tell experts to avoid all conflicts of interest because those experts could not then make a living. Society must therefore depend on such experts to disclose some conflicts (those that considerations of confidentiality allow them to disclose), to manage others, and to decline to exercise judgment where they can so decline without too much loss to those they serve. For that reason, the trend in codes of ethics in engineering and science has been away from flat prohibition of conflict of interest and toward more nuanced provisions. For example, the Code of Ethics of the Institute for Electrical and Electronic Engineers (1990) now urges members not only "to avoid real or perceived conflicts of interest whenever possible" but also "to disclose them to affected parties when they do exist."

MICHAEL DAVIS

SEE ALSO *Engineering Ethics; Professions and Professionalism.*

BIBLIOGRAPHY

Carson, Thomas L. (1994). "Conflicts of Interest." *Journal of Business Ethics* 13: 387–404. Argues that in the late twentieth century conflict of interest is endemic rather than rare.

Davis, Michael. (1993). "Conflict of Interest Revisited." *Business and Professional Ethics Journal* 12(Winter): 21–41. Classic statement of what has become the standard view.

Davis, Michael, and Andrew Stark, eds. (2002). *Conflict of Interest in the Professions*. New York: Oxford University Press. Especially Introduction, Chapter 6 (Engineering), Chapter 9 (Anthropology), Chapter 13 (Medicine), and Epilogue. A good survey of the state of thinking about conflict of interest across a broad range of professional and business activities.

Luebke, Neil R. (1987). "Conflict of Interest as a Moral Category." *Business and Professional Ethics Journal* 6(Spring): 66–81. Shows that the term conflict of interest developed its present meaning only in the middle of the twentieth century.

Krimsky, Sheldon, and L. S. Rotherberg. (2001). "The Conflict of Interest Policies in Science and Medical Journals: Editorial Practices and Author Disclosures." *Science and Engineering Ethics* 7: 205–218. Useful discussion of journal policies designed to manage conflict of interest.

Krimsky, Sheldon, and Ralph Nader. (2003). *Science in the Private Interest: Has the Lure of Profits Corrupted Biomedical Research?* Lanham, MD: Rowman and Littlefield. A study of conflict of interest in science, combining Krimsky's research with Nader's Jeremiad.

May, Larry. (1996). "Conflict of Interest." In *The Socially Responsive Self*. Chicago: University of Chicago Press. An interesting alternative to the standard approach to conflict of interest.

Pritchard, Michael. (1996). "Conflict of Interest: Conceptual and Normative Issues." *Academic Medicine* 71: 1305–1313. A thoughtful critique of the standard view.

Rodwin, Mark A. (1993). *Medicine, Money and Morals: Physicians' Conflicts of Interest*. New York: Oxford University Press. A grim indictment of a range of conflicts of interest in medicine.

Spece, Roy G., Jr.; David S. Shimm; and Allen E. Buchanan, eds. (1996). *Conflicts of Interest in Clinical Practice and Research*. New York: Oxford University Press. A collection of papers on conflict of interest in medicine covering many of the same topics as Rodwin but from a variety of viewpoints, much more nuanced.

CONFUCIAN PERSPECTIVES

• • •

Confucianism originated more than 2,000 years ago in China in the thoughts of Confucius, or Kong Zi (Master Kong, 551–479 B.C.E.). Kong Zi lived during one of the formative periods of Chinese culture, when numerous philosophical schools, such as Daoism (Taoism) and Mohism, vied for social influence. Other major early thinkers in the Confucian tradition include Mencius, or Meng Zi (371–289 B.C.E.), and Xun Zi (298–238 B.C.E.). Confucianism was established as the state ideology during the Han dynasty (206 B.C.E.–220 C.E.). As an original thinker, a powerful persuader, and a successful educator, Kong Zi became the defining philosopher of Chinese culture and one of the most influential cultural philosophers in East Asia and beyond. In the early twenty-first century Confucianism stands for a distinctive voice in global dialogues on issues that range from human rights to gender equality. As a living tradition, Confucianism also provides a unique perspective on science, technology, and ethics.

Confucian Foundations

The primary text of Kong Zi's thought that is still in existence is the *Analects* (*Lun Yu*), a posthumous collection of his sayings and his disciples' reflective remarks on his teachings. Other major Confucian classics include *The Book of Meng Zi*, *The Book of Change*, *The Book of History*, *The Odes*, *The Book of Rites*, and *The Spring and Autumn*. Although the precise dates of these works cannot be ascertained, scholars generally believe they were compiled during the Spring–Autumn and Warring States period (770–221B.C.E.). The development of Confucianism usually is divided into three phases. Classical Confucianism was developed by Kong Zi and other early thinkers. Neo-Confucianism was developed during the Song (960–1276) and Ming (1368–1644) dynasties by thinkers such as Zhu Xi (1130–1200) and Wang Yangming (1472–1529). The third phase is contemporary New-Confucianism, represented by thinkers such as Xiong Shili (1885–1968) and Mou Zongsan (1909–1995).

Historically, however, Han-Confucianism as it developed during the Han dynasty is also an important episode not only because that was the period when the tradition first became dominant in China but also because Han-Confucians extensively incorporated the notions of *yin-yang* and the Five Phases (Water, Fire, Wood, Metal, and Earth) into Confucianism. Those notions later had a great influence on the relationship of Confucianism to science and technology.

Confucianism is primarily a moral philosophy with ethics as its core. Confucian ethics has been characterized as virtue ethics. It is concerned with developing a virtuous person rather than emphasizing the following of ethical principles. Confucians see human life as a journey toward the goal of forming a virtuous character in the context of the family and other interpersonal relationships.

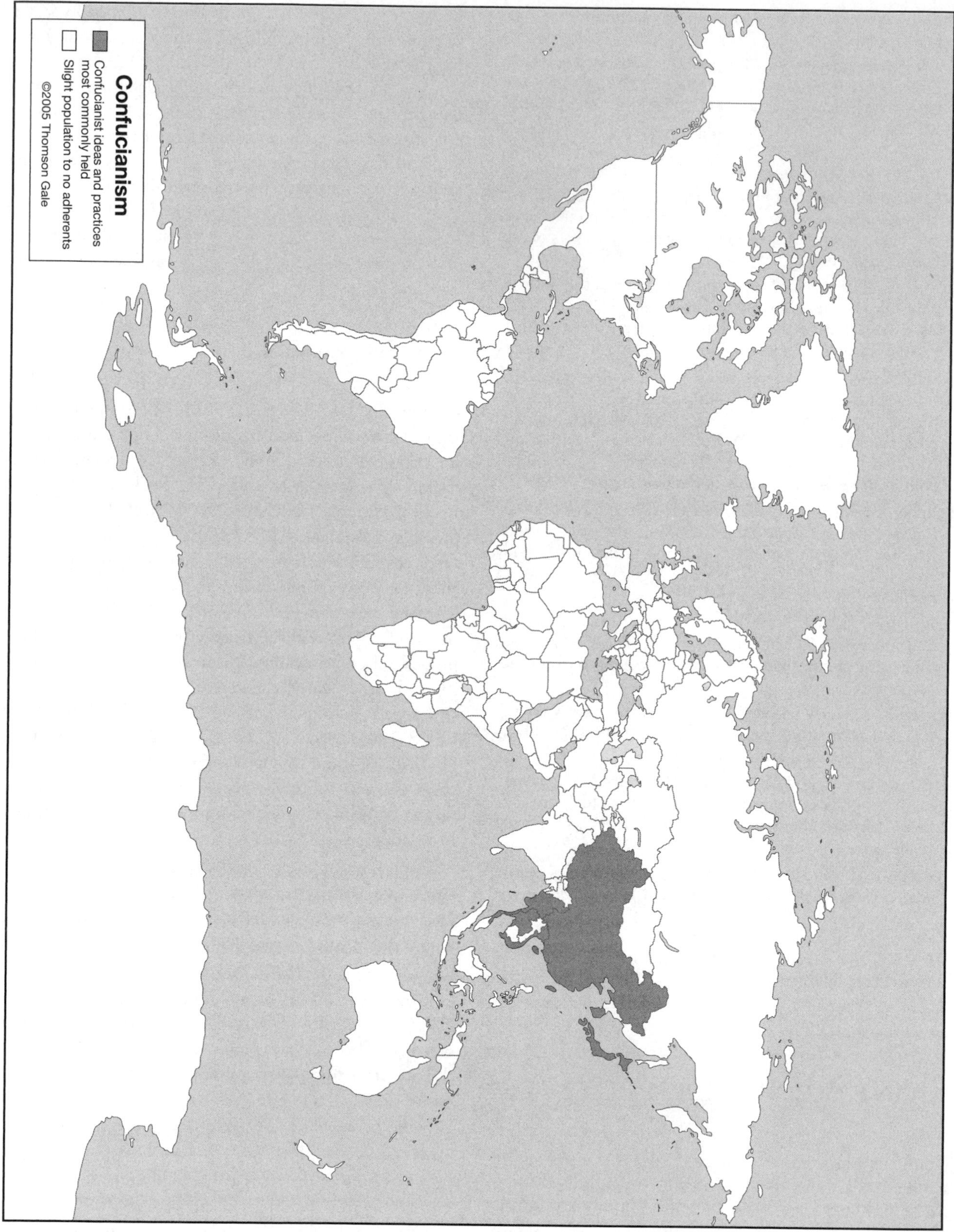
Confucianism
Confucianist ideas and practices most commonly held
Slight population to no adherents
©2005 Thomson Gale

Confucianism is not a theism. Its moral philosophy does not rest on a god or a divine being. However, it holds that there is a cosmic moral order that serves as the foundation of the moral life. This order is not carved in stone and is not a static entity; it has to be sought through human endeavors and realized through human activities. In comparison with Daoism, Confucianism places more emphasis on a person's accomplishments in society and on the positive consequences of moral edification. Whereas Daoism leaves room for supernatural forces, Confucianism is focused firmly on the earthly world and its mundane affairs. Over a long period of history Confucianism and Daoism formed a unique complementary relationship in Chinese society.

Key Concepts

Key concepts of Confucian ethics include *dao, de, ren, li,* and *yi*. The first of these concepts, *dao*, or the Way, defines the cosmic moral order. Confucians understand the cosmos as a triadic unity of Heaven, Earth, and Humanity. The *dao* is found and realized in a harmonious interaction among these three components. When it is realized, the entire world goes smoothly and the myriad things in it thrive.

Human beings participate in the realization of the *dao* by developing their *de*, or virtues. Confucians have what may be called a "person-making" ethics: One makes one's own person through learning and by extending one's knowledge and social skills. Every person is born with the potential to become a sage. Whether a person realizes his or her moral potential depends on that person's own effort. A good person is one who realizes his or her moral potential and develops into a virtuous person, one with a good character.

Whereas *de* points to particular virtues in various aspects of human life, *ren*, or humanity, as the Confucian moral ideal, stands for holistic human excellence. A *ren* person is a fully developed and well-rounded individual. Kong Zi said that a *ren* person is one who can achieve five virtues: earnestness, consideration for others, trustworthiness, diligence, and generosity.

The meaning of *li* is complex. It has been translated into English as *rites, rituals, propriety*, and *rules of proper conduct*. In the Confucian moral life *li* is the social grammar, providing guidelines for socially appropriate behavior. Unlike *ren*, *li* is tangible in that it tells people what to do in specific circumstances. For example, it is *li* to yield a seat on the bus to an elderly person and not to speak loudly in the library. Learning *li* is a necessary step for a person to develop moral virtues and become *ren*. Observance of *li* is the natural path for a person of *ren*. A society without *li* is chaotic and uncivilized; an un-*li* person is socially retarded and barbarous. Confucians, however, do not take *li* to be absolute. Recognizing the complexity and the dynamic nature of social life, Confucians value the ability to determine a course of appropriate action in complex situations.

The concept of *yi* focuses principally on what is right and fitting in particular circumstances. It calls for sound judgment and reasonableness. At times *yi* may require people to forgo personal advantages in order to do what is right. A person of *yi* demonstrates moral maturity. Other important Confucian virtues include *xiao* (filial piety), *xue* (learning), and *zhi* (wisdom).

Applications to Science, Technology, and Ethics

As a complex philosophical tradition with a long history, Confucianism has a twofold relationship to science and technology. First, as a secular philosophy Confucianism has a natural affinity to science because it includes no superstitions and does not recognize supernatural forces. When asked, Kong Zi refused to speculate about gods, ghosts, and supernatural phenomena. His focus was entirely on this world and on things that can be known. In this respect Confucianism is not opposed to science and technology.

In ancient China technology had more to do with handicrafts than with science. The Confucian classic *Rites of the Zhou* (*Zhou Li*), which was compiled during the Warring States period, contains a chapter on various types of craftsmanship in society. It attributes to early sages the invention of various handicrafts, such as the making of knives and scissors, pottery, carriages, and boats, and explicitly recognizes the important role of handicrafts in society. The chapter maintains that excellence in craftsmanship requires an integration of four things:, good timing of the season, flourishing *qi* (cosmic energy) on earth, excellent material, and superior skills. From the Confucian perspective craftsmanship is not merely a matter of technique or skill but is understood holistically in the context of the Confucian cosmology. Whereas Daoism appeared to be antagonistic to handicraft, as indicated in the *Dao De Jing*, Confucianism was receptive to it because handicrafts can be instrumental to the prosperity of the family, which Confucianism values highly.

The affinity between Confucianism and science and technology has been evidenced by historical figures such as Shen Kuo (1031–1095), who was a prominent scientist in research, a successful technocrat in civic service, and a committed Confucian in his family life. His *Brush Talks from Dream Brook* is one of the most remarkable documents of early science and technology in China. Shen not only wrote commentaries on Confucian classics, a common practice among ancient Confucian scholars, but also in his theoretical discussions of scientific topics used philosophical concepts such as *yin-yang*, the Five Phases, and *qi*, which were shared by other Confucian scholars during his time. In Shen's eyes there is no contradiction between Confucianism and science and technology.

Traditional Chinese medicine has a close connection to Confucian cosmology. The *Yellow Emperor's Inner Chapters* (*Huang Di Nei Jing*), the primary ancient text of Chinese medical science and techniques, is consistent with Confucian cosmology. The fundamentals of the entire Chinese traditional medicine are rooted in the philosophical notions of *yin-yang*, the Five Phases, and *qi*. Although these notions also can be found in Daoism, Confucians embrace them profoundly, and they are the converging points of Confucianism and Daoism. Acupuncture, for example, is based on the belief that human health depends on the smooth flow of *qi* and a good balance of *yin-yang*. The philosophy of the Five Phases provides the foundation for Chinese herbal medicine in its belief that the myriad things in nature have various combinations of the Five Phases and that the balance of the Five Phases is instrumental to the balance of *yin-yang* and the nurturing of *qi*. For example, when someone's body has too much *yin* and is short of *yang*, a herb rich in Fire may boost that person's *yang* to restore the balance.

Confucius, 551 B.C.–479 A.D. Confucius founded his school of philosophy on the concepts of benevolence, ritual, and propriety. *(Source unknown.)*

Second, Confucianism is principally a moral philosophy and places the moral life above all other aspects of human activities. For Confucians the ultimate value of human activities depends solely on their contribution or lack of a contribution to the good moral life of humanity (*ren*). In other words, apart from its contribution to the good moral life, an activity does not possess any value.

This moral view has been subjected to narrow interpretations and at times has devalued science and technology. In particular, making too direct a connection between science and the moral life may not leave room for science to grow independently, which is often a necessary condition for the flourishing of science. Confucianism is not free from criticisms of this sort: At the beginning of the twentieth century one of the two main criticisms of Confucianism was its alleged impediment to science (the other was its alleged impediment to democracy). Some criticisms of Confucianism for its hostility to science might have been exaggerated, but they were not entirely groundless.

Kong Zi apparently was not interested in technical knowledge about the natural world. When a student asked him about agricultural knowledge and skills, his reaction was negative. Xun Zi was probably the only early Confucian who had a tendency to naturalize Confucianism, a viewpoint that could have assigned natural science a larger role in the Confucian value system if it had had a broader influence. Xun Zi believed that it is human nature to learn and to know and that what people learn and know is the nature of things. However, mainstream Confucian thought has always emphasized a moral worldview. That thought focuses on moral values as the core of the cosmos and centers human existence on moral existence. Kong Zi explicitly defined true knowledge as knowledge about human affairs rather than about the natural world. This attitude was reflected in the neo-Confucian Zhang Zai's (1020–1078) formulation of the contrast between "moral knowledge" and "knowledge of the senses" and his assertion that moral knowledge cannot grow out of knowledge of the senses. Placing these two kinds of knowledge in sharp contrast or even opposition further diminishes the importance of knowledge of the natural world in comparison to the importance of moral knowledge.

Zhu Xi was the second major figure after Xun Zi to offer a chance to elevate the status of knowledge about the natural world through his interpretation of *gewu zhizhi*, an ancient concept found in the *Daxue* chapter of the Confucian *Book of Rites*. He interpreted *gewu zhizhi* to mean the investigation of things and the expansion of knowledge. According to Zhu, things in the world have their reason or principle, which can be known through empirical observation. Zhu Xi evidently had a holistic view of the world and saw a direct connection between empirical knowledge of the natural world and moral knowledge. For him the purpose of *gewu zhizhi* is to improve people's moral knowledge. Because his notion of *gewu* includes the empirical study of the natural world, he opened a door to scientific knowledge. Presumably, the investigation of things could lead to scientific knowledge about the natural world.

Unfortunately, Zhu Xi's course was reversed by another major neo-Confucian thinker, Wang Yangming. Wang initially tried to act on Zhu's idea of *gewu zhizhi* by attempting to investigate the bamboo in his yard. However, he failed miserably because he could not get any meaningful knowledge through his diligent observation of the bamboo. Wang then changed course and claimed that all useful knowledge is to be found within the heart-mind (*xin*); there is no need to look outside the heart-mind. Wang's judgment inflated to an extreme the Confucian conviction that a person's primary mission in life is to develop his or her humanity and failed to assign adequate value to the pursuit of the knowledge of the natural world. This tendency lasted till the twentieth century.

Contemporary Discussions

As science started to gain ground in Chinese society in the early twentieth century, Confucian thinkers tried to preserve the territory of moral philosophy by separating science and philosophy into two distinct realms. They argued that whereas science deals with the physical world, (Confucian) philosophy deals with the metaphysical and moral realms; therefore, the two do not conflict. After the founding of the People's Republic of China in 1949, Confucianism was subjected to severe criticisms and at times brutal repression in mainland China, although it had a significant revival during the last two decades of the twentieth century.

However, Confucianism never stopped developing. Mou Zongsan, who lived his most productive years in fHong Kong during the second half of the twentieth century, articulated a new Confucian stance on science and greatly expanded the room within Confucianism for scientific knowledge. He maintained that traditional Confucian culture failed to give adequate recognition to the form of knowledge called *zhi xing* (formal, logical thinking) and argued that to embrace both science and democracy, the spirit of Chinese culture needed "to negate itself into" the mode of *zhi xing*. Mou's philosophy marked a turning point in the long debate among Confucian thinkers about the role science and technology play in the good life and was an important stage in the development of Confucianism. After Mou the importance of science and technology was no longer an issue for Confucians.

Some scholars have attempted to interpret the history of Confucianist interactions with science and technology in a different light, arguing that Confucianism has not been as unfriendly to science and technology as sometimes is alleged. They cite the fact that science and technology in early China under Confucianism flourished and that many ancient Confucian scholars were also great scientists and technological innovators. For example, it was during the Han dynasty, when Confucianism was made the state ideology, that the basic Chinese sciences were established. Those sciences included mathematics, mathematical harmonics, mathematical astronomy, and medicine. It is possible that the attitude of Confucianism toward science and technology varied at different times, affected by specific social circumstances and influenced by individual Confucian thinkers' personal beliefs. Confucianism might have been more congenial to science and technology at certain times. It is also true that within Confucianism there is a full range of opinions on issues related to science and technology, with some being more liberal and others more conservative.

Contemporary Confucians recognize the importance of science and technology in society and in moral philosophy. Because Confucianism advocates a virtue ethics and is concerned with the full development of the holistic person, it recognizes the indispensability of ethics in science and technology in achieving that goal. Furthermore, because the goal in Confucianism is to make the *ren* person, achieve a *ren* society, and generate a harmonious world, all human activities, including science, technology, and ethics, are to serve that purpose directly or indirectly. Kong Zi said that a good person should not be a mere tool. A committed Confucian does not engage in science for the sake of science or promote technology for the sake of technology. In addition to "Is it true?" or "Does it work?" a Confucian would ask questions such as "What purpose does it serve?" "How does it contribute to the good society?" and "Does it make the world a better place?"

A case can be made that Confucianism may be more receptive to contemporary medical research, such as embryonic stem cell research. Without a doctrine of the divinely created soul, Confucians believe that a person is not born with moral worth and has to earn it through moral cultivation. Therefore, strictly speaking, the human embryo or the fetus is merely a potential human person, not yet a moral entity. Drawing on this notion, Confucians may not see embryonic stem cell research, which requires the destruction of the embryo, as morally problematic. After all, cracking an acorn is not the same as destroying a giant oak tree even though an acorn could grow into a giant oak tree.

Although Confucianism is not opposed to the development of technology, with the rapid technological advancement in the early twenty-first century, Confucians are concerned with its negative impact on the environment, its harmful effects on a harmonious world where humans and nature are closely integrated. If one uses the word *ethics* broadly to encompass the Confucians' goals of the moral life, ethics remains the primary concern for Confucians; science and technology are important tools that serve these purposes.

CHENYANG LI

SEE ALSO *Acupuncture; Chinese Perspectives; Virtue Ethics.*

BIBLIOGRAPHY

Cheng, Chung-ying, ed. (2002). *Journal of Chinese Philosophy* 29(4). Special Issue on Science, Technology, and Chinese Philosophy. Includes several recent articles on Confucianism, science, and technology.

Fan, Dainian, and Robert S. Cohen, eds. (1996). *Chinese Studies in the History and Philosophy of Science and Technology.* Boston: Dordrecht; London: Kluwer Academic. A collection of numerous articles on Chinese philosophy, science, and technology from historical perspectives.

Fan, Ruiping, ed. (1999). *Confucian Bioethics.* Boston: Dordrecht; London: Kluwer Academic. Includes articles with Confucian views on issues such as experimenting with human subjects.

Le, Aiguo. (2002). *Confucian Culture and Ancient Science and Technology in China (Rujia Wenhua yu Zhongguo Gudai Keji).* Beijing: China Books (*Zhonghua Shujiu*). A recent work focused on the affinity between Confucianism and science and technology, with rich information and superb scholarship.

Needham, Joseph. (1954). *Science and Civilization in China.* Multiple volumes. Cambridge, UK: Cambridge University Press. The most massive study of science in ancient Chinese society.

Sivin, Nathan. (1995). *Science in Ancient China: Researches and Reflections.* Brookfield, VT: Variorum. A collection of eight studies of specific topics in science in ancient China.

Yosida, Mitukuni. (1979). "The Chinese Concept of Technology: A Historical Approach," *Acta Asiatica* 36: 49–66. Offers useful information on ancient views on technology in China.

CONSCIOUSNESS

Something is conscious if there is *something that it is like to be* that thing. This widely accepted definition, proposed by philosopher Thomas Nagel (1974, reprint 1997 p. 519), emphasizes the *subjective character* of conscious experience, which is the fundamental obstacle to its scientific investigation. Scientists have no objective access to conscious states (even their own) so consciousness can only be studied scientifically by indirect means, and some believe that a complete scientific description of the world can and should be made without reference to consciousness at all. However to exclude conscious decisions from the causal chain of events would undermine all ethical and legal systems based on personal responsibility for consciously willed actions.

In the 1980s, neurophysiologist Benjamin Libet showed that when subjects were asked to make a voluntary movement at a time of their own choosing, brain activity initiating the movement (the *readiness potential*) routinely preceded by about half a second the conscious decision to make the action. Many people interpreted this as scientific proof that conscious choice and freewill are illusory, which would fit with the view that the physical universe is causally closed and deterministic. Libet himself safeguards personal freedom of action by arguing that although the brain's non-conscious readiness potential initiates an action, there is still time for the conscious mind to monitor and abort the process before the action is carried through.

Libet's work was an early example of scientific research into consciousness that combines objective information about brain activity with subjective reports from experimental subjects concerning their conscious states. Earlier generations had been handicapped by the need to choose between subjective and objective methods. Typical of these were *introspectionism*, pioneered by German psychologist Wilhelm Wundt (1832–1920), which depended on individuals analyzing their subjective thoughts, feelings, and perceptions into thousands of basic mental sensations, and the *behaviorism* of John Watson (1878–1958) and his successor B. F. Skinner (1904–1990). Watson rejected introspection, maintaining that if psychologists wanted to be real scientists they must study objective, verifiable data, which meant observable behavior. Such was his influence that consciousness was effectively banned from psychology for half a century in the mid-1900s.

The scientific study of consciousness was rehabilitated in part by new technologies that allowed the working of the brain to be objectively studied while mental processes were being carried out. The *electroencephalogram* (EEG), recording electrical activity in the brain, was available from the 1930s and used by Libet among others. Brain scanning techniques such as *positron emission tomography* (PET) and *functional magnetic resonance imaging* (fMRI), developed in the 1980s and 1990s, enabled detailed observation of active areas of the brain at work and confirmed the hypothesis that mental states are closely related to the physical condition of nerve cells (neurons). Neuroscientists were now able to observe the areas of neural activity associated with particular conscious experiences reported by human subjects, or deduced from the behavior of animals such as monkeys. Various systems in the brain were investigated, from individual cells to large networks and pathways of interconnected neurons, in the quest to identify possible *neural correlates of consciousness* (NCCs).

The exact relationship between conscious experience and the physical brain, and how and why some brain processes are conscious at all, is the core dilemma. David Chalmers, Director of the Center for Consciousness Studies at the University of Arizona at Tucson, has dubbed it the *Hard Problem*. In the mid-twentieth century the influential Oxford philosopher Gilbert Ryle (1900–1976) dismissed Descartes's dualist concept of mind-body relation as the *ghost in the machine*, and opened the way for various materialist accounts of consciousness. By the turn of the millennium most consciousness researchers embraced some form of *non-reductive materialism*, which holds that mental states are wholly caused by the physical brain, but have some quality over and above the sum of their molecular components. Variations on this theme include *property dualism* (mental states exist as properties of underlying physical states), *dual aspect monism* (the mental and the physical are two ways of looking at a single underlying reality), *emergentism* (consciousness emerges at a certain level of complexity), and *panpsychism* (every material object has an actual or potential degree of consciousness).

Treating consciousness as a real aspect of the physical world brings it back into the realm of scientific inquiry and removes the suggestion that it is an *epiphenomenon*, lying outside the causal nexus of the universe. But it does not automatically refute the claim that free choice and moral responsibility are delusions. The physical world of which consciousness is a part still appears to be deterministic, at least according to classical physics. Researchers into artificial intelligence, for instance, have drawn parallels between neuronal activity in brains and the processing of information in computers The question of whether the conscious mind itself is computational, that is, completely describable mathematically and therefore in deterministic terms, is hotly disputed.

Deterministic views are challenged within science by evidence from quantum physics, although its relevance is disputed and some of the claims speculative. For example, Oxford mathematician Roger Penrose proposes that in certain special conditions, found in the microtubules within brain cells, quantum systems provide the physical mechanism that brings about noncomputational conscious events. From a different starting point, Berkeley physicist Henry Stapp argues that quantum theory can explain how consciousness plays a creative role in shaping events and creating the world as humans know it. These views are frequently criticized, but at the very least, quantum theory puts a large question mark over the old assumption that the universe is a collection of objective facts that are (in theory at least) completely knowable.

Consideration of the ethical questions posed by the investigation and manipulation of consciousness falls under the sub-discipline of neuroethics. But the challenge to produce an account of conscious experience that provides an adequate basis for morality at all, and is at the same time both philosophically and scientifically robust, lies at the heart of all consciousness studies.

ANTHONY FREEMAN

SEE ALSO *Artifical Intelligence; Decartes, René Emotion; Emotional Intelligence; Neuroethics; Robot Toys.*

BIBLIOGRAPHY

Chalmers, David J. (1996). *The Conscious Mind: In Search of a Fundamental Theory*. New York: Oxford University Press. A technical book, written by a philosopher making use of scientific research, mostly accessible to the general reader. The stance is non-reductive physicalism.

Dennett, Daniel. (2003). *Freedom Evolves*. New York: Viking Press. A representative title from many by this popular philosopher of mind. Written from a reductive viewpoint for both professional and general readers.

Freeman, Anthony. (2003). *Consciousness: A Guide to the Debates*. Santa Barbara, CA: ABC-CLIO. A non-technical introduction that includes extracts from classic texts.

Libet, Benjamin; Anthony Freeman; and Keith Sutherland, eds. (1999). *The Volitional Brain: Towards a Neuroscience of Freewill*. Exeter, UK: Imprint Academic. Collected essays by scientists and philosophers.

Nagel, Thomas. (1974). "What Is It Like to Be a Bat?" *Philosophical Review* 83(4): 435–450. Reprinted in the 1997 book *The Nature of Consciousness: Philosophical Debates*, eds. Ned Block, Owen Flanagan, and Güven Güzeldere. Cambridge, MA: MIT Press. A seminal paper raising the need for philosophers to address subjectivity.

Popper, Karl, and John Eccles. (1983). *The Self and Its Brain: An Argument for Interactionism*. London: Routledge. Two eminent authors make the case for the unfashionable dualist approach to mind-body interaction.

CONSENSUS CONFERENCES

Consensus conferences are one of several practices (including citizen juries, scenario workshops, and deliberative polls, among others) intended to enhance deliberative public involvement in shaping social decision making about science and technology. Because public issues increasingly include complex scientific and technological components, and because the general public lacks the needed scientific knowledge, the management of those issues seems inevitably to slip out of the hands of ordinary citizens. Democratic governance, however, rests on the informed consent of ordinary people, and many observers worry that in numerous areas ordinary citizens are becoming less able to shape public policies.

Basic Issues

The basic concept behind consensus conferences is that public policies about science and technology will be improved significantly if policy makers can hear informed, deliberative public perceptions, concerns, and recommendations as they consider the choices they face. Informed and thoughtful public participation may also help to blunt two features of contemporary policy making about science and technology: intense and acrimonious partisan advocacy by both proponents and opponents of specific scientific and technological projects, and local Not-In-My-Backyard (NIMBY) campaigns based in communities likely to be directly affected by those projects. In the first case, proponents and opponents of specific science and technology projects make sensationalized and exaggerated claims about the wisdom and foresight of their perspective and the mean-spirited and hysterical positions of their antagonists. All too often, ordinary citizens (who must live with the consequences of the policy decision) are unable to sort through the conflicting claims and counterclaims. In NIMBY situations, local citizens—often frustrated by the blare and noise of partisan bickering, and distrustful of all sides in the controversy—organize to oppose, delay, and obstruct projects desired by others.

Both processes result in political and policy paralysis, the spread of cynicism and apathy, and delay in addressing pressing public needs. Consensus conferences seek to address both problems by providing a group of average, non-expert citizens with the opportunity and the resources to conduct an informed and deliberative investigation of specific technologies, to develop policy recommendations they can all endorse, and to deliver those recommendations to policy makers and the public. In this way, consensus conferences allow the deliberating citizens to confront partisan advocates with reliable information rather than sensationalism, and also help to dissipate cynicism about governmental decision making that contributes to NIMBYism.

Danish Model

The Danish Board of Technology (BOT), a research arm of the Danish Parliament, developed the basic model of a consensus conference. Several months before the parliament must address an issue with significant science and technology elements, members of the parliament may ask the BOT to conduct a consensus conference on the issue. The lead time helps assure that citizen evaluations and recommendations are available to legislators in time to help shape parliamentary debates.

The BOT takes several steps to implement a consensus conference:

- It assembles an Oversight Committee, made up of experts and stakeholders in the specific technology under inspection.
- It develops background information about the technology and its probable social, economic, political, and ethical implications.
- It recruits twelve to fifteen Danish citizens to serve as the citizen panel. The citizens are paid a stipend to cover the costs of participation.
- And, finally, it conducts the consensus conference and makes the results available to parliament, the press, and the public.

The Oversight Committee serves to guide development of the background materials that will be given to citizen panelists. Because the Oversight Committee is composed of individuals reflecting the full spectrum of opinions about the technology in question, the Committee helps to assure that background materials are fair, accurate, and accessible to ordinary people. The Oversight Committee also monitors recruitment and selection of the citizen-panelists. In a broad sense, the Oversight Committee serves to keep the entire process honest, to prevent intentional or unintentional partisan slanting of background materials or of makeup of the panel.

The actual work of the consensus conference typically takes place over three weekends, about one month apart. This marks the consensus conference as one of the most intense public participation techniques, because most other practices last only one or two days, or even two or three hours.

During the first weekend, the panelists get acquainted with each other, with the staff facilitating the sessions, and with the processes and goals of the conference. They read and discuss the background materials, and are encouraged to raise whatever issues or concerns are important to them. In this sense, the consensus conference differs from a traditional focus group in which the panelists are asked their reactions to issues raised by the focus group sponsors. In effect citizens are given control of the agenda in a consensus conference. During the second weekend, the citizen members continue to discuss the technology and the background materials, and to sharpen their issues and concerns. They also begin to develop a series of follow-on questions for content experts who will attend during the final sessions.

During the final weekend—the actual Consensus Conference—three things occur. On the first day, a series of content experts, who reflect the spectrum of opinions within the expert community, provide responses to the follow-on questions the panelists raised earlier. This is followed by an open-ended question-and-answer session, with all the experts and panelists present. The panelists can thus ask any remaining questions, probe earlier responses, and seek clarifications.

After this, panelists withdraw (along with a facilitator) to deliberate. Their goal is to arrive at a common set of policy recommendations that express their collective judgment about how best to manage the technology. This task often lasts into the early hours of the morning.

The panel's report is submitted to the content experts to catch any remaining technical errors, but the experts do not comment on the policy recommendations. The report is then delivered to parliament and the public at a press conference. The staff of the BOT point to the frequency with which contending policy constituencies refer to consensus conference reports during parliamentary debates as evidence that consensus conferences help to shape policy outcomes.

In Denmark consensus conferences have addressed an array of science and technology issues, such as genetically modified foods, infertility, the human genome project, teleworking, and transgenic animals. Consensus conferences have been organized in several other countries—including Argentina, Australia, Austria, Canada, France, Germany, Israel, Japan, the Netherlands, New Zealand, Norway, South Korea, Switzerland, and the United Kingdom—although no other country has adopted the practice as thoroughly as the Danes. Consensus conferences have been conducted about fifteen times in the United States as parts of public deliberation research.

Consensus conferences in the United States have been held at the University of Massachusetts ("Telecommunications and the Future of Democracy"), the University of New Hampshire ("Genetically-Engineered Foods"), and ten times at North Carolina State University (two conferences dealing with "Genetically Modified Foods," six Internet-based conferences dealing with "Global Warming," and two conferences dealing with "Nanotechnology"). The North Carolina State conferences were part of a National Science Foundation supported research project dealing with public deliberations about science and technology.

Further Developments

The literature about public deliberations points to a number of concerns or problems that arise when citizens deliberate together. Groups of average citizens, when deliberating, employ a variety of *decision heuristics* which, observers worry, may introduce distortions into their thinking. Ordinary citizens, for instance, seem to focus on the *risk of the month*, shifting concern from one kind of risk to another based on which risk is currently receiving the most public discussion or which has been in the news (the *availability heuristic*). Similarly they seem to draw conclusions about the dangers of specific products through *mental shortcuts*that can lead to factual errors about actual risks (*intuitive toxicology*). Risks or dangers that are exceptionally vivid also seem to gain greater public awareness, regardless of actual statistical probability (the *affect heuristic*). Critics also point to various *social cascades* in which unsubstantiated beliefs gain credibility simply because they are constantly repeated. Group polarization is another feature of some pubic deliberations. This involves the tendency for a group's final conclusions to support the group's original position, rather than a more centrist or moderate one.

The majority of studies pointing to such cognitive problems among ordinary citizens, however, focus on unfacilitated public deliberations. The current research suggests that many of these cognitive problems can be adequately addressed if professional and well-trained facilitators lead the public deliberations. Effective facilitation can, for instance, ameliorate the influence of strong-willed or domineering personalities, insure that citizen panelists are exposed to a wide argument pool, and detect and correct inappropriate decision heuristics. Consensus conferences, in particular, provide ample room for the beneficial effects of good facilitation, and provide sufficient time for the panelists to acquire substantial background information and to interact with a range of content experts. While these steps may not correct all cognitive and process issues in public deliberations, they can successfully address the most egregious problems.

Supporters of consensus conferences hope that the technique can be used wherever democratic governance of new technologies is pursued. While the outcomes of informed, deliberative citizen consideration of new technologies cannot substitute for the procedures of democratically elected government, consensus conferences may provide a mechanism for greater influence by ordinary citizens in the shaping of public policies concerning technologies that all must live with, and thereby create an enhanced level of democratic credibility for governmental decisions.

PATRICK W. HAMLETT

SEE ALSO *Constructive Technology Assessment; Discourse Ethics; Science Shops.*

BIBLIOGRAPHY

Andersen, Ida-Elizabeth, and Birgit Jæger. (1999). "Scenario Workshops and Consensus Conferences: Towards More Democratic Decision-Making." *Science and Public Policy* 26(5): 331–340.

Fishkin, James S. (1991). *Democracy and Deliberation: New Directions for Democratic Reform*. New Haven, CT: Yale University Press. Asserts both theretical and empirical arguments in favor of enhanced public deliberation about public policy in general.

Hamlett, Patrick. (2002). "Citizens' Consensus Conferences: Learning and Public Confidence." Conference presentation, Annual Meeting of the American Association the Advancement of Science (AAAS), Boston.

Joss, Simon, and John Durant, eds. (1995). *Public Participation in Science: The Role of Consensus Conferences in Europe*. London: Science Museum. Contains general discussions of the origins of the Danish consensus conference process, developments and applications in Europe, and evaluations of the use of the technique.

Mayer, Igor. (1997). *Debating Technologies: A Methodological Contribution to the Design and Evaluation of Participatory Policy Analysis*. Tilburg, The Netherlands: Tilburg University Press. Examines the theoretical claims made in favor of enhanced participatory public policy analysis involving ordinary citizens and examines the impacts of consensus conferences.

Mayer, Igor S., and Geurts, Jac L. (1996). "Consensus Conferences as Participatory Policy Analysis, a Methodological Contribution to the Social Management of Technology." In *1996 International Symposium on Technology and Society: Technical Expertise and Public Decisions Proceedings*. Danvers, MA: IEEE Society on Social Implications of Technology.

Mendelberg, Tali. (2002). "The Deliberative Citizen: Theory and Evidence." In *Research in Micropolitics*, vol. 6, eds. Michael Delli Carpini, Leonie Huddy, and Robert Y. Shapiro. New York: Elsevier Press. Provides a comprehensive review of existing empirical literature on public deliberations.

Renn, Ortwin, Thomas Webler, and Peter Wiedemann, eds. (1995). *Fairness and Competence in Citizen Participation: Evaluating Models for Environmental Discourse*. Boston: Kluwer. Assesses the implementation and impacts of several kinds of participatory public policy and public deliberation techniques.

Sunstein, Cass R. (2002). *Risk and Reason: Safety, Law, and the Environment*. New York: Cambridge University Press. Examines analytical approaches to regulating technological risks, including critiques of both affective and cognitive problems associated with public deliberation.

CONSEQUENCES

SEE *Unintended Consequences*.

CONSEQUENTIALISM

As a general category of ethical or moral theories, consequentialism refers to theories that evaluate rightness or wrongness based *exclusively* on the consequences or effects of an act or acts. Consequentialist theories may differ over what kinds of consequences matter, while agreeing that the rightness or wrongness of actions cannot be based on motives or intentions of those who act, nor on the conformity of the act to duty, virtue, piety, moral rules, or the law. Consequences are all that matter for ethics, on this view. According to consequentialists, some murders might turn out to be morally right, while some acts of sincere generosity might be wrong.

Consequentialism is the ethical theory most compatible with the empirical and quantitative focus of much of science and technology. When a consequentialist studies ethical issues in science and technology, an act is usually understood broadly to include national and local policies, programs, distributions of resources, implementations of new technologies, and the like. Consequentialism seems particularly well suited to evaluate these kinds of complex acts, because it shares with modern, positivistic science an emphasis on observation. Just as one might form and test a hypothesis about electromagnetic radiation, so too could one test an act or policy that one believes to be right. In both cases, one looks to results in the real world in order make an evaluation.

Also, consequentialist theories take into account short- and long-term effects, and hence can evaluate developments such as nuclear power, where the immediate good effects (electricity without air pollution) may be outweighed by later harmful effects (radioactive waste, illness). In focusing on observable effects over time, consequentialists seem to look in the obvious places for answers to ethical questions concerning emerging technologies. To evaluate such complicated developments as genetic engineering, nanotechnology, the Internet, or even automobile transportation, where else would one look but to the effects?

Despite the intuitive appeal of consequentialism for such ethical inquiries, the view has faced serious opposition, especially from philosophers, as its proponents try to specify which consequences are relevant to moral evaluation. The historical development of consequentialism shows a constant struggle to identify the morally relevant effects of acts and to measure them. Consequentialists have sought to elucidate a scientific ethical theory for difficult contemporary issues, but with mixed results.

The Classical View: Act Utilitarianism

The most influential version of consequentialism is known as *utilitarianism*. The basic idea behind this view is quite simple. One consequence that almost anyone would want from an act is an increase in happiness, because happiness is undeniably a good. This is the conception of the good from which utilitarianism begins, and further developments in utilitarian theory almost always get back, in some way, to the content and measure of happiness.

Utilitarians are not merely interested in their own happiness; they advocate the "greatest happiness of the greatest number." According to the founder of classical or *act utilitarianism*, Jeremy Bentham (1748–1832), an act is right if its overall tendency is to increase the proportion of happiness (or pleasure) to pain.

If one has a choice between several acts in some situation, one ought to choose the act with the best *net* effect on utility. In some cases this will be the act that increases everyone's utility. In other cases, the best act would do no more than decrease everyone's pain. For most complex acts and policies, though, the result is complicated; the same act may include both some utility and some disutility. Hence Bentham realized that he would need a quantitative method for calculating the best utilitarian act. He proposed a "felicific calculus" that attempted (unsuccessfully) to supply cardinal measurements for the utility of an act based on its intensity, duration, certainty, and similar factors. By summing measurements for every act over all those who would be affected, utilitarians could instruct society on how incrementally to increase the amount of utility its members enjoyed. Act utilitarianism, if carried out rigorously, promised a program of social reform. For individuals who used the theory to evaluate their acts, the calculus required them to count the happiness of others as though it were their own. In principle, it provided an argument for an impartial and equitable distribution of the fruits of the new industrial revolution.

Another significant aspect of Bentham's view is that his principle of utility seeks, in the long run, to maximize the utility of all sentient beings—every being that can feel pleasure or pain. In this way his theory grants moral status not just to humans, who alone can reason and talk, but also to *any* animal that can feel or suffer. Bentham argued that the pain of non-human animals must count in the felicific calculus; his view would inspire later animal rights advocates and contemporary utilitarians such as Peter Singer. Utilitarianism thus became the first modern moral theory to take seriously the harm done by humans to other animals.

Despite its progressive social and political tendencies, act utilitarianism faced major problems. Even if individuals could calculate a cardinal measurement of personal utility from a particular act, they could not be sure that this measurement was on the same scale as a measurement for another person. But the theory requires the summing of utilities over the class of those affected by the act. Utilitarianism requires cardinal *interpersonal* measurements of utility—numbers on the same scale, valid for everyone. Supposing that the theory could provide such a scale, it then seemed to *demand* constant calculation for every act, because what is required morally is to come up with the greatest sum of utility. Every option in acting would have to be considered, and such exhaustive calculations might lead to paralysis.

Finally, act utilitarianism seemed to embrace a brutish theory of the good; the pleasure of thousands of cows, chewing their cud, might outweigh the utility of a college education for one person. If there were tradeoffs to be made—and the emerging free markets of Bentham's time made those tradeoffs possible—one might end up with many satisfied cows instead of a few educated people. Worse still, act utilitarianism might ask a sacrifice of the rights of some for the utility of others. Because every good was to be reduced to utility, even future commitments of justice seemed to be beholden to the arithmetic of maximization.

Rule Utilitarianism

Bentham's protégé, John Stuart Mill (1806–1873), addressed some of the shortcomings of act utilitarianism by proposing three changes. First, he found Bentham's ethical calculations too cumbersome, and proposed instead that society adopt and enforce a set of rules which, when followed, were likely to produce the highest overall utility. The best way to be a utilitarian, on this view, would be to act according to a rule that, in conjunction with other rules, prescribed behavior that maximized total social utility. One rule could replace another in the set, provided that the change would contribute to greater overall utility. But absent now in Mill's *rule utilitarianism* was the requirement—or even the possibility—of a quantitative calculus for determining which acts to choose. Second, Mill introduced a qualitative distinction between higher and lower pleasures, thus undermining the notion of a common scale for ethical measurement, and implicitly relegating the happiness of non-human animals to insignificance. Finally, Mill argued that certain rules, what he called the rules of justice, were so important to the long-term security (and hence happiness) of society that they must be considered practically inviolable. The results of these changes made the application of rule utilitarianism less scientific but much more in line with common sense morality. Mill's theory still shared the goal of Bentham's original utilitarianism, but it allowed notions such as duties, rights, and virtues to be means to the end of increased social utility.

Market Consequentialism

By the early twentieth century, utilitarian moral philosophers and economists became interested in market activity as a replacement for the direct measurement of the consequences of an act. They saw preferences, revealed in market supply and demand, as an approximate (though indirect) indication of the utility that a single person gains by a market "act." They also were able to represent mathematically an individual's preferences over bundles of goods, and to prove some interesting theorems about these "utility functions" of individuals. By analyzing market preferences, economic consequentialists could provide quantifiable evidence of what made consumers happy. To be a consequentialist about market preferences meant to choose the act or policy that allowed all persons their highest-ordered preferences, given what an economy could supply.

The economic version of utilitarianism was made even more sophisticated by the addition of a formal theory of individual choice under uncertainty, introduced by John von Neumann and Oskar Morgenstern (1944). Their theory generated cardinal measurements of expected utilities for strategic individual choices, given plausible assumptions about an individual's utility function. Working from the results of von Neumann and Morgenstern, John Harsanyi (1955) would later provide a complementary justification of utilitarianism for social choice by employing the notion of a "social welfare function." By the end of the twentieth century, economists had transformed ethical questions over how to reach the best consequences into economic questions over how to increase market activity, trade, social welfare, and global production.

The economic consequentialists have influenced many other fields. In jurisprudence, a theory known as the economic analysis of law has advocated the interpretation of legal concepts so as to maximize wealth. In business and public policy, the cost-benefit analysis has been introduced as a decision procedure for large-scale projects. A question such as where to dump toxic waste, when addressed by the cost-benefit analysis, provides a utilitarian solution to disputes by reference to the hypothetical willingness to pay of the interested parties affected by the possible outcomes of the decision. It is not surprising then that hypothetical willingness to pay is affected by the actual ability to pay, and so the fact that dump sites end up in poor neighborhoods is explained by this "ethical" decision procedure. The Nobel laureate Amartya Sen (1970, 1985) has been the most important critic of utilitarian economics on these issues. His contributions to the debate have focused on poverty, development, and the measurement of "capability" (as opposed to raw utility) in accounting for the bases of social choice.

Pluralist Conseqentialism

Many contemporary philosophers worry that developments in utilitarian theory have undermined the spirit of consequentialism. They point out that the everyday conception of human flourishing is not as thin as wealth maximization. The British philosopher G.E. Moore even advocated an "ideal" utilitarianism that got rid of the notion of pleasure as the good, and replaced it with the good of aesthetic experience and friendship (1993). It now seems clear that, while utility may be a good, and wealth one approximation of it, there are many other goods that do not reduce to either utility or wealth. By adopting a pluralist conception of goods, critics of utilitarianism allow into the ethical decision process notions like interests, rights, human freedom, biodiversity, sustainability, and other non-economic values.

The pluralists continue to maintain that what is right to do is decided by reference exclusively to consequences—but now the list of goods in the accounting is much broader than utility. Here talk of maximization no longer makes sense; the goal is to optimize the plural goods that result from acts or policies. Stakeholder theory is one such form of consequentialism, because it tries to tailor corporate decisions to the interests of all those who have a stake in the workings of the company, and not merely to those who hold stock in it.

Special Challenges

How useful is consequentialism when one morally evaluates technologies? A particular area of ethical concern is the effect of current and near-term technologies on future generations. Nuclear power, genetic engineering, human cloning, genetic modification of food, and other momentous programs will all have effects far into the future. Some versions of consequentialism would require a counting of the effects on those who are not yet alive, even though their preferences cannot be known, and actions and choices have not yet had an impact on them. It may be assumed that, if they live, they will want clean air to breathe, clean water, safe food, and other such necessities. Harms to distant generations may be discounted by some factor, but should not be neglected entirely, for then all the consequences of acts and policies are not taken into account.

Beyond the uncertainty of how much to discount, there is a deep problem for consequentialism that has been called by Derek Parfit (1984) the "non-identity problem." One assumes that the broad technological choices that are made now could harm particular people in future generations. But a consequentialist in some future generation could not complain that current policies and choices made his or her life worse off, because the things done now will affect *who* is actually born. That person will not exist, unless people currently living do exactly the good or bad things that they end up doing. Changes in manufacturing, travel, city planning, leisure, and work will determine which future people will meet and partner, and at what point in time they will produce children. The same is true for changes in technology. Similarly, for actual persons alive in the early twenty-first century, it is extremely unlikely that *they* would have been conceived were it not for the transportation systems, migration patterns, world wars, and other life aspects of their parents.

Philosophical debate over consequentialism is likely to persist. Nonetheless, its focus on observable results and effects will keep it in the center of ethical inquiries where science and technology are concerned.

THOMAS M. POWERS

SEE ALSO *Economics and Ethics; Engineering Ethics; Mill, John Stuart.*

BIBLIOGRAPHY

Bentham, Jeremy. (1989 [1789]). *On the Principle of Morals and Legislation*. New York: Oxford University Press. A foundational text in moral philosophy and political economy.

Harsanyi, John. (1955). "Cardinal Welfare, Individualistic Ethics, and Interpersonal Comparisons of Utility. *Journal of Political Economy* 63: 309–321. Essay connecting classical utilitarianism and the social welfare function in economics.

Mill, John Stuart. (1998 [1863]). *Utilitarianism*. Oxford, UK: Oxford University Press. Classic text in moral philosophy, arguing for utilitarianism in new dress, with a "proof" of the general principle of utility, a defense of rules (especially rules of justice), and novel arguments for qualitative distinctions in pleasures.

Moore, G.E. (1993 [1903]). *Principia Ethica*. Cambridge, UK: Cambridge University Press. Features a criticism of utilitarianism from within the tradition of analytic philosophy, and offers a reformulation in terms of "ideal utilitarianism."

Parfit, Derek. (1990 [1984]). *Reasons and Persons*. Oxford, UK: Oxford University Press. Includes a sophisticated analysis of utilitarianism and the problem of future generations.

Sen, Amartya Kumar. (1970). *Collective Choice and Social Welfare*. San Francisco: Holden-Day Inc. Sophisticated overview of work in utilitarian economic theory.

Sen, Amartya Kumar. (1985). *Commodities and Capabilities*. Amsterdam: Elsevier Science Publishers. Short but important book—from Sen's Hennipman lecture in Amsterdam—which introduces the "capability" measure into utilitarian social welfare calculations.

Singer, Peter. (1990). *Animal Liberation*, 2nd edition. New York: New York Review of Books. Widely read attack on "speciesism" and defense of the consideration of animal welfare on utilitarian grounds.

Smart, J.J.C., and Williams, Bernard. (1990). *Utilitarianism: For and Against*. Cambridge, UK: Cambridge University Press. Offers an outline and a critique of a new system of utilitarian ethics.

von Neumann, John, and Morgenstern, Oskar. (1944). *Theory of Games and Economic Behavior*. Princeton, NJ: Princeton University Press. Classic text in the formal theory of choice under uncertainty.

CONSERVATION AND PRESERVATION

Ideas of *conservation* and *preservation* play central roles in ethical discussions of science and technology, especially in relation to nature and the environment. The terms also figure prominently in museum and historical work, where programs of conservation (not losing) and preservation (protecting from deterioration) are associated with specialized sciences and technologies. With regard to environmental issues, the concepts appear more closely related, both implying respect for nature.

John Muir versus Gifford Pinchot

Since the early-twentieth-century break between Gifford Pinchot (1865–1946), first director of the U.S. Forest Service, and John Muir (1838–1914), founder of the Sierra Club, conservation and preservation have sometimes served as technical concepts with different connotations. In this context, conservation signals rational human use, preservation a protection from human use.

Although originally allies in creating Yellowstone, the first national park, in 1872, Pinchot and Muir took opposed positions in the debate, which lasted from 1909 to 1913, over building a dam in the Hetch Hetchy valley of Yosemite National Park in order to supply water to a growing San Francisco. Pinchot believed that "The first great fact about conservation is that it stands for development" (Pinchot 1910, p. 42); the only question was what kind of development, and whether for short-term single-focus exploitation or long-term multiple public use. For Muir, by contrast, national parks were to be preserved in their original form. "Dam Hetch Hetchy! As well dam for water-tanks the people's cathedrals and churches, for no holier temple has ever been consecrated by the heart of man" (Muir 1912, chap. 15).

Out of this debate, which Muir and the Sierra Club lost, began a tension in the environmental movement between those who seek to conserve and those who seek to preserve nature. Conservationists sometimes accuse preservationists of failing to appreciate human needs. Preservationists accuse conservationists of being too willing to compromise the intrinsic value of nature when faced with economic or political interests. The issue, in these terms, will only grow sharper as world population races toward doubling by 2050.

The Preservation-Conservation Spectrum

But the distinction between conservation and preservation is not always clear, and in fact environmental policies may often line up along a spectrum from protection of nature or ecosystems for their own sake to libertarian exploitation. The spectrum also to some degree parallels that between ecocentric (nature centered) and anthropocentric (human centered) environmental ethics. The extreme protectionist position, evident in wilderness preservation slogans and policies, and exemplified by *Earth First!* direct action, views natural systems as possessing intrinsic value independent of human use and as better off if protected from human interventions of any kind. Conservation would fall not necessarily on the other extreme, in which nature is presented as devoid of intrinsic value except insofar as it is available for obligatory human exploitation, but somewhere in the middle.

The spectrum is slightly complicated by self-defined conservationists such as those identifying with the *Wise Use* movement, which is especially hostile toward *radical environmentalists*. According to Wise Use advocates, the pastoral ideal was kidnapped by urban wilderness ideologues who lack the living relation to the land found among farmers and ranchers and thus fail to appreciate the value of the human transformation of the earth (Arnold 1996, 1998). But given its stress on the rights of property owners to develop land in virtually any way they see fit, Wise Use is perhaps more concerned with libertarian free enterprise than with the environment.

Nevertheless conservationists do tend to stress the importance of human interests, needs, and wants over any intrinsic values nature or the environment may be thought to possess. Yet this emphasis is easily combined with various gradations emphasizing high to moderate degrees of preservation of nature from human use and with a range of balances between natural and human needs in relation to natural exploitation.

Furthermore the spectrum need not be considered simply linear. Robert Paehlke (1989) argues that preservationist and conservationist views are distributed on a grid of two axes, with the left-right political spectrum crossed by a vertical axis running from environmentalism to anti-environmentalism. The point is that environmentalists and their opponents, on ethical as well as political grounds, use terms such as conservation and preservation—along with related terms such as sustainable development and restoration ecology—in myriad and often idiosyncratic ways. Careful analysis in conjunction with accurate observation of real-world practices is necessary to know what individual groups actually mean.

Practical Applications

The implications of these controversial word uses for science and technology may not always be obvious either. Certainly strong preservation environmentalists view major technological exploitations in nature (oil drilling and pipelines, for example) as wholly negative, whereas extreme opponents believe in a technological fix for any natural shortfall, even the extinction of species or ecosystems (through DNA rather than whole species preservation), while conservationists tend to be open to a modulated range of technological interventions, including the techniques of restoration ecology.

Radical preservationists sometimes oppose further scientific examination of nature, arguing instead for the sufficiency of existing research and for more aesthetic or experiential appreciation of nature. Their opponents, by contrast, often demand something close to scientific certitude concerning problems—as in the global climate change debate—to justify any change in exploitation patterns, and thus defend making more public funds available for environmental research. Such critics view radical preservationists as too willing to accept the flimsiest of scientific evidence.

In still one more somewhat ironic comparison, those who would protect the environment from human degradation often advocate advanced technologies that pollute less and promote high-tech gear to assist individuals in the noncontaminating exploration of wilderness. Such technologies may even include photographs and IMAX presentations designed to cultivate the aesthetic appreciation of nature as something good and beautiful in itself among those who may never have any direct wilderness experience. In opposition, those who would promote diversified human utilization sometimes find themselves apologizing for whatever technologies exist and denigrating innovations that could both improve exploitation and protect nature. One example might be defending personal automobile and snowmobile use in national parks when light rail or other innovations could enhance accessibility for all, including some such as the handicapped, who have previously been excluded. Diverse assessments of ecotourism have also been known to conflate expected conservation and preservation divides.

PAUL T. DURBIN

SEE ALSO *Environmental Ethics; National Parks.*

BIBLIOGRAPHY

Arnold, Ron. (1998). *Ecology Wars: Environmentalism as if People Mattered*. Bellevue, WA: Merril Press. Reprint of a book first published in 1987, with an introduction by Alan M. Gottlieb, founder of the Wise Use movement. See also Arnold's "Overcoming Ideology," in Philip D. Brick and R. McGreggor Cawley, eds., *A Wolf in the Garden: The Land Rights Movement and the New Environmental Debate* (Lanham, MD: Rowman and Littlefield [1996]).

Foreman, Dave. (2004). *Rewilding North America: A Bold, Hopeful Vision for Conservation in the 21st Century*. Covelo, CA: Island Press. By the founder of Earth First!

Muir, John. (1912). *The Yosemite*. New York: Century.

Paehlke, Robert C. (1989). *Environmentalism and the Future of Progressive Politics*. New Haven, CT: Yale University Press.

Pinchot, Gifford. (1910). *The Fight for Conservation*. New York: Doubleday.

VanDeVeer, Donald, and Christine Pierce, eds. (1994). *The Environmental Ethics and Policy Book: Philosophy, Ecology, Economics*. Belmont, CA: Wadsworth.

CONSERVATISM

An assessment of conservative ideas about the relationship between ethics, science, and technology must begin with a brief discussion of conservatism itself. Unlike liberalism, fascism, or communism, conservatism cannot be identified with a particular conception of the ideal society. In its broadest meaning, conservatism means simply "adherence to the old and tried, against the new and untried," as Abraham Lincoln put it in his Cooper Institute speech (Lincoln 1989, p. 122). If this definition is accepted, one can be "conservative" about almost anything that has lasted a long time.

In Europe and North America over the last few centuries, however, conservatism has been associated with a defense of classical liberalism in politics and economics against first the radicalism of the French Revolution and then against socialism, Communism, fascism, and Nazism. In making this defense, conservatism has also accepted and supported the achievements of science and technology so closely identified with liberalism and capitalism. European and North American conservatism since the French Revolution is thus an inherently paradoxical enterprise, because some of the key institutions it seeks to conserve, including science and technology, are themselves generators of change. Conservatives primarily interested in economics are more likely to welcome such change than religious and cultural conservatives. Conservatism nevertheless sharply differs from the philosophical liberalism of thinkers such as John Dewey or John Rawls in that all conservatives, whatever their primary interest, insist there are sources of moral authority beyond the liberal consensus. These include revealed religion, natural law, and the insights derived from humanistic study. Science, conservatives believe, cannot answer fundamental questions about the meaning of life, nor can technology resolve the most important ethical dilemmas.

Limited Criticism of Science and Technology

Because of its emphasis on the limits of knowledge that science can make available and the benefits technology may confer, conservatism is often mistakenly associated with the wholesale condemnation of technology associated with Romanticism and also promoted by radical theorists such as Herbert Marcuse who, in *One-Dimensional Man* (1964), views technology as a form of social control and domination. The Southern Agrarians, a group of poets and writers who defended the traditions of the U.S. South, including racial segregation, in *I'll Take My Stand* (1930), were writing as romantics rather than conservatives when they objected to technology itself, as when Andrew Lytle proclaimed "a war to the death between technology and the ordinary human functions of living" (p. 202) and argued that the South "should dread industrialism like a pizen snake" (p. 234). The most influential heir of the Agrarians, Richard Weaver (1910–1963), adopted a more representative conservative viewpoint when, in *Visions of Order* (1964), he criticized not science itself but "barbarism nourished by . . . scientistic fallacies" (p. 151) and "pseudoscientific images of man" (p. 153).

The Spanish philosopher José Ortega y Gasset (1883–1955) went further in defending science when he asserted in *The Revolt of the Masses* (1930) that "liberal democracy based on technical knowledge is the highest type of public life hitherto known" (p. 52). Yet mainstream twentieth century conservatives in England and the United States shared his belief that the key issue was to find a way to maintain the real achievements of liberal democracies in the face of totalitarianism, just as conservatives in the twenty-first century seek to guard those achievements against the threats posed by new political and religious fanaticisms. Ortega believed that totalitarian regimes were made possible by the rise of the "mass-man" who felt only "radical ingratitude" toward the developments in science and technology that "has made possible the ease of his existence" (p. 58).The masses do not grasp that the devices they take for granted are really "marvels of invention and construction which can only be maintained by great effort and foresight" (p. 60). Ortega believed scientists could scarcely avoid becoming mass-persons themselves, because the specialization required by modern science made it impossible for individual scientific workers to understand science as a whole and thus achieve comprehensive vision of the universe. At the same time Ortega warned that attempts to return to a pre-industrial way of life would be suicidal.

Limited Authority of Science and Technology

Western conservatism has accepted the authority of the physical and biological sciences within their own sphere, but has sharply questioned the application of the methods of the natural sciences to the study of human beings. Edmund Burke's description of the moving spirits of the French Revolution in his *Reflections* (1790) indicts not scientists but pseudo-scientists: "sophisters, oeconomists, and calculators" (p. 170). An American admirer of Burke, Irving Babbitt (1865–1933), based his New Humanism on the distinction between what Ralph Waldo Emerson called "law for man, and law for thing" in his "Ode, Inscribed to W. H. Channing." The neglect of that distinction, Babbitt argued in his first book *Literature and the American College* (1908), leads to an intellectual climate in which "Man himself and the products of his spirit, language, and literature, are treated not as having a law of their own, but as things; as entirely subject to the same methods that have won for science such triumphs over phenomenal nature" (p. 86). Babbitt summed up his views in a short 1930 essay, "What I Believe." Although he objects when "the pseudo-scientist claims for physical science a hegemony to which it is not entitled" (p. 11), he also disclaims the romantic condemnation of intellect itself. For Babbitt the exaltation of feeling unrestrained by thought and the exaltation of mechanical efficiency for its own sake are merely two sides of the same coin. He counters what he considers the dominant trend of the age with a call for a "positive and critical humanism" (p. 14) based on a reaffirmation of "the truths of the inner life" (p. 18).

Contradictions

George Santayana (1863–1952) argued in "The Genteel Tradition at Bay" (1931) that Babbitt's New Humanism was only the last gasp of a *genteel tradition* that neither expressed nor understood what was truly dynamic in American society. Santayana had described the United States in "The Genteel Tradition in American Philosophy" (1911) as a "country with two mentalities, one a survival of the beliefs and standards of the fathers, the other an expression of the instincts, practice, and discoveries of the younger generations" (p. 39). Scientific and especially technological developments were an expression of the younger generation, while religion, philosophy and the arts were under the control of the "hereditary spirit" (p. 39) of the genteel tradition. The contrast between the two mentalities could be symbolized by the difference between two characteristic products of American architecture: the "sky-scraper" (p. 40) and the "reproduction of the colonial mansion" (p. 40). A philosopher, Santayana intimated, should understand that the new society could not be judged according to the criteria of the genteel tradition but must be accepted on its merits and judged on its own terms.

In *Reason in Science* (1906) Santayana criticized the "school of political conservatives" (p. 307) who insist on retaining the language of "theology and metaphysics"(p. 307) rather than that of science because of the loss of social stability that might ensue. Such "sensitive conservatism" (p. 307) is "entangled in a pathetic delusion" (p. 307) ; it is "conservatism in a shipwreck" (p. 307). Santayana himself was more than ready to acknowledge the validity of science, which he considered "common knowledge extended and refined" (p. 393). He criticized the critique of science by idealist metaphysicians around the beginning of the twentieth century on grounds that seem applicable to the postmodernist critique of science at the beginning of the twenty-first. It is hardly convincing, observes Santayana, "when science is systematically disparaged in favour of a method that is merely disintegrating and incapable of establishing a single positive truth" (p. 312).

Russell Kirk (1918–1994) admired both Babbitt and Santayana and included both in his seminal *The Conservative Mind* (1953). In an essay on "Civilization Without Religion" (1996) Kirk goes further than Babbitt and disagrees with Santayana in arguing that the decline of European and North American civilization could be averted only by a "restoration of religious teachings as a credible body of doctrine" (p. 15). Even Kirk, however, is careful to criticize not science but rather a *scientistic* misunderstanding of the implications of science. According to Kirk, "the principal cause of the loss of the idea of the holy is the attitude called scientism" (p. 11). It is scientism, not science, that takes it as proved that "men and women are naked apes merely; that the ends of existence are production and consumption merely; that happiness is the gratification of sensual impulses; and that concepts of the resurrection of the flesh and the life everlasting are mere exploded superstitions" (p. 11). In an essay titled "Humane Learning in the Age of the Computer" (1996), Kirk argues that technology can never replace the flesh-and-blood teacher, but he does so in the name not only of the humanities but also of science, worrying that "if facility in operating computers tends to be emphasized at the expense of serious study of physics and mathematics, the springs of the scientific imagination may dry up" (p. 122).

Critique of Scientism

In "Science and the Studies of Man," a contribution to an anthology on *Scientism and Values* (1960), Eliseo Vivas makes a representative conservative argument when he criticizes the "so-called behavioral sciences" (p. 50) for attempting to adopt the methods and assume the prestige of chemistry and physics. Vivas does not deny and indeed insists on the validity of scientific method when applied in physics and biology, but he rejects the idea that "the only valid knowledge is scientific" (p. 50). Like most other conservatives, Vivas believes that "there is philosophical knowledge of a substantive nature and that there is moral and religious knowledge and, in a qualified sense, even aesthetic knowledge" (p. 50). Vivas argues that the attempt to study human beings and their institutions according to the methods of the natural sciences results not in science but in *scientism*.

The distinction between science and scientism was not, of course, noted only by conservatives. The prestige of science among radicals and militant reformers, however, made it difficult for them to draw a line with the clarity and firmness of conservatism, even when they wanted to do so. The appeal of Dewey's pragmatism, for example, was closely linked to his proposals to use scientific techniques to reform human society. Likewise two of the outstanding examples of scientism in the twentieth century, Marxism and Freudian psychoanalysis, appealed to those who wished to either radically change or destroy bourgeois society. Both used the vocabulary of science, and both attracted adherents by claiming the authority of science.

Though the influence of Marxism was vastly more destructive, both used their prestige to challenge and undermine the traditional moral principles at the heart of conservatism. By the twenty-first century the fraudulence of both has been revealed for all but the willfully blind to see. Other versions of scientism remain, however, including the attempt to use the prestige of the theory of biological evolution to shape a secularist philosophy of human nature and view of the universe.

The Conservative Middle Ground

In opposing repudiation of the concept of truth by postmodernist skepticism, conservatism in the twenty-first century has made common cause with the natural sciences in defending knowledge that is objective and universally true. Conservatives have opposed attempts to formulate a *feminist science* or any version of science based on ethnicity. Likewise conservatives have criticized the characterization of technology as in itself demonic as claimed by some environmental radicals.

In response to the development of biotechnology, however, conservatives such as Leon Kass have continued to be guided by traditional moral principles such as the sanctity of innocent human life and human dignity. Sometimes this has led them to oppose some new uses of medical technology, such as those involved in stem cell research. The same principle of the sanctity of life has also led conservatives to object to the withdrawal of technological support from patients without their consent, whether at the behest of the state or others. Conservatism in the twenty-first century, as earlier, continues to affirm the relevance and validity of traditional ethical principles in evaluating the moral implications of new developments in science and technology, whatever those might be.

In 1932 Winston Churchill observed that "while men are gathering knowledge and power with ever-increasing and measureless speed, their virtues and their wisdom have not shown any notable improvement" (p. 279). As a true conservative, however, Churchill believed that what was required was not "progress" in thought but rather he believed it "above all things important that the moral philosophy and spiritual conceptions of men and nations should hold their own amid these formidable scientific evolutions" (p. 279).

JAMES SEATON

SEE ALSO *Agrarianism; Capitalism; Democracy; Liberalism; Marxism; Ortega y Gasset, José; President's Council on Bioethics; Scientism; Totalitarianism.*

BIBLIOGRAPHY

Babbitt, Irving. (1981 [1930]). "What I Believe." In *Irving Babbitt: Representative Writings*, ed. George A. Panichas. Lincoln: University of Nebraska Press.

Babbitt, Irving. (1986 [1908]). *Literature and the American College: Essays in Defense of the Humanities*. Washington, DC: National Humanities Institute. Irving Babbitt, with his colleague Paul Elmer More, founded the "new humanism," an attempt to reform American culture in the light of principles derived from the classics of literature, thought and religion.

Burke, Edmund. (1968 [1790]). *Reflections on the Revolution in France*. London: Penguin. Burke's reflections, according to Russell Kirk and others, marks the beginning of modern conservatism in Europe and North America.

Churchill, Winston. (1932). "Fifty Years Hence." In his *Thoughts and Adventures*. London: Thornton Butterworth.

Kass, Leon. (2002). *Life, Liberty, and the Pursuit of Dignity*. San Francisco: Encounter Books.

Kirk, Russell. (1953). *The Conservative Mind, from Burke to Santayana*. Chicago: H. Regnery Company. This book, more than any other single work, may be credited with the revival of conservative thought in the United States.

Kirk, Russell. (1996). "Civilization Without Religion." In *Redeeming the Time*, ed. Jeffrey O. Nelson. Wilmington, DE: Intercollegiate Studies Institute.

Kirk, Russell. (1996). "Humane Learning in the Age of the Computer." In *Redeeming the Time*, ed. Jeffrey O. Nelson. Wilmington, DE: Intercollegiate Studies Institute.

Lincoln, Abraham. (1989). "Address at Cooper Institute, New York City, February 27, 1860." In *Abraham Lincoln: Speeches and Writings 1859–1865*, ed. Don E. Fehrenbacher. New York: Literary Classics of the United States.

Lytle, Andrew. (1977 [1930]). "The Hind Tit." In *I'll Take My Stand: The South and the Agrarian Tradition, by Twelve Southerners*. Baton Rouge: Louisiana State University Press. The "twelve southerners" who contributed to this manifesto included such important literary figures as John Crowe Ransom, Robert Penn Warren, and Allen Tate.

Marcuse, Herbert. (1964). *One-Dimensional Man: Studies in the Ideology of Advanced Industrial Society*. Boston: Beacon Press. Marcuse was one of the leading members of the Frankfurt school.

Ortega y Gasset, José. (1957 [1930]). *The Revolt of the Masses*. New York: Norton. This is the most influential book by one of the most important philosophers of the twentieth century.

Santayana, George. (1983 [1906]). *Reason in Science*. New York: Dover Books. This is the fifth and last volume of *The Life of Reason*, the series that established Santayana's philosophical reputation.

Santayana, George. (1998 [1911]). "The Genteel Tradition in American Philosophy." In *The Genteel Tradition: Nine Essays by George Santayana*, ed. Douglas L. Wilson. Lincoln: University of Nebraska Press, pp. 37–64. Although the title refers only to "American Philosophy," this essay is one of the most insightful and influential essays on American culture ever written.

Santayana, George. (1998 [1931]). "The Genteel Tradition at Bay." In *The Genteel Tradition: Nine Essays by George Santayana*, ed. Douglas L. Wilson. Lincoln: University of Nebraska Press. In this essay Santayana invokes the concept of the "genteel tradition" to criticize the "new humanism" of Irving Babbitt and Paul Elmer More.

Vivas, Eliseo. (1960). "Science and the Studies of Man." In *Scientism and Values*, eds. Helmut Schoeck and James W. Wiggins. Princeton, NJ: Van Nostrand.

Weaver, Richard M. (1964). *Visions of Order: The Cultural Crisis of Our Time*. Baton Rouge: Louisiana State University Press.

CONSTRUCTIVE TECHNOLOGY ASSESSMENT

The core idea of constructive technology assessment (CTA) is that the social problems surrounding technology can and must be addressed through the inclusion of a large diversity of actors in technological design and implementation processes, including especially social actors. Social actors are those who experience and/or articulate and define health, environmental, or other value-laden effects of evolving technologies but are not directly engaged in technological developments. They may be consumers, citizens, employees, corporations, social groups, and more. CTA activities thus depart from traditional technology assessment (TA), which limits itself to charting the effects of given technological options, and does not attempt directly to influence or broaden the design process.

Historical Background

During the last two decades of the twentieth century, TA was widely adopted in several countries in Europe and in the United States. At first mainly conducted by technical experts, it developed toward a more participatory mode, bringing public values and opinions into the assessment of new technologies (Grin and de Graaf 1996, Vig and Paschen 2000). Both conventional expert impact assessment and various forms of participatory TA focus on shaping public policies related to technical change. TA policies have often been institutionalized in separate organizations such as the U.S. Office of Technology Assessment and the Netherlands Organization for Technology Assessment (renamed in Rathenau Institute), which serve legislatures and try to inform the broader public.

The Rathenau Institute was also heavily involved in developing the theory and practice of CTA. Since its founding, CTA practices have been taken up by many organizations, including corporations, nongovernmental organizations (NGOs), and government agencies, although not necessarily in the same way and often not under this label. These actors face different opportunities and constraints depending on their position in the innovation process. They share, however, the insight that negotiation among all stakeholders is necessary in order to deal with social problems that come with technical change.

CTA activities can take the form of dialogue workshops, consensus conferences (public debates), scenario workshops, or citizen reports. These are methods that can be used to organize structured discussions between social actors and designers (or technological actors). They only become CTA practices, however, when they focus on influencing design and technical change (Schwarz and Thompson 1990, Misa et al. 2003, Schot and Rip 1998, Schot 2001, Sørensen and Williams 2002).

Because CTA addresses innovation, it becomes a form of technology policy, although regular technology policies are not aimed at the integration of societal aspects into technical change. Some organizations and authors have called for such integration. They have argued that technology policies should aim at promoting those technologies that promise positive societal effects or externalities, as economists would term them (Carnegie Commission on Science 1992, Freeman and Soete 1997).

CTA Perspective

From a broad historical perspective, CTA practices may be viewed as a new form of management, replacing a problematic modernist way of managing technology (Misa et al. 2003). The core of modernist management lies in the separation of technology and its social effects. The lack of what may be called negotiating space between the actors involved in the design process and spokespersons for actors who are directly affected by the technology is a feature of the modernization process as it has manifested itself until the beginning of the twenty-first century.

In the modern regime of technology management, two tracks are apparent: promotion and regulation. On the one hand, there have emerged separate sites—called laboratories—where designers are given plenty of room to tinker with new technologies without having to think about the effects, because creativity might suffer. After they have been tried and tested, the black boxes are sent off into the world to bring about welfare and progress. This model encourages just plugging the technology in; playing with the technology is even considered dangerous. On the other hand, there has emerged a regulatory arena to mitigate the appearance of negative effects. Regulation does not concern itself with steering the scientific and technical developments, but rather with setting limits to their application.

Beginning in the 1970s, more and more problems and limitations became associated with this dual-track approach. Problems cropped up and so-called negative side effects of existing technologies were not easily solved through ex post facto regulation. They only worsened. Environmental problems are good examples. Since the 1980s there has been an explosion of new governmental regulations including the use of economic instruments as well as great increases in knowledge of environmental problems and solutions. Environmental advisory agencies have flourished. Yet many environmental problems have not been solved. Chimney filters and catalytic converters appear unsatisfactory. It has become clear that environmental problems must be addressed through a drastic reduction of energy and resource use. Another form of production and consumption is required. This will not come about through government regulation only, also not if it would focus on creating new market mechanisms.

An alternative form of production and consumption implies not only making environmentally-friendly technologies, but also an alternative form of making technology. The character of the technology design and implementation process is in need of change. It must be broadened to include social aspects and actors. Ultimately such a broadening could lead to a change in the current pattern of technology management (the dual-track approach). New institutions should emerge that will become platforms for the constructive integration of technology and society. It is constructive not in the sense of conflict avoidance, but in the sense that all affected are in a position to take responsibility for the construction of technology and its effects.

Features of CTA

The view that design and implementation processes must be broadened is based on the presumption that social effects are present in the form of (sometimes implicit) assumptions about the world in which the product will function. Thus, when technologies are designed, assumptions are made about users, regulations, available infrastructures, and responsibilities between various actors.

In technology studies, the notion of *scripts* is used to refer to this set of assumptions (Akrich 1992). The effect of broadening (and thus of the application of CTA) is that the designers' scripts are articulated and laid out as early as possible to the users, governments, and other interested parties, all of whom have their own scripts, and who will feel the effects of the technology. From the point of view of CTA, it is important to make room for such an early and more regular confrontation and exchange of all the scripts. Thus CTA processes acquire their three normative beneficial features: (1) anticipation, (2) reflexivity, and (3) social learning.

ANTICIPATION. Whenever users, social groups, and citizens take part in the design processes, they are more likely to bring in social aspects at an early stage than are designers. Designers rarely anticipate social effect; they have a hard enough time anticipating market conditions in a timely fashion. They react to market signals and social effects only when they occur, which leads to ad hoc problem solving. In the field of management studies, this lack of sensitivity toward user needs has been identified as a barrier for successful innovation.

Despite the emphasis on anticipation, there is no presumption that all social effects can be predicted. On the contrary, it must be assumed that technological development is nonlinear and unpredictable. During development all kinds of unexpected side roads and branching emerge. The given unpredictability of technological development has two implications. First, anticipation must be organized into a regular activity, including during the phase of implementation. That is when unforeseen effects emerge by way of new interactions and applications. Owing to the importance of anticipating social effects as early as possible, corporations and other technology actors can be advised to organize a trajectory to develop scenarios for coping with social effects alongside product development trajectories. Second, the technology development process should be flexibly structured so that choices can be deferred or altered.

REFLEXIVITY. Broadening the design process results in being able to notice earlier and more clearly that social effects are coupled to specific technical options and that designers design not only technological but social effects. Scripts can no longer remain hidden. The effects that emerge are dependent not only on the designers' scripts but also often on the outcomes of complex interactions between designers, users, third parties, and the context in which these actors operate.

CTA activities aim to stimulate actors to take account of the presence of scripts and realize that technological developments and social effects are coproduced. Actors thereby become reflexive. They must integrate technology and its effects into their thoughts and actions. Consensus may be reached, but controversies could very well occur as CTA exposes hidden scripts and places them next to one another. This need not be such a great problem in societies where controversies are a routine and normal part of the process of technology development. Analyses of controversies have shown that attempts often are made to suppress reflexivity. Attempts are made to separate technical facts from assumptions about the social reality in which the technologies function. Controversies subsequently take the unproductive course of the dual-track regime, either emphasizing promotion or regulation of new technologies.

SOCIAL LEARNING PROCESSES. Learning may occur on two levels. First-order learning leads to developing a better ability to specify and define one's own design. Second-order learning means learning about one's own assumptions and scripts, learning that one is creating new couplings and demands. CTA relates to both forms of learning. It is important to embed technological development in social learning processes as early as possible so that users, designers, and third parties have the opportunity to scrutinize their own presumptions and come to new specifications. In practice, design processes then become more symmetrical from the beginning. As much attention is paid to technical as market and social issues. Design processes become open (so actors are ready to partake) and space is made for experimentation, for trying out various couplings and problem definitions.

Changing the Design Process

CTA activities are not directed in the first instance at such substantive goals as the reduction of environmental pollution, the defense of privacy, or other such social goals. Thus, for instance, the development of wind energy or a security system to guard against bank fraud cannot be automatically labeled CTA. The purpose of CTA is to shape technological development processes in such a way that social aspects are symmetrically considered.

When design processes assume the character of CTA, fewer undesired and more desired effects will result. Such a claim is based on two arguments: (1) By incorporating anticipation, reflexivity, and social learning, technology development becomes more transparent and more compliant to the wishes of various social actors. (2) In a society where CTA processes have become the norm, technology developers and those likely to be affected by the technology will be in the position to negotiate about the technology. An ability to formulate sociotechnical critique and contribute to design will become widespread. Resistance to specific social aspects will not be viewed as technophobia, but as an opportunity to optimize the design (or achieve a better fit in society).

The effect of CTA will not be to bring technology under control so that it plays a less dominant role in society. Rather, it aims to change the form of control and how technology development is played out. The goal is to anticipate earlier and more frequently, to set up design processes to stimulate reflexivity and learning, and thus to create greater space for experimentation. Possible technologies should be made more open and flexible so users easily can have control over them. Technological development will also become more complex. More coordination and new competencies will be required. In some cases the processes will slow. New institutions will emerge to encourage negotiation between developers, users, and third parties. Should design processes acquire the character of CTA, technologists will not suddenly see their work disappear or have it constantly evaluated by new bureaucracies. Almost all of the incremental design changes will not require negotiation. In the program of requirements, allowance routinely will have been made for social aspects (including flexibility). However, the variety of technological designs probably will increase, as more groups will be involved in their capacities as knowledge producers and technology developers.

The three quality criteria for CTA processes make apparent that broadening the design process is not an end in itself, and that "broader" does not necessarily mean "better." Broader is better only in those design processes where space has been created for anticipation, reflexivity, and learning. That provides some guarantee that processes should result in better technology, which is to say technology with more positive and fewer negative effects. These three criteria also allow existing CTA activities to be evaluated, and suggest directions for improvement.

JOHAN W. SCHOT

SEE ALSO *Consensus Conferences; Discourse Ethics; Expertise; Office of Technology Assessment.*

BIBLIOGRAPHY

Akrich, Madeleine. (1992). "The De-Scription of Technical Objects." In *Shaping Technology/Building Society: Studies in Sociotechnical Change*, ed. Wiebe E. Bijker and John Law. Cambridge, MA: MIT Press. Classic article on how technical objects are turned into black boxes, become stable, naturalized, and depoliticized. The process of their making is concealed. The article is informed by actor—network theory.

Carnegie Commission on Science. (1992). *Enabling the Future: Linking Science and Technology to Societal Goals*. New York: Carnegie Corp. A report that integrates social dimensions in a proposal for a new kind of technology policy.

Freeman, Chris, and Luc Soete. (1997). *The Economics of Industrial Innovation*, 3rd edition. Cambridge, MA: MIT Press. A highly influential yet stylized history and broad review of theories of innovation.

Grin, John, and Henk van de Graaf. (1996). "Technology Assessment as Learning." *Science, Technology, and Human Values* 20: 349–366. An excellent exploration of the nature of learning processes involved in technology assessment.

Misa, Thomas J.; Philip Brey; and Andrew Feenberg, eds. (2003). *Modernity and Technology*. Cambridge: MIT Press. A truly interdisciplinary exploration of the connections between technology and the modern world. The ambition is to lay the foundations for a new field, hence it includes discussions on theory and method, next to a wide range of empirical studies.

Rip, Arie; Thomas J. Misa; and Johan Schot, eds. (1995). *Managing Technology in Society*. London: Pinter. An analysis of the underdevelopment of innovative potential for achieving societal goals. Factors that block change, as well as conditions for successful learning, are identified in a wide-ranging selection of cases that encompass biotechnology, clean technologies, information and medical technologies.

Schot, Johan. (2001). "Constructive Technology Assessment as Reflexive Technology Politics." In *Technology and Ethics. A European Quest for Responsible Engineering*, ed. Philippe Goujon and Bertrand Hériard Dubreuil. Leuven, Belgium: Peeters. A history of the emergence of Constructive Technology Assessment and an analysis of experiences leading to suggestions for a new research agenda and practice.

Schot, Johan, and Arie Rip. (1998). "The Past and Future of Constructive Technology Assessment." *Technological Forecasting and Social Change* 54: 251–268. An attempt to define criteria for a good Constructive Technology Assessment practice.

Schwarz, Michael, and Michael Thompson. (1990). *Divided We Stand: Redefining Politics, Technology, and Social Choice*. Philadelphia: University of Pennsylvania Press. An introduction to a cultural theory of risk leading to a plea for constrained pluralism in technology introduction and adoption discussion and practices.

Sørensen, Knut H., and Robin Williams, eds. (2002). *Shaping Technology, Guiding Policy: Concepts, Spaces, and Tools*. Cheltenham, UK: Elgar. An overview of recent developments in technology studies. It introduces a range of new concepts and ideas, and it highlights both the policy implications of these concepts and explores new possibilities for intervention by government, policymakers, managers and the public.

Vig, Norman J., and Herbert Paschen, eds. (2000). *Parliaments and Technology: The Development of Technology Assessment in Europe*. Albany: State University of New York Press. An informative and historical country based overview of technology assessment.

CONSUMERISM

Consumerism is a way of life combining high levels of material affluence with an emphasis on symbolic and emotional meanings associated with shopping and possessions. The United States continues to lead the way, but the phenomenon increasingly is of global scope. Consumerism can be interpreted positively as a means of stimulating the economy while facilitating people's liberties to shape their identities and subcultures. In contrast, critics perceive consumerism as a manipulated and environmentally destructive habit leading to too many units of stuff being designed, produced, advertised, sold, and discarded (Rosenblatt 1999, World Watch Institute 2004). All may agree that "The one unambiguous result of modern capitalism, of the industrial revolution, and of marketing ... is: In the way we live now, you are not what you make. You are what you consume" (Twitchell 2002, p. 1).

Infrastructure of Consumption

Consumerism involves not just the conventional *shopaholic*, but a complicated set of organizations, relationships, and ethically problematic practices involving science and technology. Product designers, manufacturing engineers, solid state physicists, and those trained in just about every other scientific and technical specialty have participated directly or indirectly in the development and spread of consumer society. Chemists created synthetic pesticides, PCBs, and PVC plastics, enabling businesses to produce and consumers to purchase products that inadvertently scattered billions of pounds of toxic compounds across the landscape. Civil engineers paved and built, making possible an automobile-centered way of life, that enhanced mobility while creating urban sprawl. Agricultural scientists helped construct the modern diet, combining unprecedented variety and nutrition with an obesity epidemic. Computer engineers' amazing achievements also were crucial in spreading pornography via the Internet, even though it was not the engineers themselves who produced or downloaded it.

Technologists are joined by government in fostering consumerism. The basic science integrated into leading-edge technologies such as carbon nanotubes derives partly from taxpayer-funded research and other government subsidies. Transport, electricity, communications, agriculture, and other infrastructure of consumer society all benefit from advantageous tax treatment or outright subsidy, a favorable legal environment, and government stimulation of the economy by means of monetary and fiscal policy. Military research and development (R&D) also has been indispensable; for example billions of aluminum beverage cans annually derive from aluminum smelting procedures developed for aircraft construction during the World War II.

Drawing in part on ideas developed via government-sponsored R&D, business executives search for market niches while hiring experts to deploy technological innovation as a competitive strategy. Franchises and fast food restaurants, *big box* stores and malls, cruise ships, theme parks, sports and musical performance arenas, resorts, and casinos all depend on technologically enabled data processing, communication, and transport of customers, merchandise, food, and drink from all over the globe. These and other forms of consumerism are reshaping everyday life worldwide by a process that some sociologists refer to as *McDonaldization*. The quest for efficiency, calculability, predictability, and "control through nonhuman technology" achieves amazing results, but cumulatively may constitute "the irrationality of rationality" (Ritzer 2004, p. 15–16).

Public Receptivity

As indispensable as technologists, business executives, and government officials have been in development of consumerism, they could not have done it without a receptive audience. If there is a dividing line between purchasing and consumerism, it perhaps occurs when purchasing becomes more about shopping and its psychosocial benefits than about actual use of the purchased items. Friends may prescribe a shopping trip for someone who is depressed; *bargains* and *sales* are avidly sought, even though the total expended is certain to be higher when one goes shopping than when one does not; and somewhere in many shoppers' minds is an expectation of approving looks or words that may be evoked by a new garment or tool. The symbolic, emotional, and interpersonal elements of consumerism are difficult to overstate.

That is not to deny that consumers exercise choice; of course they do, in part because the variety of possible purchases is so great that choice is inescapable. Nevertheless just as families come voluntarily to Disney World and then are channeled into preformed experiences, so more generally is consumer behavior in some respects channeled for the convenience and profitability of business. To attract customers, merchandisers play on consumers' envy, shame, and pride, expending 1 trillion dollars annually worldwide on advertising, attractive packaging, and other selling techniques. A small army of psychologists and statisticians conduct market research to learn how to stimulate sales, "constantly gaining more precision in pinpointing the demographic and lifestyle trends of consumer segments, employing new tools such as Internet *cookies* to monitor the *click-streams* of e-shoppers" (Cohen 2003, p. 402). The finance industry brilliantly stimulates the borrowing necessary to keep spending high.

Criticisms and Rejoinders

The disposable income required to purchase a growing array of goods and services is of course far more available to the affluent, who are located mainly in North America, Japan, and Europe. At the other end of the spectrum are approximately 1 billion persons who live in absolute poverty, about as many humans as the total number alive prior to the Industrial Revolution. To families without toilets or clean drinking water, television broadcasts the lifestyles of the rich and thereby stimulates consumer aspirations and helps spread consumer society across the globe. Within affluent cultures, intangible ethical consequences of consumerism appear to include deterioration of face-to-face community, increased rates of psychological depression without commensurate improvements in happiness (Lane 2000), and reduced interaction among family members as children turn increasingly to the televisions and computers in their bedrooms. Parents' long working hours sometimes come at the expense of sleep, leisure, family, and friends—a syndrome far more common in some countries (such as the United States) than in others (Schor 1998).

Consumerism is environmentally problematic in obvious ways, but also more subtly, as when distant consumers' appetites for shrimp, teak, and coffee disrupt fragile tropical ecosystems (Tucker 2002). Whether consumerism potentially can be made compatible with environmental sustainability is debatable. The formula for calculating ecological damage is roughly the total number of humans, multiplied by the amount consumed per person, multiplied by the resources utilized and toxicity released per unit of consumption. If the human population declines soon enough, and if technologists figure out how to dramatically reduce resource usage and pollution per unit produced and consumed, increasing material affluence per person might be compatible with greatly reduced environmental damage. Advocates of *natural capitalism* propose radically reconceptualized ways of providing housing, transport, and consumer products (Hawken et al. 1998, McDonough and Braungart 2002); and a few nanotechnologists believe that molecular manufacturing eventually may eliminate hazardous wastes and other side effects of production. As of the early twenty-first century, however, reductions in pollution per unit in most industries have been offset by population growth and by increased consumption per person.

Not everyone agrees with the above diagnosis. Among counterarguments, they point out that contemporary economies are organized to require an unpleasant choice: allow recession and unemployment, or stimulate the economy through ever-higher levels of consumer spending. In poorer nations, increased investment and purchasing theoretically might be devoted to basic needs including water supply systems, safe sanitation, housing, and nutrition. In the already affluent nations, however, economic growth tends to mean more elaborate barbeque grills, second homes, cosmetic surgery, and other luxuries. These are lesser evils, or not evils at all, to those who emphasize the benefits of full employment, interesting jobs, and liberty to purchase a lifestyle more of one's own choosing than previously possible for most of humanity, together with the value of technological innovation as a means of making life more diverse and more interesting (McCracken 1988).

The Challenge of Change

Few knowledgeable observers presently consider consumer trends compatible with environmental sustainability, but those concerned about unlimited consumerism face a difficult task in addressing the issue. It is easy to make products and production processes a bit greener by, for instance, creating biodegradable carpeting. But limiting the total volume of production and consumption is far more difficult, requiring people to forego some of what they have learned to want. Such a change in consumer mentality presumably would require slowing the drumbeat of messages encouraging consumption, and perhaps even a ban on advertising as well as tight restrictions on consumer credit. Such changes surely depend on ardent environmentalists and other slow-growth advocates winning more elections, which cannot happen without a different attitude among citizens. In other words, consumerism is constructed as a circle, a vicious circle in the eyes of critics.

Changed thinking among scientists, engineers, and other technically trained persons also might be necessary to intervene in the consumerist trajectory. In effect, technoscientists now gain governmental research funding by helping create weaponry, communications, transport, and other innovations helpful in military affairs and in economic activities valued by governing elites. A similar expectation leads industry to help fund scientific research, employ technoscientific consultants, and hire college graduates in chemistry, biotechnology, computer science, and other technical fields. All this makes good sense, in a way; but the partially unintended, collective consequences include the problematic aspects of consumerism.

Breaking out of the consumerist cycle would involve billions of persons over generations in evolving a commendable, interesting, high-technology, lower-consumption way of life. This, arguably, is the master challenge for human civilization—an activity so far-reaching and visionary that no one can fully imagine what would be involved. However a first step probably would require that more people begin to think of consumerism as an ethical, technological, economic, and political issue to be addressed.

EDWARD J. WOODHOUSE

SEE ALSO *Advertising, Marketing, and Public Relations; Affluence; Cosmetics; Material Culture; Materialism; Popular Culture; Population; Waste.*

BIBLIOGRAPHY

Cohen, Lizabeth. (2003). *A Consumer's Republic: The Politics of Mass Consumption in Postwar America*. New York: Basic Books. A historical study of the shift in the U.S. toward consumer-citizens who focus not on politics as a quest for the common good but on having government facilitate personal consumption.

Hawken, Paul, Amory Lovins; and L. Hunter Lovins. (1998). *Natural Capitalism: Creating the Next Industrial Revolution*. Boston: Little, Brown. Proposes changes in housing, transport, and many other aspects of production and consumption to make compatible the goals of business profitability, technological innovation, fairness, diversity, and environmental sustainability.

Lane, Robert E. (2000). *The Loss of Happiness in Market Democracies*. New Haven, CT: Yale University Press. Summarizes and interprets psychological research calling into question the hope that consumerism can lead to a permanent state of enhanced life satisfaction.

McCracken, Grant D. (1988). *Culture and Consumption: New Approaches to the Symbolic Character of Consumer Goods and Activities*. Bloomington: Indiana University Press. Materialism as a quest for intangible goods such as identity and belonging.

McDonough, William, and Michael Braungart. (2002). *Cradle to Cradle: Remaking the Way We Make Things*. New York: North Point Press. A visionary yet down-to-earth analysis of how a highly technological lifestyle could be elegantly designed to be environmentally benign.

Ritzer, George. (2004). *The McDonaldization of Society*, revised new century edition. Thousand Oaks, CA: Pine Forge Press. Sociological interpretation of the mechanisms that major organizations use to serve growing numbers of people in standardized ways, while displacing local businesses and helping change the culture of everyday life.

Rosenblatt, Roger, ed. (1999). *Consuming Desires: Consumption, Culture, and the Pursuit of Happiness*. Washington, DC: Island Press. A collection of essays by critics of consumer society.

Schor, Juliet B. (1998). *The Overspent American: Upscaling, Downshifting, and the New Consumer*. New York: Basic Books. A nonconventional economist's interpretation of the quest for status via consumption.

Tucker, Richard. (2002). "Environmentally Damaging Consumption: The Impact of American Markets on Tropical Ecosystems in the Twentieth Century." In *Confronting Consumption*, ed. Thomas Princen, Michael Maniates, and Ken Conca. Cambridge, MA: MIT Press.

Twitchell, James B. (2002). *Living It Up: Our Love Affair with Luxury*. New York: Columbia University Press. An even-handed attempt to find out what luxury means to consumers, based in part on ethnographic research in Beverly Hills and other shopping havens.

World Watch Institute. (2004). *State of the World 2004: The Consumer Society*. New York: W. W. Norton. Describes global environmental effects of consumption, and presents a plan for moving toward what the authors consider a more just and more sustainable civilization.

CONTRACEPTION

SEE *Birth Control.*

CONTRACTS

Contracts are legally enforceable agreements between persons that specify transactions or define relations between them. Either informal or written, they may concern any lawful human transaction, from purchases and loans to hiring and marriage. In engineering and science, contracts play important roles because, in both domains, practitioners do a great deal of work under some form of contract. Defining what the parties are obligated or permitted to do, contracts establish an ethical framework for engineering and scientific work, and they present ethical problems. The ethical framework has at its core one or more promises. Because the promises are legally enforceable, they involve a third actor in addition to the promisor and the promisee, government.

Contracts in Engineering

For most engineers and many scientists, the employment contract frames their professional activities. American courts apply the common law doctrine of "employment at will" when interpreting employment contracts for engineers and scientists. Under this doctrine, the employer is free to hire and fire at will, and the employee is free to take up employment and resign at will. This means that an employer may dismiss an "at will" employee, in the words of the court in an often cited case, "for good cause, no cause, or even cause morally wrong, without thereby being guilty of moral wrong" (*Payne* v. *Western and Atlantic RR*, 81 Tenn. 507, 519–20 [1884]). As a consequence, for example, engineers have reason to fear that by asking challenging questions about the safety of a project, they risk being fired. The "employment at will" doctrine is subject to limitations expressly indicated in federal and state statutes (for example, civil rights laws), to public policy exceptions courts have worked out, and to express provisions in the employment contract, such as provision for a term of one year.

The contract includes the usual terms of employment: salary, compensation, health and pension benefits, etc., but in addition may include "employment agreements" concerning intellectual property, confidentiality, and restrictions on future employment. At the time of taking employment, engineers and scientists often enter these agreements with insufficient appreciation of the implications. Sometimes, in this way, engineers or scientists unwittingly enter agreements that are so restrictive they could not be enforced. An example is an agreement that excludes future employment with a competitor to the extent of putting the engineer's future livelihood at risk.

Engineers or scientists may be surprised to discover that they are legally and ethically obligated by the employment agreement to maintain the secrecy of certain information even after changing employers. At the next job, an engineer or scientist may have to decide whether particular indirect uses of information gained from a former employer are permissible. Engineers or scientists may also have to decide whether to maintain the confidentiality of information that they believe a client or customer needs to avoid certain harms.

Some have argued that the code of ethics does or should rank as an implied element of the engineer's or scientist's employment contract. Viewed this way, the code would provide a barrier protecting engineers or scientists from being required by their employers to engage in behavior that violates the code. One way to interpret this claim is by invoking the status of an engineer as a professional: An employee trained and hired as an engineer is bound by all the standards of engineering, including ethical standards. The employment contract cannot require engineers to violate their ethical standards. Courts, however, have not been receptive to this interpretation.

Contracts bear on engineers' and scientists' work in another important way—through the contractual agreements that their employers (or they themselves) make with other business organizations, non-profit organizations, interest groups, and government agencies at every level of government. Engineers' functions—design, testing, maintenance, and operations—and their project-related dealings with purchasing agents, marketing specialists, customers, vendors, and construction contractors, as well as with other engineers, are usually associated with such contracts. The same is true of scientists when they function similarly.

Some common ethical problems for engineers typically arise from contracts of this sort. An example is the "deadline problem" that occurs, for example, when engineers discover in the course of their work that they cannot meet both the specifications for the product and the delivery date to which they originally agreed. They may have to develop options, such as working overtime or negotiating a compromise.

Yet engineers contribute to devising these sorts of contracts as well as to implementing them. They participate in defining projects and determining specifications for products even when they do not directly take part in contract negotiations. Their judgments about the time and resources needed to complete projects (a new chemical plant, for example) often help to decide the terms of contracts by which they and other engineers are bound. Strategies of preventive ethics may help engineers avoid ethical problems associated with devising contracts.

So far, this entry has focused on ethical problems and strategies from the perspective of engineers and scientists as promisors. In most cases when engineers and scientists are promisors, the promissee is a large company or firm. The company's perspective brings to the fore other ethical problems and needs for preventive ethics. For example, employers cannot easily determine whether engineers are faithfully abiding by their promise to maintain the secrecy of information at a new place of employment. Companies cannot pursue former employees by legal means if they do not have tangible evidence of, for example, the transfer of confidential information to the new employer. From this perspective, strategies of preventive ethics, such as rewards for creativity to valuable employees, may be useful. Companies can also use contracts to provide incentives to valued employees to stay with the company and to departing employees to maintain desired confidentiality, in this way protecting engineers and scientists from temptations to which they are subject.

In consulting firms (for example, environmental consulting firms) that require or permit the firm's engineers and scientists to obtain and implement contracts on their own, other problems arise. Scientists or engineers may unwittingly enter contracts with clients whose interests collide with those of other clients of the firm. Problems about the treatment of their reports may arise for engineers and scientists in these firms, and also in other contexts. The firm may object to the engineer's client using a report in a press conference or require the engineer to suppress a report altogether. An academic engineer who does independent consulting under contract may have to decide how to handle a client's decision to bury a report revealing the client's responsibility for some harm.

Contracts in Science

In scientific research and in academic engineering, contracts are pervasive in defining conditions attaching to awards of funding necessary for conducting research. In universities, investigators, students, and postdoctoral scholars have a contractual duty to abide by the institution's rules. Informal agreements in research groups under these rules are similarly binding. Graduate students are often surprised to learn that they do not own data they themselves collect. They may perceive this rule or agreement in research groups as an unfair hindrance to advancing their careers. Nevertheless, they are ethically bound to abide by these agreements unless they can negotiate other terms with the principal investigator.

The power disparity among senior investigators, graduate students, postdoctoral scholars, and junior investigators complicates the ethical situation in research groups. Their leaders have the power to make the ground rules for conducting research in their groups. Those subject to this power and dependent on their leaders for research support and recommendations for future employment are not in a position to contest rules. Because of the power disparity, safeguards should be built into the ground rules to protect the vulnerable, less powerful members of groups. Senior investigators can begin by making informal understandings explicit and open to discussion and revision.

In conducting research for commercial sector firms, scientists and engineering researchers are required to sign contracts that allow firms proprietary control not only over copyrights and patents, but usually also over data, tools, resources, and techniques. As a consequence, "virtually any piece of information or equipment used in industry-sponsored research can become company property" (Resnick 1988, p. 31). By these agreements, even chemical formulas and DNA sequences can become company property.

These contracts also commonly require scientists and engineers in the commercial sector to submit their publications or public presentations for company review and to accept delays beyond the limit acceptable in academe. In some cases, companies suppress publication altogether, in this way requiring engineers or scientists to violate professional standards. In light of the central value of open publication in science, a value that serves science and the public welfare, these requirements present ethical conflicts for scientists.

As encompassing as these agreements are, courts have upheld them. In doing so, they represent the public's interest in ongoing scientific research and development. The underlying assumption is that such proprietary control is essential for companies to gain a return on the heavy investment required for scientific research and product development. Without the assurance of a return, they will not take the risk of investing. However, some companies may suppress results and refuse to share useful tools and resources beyond the need to realize the return on their investment. Consequently, scientists and the public may fail to receive the benefits of the propagation of new knowledge and inventions.

In the interests of the promisor, the promisee, and the public, companies should allow their scientists and engineers to publish results and share resources and tools in a timely fashion. Industry-sponsored research contracts should not require scientists or engineering researchers to violate professional standards. Rather, contracts between companies and researchers should be written to strike an appropriate balance between proprietary control over information and scientists' responsibilities to publish results and share resources to the benefit of science and the public.

This overview of ethical obligations, responsibilities, and ethical problems associated with contracts in science and engineering points to the need for practitioners in these domains to be taught to pay close attention to contracts. In many circumstances, engineers and scientists can influence the terms of contracts in such a way as to reduce the likelihood of their facing ethical problems later. Promisors and promisees can become oriented to devising and using strategies of preventive ethics to avoid violating professional ethical standards.

VIVIAN WEIL

SEE ALSO *Conflict of Interest; Engineering Ethics.*

BIBLIOGRAPHY

Bennett, F. Lawrence. (1996). *The Management of Engineering: Human, Quality, Organizational, Legal, and Ethical Aspects of Professional Practice*. New York: Wiley. A reference volume as well as a textbook that considers contracting and ethical issues for engineers from the perspective of managers of engineers.

Campbell, Sally A. (1997). "Balancing the Responsibilities of the Environmental Consultant to Create a Personal Ethical System." In *Proceedings of the Air & Waste Management Association's 90th Annual Meeting & Exhibition, June 8–13, 1997*. Toronto: Air & Waste Management Association. An environmental consultant's perspective on engineers' and scientists' ethical responsibilities to the public and the scientific community when they provide estimates of environmental impacts and contribute to environmental decision making, under contract to management.

Orlans, Harold. (1973). *Contracting for Knowledge*. San Francisco: Jossey-Bass. A broad consideration of problems inherent in contracting for knowledge, including social scientists' research.

Resnick, David B. (1998). "Industry-Sponsored Research: Secrecy Versus Corporate Responsibility." *Business and Society Review* 99(1): 31–35. A philosopher's analysis of contractual constraints on industrial scientists and engineers that documents cases of excessive control and indicates how business organizations can manage such contracts to the benefit of the public and the scientific community as well as their own benefit.

Watson, Alan. (1991). *Roman Law and Comparative Law*. Athens: University of Georgia Press. A solid scholarly volume that explains contracts in the law, with an account of the origins of contracts in Roman Law.

CORRUPTION

Corruption derives from the Latin verb *corrumpere*, which means to break into pieces, destroy, defraud, falsify, seduce, or bribe. But the meanings hardly end with those. They are merely one set of a procession of definitions and interpretations amassed over the centuries, all signifying some contagiously harmful, unjust, self-serving, often repulsive divergence from moral conduct.

Definitions

Corruption defies and defiles what is generally perceived as the common good. In its malevolent extreme—such as systematic and widespread murder, torture, rape, or pillage, undertaken to maximize power—corruption can attain the dimensions of evil. At the lesser extreme, acts such as bribery, embezzlement, plagiarism, or falsifying research data, when done on a small scale and episodically, can be seen as unethical, immoral, or deranged, though not necessarily corrupt. Scope can often define corruption.

Science has its own literal definitions of corruption. Data are sometimes called corrupted. In biology, corruption is the process of living matter's decomposition. Similarly, a spoiled laboratory sample can likewise be described as having been corrupted. Terms such as rot, putrescence, and decay all serve well as descriptives for the revulsion corruption can generate. Corruption covers a multitude of sins and therefore has an almost limitless repertoire of baleful synonyms and colorful case examples.

Scholarship on corruption in science is rare, in technology (e.g., patent piracy, computer hacking) increasingly frequent. But scholarly work on corruption in governments, wherever they may be, is abundant. The challenge in the science and technology sector is to connect the hidden motivations and behavior patterns of those in the technical world to that of the political and economic spheres so that technical professionals can play stronger roles in perceiving their own relevance in stemming corruption's incessant growth.

Organizational Approaches

A handful of organizations with ambitious programs to understand and prevent corruption have attempted to establish satisfactory definitions of corruption. The World Bank, which in 1999 launched a vigorous anti-corruption program, defines corruption as "the abuse of a public position for private gain." Transparency International, long the leading body in tracking and studying corruption, defines it as "behaviour on the part of officials in the public sector, whether politicians or civil servants, in which they improperly and unlawfully enrich themselves, or those close to them, by the misuse of the power entrusted to them." Because of the global trend toward the privatization of public functions, it extends that definition to abuses in the private sector.

A third body, the Organisation for Economic Co-operation and Development (OECD), shuns any attempt to define corruption but has undertaken considerable work in gathering statistics, convening conferences, and issuing reports on such subjects as bribery, export credits, corruption in individual countries, and corruption's impact on development.

Any generalized treatment of corruption that is less than criminal or evil can entail considerable subjective judgment, thus inviting both a self-critical eye and rhetorical reflection. Often the word is used loosely in tirades against political opponents, such as a "corrupt" policy by one political party or another involving the environment, the elderly, or illicit campaign tactics. Charges of corruption can be flung when scientists rally against the packing of technical panels by a government whose political party they oppose. In the brutal give and take of politics, judgments about corruption's severity and perhaps its very existence are best done with care, case by case, even item by item, with emotions held in tow.

The Situation within Science, Engineering, and Technology

Science, engineering, and technology—technology being the useful products of engineering—are themselves fertile soils for corruption. Under the thrust of technological change, they can serve as tools (genetic engineering, virology, the computer, and digital communications as examples) to expand the range of corruption's infectivity. Thus, right at the start, the technical world can be mired within conflicting goals when business, engineering, and science comingle. Not only that, but history displays the macabre paradox of science and engineering specifically employed for evil means such as the freezing of human beings by Nazi scientists to study the process of death and the feasibility of resuscitation, or the infamous Tuskegee Syphilis Study (1932–1972) on prison inmates in Tuskegee, Alabama, as well as the radiation experiments performed on unwitting human subjects by the Atomic Energy Commission from 1944 to 1974. Further, it could easily be argued that weak implementation of occupational safety and health laws leading to worker deaths is also a form of corruption of the public good.

The values of science, which derive from philosophical and moral thought, are their own protection against any infestation of corruption. The inner character of science contains the ethical outcome of improving the lot of humankind and adhering to a strict code that imposes integrity on its practitioners. For years, the scientific community has striven to reduce the incidence of data falsification, arguing that the act of falsification erodes the honesty and openness that feeds scientific progress. Thus, science contains within itself a moral value all scientists are trained to revere. But, like other human beings, scientists can cheat, lie, and steal. The question is whether one chooses to call such flaws corruption—whether to expand the definition of corruption to include the corruption of values. At this moment in the sociology and psychology of science, divergent behavior in the technical fields rests in the discipline of ethics, broad enough in itself.

Thus, the tracking and policing of unethical behavior among technical professionals has been left to science and engineering societies, journals devoted to science/society issues and to the field of misconduct and malpractice, inspectors general for the technical agencies of government, the agencies themselves (through, for example, the Office of Research Integrity at the National Institutes of Health), and science and engineering workplaces. Corruption involving science and technology, however, does come in for significant treatment in the corruption literature because the capital transferred for development projects that involve science and engineering is often skimmed for payoffs at either the contractor or government level. Thus it is clear that those within the science and engineering community whose work engages them in development projects have a stake in corruption at the level of the Third World. Whistleblowing is one major response by technical people to perceived violations of ethical practice among their higher-ups. Unfortunately, whistleblowers are too infrequently rewarded—and often punished—for acting on their sense of outrage.

How corruption can be differentiated from immorality is an open question. If a lie is immoral, then scientific fraud—whether by plagiarizing texts or falsifying data—is immoral as well. But whether it is corruption is more a question of philosophy than practicality. Oftentimes, examples of fudging laboratory work for neater results might well be seen as advancing the cause of a research project. If the loss of research support for a worthy program, for example, is threatened by a bit of discrepant data, then the researcher might consider "tidying up" the results for the sake of saving the grant.

Trends and Outlook

Where corruption in science and engineering perhaps bears most watching is in the relatively recent marriage between corporations and universities in conducting genetic engineering research. The field itself has long presented ethical and moral dilemmas, but the risk of corruption increases in the high intellectual property stakes involved in genetic discoveries. The fear is that academic and intellectual freedom has been "corrupted" when scientists working under the support of the corporation deliberately withhold data from colleagues at competing institutions. These practices have taken place to a disturbing extent with no final consensus in view.

Corruption will always be present within the human realm. The war on it in the developing world has become vigorous and is showing success. Evidence shows that as those countries democratize and generate more internal wealth, corruption will decrease. At the same time, however, the growth of new scientific and technological tools will render corruption increasingly creative and sophisticated. The incursion into personal privacy through sensor technology applied to "protect democracy" can be seen as chilling enough. The challenge, then, is to anticipate what new forms of infectious malfeasance loom as the science behind biotechnology and nanotechnology, and the digital instruments of technology generate new ways of doing harm.

WIL LEPKOWSKI

SEE ALSO *Development Ethics; Engineering Ethics; Ethics: Overview; Office of Research Integrity; Tuskegee Experiment; Whistleblowing.*

BIBLIOGRAPHY

Caiden, Gerald E. (1988). "Toward a General Theory of Official Corruption." *Asian Journal of Public Administration* 10(1): 3–26.

Krimsky, Sheldon. (2003). *Science in the Private Interest: Has the Lure of Profits Corrupted Biomedical Research?* Lanham, MD: Rowman and Littlefield.

Rose-Ackerman, Susan. (1999). *Corruption and Government: Causes, Consequences, and Reform*. New York: Cambridge University Press.

INTERNET RESOURCE

Transparency International. Available from http://www.transparency.org.

COSMETICS

The term *cosmetics* comes from the Greek *kosmos* for order, referring in this case to the well-ordered face or appearance. Cosmetics are substances applied to the skin or hair to create a pleasing appearance. In the early-twenty-first century, they are alternately seen as the bane of modern women's existence (creating a time-consuming *third shift* [Wolf 1991] for women) and as a simple, popular tool for personal transformation. Some feminists deride the cosmetics industry as an ethically corrupt patriarchal institution that intentionally makes women feel that their natural faces are inadequate and exacerbates the identification of value with superficial appearance (Bordo 1995), whereas others cheer the liberating effects of bringing control over self-image and appearance within the grasp of every person. Ethical concerns raised in the history of the cosmetics industry remain and are exacerbated by technological innovations and the increasing consumer culture.

History of Cosmetics

The practice of painting and tattooing the body dates back to early-Neanderthal humans, when natural mud, ash, and natural dyes were used for not so much for enhancing beauty, but for camouflage, inspiring fear in others, and representation of animal gods in ritual ceremonies. In ancient Egypt, body painting focused on the eyes, with black antimony powder and green malachite lining used for protection from the sun as well as for decoration. Cosmetics and perfumes were used by both sexes in ancient Egypt and Rome. Later, in medieval Europe, strict religious norms identified cosmetics as the devil's work—a sign of vanity and deception. The Renaissance period brought cosmetics back in style, emphasizing the human ability to improve upon nature. In Elizabethan England, both sexes powdered their faces for a pale complexion, while women also used rouge and lip color, and covered the entire face with egg white for preservation. Men and women of the upper classes devoted significant amounts of money and time to maintaining an aristocratic appearance (Gunn 1973).

In *Hope in a Jar* (1998), social historian Kathy Peiss tells the story of the cosmetics industry in the United States. The American Revolution led to a rejection of the English tradition of wigs and facial powders as signs of aristocratic standing for men. Yet women's virtue continued to be linked with appearance. Women kept instructions for homemade cosmetics intermingled with potions for curing rashes and maintaining good health. Traditional family recipes (using household items such as oatmeal, lye, charcoal, and berries, among others) were commonly exchanged through social networks; for advice, one went to a friend or family member, not a pharmacist or physician. But more efficient and less risky substances were often available at the pharmacy, and soon women began buying special ingredients for their beauty concoctions. Pharmacists recognized an opportunity for packaging recipes of their own and selling them as finished products. Advertising created brand recognition and motivated women to seek the lifestyles they saw in print. Thus by the early 1930s, most women in the United States reported that *putting on a face* was a daily activity involving commercial beauty products (Peiss 1998).

Despite this increasing popularity for commercial cosmetics, early critics voiced concerns. Some questioned the monetary and time costs invested for such temporary results. Others expressed moral contempt for a practice that was viewed as an enemy of authenticity, a way to fake one's way into beauty. Early associations between cosmetics and women of low status (e.g., prostitutes and vaudeville showgirls) contributed to this distrust. Yet in a society that historically undervalued women's intellectual capacities and overemphasized their aesthetic value, the cosmetics industry flourished. Looking good was a ticket to increased social status. Even women who initially rejected cosmetics as an inappropriate solution to problems of inequality felt social pressure to use them. Similar pressures have more recently led to increased use of cosmetic surgery for women, the expansion of the cosmetics market to men's products, and biotech research into more effective and individualized cosmetics products.

As the cosmetics industry has become more dependent on science and technology, significant ethical issues have been highlighted, and termed *cosmethics*. The issues range from gender equity to safety concerns and animal testing. Codes of ethics have been formulated by the cosmetics industry to begin to address these issues as they arise in development, manufacturing, distribution, and advertising (ICMAD). In the United States, the Food and Drug Administration (FDA) does not require cosmetics safety testing prior to public sales because cosmetics are not considered drugs. However the FDA publishes guidelines for good manufacturing, and all cosmetics manufacturers must comply with the Food, Drug, and Cosmetic Act and the Fair Packaging and Labeling Act; products without substantiated safety must bear the warning, "The safety of this product has not been determined."

Equity Issues

The contemporary cosmetics industry was largely founded by women (e.g., Elizabeth Arden, Madam C. J. Walker), who recognized the opportunity to make use of recalcitrant appearance norms for their benefit (elevating women's status by turning men's weaknesses against

Various cosmetics. Use of cosmetics for the purpose of enhancing beauty dates back to ancient Egypt. *(AP/Wide World Photos.)*

them) and built on the tradition of women's home-beauty networks. Women who were overworked, underappreciated, lacked self-esteem, or simply desired attention for themselves were offered a medium through which to connect with other women, pamper each other, and share concerns. Furthermore women could experiment with new identities for themselves through the use of cosmetics. Such benefits continue to be heralded in the early-twenty-first century. Of course, for convenience, most women settle on a standard routine that best fits their sense of themselves. Thus, in order to maintain a *normal* appearance, they come to rely on regular purchases of the associated products. Consumer purchases are required simply to be oneself. How ironic that the product heralded as an opportunity for self-creation, self-care, and shared intimacy among women turns into a requirement of time, energy, and financial investment. In a society highly attuned to appearance, serious consequences ride on conforming to the norm: preservation of jobs, relationships, and self-esteem. Indeed the pernicious dynamic of commercialization and biased norms of appearance has resulted in studies showing that many contemporary women spend significant time each day applying cosmetics, find them essential to wear in a wide variety of circumstances, and believe that their attractiveness depends on cosmetics (Cash and Wunderlee 1987, Kelson et al. 1990).

This situation is problematic for several reasons. First, although emphasis on men's grooming is increasing (Bordo 2000), the value placed on appearance is still decidedly greater for women than men. For women, the use of cosmetics is tied to social status and credibility in the workplace (Dellinger and Williams 1997). Second the image of beauty proclaimed by the industry is decidedly narrow, favoring a white, Western ideal, even when models are from different racial or ethnic groups (Perlmutter 2000, Bordo 1995). This imposition of one version of beauty on all reinforces the historically unjust social status of many women of color. The Western beauty bias can also be seen in scholarship on racialized uses of cosmetic surgery (Kaw 1993).

Safety and Animal Testing

To ensure that cosmetics are safe for human use, animal testing has been employed to determine toxicity and likely reactions to chemicals in the products. The LD-50 test (lethal toxicity for 50% of the animals tested) started in 1927 (Singer 1999) and was developed to determine the strength of various drugs for medical purposes. The testing quickly spread to other applications, including ingestion of lipstick and other cosmetics. It was an industry standard until the early 1980s, when animals rights groups pressured the industries to rethink both the efficacy and ethics of the test. Given species differences and drastic disparities in the amount and time frame for ingestion, the applicability of the test for human usage was unclear at best, and half the experimental animal populations had to die to complete the test. As one activist wrote, "The test defies common sense. Does one really need to know how many bars of pure Ivory soap kill a dog?" (Singer 1999, p. 10). Following public pressure, in 1985 the cosmetics industry moved to a limited test that feeds a smaller amount of the product to a smaller group of animals, and discontinues the study if no harmful effects are found. Similarly, since the 1940s, the Draize eye test has used conscious but immobilized rabbits to ascertain effects such as redness, blistering, and blindness that might result from direct contact of a cosmetic product with the eye. Rabbits' eyes are dabbed with the product, and observed over time to record eye damage and discomfort. Pressure from animal rights activists for alternative models to ensure safety convinced the industry to contribute its own funds to research aimed at refinement, reduction, and replacement of animal use. *Animal-free testing* now has marketing appeal as well as ethical grounding. In 2002 the European Parliament banned the sales of animal-tested cosmetics produced throughout the European Union, a ban that will, in the future, apply to animal-tested cosmetics produced in other areas of the world.

Although contemporary cosmetics has advanced significantly from the heyday of animal testing and the previous dangers of unregulated and untested products (e.g., in the Elizabethan period the use of ceruse, or white lead, for complexion whitening led to toxic reactions, sometimes with deadly consequences), the risks of cosmetic use have not been eradicated. Advances in science and technology have brought the advent of *cosmeceuticals* or beauty products designed to make use of medical and pharmaceutical advances for nonmedical purposes. These include Retin A-enriched facial cream to diminish wrinkles, baldness treatments, and other cosmetic products with biologically active agents. In the United States, this rapidly growing industry (Lamas 2003) is not subject to regulation and testing by the FDA because cosmeceuticals are not considered drugs (which affect the body's structure and function). Yet this claim is difficult to confirm without the very testing that has been waived due to the categorization scheme. Cosmeceuticals are often sold in the offices of dermatologists and other physicians, and may be easily mistaken for tested medical treatments by patient-consumers. Even overlooking the likely ethical conflict of interests, one wonders whether such new and improved cosmetic treatments really advance human options or instead quietly increase burdens, as people try to keep up appearances.

SARA GOERING

SEE ALSO *Affluence; Body; Consumerism.*

BIBLIOGRAPHY

Bordo, Susan. (1995). *Unbearable Weight: Feminism, Western Culture, and the Body*. Berkeley: University of California Press.

Bordo, Susan. (2000). "Beauty (Re)Discovers the Male Body." In *Beauty Matters*, ed. Peg Brand. Bloomington: Indiana University Press. Wide-ranging, accessible coverage of issues related to gender and the body.

Cash, Thomas, and James Wunderlee Jr. (1987). "Self-Monitoring and Cosmetics Use among College Women." *Journal of Social Behavior and Personality* 2(4): 563–566.

Dellinger, Kirsten, and Christine Williams. (1997). "Makeup at Work: Negotiating Appearance Rules in the Workplace." *Gender & Society* 11(2): 151–177.

Gunn, Fenja. (1973). *The Artificial Face: A History of Cosmetics*. New York: Hippocrene Books. Coverage of the world history of cosmetics from Neanderthal times through the twentieth century.

Kaw, Eugenia. (1993). "Medicalization of Racial Features: Asian American Women and Cosmetic Surgery." *Medical Anthropology Quarterly* 7(1): 74–89.

Kelson, Tamar; Ann Kearney-Cooke; and Leonard Lansky. (1990). "Body-Image and Body-Beautification among

Female College Students." *Perceptual and Motor Skills* 71: 281–289.

Lamas, Daniela. (2003). "The Business of Beauty." *Miami Herald*, September 23, p. 11E.

Peiss, Kathy Lee. (1998). *Hope in a Jar: The Making of America's Beauty Culture*. New York: Metropolitan Books.

Perlmutter, Dawn. (2000). "Miss America: Whose Ideal?" In *Beauty Matters*, ed. Peg Brand. Bloomington: Indiana University Press.

Singer, Peter. (1999). "Henry Spira's Search for Common Ground on Animal Testing." *Cambridge Quarterly of Healthcare Ethics* 8: 9–22. Interesting discussion of one individual's contributions to the animal rights movement.

Wolf, Naomi. (1991). *The Beauty Myth: How Images of Beauty are Used against Women*. New York: Bantam/Doubleday. Popular work on beauty and the pressures to achieve it.

INTERNET RESOURCE

Independent Cosmetics Manufacturer and Distributors, Inc. (ICMAD). Code of ethics available from http://icmad.org/join/codeofethics1.pdf

COSMOLOGY

The night sky is a primal wonder whose infinite nature spurs a longing to understand human existence. The realization that they are beneath a vastness and majesty beyond their personal experience impels people to attempt to know themselves and their place in all that there is. This is a religious impulse and is also the impulse behind cosmology.

From Astronomy to Cosmology

Cosmology is, however, a uniquely modern science of the history, structure, and dynamics of the universe. Although *astronomy* is a transliteration from the Greek, the word *cosmology* is a seventeenth-century coinage from an imaginary Greek term. It thus denotes a new, uniquely scientific way to deal with primal wonder about the night sky that was designed to replace the myths that represented primordial efforts to respond to that wonder.

The myths on which traditional societies were built were inspired by and speak to the origins of humankind and its place in the universe. Because the nature of the firmament is unknowable by the direct senses, until recently those myths were untestable and therefore perennial. The birth of technology changed that situation. Tools that take advantage of natural laws and allow humankind to manipulate those laws changed what was knowable. Systematic observations of the motion of the planets that were motivated by Tycho Brahe's (1546–1601) desire to find God's perfection in the sky led Johannes Kepler (1571–1630) to devise a model of the solar system with the sun at its center. Timepieces and levers set the stage for Isaac Newton's (1643–1727) grasp of gravity and its implications for the cosmos. Newton's calculus, a kind of conceptual technology, captured physical law with a generality and precision of unprecedented scope.

Today fossil light from the beginning of time is collected by immense machines both on the earth and in space and analyzed electronically to reveal the most intimate details of the universe and its beginnings. Modern cosmology weaves a creation story that passes the tests of science. The same methodology that has laid out physical truth and made possible the ability to control nature has allowed humankind to know the extent and origin of all that there is. In the process the inevitable imperial nature of science has taken over, displacing the old myths with cold certainty and weakening the ground beneath religions, belief systems, and structures of morality. As science replaces older foundational beliefs, it becomes complicit in the moral confusion of the modern age.

Can heaven survive the heat death of the universe? Will the cherished views of earlier cultures on the origin and meaning of human existence be another casualty of modern science? As astronomers divine the mysteries of the origin and evolution of the universe, are they culpable for the elimination of worldviews that may have had legitimate purposes but did not stand up to the scrutiny of scientific methodology?

The Emergence of the Big Bang Theory

In 1929 the astronomer Edwin Hubble (1889–1953) announced that the recessional velocities of galaxies are proportional to how far away they are. The farthest galaxies were said to be receding the fastest, as measured by the Doppler shifts of their emitted light. The Doppler shift is the stretching of light waves from objects that are receding from the earth at high velocity. Hence, distant galaxies appear redder. The constant of proportionality (between distance and recession velocity) became known as the Hubble constant. The implications of this relationship are profound. The simplest explanation of it is that at some time in the very distant past all the galaxies were packed together. The reciprocal of the Hubble constant is approximately the age of the universe: about 14 billion years.

How far back in time is it possible to see? What immense, sophisticated, and expensive instruments are required to see something as esoteric as the first light of the universe? In fact, one can see the radiation from the explosion of the Big Bang in almost every living room in the United States and almost any household in the world. All that it is necessary to do is to unplug the cable from a television set and set it to a channel where there is no broadcast. Part of that chaotic, somewhat disturbing pattern known as snow is the microwave echo of the Big Bang, which was released when the universe became transparent 200,000 years after it was born.

In 1965 Arnio Penzias (b. 1933) and Robert Wilson (b. 1936) of Bell Laboratories were working on a state-of-the-art antenna for the emerging technology of satellite telecommunications. Wherever they pointed their antenna in the sky, they heard a constant hum. In one of the most serendipitous discoveries in the history science, the cosmic microwave background (CMB) radiation had been found, and at a frequency exactly in agreement with the theory of the Big Bang (Sciama 1973). (Penzias and Wilson won the 1978 Nobel Prize for their discovery.) Since the Big Bang space has been cooling as it expands. If one runs the movie of the evolution of the universe backward to the point where all the galaxies coalesce, one finds that the "primeval egg" began expanding at nearly the speed of light 14 billion years ago. From the inferno of creation to the present the science of thermodynamics predicts that space should have cooled to 2.7 degrees Celsius above absolute zero. The frequencies Penzias and Wilson heard in the CMB correspond exactly to that temperature.

Cosmology and nuclear physics began to merge when scientists started to consider the first three minutes of the universe, a point made clear in Steven Weinberg's *The First Three Minutes: A Modern View of the Origin of the Universe* (Weinberg 1977). During that time all the fundamental particles—the neutrons, protons, and electrons that make up atoms and the rest of the fundamental particle zoo—were formed. As the universe expanded and cooled, mostly hydrogen nuclei were formed, but a fraction of them teamed with neutrons to make helium, deuterium, and lithium. According to nuclear physics, the relative amounts of each of these elements are quite sensitive to the conditions of the early universe. From that period of nucleosynthesis right after the Big Bang nuclear physics predicted that the universe should have been formed with about 76 percent hydrogen, 24 percent helium, and less than 1 percent heavier elements. In an affirmation of the Big Bang theory spectroscopists have shown that wherever one looks in the universe those ratios prevail.

With the evidence provided by Hubble's observation that the universe is expanding, the measurement of the CMB, and the correct prediction of nucleosynthesis during the first three minutes of the universe the Big Bang has been accepted as the real story of the universe. However, adjustments have been made to it.

The Structure of the Universe

A map of the universe as it is currently understood is shown in Figure 1. The bottom of the chart shows the center of the earth, and the top represents the farthest that can be seen: the CMB. The scale is logarithmic so that any quarter inch on the chart represents ten times the distance of the quarter inch below it. Two populations of artificial satellites populate space immediately above the earth: low orbit satellites at about 200 miles and geostationary satellites at 23,000 miles. The planets, asteroid belt, and Kuiper belt can be seen in the bottom half of the chart. The Kuiper belt is a vast ring of large comets that orbit the sun outside Pluto. Midway on the chart is the Oort cloud, a much larger spherical shell of comets that are bound loosely to the sun. Nearby stars, galactic stars, and the center and edge of the galaxy follow as one moves outward. The Milky Way is part of the local group, a loose collection of about two dozen galaxies that are gravitationally bound. Beyond that is the large-scale structure of the universe. Galaxies fill the heavens in these vast reaches, but they are not randomly placed. Not only do they form clusters, there are coherent structures that are significant fractions of the size of the universe. The Great Wall is one such structure: a long filament of galaxies that is 300 million light-years from the earth.

In fact, the large-scale structure of the universe is foamy and filamentary, as shown in Figure 2 (Gott et al. 2004). In this figure each point represents a galaxy: The foamy nature of the universe can be seen out to 2.7 billion light-years in this diagram. The foam seems to become less dense farther from the earth or, equivalently, farther back in time. In fact, it extends as far back as can be seen. The blank wedge-shaped regions are places in the sky where it is impossible to see out of this galaxy. This is the plane of the Milky Way.

The foamy structure of the universe must be indicative of the small, quantum asymmetries that were imparted during the Big Bang. One can imagine that a perfectly spherical explosion would result in a smooth, uniform universe with no structure. However, somehow

FIGURE 1

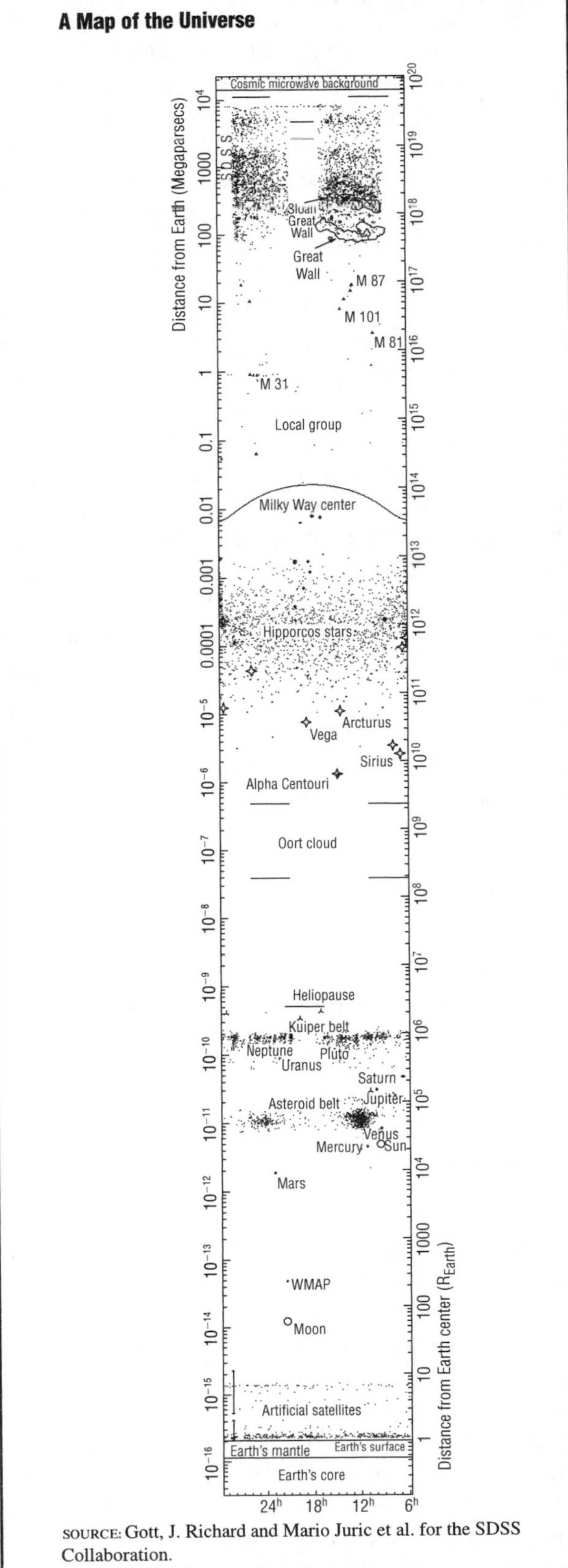

SOURCE: Gott, J. Richard and Mario Juric et al. for the SDSS Collaboration.

A map of the Universe (Gott et al., 2004). The vertical axis is distance on a logarithmic scale. At the bottom is th center of the Earth, and at the top is the most distant feature of the Universe, the Cosmic Microwave Background.

small asymmetries must have been present and were amplified by the force of gravity as the universe evolved and expanded. The structure that is seen is not consistent with the amount of matter and energy observed in the universe. There does not appear to be enough gravity to hold it all together, and this is where dark matter comes in.

Dark Matter

The direct evidence for dark matter is simple. Galaxies usually exist in gravitationally bound clusters of a few to several dozen. The motion of the galaxies around their common center, a matter of Newtonian physics, is completely inconsistent with the amount of matter that is seen. The motion of individual galaxies within a cluster can be explained only by the existence of an additional strong gravitational field. In fact, every galaxy or cluster must have a spherical halo of matter around it that is undetectable with electromagnetic radiation but is five times more abundant than the matter in the galaxies themselves. Little else is known about this mysterious cold dark matter, but its existence is generally accepted and there is an ongoing effort to detect it directly.

The Cosmic Microwave Background and Dark Energy

The microwave background also has structure. If the universe began as a microscopic primeval egg, it must have undergone vigorous quantum fluctuations in energy, shape, and even dimensionality. The imprint of those quantum fluctuations is seen in the spatial structure of the microwave background. To an incredible degree, however (about one part in a million), the microwave background is uniform. This implies that at one time the universe was small enough that it could come to thermal equilibrium but then grew rapidly, freezing in both the large-scale isotropy and the quantum fluctuations. This freezing in would have happened during an inflationary period when the universe accelerated outward at an exponential rate.

This is a decidedly nonintuitive move for a universe to make. What caused the universe to accelerate in the first place? In the old standard model of the Big Bang, without inflation, a prime mover is required, but only at the instant of creation. The explosion casts matter and

FIGURE 2

A Map of the Universe out to 2.74 Billion Light Years from the Earth

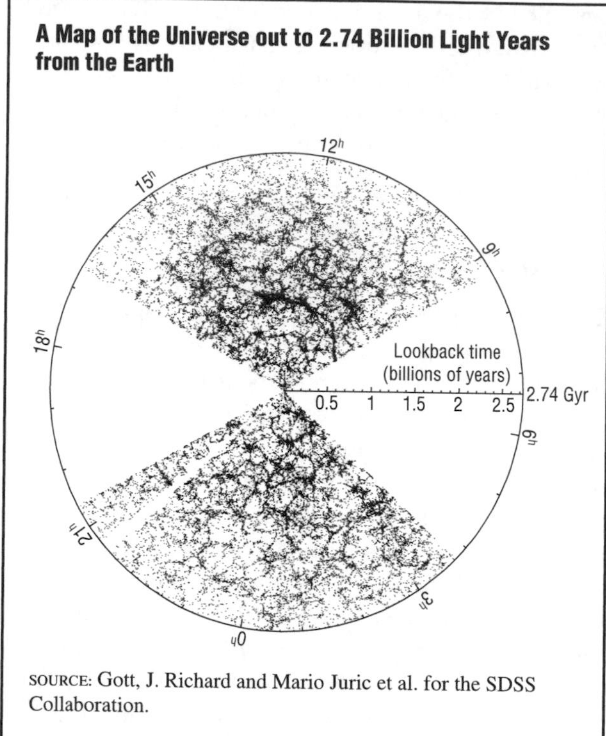

SOURCE: Gott, J. Richard and Mario Juric et al. for the SDSS Collaboration.

A map of the Universe out to 2.74 billion light years from the Earth (Gott et al., 2004). The scale is linear. Galaxies are represented by dots; the large scale, foamy, bubbly, filamentary structure of the Universe is visible. The blank wedges on the left and right are due to our lack of ability to see outside our own galaxy in these regions. They are in the plane of the Milky Way.

energy outward, expanding under this initial, unimaginable force but eventually slowing down as gravity pulls everything back to the center. The central question in cosmology at the start of the twenty-first century has been, What is the density of the universe? If the density is too low, gravity will never win and the universe will expand forever. If the density is high, beyond a critical point, the universe eventually will slow to a stop and begin to fall in on itself. The end is the Big Crunch, perhaps followed by reincarnation as the cycle begins all over again.

Neither of these scenarios appears to be the likely fate of the universe, however, based on the smooth nature of the microwave background radiation. Instead, the universe appears to exist in a state in between these scenarios, like a penny that has landed on its edge. It seems that the universe is flat, a spacetime geometry that means that the universe will continue to expand forever, although more and more slowly, approaching a stop at *t* equals infinity. The problem is that when one adds up all the mass and energy and dark matter, the universe is shy of the total amount required for a flat geometry by a factor of two.

This is where two problems are solved at once by the inflationary theory. There are quantum mechanical reasons to suspect that the vacuum itself has energy. That is, there is some underlying fabric that wildly undulates, popping fundamental particles into existence from nothing and swiftly returning them to the weave. Those particles have been observed, although the nature of the fabric and the energy it imparts to the vacuum remain mysterious. At one time the physicist Albert Einstein (1879–1955) postulated that energy, which he inserted into his equations as a cosmological constant. His goal was to produce a model of a steady-state universe, infinite and isotropic in time and space, largely because he felt that that was more aesthetically reasonable than a universe that began with a Big Bang. Although Alexander Friedman (1888–1925) showed that the Big Bang was a valid solution to Einstein's equations, Einstein abhorred that theory. However, he abhorred the ad hoc adjustment to his equations even more, and when the empirical evidence for a Big Bang could not be ignored, he declared the cosmological constant his biggest mistake. On new empirical grounds it must be included again, although a fundamental theory of its origins probably will require the achievement of a grand unified theory, a theory of everything, that string theory seems to promise for the future (Greene 2003).

This quantum vacuum energy is called the dark energy, and there is twice as much of it as there is of everything else that can be seen and measured. The dark energy has been implicated in the inflationary era of the universe and may have been the driving force for it. Still, aside from problems with identifying the quantum vacuum energy with the missing energy of the universe, the invention of the dark energy seems contrived.

There has, however, been an important recent discovery whose status has increased steadily. By very carefully measuring the red shifts, and hence the recessional velocities of galaxies deep into the universe, cosmologists have been able to map the evolution of the expansion rate of the universe. They have found that although the universe slowed down steadily after inflation, as a result of gravity, about 5 billion years ago it began to speed up again (Greene 2003). Today not only is the universe expanding, its expansion rate is increasing. The universe is accelerating, and something must be causing that. The culprit is the dark energy that permeates the vacuum.

The Story of The Creation

The newest creation story is surely not the final answer. A final theory will emerge only when there is a full understanding of how gravity is related to the other three forces and when the theories of gravitation and quantum mechanics are united. Enormous conceptual progress has been made with the development of string theory and its big brother, M (membrane) theory. String theory envisions particles as one-dimensional strings that vibrate not only in the known universe but also within six other hidden dimensions that are curled too small to be seen but that exist at every point in space (Greene 2003). A majority of cosmologists and theoretical physicists consider string theory the most promising and testable avenue for developing a true "theory of everything."

In the beginning there was an incredibly hot multidimensional nugget that was about one Planck scale (10^{-33} centimeters) in length. According to string theory, this Planckian egg is the smallest that anything can be. Squeezing it tighter makes it bigger and cooler. String theory avoids the singularity of the conventional Big Bang theory by considering the behavior of matter and energy at the very finest scales. It cannot say, however, what may have existed before this state, although this is an area of ongoing research.

The nugget had the entire mass of the universe in it, and it underwent transitions in its topography rapidly and randomly. Between 10^{-36} and 10^{-34} seconds after the start of time three dimensions suddenly broke free of their confining strings and inflated ferociously in a violent, exponential expansion. Alan Guth (b. 1947) of the Massachusetts Institute of Technology first showed that inflationary expansion of the universe represents a particular solution to Einstein's equations and can explain a deeply perplexing aspect of the CMB: its overall isotropy. The remaining dimensions stayed curled together, fundamentally influencing the nature of the particles and forces that became manifest in the three macroscopic dimensions. At one-hundred-thousandth of a second quarks began to clump into protons and neutrons.

Meanwhile, as the universe cooled, something strange was happening to the force within it. It was born with only one force, but as it cooled, it underwent phase transitions by which new forces were cleaved from the original one. Ultimately, for reasons that are not understood, the universe ended up with four forces: gravitation, electromagnetism, and the weak and strong nuclear forces. From a hundredth of a second to three minutes after the Big Bang the elements were formed. At 200,000 years the universe had cooled enough for stable atoms to form. In other words, the universe cooled from a plasma to a gas and became transparent. The photons streaming outward at that time are the blips seen on television sets.

Perhaps a billion years after the Big Bang galaxies began to form. The universe continued to expand at close to the speed of light, but the relentless action of gravity caused its expansion to slow. However, 9 billion years after the origin of the universe its expansion began to accelerate, most likely as a result of the repulsive force of the quantum vacuum energy. If this trend continues, the acceleration of the universe will cause galaxies to fly ever more rapidly away from one another. Some day even the closest galaxy will be too far away to see; the galaxies will be beyond the light horizon. Some day all the fuel for the stars will be used up, first hydrogen and then helium, carbon, and oxygen, until the last sun flickers and the universe is plunged into eternal darkness.

The Ethical and Political Dimensions of Cosmology

For many scientific disciplines the cause-and-effect relationship between scientific outcomes and the well-being of people is of great importance: Scientific results and their technological progeny are the dominant forces shaping the future of the world. The role science will play in determining the quality of life for every human being on the planet is of course determined by the elite that funds science. In this way all scientific enterprise is embedded in the greater moral problem of how individuals and groups should conduct themselves. Is it better for the powerful to channel their efforts solely for competitive self-benefit or to distribute knowledge and technology among all people? What are the consequences of pushing technologies on societies that may not want them? In some fields these issues spring directly from contemplation of the promise and implications of their projects. If it is possible to choose the human qualities of a person through genetic engineering, who will decide what those qualities will be, and to whose progeny will they go? Other subjects may be further afield, but the conceptual shift forced on science by the quantum nature of the infinitesimal in the 1920s has led to the most transforming technology in history: electronics.

Cosmology evokes a sense of the most benign and pure of sciences. The fascination of contemplating what is out there, combined with the fact that humankind cannot do anything to it, lends the study of space its alluring innocence. That of course is the old view. Cosmology is coming dangerously close to asking God rather direct questions.

To some degree scientific disciplines can be categorized by how influential ethics is thought to be in a particular field. Indeed, the ethical weight of astronomy, compared with that of genetics, lends it a kind of lightness and purity that is perceived by the people who fund it. Virtually everyone on the planet has gazed up and rested briefly in that human space where one wonders what it all is and what it all means. The pursuit of these wonders feels ennobling, partly because of the human space it comes from and partly because it is difficult to imagine how contemplation of the stars could alter the fate of humankind.

The modern science of cosmology is perhaps as far removed from the day-to-day concerns of humanity as any human endeavor can be. Futurists may conjure colorful uses for the discoveries of scientific research on the nature and origin of the universe, but this is not a matter of dealing with transistors or life-extending drugs. No one argues that cosmology is studied because of its economic impact. However, this does not mean that the study of the universe lacks an economic impact. The latest discoveries in astronomy have always depended on progress in computer, space, and detector technology (Tegmark 2002). Synergism between the astronomical sciences and industrial and military concerns is strong and growing, and both enterprises benefit.

Philosophical Issues

As self-aware beings people share a special, emergent property of the universe: consciousness. Is the quality of this aspect of nature in some way different from, say, the way space is curved as a result of the distribution of mass in the universe? What is special about the way living, replicating systems employ available resources to thrive, evolve, and produce beings that are capable of studying the deepest questions about their existence? Is mind a statistically unlikely property to have emerged from a universe with 1,000,000,000,000,000,000,000 solar systems? Or is the quality of mind ubiquitous and unifying like gravitation or other universal physical laws? Science is engaged in exploring the origin and nature of the universe as it never has before, along with the role of life and consciousness within it.

Every culture has a cosmology. Science has become the sine qua non of truth, and its revelations are taken as gospel. The insights of science into the nature of the universe therefore are assumed to or allowed to subsume all prior knowledge. It is incumbent on all scientists to ask whether their work leads to living together in harmony or interferes with that harmony. Where is the role of heart or spirit in the exploration of the cosmos or, for that matter, in any scientific endeavor? The scientific study of the origin and structure of the universe is a journey that has begun to yield answers to questions that once were the purview of religion and myth. What is done with this knowledge and what its ultimate meaning may be should be an essential component of the science of cosmology.

MARK A. BULLOCK

SEE ALSO *Astronomy*.

BIBLIOGRAPHY

Greene, Brian. (2003). *The Elegant Universe: Superstrings, Hidden Dimensions, and the Quest for the Ultimate Theory*, 2nd edition. New York: Vintage. A superb popular treatment of the physics of the ultra-small.

Gott, J. Richard; Mario Juric; David Schlegel; *et al.* (2004). *A Map of the Universe*. Princeton, NJ: Princeton University Press. Available from http://arxiv.org/abs/astro-ph/?0310571. An innovative approach to mapping the universe at all scales.

Sciama, D. W. (1973). *Modern Cosmology*. Cambridge, UK: Cambridge University Press. A classic text in cosmology.

Tegmark, Max. (2002). "Measuring Spacetime: From the Big Bang to Black Holes." *Science* 296: 1427–1433. A technical article by one of the leaders in cosmology.

Weinberg, Steven. (1977). *The First Three Minutes: A Modern View of the Origin of the Universe*. New York: Basic Books. A widely-read popular account of the role of nuclear physics in the origin of the universe.

COUNCIL ON BIOETHICS

SEE *Bioethics Commissions and Committees; President's Council on Bioethics*.

CREATIONISM

SEE *Evolution-Creationism Debate*.

CRIME

• • •

Crimes are commissions of acts that are publicly proscribed or the omissions of duties that thereby make offenders liable to legal punishment. More colloquially, a crime is any grave offense, particularly against morality, and thus something reprehensible, foolish, or disgraceful. Criminal behavior is in most cases unethical; it has also been subjected to scientific study in criminology. Technological change has in turn given rise to new forms of crime.

Legal Traditions

In some legal traditions, there is a distinction between *crimes* and *torts*. The former are offenses against the state or society that are enforced by agents of the state. The latter are offenses against specific citizens, which the machinery of the state will enforce only if victims pursue their grievances in the form of a civil suit. The boundary between these categories is fluid, as discussed below with respect to homicide's historical transition from tort to crime. In keeping with ordinary parlance, both sorts of offenses are considered here.

What qualifies as crime in both its technical and informal meanings is cross-culturally variable, because laws and norms are cross-culturally variable. Premarital sex, profanity, abortion, political dissent, alcohol use, homosexuality, littering, and remaining standing in the presence of the king are all crimes in some societies but not in others. Theories of crime are thus concerned not only with the causes of criminal behavior, but with social norms and the labeling of acts. However, the fact that what is considered crime varies between times and places does not imply that it is arbitrarily constituted. There is substantial overlap in the content of criminal codes, both written and traditional, from around the world. The acts that are most consistently criminalized are concentrated in a few principal domains: certain acts of violence, certain sexual acts, certain acts of expropriation, and certain betrayals of the collectivity to rival collectivities. In general, crime entails self-interested action that violates the interests of others.

Most crimes have identifiable victims, and for criminal sanctions to be widely accepted as legitimate and just, it is important both that the victimization was undeserved and that the offender behaved with inadequate consideration of the victim's interests. Law sometimes excludes consideration of whether a victimization was deserved when deciding an offender's guilt, but this is by no means generally true—consider the breadth of cases in which "provocation" can mitigate criminal responsibility—and even where it is true, the prevalent defense practice of "putting the victim on trial" suggests that desert is a more influential consideration than a literal reading of criminal codes might suggest. As for the offender, it is not enough in Anglo-American law that a wrongful act (an *actus reus*) was committed; there must also have been a wrongful intent (*mens rea*).

The essence of the *mens rea* criterion is that the wrong-doer was overvaluing his own interests and undervaluing those of others. The two principal justifications for criminal sanctions both demand such a criterion. If criminal sanctions constitute just moral retribution, then assigning culpability without reference to intent is wrong. Alternatively, if criminal sanctions are justified by their social utility, then punishing outcomes without regard to intentions is unlikely to deter antisocial behavior. However, the concept of *mens rea* is necessarily broader than just a specifically *malevolent* intent, because it encompasses reckless disregard for the well-being of others, thereby permitting the criminalization of acts such as drunk driving in which the perpetrator may have intended no harm to anyone but was still excessively overvaluing his own desires relative to the interests of others.

In modern nation-states, criminal offenses are considered offenses against the state and it is the state that prosecutes them. This practice has evolved historically from the "self-help" justice characteristic of traditional societies lacking professional police or judiciary, where victims or their relatives might demand material compensation or undertake retaliatory action in response to offenses against persons or property. Blood revenge in retaliation for homicide and persistent blood feuds between lineages are cross-culturally widespread manifestations of such self-help justice. The first step toward a criminal justice system occurs when a socially recognized power, such as a king or a council of elders, rules on the validity of grievances and hence the legitimacy of retaliation. Note, however, that punitive response remains in the hands of victims, with the consequence, for example, that killing someone who lacked family and friends would not be penalized.

It is only relatively recently that nation-states have assumed the responsibility (at least in principle) of punishing violations against all citizens. In Britain, for example, crimes became crimes against the state only after the Norman conquest of 1066, and even then, a murder victim's lord or kinsman might still negotiate monetary compensation from the killer or his/her kin. However, because such agreements did not affect prosecution by the crown and resultant fines, confiscation of the offender's belongings, and corporal or capital punishment, and because William the Conqueror also treated private retaliation as a crime, there was little incentive for a killer or his kinsmen to reach an accord with the victims. These practices gradually faded away, as did any central role for victims of crime other than as witnesses.

From the king's or state's perspective, blood revenge and feuds between powerful families were disruptive of social order, jeopardized the tax base, and weakened societal defensive capabilities against external threat. Why the citizenry succumbed to the rise of state authority also seems clear. An ideal of impersonal state-administered justice has been associated historically, and presumably causally, with a decline in the solidarity of kin groups and a rise in contractual relationships and individual responsibility. Impersonal justice is widely considered essential for keeping the citizenry safe from predatory victimization, and it certainly does extend the umbrella of protection to the relatively powerless. Moreover, even those with retaliatory and deterrent capability may welcome it. In the case of homicides, for example, the powerful as well as the weak may be relieved to relinquish the duty of vengeance, but only if they can trust the machinery of state to punish their enemies on their behalf.

Criminology

Although crimes always entail conflicts of interest, not all conflictual action is criminal. It follows that a general theory of crime requires both a theory of the nature of human interests and a theory of what legitimizes some, but not other, ways of pursuing self-interest at others' expense. The academic discipline of criminology arose primarily within sociology, and most theories of crime rely primarily on sociological concepts such as inequity, power, norms, legitimacy, and social control. Underlying psychological theories, in the form of assumptions about human desires, developmental susceptibilities, and social inferences, are typically more implicit than explicit, and at an even more basic level, criminological theories almost never explicitly address the origins and elements of a human being's interests, which must be identified before one can recognize violations thereof. Arguably, this question is within the domain of evolutionary biology, which provides the only relevant scientific theory, namely that the apprehension of where one's interests reside has evolved to promote Darwinian fitness within the circumstances prevailing in ancestral environments. This level of analysis is uniquely able to shed light on such questions as why rape is considered a particularly horrific violation regardless of attendant physical trauma, why men are more likely than women to respond violently to social disadvantage, why maternally perpetrated infanticide is widely considered a less heinous offense than other homicides if indeed it is an offense at all, and why adultery is a sexually asymmetrical offense defined as sexual contact between a married woman and a man other than her husband in all premodern legal codes.

Psychological science is primarily concerned with elucidating the mental and behavioral processes characteristic of a prototypical human being: how memories are laid down and retrieved, how people make probabilistic inferences, what emotions people all share, and so forth. A secondary focus of psychological science is the elucidation of how individuals differ. Both lines of inquiry are relevant to understanding crime.

At the panhuman level of analysis, psychologists investigate basic mental processes, and attempt to explain historical, cultural, and ecological variability in behavior as contingent products of a universal psychology's responses to variable circumstances and experiences. Anger, for example, is a motivational/emotional state that can be elicited in any normal person, with characteristic effects on physiology and information processing; it plays a role both in mobilizing physiological resources for violent action and in advertising one's likelihood of engaging in such action. Note that these claims entail hypotheses about the functions of being angry. A psychologist who assumes, for example, that the principal function of the psychophysiology of anger is to mobilize the organism for effective physical assaults will look for a somewhat different set of manifestations and social controls than another who instead assumes that anger functions primarily to threaten and deter so as to limit the costs of violent confrontations. Within this universalist research tradition, the reasons why people vary in their frequency and intensity of anger are to be sought in the social and material forces impinging upon them.

Notwithstanding advances in the understanding of how this universal human response operates, both centrally and peripherally, it is also evident that individuals differ in their responses to identical circumstances and stimuli. Whether these differences can be attributed to the cumulative effects of prior experiences acting on a universal human nature, or instead require a different sort of theory of individual differences, is not always apparent. Psychiatrists have identified a personality type that is disproportionately responsible for crime, especially violent crime: the "antisocial personality." Risk factors associated with the development and maintenance of antisocial personality include poverty, maleness, early maturity, poor school performance, parental criminal history, and psychopathology, implying that antisocial personality is in large part a facultative developmental response to experiential indicators of the lesser utility of developing a more "prosocial" personality. However, there is also evidence from twin and adoption studies that antisocial personality is substantially heritable, implying that individual differences in behavior are attributable to genetic differences.

Despite a large body of research on the genetics of crime, there has been relatively little consideration of this puzzle: why does genetic variability affecting phenomena such as criminal behavior exist? The reason for asking is because natural selection generally tends to eliminate genotypes with suboptimal phenotypic consequences, and one might expect that selection would have favored a panhuman phenotypic "design" with violence and other conflict behavior under appropriate contingent control. One possible answer to the puzzle is that heritable variation in antisocial behavior is a modern phenomenon and there has been insufficient time and/or fitness cost to eliminate the variability from human populations. A more interesting possibility is that antisocial personality types have social and material advantages in populations where they are rare and can exploit the trust and friendliness of the prosocial types.

Discussions of crime are often couched in the language of pathology. This is appropriate insofar as criminal acts reflect psychoses, delusions, and brain damage, but the language of pathology can mislead. Pathologies are failures of anatomical, physiological, and psychological adaptations, as a result of mishap, senescent decline, or subversion by biotic agents, such that the adaptations are no longer achieving the functions for which they evolved. The prototype of a pathology is a fracture: A broken bone can no longer perform its function. But crimes against people and property are not clearly pathological, and the term is certainly not applicable to violence in general. Violence is often well-regulated, self-interested behavior, and there are parts of the normally-functioning human brain that are dedicated to the production of controlled violence. The misconception that human violence is pathological has perhaps been reinforced by studies linking it to disadvantaged backgrounds and environments, but these associations are by no means universal. In nonstate societies, violence has been a prominent attribute of high-status men and a contributor to their social success. In modern state societies, the welfare of most people no longer depends on their own or their allies' violent capabilities, so violence is relatively rare and relatively likely to reflect psychological pathology. Nevertheless, disproportionate numbers of violent offenders are drawn from groups who lack access to the opportunities and protective state services available to more fortunate citizens, and who therefore find themselves in "self-help" circumstances much like those experienced by most people's ancestors.

Most crime is committed by men, and more specifically by young men. Criminologists and other social scientists have offered various hypotheses to explain these facts, but most of these hypotheses invoke local aspects of particular societies and thus provide no candidate explanation for the cross-cultural generality. Such consistently gendered behavior is better understood in terms of the different selection pressures confronting humanity's male versus female ancestors. There is morphological, physiological, developmental, and psychological evidence that humans evolved under chronic circumstances in which the variance in fitness was greater among males than among females: men had both a higher ceiling on their potential progeny and a higher chance of dying childless. In human beings, as in other animals, a higher variance in reproductive success has selected for a psyche that is more inclined to see life as a competitive contest with same-sex rivals, and is more willing to accept risks in the pursuit of material and social resources, including a willingness to embrace dangerous confrontations.

Criminal offenders have been characterized as lacking self-control and the capacity to delay gratification. In effect, criminal offenders tend to value the near future more highly, relative to more distant futures, than do law-abiding citizens. Discussion of these phenomena often presupposes that steep discounting of the future is dysfunctional, but an alternative view is that the human psyche has been designed by selection to adjust the discount rate ("patience") in relation to age, sex, and social and material circumstances. In this view, a short time horizon may be a rational response to information that indicates an uncertain or low probability of surviving to reap delayed benefits, and the sort of reckless, risk-accepting mindset that facilitates criminal acts may be aroused when the expected material or social profits from safer courses of action are negligible.

Variations in rates of crime are social phenomena, affected by sociological and demographic variables such as local cultural practices and the population's age structure. Elucidating exactly how and why these variables affect criminal acts by some and not other citizens is a project requiring interdisciplinary synthesis involving all social and biological sciences.

MARGO WILSON
MARTIN DALY

SEE ALSO *Death Penalty; Monitoring and Surveillance; Police; Science, Technology, and Law.*

BIBLIOGRAPHY

Gibson, Mary. (2002). *Born to Crime: Cesare Lombroso and the Origins of Biological Criminology*. Westport, CT: Praeger.

Hare, Robert. (1993). *Without Conscience: The Disturbing World of the Psychopaths Among Us*. New York: Pocket Books.

Wilson, James Q., and Richard Herrnstein. (1985). *Crime and Human Nature*. New York: Simon & Schuster.

CRITICAL SOCIAL THEORY

Critical social theory constitutes an effort to rethink and reform Marxist social criticism; it characteristically rejects mainstream political and intellectual views, criticizes capitalism, promotes human liberation, and consequently attempts to expose domination and oppression in their many forms. The extent to which science and technology may be associated with domination and oppression has been a major theme of critical theory.

Background and Method

Critical theory is not so much a particular theory as a tradition of thought historically associated with the Institute for Social Research, founded at the University of Frankfurt, Germany, in 1923. It is thus also commonly known as the Frankfurt School. The rise of Nazism forced Institute members into exile in 1933; the Institute then became affiliated with the Studies in Philosophy and Social Science program at Columbia University in New York City in 1935. The original school was reestablished in Frankfurt in 1953.

The Frankfurt School was a multidisciplinary group that included philosophers, sociologists, economists, political scientists, legal theorists, psychoanalysts, and others. Key members of the first generation were Max Horkheimer (1895–1973), Theodor Adorno (1903–1969), Erich Fromm (1900–1980), Herbert Marcuse (1898–1979), Leo Lowenthal (1900–1993), and Franz Neumann (1900–1954), with Walter Benjamin (1892–1940) as a close associate. Important members of second and third generations include Jürgen Habermas (a student of Adorno), Axel Honneth, Andrew Feenberg (a student of Habermas), Douglas Kellner, Steven Best (a student of Kellner), Albrecht Wellmer, Claus Offe, Nancy Fraser, and Martin Beck Matustik. Distributed now among institutions in the United States (Kellner is at the University of California in Los Angeles, Best at the University of Texas in El Paso, Fraser at the New School in New York) and Canada (Feenberg is at Simon Fraser University in Vancouver, British Columbia) as well as Germany, critical theorists have continued to include as part of their engagements with contemporary issues a critical dialogue with the works of Immanuel Kant (1724–1804), Georg Wilhelm Friedrich Hegel (1770–1831), Karl Marx (1818–1883), Søren Kierkegaard (1813–1855), Max Weber (1864–1920), Gyorgy Lukács (1885–1971), and Sigmund Freud (1856–1939).

The key method of critical theory is *immanent critique*, which focuses on the internal tensions of the theory or social form under analysis. Using immanent critique, critical theorists identify the internal contradictions in society and in thought, with the aim of analyzing and identifying (a) prospects for progressive social change and (b) those structures of society and consciousness that contribute to human domination. Critical theorists aim to aid the process of progressive social change by identifying not only what is, but also identifying the existing (explicit and implicit) ideals of any given situation, and analyzing the gap between what is and what might and ought to be. When applying immanent critique to science and technology, critical theorists identify both oppressive and the liberatory potentials.

Regarding science and technology, all critical theorists hold that science and technology are intertwined into a single complex or realm of human activity that in the early twenty-first century is commonly called *technoscience*. Further, they believe that technoscience is not neutral with respect to human values, but rather creates and bears value. They argue that the tools people use shape ways of life in societies where technoscience has become pervasive. Hence, how individuals do things determines who and what they are, and technological development transforms what it is to be human. But while critical theorists agree that the apparently neutral formulations of science and technology often hide oppressive or repressive interests, they differ in their ideas about whether technoscience is of necessity a force for dehumanization, and if not, why and how it might serve as a force for greater freedom.

From Hope to Dystopia: Horkheimer and Adorno

One strand of the critical theory tradition contains an initially hopeful view that technoscientific progress might inevitably drive forward human progress and contribute to the realization of greater freedom. This later gives way to a dystopian view, in which technoscience is equated with domination. In the 1920 and 1930s, many members of the Institute adopted a rather orthodox version of Marxism, arguing that the socialist revolution is a natural and inevitable outcome of the internal contradictions of capitalism. In line with this idea, Horkheimer, the second director of the Institute and the person who first named the members' work "critical theory," argues that progress in the forces of production has created objective possibilities for human liberation. These possibilities have not yet been realized because capitalism limits the progress of science and technology and thus restricts human progress. For Horkheimer, only a social and political revolution can unleash greater progress in the technosciences and harness technoscience to the cause of human liberation (Horkheimer 1972).

INSTRUMENTAL DOMINATION. While in exile in the United States during the late 1930s and 1940s, Horkheimer and Adorno reconceptualized their views on science and technology. They came to believe that the project of the European Enlightenment has turned into a mythology, and that modern reason and modern autonomy are rooted in the domination of non-human nature, other humans, and people's inner lives (Adorno and Horkheimer, 2002). They claim that the ideal of the Enlightenment is an ever-larger rational conversation about goals, values, and desires that expands the realm of human knowledge and action. Thus, they believe, the Enlightenment is an effort to increase human freedom and self-determination. But the course of reason since the Enlightenment has been increasingly to refuse to think about real alternatives. Rationality becomes, they argue, reduced to instrumental thinking: that is, to reasoning about efficient means to already given ends. This mode of thinking—instrumental reasoning—has become, they argue, the mode of thought characteristic of western culture in general, and of the technosciences in particular.

As they investigate the increasing integration of economics and politics, they find that society is ever more structured around the capitalist value of profit making and the technoscientific value of efficiency. Technological advances, including the increasing fragmentation and mechanization of work tasks, transform the work process. Work becomes more repetitive and mind numbing; workers are ever more isolated from one another, and have ever less time to critically reflect on their work or lives.

Thus, for Adorno and Horkheimer, technoscientific development brings with it increasing dehumanization. Modern institutions and ideas, including transnational organizations and democracy, are shaped and guided by instrumental rationality, and exist primarily to preserve themselves. It is no longer possible to ask about, or critically evaluate, ends; these are taken for granted. Because only questions about means can be considered by instrumental rationality, questions about ends are now considered irrational. So the progress of Enlightenment reason, restricted to instrumental rationality, contradicts the very goal sought by the Enlightenment—the increasing liberation of human beings. And modern technoscience, which should contribute to greater human freedom, increasingly becomes a cage of our own making.

CULTURE INDUSTRY. According to Adorno and Horkheimer, technology now carries the values of capitalism and of a consumer society. They coin the term "culture industry" to signify the process of the industrialization of mass-produced culture and the commercial imperatives that drive the system. The culture industry creates distractions, and the semblance of freedom (such as through the choice of which TV show to watch, or which breakfast cereal to purchase). But it offers no real alternative and only serves to distract people from careful reflection on the conditions of their lives. Adorno and Horkheimer attempt to demonstrate that the products of the culture industry commodify and mechanize everyday life, and that consumers of popular culture accept the pre-given ends of their culture and worry about how to organize their lives to acquire as many of these goods as possible. Thus the values of efficiency and instrumentality that characterize the technosciences and industrial production slowly shape the whole of society.

They further claim that in contemporary culture there is little critical awareness of technology because what is thinkable is constrained to those options considered rational under a narrow instrumental definition of rationality. Thus it is difficult for people to think of technology as a bearer of values. The technosciences appear to be value neutral, and the values of efficiency and instrumentality seem to be the only values it is rational to adopt. Hence, the dominant conception of technoscience is as something good if in the right hands. Adorno and Horkheimer argue that so long as instrumental reasoning is the dominant mode of thinking in Western culture, then human liberation will be blocked. Further, because instrumental rationality characterizes the Enlightenment and subsequent cultures at their very core, and is at the essence of technoscience, then technoscience necessarily leads to domination and dehumanization.

This increasingly dystopian view of technoscience is reinforced by the exposure of the great depths of evil that technoscience produced in the service of fascism, and in the Soviet system. By focusing only on means, many engineers, scientists, and technicians made death camps more efficient and produced propaganda and weapons for the oppression and control of people. As Horkheimer and Adorno understand things, all of this was made possible by instrumental reason that comes to see everything, even human beings, as objects of study and manipulation. They see liberal capitalism as also a system of domination because the growth of the culture industry, and the spread of technocratic thinking, only spreads domination over inner and outer nature. This process is all the more insidious because it does not appear as domination, but rather as entertainment, or simply as reality.

AESTHETIC LIBERATION. There is, however, one sphere of culture, they argue, that resists instrumentalization, and this is the fine arts. The great artists have, in their works, preserved and exemplified autonomy, thereby resisting merely instrumental concerns. In his last great work Adorno develops a complex theory of aesthetic resistance as maintaining a critical function, and as preserving the last vestige of humanness in an increasingly technological and inhumane world (Adorno 1998).

There are many questions and responses to this version of critical theory and its dystopian view of technoscience. American pragmatists, especially John Dewey and Larry Hickman, develop a version of instrumentalism that, rather than rejecting critical reflection on the ends of activity, requires it. Pragmatists have further criticized Adorno and Horkheimer for their increasing disengagement from any projects of real social change. Another criticism is that the work of Adorno and Horkheimer is elitist and escapist, especially in recommending the highly formal and abstract work of artists such as Arnold Schönberg (1874–1951). Such a detached view fails to live up to the goal of decreasing oppression. From within critical theory, Benjamin, Marcuse, Habermas, and Feenberg all break with dire pessimism and offer theories of technoscience as potentially aiding human liberation.

Liberatory Possibilities

There is another strand of thinking about technoscience within critical theory, composed of those who reject the pessimism of Horkheimer and Adorno and who maintain that technoscience can be useful in fighting domination. As with critical theory as a whole, this tradition contains multiple particular positions, some of which are at odds with each other. All maintain, however, the method of immanent critique, and the commitment to a critical analysis of culture with the aim of aiding human liberation. The four strands of critical theory that identify liberatory possibilities in technoscience are:

(1) the idea that technological change will sweep away old and oppressive cultural forms (Benjamin);

(2) that technoscience is oppressive under capitalism, but might be otherwise under a different social order, and hence might embody different values (Marcuse);

(3) that technoscience has an internal logic appropriate to its own realm, but that it must be restrained or all of life will fall under its sway (Habermas);

(4) that technoscience always contains internal contradictions, and thus always contains potentials both for oppression and liberation (Feenberg, Kellner, and Best).

WALTER BENJAMIN. The idea that technological change might sweep away oppressive aspects of culture is most clearly stated by Benjamin. For him, there are progressive possibilities in new technologies of cultural production, especially film, radio, and photography. Traditional forms of art maintain their cultural power through the aura of the authentic original. This gives the great works of art a mythic status that has served to present, maintain, and further the power of some, such as the church, the wealthy, and the state, over others.

Benjamin argues that the technologies of mechanical reproduction break down the aura and shatter the myth of authenticity. For example, not only is it difficult to determine which, if any, photographic print is the original, but also mechanical reproduction allows people to replicate the great works from history. Thus high culture loses its mystifying power. Further, media culture could cultivate individuals better able to judge and analyze their culture. By processing the flow of images in film, people develop the ability to better parry and comprehend the erratic and powerful flow of experiences in industrialized, urbanized societies. For Benjamin, the buildings, pictures, and stories of avant-garde artists, work that was often highly dependent on technology, was a form in which humanity was preparing itself to survive even the darkest night of fascism.

HERBERT MARCUSE. The position that technoscience is oppressive under capitalism, but might be otherwise, is clearly articulated in the work of Marcuse. Unlike Adorno and Horkheimer, who see technoscience as having a necessarily oppressive essence, Marcuse believes it is possible to identify and understand the specific historical and social forces that lead to oppressive technoscience.

Under capitalism, Marcuse argues, technology produces a mass culture that habituates individuals to conform to the dominant patterns of thought and behavior, and thus provides powerful instruments of social control and domination. This is so, he claims, because under capitalism, technology reflects particular class interests in what he calls a "one-dimensional society" (Marcuse 1964). Consumer culture, which is made possible by the rapid advances of the technosciences, is seductive, and sexually charged, while work is ever longer and more soul-killing. Rather than the sublimation of desire discussed by Freud, which leads to the great and meaningful products of human culture, Marcuse identifies a process of repressive desublimation in which everything becomes sexualized, but meaning and satisfaction are ever more elusive.

However, for Marcuse, technology could, through its advance and transformation, mechanize most socially necessary work, and thus free human beings for greater creative self-expression and social experimentation. Technology would cease to be autonomous, as it is in the one-dimensional society, and would become subordinate to a substantive notion of the good life, one that is fundamentally aesthetic in nature. Marcuse has an aesthetic model of human beings as free, self-creative beings. He believes that only spontaneous creative activity could break out of the one-dimensionality of life under capitalism. Hence, a new form of technoscience, one that embodies not mere instrumentality, but also allows for spontaneity and creativity, might further human liberation. Because of the centrality of one-dimensional intrumental rationality in modern society, Marcuse hypothesized that the likely sources of the ideas and energies for radical social change, including new forms of science and technology, would come not from the working class as traditionally conceived, but would be found in those most marginalized in society—people of color, women, and the disenchanted young. Among others, Angela Davis was both inspired by, and inspiration for Marcuse's work.

Critics rightly note that this alternative is highly speculative and underdeveloped. In his development of still another strand of critical theory that sees technoscience in a potentially positive light, Habermas criticizes Marcuse's position as hopeless romanticism, and one that dangerously will restrict the careful use of instrumental reasoning in the areas where it is appropriate to use it.

JÜRGEN HABERMAS. The third version of critical theory that views technoscience as having some liberatory potential is exemplified in the work of Habermas. He argues that technoscience brings great benefits to humans in modern cultures, and that insofar as it is concerned with technoscientific questions it should remain true to its own internal values. A problem arises when individuals allow technoscience and technoscientific values to take over other realms of human life that should not be organized around values of productivity and efficiency. Habermas criticizes the tendency of modern societies to subject all areas of human life to instrumental reasoning. For example, the sorts of thinking best suited to determining how to build a bridge are not the same as those best suited to nurturing friendship, neither are the skills and modes of thinking that characterize consumption those best suited to responsible citizenship. Habermas claims that it is dangerous to allow the values of either realm to seep into the other. On the one hand, the result is dehumanization of human relationships, and many of the destructive possibilities identified by other critical theorists. On the other, the consequence is bad science, and the pursuit of technical knowledge will be subordinated to ideology. Thus, technoscience, properly constrained, is necessary to human liberation, and to decreasing suffering and oppression.

Some critics argue that his position offers no concrete criteria for changing technology. Others claim that his position is hopelessly naïve, and that the technosciences cannot be constrained in the manner he suggests, so that Habermas's theory is actually a justification of the status quo.

ANDREW FEENBERG. The most recent work in critical theory of technology adopts a fourth position and argues that technoscience always contains contradictory possibilities. This is so because there are many dimensions to technoscience, many of which traditional accounts fail to identify. For this reason Feenberg argues that technology should be reconceived of through instrumentalization theory. This theory distinguishes between the understanding of technology by technical experts and philosophers of technology, and the understanding of technology within a specific social context by those who use it and are affected by it. Users of technology often deploy it in unintended and often unanticipated but imaginative ways. These uses often challenge existing technological systems and social orders. By better understanding and developing these contradictory potentials, he argues, the critical theorist can further the goal of assisting the cause of human liberation. Feenberg continues the Frankfurt school interest in popular culture, but is more sensitive to the political complexity of contemporary culture, and thus to the ambipotent nature of technological change. His work engages not only theorists such as Habermas and Heidegger, but included empirically rich case studies of French communications technologies, Japanese conceptions of technology, science fiction, and film. Feenberg returns the tradition of critical social theory to its multi-disciplinary roots, and is active in empirical research on the development and uses of technology, especially educational technologies.

DOUGLAS KELLNER AND STEVEN BEST. Kellner and Best bring critical theory into dialogue with postmodern and poststructuralist thinkers such as Jean Baudrillard, Michel Foucault, and Arthur Croker. Along with Feenberg, they also bring critical theory into dialogue with the pragmatist tradition. Kellner and Best also continue and revitalize the tradition of culture industry critique. However, unlike Adorno, they work to identify the contradictory potentials present in popular culture. Kellner has long explored the oppositional possibilities within technology, especially in alternative media and education. Best is also expanding critical theory into environmental philosophy.

Assessment

Contemporary critical theorists agree that there are liberatory possibilities in technoscience, but only the careful use of human will and consciousness can bring these to fruition. The future of critical theory promises an ever-greater dialogue with other applied traditions in philosophy, especially with pragmatism. Although some, such as Larry Hickman, have argued that critical theory is still too tied to an anti-technology paradigm that limits its practical usefulness, critical theorists are becoming more involved in concrete issues, from the alternative media work of Kellner to the work on computer-based learning of Feenberg, and this trend too promises to make critical theory more empirically rich, and thus better able to work toward the goal of increasing the realm of human freedom.

J. CRAIG HANKS

SEE ALSO *Autonomous Technology; Capitalism; Efficiency; Fascism; Freedom; Habermas, Jürgen; Marcuse, Herbert; Marxism; Marx, Karl; Neutrality in Science and Technology; Popular Culture; Socialism; Utopia and Dystopia; Work.*

BIBLIOGRAPHY

Adorno, Theodor. (1983). *Negative Dialectics*. New York: Continuum. Explores the limits of conceptualization, and hence of any instrumental understanding.

Adorno, Theodor. (1998). *Aesthetic Theory*, trans. Robert Hullot-Kentor. Minneapolis: University of Minnesota Press. Adorno's account of how the aesthetic dimension offers the possibility of resistance to instrumental rationality.

Adorno, Theodor, and Max Horkheimer. (2002). *Dialectic of Enlightenment*, ed. Gunzelin Schmid Noerr, trans. Edmund Jephcott. Stanford, CA: Stanford University Press. An account of how enlightenment rationality turns in on itself and leads to an irrational society.

Benjamin, Walter. (1968). *Illuminations*, ed. Hannah Arendt, trans. Harry Zohn. New York: Harcourt, Brace & World. Collected essays, including his important work on art in the age of mechanical reproduction.

Feenberg, Andrew. (1995). *Alternative Modernity: The Technical Turn in Philosophy and Social Theory*. Berkeley: University of California Press.

Feenberg, Andrew. (1999). *Questioning Technology*. New York: Routledge. A critical account of philosophy of technology and of the possibilities of democratic interventions in the course of technology.

Feenberg, Andrew. (2002). *Transforming Technology*. New York: Oxford University Press. An examination of the multivalent possibilities of technological change and its effects on labor, education, the environment, and politics.

Habermas, Jürgen. (1970). *Toward a Rational Society: Student Protest, Science, and Politics*, trans. Jeremy J. Shapiro. Boston: Beacon Press. An early explication of Habermas's account of positive possibilities of science and technology, offered in the context of a discussion of education, science, communication, and political action.

Habermas, Jürgen. (1971). *Knowledge and Human Interests*, trans. Jeremy J. Shapiro. Boston: Beacon Press. Habermas discusses the quasi-transcendental anthropological interests in communication, survival and control, and liberation, and how these interests shape our quest for knowledge.

Habermas, Jürgen. (1984–1987). *The Theory of Communicative Action*, 2 vols., trans. Thomas McCarthy. Boston: Beacon Press. Habermas works out implications of, and revises, his account of communicative rationality, and thus the promises and limits of science and technology.

Habermas, Jürgen. (2003). *The Future of Human Nature*. Cambridge, UK: Polity. A critique of biotechnology.

Horkheimer, Max. (1947). *The Eclipse of Reason*. New York: Continuum. An extended argument that objective reason has been overcome by partial, instrumental, and subjective reason.

Horkheimer, Max. (1972). *Critical Theory: Selected Essays*, trans. Mathew J. O'Connell. New York: Herder and Herder. Critique of instrumental rationality, and his important account of how critical theory differs from traditional theory.

Kellner, Douglas, and Steven Best. (2001). *The Postmodern Adventure: Science, Technology, and Cultural Studies at the Third Millennium*. New York: Guilford Press. A critique not only of post-modernist celebrations of science and technology, but also a critique of wholesale rejections of science and technology.

Marcuse, Herbert. (1955). *Eros and Civilization: A Philosophical Inquiry into Freud*. Boston: Beacon Press. Marcuse brings Freud and Marx together and examines ways in which contemporary society deals with desire.

Marcuse, Herbert. (1964). *One Dimensional Man*. Boston: Beacon Press. A discussion of how an affluent capitalist society can buy off, or co-opt, its members and thus diffuse opposition. Central is an account of the force of instrumental rationality and how contemporary science and technology restrict thinking and creativity.

Marcuse, Herbert. (1978). *The Aesthetic Dimension*. Boston, Beacon Press. Marcuse's argument that art can transform how people see the world, and offers help in imagining new and better possibilities.

CRITICAL THEORY

SEE *Critical Social Theory*.

CULTURAL CRITICISM

SEE *Anglo-Catholic Cultural Criticism*.

CULTURAL LAG

• • •

The U.S. sociologist William F. Ogburn (1886–1959) developed the concept of cultural lag, which occurs when unequal rates or degrees of change between interdependent parts of culture leads to "maladjustment" (1922). According to Ogburn, as new inventions are introduced into society, a maladjustment occurs and a period of adjustment is required. Most often these inventions are technological in nature, and are part of what he termed "material culture." However, Ogburn noted that "non-material culture" can also drive change. For example, he cites India in the early years of Buddhism as a case where religion was driving change in other areas of culture (1964).

Ogburn's classic description of technologically-driven cultural lag was the period required for society to adapt to the speed of the automobile (1964). It took some time for the social institutions and customs of road building to adapt to the ability of new cars to travel much faster than horses and older car models. A more pressing example is provided by the advent of nuclear weapons, which represent an enormous leap in scientific knowledge without a complimentary advance in political institutions capable of regulating and using that knowledge wisely. Another example is provided by the rapid advances in biomedical technologies and the ability of institutionalized ethics committees, such as Institutional Review Boards (IRBs) and Institutional Biosafety Committees (IBCs), to adapt to those changes and make wise decisions. The depletion of natural resources, especially oil, represents a broader interpretation of cultural lag, where changes in the material environment may outpace the cultural response to those changes.

FIGURE 1

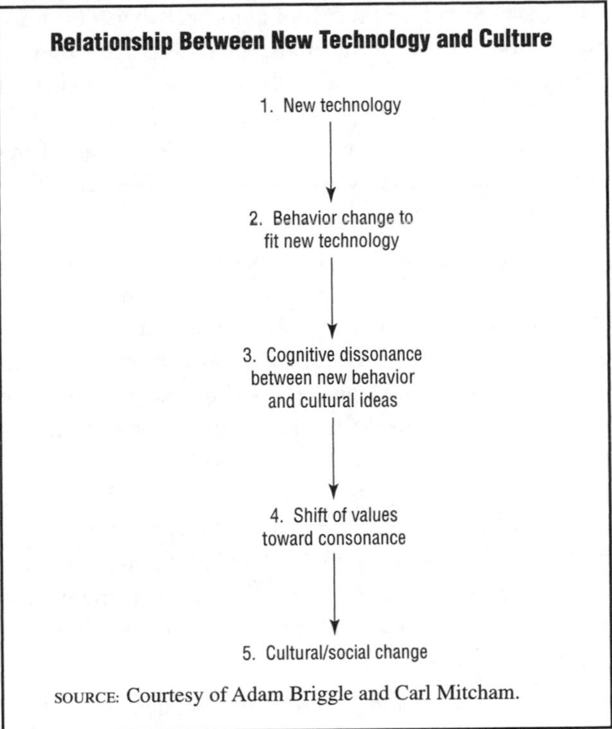

SOURCE: Courtesy of Adam Briggle and Carl Mitcham.

Numerous other cases exist where science and technology have advanced more rapidly than the spiritual, social, or political aspects of culture. Indeed, the anthropological studies collected by Edward H. Spicer (1952) and H. Russell Bernard and Pertti J. Pelto (1987) document examples of a relationship that Bernard and Pelto simplify as shown in Figure 1. Such maladjustment can prove socially harmful.

However, the concept of cultural lag must be interpreted and applied carefully in order to avoid dubious assumptions about progress. First, it must be recognized that culture can also lead rather than follow. Many historical analyses of how modern science and technology arose in Europe after the 1500s, such as those by Max Weber (1904), Lynn White, Jr. (1978), and others, have argued that cultural change preceded technological change. Second, it need not follow that "lagging" aspects of culture must simply be altered in order to "catch up" with more rapidly changing elements. If applied interculturally, the concept can also promote Eurocentric assumptions about "underdeveloped" parts of the world, and lead to irresponsible transfer and application of technologies.

Several evaluations of cultural lag exist in terms of its ability to describe and predict cultural change (Brinkman and Brinkman 1997). More important, however, is the need to deconstruct any bias toward an inadequate notion of *progress* within the metaphor of cultural lag. It is intuitive that various parts of culture change at different rates and thus no longer fit together smoothly. Yet this does not necessarily mean that one part now "lags behind" another. The metaphor of *cultural lag* easily connotes the "failure" of different cultures or parts of culture to adjust to change, as if there were no agency or choice outside of simply running along the treadmill of material change.

In other words, as Alvin Toffler argues, cultural lag needs a balancing term of "future shock," which describes "the shattering stress and disorientation that we induce in individuals by subjecting them to too much change in too short a time" (1970, p. 4). Building directly off of Ogburn's concept, Toffler explains, "The concept of future shock … suggests that there must be balance, not merely between rates of change in different sectors [of society], but between the pace of environmental change and the limited pace of human response. For future shock grows out of the increasing lag between the two" (p. 5).

He makes the argument that rapid change is neither indisputably good nor out of one's control to shape and sometimes slow down. The future can arrive too soon for society's own good. This highlights the central idea within cultural lag of proportionality, equilibrium, and harmony (the right adjustment) among the parts of culture. As Toffler argues, "The only way to maintain any semblance of equilibrium … will be to meet invention with invention—to design new personal and social change-regulators. Thus we need neither blind acceptance nor blind resistance, but an array of creative strategies for shaping, deflecting, accelerating, or decelerating change selectively" (p. 331). Achieving this selective change is not a simple, technical matter of "catching up," but rather a series of decisions about the meaning of the good life and the ideal society.

ADAM BRIGGLE
CARL MITCHAM

SEE ALSO *Double Effect and Dual Use; Science, Technology, and Society Studies; Social Theory of Science and Technology; Unintended Consequences.*

BIBLIOGRAPHY

Bernard, H. Russell, and Pertti Pelto, eds. (1987). *Technology and Social Change*, 2nd edition. Prospect Heights, IL: Waveland Press.

Brinkman, R.L., and J.E. Brinkman. (1997). "Cultural Lag: Conception and Theory." *International Journal of Social Economics* 24(6): 609–627. Surveys and evaluates the theory of cultural lag and some of its applications.

Ogburn, William F. (1922). *Social Change with Respect to Nature and Original Culture*. New York: Viking. Ogburn's first systematic formulation of cultural lag as a part of his theory of social change, which is presented in the chapter "The Hypothesis of Cultural Lag."

Ogburn, William F. (1964). "Cultural Lag as Theory." In *On Culture and Social Change*, ed. Otis Dudley Duncan. Chicago: University of Chicago Press. A concise overview of the concept with several examples where lags accumulate as the result of rapid technological change. Also provides some background to its development and addresses criticisms.

Ogburn, William F. (1964). *On Culture and Social Change*, ed. Otis Dudley Duncan. Chicago: University of Chicago Press. A posthumous collection of 25 essays by Ogburn. Includes works on social evolution, social trends, short-run changes, and methods.

Spicer, Edward H., ed. (1952). *Human Problems in Technological Change: A Casebook*. New York: Russell Sage. Fifteen case studies, including the often cited "Steel Axes for Stone Age Australians" by Luriston Sharp.

Toffler, Alvin. (1970). *Future Shock*. New York: Random House. A 505-page book that argues that there can be too much change in too short of a time, and therefore peoples need to improve their ability to wisely regulate, moderate, and apply technology to serve human ends.

Weber, Max. (2001 [1904]). *The Protestant Ethic and the Spirit of Capitalism*, trans. Stephen Kalberg. Chicago: Fitzroy Dearborn.

White, Lynn Jr. (1978). *Medieval Religion and Technology: Collected Essays*. Berkeley: University of California Press.

CYBERCULTURE

In a restricted but popular sense, cyberculture denotes the hacker subculture along with various social and artistic manifestations; as such it references feedback loops, computer slang, video games, the Internet, hypertext, virtual communities, and more. In a wider and more argumentative sense, cyberculture refers to contemporary culture in its totality, insofar as it has been influenced by cybernetic technology and its creative ideas. In both senses cyberculture has become a new scientific and technological context that stimulates ethical reflection.

Historical Development

The term *cyberculture* appeared in the 1980s but is ultimately dependent on Norbert Wiener's creation of the science of "cybernetics" (1948). An initial cyberculture emerged before the term itself when the scholarly community attempted to apply cybernetics to the interpretation of phenomena in psychology, economics, politics, anthropology, and education. The work of Gregory Bateson (1972) and Heinz von Foerster (1984) in the development of "second-order cybernetics" was central to this development, as was the promotion of information and systems theory. In the Soviet Union cybernetics, after initially being rejected under late Stalinism as another form of bourgeois ideology, also exercised a special attraction as a possible means to reconcile central planning with the increasing complexities of large-scale systems that were straining under top-down management inefficiencies (Gerovitch 2002). Cyberculture in these senses was never so named, and was never more than an issue among specialist intellectuals.

A second-stage cyberculture emerged in science fiction from the mid-1980s. Bruce Bethke (in his 1983 short story "Cyberpunk"), William Gibson (in 1984's *Neuromancer*), and others developed a new form of science fiction; in opposition to classical science fiction, which had become somewhat domesticated, such authors introduced raw (punklike) elements and expressed a negative vision of the short-term future. Bruce Sterling (1986) provides a general introduction to this form of cyberculture. Promoted in part simply by the linguistic accident that *cyber* could be easily prefixed to anything from space to sex, cyberculture experienced a rapid inflationary moment in *cyburbia* and *cyberia*, *cyberphilia* and *cyberphobia*.

Science-fiction writer Neal Stephenson justified this inflation by declaring: "Our concept of cyberspace, cyberculture, and cyber-everything is . . . a European idea, rooted in Deuteronomy, Socrates, Galileo, Jefferson, Edison, Jobs, Wozniak, glasnost, perestroika, and the United Federation of Planets" (1994, p. 100). In this sense, cyberculture includes everything from science and technology to politics and literature as it has been altered by the mediation of computers, digital interactivity, and "hacktivism" (Himanen 2001). From such an amplified perspective, cyberculture is simply that culture which emerges through symbiosis with cybernetic or information technology, itself understood as the fulfillment of technoscience, after the manner of Martin Heidegger's identification of cybernetics as the ultimate stage of metaphysics (Heidegger 1972). Indeed, the methods of experimentation and logical analysis that are central to science have now been supplemented with simulation modeling that introduces something such as cyber-experimentation into science.

Using a distinction between culture (of and related to nature or the body) and civilization (of or related to politics and rationality), cyberculture may also be thought of as constituted primarily by those human interactions with the material world of advanced technological artifice that are replacing nature as the basic context for human experience. Cybertechnology in some form has come to exist in the background of all new political orders and rational discourse, and even encourages human beings to consider the ways in which they are becoming cyborgs (Haraway 1991) or posthumans (Hayles 1999).

The general examination of cyberculture in these disparate senses is found in cyberculture studies, which includes the more focused field of cyborg studies. According to David Silver, director of the Resource Center for Cyberculture Studies, this kind of activity has passed from popular promotion based on the image of a "cybernetic frontier" through an initial scholarly concern for sociological (virtual communities) and psychological (online identity transformation) implications, to what he terms "critical cyberculture studies." In critical cyberculture studies the ethical issues implicit in such works as Howard Rheingold's *The Virtual Community: Homesteading on the Electronic Frontier* (1993) and Sherry Turkle's *Life on the Screen: Identity in the Age of the Internet* (1995) become explicit themes.

Ethical Issues

The shift from description to critical assessment has taken place around four overlapping themes. First, questions are raised about the personal and environmental safety of cybernetic hardware. The silicon chip and carbon-zinc battery industries are not as obviously polluting as steel mills and chemical plants; they nonetheless present major challenges to worker safety and environmental contamination in both the production and disposal cycles. Safety and ergonomic issues are further associated with the use of screens (eyestrain) and hands (keyboard and mouse strain).

Second, critical issues are further associated with economic and political discussions of dot-com cyber-industries. Concerns for the economic and political impacts of automation extend into discussions about cybernation, cybercrime, accounting fraud, marketing hype, treatment of labor, and concentrations of wealth and power in the networked society. Debates about a possible digital divide also fit in this category. At the same time, Pekka Himanen (2001) has argued that a distinctive cyber-economics is growing out of the "hacker ethic" applied to business affairs using open-source software. Finally, questions of cyberpower have been posed in relation to adaptions of the Internet to enhance democracy, to plot or practice criminal and terrorist communications (including venial hacking or "cracking" and the launching of viruses), and to police those same communications.

Third, detailed historical, sociological, and psychological studies have attempted to contextualize the practices characteristic of cyberculture. Empirical case studies qualify both promotional hype and jeremiad alarms. Cybersex is not unexpectedly one of the most written about topics (see, for example, Ben-Ze'ev 2004). But cyberculture is revealed as not so much cut loose from culture as culture in a new form, full of subtle negotiations taking place between online and off-line worlds, yet still with persistent dangers. The standards of acceptable behavior in cyberspace—for online communications, for instance—are constructed in ways that mirror what happens in playgrounds or offices.

Fourth, the narratives of cyberculture call for aesthetic and literary criticism. What are the distinctive structures of motion pictures of the cyberfuture such as *Blade Runner* (1982), *The Terminator* (1984), and *The Matrix* (1999)? Is cyberart a distinctive form that enhances—or does it only exploit and entertain? Can computers write poetry? In what ways do such stories and productions inform or obscure the phenomena they both use and challenge? What distinctive roles do violence, glamour, sex, and speed play in cyberspace? The mass production of virtual pornography, including bestiality and pederasty, poses special questions for cultural criticism.

These four themes, along with issues of ethical responsibilities among cyberprofessionals and questions about the ontological status of cyberrealities, are included in an increasing number of books focused on cyberethics. (The Association for Information Systems nevertheless restricts "cyberethics" to information system ethics.) Although all these themes appear in other encyclopedia articles, their relations deserve to be highlighted here to emphasize synergies and interactions among the various dimensions of coming to ethical terms with the new life human beings are creating for themselves through cyberculture, whether narrowly or broadly defined.

ANDONI ALONSO
IÑAKI ARZOZ
CARL MITCHAM

SEE ALSO *Cybernetics; Information Overload; Science, Technology, and Literature.*

BIBLIOGRAPHY

Bateson, Gregory. (1972). *Steps to an Ecology of Mind: Collected Essays in Anthropology, Psychiatry, Evolution, and Epistemology*. San Francisco: Chandler.

Ben-Ze'ev, Aaron. (2004). *Love Online: Emotions on the Internet*. Cambridge, UK: Cambridge University Press.

Bethke, Bruce. (1983). "Cyberpunk." *Amazing Science Fiction Stories* 57(4): 94–105.

Gerovitch, Slava. (2002). *From Newspeak to Cyberspeak: A History of Soviet Cybernetics*. Cambridge, MA: MIT Press.

Gibson, William. (1984). *Neuromancer*. New York: Berkeley.

Haraway, Donna J. (1991). *Simians, Cyborgs, and Women: The Reinvention of Nature*. New York: Routledge.

Hayles, N. Katherine. (1999). *How We Became Posthuman: Virtual Bodies in Cybernetics, Literature, and Informatics*. Chicago: University of Chicago Press.

Heidegger, Martin. (1972). "The End of Philosophy and the Task of Thinking." In *On Time and Being*, trans. Joan Stambaugh. New York: Harper and Row. Originally published, 1964. For interpretation, see Herbert L. Dreyfus, "Cybernetics as the Last Stage of Metaphysics." In *Proceedings of the XIVth International Congress of Philosophy*, Vol. 2. Vienna: Herder.

Himanen, Pekka. (2001). *The Hacker Ethic, and the Spirit of the Information Age*. New York: Random House.

Jones, Steven G. (1997). *Virtual Culture: Identity and Communication in Cybersociety*. Thousand Oaks, CA: Sage.

Rheingold, Howard. (2000). *The Virtual Community: Homesteading on the Electronic Frontier*, rev. edition. Cambridge, MA: MIT Press.

Silver, David, and Donald Snyder. (2003). "Cyberculture and Related Studies." In *Education and Technology: An Encyclopedia*, Vol. 1, ed. Ann Kovalchick and Kara Dawson. Santa Barbara, CA: ABC-CLIO.

Stephenson, Neal. (1994). "In the Kingdom of Mao Bell; or, Destroy the Users on the Waiting List!" *Wired*, no. 2.02: 98–103, 128–132.

Sterling, Bruce, ed. (1986). *Mirrorshades: The Cyberpunk Anthology*. New York: Arbor House.

Turkle, Sherry. (1995). *Life on the Screen: Identity in the Age of the Internet*. New York: Simon and Schuster.

von Foerster, Heinz. (1984). *Observing Systems*. Seaside, CA: Intersystems Publications.

Wiener, Norbert. (1948). *Cybernetics; or, Control and Communication in the Animal and the Machine*. New York: Wiley. 2nd edition, Cambridge, MA: MIT Press, 1961.

INTERNET RESOURCE

Resource Center for Cyberculture Studies. Available from http://www.com.washington.edu/rccs/.

CYBERNETICS

Cybernetics is defined classically as the study of "control and communication in the animal and the machine" (Wiener 1948). After the decline of classical cybernetics, the field underwent a rebirth as "second-order cybernetics" in the early 1970s. Second-order cybernetics is more closely and more obviously involved with ethics than classical cybernetics (and certainly promotes a radically different worldview), but both have important contributions to make to reflections on science, technology, and ethics. Cyberculture, an increasingly important phenomenon that includes elements as diverse as email and chat rooms, electronic commerce and gaming, virtual reality and digital politics, has its origins not just in computers but also in the lesser known field of cybernetics (from which it takes its name).

Cybernetics

Cybernetics was originally promoted by the mathematician Norbert Wiener (1894–1964) in his 1948 book of that name (although W. Ross Ashby's 1956 book, *An Introduction to Cybernetics*, is considered the classic introductory text). The terms of cybernetics (including *goals* and *purposiveness*, *feedback*, and *mechanism as metaphor*) had been previously used, as was the concept of control as attaining and maintaining desired states, rather than restricting the actions of others—but not as concepts forged into a coherent field. In the development of cybernetics, two groups were particularly important: the informal association of Wiener, Arturo Rosenblueth (1900–1970), and Julian Bigelow (1913–2003) at the Massachusetts Institute of Technology (MIT); and the Josiah Macy Jr. Foundation meetings on "Circular, Causal, and Feedback Mechanisms" (which assumed the supertitle "Cybernetics" after the publication of Wiener's book), which included Warren McCulloch (1898–1969), Walter Pitts (1923–1969), Margaret Mead (1901–1978), Gregory Bateson (1904–1980), Heinz von Foerster (1911–2002), and Wiener and Rosenblueth.

The term *cybernetics* was derived from the Greek *kybernetes*, meaning "helmsman," and the field initially examined the behavior of (often complex) systems to develop models for improving system performance. The models were based on a notion of universally applicable mechanism: No essential differentiation was made between animate and inanimate systems. Examination of behaviors meant that systems which seemed impossibly complex or obscure no longer needed to remain so. If cyberneticians could not see what constituted a system, they could treat the system as a black box, which, through careful study of the inputs and consequent outputs, could be notionally "whitened" to the point that a viable mechanism relating input and output could be imagined, even if the actual mechanism remained unknown.

The intention was that systems would become controllable or better able to achieve the aims for which they were intended. The systems that cyberneticians studied were assumed to have observer-defined goals. Potential for error was understood to be omnipresent. To correct an aberration in the behavior of a system, differences between the (hypothesized) goal and behavior were examined, and the system adjusted to compensate for any difference (error). The process of error determination and correction continued until the system began to attain (and continue to attain) its goal.

Although the physical systems initially considered by cyberneticians were military and mechanical (starting with antiaircraft guns and developed through W. Grey Walter's electronic "tortoise" and Ashby's "homeostat," as much as through the computer and the robot), the animate quickly grew to be of equal significance. Application to social, anthropological, and psychological issues was pursued by Mead and Bateson (Bateson 1972a), especially in regard to mental health issues—a concern that Bateson shared with Ashby, also a psychologist. Management cybernetics was born of Stafford Beer (1926–2002) in the 1960s, and Gordon Pask (1928–1996) began cybernetic studies of teaching and learning in the 1950s.

There are many similarities between classical cybernetics and the slightly later mathematical theory of communication, or information theory, of Claude Shannon and Warren Weaver (1949); and general systems theory and its siblings, such as systems science, as developed by Ludwig von Bertalanffy (1950), making differentiation between these approaches difficult. Which term is used is frequently no more than a personal preference or historical accident. All of these approaches made notable contributions to such scientific and technological understandings and developments as the relationship between wholes and parts, automated control systems, approaches to complexity, developments in computing and communications hardware and software, and homeostasis in biological systems—to list but a few.

Early on, Wiener recognized ethical dangers in the cybernetic approach. The conjunction of animal and machine, even used metaphorically, has ethical implications—especially when the metaphor is predominantly of the animal as machine rather than the machine as animal. Another typical (and well-known) danger is that associated with the power of the machine, as exemplified, for example, in Isaac Asimov's "Three Laws of Robotics," from his science-fiction writings, which read:

First Law: A robot may not injure a human being, or, through inaction, allow a human being to come to harm.

Second Law: A robot must obey orders given it by human beings, except where such orders would conflict with the First Law.

Third Law: A robot must protect its own existence as long as such protection does not conflict with the First or Second Law. (Asimov 1942)

Wiener's *Human Use of Human Beings* (1950) is his attempt to come to terms with the most important of these dangers. He was not alone in this awareness. These ethical considerations, however, are not peculiar to cybernetics.

Second-Order Cybernetics

The initial promise of cybernetics was more than could be delivered, and the subject fell out of favor. By 1970 its funding base had eroded (with assistance from the Mansfield Agreement, a U.S. law introduced to prevent the military from funding any speculative research, or research that might not lead to an immediate military outcome). For some cyberneticians this indicated retrenchment, for others reconsideration leading to a new beginning: second-order cybernetics. The critical insight differentiating second-order cybernetics from classical (first-order) cybernetics is that second-order cybernetics takes cybernetic circularity more seriously.

Classical cybernetics exists within a worldview in which energy considerations reign paramount. The feedback loop is understood as requiring insignificant amounts of energy, thus creating a hierarchy. The controller, using relatively (and ignorably) little energy, controls the controlled, which is the big energy using part of the system. In second-order cybernetics, form and information are considered in preference to energy. In a second-order cybernetic control loop, the information passed between controller and controlled is understood to be of equal status. First-order hierarchy disappears. Each component in the loop contributes to the control of the whole. In effect, each component controls the other and the controller/controlled distinction is seen as a matter of role. The circular form of the cybernetic system is no longer disguised.

The difference was not initially presented this way. The originator of second-order cybernetics, von Foerster, made the following distinction on the frontispiece of his compilation "The Cybernetics of Cybernetics" (1975):

> First order cybernetics—the cybernetics of observed systems / Second order cybernetics—the cybernetics of observing systems.

These two characterizations, however, appear similar if one treats *observe* and *control* as interchangeable verbs, and remembers that the observing/controlling system is observing/controlling the observed/controlled system in order to develop understanding, which requires feedback. Furthermore, these concerns are similar to those expressed in the involved observer of Ernst von Glasersfeld's *Radical Constructivism* (1987).

The circular systems of Second Order Cybernetics are essentially autonomous. Their stability derives from their (internal) maintenance of their circular processes. To an external observer they may appear to veer wildly. An example is the Autopoietic system of Humberto Maturana, Francisco Varela and Ricardo Uribe. This system constructs and then maintains itself, providing a model of "life"—or, rather, "living." Such systems are said to be *organisationally closed* but *informationally open*: the form of the system maintains (distinguishes) itself, is in this manner autonomous (Maturana and Varela 1992). Information enters, passes through (is processed by) and exits it. The system distinguishes itself as itself. Because these systems are autonomous, any meaning the information passing through them may have is unique, private to each system. Communication between these systems cannot be by transmission of meaning because each system builds its own meaning: Meanings are not communicated. Uncoded communication may, however, occur through conversation. Pask's conversation theory (a formalized version of everyday conversation developed, initially, to support communication in learning environments) provides a structure to sustain communication that is formally equivalent to the other circular systems of second-order cybernetics (Glanville 1996).

Admitting autonomy and conversation requires a system that accepts that, individually, one sees differently and understand uniquely, while acting as though one believes the objects one observes are the same. Otherwise, one's relativism would lead to isolation because one has nothing communicable and there is no one to communicate with. Ranulph Glanville's theory of "Objects" (1975) provides the framework that allows individuals to believe they each make different observations of the world, yet can act as if observing the same "Object"—the essential conceptual basis making second-order cybernetics and its ethical implications viable.

Second-order cybernetics has made notable contributions in such areas of human understanding as learning, conversational communication, and the emergence of the unanticipated (often through conversational processes). In particular, through the concepts and mechanisms of autopoiesis, it has aided in the understanding of how social systems acquire stability. Nevertheless, second-order cybernetics is probably better thought of more as a way of understanding than as a technology.

Ethics

There are those who would argue that, perhaps more than any other scientific or technological field, second-order cybernetics constitutes an effort to develop a scientific basis for ethics. As such it constitutes an important contribution to any discussion concerned with science, technology, and ethics. This section sketches the basis of this contribution.

Second-order cybernetics' circular systems are autonomous—the starting point for the ethical implications of second-order cybernetics. Von Foerster was among the first to register the ethical dimension in his essay, originally published in 1973, titled "On Constructing a Reality" (von Foerster 2003a); even more relevant was his 1992 essay, "Ethics and Second-Order Cybernetics." (Von Foerster's 1993 German book *KybernEthik* originated the term *CybernEthics*.)

Von Foerster proposed two imperatives:

> Ethical imperative:/Act always so as to increase the number of choices. / Aesthetical imperative: / If you desire to see, learn how to act.

The ethical imperative insists that cybernetics has a dimension in ethics. Cybernetics implies generosity, increasing options. Von Foerster contrasted the essential meanness of morality (restrictions applied to others) to the generosity of ethics (which comes from within.)

The origin of this ethical concern can be seen to lie in the age-old question of what reality, if any, we can know independent of our knowing (i.e., is there a mind-independent reality [MIR]?). Although making a strong assumption of MIR is now commonplace, the question is in principle undecidable. Von Foerster remarked, "only we can decide the undecidable," leaving responsibility for answering this question (and, hence, for determining how we act) with each individual: one pursues whichever option one chooses. One's approach to one's world starts from this choice, which can be made once, or remade at will.

In second-order cybernetics, one's understanding of the world may be said to derive from a position of essential ignorance. The black box provides a mechanism for this. The understanding an observer builds through interacting with experience is (in the black box model) tentative: A reliable description of behavior emanating from the box may suggest it has been whitened, but nothing about the black box and our relationship to it has changed. It remains unopened (and unopenable)—provisional, as black as ever. Knowledge gained from using this model is based in profound ignorance. One cannot, therefore, insist on rightness and should tread warily, respecting the different views of others. The ethical implication of ignorance is respect for the views of others since one can never be certain, oneself. The views of others are considered as equal in stature to one's own—which does not mean theirs—or one's own—are either correct or viable.

Furthermore, the relationship between the behaviors (or signals), that is, the input and the output that black boxes are taken to act on—causing input to become output—results from interaction between observers and their own black boxes. Causality and its legal counterpart, blame, are seen to arise not from mechanism but from patterns observed by observers. The value of this understanding in how one acts cannot be overemphasized, and is confirmed in many psychotherapies that depend for their effectiveness on persuading people that the blaming causality they see is their construction and responsibility. It is not what happens to one that matters, but how one responds to it.

The black box model requires that one distinguishes: If there is no distinction between behaviors there is nothing to experience. In essence, why distinguish myself if I am alone? Distinguishing myself, I distinguish myself also from another. This act of distinguishing brings into being and implies mutualism:

whatever qualities may be attributed to one side of the distinction may (but need not) be attributed to the other. What I take for myself I may give you—this is von Foerster's ethical imperative again.

Distinctions, made in observing, can be considered a basis upon which observers construct experience, including experience of themselves. In order to assume experience is not solipsistic we assume that the other constructs (its experience of) itself (and us) in a reciprocal manner—another form of mutuality. Self-construction and maintenance indicate organizational closure: There is a boundary (it distinguishes its self) and the system is autonomous. An autonomous system is responsible. It has built itself, maintains itself (is organizationally closed), while it remains informationally open (communicates with its environment, thus substantiating the claim that, in distinguishing, one both distinguishes and distinguishes from). Bateson brings these ideas together when he uses the notion of difference (distinction) to define information: the difference that makes a difference (Bateson 1972b). The acceptance of responsibility grows out of autonomy (von Foerster 2003b): Autonomous systems are responsible for their actions. Here is the source of the aesthetical imperative.

There remains communication—that is, conversation. When communication is understood as individual construction of—and responsibility for—meaning and understanding by each participant (rather than the transmission of meanings and understandings), one can see that to understand the other one trusts the other's goodwill, acting with generosity, trust, honesty, and openness to build the understandings one will map onto each other's. This is an interaction. Teaching and learning (and much else beside) are interactive—the reason Pask developed conversation theory.

In turn, this understanding reveals that all one knows requires an observer's (knower's) presence, an understanding crucial in how one treats learning. Maturana said, "Everything said is said by an observer." Von Foerster retorted, "Everything said is said to an observer" (Von Foerster in Krippendorf 1979, p. 5). Respecting the observer is an ethical behavior.

Conclusion

Second-order cybernetics implies individuals are willing to treat each other, and (other, second-order) cybernetic systems, with a goodwill and generosity that can and should be understood as ethical implications. These go against some of the meaner understandings people currently and fashionably hold about their position in the world. Second-order cybernetics provides, in the ethical arena, hope and delight: those behaviors that are often considered higher, more civilized, and better are assumed and sustained in this way of understanding—a better-than-good reason for taking its lessons seriously.

RANULPH GLANVILLE

SEE ALSO *Automation; Cyberspace; Posthumanism; Science, Technology, and Literature; Wiener, Norbert.*

BIBLIOGRAPHY

Ashby, W. Ross. (1956). *An Introduction to Cybernetics*. London: Chapman and Hall. The most complete textbook introduction to cybernetics presented with simple profundity.

Asimov, Isaac. (1942). "Runaround." Reprinted in his *Robot Dreams*. London: Victor Gollancz, London, 1989.

Bateson, Gregory. (1972a). *Steps to an Ecology of Mind: Collected Essays in Anthropology, Psychiatry, Evolution, and Epistemology*. San Francisco: Chandler. Collected essays and commentaries from a multidisciplinary scholar who was one of the founding fathers of cybernetics, and whose writings can be seen to anticipate second-order cybernetics in several respects.

Bateson, Gregory. (1972b). "Pathologies of Epistemology." In *Steps to an Ecology of Mind*. An exploration of why we need the sort of shift that second order cybernetics embodies.

Glanville, Ranulph. (1975). *A Cybernetic Development of Theories of Epistemology and Observation, with reference to Space and Time, as seen in Architecture*. Ph.D. Thesis, Brunel University, 1975. Also known as *The Object of Objects, the Point of Points,—or Something about Things*.

Glanville, Ranulph. (1996). "Communication without Coding: Cybernetics, Meaning and Language (How Language, Becoming a System, Betrays Itself)." Invited paper in *Modern Language Notes* 111 (3): 441–462. Explores communication, language and meaning when the mechanism is conversation rather than code.

Glanville, Ranulph. (2002). "Second-Order Cybernetics." In *Systems Science and Cybernetics: The Long Road to World Sociosystemicity*, ed. F. Para-Luna. Oxford: Encyclopaedia of Life Support Systems (EOLSS) Publishers. One of the clearest explanations of a second-order cybernetic position and its epistemology.

Maturana, Humberto R., and Francisco J. Varela. (1992). *The Tree of Knowledge: The Biological Roots of Human Understanding*, rev. edition, trans. Robert Paolucci. Boston: Shambala.

Shannon, Claude, and Warren Weaver. (1998 [1949]). *The Mathematical Theory of Communication*. Urbana: University of Illinois Press. The classic text introducing what came to be called Information Theory.

Von Bertalanffy, Ludwig. (1950). "An Outline of General Systems Theory." *British Journal for the Philosophy of Science*

1: 134–165. One of von Bertalanffy's earlier papers discussing his proposals for a General Systems Theory.

Von Foerster, Heinz. (1979). "Cybernetics of Cybernetics." In *Communication and Control in Society*, ed. Klaus Krippendorf. New York: Gordon and Breach. Reprinted in *Understanding Understanding: Essays on Cybernetics and Cognition*, ed. Heinz Von Foerster. New York: Springer (2003).

Von Foerster, Heinz. (1991). "Ethics and Second-Order Cybernetics." In French in *Systemes, Ethique, Perspectives en therapie familiale*, ed. Yveline Rey and Bernard Prieur. Paris, ESF Editeur. Also in *Understanding Understanding: Essays on Cybernetics and Cognition*, ed. Heinz Von Foerster. New York: Springer (2003).

Von Foerster, Heinz. (1993). *Kybernethik*. Berlin: Merve Verlag.

Von Foerster, Heinz. (1995 [1975]). *The Cybernetics of Cybernetics*. Minneapolis: Future Systems Inc. A collection of cybernetic papers treated to cybernetic analysis providing a starting point for second order cybernetics. Von Foerster compiled and edited this collection, which was assembled by students at the University of Illinois and published by the publisher of the *Whole Earth Catalogue*, Stewart Brand.

Von Foerster, Heinz. (2003a). "On Constructing a Reality." In his *Understanding Understanding: Essays on Cybernetics and Cognition*. New York: Springer. Originally published in *Environmental Design Research*, Vol. 2, ed. Wolfgang F. E. Preiser (Stroudsburg, PA: Dowden, Hutchinson and Ross, 1973).

Von Foerster, Heinz. (2003b). "Responsibilities of Competence." In his *Understanding Understanding: Essays on Cybernetics and Cognition*. New York: Springer. Originally published in *Journal of Cybernetics* 2, no. 2 (1972): 1–6. The first venture into ethics by the father of second-order cybernetics.

Wiener, Norbert. (1948). *Cybernetics; or, Communication and Control in the Animal and the Machine*. New York: Wiley. 2nd edition, Cambridge, MA: MIT Press, 1961.

Wiener, Norbert. (1950). *The Human Use of Human Beings: Cybernetics and Society*. Boston: Houghton Mifflin.

INTERNET RESOURCE

Ashby, W. Ross. *An Introduction to Cybernetics*. Available from http://pespmc1.vub.ac.be/ASHBBOOK.html.

CYBERSPACE

• • •

Cyberspace is a term used to describe a new kind of "space" that has been made possible by the Internet. The word has a short but complex history with obscure and shifting meanings and constitutes a context for ethical issues related to science and technology.

In everyday life the notion of space is self-evident and denotes that, along with time, "in which" people live. In mathematics it refers to a collection of elements, such as points, that satisfy certain mathematical postulates. In both cases space is more given than created. In the first case, space is given, while in the second case it is a created, abstract space that people can understand conceptually but cannot directly experience.

The term *cyberspace* gained notice after William Gibson's use of it in his science fiction novel *Neuromancer* (1984). Through one of the novel's characters Gibson speaks of cyberspace as "consensual hallucination experienced daily by billions" of people, thus referring to a "non-real" space that is common to all. More specifically, he speaks about a "graphical representation of data" that emerges by abstraction from "every computer." One comes to be in cyberspace by turning a switch "on" and thus producing an instantaneous transition to it. Once there, people can enjoy the "bodiless exultation of cyberspace." Although they are somewhat confusing, these are powerful characterizations.

Background

The prefix *cyber* derives from *cybernetics*, a term coined by the mathematician Norbert Wiener (1894–1964) in 1948 to denote the study of control processes in machines and animals. That term was derived from the Greek *kubernetes*, meaning "governor" or "pilot." Cyberspace, then, is a kind of "controlled," humanly produced space.

Different Senses

In one of its senses cyberspace refers to the "spaces" associated with virtual reality, an advanced computer-based technology in which people wear headsets with stereoscopic displays, carry trackers that sense their motion, and use special input devices. With the help of those devices people navigate in "simulated" spaces, typically graphical representations of three-dimensional mathematical spaces. The integrated use of these devices creates an experience of immersion in a "virtual" reality, thus realizing an important aspect of Gibson's vision: that it is possible to enter into cyberspace, leaving the body behind.

In another sense, which became predominant in the mid-1990s, cyberspace refers to the integrated "space" made possible by the Internet, which is populated by large numbers of entities of various kinds and in which people perform multiple activities. Although this space does not support immersion, it brings to life another important ingredient of Gibson's cyberspace: the fact that it is common to all.

In *City of Bits,* William Mitchell approaches the Internet from the perspective of space and place and suggests that "the worldwide computer network—the electronic agora—subverts, displaces, and radically redefines our notions of gathering place, community, and urban life" (1995, p. 8). Mitchell proposes that the Internet is *antispatial* in the sense that it is "nowhere in particular but everywhere at once" and that it is *noncorporeal* because people's identity in it is "electronic" and disembodied. In addition, because of this disembodiment, the constructions others make of people in an effort to give those people an identity are *fragmented.* Also, the Internet favors *asynchronic* communication.

Increasingly, the word *Internet* is being invested with a broad meaning to encompass the notion of cyberspace in the second sense discussed above. For this reason ethical issues arising in cyberspace are covered under the entry "Internet." Other ethical issues are discussed in the entries "Cyberculture" and "Computer Ethics."

AGUSTIN A. ARAYA

SEE ALSO *Cybernetics; Internet; Space; Virtual Reality.*

BIBLIOGRAPHY

Gibson, William. (1984). *Neuromancer.* New York: Ace Books. This is a science fiction novel that became very popular among science fiction readers. It was also a source of inspiration for people working on certain advanced computer technologies.

Halbert, Terry, and Elaine Ingulli. (2002). *CyberEthics.* Cincinnati: West-Thomson Learning. A textbook on ethical issues that arise in the Internet and cyberspace. It combines analysis by the two authors with extensive reprints from works of several well-known authors in the area. Emphasizes a legal perspective and strives to present a balanced view of the issues by examining them from multiple viewpoints.

Mitchell, William J. (1995). *City of Bits: Space, Place and the Infobahn.* Cambridge, MA: MIT Press. Perceptive analysis of several dimensions of the Internet and cyberspace, made partly from the point of view of architecture, with its emphasis on space and place, and partly from the author's early experience with the Internet.

CYBORGS

A cyborg is a crossbreed of a human and a machine. The cyborg metaphor was coined by the astronautics researcher Manfred Clynes and the psychiatrist Nathan Kline (Clynes and Kline 1960, pp. 26–27), who argued that space travel required the development of "self-regulating human-machine systems." Such systems were termed cyborgs, from *cyb*ernetic technology and *org*anism. However, the term is not restricted to astronautics. Robotic beings that blur the distinction between humans and machines inhabit myriad science fiction novels and films, such as *Star Trek* (1979), *Robocop* (1987), *Blade Runner* (1982), and *Terminator* (1984). Above all, *cyborg* derives its intellectual influence from Donna Haraway's "Cyborg Manifesto (1985).

This manifesto rang in Haraway's presence as a leading theorist in the field broadly defined as science and technology studies. Haraway was educated as a primatologist, philosopher and historian of science and technology. In the early twenty-first century she teaches as a professor of the history of consciousness at the university at Santa Cruz, United States. In addition to a long list of essays, Haraway is the author of *Crystals, Fabrics and Fields* (1976), *Primate Visions* (1989) and most recently, the *Companion Species Manifesto* (2003), in which she revises her view of cyborgs by arguing that dogs are more important.

The Cyborg Manifesto is a complex, ironic, cacophonous text. Although it initially was addressed to feminist thinkers, it has had a considerable impact in the broader field of science and technology studies. It moves from reflection on the human condition in technological culture to a critique of politics and power relations. Haraway's critique includes current feminist strategies, which she describes as an extension of "identity politics" that defends fixed identities by victimizing the excluded. The manifesto argues for the pleasure of confusing identities. It invites feminists to play with ideas as hybridization and crossing boundaries.

People ceaselessly strive for an ordered world. Science and technology are considered as means to improve that ordering. But at the same time, they unwillingly destroy the ordering principles. As a result of findings in science, technology, and medicine, traditional binary oppositions between human and animal, organism and machine, nature and culture, man and woman, fact and fiction, body and mind, and subject and object increasingly have been blurred. Humans and animals more and more resemble cyborgs, with their bodies being equipped with pacemakers, dental prostheses, implants, and xenotransplants or modified by genetic engineering or cloning. Outside the body the dependency between living beings and machines has increased too.

The cyborg is not only a descriptive category. According to Haraway, the blurring of borders should be actively pursued. "By the late twentieth century, our time, a mythic time, we are all chimeras, theorized and fabricated hybrids of machine and organism; in short we are cyborgs. The cyborg is our ontology; it gives us our politics" (Haraway 1994, p. 150).

Cyborgs not only disrupt orderly power structures and fixed interests but also signify a challenge to settled politics, which assumes that binary oppositions or identities are natural distinctions. Actually those oppositions are cultural constructions. Haraway underlines the critical function of the cyborg concept, especially for feminist politics. The current dualistic thinking involves a "logic of dominance" because the parts of the dualisms are not equivalent. Thus, the logic produces hierarchies that legitimize men dominating women, whites dominating blacks, and humans dominating animals.

Instead, Haraway suggests that people should undermine these hierarchies by actively exploring and mobilizing the blurring of borders. "Perhaps, ironically, we can learn from our fusions with animals and machines how not to be man, the embodiment of western logos"(Haraway 1991b, p. 173).

This might suggest that Haraway simply reinforces what science and technology already do: blurring boundaries. But Haraway wants to make explicit the assumed identities and boundaries, whereas science and technology blur them in an implicit and unintended way in their strive for control of nature and order. This unintended blurring has also been articulated by the French philosopher Bruno Latour in "we have never been modern" (Latour 1993). Latour speaks about *hybrids*, which are mixtures of humans and nonhumans, like cyborgs. According to Latour, modern science and technology have caused a "proliferation of hybrids." Cyborg politics tries to escape the logic of dominance and its inherent essentialism: "Queering what counts as nature is my categorical imperative. Queering specific normalized categories is not for the easy frisson of transgression, but for the hope of lovable worlds" (Haraway 1994, p. 60).

The virtue of cyborg politics is that as soon as individuals acknowledge their identities and boundaries to be culturally constructed, they can reconstruct them in a more thoughtful way. And as soon as people acknowledge that their identity and that of others is necessarily fragmented, they can no longer dominate others, neither be dominated, Haraway asserts. Thus, the ironical play with boundaries is not without obligations. Players should take responsibility in reconstructing them (Haraway 1991a). The model of dominance should be replaced for a model of responsibility for other people as well as for machines. Like people, machines have no singular identity: "the machine is us, our processes, an aspect of our embodiment. We can be responsible for machines, *they* do not dominate or threaten us. We are responsible for boundaries, we are they" (Haraway 1991b, p. 180). However, how this responsibility towards machines and boundaries should be shaped in practice, remains unsettled in Haraway's work.

The philosophical importance of cyborg politics is not situated entirely in its anti-essentialism, for this is a common philosophical theme (Munnik 2001). Its importance is in the focus on the political potencies and challenges of technology crossing fundamental boundaries. Cyborg politics distinguishes itself from most critical approaches by not one-sidedly stressing the fearful risks of new technologies. By emphasizing peoples' responsibility of reconstructing identities, cyborg theory offers a radical and original approach toward the philosophy of technology.

MARTIJNTJE W. SMITS

SEE ALSO *Androids; Posthumanism; Robots and Robotics.*

BIBLIOGRAPHY

Clynes, Manfred, and Nathan Kline. (1960). "Cyborgs and Space." *Astronautics* 13: 26–27, 74–75.

Gray, Chris Hables, ed., with assistance of Heidi J. Figueroa-Sarriera and Steven Mentor. (1995). *The Cyborg Handbook.* New York: Routledge. Forty-three original and reprinted articles on the origin of the cyborg concept; cyborgs in engineering, medicine, and fiction; and cyborg anthropology and politics. The foreword is by Donna Haraway.

Haraway, Donna. (1991a). "A Cyborg Manifesto: Science Technology and Socialist-Feminism in the Late Twentieth Century." *Socialist Reviews* 8: 65–108. Included in Haraway's *Simians, Cyborgs and Women: The Reinvention of Nature*. New York: Routledge, 1991.

Haraway, Donna. (1991b). *Simians, Cyborgs and Women: The Reinvention of Nature*. New York: Routledge, 1991.

Haraway, Donna. (1994). "A Game of Cat's Cradle: Science Studies, Feminist Theory, Cultural Studies." *Configurations* 2(no. 1): 59–71.

Latour, Bruno (1993). *We Have Never Been Modern*. New York, Harvester Wheatsheaf. Translation of *Nous N'avons Été Modernes* (Paris: La Découverte, 1991).

Levidow, Les, and Kevin Robins, eds. (1989). *Cyborg Worlds: The Military Information Society*. London: Free Association Books.

Munnik, René. (2001). "Donna Haraway: Cyborgs for Earthly Survival?" In *American Philosophy of Technology. The Empirical Turn*, ed. Hans Achterhuis. Bloomington: Indiana University Press.

Smits, Martijntje. (2002). *Monsterbezwering*. Amsterdam: Boom.